Springer Series in Optical Sciences   Volume 54

Springer Series in Optical Sciences

Volumes 1–41 are listed on the back inside cover

Lasers, Spectroscopy and New Ideas

A Tribute to Arthur L. Schawlow

Editors:
W. M. Yen and M. D. Levenson

With 161 Figures

Springer-Verlag Berlin Heidelberg New York
London Paris Tokyo

Professor William M. Yen

Department of Physics and Astronomy, The University of Georgia,
Athens, GA 30602, USA

Dr. Marc D. Levenson

K69/803 (E), IBM Almaden Research Center, 650 Harry Rd.,
San Jose, CA 95120, USA

ISBN 3-540-18296-9 Springer-Verlag Berlin Heidelberg New York Tokyo
ISBN 0-387-18296-9 Springer-Verlag New York Heidelberg Berlin Tokyo

Library of Congress Cataloging-in-Publication Data. Lasers, spectroscopy, and new ideas. (Springer series in optical sciences '; v. 54) Includes index. 1. Lasers. 2. Spectrum analysis. 3. Schawlow, Arthur L., 1921-. I. Levenson, Marc D. II. Yen, W.M. (William M.) III. Schawlow, Arthur L., 1921-. IV. Series. QC688.L37 1987 535.5'8 87-23332

Printing: Druckhaus Beltz, 6944 Hemsbach/Bergstr.
Binding: J. Schäffer GmbH & Co. KG, 6718 Grünstadt.
2153/3150-543210

To Do Successful Research,
You Don't Need to Know Everything.
You Just Need to Know of
One Thing That Isn't Known.
A.L. Schawlow

Arthur L. Schawlow

Foreword

This volume originated in a happy event honoring Arthur Schawlow on his 65th birthday. As a research physicist, Schawlow has been a major influence on the present nature of physics and of high technology. He has also had a role, through the American Physical Society and other organizations, in shaping policy for the world of physicists. Important as these professional activities have been, the contributions to this volume were not prepared just for these reasons, but more for Art Schawlow the friend, colleague, and teacher. I am one who has had the privilege of knowing and collaborating with Art, probably over a longer period of time than others participating in this volume, and in a number of different enterprises; his friendship and stimulating scientific abilities are a very significant part of my own life. It is hence a pleasure to take part in this volume celebrating his contributions to science and to scientists.

Schawlow's career has been geographically centered at the University of Toronto, Columbia University, the Bell Telephone Laboratories, and Stanford University. But, as is illustrated by the papers of this volume, its effects and his personal influence have diffused widely. In his own work, Art Schawlow is noted for thoughtful imagination, keen physical intuition, and what might be thought an interest in gadgets – not just any gadgets, but beautiful and innovative mechanisms or new techniques in which he characteristically recognizes important potentials. One can say that he has always been a spectroscopist – from a thesis at Toronto in optical spectroscopy, to his work and book on microwave spectroscopy, the first publication on the laser, and most recently his part in the inventive development of laser spectroscopy to remarkable refinement, precision, and power. From this brief list it is obvious that his work has also had great breadth and touched many fields.

Art has also touched many people, and always with consideration, friendship, a delight in scientific discovery, and an infectious sense of humor. I believe the inspiration and motivation for this volume spring largely from these latter warm personal qualities. The contributions it contains will illustrate some of the many fields and individuals indebted to Arthur Schawlow, and it is hoped that here and there the authors will have captured at least some approximation of his almost inimitable humor.

Berkeley, California   March, 1987   *Charles H. Townes*

Preface

It occurred to some of us as the occasion of Art Schawlow's sixty-fifth birthday approached in 1986 that we needed to make an appropriate gesture to honor this man, not only for his well-recognized scientific contributions, but also for the personal legacy which he is leaving to, and the influence he has exerted on, everyone who has had the privilege of coming into contact with him through the years. After some false starts, it was decided that a collection of articles and reminiscences would serve as an appropriate vehicle for such a tribute, and it is thus that this venture came into existence.

To quantify or enumerate Art's contributions to the scientific literature is a relatively easy task, and his articles and reviews, many of which are classics, are clear, concise and numerous. He has co-authored papers with almost one hundred different people and he has worked with nearly seventy-five collaborators with varying functions during his period at Stanford. The range of topics to which his contributions are addressed is impressively wide-ranging and spans subjects as diverse as Doppler-free atomic spectroscopy and the properties of xenon flash lamp discharges. Needless to say, this volume of scientific work has had undeniable influence and impact in a number of areas of scientific and technical importance, which we need not belabor here. The nature of Art's influence is sampled in the articles presented here and is evinced by the accomplishments of the many researchers he has trained.

It is much more difficult to provide an adequate measure of Art's other contributions, especially those concerned with the fostering of scientific ideas and scientific talent and attitudes. Indeed, it is because of the very positive influence he exerted on many of us with respect to our professional growth that we decided to organize this celebratory volume. For those of us who have had the privilege of falling under his tutelage, it is generally agreed that he attempted to teach us (sometimes successfully, many times not) that very simple concepts are normally sufficient to explain even the most complex observations. This principle has served us all well in our subsequent careers. In addition to developing and encouraging new scientific ideas and approaches, Art has always provided a personal touch in his interactions; in these he reveals without fail his patience, his intrinsic kindness, his humanism, and his humor. This touch was most welcome as it nurtured self-confidence in the many raw and inexperienced graduate students and postdocs that joined his effort at Stanford, including the two editors of this volume.

It was the humanistic side of his influence that led us to choose the general tone of this collection of writings. The authors who graciously agreed to participate in this effort represent a sample of the many scientific areas in which Art has left a legacy or made an impact. We suggested to all the contributors that they write their articles in such a way as to include not only some description of some phase of their present area of scientific endeavors but also to include impressions as to how their personal attitudes and development were affected by interactions with Art. Some of the contributions describe work in which the authors are currently engaged, while others are archival, as they are concerned with the evolution of areas in which Art has made seminal contributions. By and large, we are pleased by the results of the effort, and we believe that in this collection a number of the contributions will remain relevant well into the future, especially those which were designed to be historical. We have incorporated, between parts, anecdotes and other items which address only the humanistic side and are exemplary of the joy and humor which normally prevail in any association with Art. Indeed, we would also have liked to provide recordings of his jazz clarinet playing dating from his graduate school days, but unfortunately he would not allow their release for circulation.

The volume is organized as follows: The contributed articles are divided into four areas. The first three parts include material devoted to areas in which Art has had an undeniable role, either in establishing a field of endeavor or in exercising exceptional leadership. These are, in sequence, lasers and laser spectroscopy, spectroscopy of atomic and molecular systems, and spectroscopy in the condensed phases. Each of these parts contain four to six papers from authors who have made recognizable contributions in each of the respective areas and who, following their contact with Art, have gone on to distinguished careers of their own. The fourth part consists of three contributions which are illustrative of areas where Art has had an indirect influence, in these cases by training a cadre of scientists who have advanced other frontiers by utilizing those attitudes which are so characteristic of "The Boss". The picture we have succeeded in presenting in this sampling does not totally summarize all the accomplishments of Art Schawlow. Many of us are cognizant of the fact that Art made a pioneering attempt at laser isotope separation in the early 1960s, that he played a principal role in interpreting the spectra of magnetically ordered materials, and that methods to induce cooling in atoms with lasers were suggested by him in the early 1970s. Regardless of the shortcomings of this collection, for which we, the editors, assume full responsibility, we believe that each of the contributions has its own worth; in some instances the articles are important reviews in their own right, albeit softened somewhat from the usual austere scientific format because of the nature of this enterprise. The advantage in return is that the majority of the contributions are eminently readable and will be understood by a wide range of readers not directly involved in the specific areas of scientific endeavor.

It is always difficult to take time out from the many pursuits which normally engage our time to participate in extracurricular ventures. It is indeed gratifying that so many people readily agreed to contribute to this volume and, for the most part, produced their manuscripts on time. The editors would thus like to take this opportunity to express their thanks to all who participated in this worthwhile cause and also to Dr. Helmut Lotsch and Springer-Verlag for their cooperation, which made this volume possible. Ms. Nancy Bachman of the University of Georgia is thankfully acknowledged for her assistance in sundry editorial tasks. And, of course, Mrs. Fred-a Jurian is acknowledged to be the true "boss of bosses" of the operations at Stanford, and she bears direct responsibility for many of us having survived the vicissitudes of our youth, perhaps at times to her regret. Indeed, this volume is also a tribute to her wisdom, concern and kindness.

San Jose, California   *W.M. Yen*
April 1987   *M.D. Levenson*

Contents

Part III   Solid State Spectroscopy

Part IV   Miscellaneous Ideas

Part I

Lasers and Laser Spectroscopic Techniques

Sign on entrance to Schawlow's Stanford laboratory

From (Incr)edible Lasers to New Spectroscopy

T.W. Hänsch

Sektion Physik, Universität München, and Max-Planck-Institut für Quantenoptik, D-8046 Garching, Fed. Rep. of Germany

Arthur L. Schawlow was already a very famous man when I first met him at a summer school in Scotland in 1969, a few months after I had received my doctorate from the University of Heidelberg in Germany. Immediately captivated by his personality, his quick and sharp mind, and his warm humor, I wrote to him, asking if he would take me on as a postdoc for a year or two. Fortunately, he agreed, and I arrived at his laboratory in May 1970 with a NATO Fellowship. Little did I know that I would join the faculty of the Physics Department at Stanford University two years later, and that I would be able to enjoy a close association and friendship with Art Schawlow for the next 16 years.

The early years of this period were most exhilarating, since we found ourselves at the heart of a revolution in laser spectroscopy. Many accounts have been written of the research at Stanford during this time [1–3]. Here I hope to add some personal impressions and anecdotes which capture a little bit of the human side of this great scientist.

The Incredible Laser

After arriving at Stanford, I was fascinated by a special "magic" atmosphere in Art Schawlow's laboratory. Walking down the hallway on the second floor of the Varian Physics building, a futuristic poster on one of the doors caught my eye. It showed an enormous laser gun blasting at some attacking rockets in the sky. The caption in bold letters read: "The Incredible Laser". In smaller letters below, someone had written "for credible lasers, see inside".

There could be no doubt that to Art Schawlow science is great fun. Despite his extremely busy schedule, he would often find the time to treat visitors to some amazing and entertaining demonstrations. Rummaging in a huge briefcase, he would for instance pull out his famous red toy laser gun, into which his technician has skillfully installed a real small ruby laser. With serious voice at first, he would begin to explain: "We found the whole idea of the laser was some kind of a death ray. So Mr. Sherwin built us this ray gun. Having a weapon, we had to do a little hunting and went looking for some animals. Around Stanford, the only place you find animals is in the zoo. But at the zoo the animals were all rather big and fierce. So we just bought a balloon for the kids."

At this point he would begin to noisily inflate a large clear balloon. "But when we looked at the balloon there was something funny about it - there was something inside it." Gradually, a blue balloon begins to appear inside, with big ears like Mickey Mouse. "There was a mouse inside that balloon. You know it is terrible the way mice get into everything! So we had to get our more or less trusty laser and dispose of it." With a pull of the trigger, the ray gun flashes, and the inner balloon bursts with a loud pop while the outer balloon remains unharmed. This instant is captured in Fig. 1.

Figure 1. Art Schawlow using his ruby laser ray gun to dispose of a mouse.

"Now this is a very serious experiment. It works because the outer balloon is clear so that the red light flash of the laser passes through it without being absorbed. The inner balloon is dark blue and absorbs red light so that a hot spot was formed on the surface. This illustrates how, with lasers, light is no longer something to look with, it is something you can do things with, and you can do them at places where you can see but not touch, as for instance at the retina of the eye. One of the very first applications of lasers was for surgery inside the eye, to prevent blindness from either a detached retina or leaky blood vessels." Art Schawlow likes to add that he had never heard of such diseases when he started his work on lasers. And if he had set out to find a treatment, he certainly would not have been fooling around with atoms and stimulated emission of radiation.

Sometimes the laser ray gun fails to work, but Art Schawlow is ready for such mishaps. By pulling another trigger, he can fire a small spring-loaded arrow with a rubber tip. "This is our second-strike capability."

Another of Art Schawlow's favorite demonstrations is his "laser eraser", a small flashlamp pumped Nd:glass laser which can be used to evaporate the ink from type-written or printed paper. With an impish twinkle in his eye, he would ask an unsuspecting visitor for a one-dollar bill. Pointing the eraser at the eye of George Washington, he innocently asks his victim to push a button. With a pop the eye vanishes so that Washington now appears to wear a monocle. "What have you done?", Art Schawlow would exclaim. "You know that it is illegal to deface US-, currency!" Quite a few of these dollar bills are probably still being treasured as souvenirs, even though the future of the laser eraser may have become somewhat clouded by the success of personal computers and word processors.

Art Schawlow has little patience with abstract theory or tedious mathematical derivations. But he combines a vast range of knowledge and interests with a brilliant intuition. He has a unique gift of seeing the significance of new discoveries, and he can explain complex ideas in the most simple and lucid terms. To make a point in a public lecture, he can draw on all the skills of a good comedian. To give an example, I will never forget his explanation of "coherence":

"Now, nearly everybody knows what is meant by 'coherent'. But in case there is a chemist in the audience, let me explain: you see from the beginning of time until the last twenty years or so, all light sources have been essentially hot bodies, whether it is the sun, or a tungsten filament in a lamp, a flame, or the atoms in a neon sign. The atoms are jostled around by the thermal agitation, and as an atom gets struck, for a moment it stores a little bit of energy. But in a millionth of a second or so, it releases it and sends out another light wave. Now, there are a lot of these atoms and you can picture this process as being like raindrops falling into a still pond. As each drop lands - or as each atom emits - the waves spread out in ever widening circles." With movements of his arms and hands he would illustrate this spreading of waves. "So you have atoms here going ping, and ping - the big ones go pong - ..." Gesticulating with ever increasing speed, he bursts into a wild and hilarious jumble of "ping, ping, pong, pong, ping..." Suddenly he would stop and explain, after a measured pause: "Now, for some reason, people call this kind of light 'incoherent'..."

Art Schawlow has not only made immense contributions to the public understanding of lasers and optical science. With his keen interest in fundamental physics and his contagious enthusiasm, he has a rare ability to inspire students and co-workers to high achievements. Sometimes, he would visit a young graduate student in his laboratory and ask: "What have you discovered?" To most, the thought that they were there to discover something new came almost as a revelation. But how does one do that? Art Schawlow gives very important encouragement when he emphasizes that one does not have to study everything that is known about a subject in order to discover something new. One only has to find one thing that was not known.

Edible and Other Dye Lasers

After the "Sputnik-shock" in the sixties, quite a lot of money had flown into university laboratories in the United States, and Art Schawlow had managed to accumulate an enviable collection of instruments and expensive components. "I have been poor, and I have been rich," he sometimes quipped, "and let me tell you, rich is better! As experimentalists, we always can find something to do, even if we have to work with string and sealing wax. But then, a lot of talent, time and effort gets wasted. One problem is, one never knows what remains undiscovered simply because the right equipment is not there at the right time."

I soon started to enjoy my work in the laboratory tremendously, since Art Schawlow left me complete freedom in my research while giving me access to all his treasures. He even agreed to let me purchase an AVCO nitrogen laser which I was planning to use as pump source for a tunable dye laser. Laser action in dyes had been discovered a few years earlier by Peter Sorokin and independently by Fritz Schäfer. I felt that it should somehow be possible to make such a dye laser so highly monochromatic that it would permit us to study spectral lines of free atoms and molecules by the powerful new methods of Doppler-free saturation spectroscopy which Peter Toschek and I had begun to develop during my thesis work at Heidelberg.

The new nitrogen laser turned out to be a marvellous toy. By simply focusing its ultraviolet output beam with a cylindrical lens into a glass cell filled with an organic dye solution, we could produce spectacularly colorful intense beams of laser-like amplified spontaneous emission. By adding a diffraction grating and a mirror to form an optical cavity, the color of the output beam could be changed at will. At one time I focused the blue light of such a dye laser into a single drop of watery solution of fluorescein. This drop then became a dye laser all of its own, emitting an intense beam of green light. The laser cavity was simply formed by the surfaces of the liquid.

Observing this droplet laser with obvious delight, Art Schawlow postulated that "anything will lase if you hit it hard enough." Thinking about challenges to prove such a claim, we wondered if the colorful gelatine desserts popular with children would show laser action when pumped with the nitrogen laser. The next morning, Art Schawlow came to work waving a package with twelve different flavors of "Knox Jello". Using the hot water supply in the darkroom, we prepared two of the desserts in plastic cups, following the manufacturer's instructions. After they had begun to gel, we took them to the lab and focused the nitrogen laser beam to illuminate a line on the flat surface of the wobbly substance. There was distinct fluorescence, but no laser action. In resignation, Art Schawlow would return to his office and enjoy the obstinate experiment as a snack. This ritual was repeated every morning for a week until we had tried all twelve flavors without luck.

Determined to demonstrate an edible laser, we finally mixed up a packet of clear, flavorless gelatine and added some sodium fluorescein. This experiment was an instant success. With a knife we could cut the new laser material into rods or other shapes. The paper describing this laser would soon be posted on many bulletin boards [4]. A few people considered this experiment a frivolity, but it actually led to some rather important technical developments. Soon afterwards, Kogelnik and Shank [5] exposed a dichromated gelatine film to the interference pattern of two ultraviolet laser beams and demonstrated the first distributed feedback laser. Today, distributed feedback plays an increasingly important role in semiconductor diode lasers for optical communications.

From the beginning, Art Schawlow was very enthusiastic about my plans to develop a highly monochromatic tunable dye laser. Such a tool would open many exciting new possibilities for studying the structure of atoms, molecules, and solids. The nitrogen laser appeared as a particularly attractive pump source, because many dyes spanning the visible spectrum could be pumped with good efficiency at high pulse repetition rates. Past attempts to achieve narrow linewidths with a nitrogen-pumped dye laser had remained unsuccessful. But I was hopeful that it should be possible to isolate a single axial mode with the help of a holographic diffraction grating and a birefringent Lyot filter. After encouraging preliminary experiments, I submitted a paper for presentation at the APS Meeting at Stanford University in December 1970 [6].

As the conference approached, I became rather panicky, because the envisioned scheme did not work reliably. Almost in desperation, I tried an entirely different approach. I moved the grating far away from the dye cell, took out the Lyot filter, and inserted into the cavity a small Zeiss telescope which I happened to carry in my pocket so that I could read the slides during a lecture from the backrow. The idea was that a beam-expanding telescope would illuminate more lines at the grating and so improve the resolution. This little trick worked well beyond my expectations [7]. With a larger telescope, the linewidth could be reduced to a few hundredths of an angstrom. Even narrower lines could be achieved by inserting a tilted Fabry-Perot etalon into the cavity. With this suprisingly simple scheme, we could now produce coherent light of a spectral purity and brightness that was previously only available from gas lasers with very limited tuning range.

Spectroscopy without Doppler Broadening

During my thesis work at the University of Heidelberg I had become intrigued by the potential of lasers for high resolution spectroscopy. By exploiting the spectral hole-burning effect discovered by Willis Lamb, Bill Bennett and Ali Javan, I was able to develop an early form of saturation spectroscopy [8] which could circumvent the Doppler broadening of spectral lines. With Peter Toschek, we used this method to study collision processes and nonlinear optical phenomena in neon discharges. When I came to Stanford, I collaborated with Peter Smith, then on sabbatical at Berkeley. We demonstrated a simple new method of saturation spectroscopy, using just one single He-Ne laser and an external gas cell [9]. A very similar technique was developed independently by Christian Bordé in Paris [10].

In Art Schawlow's laboratory, Marc Levenson, then a graduate student, was working with a krypton ion laser. By placing a prism and etalon into the cavity, he could make the laser work in a single axial mode, manually tunable over a few gigahertz. During a visit of Peter Toschek in the summer of 1970, we decided jointly to try and use this laser for saturation spectroscopy of iodine vapor, since it was known that the diatomic iodine molecule has several absorption lines in accidental coincidence with krypton and argon ion laser wavelengths. The experiment was at first unsuccessful. A few weeks later, Marc Levenson found out that we had worked with a contaminated iodine cell. With a new cell, we soon obtained very pretty spectra, showing all 21 hyperfine components of one line completely resolved [11].

Art Schawlow was delighted with these results. Even then, he was so much in demand as a public speaker that he sometimes defined "genius" as "an infinite capacity to take planes." In his lectures, he would describe the new experiment with his unique clarity and simplicity by first reminding people of the Doppler effect: "If an object is moving towards you - like a train - the emitted sound goes up in pitch." While making this point he would walk towards the audience and his voice assumed a funny high pitch. Next he would walk backwards and explain in a deep bass voice: "...and if it is moving away from you, it goes down in pitch." Returning to his normal voice: "Now the same is true for light. So in a gas, where the molecules are moving in all directions, the spectral lines appear blurred and spread out by the Doppler effect."

Then he would go on to explain the method of Doppler-free saturation spectroscopy. Showing a drawing of the apparatus as in Fig. 2, he would point out: "The light from a tunable laser is divided by a beam splitter into a strong saturating beam and a weaker probe beam. These light beams are sent through the absorbing gas in nearly opposite directions. When the saturating beam is on, it bleaches a path through the cell, and a stronger probe signal is received at the detector. As the saturating beam is alternately stopped and transmitted by the chopper, the probe signal is

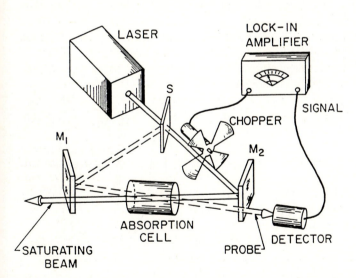

Figure 2. Apparatus for saturation spectroscopy without Doppler broadening.

modulated. However, this happens only for those wavelengths that interact with atoms that are standing still, or at most moving sideways. Because from a moving atom, the two light beams appear Doppler-shifted in opposite directions, so that they cannot both be in resonance at the same time."

Despite such powerful new techniques, the situation was at first quite frustrating to laser spectroscopists. Although there were hundreds of different lasers, they could only be tuned over very small wavelength ranges, and laser spectroscopy was limited to the laser transitions themselves or to a few molecular transitions in accidental coincidence with known laser lines. Lamenting this fact, Art Schawlow would recount how, as a graduate student, working in atomic spectroscopy, he discovered the true definition of a diatomic molecule: "It is a molecule with one atom too many!"

The whole situation changed completely with the advent of highly monochromatic tunable dye lasers. One of the first experiments with our new laser was the realization of an old dream. Jointly with Issa Shahin, one of Art Schawlow's graduate students, we recorded Doppler-free saturation spectra of the yellow D-lines of atomic sodium [12]. One such spectrum is shown in a little cartoon (Fig. 3) which I left on Art Schawlow's desk after an exhilarating night.

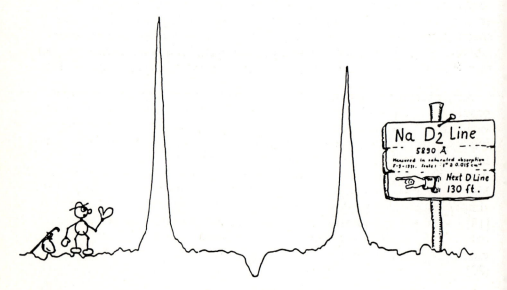

Figure 3. Cartoon dated August 9, 1971, with a Doppler-free saturation spectrum of the yellow sodium D_2 line

When Art Schawlow saw these results, he immediately urged: "You have to do the same with the red Balmer-α line of atomic hydrogen!" As the Rosetta Stone unlocked the secret of Egyptian hieroglyphics, the Balmer spectrum of the simple hydrogen atom opened up the laws governing atoms, and eventually molecules, liquids and solids [13]. It inspired the pathbreaking discoveries of Bohr, Sommerfeld, de Broglie, Schrödinger, and Dirac. For more than a decade, in the thirties and fourties, the line profile of the Balmer-α line presented one of the central problems of atomic spectroscopy. There appeared to be some discrepancies between observations and the predictions of the Dirac theory. But Doppler broadening is particularly large for the light hydrogen atoms, and classical spectroscopists never succeeded in resolving single fine structure components. In the 1940s, Lamb and Retherford used a radiofrequency method to reveal a new fine structure in the lower level of the transition that produces the Balmer-α line. The discovery of this Lamb shift led to the development of quantum electrodynamics by Feynman, Schwinger, and Tomonaga.

Within a few weeks, Issa Shahin and I set up an old-fashioned Wood-type gas discharge tube in order to produce hydrogen atoms in the excited n=2 state. With a simple change of the dye solution, our laser was ready for saturation spectroscopy of the Balmer-α line. Even the first Doppler-free spectra revealed single fine structure components [14]. We were thrilled that we could observe the Lamb shift directly in the optical spectrum, as illustrated in Fig. 4. This experiment set the beginning of an exploration of the simple hydrogen atom by high resolution laser spectroscopy that has grown into a fairly big enterprise in numerous laboratories [15]. Absolute wavelength measurements have meanwhile led to a 300-fold improvement in the accuracy of the Rydberg constant [16].

At Stanford, jointly with a group of bright graduate students and postdoctoral visitors, we went on to develop and demonstrate different variants of Doppler-free saturation spectroscopy. I remember a discussion with Art Schawlow at a conference in Esfahan in 1971, during which he invented intermodulated fluorescence spectroscopy, soon to be demonstrated by his student Mike Sorem [17]. Later, Art Schawlow and his group were very successful in applying saturation techniques to the simplification of complex molecular spectra by selective level labeling [18]. With the new tunable dye lasers, we could explore other fascinating coherent light techniques such as ultrasensitive detection of atoms and molecules by intracavity spectroscopy [19] or laser-excited fluorescence [20].

Much to our later chagrin, we long ignored a particularly elegant possibility for high resolution spectroscopy, the method of Doppler-free

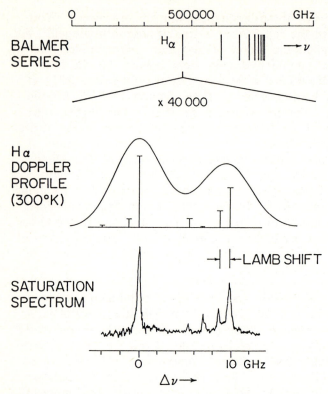

Figure 4. Balmer series of the simple hydrogen atom and fine structure of the red line H-α, resolved by saturation spectroscopy.

two-photon spectroscopy, as proposed in 1970 by Veniamin Chebotayev and coworkers [21]: If two counterpropagating light beams from a laser excite two-photon transitions in a gas, the two first-order Doppler shifts cancel, and a narrow resonance is expected, without any need to select slow atoms. This prediction was already implicitly contained in a theory of three-level gas atoms interacting with two laser beams which Peter Toschek and I had worked out [22]. However, we did not have widely tunable monochromatic lasers at that time, and we did not think seriously about exciting atoms with two equal photons.

At Stanford, Art Schawlow and I became aware of Chebotayev's proposal only in 1974 during a visit of Vladilen Letokhov. It was immediately obvious that such an experiment must work and that it should be very easy with our tunable dye lasers. Letokhov went on to visit Boston, where Marc Levenson was meanwhile working with Nico Bloembergen at Harvard. Soon afterwards, Marc succeeded in demonstrating pulsed Doppler-free two-photon spectroscopy in sodium vapor, almost simultaneously with an independent experiment of Bernard Cagnac's group in Paris. Our own paper [23] was unfortunately submitted a few weeks later, but we observed clean, well-resolved spectra with a cw dye laser of low power. This experiment was so simple that Art took delight in showing a color slide with a Mickey Mouse in front of the apparatus. "I have come to realize that I am really rather stupid", he would remark. "The only thing that saves me is that everybody else is rather stupid too - sometimes."

Doppler-free two-photon spectroscopy of atomic hydrogen holds particularly intriguing challenges, since the two-photon transition from the $1S$ ground state to the metastable $2S$ state has a natural linewidth of only about 1 Hz [21]. Unfortunately, such an experiment requires ultraviolet radiation in the difficult wavelength region near 243 nm. The first experiments had to resort to frequency-doubled pulsed dye lasers with rather large instrumental linewidths. But recent cw experiments at Stanford have measured the Lamb shift of the hydrogen ground state to within about 1 MHz [24]. Future experiments now being set up in our laboratories in Garching promise dramatic further improvements in spectral resolution [15]. They will permit tests of basic physics laws with unprecedented accuracy, and if past experience is any guide, the biggest surprise would perhaps be if we found no surprise.

Cooling with Laser Light

By 1974, our group at Stanford had become quite used to manipulating gas atoms with the help of intense, monochromatic laser light. In saturation spectroscopy, we were routinely selecting and exciting slow atoms with the pump beam. At about the same time, Art Ashkin and coworkers at Bell Laboratories reported on a series of interesting experiments levitating microscopic objects with laser radiation pressure. In a related earlier experiment, even before the advent of lasers, a beam of sodium atoms had been deflected by resonant radiation pressure.

Talking about these experiments, it suddenly became obvious to Art Schawlow and me that it should be possible to cool an atomic gas by resonant radiation pressure. The idea was to illuminate a gas sample with laser light tuned just below the atomic resonance frequency. Even if the light were propagating isotropically in all directions, it should exert a "viscous" braking force on any atom. Because of the Doppler effect, a moving atom will absorb preferentially those photons which are propagating in the opposite direction. Since the reemission is isotropic, each scattered photon will, on the average, reduce the momentum of the atom by an amount $\hbar k$. Some simple back-of-the-envelope calculations showed that atomic gases could be cooled very quickly to very low temperatures in this way.

Once we thought of it, the method appeared so obvious to me that I doubted we could publish it without demonstrating laser cooling in an actual experiment. But Art Schawlow called in a few colleagues and began to explain our idea. To my surprise, most shook their heads in disbelief and said something like "Cooling a gas by shining in laser light? You must be crazy!" After the visitors had left, Art said to me: "Now I am sure we have got something!" and we proceeded to write up our paper [25].

This scheme has been demonstrated recently with spectacular success by Steve Chu et al. [26] in their "optical molasses" experiments with sodium atoms. Some special tricks were required to avoid optical pumping between hyperfine levels. So Chu and his coworkers had the good sense to wait with this experiment until optical communications engineers had developed proper tools for modulating and controlling laser light. The ability to cool atoms by laser radiation pressure quickly to millikelvin temperatures and below is opening a whole new world to atomic physicists. It has become possible to trap cold neutral atoms magnetically or optically, to manipulate them, and to look for new quantum phenomena in their interactions with each other or with surfaces. Cold trapped atoms or ions may also lead to new incredibly precise atomic clocks.

More than once, Art Schawlow has uncovered new concepts which have opened up entire new fields of scientific endeavor, even though in hindsight some of his ideas may appear simple and almost obvious. The most dramatic example is of course the laser. Art Schawlow was the first to realize that, at optical wavelengths, just two mirrors would produce a laser resonator. With his role in the invention of the laser, he has reshaped modern science and technology to an extent not matched by many others. What a wonderful inspiration for future researchers when this great magician of science asserts that there must be many other beautiful and simple concepts lying around, ready to be discovered!

Acknowledgements

I am grateful to Peter Toschek for critically reading the manuscript and suggesting several changes. This article could not have been completed in time without the spirited help of Sigrid Oetjen, my secretary.

REFERENCES

1. A.L. Schawlow, J. Opt. Soc. Am. 67, 140 (1977)
2. T.W. Hänsch, Physics Today 30, 34 (1977)
3. A.L. Schawlow, Rev. Mod. Phys. 54, 687 (1982)
4. T.W. Hänsch, M. Pernier, and A.L. Schawlow, IEEE J. Quant. Electr. QE-7, 45 (1971)
5. H. Kogelnik and C.V. Shank, Appl. Phys. Lett. 18, 152 (1971)
6. T.W. Hänsch and A.L. Schawlow, Bull. Am. Phys. Soc. 15, 1638 (1970)
7. T.W. Hänsch, Appl. Opt. 11, 895 (1972)
8. T.W. Hänsch and P. Toschek, IEEE J. Quant. Electr. QE-4, 467 (1968)
9. P.W. Smith and T.W. Hänsch, Phys. Rev. Lett. 26, 740 (1971)
10. C. Bordé, C.R. Acad. Sci. Paris, 271, 371 (1970)
11. T.W. Hänsch, M.D. Levenson, and A.L. Schawlow, Phys. Rev. Lett. 26, 946 (1971)
12. T.W. Hänsch, I.S. Shahin, and A.L. Schawlow, Phys. Rev. Lett. 27, 707 (1971)
13. T.W. Hänsch, A.L. Schawlow, and G.W. Series, Scientific American 240, 94 (1979)
14. T.W. Hänsch, I.S. Shahin, and A.L. Schawlow, Nature 235, 63 (1972)
15. T.W. Hänsch, R.G. Beausoleil, B. Couillaud, C.J. Foot, E.A. Hildum, and D.H. McIntyre, Laser Spectroscopy VIII, S. Svanberg and W. Persson, Eds., Springer Series in Optical Sciences, Springer-Verlag, New York, Heidelberg, 1987
16. P. Zhao, W. Lichten, H.P. Layer, and J.C. Bergquist, Phys. Rev. Lett. 58, 1293 (1987)
17. M.S. Sorem and A.L. Schawlow, Opt. Comm. 5, 148 (1972)
18. R. Teets, R. Feinberg, T.W. Hänsch, and A.L. Schawlow, Phys. Rev. Lett. 37, 683 (1976)
19. T.W. Hänsch, A.L. Schawlow, and P. Toschek, IEEE J. Quant. Electr. QE-8, 802 (1972)
20. W.M. Fairbank, Jr., T.W. Hänsch, and A.L. Schawlow, J. Opt. Soc. Am. 65, 199 (1975)
21. L.S. Vasilenko, V.P. Chebotayev, and A.V. Shishaev, JETP Letters 12, 113 (1970)
22. T.W. Hänsch and P. Toschek, Z. Physik 236, 213 (1970)
23. T.W. Hänsch, K.C. Harvey, G. Meisel, and A.L. Schawlow, Opt. Comm. 11, 50 (1974)
24. R.G. Beausoleil, D.H. McIntyre, C.J. Foot, B. Couillaud, and T.W. Hänsch, Phys. Rev. A35, 4878 (1987)

25. T.W. Hänsch and A.L. Schawlow, Opt. Comm. 13, 68 (1975)
26. S. Chu, L. Hollberg, J. Bjorkholm, A. Cable, and A. Ashkin, Phys. Rev. Lett. 55, 48 (1985)

High-Power Solid State Lasers

J.F. Holzrichter

Lawrence Livermore National Laboratory, Livermore, CA 94550, USA

In 1958 SCHAWLOW and TOWNES [1] described the build-up of coherent light in an optical resonator, or laser. They noted that amplified coherent light would possess two desirable properties: precise, narrow wavelengths, and high brightness or focusability. The first laser, demonstrated by MAIMAN et al. [2] in 1961, produced a peak power of approximately 3 kW $(3 \times 10^3 \mathrm{W}$; i.e., 0.5 J of light in a pulse a few hundred microseconds long) and verified many of the Schawlow/Townes predictions.

Fig. 1. The Nova laser provides up to 100 TW (100 kJ in 10^{-9} s) at 1.05 μm. It can be harmonically converted to 0.53 or 0.35 μm with 60% efficiency. Most fusion experiments are conducted using 0.35-μm light

Today, high-power lasers (Fig. 1) are capable of more than 100 TW (10^{14} W). The high irradiances attainable with such lasers are making possible many technological advances, of which recent and important examples are the demonstration of inertial-confinement fusion (ICF) [3] and of laser gain at soft x-ray wavelengths [4].

Between 1959 and 1961, COLGATE et al. [5] suggested that high-power lasers could provide the irradiance required to compress small quantities of deuterium and tritium to high densities, initiating fusion and releasing thermonuclear energy.

In 1958, SCHAWLOW and TOWNES [6] wrote of the difficulty of producing lasers at x-ray wavelengths:

> "... in the ultraviolet region at $\lambda = 1000$ Å, one may expect spontaneous emissions of intensities near ten watts.... Another decrease of a factor of 10 in λ would bring the spontaneous emission to the clearly prohibitive value of 100 kilowatts.... unless

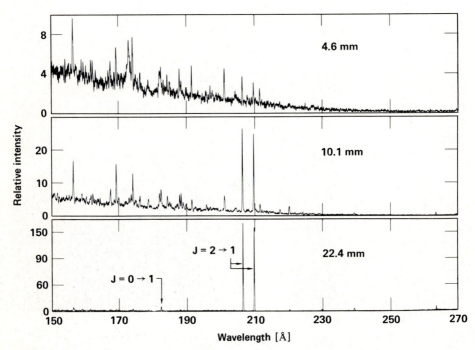

Fig. 2. **Variation in line intensity with the length of the amplifying neon-like selenium plasma confirms lasing at 206 and 209 Å**

some radically new approach is found, [maser systems] cannot be pushed to wavelengths much shorter than those in the ultraviolet region."

In 1985, MATTHEWS, ROSEN, et al. [4] observed laser gain in neon-like selenium at 206 Å (Fig. 2); recently, they observed gain at 66 Å in nickel-like europium [7]. In both experiments, the pumping source was the Novette laser, a 10-TW (10^{13}-W), harmonically converted, two-beam glass laser at the Lawrence Livermore National Laboratory (LLNL).

1. The 1960s at Stanford and Elsewhere

Schawlow and his associates and students demonstrated many of the laser technologies now being pursued in university, industrial, and national laboratory programs. These included the technology of high-power pulsed lasers with single temporal and spatial modes, frequency stabilization, wavelength tunability, the spectroscopy of laser media, and the focusing and diagnostic techniques necessary for the effective use of lasers in experimentation.

One of the more unusual experiments of the period—possibly one of the earliest ICF experiments—is shown in Fig. 3. The outer layer of a spherical object (a potato!) was irradiated with a 100-MW laser, one of the most powerful lasers available at the time. The ablation of the skin caused a shock wave to converge inward. Emmett and Schawlow are reported to have said at the time that if a powerful enough laser (the LLNL Nova laser would suffice) were to spherically irradiate a small enough potato, the skin would be instantly re-

Fig. 3. J. L. Emmett and A. L. Schawlow align an early "fusion" experiment. Laser ablation of an Idaho potato "fusion-fries" the inside and removes the skin at the same time. Patent pending!

moved and the potato would be heated to near-stellar temperatures, creating a "fusion fry."

Experiments at Stanford and elsewhere (notably at Harvard and Michigan) showed that laser beams focused to high irradiance ($>10^9$ W/cm^2) produced electric fields strong enough ($>10^6$ V/cm) to modify the electron distribution around the nucleus. An intense laser field mixes high-lying electronic wave functions into the ground-state wave functions that are summed in calculating the susceptibility tensor of a material, thus making the optical response of the material to the laser field a nonlinear function of the field strength [8]. Through this interaction, an intense beam can increase the index of refraction of the medium (solid, liquid, or gas) in which the beam is propagating. This causes self-focusing of the laser beam, because wherever the beam is most intense the refractive index increases, causing a local lens-like action that further increases the intensity of that portion of the beam [9]. The nonlinear response to the electric field strength (a result of anharmonic electron motion) causes waves to be generated at harmonics of the frequency of the incoming beam [10]. This is used to efficiently ($>90\%$) convert the laser light to 2nd, 3rd, or 4th harmonic light [11].

In what are called three-wave interactions, incoming laser light can couple through virtual electronic transitions to elementary excitations in a material. The difference wave is Brillouin light if sound waves are the elementary excitation, or Raman light if vibrational transitions are the elementary excitations [12]. This difference wave can grow to high amplitude, and can ultimately carry more than 70% of the incoming laser energy.

High intensity can also cause complex multiphoton absorption processes, which can lead to ionization or other effects. A study [13] with a picosecond UV laser showed that at irradiances of roughly 10^{12} to 10^{13} W/cm^2, an atom such as uranium can lose its entire outer electron shell (8 electrons), corresponding to the absorption of nearly 100 photons from the laser field.

Most of these early experiments were performed with ruby lasers several millimeters in diameter, which delivered 10- to 100-MW beams (0.1 J in 10^{-8} s) that were focused (with some difficulty) to irradiances of 10^9 W/cm^2. A remarkable number of nonlinear processes were observed and quantified in the 1960s. Today, high-power fusion lasers such as Nova propagate 0.7-m-diameter beams at irradiances greater than 10^9 W/cm^2 over many meters before they are focused onto their targets. The irradiances of these beams are high enough without focusing to cause efficient (70%) harmonic conversion in 1-cm-thick plates of crystalline potassium dihydrogen phosphate (KDP); they are, in fact, high enough to push the limits of stable beam propagation in all media in which they travel. Propagation is limited by stimulated rotational Raman scattering and nonlinear self-focusing in air, and by transverse Raman or Brillouin scattering in transparent laser materials such as KDP crystals, fused silica, and even laser glass itself.

The 1960s were a time of great excitement in the laser field. Nonlinear optics was developed; in 1981 BLOEMBERGEN [14] received the Nobel Prize for his pioneering work in this field. The tunable dye laser was invented by SOROKIN [15], developed as a cw laser by SNAVELY [16], and developed to a high art form by HÄNSCH and his coworkers in Schawlow's laboratory at Stanford [17]. A renaissance in optical spectroscopy began in the 1960s and continues to this day. SCHAWLOW [18] received the Nobel Prize in 1981 for his leadership in this field. Being a student in Schawlow's laboratory during this period was a special experience. His attitudes and approaches to scientific investigation strongly influenced our work at Livermore and that of his students elsewhere.

The 1960s saw the development of a host of other important laser technologies. The invention of the semiconductor laser, together with glass fiber technology, has led to high-bandwidth communication by means of light waves. Arrays of semiconductor lasers are being used as efficient (>50%), high-power pump sources for solid state lasers and are leading to remarkable improvements in solid state laser technology. Laser-driven ICF and laser-induced isotope separation began; an important experiment by TIFFANY and SCHAWLOW [19] contributed to progress in isotope separation. Industrial and medical applications based on the welding, cutting, and photoselective properties of the laser were demonstrated (see Fig. 4 for an amusing

Fig. 4.   A. L. Schawlow demonstrates a laser technique relying on photoselectivity to eradicate a giant blue mouse using a red (ruby) laser beam. He also demonstrates the "action-at-a-distance" properties of the laser that make it so useful as a scientific probe, welding tool, etc.

example). Physicists, chemists, engineers, and others began to use the laser as a noninvasive probe for a wide variety of physical and chemical measurements that have contributed greatly to our understanding of the physical world.

2. Inertial-Confinement Fusion

In the 1970s, large national programs were begun to examine the possibility and practicality of ICF as a source of commercial electrical energy. This work grew out of pioneering work on high-power lasers in the 1960s at the Centre de l'Energie Atomique at Limeil by Bobin, Floux, and their coworkers; at the Lebedev institute by Basov and Prokhorov; and at LLNL by Nuckolls, Kidder, and others [20]. In 1970, laser power was about 0.01 TW (10^{10} W). Laser power of 10 to 100 TW (10^{13} to 10^{14} W) was needed to uniformly irradiate small fusion targets (0.1 to 1 cm in diameter) at $>10^{14}$ W/cm^2 [21]. This irradiance produces pressures greater than 10 Mbar (10 million atmospheres), enough to compress the most easily ignited fusion fuel, an equimolar mixture of deuterium and tritium, to fusion conditions—ion densities greater than 200 g/cm^2 (1000 times the density of solid DT) and ion temperatures greater than 5 keV (50 million degrees Celsius) [21]. Experiments at LLNL [3] recently confirmed that these conditions can be attained.

The development of the laser from its first demonstration in 1961 to the 100-TW fusion systems of today (see Fig. 5) has been made possible by con-

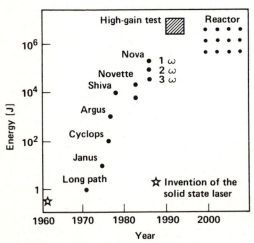

Fig. 5. Laser performance has increased dramatically since the first demonstration by Maiman in 1961. The three "Nova" points represent (top to bottom) performance at 1.05, 0.52, and 0.35 μm. "High gain test" and "Reactor" are laser systems contemplated for future demonstrations

tributions from many laboratories; see Ref. 22 for a review. This laser power gives us access in the laboratory to high temperatures and pressures that heretofore existed only in astrophysical environments (stars) or in nuclear explosions.

3. Physics of High-Power Lasers

A high-power laser designed for fusion and x-ray laser applications uses a small laser oscillator to generate a millijoule pulse in a single spatial and temporal mode, and then amplifies that pulse to thousands of joules (a gain of 10^6 to 10^8) [22]. Typical pulse durations are 10^{-9} s. The beam (a pulse of light about 30 cm long) propagates through the laser components at an irradiance greater than 10^9 W/cm^2.

Under these conditions, longitudinal and transverse nonlinear phenomena limit the irradiance of the pulse and the path length over which the pulse can propagate. Catastrophic breakdown of an optical material, associated with the formation of an absorbing, obscuring plasma, is the ultimate propagation limit and occurs at fluences of roughly 5 to 20 J/cm^2. The total power that can be carried by a single beam then depends on the irradiance and fluence of the beam and on the cross-sectional area to which the beam can be expanded before breakdown or nonlinear self-focusing occurs.

Nonlinear self-focusing [23] leads to near-field amplitude noise N and to far-field (focal) spread in a beam. The near-field noise grows as

$$N \simeq N_{\text{init}} \exp\left(\frac{2\pi}{\lambda} \int \gamma I \, dl\right) \equiv N_{\text{init}} \exp(B), \tag{1}$$

where N_{init} is the initial beam noise perturbation; λ is the laser wavelength; γ is the nonlinear retardation of the beam, caused by local increases in the refractive index of the material through which the beam propagates; and I is the beam irradiance. (Note that the beam noise grows more rapidly for shorter wavelengths.) The integration in (1) is carried out along the propagation path. The resulting so-called B-integral, giving the noise growth "gain," is also equal to the phase front retardation, in radians. (One wave of retardation corresponds to $2\pi = 6.28$ radians or to $B = 6.28$.) Typical near-field B limits are 2.5 to 3, which lead to near-field noise gain $\exp(B) = 12$ to 20. Far-field noise (beam spreading) grows as $\exp(2B)$, because both phase and amplitude noise contribute to the far field.

Given the cost of optical materials and the costs of construction generally, it is desirable to maximize the irradiance and fluence delivered by a given laser system. Progress in high-power lasers has resulted from efforts to minimize all parameters other than the irradiance (1), which we maximize subject to a fixed N:

- Lower-γ materials, e.g., fluoride-based glasses.
- Higher laser-gain coefficients and thinner optical components, which decrease the beam path length in the laser materials.
- Control of the spatial and temporal uniformity of I to prevent hot spots, which are exponentially amplified. Production of bandwidth-limited temporal pulses with no time spikes, and smooth, single-phase, noise-free transverse beams.
- Attainment of systems with low noise amplitude N_{init} by super-cleaning optical surfaces, by control of diffraction ripples with imaging apertures and apodizing apertures, and (most importantly) by periodically reducing the noise amplitude using spatial filtering.

Limits on the fluence ($\int I dt$) are different from irradiance limits because they occur when defects in optical elements absorb enough laser energy that an absorbing plasma develops, resulting in damage to the laser material. Low-loss, inclusion-free coatings, surfaces, and bulk materials permit optical elements to sustain higher fluences.

Higher fluences naturally occur for beams of fixed irradiance, defined in (1) above, as the laser pulse duration is increased from 1 to 10 ns. The current evolution of fusion lasers toward longer pulse duration for larger targets makes the development of high-fluence materials more and more important.

Large-area (high-Fresnel-number) systems are desirable because it is cost effective to maximize the power from a single beam aperture before paying the price of additional beams. Such systems have been made possible by advances in the production of low-cost, large-area laser materials, notably the development of new glasses, continuous polishing and lapping, and sol-gel coatings, and by the invention of imaging techniques (the Hunt relay [22]) that make it possibe to operate a laser with maximum fluence over the full optical aperture.

Beam area is ultimately limited by parasitic transverse laser oscillation in the gain medium or by transverse stimulated processes (e.g., Raman). If the transverse dimension of the medium is great enough, a sufficient laser gain-length or transverse stimulated gain-length can develop. This limits the gain coefficient of the laser host or the intensity of the laser beam.

To obtain laser power higher than that available from a single optimized beam, multiple beams are needed. The engineering and control of long, multibeam systems has been demonstrated. We have shown that multiple beams can be focused satisfactorily on a common target.

The output of glass laser systems can be harmonically converted from 1.05 μm to 0.52, 0.35, or 0.26 μm with efficiencies of up to 70%. This has made possible the generation of ICF plasmas free from suprathermal ("hot") electrons [3]. Hot electrons, which preheat the fusion fuel and greatly reduce the compression, are produced when plasma waves are stimulated by Raman and other processes. Laser light with wavelengths of 1 μm and longer is absorbed in a plasma at low electron densities ($n_e < 10^{21}$ cm^{-3}), at which electron-ion collision rates are too low to damp the plasma waves. But for $\lambda \leq 0.5\,\mu$m, absorption takes place at $n_e \geq 4 \times 10^{21}$ cm^{-3}; damping is then strong enough that at fusion irradiances ($>10^{14}$ W/cm^2), the plasma near the target is Maxwellian and free of hot electrons. Under these conditions, fusion implosions proceed satisfactorily [3].

4. Future of Solid State Lasers

Large investments in materials, research, computer models, and special fabrication technologies, together with new ideas and technologies, are leading to rapidly expanding applications of solid state lasers. Our experiments indicate that a demonstration of high-gain fusion in the laboratory will require a 5- to 10-MJ, short-wavelength laser. This represents roughly a fiftyfold increase in pulse energy with respect to the Nova system. Analysis of new laser architectures, new high-fluence materials, and lower-cost production technologies leads us to believe that such a system can be built at a cost only 2 or 3 times that of Nova, or roughly $300 to $600 million in 1987 dollars.

Recent calculations and experiments [24] show that crystal and (surprisingly) glass solid state laser media can be cooled well enough to generate high-optical-quality outputs of over 1000 W per laser plate. Multiplate, multibeam systems should be able to generate over a megawatt of average power. The ability of solid state laser systems to amplify nanosecond pulses makes it possible to convert the fundamental laser wavelength to almost any wavelength from the UV to the near IR using harmonic, parametric, or Raman processes. This can occur under average power conditions [24]. With the advent of GaAlAs laser diode arrays, which generate 810-nm light with an electricity-to-light conversion efficiency of over 60%, it is possible to contemplate solid state laser systems with overall efficiencies of 10 to 20%. A 1-cm^2 GaAlAs chip with 100 W/cm^2 pump output might be used to excite a 1-cm^2 solid state host at an efficiency of 30 to 40%; the 1-μm output of this system could be harmonically converted to 0.5 μm or shorter wavelengths with an overall system efficiency of 10 to 20%. Such a system, occupying a package a few cubic centimeters in volume, could operate on 110-V electrical energy, could be air-cooled, and could deliver up to 20 W of 0.5-μm light. Scaled up, this technology can be expected to lead to megawatt fusion systems capable of operating at efficiencies greater than 10%.

5. Summary

Work begun by Schawlow at Bell Laboratories in the 1950s on the optical properties of solids contributed to the demonstration of the laser, to a remarkable series of scientific achievements in the 1960s, and to applications in the 1970s. In the 1980s, solid state technologies are beginning to provide powerful, efficient, multiwavelength lasers for scientific, energy, defense, industrial, and medical applications, and they are likely to become the technology of choice for visible-laser applications in the 1990s.

Acknowledgments

While I have had the privilege of writing this article, which relates our work on solid state lasers at Livermore to Schawlow's work at Stanford, my Livermore colleagues have in fact done most of the work. In particular I acknowledge J. L. Emmett, Schawlow's second graduate student, who came to Livermore in 1972 to direct the Inertial Fusion Program and who began the Laser Isotope Separation Program in 1974. W. F. Krupke and A. C. Haussmann worked closely with Emmett from the beginning to develop these programs to their present levels of success. It has been a pleasure working with these people, and with my other Livermore colleagues; I have inadequately referenced their contributions, which can be found in the Lawrence Livermore National Laboratory *Laser Program Annual Reports* for 1972 to the present. I thank P. W. Murphy for editing this article.

Work performed under the auspices of the U.S. Department of Energy by the Lawrence Livermore National Laboratory under Contract W-7405-Eng-48.

References

1. A. L. Schawlow and C. H. Townes, "Infrared and Optical Masers," *Phys. Rev.* **112** (6), 1940–1949 (1958).
2. T. H. Maiman, R. H. Hoskins, I. J. D'Haenens, C. K. Asawa, and V. Evtuhov, "Stimulated Optical Emission in Fluorescent Solids, II. Spectroscopy and Stimulated Emission in Ruby," *Phys. Rev.* **4** (123), 1151–1157 (1961).
3. J. F. Holzrichter, E. M. Campbell, J. D. Lindl, and E. Storm, "Research with High-Power Short-Wavelength Lasers," *Science* **229**, 4718 (1985).
4. D. L. Matthews, P. L. Hagelstein, M. D. Rosen, M. J. Eckart, N. M. Ceglio, A. U. Hazi, H. Medecki, B. J. MacGowan, J. E. Trebes, B. L. Whitten, E. M. Campbell, C. W. Hatcher, A. M. Hawryluk, R. L. Kauffman, L. D. Pleasance, G. Rambach, J. H. Scofield, G. Stone, and T. A. Weaver, *Phys. Rev. Lett.* **54**, 110 (1985); M. D. Rosen, P. L. Hagelstein, D. L. Matthews, E. M. Campbell, A. U. Hazi, B. L. Whitten, B. MacGowan, R. E. Turner, R. W. Lee, G. Charatis, Gar. E. Busch, C. L. Shepard, P. D. Rockett, and R. R. Johnson, *Phys. Rev. Lett.* **54**, 106 (1985). In recent work, x-ray mirrors with >50% normal-incidence reflectivity near 200 Å have been made [T. W. Barbee, Jr., *AIP Conf. Proc.* **119**, 311 (1984); E. S. Spiller, *AIP Conf. Proc.* **119**, 312 (1984)].
5. S. Colgate, R. E. Kidder, J. H. Nuckolls, R. F. Zabawski, and E. Teller, Lawrence Livermore National Laboratory, Livermore, CA, unpublished calculations (1961).
6. [1], p. 1949.
7. D. L. Matthews, Lawrence Livermore National Laboratory, Livermore, CA, private communication (1987).
8. N. Bloembergen, *Nonlinear Optics* (Benjamin, New York, 1965); Y. R. Shen, *The Principles of Nonlinear Optics* (Wiley, New York, 1984).
9. R. L. Carman, R. Y. Chiao, and P. L. Kelley, "Observation of Degenerate Stimulated Four-Photon Interaction and Four-Wave Parametric Amplification," *Phys. Rev. Lett.* **17**, 1281 (1966); Y. R. Shen, "Self-Focusing: Experimental," and J. H. Marburger, "Self-Focusing: Theory," in *Progress in Quantum Electronics*, ed. by J. H. Sanders and Stenholm, Vol. 4, Part 1 (Pergamon, New York, 1975).
10. P. A. Franken, A. E. Hill, C. W. Peters, and G. Weinreich, *Phys. Rev. Lett.* **7**, 118 (1961).
11. D. Eimerl, "Thin-Thick Quadrature Frequency Conversion," Lawrence Livermore National Laboratory, Livermore, CA, UCRL-92087 (1984): also in *Proceedings of the International Conference on Lasers '84* (November 26–30, 1984); D. Eimerl, "Quadrature Frequency Conversion," Lawrence Livermore National Laboratory, Livermore, CA, UCRL-95424 (1986): to be published in *IEEE J. Quantum Electron.*

12. A. Denzhofer, A. Laubereau, and W. Kaiser, "High Intensity Raman Interactions," *Prog. Quantum Electron.* **6**, 55–140 (1979).

13. T. S. Luk, H. Pummer, K. Boyer, M. Sahidi, H. Egger, and C. K. Rhodes, *Phys Rev. Lett.* **51**, 110 (1983).

14. N. Bloembergen, "Nonlinear Optics and Spectroscopy," *Science* **216**, 4550 (1982).

15. P. P. Sorokin and J. R. Lankard, *IBM J. Res. Dev.* **11**, 148 (1967).

16. B. B. Snavely, O. G. Peterson, and R. F. Reithel *Appl. Phys. Lett.* **111**, 275 (1967).

17. T. W. Hänsch, "High Resolution Laser Spectroscopy," in *Advances in Laser Spectroscopy*, F. T. Arecchi, F. Strumia, and H. Walther, Eds. (Plenum Press, New York, 1983), p. 127; T. W. Hänsch, "Sub-Doppler Spectroscopy," in *Atomic Physics 8*, L. Lindgren, A. Rosen, and S. Svanberg, Eds. (Plenum Press, New York, 1983), p. 55.

18. A. L. Schawlow, "Spectroscopy in a New Light," *Science* **217**, 4554 (1982).

19. W. B. Tiffany, H. W. Moos, and A. L. Schawlow, "Selective Laser Photocatalysis of Bromine Reactions, *Science* **157**(2784), 40–43 (1967); W. B. Tiffany, "Selective Photochemistry of Bromine using a Ruby Laser," *J. Chem. Phys.* **48**(7), 3019–3031 (1968).

20. *Laser Interaction and Related Plasma Phenomena*, H. J. Schwarz and H. Hora, Eds. (Plenum Press, New York, 1971–1984), vols. 1–6.

21. J. H. Nuckolls, L. L. Wood, A. R. Thiessen, and G. B. Zimmerman, *Nature (London)* **239**, 139 (1972); J. L. Emmett, J. H. Nuckolls, and L. L. Wood, "Fusion Power by Laser Implosion," *Sci Am.*, pp. 24–37 (June 1974). For detailed descriptions of laser-plasma investigations, see the annual reports of the following laboratories: CEA, Limeil, France; KMS Fusion, Inc., Ann Arbor, MI; Naval Research Laboratory, Washington, DC; Osaka University, Osaka, Japan; Sandia Laboratories, Albuquerque, NM; the Lebedev Physical Institute, Moscow, USSR; the Max Planck Institute for Plasma Physics, Garching, Germany; the Rutherford Laboratory, Didcot, England; Lawrence Livermore National Laboratory, Livermore, CA; Los Alamos National Laboratory, Los Alamos, NM; University of Rochester (Laboratory for Laser Energetics), Rochester, NY.

22. J. F. Holzrichter, D. Eimerl, E. V. George, J. B. Trenholme, W. W. Simmons, and J. T. Hunt, *J. Fusion Energy* **2**, 5 (1982). Available as Lawrence Livermore National Laboratory, Livermore, CA, UCRL-52868-1. See also *Laser Program Annual Report*, Lawrence Livermore National Laboratory, Livermore, CA, UCRL-50021-73 (1974) to UCRL-50021-84 (1985).

23. J. B. Trenholme, in [22], p.28.

24. J. L. Emmett, W. F. Krupke, and J. B. Trenholme, *The Future Development of High-Power Solid State Laser Systems* (*Physics of Laser Fusion*, vol. 4), Lawrence Livermore National Laboratory, Livermore, CA, UCRL-53344 (1982).

From Micromasers to Antimasers: When One Photon in a Cavity May Be One Too Many...

S. Haroche

Ecole Normale Supérieure, Paris, France and
Yale University, New Haven, CT, USA

The direction of physics is sometimes unpredictable. When Arthur Schawlow and the other pioneers of lasers were trying to excite as many atoms as possible in an optical cavity to get the first lasers going, they would hardly have guessed that some twenty years later an active domain of quantum optics would be concerned with maser or laser systems in which the cavity is empty – or contains only one atom at a given time, or with cavities unable to sustain a single photon emitted by the atomic medium... This domain of quantum optics is now called "Cavity Quantum Electro-dynamics" [1, 2]. Although theory in this field can be traced back to articles [3] and short notes [4] published a long time ago, the development of cavity Q.E.D. experiments has recently come as a natural extension of the research on Rydberg atom radiative properties [1, 2, 5, 6]. In this paper, written as a token of my friendship with and admiration for Art Schawlow, I review some aspects of this research which I have been interested in over the last seven or eight years at Ecole Normale Supérieure in Paris, as well as more recently at Yale University.

1. Rydberg Atom Transient Masers : A Case Study in Superradiance

A sample of Rydberg atoms – i.e. of atoms excited in an energy level close to the atomic ionization limit – is an ideal active medium for a laser or a maser : there are a very large number of inverted transitions down to more bound states. The transition matrix elements towards nearby levels are huge (they scale as the size of the atom, i.e., as n^2 when n is the principal quantum number of the Rydberg state). Moreover, the relaxation times of these atoms are quite long (in the millisecond to microsecond range for n ~ 30, if the atoms are not too perturbed by external fields). As a result, superradiant emission without mirrors occurs with Rydberg samples containing typically about 10^6 atoms [7]. If the sample is prepared in a cavity with a moderate quality factor Q (in the range 10^3 to 10^4) a few hundred to a few thousand atoms are enough to reach the emission threshold [7,8]. This emission usually occurs at microwave frequencies (centimeter or millimeter wavelengths) corresponding to transitions between nearby levels in the Rydberg spectrum.

In fact, we observed our first Rydberg masers somewhat by chance, when performing microwave spectroscopy in very excited states of sodium. The setup we were using is sketched in Fig. 1a : an atomic beam of Na was excited by a short laser pulse into a Rydberg state ns or nd (n ~ 20 to 40) inside a Fabry-Perot cavity into which we fed microwaves through a waveguide. After crossing the cavity, the atoms were analyzed with a Rydberg state selector (Fig. 1b) : a ramp of electric field was applied between two condenser plates. This field reached the ionization threshold for the two levels implied in the Rydberg-Rydberg transition at slightly different times (the lower more bound state requiring a somewhat larger ionizing field than the upper one). This resulted in two time resolved electron peaks detected by an electron multiplier. The resonance appeared as a change in the relative weights of the peaks associated with the initial and final states of the transition as the microwave frequency was scanned. We soon noticed that resonances could in fact be observed without applying any microwaves to the cavity. When the mirror distance L was tuned to an integer number of half-wavelengths of an atomic transition between the initially prepared Rydberg level and a lower state, we observed a strong and fast transfer of population to this latter state, demonstrating a transient maser or laser action in the system. In this way, we have observed hundreds of Rydberg transitions in Na, Cs [8] and more recently in lithium, corresponding to $\Delta n = 1, 2, 3$ transitions. Cascading transitions sharing a common level and masing successively or in competition have also been observed [8]. The thresholds - even with moderate Q cavities - were small (N ~ 10^2 to 10^4) and it was in fact difficult to avoid this maser action when it was a nuisance in other kinds of experiments. The first direct use of this effect has been to perform cheap and convenient Rydberg atom spectroscopy without a microwave source. The Rydberg-Rydberg frequency was merely determined by the lengths of the cavity corresponding to the maser action between the levels, a variant of the "spectroscopy with a ruler", type of experiment often advocated by Art Schawlow.

We then realized that these transient maser systems were ideal to test the theory of superradiance [9,10], first developed by Dicke [11],

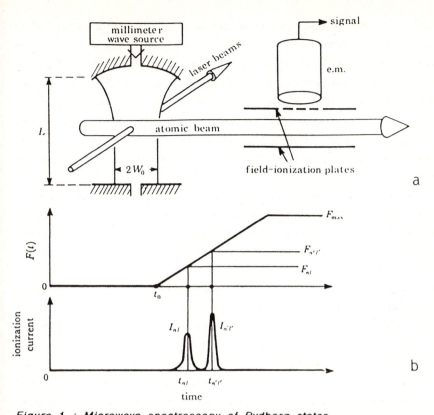

Figure 1 : Microwave spectroscopy of Rydberg states.
a) Sketch of setup showing atomic beam excited by laser pulses inside the Fabry–Perot millimeter wave cavity. Rydberg atoms interact with the microwave during their propagation through the cavity waist (w_0), then drift into the detection zone (field ionization plates). b) Time evolution of electric field produced by the plates and of resulting ionization current. The delay of the field ramp is set so that atoms excited at t=0 in the cavity have reached the detector at time t_0. Times $t_{n\iota}$ and $t_{n'\iota'}$ correspond respectively to the ionization threshold fields $F_{n\iota}$ and $F_{n'\iota'}$ of the Rydberg levels $n\iota$ and $n'\iota'$ involved in the transition. If the upper level $n\iota$ is initially prepared, resonant radiative transfer to level $n'\iota'$ results in a change of the ionization peak shape from $I_{n\iota}$ to $I_{n'\iota'}$. Maser action is easily observed without any external microwave when cavity length L is tuned into resonance.

which, until the development of Rydberg atom experiments, had been very difficult to check experimentally in detail. The ideal textbook superradiance situation deals with two-level atoms symmetrically coupled to a damped mode of the electromagnetic field. The symmetrical coupling entails a fast radiative deexcitation of the atomic system, on a time scale inversely proportional to the number N of atoms involved in the process. This situation is quite easily achieved with a sample of Rydberg atoms in a microwave cavity : only the two levels connected by the transition resonant with the cavity are relevant during the time of the experiment since the rate of emission to other states is usually very small; the symmetric coupling to the field is realized by preparing the atoms at an antinode position in the cavity (a millimeter size sample fulfils this condition). The mode structure of the field in the cavity is very simple and usually one mode only is coupled to the atoms.

In a series of experiments performed in 1982–1984, we made an extensive check of the Dicke theory. We studied the evolution of the average atomic energy during a superradiant pulse and the fluctuations around this average [8, 12]. We showed also that the collective behavior of the atomic system was manifest not only in emission processes, but also in absorption and we demonstrated an effect of collective absorption of thermal radiation by a sample of Rydberg atoms in a resonant cavity [13].

A simple way to analyze these experiments consists in describing the symmetrical ensemble of N two-level atoms as an angular momentum J=N/2 evolving in an abstract space [1, 8] (Bloch vector). This vector evolution obeys a pendulum-like equation, the radiation reaction of the atoms being analogous to the pendulum gravity restoring couple and the cavity damping playing the role of a viscous drag force. In this model, the potential and the kinetic energies of the pendulum are respectively associated to the atomic and field energies in the cavity and the vacuum fluctuations and thermal field noise correspond to Langevin-type forces responsible for the Brownian motion of the pendulum. In other words, transient Rydberg maser experiments can be viewed as exploring the dynamics of a pendulum starting either from its unstable equilibrium position (emission) or from its stable one (absorption) and being triggered away from this position by Brownian forces. The Dicke superradiant regime corresponds in this point of view to the overdamped evolution of the pendulum starting from its unstable position (pendulum falling with a strong viscous drag force). If the cavity damping is small enough, the pendulum evolution becomes oscillatory, corresponding to a quasi-reversible exchange of energy between the atom and the field with a pseudofrequency proportional to the square root of the atom number in the cavity. This regime too had been predicted long ago in superradiance theories [14], but had never been clearly observed before in quantum

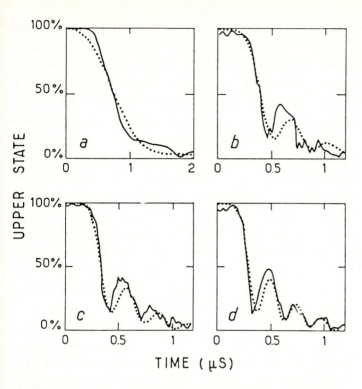

Figure 2 : *Time evolution of a transient Rydberg atom maser for increasing atom number N in the active medium (from [15]). The relative population of the upper Rydberg state, averaged over a few hundred pulses, is recorded as a function of the interaction time between the atoms and the cavity (t=0 corresponds to the initial preparation of the system). Traces a, b, c, d correspond respectively to N = 2000, 12000, 27000, 40000 atoms. The regime is overdamped in trace a. Traces b, c, d clearly exhibit the pendulum-like behavior of the system.*

optics. We were able to study it in detail with our Rydberg atom masers [15]. Figure 2 shows recordings of the mean atomic energy evolution, corresponding to increasing atom number in the sample. The pendulum-like oscillations of the system are clearly demonstrated in this experiment.

2. Microscopic Masers : One Rydberg Atom in the Cavity

In all the experiments mentioned above, N was in the range 10^3–10^5. A natural extension of these studies was to try to decrease the atom number in the resonator while increasing the cavity Q. The goal was eventually to achieve a Rydberg maser operating with a single atom in the cavity. This concept deserves some discussion since masers or lasers are usually supposed to operate as collective systems. In fact, a single atom maser starts as an atom-cavity coupled system in which the atom is induced to radiate a photon much faster than in free space. The cavity around the atom modifies the mode density of the electromagnetic field and, if it is resonant with the atomic transition, increases the density of photon final states open for the atomic decay, so that the spontaneous emission rate is enhanced [1,2]. Alternatively, one can say that the atom interacts collectively with its electric images in the cavity walls and radiates in this way faster than in free space. Here again, theorists had predicted this effect long ago, but Rydberg atoms gave the first experimental opportunity to observe it directly. Replacing our ordinary copper cavity mirrors by niobium superconducting ones, we were able to increase the Q value to about 10^6 and to observe in 1983 the enhancement of the spontaneous emission rate of a single Rydberg atom in the cavity [16]. The atomic flux was reduced so that most of the time only one Rydberg atom per laser pulse crossed the cavity. This atom was transferred to the lower state of the transition during the cavity crossing time when the cavity was tuned to resonance whereas it stayed excited in the initial level if the cavity was off-resonant (see Fig. 3). This experiment was the one-atom limit of the overdamped regime of a

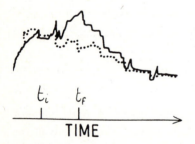

t_i t_f

TIME

Figure 3 : *Time resolved ionization signal demonstrating the one-atom transient maser action. The signal is the ionization current recorded during the ramp of ionizing field following each laser pulse (see Fig.1b). The signal is averaged over a few hundred laser pulses. Times t_i and t_f correspond respectively to the ionization thresholds of the upper level i and lower level f of the transition ($23S_{1/2}$ and $22P_{1/2}$ states of Na). The laser intensity is reduced so that for most pulses, one or two atoms are excited in the cavity. The dotted and full line curves correspond respectively to off and on-resonance cavity. The increase of signal around t_f in the resonant case demonstrates the spontaneous emission enhancement effect (from [16]).*

transient Rydberg maser : the emitted photon was rapidly damped in the cavity walls.

In order to observe the one-atom maser oscillating regime – analogous to the oscillatory behavior mentioned above in the collective case – it was necessary to increase the cavity Q by at least two orders of magnitude. The emitted photon could then be stored long enough for the atom to be able to reabsorb it at a later time giving rise to a quasi reversible single photon exchange between the atom and the cavity mode. We described this ideal situation in several publications [5,8,16] and went on developing cavities for its experimental realization. But the Fabry-Perot resonators with open structures we were using at that time were mechanically too unstable to achieve the required high Q's. The solution to the cavity problem has been found by Walther, Meschede and Muller in Munich who have developed closed and rigid superconducting cylindrical microwave cavities operating in a low order with a Q in the 10^8–10^9 range. Using these cavities, in 1984 they observed for the first time a single atom maser with a long photon storing time [17]. In their experiment, the atoms (rubidium in 63P state) were prepared by continuous wave laser excitation, with a flux of the order of 10^3–10^4 atom per second, so that successive radiators entered the cavity before the photons radiated by preceding atoms had disappeared. In steady state operation, this system thus realizes a maser with less than one atom on average in the medium and only a few photons in the cavity. Although it has not yet been observed, the one atom-one photon exchange regime mentioned above is now within reach of experimental investigation.

Rydberg atom micromasers are now a class of quantum optics systems on their own. The statistical properties of the electromagnetic field radiated by these systems are quite interesting. Non-classical field statistics, squeezing, generation of pure Fock states corresponding to a well defined number of photons in the cavity have been predicted theoretically [18,19] and are now being checked in difficult experiments involving very high Q cavities at very low temperatures and well controlled atomic beams (velocity selection of atoms is very important in these experiments) [20].

3. Two-Photon Rydberg Atom Micromasers

The combination of Rydberg atom and high Q cavities also offers ideal opportunities to realize other kinds of new quantum optics devices. At Ecole Normale Supérieure, we are presently trying to build a Rydberg atom micromaser operating in a cw regime on a two-photon transition. Two photon lasers or masers have been for a long time the most elusive systems in quantum optics. Although the possibility of amplifying the electromagnetic field on a two-photon degenerate or non-degenerate transition between two atomic levels of the same parity was pointed out in the early days of lasers [21,22] and considered since in a large body of theoretical papers [23], these devices have never been demonstrated so far in cw operation (one report does exist however of two photon amplification in pulsed regime [24]). On ordinary transitions, the two photon gain is indeed very low. Furthermore, a two photon laser generally requires a triggering field in order to build up an oscillation large enough so that the two-photon quadratic gain overcomes the linear losses of the cavity. The triggering field is likely to produce unwanted competing effects (Raman processes, multiple wave mixing), depleting the pumped level before two-photon amplification can occur. All these difficulties vanish with Rydberg atoms in super-conducting cavities [25]. Very high gain is achievable due to the large intrinsic atomic dipoles of Rydberg levels and most importantly to the existence of near coincidences in the Rydberg spectrum with a relay level almost midway between the initial and final state of the two-photon transition (see Fig. 4). Under optimum conditions (40S → 39S transition in ^{85}Rb

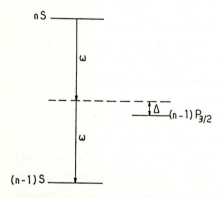

Figure 4 : *Energy diagram relevant to the two-photon Rydberg atom maser. The cavity is tuned at frequency ω, corresponding to a two-photon degenerate transition between the $nS_{1/2}$ and $(n-1)S_{1/2}$ levels. The $(n-1)P_{3/2}$ level is very close to the middle of the energy interval, which considerably enhances the two-photon amplitude. The most favorable case is n = 40 in ^{85}rubidium ($\Delta \simeq 39MHz$ only). The two photon maser action will be demonstrated by recording the population transfer from $nS_{1/2}$ to $(n-1)S_{1/2}$, while the $(n-1)P_{3/2}$ population should remain zero.*

atoms) with a niobium superconducting cavity having a Q ~ 2.10^8 at ν = 68 GHz, we expect that the Rydberg maser should operate on the two photon transition with only about one atom at a time and a few tens of microwave photons in the cavity. Excitation of the atoms in the 40s state of rubidium is achieved by a four step cw process ($5S_{1/2} \to 5P_{3/2} \to 5D_{5/2} \to 40P_{3/2}$ transitions are induced by three laser diodes at λ_1 = 7802Å, λ_2 = 7759Å and λ_3 = 1.26 μm respectively and the final $40P_{3/2} \to 40S_{1/2}$ transition by a microwave source at 62 GHz). We have achieved in this way Rydberg atom fluxes in the range 10^6-10^{7}atoms/s, corresponding to tens of atoms in the cavity at a time. The fine cavity tuning will be achieved, as in [17], by mechanical pressure on the cavity walls and the maser action will be detected by monitoring the atomic populations in the 40S and 39S states in atoms emerging from the cavity.

We have performed a complete theoretical analysis of these systems [26]. Here again, the microscopic character of the device leads to interesting new effects to be studied. We have shown that the distribution of the photon number n_φ in this maser diffuses according to a Fokker-Planck equation, with an effective potential $V(n_\varphi)$ presenting typically several minima. Multistable operation and quantum diffusion between these minima are predicted and it will be interesting to investigate these experimentally. The passage time from one minimum to a neighboring one is shown to increase very quickly with the atom and photon numbers in the cavity, which means that such effects would be unobservable with "classical" macroscopic masers containing very large photon numbers. Even the initiation of the maser action is predicted to be different in the two-photon micromaser. In fact, spontaneous noise can be large enough to trigger the passage of the system in a finite time from the equilibrium around n_φ = 0 to stable operation with n_φ > 0, making external triggering unnecessary.

It is possible that the study of these new types of quantum systems will lead to situations analogous to the ones now being explored in the experiments monitoring single atoms or ions : the passage from one equilibrium state to another is in particular reminiscent of quantum jump behavior.

4. "Antimaser or Antilasers" : Inhibition of Spontaneous Emission in a Cavity

As discussed above, the basic process in the dynamics of a microscopic maser is the coupling of a single atom to a mode of the electromagnetic field sustained by a cavity. In this respect, micromaser physics is directly related to cavity Q.E.D., the domain of Q.E.D. studying how the cavity boundary conditions modify the spontaneous atomic radiative processes. In a micromaser, the cavity is resonant with an atomic transition, which corresponds to an enhancement of the field mode density at atomic frequency with respect to its free space value and to an acceleration of one-photon as well as two-photon spontaneous emission processes. The opposite effect – inhibition of spontaneous emission in an off-resonant cavity – has been predicted [27] and also observed in Rydberg atom physics : if the atom is excited in a cavity below cutoff for the atomic transition, there is no mode available for the photon decay and the atom should remain excited much longer than in free space.

The simplest cavity geometry corresponding to this situation is the biplanar configuration with two plane parallel metallic surfaces separated by a gap d. Figure 5a shows the mode density $\rho(\omega)$ versus frequency ω in this structure, for modes which at the midgap position $z = d/2$ are polarized either parallel (σ) or perpendicular (π) to the surfaces ($\rho_\sigma^{(cav)}$ and $\rho_\pi^{(cav)}$ respectively). The dotted curve indicates for comparison the free-space mode density $\rho_0(\omega) \sim (\omega^2/3\pi^2c^3)$. The striking feature of these distributions is the complete cutoff of field modes for $\omega < \omega_0 = (\pi c/d)$ in σ-

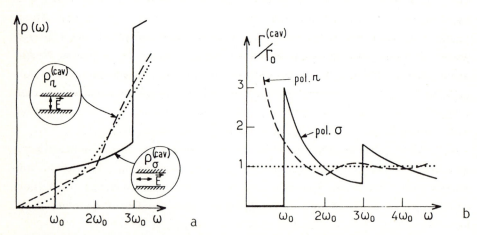

Figure 5 : a) Mode density $\rho(\omega)$ versus ω in a cavity made of two plane parallel mirrors separated by a gap d. Density evaluated at midgap position $z = d/2$. Full line : σ-polarization. Dashed line : π-polarization. For comparison, the free space mode density is represented by the dotted line. b) Ratio $\Gamma^{(cav)}/\Gamma_0$ versus frequency ω. Full line : spontaneous emission in polarization σ; dashed line : spontaneous emission in polarization π.

polarization whereas modes exist down to $\omega = 0$ in π-polarization (corresponding to the electrostatic limit with a static electric field perpendicular to the condenser plates). The spontaneous emission rate being proportional to the mode density, the rate in the cavity $\Gamma^{(cav)}$ is merely equal to the rate in free space Γ_0, multiplied by the ratio $\rho^{(cav)}/\rho_0$ (see Fig. 5b). In order to inhibit the atomic emission in such a structure, it is thus necessary to prepare an excited state corresponding to an electric dipole parallel to the mirrors and to have a plate separation $d < \lambda/2$. These conditions have been achieved by Hulet, Hilfer and Kleppner [28] who have sent Rydberg atoms in high angular momentum "circular" states between two aluminum mirrors separated by a gap $d = 0.23$ mm. The circular Rydberg states correspond to a toroïdal electronic orbital lying in a plane perpendicular to the atomic angular momentum. These states are prepared by feeding angular momentum into the Rydberg atoms by absorption of a large number of circularly polarized microwave photons [29]. The atoms are subjected to an electric field along the direction of this angular momentum and the microwave photons induce transitions between the Stark sublevels in this electric field. By aligning this directing field along the normal to the mirror, the Rydberg atoms are thus prepared with their orbit parallel to the conductor plates and the electric dipole corresponding to transitions to the adjacent lower circular state is parallel to the mirrors. The inhibition of spontaneous emission was observed by monitoring the flux of excited atoms having survived spontaneous decay in the structure, after a crossing time of the order of the natural lifetime of the transition (~ 1 ms). Evidence for a large suppression of spontaneous emission has been obtained in this experiment. This experiment had been in fact preceded by a demonstration of radiative decay inhibition in the cyclotron resonance of a free electron in a Penning trap [30]. The cavity was made by the electrodes of the trap. Due to the much more complex geometry of this cavity, the interpretation of this experiment and the quantitative check of cavity Q.E.D. theory were more delicate in this case.

In all these microwave experiments, the spontaneous processes to be inhibited were quite slow (natural spontaneous emission rates Γ_0 in the 10 s^{-1} to 10^3 s^{-1} range). We have very recently shown at Yale [31] that it is possible to extend these experiments in the optical domain, where spontaneous emission processes are intrinsically much stronger ($\Gamma_0 \sim 10^6$ s^{-1}). The atoms (which are now in strongly bound excited states and no longer in Rydberg levels) must be sent in micrometer-sized cavities. We have realized such microscopic cavities [32] by pressing two $\lambda/10$ flat mirrors against each other with 1μm-thick metal foils used as spacers between them. We were able in this way to build mirror tunnels with a gap $d\sim 1\mu$ and a length $\iota=8$mm, subtending an angle of only $\sim 10^{-4}$ rad.

The relevant energy levels for this experiment are shown in Figs. 6a and 6b and the experimental setup is sketched in Fig. 6c. Exciting the Cs atoms in the $5D_{5/2}$ level and sending these atoms across the mirror tunnel, we were able to observe the inhibition of the $5D_{5/2} \rightarrow 6P_{3/2}$ transition at $\lambda=3.49\mu$m. The natural lifetime of this transition – which has a branching ratio 1 – is $\tau=1/\Gamma_0=1.6$ μs. One laser system (laser 1) was used to prepare the atoms in the $5D_{5/2}$ level before they entered the cavity : the atoms were excited into the $7P_{3/2}$ state ($6S_{1/2} \rightarrow 7P_{3/2}$ transition at $\lambda_1=0.4555$ μm) from which $\sim 10\%$ cascade down very rapidly (~ 150 ns) into the $5D_{5/2}$ state. This state has a hyperfine structure (F=1 to 6 where F is the total system angular momentum). Only levels F=4, 5 and 6 are prepared by our excitation scheme (see Fig. 6b). A second laser system (laser 2) was used to detect the atoms remaining in the $5D_{5/2}$ level after crossing the cavity (in a time $\sim 20\mu$s). This laser was tuned across the transition from the $5D_{5/2}$ level into the 26f Rydberg state ($\lambda_2 \sim 6010$ $\overset{\bullet}{A}$). The Rydberg atoms were subsequently field-ionized. Counting the resulting electrons as a function of the frequency ν_2 of laser 2 provided a direct measure of the number of atoms remaining in the F=4, 5, 6 hyperfine levels of the $5D_{5/2}$ state after the cavity crossing (these numbers being proportional to the intensity of lines # 1, 2 and 3, respectively, in the energy diagram of Fig. 6b). We observed a large detection signal corresponding to peak # 3, providing a clear demonstration of spontaneous emission inhibition for atoms in the $5D_{5/2}$ F=6 state. The detected atoms had survived the gap crossing without decaying, remaining excited during about 13 natural lifetimes... The probability of such a long survival in free space would be quite negligible. The absence of peaks # 1 and 2 in the detection signal is easy to understand [31]. Among the atoms in the $5D_{5/2}$ level, only those which were in substates with electric dipoles radiating parallel to the surfaces had their lifetime lengthened to the extent that they could survive the cavity crossing. These substates correspond to the maximum possible value of the angular momentum component along the normal to the mirrors (F=6, $M_F=\pm 6$ states). A mere inspection of Fig. 6b shows that only these substates radiate by pure σ-emission towards the $M_{F'}=\pm 5$ substates of the $6P_{3/2}$ F'=5 final level. All the other levels – and in

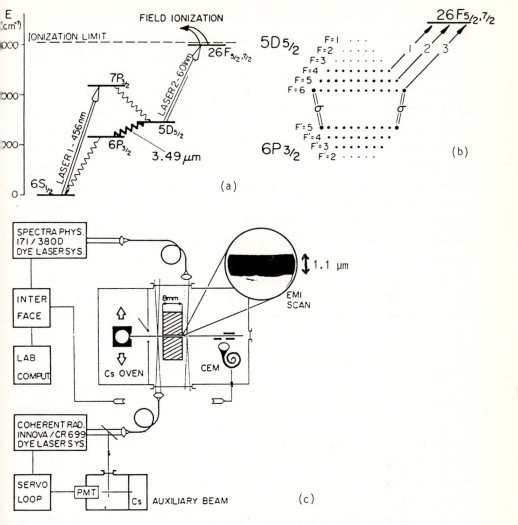

Figure 6 : a) Cesium energy level relevant to the optical spontaneous emission inhibition experiment. b) Close-up showing the hyperfine structure of the $5D_{5/2}$ and $6P_{3/2}$ states. The extreme $F = 6$, $M_F = \pm 6$ states of the $5D_{5/2}$ level are the only ones to radiate by pure σ-emission (to the $F' = 5$, $M_{F'} = \pm 5$ substates of the $6P_{3/2}$ level). Hence, only the hyperfine component # 3 remains in the absorption from level $5D_{5/2}$ after the atoms have crossed the cavity. c) Sketch of experimental setup. Laser 1 (at bottom) is locked on the $6S_{1/2} \rightarrow 7P_{3/2}$ resonance line of an auxiliary cesium beam. Laser 2 (at top) is scanned across the $5D_{5/2} \rightarrow 26F$ absorption line. Laser 1 excites the atoms upstream the mirror cavity; laser 2 detects them downstream. The Rydberg atoms are detected by field ionization and counting the resulting electrons with the channel electron multiplier (CEM). The insert shows an electron microscope picture of the mirror gap exit.

particular the F=5 and F=4 ones – can also decay via π–polarized ΔM=0 channels which are not cut–off in the cavity (see Fig. 5). Hence, only peak # 3 appears in the absorption spectrum of the excited atoms after the cavity crossing. This provides evidence of the anisotropy of the spontaneous emission inhibition effect in the cavity. We have also demonstrated this anisotropy in a very dramatic way by studying the magnetic field dependence of the excited atom transmission signal. By applying a magnetic field with a component parallel to the mirrors, we were able to mix the M_F = ±6 long lived states with other $|M_F|$ ≤ 5 states able to radiate π–polarized photons in the cavity. This mixing thus resulted in a quenching of the excited state in the mirror structure and in a decrease of the excited state transmission. Figure 7 shows the result of this experiment : the excited atomic transmission through the tunnel is plotted as a function of the angle θ between the direction of an applied magnetic field $\vec{B}$ and the normal to the mirrors : the points are experimental and the full line curve is theoretical. The fair agreement between them demonstrates that the cavity Q. E. D. processes are well understood in this case. This experiment clearly shows

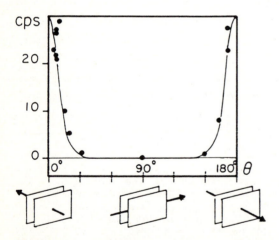

Figure 7 : *Spontaneous emission inhibition experiment at optical frequency : Excited state transmission through the tunnel versus the angle θ between an applied magnetic field and the normal to the mirrors. For θ = 0 and θ = 180°, a large transmission is observed, demonstrating the spontaneous emission inhibition effect for F = 6, M_F = ±6 states. When θ is different from zero or 180°, the transverse field components mix the long lived M_F = ±6 states with shorter lived $|M_F|$ ≤ 5 ones and the excited states are quenched before emerging from the cavity. The points are experimental and the solid line curve is given by theory.*

that it is possible to "manipulate" the vacuum fluctuations in a confined space and to radically alter the spontaneous emission processes induced by these fluctuations. The vacuum field, obviously isotropic in free space, becomes anisotropic in the cavity and this leads to an anisotropy of the spontaneous emission properties of the excited atomic state in this structure. It is tempting to call "antimaser" or "antilaser" such systems made of excited atoms crossing a cavity in which they are forbidden to emit a single photon... The reduction of the quantum noise in these devices bears some similitude with the effect known as "squeezing" [33]. In cavity Q.E.D., as in squeezing, one is able to decrease the effect on some carefully chosen physical observables of the vacuum field fluctuations.

The above discussion immediately raises the question of the possible use of this spontaneous emission inhibition effect for improving the ultimate resolution of spectroscopic experiments : since one can greatly enhance the natural lifetime of an excited atomic state, can one reduce in the same proportion the natural linewidth of a spectral line originating from this state ? The answer to this question is unfortunately not simple. In fact, the effect of the cavity is not only to modify the excited levels lifetime, but also to shift the atomic energies through the atom–metal surface Van der Waals interaction [34,35]. This latter effect cannot be separated from the lifetime enhancement phenomenon. The change of the radiative lifetime in the cavity is due to the atom coupling with the modes resonant with the atomic transition (dissipative atom–field interaction). The atomic system also interacts with the non-resonant modes (dispersive atom–field coupling). This coupling can be understood in terms of virtual photon exchanges and is responsible for Lamb–shift type of effects... The presence of the cavity changes these processes as it changes the resonant ones (virtual photons, as real ones, can only be emitted in modes compatible with the cavity geometry). The cavity-induced change in the atomic energy levels depends upon the atomic position in the cavity. It results in inhomogeneous level shifts which could be measured in spectroscopy experiments involving the atoms in the cavity. Furthermore, the derivative of these shifts with respect to the atomic position corresponds to the Van der Waals force pulling the atoms to the mirrors. In our 8mm long cavity, about 80% of the Cs atoms are deflected onto the walls by the force and the spontaneous emission inhibition effect is observed on the ~ 20% escaping collision with the mirrors. The lifetime enhancement ratio obtained in our experiment (~ 13) is thus a limit difficult to improve without loosing all the atoms to the walls of the cavity. The above discussion shows that while suppressing the atom coupling to the field resonant modes, we have in fact strongly coupled it to the cavity via the dispersive part of the atom-field interaction. Thus, if one tried to take advantage of the spontaneous emission inhibition to improve spectroscopic resolution beyond the natural width, one would be very quickly limited by the existence of inhomogeneous Van der Waals shifts. We would thus have replaced the "atom + vacuum field" system by the "atom + cavity" one. The experimental study of atomic energy level shifts in a cavity is certainly another interesting field of cavity Q.E.D. that we intend to study in the future.

The experiments in which I have been involved and which are reviewed here have been carried out in collaboration with colleagues and students in two research groups. In Paris, I have worked over a long period of time with C. Fabre, Philippe Goy, M. Gross and J.M. Raimond along with several graduate students (G. Vitrant, Y. Kaluzny, J. Liang, M. Brune, J. Hare) and foreign visitors (L. Moi). In Yale, my coworkers are E. Hinds, D. Meschede and again L. Moi with two graduate students, A. Anderson and W. Jhe... I was also very fortunate to be associated with Art Schawlow and his group at an early stage of my career. Many of the ideas and techniques I have developed later in the research reviewed in this paper owe much to the experience I gained in his laboratory. I am very glad to have the opportunity to acknowledge it in this article.

References

[1]   S. Haroche, J.M. Raimond: Adv. At. Mol. Phys. $\underline{20}$, 347 (1985)

[2]   S. Haroche: in Proceedings of Symposium Alfred Kastler, Ann. Phys. (Paris) $\underline{10}$, 811 (1985); P. Dobiasch, H. Walther: ibid. $\underline{10}$, 825 (1985)

[3]   H.B. Casimir, D. Polder: Phys. Rev. $\underline{73}$, 360 (1948)

[4]   E. Purcell: Phys. Rev. $\underline{69}$, 681 (1946)

[5]   S. Haroche, P. Goy, J.M. Raimond, C. Fabre, M. Gross: Philos. Trans. R. Soc. London, Ser. A $\underline{307}$, 659 (1982); S. Haroche: In New Trends in Atomic Physics, Les Houches Summer School, Session 38, ed. by G. Grynberg, R. Stora (North-Holland, Amsterdam 1984)

[6]   T. F. Gallagher: In Rydberg States of Atoms and Molecules, ed. by
         R. F. Stebbings, F. B. Dunning (Cambridge University Press, New
         York 1983); J. A. C. Gallas, G. Leuchs, H. Walther, H. Figger:
         Adv. At. Mol. Phys. $\underline{20}$, 413 (1985)

[7]   M. Gross, P. Goy, C. Fabre, S. Haroche, J. M. Raimond: Phys.
         Rev. Lett. $\underline{43}$, 343 (1979)

[8]   L. Moi, P. Goy, M. Gross, J. M. Raimond, C. Fabre, S. Haroche:
         Phys. Rev. A $\underline{27}$, 2043 (1983)

[9]   R. Bonifacio, R. Schwendimann, F. Haake: Phys. Rev. A $\underline{4}$, 302,
         854 (1971); F. Haake, H. King, G. Schröder, J. Haus, R. J.
         Glauber: Phys. Rev. A $\underline{20}$, 2047 (1979); F. Haake, J. Haus,
         R. J. Glauber: Phys. Rev. A $\underline{23}$, 3255 (1981)

[10]   M. Gross, S. Haroche: Phys. Rep. $\underline{93}$, 302 (1982)

[11]   R. H. Dicke: Phys. Rev. $\underline{93}$, 99 (1954)

[12]   J. M. Raimond, P. Goy, M. Gross, C. Fabre, S. Haroche: Phys.
         Rev. Lett. $\underline{49}$, 1924 (1982)

[13]   J. M. Raimond, P. Goy, M. Gross, C. Fabre, S. Haroche: Phys.
         Rev. Lett. $\underline{49}$, 117 (1982)

[14]   R. Bonifacio, L. A. Lugiato: Phys. Rev. A $\underline{11}$, 1507 (1975); ibid.
         $\underline{12}$, 587 (1975)

[15]   Y. Kaluzny, P. Goy, M. Gross, J. M. Raimond, S. Haroche: Phys.
         Rev. Lett. $\underline{51}$, 1175 (1983)

[16]   P. Goy, J. M. Raimond, M. Gross, S. Haroche: Phys. Rev. Lett.
         $\underline{50}$, 1903 (1983)

[17]   D. Meschede, H. Walther, G. Müller: Phys. Rev. Lett. $\underline{54}$, 551
         (1985)

[18]   P. Filipowicz, J. Javanainen, P. Meystre: Phys. Rev. A $\underline{34}$, 3077
         (1986)

[19]   P. Filipowicz, J. Javanainen, P. Meystre: Optics Commun. $\underline{58}$, 327
         (1986)

[20]   G. Rempe, H. Walther, N. Klein: Phys. Rev. Lett. $\underline{58}$,
         353 (1987)

[21]   P. P. Sorokin, N. Braslau: IBM J. Res. Dev. $\underline{8}$, 177 (1964)

[22]   A. M. Prokhorov: Science $\underline{149}$, 828 (1965); V. S. Letokhov: Pis'ma
         Zh. Eksp. Teor. Fiz. $\underline{7}$, 284 (1968) [JETP Lett. $\underline{7}$, 221 (1968)]

[23]   L. M. Narducci, W. W. Eidson, P. Furcinetti, D. C. Eteson: Phys. Rev. A 16, 1665 (1977); R. L. Carman: Phys. Rev. A 12, 2048 (1975); H. P. Yuen: Phys. Rev. A 13, 226 (1976); N. Nayak, B. K. Mohanty: Phys. Rev. A 19, 1204 (1979)

[24]   B. Nikolaus, D. Z. Zhang, P. E. Toscheck: Phys. Rev. Lett. 47, 171 (1981)

[25]   M. Brune, J. M. Raimond, S. Haroche: Phys. Rev. A 35, 154 (1987)

[26]   L. Davidovich, J. M. Raimond, M. Brune , S. Haroche: to be published

[27]   D. Kleppner: Phys. Rev. Lett. 47, 233 (1981)

[28]   R. G. Hulet, E. S. Hilfer, D. Kleppner: Phys. Rev. Lett. 55, 2137 (1985)

[29]   R. G. Hulet, D. Kleppner: Phys. Rev. Lett. 51, 1430 (1983)

[30]   G. Gabrielse, H. Dehmelt: Phys. Rev. Lett. 55, 67 (1985)

[31]   W. Jhe, A. Anderson, E. A. Hinds, D. Meschede, L. Moi, S. Haroche: Phys. Rev. Lett. 58, 666 (1987)

[32]   A. Anderson, S. Haroche, E. A. Hinds, W. Jhe, D. Meschede, L. Moi: Phys. Rev. A 34, 3513 (1986)

[33]   R. E. Slusher, L. W. Hollberg, B. Yurke, J. C. Mertz, J. F. Valley: Phys. Rev. Lett. 55, 2409 (1985); L. A. Wu, H. J. Kimble, J. L. Hall, H. Wu: Phys. Rev. Lett. 57, 2520 (1986)

[34]   J. E. Lennard-Jones, Trans. Faraday Soc. 28, 336 (1932)

[35]   C. Lütken, M. Raindal, Phys. Rev. A 31, 2082 (1985); Phys. Scr. 28, 209 (1983)

Laser Glass: An Engineered Material

S.E. Stokowski

Lawrence Livermore National Laboratory, Livermore, CA 94550, USA

Foreword

In 1958 SCHAWLOW and TOWNES [1] proposed that an optical maser could be made with the right combination of spectroscopic properties of an excited atom, ion, or molecule in a gas or solid, an optical pumping source, and a resonant cavity. The key to their proposal was their choice of a Fabry-Perot cavity as the resonator. This proposal stimulated several efforts to make an optical maser. In 1960 MAIMAN [2] demonstrated coherent emission from a ruby crystal, and the laser era had begun.

In the first years of its existence the laser was a curiosity; many people said that it was an interesting device looking for an application. Looking back from the perspective of a quarter century it is easy to deride such a comment. However, that is how a completely new and unexpected invention is usually received. Until people have a chance to store the unique characteristics of a new invention in their memories, they have difficulty in thinking of applications.

But applications for the laser have come in droves, ranging from the commonplace, such as supermarket scanners and surveying instruments, to the unusual, such as nuclear fusion drivers; from medical and surgical instruments for healing to weapons, such as missile designators and destroyers; from communications to compact disk players; from toys to scientific instruments used on the frontiers of science. We now take the laser for granted. In a little more than a quarter century the result of Schawlow's and Townes's proposal has made a substantial impact on our lives. Future laser applications are likely to make present ones pale in comparison.

Spectroscopy has played a key role in laser development. Schawlow, a spectroscopist by training and inclination, understood this from the beginning and has always been at the forefront of laser spectroscopy. My own interest in spectroscopy stems from Schawlow's excitement and enthusiasm for the new world of spectroscopy that opened up with the advent of the laser. When I was a serious graduate student, Schawlow, through his infectious joy and his ability to have fun with science, taught me that we probably do our best work when we approach nature with the curiosity and openness of a child. If we can work in this way, we are not burdened with prejudices that are obstacles to new insights.

This paper describes how glass has been made useful in lasers. My main theme is that laser glass has been "engineered" to meet a variety of applications. Engineering is defined in a dictionary as "a science by which the properties of matter and sources of energy are made useful to man in structures, machines, and products." In our search for better engineering materials, such as laser glass, we must understand materials science. Science and engineering are necessarily coupled; they are, in a sense, two sides of the same coin. When this coupling is effective, advances occur rapidly.

I acknowledge the many scientists who have contributed tremendously to laser glass development and with whom I have had the privilege of collaborating over the years, in particular, C. F. Cline, L. M. Cook, K.-H. Mader, T. Izumitani, S. D. Jacobs, H. E. Meisner, J. D. Myers, C. F. Rapp, E. Snitzer, H. Toratani, and M. J. Weber.

1. Introduction

The first laser material was a crystal, ruby. MAIMAN [2] used it to generate coherent light for the first time on May 15, 1960. Soon after that JAVAN et al. [3] demonstrated lasing action in a helium–neon gas mixture. Having realized that neodymium in a glass should produce laser light if the emission cross section was not too low because of the inhomogeneous line broadening (that is, broadening arising from inhomogeneity of the host glass), in 1961 SNITZER [4] demonstrated a neodymium-doped glass (Nd:glass) laser using barium crown glass. After this, other rare earth–doped glasses were made into lasers; dopants included Yb, Ho, Er, Tm, and Tb [5–9]. In 1963 YOUNG [10] made a cw laser from Nd:glass. By 1963 neodymium-doped yttrium aluminum garnet (Nd:YAG) had come into the marketplace, and rare earth–doped laser glass did not seem to be a competitor to crystalline YAG. Laser glass has lower thermal conductivity than YAG, which prevents it from being used in a cw or high-repetition rate mode. However, it was clear that laser glass has properties that complement those of laser crystals.

In their 1968 review, SNITZER and YOUNG [11] pointed out the principal advantages of laser glass: it can be made in a variety of sizes and shapes, with excellent homogeneity and low birefringence; it is less expensive than crystals; and it has good coupling to broadband pump sources, such as flashlamps, and is thus capable of high stored energy density.

Glass is described as an amorphous solid or as a frozen liquid. A crystal, on the other hand, has a unit structure that repeats itself over a long range. In a crystal this translational symmetry means that the laser-active ions reside in sites that are either identical throughout the crystal or can be reduced to a few equivalent sites. Thus transitions in crystals have narrow line widths. Glass by its nature has a variety of sites that differ from each other in the number and position of the surrounding anions. Thus transitions in glass are inhomogeneous and relatively broad. This basic structural difference between glasses and crystals leads to an important consequence: emission cross sections for ions in crystals are considerably higher than those for ions in glasses. The lasing threshold depends on the emission cross section, so why would laser glasses be of any interest?

Early researchers realized that emission cross section is not the whole story, because the laser gain coefficient g is equal to the product of the cross section σ and the excited state inversion density N^*:

$$g = \sigma N^* . \tag{1}$$

The broader emission lines and generally lower cross sections of glass are compensated by its broader absorption bands, which allow glass to couple more efficiently to broadband pump sources than do crystals.

Laser glass can store large amounts of energy because it can be produced in large sizes and because of its low cross section. Low cross sections are helpful in large laser systems because the obtainable gain is limited by amplified spontaneous emission and parasitic oscillations [12–14]. The rule of thumb is that the medium becomes harder and harder to pump when the product of the gain coefficient and the largest dimension of the laser material rises above 3 or 4. Cross sections in Nd:YAG crystals are about 10 times greater than in glasses: thus laser glasses can store energy at a density about 5 to 10 times that in Nd:YAG crystals.

Laser glass has an additional advantage over laser crystals in lasers operating in the short-pulse (picosecond) regime. The broad fluorescence line width of glass allows for shorter, transform-limited pulses. This property was first demonstrated in 1965 by MOCKER and COLLINS [15].

Glass has always had a fascination for materials scientists because of its variety of compositions, which give them a certain amount of control over its properties. In a sense you have to take what you get with crystals, but glass can be "engineered."

Making glass has been likened to making soup. In fact, early glass melters treated glass melts as a soup and took the empirical approach to discovering new glasses or better methods for making glass. In creating a soup, you decide on the base—for example, chicken, beef, or pork. Then you add modifiers (vegetables)—corn, carrots, onions, etc. Some people like their soup thick, so you may add thickeners, such as corn starch. Spices added in small amounts flavor the soup, and make the difference between a culinary delight and a bland, unappetizing mix.

The analogy between soup-making and glass-making is as follows: the base, or glass former, is usually the major constituent; SiO_2, P_2O_5, B_2O_3, and BeF_2 are examples. The glass formers form the backbone structure of the glass, which is based on SiO_4, PO_4, or BeF_4 tetrahedral units or BO_3 planar, triangular units. The modifiers do not form glasses by themselves, but open up the glass structure by breaking bonds. Examples of modifiers are the alkali metals Li, Na, and K, and the alkaline earths Mg, Ca, and Ba. Thickeners are the marginal glass formers or intermediates, which are incorporated into the backbone structure of the glass and which change the viscosity and improve durability of the glass. Examples of these are Al_2O_3, AlF_3, TiO_2, and TeO_2. Mixes of intermediates commonly form glasses. The spices are the minor additions that can make the difference between a usable commercial glass and one that never gets out of the research laboratory. Such additives as Sb_2O_3, Nb_2O_5, F_2, and As_2O_3 keep the glass free of bubbles, lower the water (OH^-) content, or prevent radiation-caused absorption (solarization).

In keeping with the soup analogy, the best soups, except for consommés, usually have the greatest number of ingredients. Practical glasses have a great number of components, primarily because of what we call the "confusion factor." The confusion factor solves the greatest problem in glass manufacture, crystallization. The more components in the glass melt, the more difficult it is to crystallize, as it is less likely that a critical nucleus will form and grow because components not incorporated into the crystals must be rejected.

SUN [16] categorized oxide compounds as glass formers, intermediates, or modifiers on the basis of their calculated bond energies, which are shown in Fig. 1. Glass formers generally have bond energies above 350 kJ/mole; intermediates have bond energies of 250 to 350 kJ/mole; and modifiers have bond energies below 250 kJ/mole. Neodymium oxide has a bond energy of 213 kJ/mole for 8-fold coordination, which places it near the high end of the modifier category.

The first laser glasses were silicates because they were the most common and the easiest to make. Initially, small-diameter (12 mm) laser glass rods were made and tested by many laboratories; rods of this diameter and 150 mm long were capable of 70-J output per pulse, with pulse durations of 0.6 ms. As the glass melters learned how to melt laser glass with greater

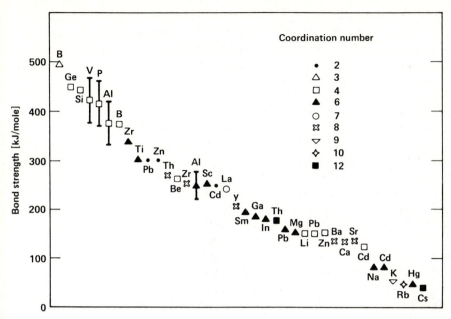

Fig. 1. Energies of cation-oxygen bonds calculated by SUN [16] for the coordination numbers indicated; in order of decreasing bond energy (abscissa arbitrary)

homogeneity and less stress, laser glass rods increased in diameter and length. By 1966, the energy output reached 5 kJ in a 3-ms-long pulse, from 30-mm-diam, 950-mm-long rods [17].

To increase coupling efficiency (the efficiency of coupling between the emission spectrum of the pump, typically a flashlamp, and the absorption spectrum of the Nd:glass), researchers from the beginning considered the use of sensitizers: that is, of an ion that absorbs flashlamp energy not absorbed by a lasing ion and transfers it to the lasing ion. Many ions have been tried for sensitizing Nd^{3+}, including Ce^{3+}, Mn^{2+}, UO_2^{2+}, Tb^{3+}, Bi^{3+}, Eu^{3+}, and Cr^{3+} [18–31]. Only Ce^{3+} proved to be of any practical use, and even Ce^{3+} is of marginal benefit because its absorption is in the ultraviolet.

Laser glasses based on glass formers other than silica were investigated: borates and germanates were first reported in 1963 [32], fluoroberyllates in 1966 [33], fluorophosphates in 1966 [34], and phosphates in 1967 [35]. The interest in laser glass studies increased significantly after BASOV and KROKHIN [36] initiated work on laser-produced plasmas, which was followed by suggestions that lasers could be used to compress and heat deuterium–tritium fuel for generating fusion energy [37,38]. In 1972 BASOV et al. [39] demonstrated neutron generation from a laser-irradiated target. The high energies and powers required for these lasers meant that larger laser glass components were needed [40]. The challenge of manufacturing large pieces of optically homogeneous laser glass was taken up by Corning, Hoya, Kigre, Owens-Illinois, and Schott. As larger and larger laser drivers for fusion were built, the melting technology developed by these companies kept pace with the requirements. By the late 1970s individual laser glass components had reached volumes of 7 liters [41,42].

In the last ten years, improved laser glasses have emerged from many compositional studies; these glasses are more efficient, have better energy storage, and are more robust. In this paper I describe some of the considerations that went into laser glass development. I hope I can give you a flavor of the research in this area.

2. What are the Important Laser Glass Properties?

Soon after the invention of the Nd:glass laser, researchers were measuring the variation in neodymium spectroscopic properties with glass composition [43,44]. However, it became clear very quickly that given the high number of possible glass compositions, one had better concentrate on those that had the greatest chance of providing the desired properties. But what properties made for a good laser glass? That question has several answers depending on the application.

In the early days of lasers the performance characteristics desired were high gain, high stored energy, and low loss. Gain and stored energy depend on stimulated emission cross section, fluorescence lifetime, and coupling efficiency.

In small oscillator systems, gain is the important parameter; therefore, according to (1), high stimulated emission cross section is desirable. On the other hand, in Q-switched systems both gain and stored energy are important. For large apertures the gain coefficient is limited by amplified spontaneous emission and perhaps by parasitic oscillations. Lower cross sections are favored to increase stored energy, but very low cross sections are undesirable because of the high fluences required to extract the stored energy and because of the need for very low material absorption. Energy extraction from a laser material becomes efficient when the laser fluence in the material reaches at least twice the saturation fluence

$$
E_s = \frac{h\nu}{(1 + g_u/g_l)\,\sigma}, \tag{2}
$$

where g_u and g_l are the degeneracies of the upper and lower energy levels, respectively. High fluence is then needed for materials of very low cross section, in which case the damage-fluence threshold of the material may be exceeded. Thus the optimum cross section depends on the specific system.

In all laser systems the coupling between the pump and the active lasing ion must be maximized for best efficiency. Both a spectral and temporal match between the pump and material must be achieved. The most common pump source is the xenon flashlamp, which is an efficient ($>80\%$), high-brightness source [45–47]. Xenon is primarily a broadband emitter with a blackbody temperature of about 10 000 K. The flashlamp pulse duration for most efficient lamp operation is 200 to 600 μs. The fluorescence lifetime of the active ion should be longer than the pump pulse. The lifetime of the $^4F_{3/2}$ state of Nd^{3+} varies between 10 and 1000 μs, depending on the glass composition and on the concentrations of neodymium and of quenching impurities.

Long fluorescence lifetimes are usually associated with low cross sections, through the relation

$$\sigma \tau_R = \frac{\lambda_p^4 \beta}{8\pi c n^2 \Delta \lambda_{eff}},$$

(3)

where τ_R is the radiative lifetime, λ_p is the wavelength at the fluorescence peak, β is the branching ratio (the ratio of the fluorescence intensity for an individual transition to the total fluorescence intensity), n is the refractive index, and $\Delta \lambda_{eff}$ is the effective line width (determined by integrating the fluorescence intensity over wavelength and dividing by the intensity at the peak). For different neodymium glasses, λ_p and β do not vary substantially; therefore, the product $n^2 \sigma \Delta \lambda_{eff}$ is linearly related to τ_R^{-1}, as can be seen in Fig. 2.

For maximum coupling to the flashlamp pump we want a material with long fluorescence lifetime and high absorption. For high absorption we need high neodymium concentrations and thick materials. However, high neodymium concentrations decrease the fluorescence lifetime and make the laser material harder to pump. The thickness of the laser material and the neodymium concentration are then determined by a competition between maximizing absorption and fluorescence lifetime. As an example of this phenomenon, Fig. 3 shows the calculated gain coefficient vs neodymium concentration and lamp pulse width [48]. For low neodymium concentrations the lifetime is long, but absorption is low, thus requiring thick laser glass; at high neodymium concentrations absorption is good, but the lifetime is too short. For short pump pulses, the flashlamp is not efficient because of lamp opacity, or reabsorption; for long pulses, too much energy drains out of the excited state during the pump pulse.

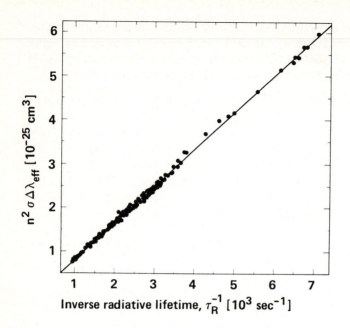

Fig. 2. The product $n^2\sigma\,\Delta\lambda_{\text{eff}}$ vs the inverse of the calculated radiative fluorescence lifetime, τ_R^{-1}, for 300 glasses

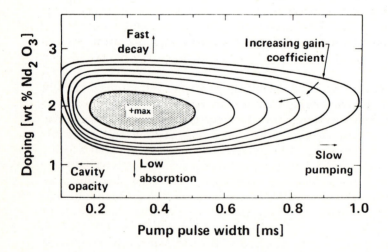

Fig. 3. Calculated gain coefficient contours for various neodymium concentrations and pump pulse durations

Soon after the advent of the laser, scientists found that its high fluence had an undesirable side effect, laser-induced damage of the laser material or of the optical material through which the beam passed [49]. Initially, filamentary "angel hair" tracks were seen in laser materials; these were later explained as due to self-focusing of the laser beam as a result of intensity-dependent refractive index changes [50]. This nonlinearity in the refractive index is critically important in designing laser oscillators and amplifiers. Even at laser fluences below those at which self-focusing occurs, refractive index nonlinearities lead to exponential growth of a small-scale irregularities in the beam profile. Thus diffraction fringes from small-particle obscurations in the beam grow as the laser beam passes through an optical material, ultimately leading to very large variations in the beam intensity. The fluctuations grow as $\exp(B)$, where [51,52]

$$ B = \frac{2\pi}{\lambda} \int \gamma I \, dz , \tag{4} $$

where λ is the wavelength, γ is the nonlinear refractive index coefficient, and I is the optical intensity. The coefficient γ is defined by $\Delta n = \gamma I$, where Δn is the change in refractive index. Alternatively we have $\Delta n = n_2 \langle E^2 \rangle$, where n_2 is the nonlinear refractive index and E is the amplitude of the optical electric field; thus n_2 and γ are related by

$$ n_2 = \frac{cn}{40\pi} \gamma , \tag{5} $$

where n_2 is in esu and γ is in m^2/W.

Nonlinear refractive indices are very difficult to measure because high-power lasers are required and the measurement samples must be of good optical quality [53–58]. Fortunately BOLING, GLASS, and OWYOUNG [59] formulated an empirical relation for predicting nonlinear refractive indices in optical solids from the linear index and the dispersion, which are relatively easy to measure. The dispersion is characterized by the Abbe number v, defined by

$$ v = \frac{n_D - 1}{n_F - n_C} , \tag{6} $$

where n_F, n_D, and n_C are the linear refractive indices at 486.1, 589.3, and 656.3 nm, respectively. In terms of n_D and v, we have [59]

$$ \gamma = \frac{K \left(n_D - 1 \right) \left(n_D^2 + 2 \right)^2}{n_D v \left[1.52 + \left(n_D^2 + 2 \right) \left(n_D + 1 \right) v/6 n_D \right]^{1/2}} , \tag{7} $$

where $K = 2.8 \times 10^{-10}$ m²/W is obtained from fitting experimental values for γ at 1064 nm. Equations (7) and (5) are extensively used to predict γ and n_2, but (7) must be used with caution for high values of n_D, for which it overestimates γ.

Figure 4 is frequently used to illustrate the dependence of γ on n_D and Abbe number v; a similar figure was first published by GLASS [60] in 1975. Figure 4 also shows that fluoride glasses have much lower linear and nonlinear indices than oxide glasses, so that fluoride laser glasses are desirable for laser systems that deliver very high fluences.

Thus we see that emission cross section, the absorption spectrum and strength, the fluorescence lifetime as a function of neodymium concentration, and the nonlinear index of refraction are the important intrinsic properties for evaluating the utility of a glass for laser applications. Other properties are generally less important, but may be critically important in some applications. These properties include thermal expansion α and the change of refractive index with temperature $\delta n/\delta T$, which together affect the thermal change of optical path length s through the relation

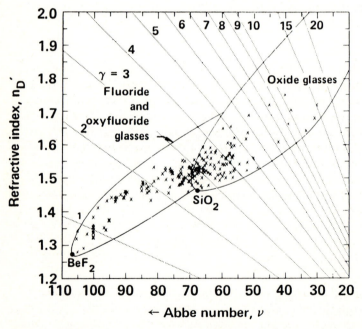

Fig. 4. Refractive index n_D and Abbe number v of optical glasses. Dotted lines of constant γ (intensity-dependent index change in 10^{-20} m²/W) calculated from (7)

$$\frac{\delta s}{\delta T} = \frac{\delta n}{\delta T} + \alpha(n - 1). \tag{8}$$

Athermal glasses (those for which $\delta s/\delta T = 0$) are desired for applications in which large thermal gradients exist, particularly in laser oscillators or in amplifiers with high-average-power output.

The stress-optic coefficients of glasses determine the amount of birefringence induced by thermal stresses or by residual stresses from the melting process. Thermomechanical properties are important for high-average-power operation. The power available from a material is proportional to the thermomechanical figure of merit [61]

$$R_T = \frac{\kappa (1 - \nu) S_T}{\alpha E}, \tag{9}$$

where κ is the thermal conductivity, ν is now Poisson's ratio, S_T is the tensile strength, and E is Young's modulus.

Some characteristics of laser glass are extrinsic, in that they depend on foreign or undesired material in the glass. These characteristics are susceptibility to laser-induced damage, absorption at the laser frequency, and solarization. Much of the utility of a laser glass depends on finding ways to avoid these effects.

Laser-induced damage in laser glasses is in most cases due to the presence of metallic inclusions [62]. Most glasses are melted and refined in platinum crucibles, so platinum particles of sizes 1 to 100 μm are commonly seen as inclusions. Laser fluences of about 2 J cm^{-2} can heat the surface of such small platinum particles above their vaporization temperature [63]. The pressure of the resulting platinum vapor is greater than the bulk strength of the glass, which therefore fractures. Glass manufacturers attempt to prevent the introduction of metallic platinum into the melt, which occurs as a result of vapor transport or of mechanical damage to the crucible. If it is present, platinum can be dissolved into the melt and its susceptibility to damage thus eliminated by refining the glass in an oxidizing environment. On the other hand, a high concentration of ionic platinum in the glass will absorb pump photons in the blue and ultraviolet and thereby lower pumping efficiency.

Improving laser glass thus depends on an understanding of the variation of spectroscopic, optical, and thermomechanical properties with composition. Because of the large number of possible glass compositions, this approach ensures job security for those who search for the Holy Grail of laser glass. I now describe these properties of Nd:glasses and give some background on the physical mechanisms for the observed variations.

3. What Determines Laser Glass Properties?

3.1 Spectral Inhomogeneity

Glass has no long-range spatial order. Thus, each neodymium ion in glass sees a different environment, resulting in substantial spectral inhomogeneity.

Within a given glass, what are the ranges of values of energy levels, transition strengths, electron-phonon coupling coefficients, and homogeneous line widths? Fluorescence line-narrowing (FLN) experiments [64,65], in which a spectrally narrow region can be explored by excitation with a narrow-band laser, have answered these questions for a variety of glasses. Typically the homogeneous width of the $^4F_{3/2} \rightarrow {^4I_{11/2}}$ transition in Nd^{3+} is 20 to 30 cm^{-1} [66–68], whereas the inhomogeneous width is about 250 cm^{-1}.

One indication of the spectroscopic inhomogeneity of a glass is in the deviation of the fluorescence decay from a pure exponential. The different decay rates of individual ions result in a nonexponential decay, even in the absence of energy migration or quenching. Typical e-folding times for fluorescence decays of Nd^{3+} in different glasses are given in Table 1. Of particular note is the substantial nonexponential nature of silicate glass decays compared with the nearly exponential decays of phosphates, which thus appear to be more homogeneous than silicates. This difference was confirmed, using FLN techniques, by BRECHER et al. [64,65], who found decay

Table 1. Nd^{3+} $^4F_{3/2}$ fluorescence decay e-folding times in various glasses for low neodymium concentration, where neodymium self-quenching and Nd-Nd energy migration are absent. Percentage changes in these times indicate the inhomogeneity of the transition strengths

| Glass | Glass type | e-folding times (μs)[a] | | | % change from τ_1 to τ_3 |
		1st	2nd	3rd	
LG-670	Silicate	370	385	395	7
LG-660	Silicate	420	560	580	38
LG-650	Silicate	690	860	1035	50
LG-750	Phosphate	390	390	385	1
UP-16	Phosphate	343	360	368	7
P-101	Phosphate	376	399	430	14
LG-810	Fluorophosphate	540	576	583	8
B-101	Fluoroberyllate	666	685	702	5
K-1261	Tellurite	175	176	181	3

[a] The nth e-folding time is the time taken for the fluorescence intensity to decay from e^{-n+1} to e^{-n} of its initial intensity.

rates in a silicate glass between 2400 and 3300 s^{-1} (a 38% variation) depending on the excitation wavelength within the $^4I_{9/2} \rightarrow {}^2P_{1/2}$ transition. For a phosphate glass the rates vary from 2600 to 3050 s^{-1} (a 17% variation).

Spectral inhomogeneities can significantly decrease the extractable energy of Nd:glass amplifiers and oscillators. In a homogeneous system the saturation fluence E_s is given by (2). In an inhomogeneous system the satura-

Table 2. Measured saturation fluences of neodymium-doped laser glasses [69,70]

Glass	Supplier	Cross section (pm^2)		Saturation fluence (J/cm^2)[a] ± 0.3 J/cm^2	$h\nu/\sigma$ (J/cm^2)
		1053 nm	1064 nm		
Phosphates					
P-101	Hoya	3.0	—	5.3	6.2
LHG-8	Hoya	4.0	—	4.3	4.6
Q-88	Kigre	4.0	—	4.1	4.6
Q-94	Kigre	3.8	—	4.2	4.9
Q-98	Kigre	4.1	—	4.0	4.5
LG-750	Schott	4.0	—	4.4	4.6
EV-4	Owens-Illinois	4.2	—	3.8	4.9
Silicates					
ED-2	Owens-Illinois	—	2.6	4.5	7.1
LSG-91H	Hoya	—	2.3	4.9	8.1
LG-56	Schott	0.84	0.8	9.0[b]	23
OI-H9	Owens-Illinois	—	1.45	6.7	12.8
Fluoro-phosphates					
LG-810	Schott	2.5	—	5.1	7.4
LG-800	Schott	2.7	—	4.0	6.9
LHG-10	Hoya	2.6	—	4.9	7.1
E-309	Owens-Illinois	2.4	—	5.0	7.7
Fluoro-beryllate					
B-101	Corning	2.6	—	5.5	7.1

[a] For 20-ns pulse length and an output fluence of 5 J/cm^2; saturation fluence for 1-ns pulses is about 5% lower.
[b] 8.2 $\pm$ 1.0 J/cm^2 at E_{out} = 10 J/cm^2, 1053 nm.

tion fluence will vary with the input energy because the excited ions with high cross sections will give up their energy more rapidly than those with low cross sections. Spectral hole-burning can also occur. MARTIN and MILAM [69] and YAREMA and MILAM [70] measured E_s for phosphate, silicate, fluorophosphate, and fluoroberyllate glasses. Their values, listed in Table 2, show that in phosphates the E_s values are close to $h\nu/\sigma$. However, in silicates E_s is about 50% lower than $h\nu/\sigma$, indicating that many ions are not in resonance with the extracting laser pulse.

3.2 Emission Cross Section and Absorption Spectra

The emission cross section, the absorption spectrum, and the absorption strength are needed to predict the performance of a laser glass. The absorption strength can be determined once the neodymium ion density is known. However, the emission cross sections for the $^4F_{3/2} \rightarrow {}^4I_{11/2}$ and $^4F_{3/2} \rightarrow {}^4I_{13/2}$ transitions cannot be determined from the absorption transitions because the terminal states are 2000 cm^{-1} and 4000 cm^{-1}, respectively, above the ground state. These states have very low populations and very weak, usually unobservable, absorptions.

Several spectroscopic means of estimating the emission cross section have been used; the most convenient and consistent way is through the application of the JUDD-OFELT [71–73] treatment of spectral intensities of the rare earths. This theory was first applied by KRUPKE [74] to calculating emission cross sections in 1974. In the Judd-Ofelt treatment, the line strength S of a transition between two J states is given by

$$S(aJ:bJ') = \sum_{t=2,4,6} \Omega_t |\langle aJ| |U^{(t)}| |bJ'\rangle|^2 \qquad (10)$$

where a and b denote other quantum numbers specifying the eigenstates. The reduced matrix elements of the unit tensor operator $U^{(t)}$ are calculated in an intermediate coupling approximation; KRUPKE [74] gives their values.

The Judd-Ofelt parameters vary from glass to glass and are determined from a least-squares fit of the integrated per-ion absorption band intensities. Once the Judd-Ofelt parameters are known, the peak emission cross section can be calculated from

$$\sigma(\lambda_p) = \frac{8\pi^3 e^2}{27hc\,(2J+1)} \frac{\lambda_p}{\Delta\lambda_{eff}} \frac{(n^2+2)^2}{n} S(aJ:bJ'), \qquad (11)$$

where λ_p is the peak fluorescence wavelength, $\Delta\lambda_{eff}$ is the effective line width, given by

$$\Delta\lambda_{\text{eff}} = \int \frac{I(\lambda)\, d\lambda}{I(\lambda_p)},$$

and n is the refractive index. The cross section thus calculated is accurate to $\pm 10\%$ if the Nd^{3+} concentration is accurately measured.

The Judd-Ofelt formalism predicts the strengths of individual absorption bands to better than 10%, except for a systematic 10 to 20% overestimation of the $^4I_{9/2} \rightarrow {}^4F_{3/2}$ transition strength. Measurements [75,76] of the branching ratios of the $^4F_{3/2} \rightarrow {}^4I_J$ transitions (J = 9/2, 11/2, 13/2, 15/2) have shown that they are very close to those predicted with the Judd-Ofelt formalism. Therefore, the calculated cross sections of all the $^4F_{3/2} \rightarrow {}^4I_J$ transitions may be too high by 10 to 20%.

Evidence for this overestimation comes from comparisons of the calculated radiative lifetime and the measured fluorescence lifetime at low neodymium concentrations, particularly for phosphate glasses. I find that the average calculated lifetime is about 20% lower than the measured lifetime in glasses. To maintain consistency in comparing different glasses, I have used the calculated Judd-Ofelt parameters in estimating emission cross sections, keeping in mind that some adjustment may be necessary if an improved calculation of rare earth spectral intensities becomes available.

3.3 Fluorescence Lifetimes

For good efficiency, fluorescence lifetimes of Nd:glasses should be longer than the pump pulse. Flashlamps driven at highest efficiency have pulse durations of 100 to 600 μs. In glass, Nd^{3+} has fluorescence lifetimes in the same range. POWELL, MURRAY, and JANCAITIS [77] found that the efficiency of pumping neodymium with flashlamps increases approximately as the square root of the lifetime if the fluorescence lifetime is about the same as the pump pulse duration.

The maximum fluorescence lifetime is set by the radiative lifetime. Non-radiative processes, if present, will shorten, or quench, this lifetime. These fluorescence-quenching processes, shown in Fig. 5, include (a) Nd–Nd self-quenching, (b) quenching induced by impurities (e.g., transition-metal ions), (c) nonradiative multiphonon relaxation, and (d) multiphonon energy transfer processes involving glass matrix vibrations or molecular impurities such as water.

Radiative lifetimes are related to the emission cross section through (3). Longer radiative lifetimes therefore generally mean lower cross sections (Fig. 2). Narrower effective line widths can offset this trend. The glass refrac-

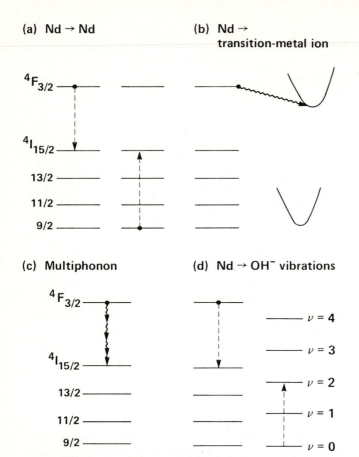

Fig. 5. Fluorescence quenching processes for Nd^{3+}: (a) Nd–Nd self-quenching, (b) quenching induced by impurities (e.g., transition-metal ions), (c) nonradiative multiphonon relaxation, and (d) energy transfer to OH^- (water) vibrational energy

tive index also plays a role, in that fluoride glasses generally have longer lifetimes because of the n^{-2} dependence in (3).

For Nd^{3+} the large energy gap between the $^4F_{3/2}$ and $^4I_{15/2}$ states makes the multiphonon relaxation rate between these states generally small compared with the radiative rate. LAYNE, LOWDERMILK, and WEBER [78,79] measured multiphonon rates W_{NR} for rare earths in a variety of glasses and found that they can be empirically expressed as

$$W_{NR} = W_0(T) \exp(-\alpha \, \Delta E), \tag{12}$$

where $W_0(T)$ depends on the host glass and temperature, α depends on the host glass, and ΔE is the energy gap between neighboring states. The rates are proportional to the highest-energy vibrational frequency in the different base glass types. For the $^4F_{3/2}$ state of Nd^{3+}, the results [78,79] are summarized in Table 3. Borate glasses, because of their high vibrational frequency, have high nonradiative rates and thus low quantum efficiency.

Table 3. Multiphonon decay rates of rare-earth ions in various glass types, expressed as $W_{NR} = W_0(T) \exp(-\alpha \, \Delta E)$ with the listed parameters. The highest-frequency optical phonon in each glass type is listed, along with the predicted multiphonon decay rate ω_{max} for the Nd^{3+} $^4F_{3/2} \rightarrow$ $^4I_{15/2}$ transition ($\Delta E \simeq 5200$ cm^{-1}) (data from [78, 79])

Glass type	W_0 (s^{-1})	α (cm)	ω_{max} (cm^{-1})	W_{NR} $(s^{-1})^a$
Borate	5×10^{12}	3.94×10^{-3}	1350	6300
Phosphate	4×10^{12}	4.61×10^{-3}	1300	150
Silicate	3×10^{12}	4.95×10^{-3}	1100	20
Germanate	2×10^{11}	4.89×10^{-3}	875	2
Tellurite	1×10^{11}	4.91×10^{-3}	775	<1
Fluoroberyllate	9×10^{11}	4.98×10^{-3}	1050	5

$^a \Delta E = 5200$ cm^{-1}.

GAPONTSEV [80] has also studied nonradiative rates in many glasses and relates them to the measured infrared multivibrational spectrum. Although the nonradiative rates follow the general exponential trend described by Layne, Lowdermilk, and Weber, the rates can increase significantly if ΔE is in resonance with overtone or combination vibrational frequencies of the host glass matrix.

An intrinsic nonradiative quenching mechanism is associated with neodymium pairs [Fig. 5(a)]. One excited Nd^{3+} ion can nonradiatively decay to the $^4I_{15/2}$ state and raise a neighboring neodymium ion to the $^4I_{15/2}$ or $^4I_{13/2}$ state with the emission or absorption of vibrational energy. This self-quenching increases as the average neodymium ion distance decreases with increasing neodymium concentration. This so-called concentration quenching is intrinsic to a given glass host. Many researchers have studied this phenomenon in an attempt to establish empirical rules that might help in finding glasses with low concentration quenching.

Energy transfer between ions is generally described on the basis of a dipole-dipole interaction, as first put forward by FÖRSTER [81] and DEXTER

[82,83]. The transfer rate W_{DA} from a donor D to an acceptor A can be written as

$$W_{DA} = \frac{9\chi^2}{128\pi^5 N_A n^4 \tau_{0D} R_{DA}^6} \int \frac{g_D(\nu) K_A(\nu)\, d\nu}{\nu^4},$$ (13)

where

χ^2 = a factor that takes into account the orientational averaging of the dipoles,

N_A = acceptor concentration,

n = refractive index,

τ_{0D} = donor radiative lifetime,

R_{DA} = donor-acceptor distance,

$g_D(\nu)$ = shape factor of the fluorescence band,

$K_A(\nu)$ = absorption coefficient of the acceptor,

ν = frequency on a wave number scale.

Reduced energy transfer, and thus reduced fluorescence quenching, can occur under the following (independent) conditions: (a) the average distance between the ions is large, (b) the transition strengths are low, or (c) the overlap between the fluorescence and absorption spectra is small.

Concentration quenching of the neodymium system has been studied extensively. Figure 6 shows the fluorescence decay rates of several glasses as

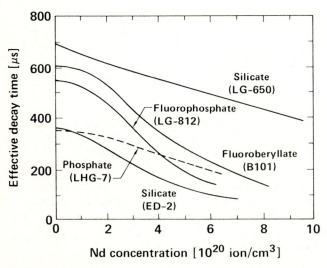

Fig. 6. Time required for the Nd^{3+} $^4F_{3/2}$ fluorescence to decay to $1/e$ of its initial value vs Nd^{3+} ion density (from STOKOWSKI [84])

a function of the neodymium concentration. In general, phosphate glasses (and particularly ultraphosphates) do not have strong concentration quenching. Most researchers believe that this weaker quenching is due to the greater average distance between neodymium ions in the phosphates. This view is supported by the low concentration quenching observed in ultraphosphate crystals, in which the smallest Nd–Nd distances are greater than 6 Å. Thus phosphate glasses have a decided advantage over other types of glass, particularly when high neodymium concentrations are needed in small systems.

The effect of transition strength on quenching can be seen in the silicate glasses LG-670 (ED-2), LG-660, and LG-650. The neodymium concentration for which the lifetime is reduced by half is listed in Table 4 and compared with the calculated absorption strength of the $^4I_{9/2} \rightarrow {}^4I_{15/2}$ transition. Lower line strengths result in lower concentration quenching.

Table 4. Fluorescence decay characteristics of three neodymium-doped silicate glasses: fluorescence decay time T_0 (first e-folding time), neodymium concentration $\rho_{1/2}$ at which the first e-folding time is one-half that for low neodymium concentrations, and peak emission cross section σ for the $^4F_{3/2} \rightarrow {}^4I_{11/2}$ transition.

Glass	T_0 (μs)	$\rho_{1/2}$ (10^{20} cm^{-3})	σ (pm^2)
LG-670	370	4	2.7
LG-660	420	6	1.8
LG-650	690	10	1.1

The neodymium self-quenching rate increases as the square of the neodymium concentration, because of the R^{-6} dependence of the energy transfer rate given by (13). However, in some ultraphosphate glasses the quenching rate increases linearly with neodymium concentration; in LG-650, the dependence is cubic (Fig. 7).

To investigate the effect of the overlap of the fluorescence and absorption transitions on self-quenching in the neodymium system, we measured the $^4F_{3/2} \rightarrow {}^4I_{15/2}$ fluorescence transition and the $^4I_{9/2} \rightarrow {}^4I_{15/2}$ absorption transition for different glasses and calculated the overlap integral. The overlap varies by less than 5% from glass to glass, and thus is not an important factor.

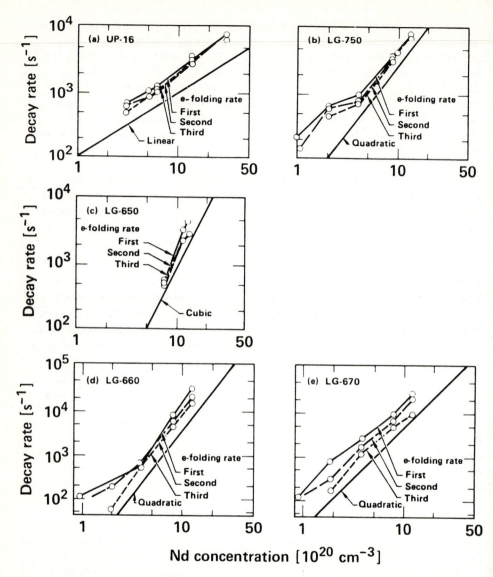

Fig. 7. Nonradiative decay rates vs neodymium concentration in (a) UP-16 phosphate, (b) LG-750 phosphate, (c) LG-650 silicate, (d) LG-660 silicate, and (e) LG-670 silicate laser glasses

Transition-metal and rare earth impurities can quench the neodymium fluorescence. STOKOWSKI and KRASHKEVICH [85] have measured the effect of these impurities on the neodymium fluorescence lifetime. Their results are listed in Table 5.

Molecular impurity vibrations can also lead to nonradiative quenching of the neodymium fluorescence. The only molecular impurity that has been studied to any extent is OH^-, whose high vibrational frequency gives it a strong effect on the nonradiative decay of neodymium. The quenching rates of OH^- are shown in Table 6. These rates are expressed in terms of the absorption coefficient at the OH^- fundamental vibrational frequency, which varies with glass type, as shown in Table 6. The water content of phosphate glasses can be very high because the OH^- enters the structure easily by converting a $P{=}O$ double bond to a single $P{-}OH$ bond. Silicate glasses are less susceptible to water contamination, but they sometimes contain enough water to affect the neodymium fluorescence lifetime. Commercial laser glass manufacturers have reduced the water content of their glasses to levels that result in an absorption coefficient of less than 2 cm^{-1} at the OH^- IR band peak.

Table 5. **Nonradiative quenching rate of neodymium fluorescence for 300 wt. ppm of various transition-metal impurities for a neodymium concentration of 5×10^{19} cm^{-3}. Data from STOKOWSKI and KRASHKEVICH [85].**

| Transition metal | Quenching rate (s^{-1}) | |
	Phosphate glass (UP-91)	Silicate glass (LG-660)
V	900	—
Cr	250	—
Fe	900	880
Co	1500	1740
Ni	1080	1050
Cu	1950	300

Table 6. **Nonradiative quenching rate of neodymium fluorescence for various neodymium concentrations in three laser glasses assumed to contain water. Data from STOKOWSKI and KRASHKEVICH [85].**

| Glass and glass type | Decay rate for pure sample (s^{-1}) | Vibrational frequency of OH^- (cm^{-1}) | Quenching rate (s^{-1}) per cm^{-1} coefficient at OH^- peak | | |
			0.5×10^{20} ions/cm^3	5×10^{20} ions/cm^3	10×10^{20} ions/cm^3
UP-91 phosphate	2650	2750	61	163	287
UP-16 phosphate	2565	2750	59	154	303
LG-660 silicate	2060	2880	70	162	319

3.4 Compositional Effects

The spectroscopic, optical, and thermomechanical properties of neodymium-doped glasses are sensitive to the glass composition. Variations in these properties are largely determined by the glass structure, which in turn depends on the glass former and on the glass modifiers.

To develop empirical rules to represent the variation of Nd:glass properties with composition, early researchers melted and characterized a large variety of glasses. Compositional studies of Nd:glass have been going on for over 25 years and will probably continue for a long time. The results of these studies are published in many articles [86–96]. The most extensive compilation of data on Nd:glass is that of STOKOWSKI, SAROYAN, and WEBER [97].

To summarize the results of a large amount of work, I describe below how the glass former affects Nd:glass properties in general, and then consider the effects of the typical glass modifiers, the alkali metals and alkaline earths.

Figure 8 shows the ranges of the spectroscopic properties found in the different major glass types. Phosphate glasses have higher cross sections than silicate or fluoroberyllate glasses. Differences in glass structure can account for this: silicate and fluoroberyllate glasses form a three-dimensional structure based on the SiO_2 or BeF_2 tetrahedra, whereas phosphate glasses consist of chains of PO_4 tetrahedra. In phosphate glasses, however, the average chain length depends on the P_2O_5 content. At low P_2O_5 content the chains are short. As the P_2O_5 content increases the chains get longer and theoretically are infinitely long at the metaphosphate composition (Fig. 9), assuming no cross-linking. In glasses with still higher P_2O_5 content (the ultraphosphates), cross-linking occurs and the structure is more three-dimensional. Borate glasses are based on the BO_3 planar molecular unit and thus constitute a third structural type.

Consider how a neodymium ion may integrate itself into the glass structure. The bonding of neodymium is primarily electrostatic, because the shielded f electrons do not participate strongly in covalent bonding. Given that the ionic radius of neodymium is about 1.08 Å, the closest packing that oxygen and fluorine ions can have with neodymium is about 8-fold coordination. The neodymium-oxygen bond energy of 213 kJ per mole with this coordination means, however, that neodymium is not a strong modifier of glass structure (Fig. 1). In fact, stable phosphate glasses can be made with neodymium concentrations of up to 4×10^{21} cm^{-3}.

The transition energies, transition strengths, and inhomogeneous line widths can tell us much about the local neodymium ion environment and its variation.

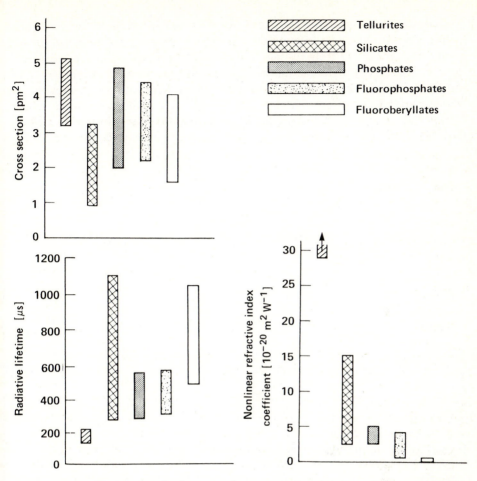

Fig. 8. Variation of Nd^{3+} $^4F_{3/2} \rightarrow {}^4I_{1/2}$ emission cross section, calculated $^4F_{3/2}$ radiative lifetime, and nonlinear refractive index coefficient γ for tellurite, silicate, phosphate, fluorophosphate, and fluoroberyllate glasses

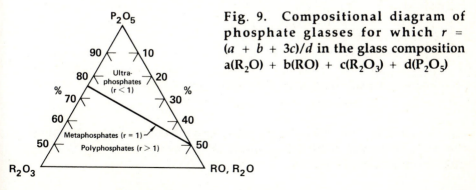

Fig. 9. Compositional diagram of phosphate glasses for which $r = (a + b + 3c)/d$ in the glass composition $a(R_2O) + b(RO) + c(R_2O_3) + d(P_2O_5)$

The transition energies are determined by the splitting of the $4f^3$ states by the electron-electron interaction. The more covalent the neodymium-ligand bond, the more time the neodymium electrons spend apart; therefore, the electron-electron interaction is reduced and the energy differences and transition energies are reduced. This effect is known as the nephelauxetic or "cloud-expanding" effect. Thus, the $^4F_{3/2} \rightarrow {}^4I_{11/2}$ transition occurs at lower energy (i.e., at longer wavelengths) in oxide glasses than in fluoride glasses (Fig. 10). Further, the transition in phosphate glasses is at shorter wave-

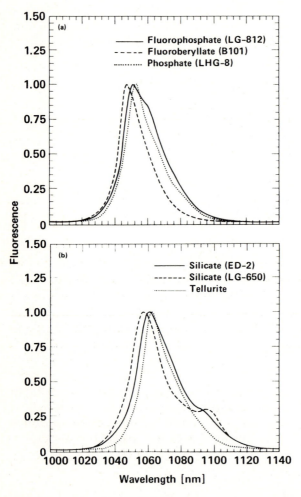

Fig. 10. Relative Nd^{3+} $^4F_{3/2}$ fluorescence spectra: (a) in a fluorophosphate, a fluoroberyllate, and a phosphate glass; (b) in two silicates and a tellurite glass (from STOKOWSKI [84])

lengths than in silicate glasses, implying that the neodymium in silicates participates more in bonding.

The transition strengths (or the Judd-Ofelt Ω parameters) are determined by the noncentrosymmetric local field surrounding the neodymium ion. This field mixes the higher-energy d-electronic states with the f states, thereby allowing electric dipole transitions to occur. The strength of this local field is determined by the distortion of the neodymium-anion ligands as a result of the bonding and structural requirements of the glass-former ion in the neighborhood of the neodymium ion. The transition strengths affect the radiative lifetime of the $^4F_{3/2}$ state. This lifetime depends on Ω_4 and Ω_6 and on the local field-correction factor $n(n^2 + 2)^2$. The calculated radiative lifetime is therefore a good indicator of the strength of the noncentrosymmetric field.

The inhomogeneous width of transitions is affected by variations in the local neodymium field that arise from structural variations in the glass. Large inhomogeneous line widths mean that the neodymium ion has less influence on its environment, which is determined more by the glass former. The site-to-site variations in glass structure are also determined by the temperature at which they are frozen in. Thus, we would expect greater variations in local structure for glasses that have a higher transformation temperature. The effective line width of the $^4F_{3/2} \rightarrow {}^4I_{11/2}$ fluorescence transition is determined by both the inhomogeneous width and the splitting of the six Kramer doublet states by the crystalline field.

Because of its importance in determining the emission cross section for the lasing transitions, we investigated the effective line width as a function of composition. Figures 11 through 14 illustrate the effects of different modifier ions on the calculated radiative lifetimes, emission cross sections, and effective line widths of neodymium in various glasses. Of particular note is the variety of dependences on the modifier. For instance, silicate and fluoroberyllate glasses, whose structures are both based on SiO_4 or BeF_4 tetrahedra, have opposite slopes with respect to changes in the alkali metal modifiers. The main difference between the glasses is that silicates are primarily covalent glasses, whereas fluoroberyllates are more ionic. The larger alkali ions have lower electrostatic field strengths, which explains the behavior of silicate glasses. We would expect the fluoroberyllate glasses to have the same trend; the fact that they do not indicates that the smaller alkali ions must be distributed around the neodymium ion in a more symmetrical way in the fluoroberyllates.

In general the effective line width depends little on the alkali ion and decreases somewhat as the alkaline earth varies from Mg to Ba in all the

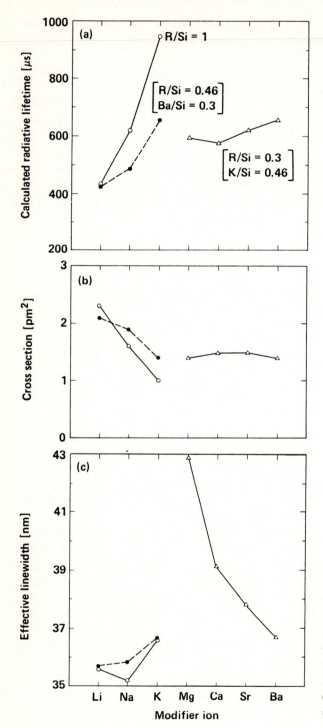

Fig. 11. Effects of modifier ions in silicate glasses on (a) calculated radiative lifetimes, (b) $^4F_{3/2} \rightarrow {}^4I_{11/2}$ emission cross sections, and (c) effective line

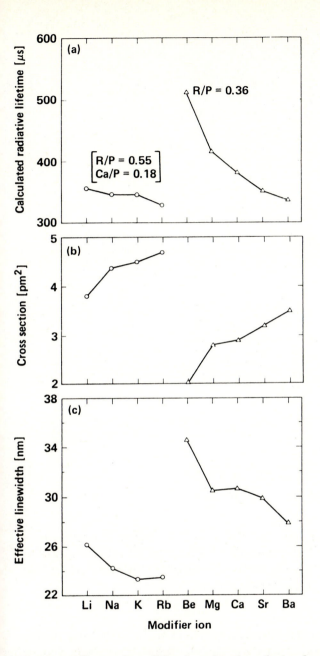

Fig. 12. Effects of modifier ions in phosphate glasses on (a) calculated radiative lifetimes, (b) $^4F_{3/2} \rightarrow {}^4I_{11/2}$ emission cross sections, and (c) effective line widths of the of the $^4F_{3/2} \rightarrow {}^4I_{11/2}$ fluorescence

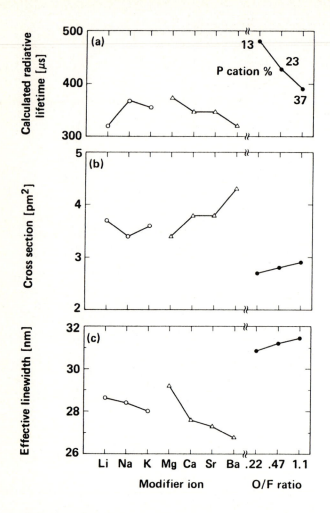

Fig. 13. Effects of modifier ions and the ratio O/F of ionic concentrations of oxygen and fluorine in fluorophosphate glasses on (a) calculated radiative lifetimes, (b) $^4F_{3/2} \rightarrow {}^4I_{11/2}$ emission cross sections, and (c) effective line widths of the $^4F_{3/2} \rightarrow {}^4I_{11/2}$ fluorescence

glass types. However, I noted an interesting correlation between the effective line width and the glass transformation temperature in a series of silicate glasses of silica content from 60 to 70 mole % with various modifiers. This correlation, shown in Fig. 15, lends support to the idea that the inhomogeneities in the local neodymium environment are greater when the glass structure is frozen in at higher temperatures, at which larger fluctuations in glass structure are present.

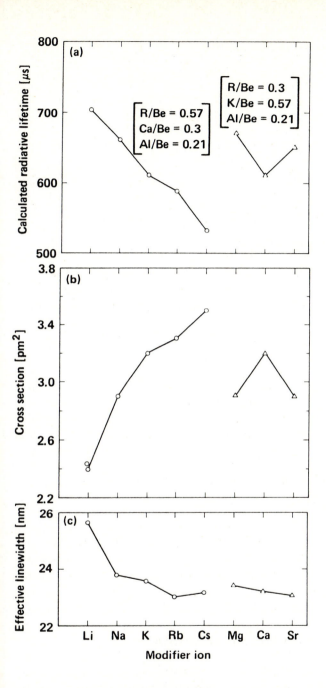

Fig. 14.   Effects of modifier ions in fluoroberyllate glasses on (a) calculated radiative lifetimes, (b) $^4F_{3/2} \rightarrow {}^4I_{11/2}$ emission cross sections, and (c)
effective line widths of the $^4F_{3/2} \rightarrow {}^4I_{11/2}$ fluorescence

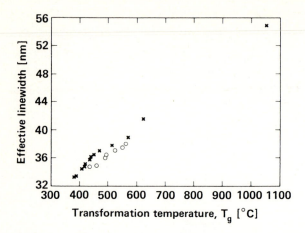

Fig. 15. Effective line width of the $^4F_{3/2} \rightarrow {}^4I_{11/2}$ transition for a series of Schott silicate glasses (crosses) vs measured transformation temperature. Open circles are for other silicate and borosilicate glasses. Cross in upper right corner is for fused silica

Phosphate glasses have proven to be the most useful of all the possible laser glass hosts. Their utility comes from their generally higher cross sections (which give higher gain), their lower concentration quenching (which allows more neodymium in the glass, giving higher efficiency), and their more homogeneous spectral characteristics (which allow for more efficient energy extraction).

In 1983 and 1984 LLNL, Schott, and Hoya investigated new phosphate compositions, particularly in the ultraphosphate region (Fig. 9), where low-quenching glasses were first found by VORONKO et al. [98]. As a result of these studies [99], two compositional regions stand out. The first group, alkali-lanthanide-phosphates, such as K_2O–La_2O_3–P_2O_5 or the lithium-sodium equivalents, have lower concentration quenching. For instance; the 210-μs fluorescence lifetime of UP-16 is the longest, for a neodymium concentration of 10^{21} ions cm^{-3}, of any known glass. The second group, Al_2O_3-containing ultraphosphates, have somewhat higher concentration quenching than the first group, but their thermomechanical characteristics are better.

In the aluminum-free, alkali-ultraphosphate glasses, there is a range of compositions in which concentration quenching is minimal, as shown in Fig. 16. This compositional range seems to be correlated with the appearance of a stable glass structure, because, at high K_2O/P_2O_5 ratios, the glass melt crystallizes; at low ratios, the P_2O_5 volatilizes excessively. UP-16 is in the

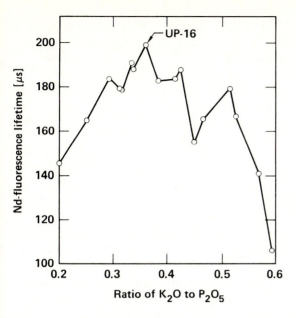

Fig. 16. Neodymium fluorescence lifetime for neodymium concentration of 13×10^{20} ions cm^{-3} in K$_2$O–Ln$_2$O$_3$–P$_2$O$_5$ glasses

center of this region and makes a good glass. Adding more than 5 mole % of alkaline earths (Mg, Ca, or Ba) or alumina to UP-16 results in increased concentration quenching, but the poor thermomechanical characteristics of UP-16 are improved by adding alumina.

We have found that adding up to 10 mole % of B$_2$O$_3$ can increase fluorescence lifetimes by 10% at high neodymium concentrations. Normally, B$_2$O$_3$ strongly quenches the neodymium fluorescence because of the high-frequency vibrational modes of the BO$_3$ group. However, B$_2$O$_3$ in concentrations of less than 10 mole % in ultraphosphates incorporates itself into the glass structure as BO$_4$ side groups on the phosphate chains, so that BO$_3$ chain terminators are absent.

3.5 Thermomechanical Properties

Thermomechanical properties of laser glasses are critically important in lasers for high-average-power operation. In such lasers, heat-removal rates and thermal stresses determine the maximum power that the laser medium can sustain. The thermomechanical figure of merit R_T [61] for a laser material designed to run at its fracture limit is given by (9). In brittle materials the fracture strength S_T, which appears in (9), depends on the critical-fracture toughness K_c, which characterizes the energy required to propagate a crack, and on the maximum surface flaw size a:

$$S_T = YK_c a^{-1/2}, \tag{14}$$

where Y is a dimensionless parameter that depends on the location and geometry of the flaw.

We have looked for correlations between thermomechanical properties and glass compositions. Figure 17 shows that thermal conductivity κ and the thermal expansion coefficient α are strongly correlated. The inverse relation between κ and α arises from the dependence of both on the anharmonics of glass structural vibrations. Large anharmonicities result in low thermal conductivities and high thermal expansions.

Young's modulus and the thermal expansion coefficient are also inversely related for phosphate glasses; their product αE has a value of 0.56 ± 0.14 MPa K^{-1}.

Poisson's ratio v for glasses is in the range of 0.25 ± 0.03, a relatively small variation.

The fracture toughness varies with the glass former and the modifier ion over the range 0.3 to 0.9 MPa m$^{1/2}$, with phosphate glasses at the lower end

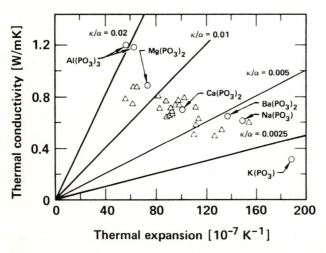

Fig. 17. Thermal conductivity and thermal expansion are inversely correlated in phosphate glasses. Lines of constant κ/α are shown. Unlabeled data points represent multicomponent glasses

and silicates at the upper end. Within a given glass type, K_c can vary by about a factor of 2.

As a result of variations, from glass to glass, in the material properties κ, ν, α, and E in (9), and of variations in S_T insofar as it is affected by the material property K_c, thermomechanical figures of merit differ from one another by factors of no more than 2 or 3. In high-average-power lasers, we would like to have thermomechanical figures of merit an order of magnitude higher.

Such a large increase in R_T can be obtained only by going outside the material properties: this means increasing the glass fracture strength S_T through reductions in the maximum surface flaw size a, which is determined when the component is fabricated. Grinding and finishing are therefore critically important to obtaining high-strength laser glass components. MARION [100] has found that grinding leaves deep, but optically invisible subsurface cracks. This damage can be removed in subsequent grinding or polishing by removing more material than is typical in commercial processes. Factor-of-4 increases in S_T have been observed. Greater improvements (up to a factor of 15) can be obtained by etching the glass.

In use, the strength of a glass component will be degraded by handling, laser damage, and attack by atmospheric water vapor. To make the glass more durable, fractures generated by surface flaws under stress must be prevented from propagating by a compressive surface layer. Compressive layers can be formed by ion-beam–sputtered coatings [101] or ion-exchanged layers [102–104]. Ion-exchanged layers have made a factor-of-3 improvement in the fracture strength of laser glasses [104].

4. Future Possibilities

What further improvements might be made in laser glass? For large-aperture lasers such as those used in fusion research, the system cost per output joule is inversely proportional to the stored energy density in the material. Higher energy density will require higher neodymium concentrations. But high neodymium concentrations generally decrease fluorescence lifetimes, making flashlamp pumping more difficult. However, there is still some hope that glasses might be found that have very low concentration quenching. Our knowledge of the quenching mechanisms leads us to expect that these glasses will have large Nd–Nd distances and low transition strengths. The ultraphosphate compositional region should be investigated further—it may still have some surprises for us.

Increased energy density can also be obtained by improving pumping efficiency. Sensitization remains a possibility, because many sensitization schemes remain to be investigated, but I do not expect this approach to be successful.

The development of larger and more efficient arrays of laser diodes makes them increasingly attractive for pumping Nd:glass. The near-resonance pumping possible with laser diodes also means lower heat input to the laser glass, which makes diodes particularly desirable for pumping high-average-power lasers. However, the packing density of diode arrays must increase—and their cost must decrease considerably—before they will be used extensively.

Another way to increase energy density is to reduce the losses caused by amplified spontaneous emission. A saturable absorber in laser glass might be one way to reduce this loss. The search for such an absorber has so far been very limited; I believe that this area warrants more attention.

High laser fluences will be needed to extract energy stored at high densities. These high fluences cause damage in current optical and laser materials, primarily because of metallic inclusions. The only sure way to avoid these inclusions, which come from the metal crucibles commonly used to contain the melt, is to melt glass in nonmetallic containers. The optical glass manufacturers have begun work on using vitreous carbon, a highly promising future crucible material, for glass melts. The availability of metallic-inclusion-free laser materials will be critically important for future laser systems.

High laser fluences also mean that materials with lower nonlinear refractive indices are necessary to prevent self-focusing. Fluoride glasses have substantially lower nonlinear indices than oxide glasses. With the present high level of interest in fluoride glasses for optical fibers, research in this area will probably result in new practical laser glasses.

Research on composite laser materials promises to open up new territory. Marion and Stokowski have suggested that a composite laser material might be made consisting of a glassy phase and a crystalline phase. These composites can be formed by mixing small crystalline particles in a glass melt, which is then processed similarly to normal glasses. The two phases must match in refractive index and thermal strain to prevent high scattering losses. These composites are to be contrasted with glass ceramics, which are partially crystallized glasses in which the refractive-index mismatch between the two phases leads to high scattering losses. Marion and Stokowski have initiated an experimental program to find and make a suitable material.

5. Acknowledgment to Arthur Schawlow

Arthur Schawlow's scientific accomplishments are well known, particularly those connected with the laser and with high-resolution spectroscopy of the hydrogen atom. Those accomplishments are well documented in many publications. Schawlow's career as a university professor is not as well known, but his influence can be seen in the accomplishments of his students. I particularly remember the humor (irreverent at times) with which he emphasizes an insight into the physical world. His humor is an effective teaching technique; it's hard to forget the props, such as his "Mickey Mouse" balloon and his "Jell-o" laser, that he used to get across his points. But these amusing items are an example of his great imagination and intuitive thinking. In this way he taught me to step out of the constraining world of existing theory and calculations and to imagine new worlds of possibilities. I remember a discussion of the second-order phase transition in $SrTiO_3$, in which the crystal structure goes from cubic to tetragonal. We were comparing this transition to the ordering of the spin dipole (a first-rank tensor) in ferromagnets and antiferromagnets. Schawlow suggested that the transition we were observing in $SrTiO_3$ was analogous to an ordering in the orientation of a second-rank tensor (the lattice strain), and this was later confirmed.

I thank Arthur Schawlow for taking me on as a research student, for the opportunity to work with him, and for imparting to me his excitement about discovering the mysteries of the physical world.

Work performed under the auspices of the U.S. Department of Energy by the Lawrence Livermore National Laboratory under Contract W-7405-Eng-48.

References

1. A. L. Schawlow and C. H. Townes: "Infrared and Optical Masers," *Phys. Rev.* **112**, 1940–1949 (1958).
2. T. H. Maiman: "Stimulated Optical Radiation in Ruby," *Nature* **187**, 493–494 (1960).
3. A. Javan, W. R. Bennett, Jr., and D. R. Herriot: "Population Inversion and Continuous Optical Maser Oscillation in a Gas Discharge Containing a Ne–Ne Mixture," *Phys. Rev. Lett.* **6**, 106 (1961).
4. E. Snitzer: "Optical Maser Action of Nd^{+3} in a Barium Crown Glass," *Phys. Rev. Lett.* **7**, 444 (1961).
5. H. W. Etzel, H. W. Gandy, and R. J. Ginther: "Stimulated Emission of Infrared Radiation From Ytterbium-Activated Silicate Glass," *Appl. Opt.* **1**, 534 (1962).

6. H. W. Gandy and R. J. Ginther: "Stimulated Emission from Holmium Activated Silicate Glass", *Proc. IRE* **50**, 2113 (1962).

7. E. Snitzer and R. Woodcock: "Yb^{3+}-Er^{3+} Glass Laser," *Appl. Phys. Lett.* **6**, 45 (1965).

8. H. W. Gandy, R. J. Ginther, and J. F. Weller: "Stimulated Emission of Tm^{3+} Radiation In Silicate Glass," *J. Appl. Phys.* **38**, 3030 (1967).

9. S. I. Andreev, M. R. Bedilov, G. O. Karapetyan, and V. M. Likhachev: "Stimulated Radiation of Glass Activated By Terbium," *Sov. J. Opt. Tech.* **34**, 819 (1967): *Opt.-Mekh. Promst.* **34**, 60 (1967).

10. C. G. Young: "Continuous Glass Laser," *Appl. Phys. Lett.* **2**, 151 (1963).

11. E. Snitzer and C. G. Young: "Glass Lasers," in *Advances in Lasers*, ed. by A. Levine, vol. 2 (Dekker, New York, 1968).

12. J. M. McMahon, J. L. Emmett, J. F. Holzrichter, and J. B. Trenholme: "A Glass-Disk-Laser Amplifier," *IEEE J. Quantum Electron.* **QE-9**, 992 (1973).

13. D. C. Brown: "Parasitic Oscillations In Large Aperture Nd^{3+} Glass Amplifiers Revisited," *Appl. Opt.* **12**, 2215 (1973).

14. J. A. Glaze, S. Guch, and J. B. Trenholme: "Parasitic Suppression In Large Aperture Nd;Glass Disk Laser Amplifiers," *Appl. Opt.* **13**, 2808 (1974).

15. H. W. Mocker and R. J. Collins: "Mode Competition and Self-Locking Effects in a Q-Switched Ruby Laser," *Appl. Phys. Lett.* **7**, 270 (1965).

16. K. H. Sun: "Fundamental Condition of Glass Formation," *J. Am. Ceram. Soc.* **30**, 277 (1947).

17. C. G. Young: "Glass Laser Delivers 5000-Joule Output," *Laser Focus* **3**, 36 (February, 1967).

18. H. W. Gandy, R. J. Ginther, and J. F. Weller: "Energy Transfer in Silicate Glass Coactivated with Cerium and Neodymium," *Phys. Lett.* **11**, 213 (1964).

19. R. R. Jacobs, C. B. Layne, M. J. Weber, and C. Rapp: "$Ce^{3+} \rightarrow Nd^{3+}$ Energy Transfer in Silicate Glass," *J. Appl. Phys.* **47**, 2020 (1976).

20. S. Shionoya and E. Nakazawa: "Sensitization of Nd^{3+} Luminescence by Mn^{2+} and Ce^{3+} in Glasses," *Appl. Phys. Lett.* **6**, 117 (1965).

21. N. T. Melamed, C. Hirayama, and E. K. Davis: "Laser Action in Neodymium-doped Glass Produced Through Energy Transfer," *Appl. Phys. Lett.* **7**, 170 (1965).

22. N. T. Melamed, C. Hirayama, and P. W. French: "Laser Action in Uranyl-Sensitized Nd-Doped Glass," *Appl. Phys. Lett.* **6**, 43 (1965).

23. H. W. Gandy, R. J. Ginther, and J. F. Weller: "Energy Transfer in Triply Activated Glasses," *Appl. Phys. Lett.* **6**, 46 (1965).

24. J. C. Joshi, N. C. Pandey, B. C. Joshi, and J. Joshi: "Energy Transfer from $UO_2 \rightarrow Nd^{3+}$ in Barium Borate Glass," *J. Luminescence* **16**, 435 (1978).

25. A. Y. Cabezas and L. G. DeShazer: "Radiative Transfer of Energy Between Rare-Earth Ions in Glass," *Appl. Phys. Lett.* **4**, 37 (1964).

26. R. Reisfeld and Y. Kalisky: "Energy Transfer Between Bi^{3+} and Nd^{3+} in Germanate Glass," *Chem. Phys. Lett.* **50**, 199 (1977).

27. E. J. Sharp, M. J. Weber, and G. Cleek: "Energy Transfer and Fluorescence Quenching in Eu- and Nd-Doped Silicate Glasses," *J. Appl. Phys.* **47**, 364 (1976).

28. G. O. Karapetyan, V. P. Kovalyov, and S. G. Lunter: "Chromium Sensitization of the Neodymium Luminescence in Glass," *Opt. Spectrosc. USSR* **19**, 529 (1965); *Opt. Spectrosk.* **19**, 951 (1965).

29. G. Dauge: "Nonradiative Energy Transfer in Silicate Glass," *IEEE J. Quantum Electron.* **QE-2**, lviii (1966).

30. J. G. Edwards and S. Gomulka: "Enhanced Performance of Nd Laser Glass by Double Doping With Cr," *J. Phys. D.* **12**, 187 (1979).

31. A. G. Avanesov, Yu. K. Voron'kov, B. I. Denker, G. V. Maosimova, V. V. Osiko, A. M. Prokhorov, and I. A. Shcherbakov: "Nonradiative Energy Transfer from Cr^{3+} to Nd^{3+} Ions in Glasses with High Neodymium Concentrations," *Sov. J. Quantum Electron.* **9**, 935 (1979); *Kvantovaya Elektron.* **6**, 1583 (1979).

32. R. D. Maurer: "Nd^{3+} Fluorescence and Stimulated Emission in Oxide Glasses," in *Proceedings of the Symposium on Optical Masers*, Microwave Research Institute Symposia Series Vol. XIII (Polytechnic Press, Brooklyn, NY, 1963) p. 435.

33. G. T. Petrovksii, M. N. Tolstoi, P. P. Feofilov, G. A. Tsurikova, and V. N. Shapovalov: "Luminescence and Stimulated Emission of Neodymium in Beryllium Fluoride Glass," *Opt. Spectrosc. USSR* **21**, 72 (1966); *Opt. Spektrosk.* **21**, 126 (1966).

34. F. Auzel: "Emission Stimulée de Er^{3+} dans un Verre Fluorophosphate," *C. R. Acad. Sc. Ser. B* **263**, 765 (1966).

35. G. Deutschbein, C. Pautrat, and I. M. Svirchevsky: "Phosphate Glasses, New Laser Materials," *Rev. Phys. Appl.* **1**, 29 (1967).

36. N. G. Basov and O. N. Krokhin: "Conditions For Heating Up of a Plasma by the Radiation from an Optical Generator," *Sov. Phys. JETP Lett.* **19**, 123–125 (1964).

37. R. E. Kidder: "Applications of Lasers to the Production of High-Temperature and High Pressure Plasma," *Nucl. Fusion* **8**, 3–12 (1968).

38. J. Nuckolls, L. Wood, A. Thiessen, and G. Zimmerman: "Laser Compression of Matter to Super-High Densities: Thermonuclear (CTR) Applications," *Nature* **239**, 139–142 (1972).

39. N. G. Basov, Yu. S. Ivanov, O. N. Krokhin, Yu. A. Mikhailov, G. V. Sklizkov, and S. I. Fedotov: "Neutron Generation In Spherical Irradiation of a Target by High-Power Laser Radiation," *Sov. Phys. JETP Lett.* **15**, 417–419 (1972).

40. J. E. Swain, R. E. Kidder, K. Pettipiece, F. Rainer, E. D. Baird, and B. Loth: "Large-Aperture Glass Disk Laser System," *J. Appl. Phys.* **40**, 3973 (1969).

41. S. E. Stokowski, W. H. Lowdermilk, F. T. Marchi, J. E. Swain, E. P. Wallerstein, and G. R. Wirtenson: "Advances in Optical Materials for Large Aperture Lasers," in *Proceedings of Electro-Optics/Laser '81*, Anaheim, CA, Nov. 17–19 (Industrial & Scientific Conf. Management, Chicago, 1981), p. 203.

42. W. W. Simmons and R. O. Godwin: "Nova Laser Fusion Facility—Design, Engineering, and Assembly Overview," *Nucl. Technol. Fusion* **4**, 8–24 (1983).

43. D. W. Harper: "Assessment of Neodymium Optical Maser Glass," *Phys. Chem Glasses* **5**, 11 (1964).

44. C. Hirayama and D. W. Lewis: *Phys. Chem. Glasses* **5**, 44 (1964).

45. J. B. Trenholme and J. L. Emmett: "Xenon Flashlamp Model for Performance Prediction" in *Proceedings of Ninth International Conference on High Speed Photography*, ed. by W. G. Hyzen and W. G. Chase (Society of Motion Picture and Television Engineers, New York, 1970), p. 299.

46. A summary of the Trenholme-Emmett model and a review of flashlamp pumping of Nd:Glass lasers is found in: D. C. Brown: *High-Peak-Power Nd:Glass Laser Systems* (Springer, Berlin, 1981), Ch. 3.

47. H. T. Powell, A. C. Erlandson, and K. S. Jancaitis: "Characterization of High Power Flashlamps and Application to Nd:glass Laser Pumping," *Proc. SPIE Conf. on Flashlamp Pumped Laser Technology* **609**, 78 (1986).

48. J. B. Trenholme: Lawrence Livermore National Laboratory, Livermore, CA, private communication (1987).

49. M. Hercher: *J. Opt. Soc. Am.* **54**, 563 (1964).

50. R. Y. Chiao, E. Garmire, and C. H. Townes: "Self-Trapping of Optical Beams," *Phys. Rev. Lett.* **13**, 479 (1964).

51. J. B. Trenholme: "Small-Scale Instability Growth: Review of Small Signal Theory," in *Laser Program Annual Report 74*, Lawrence Livermore National Laboratory, Livermore, Calif., UCRL-50021-74 (1975), p. 179.

52. E. S. Bliss, J. T. Hunt, P. A. Renard, G. E. Sommargren, and H. J. Weaver: "Effects of Nonlinear Propagation on Laser Focusing Properties," *IEEE J. Quantum Electron.* **QE-12**, 402 (1976).

53. E. S. Bliss, D. R. Speck, and W. W. Simmons: "Direct Interferometric Measurements of the Nonlinear Refractive Index Coefficient n_2 in Laser Materials," *Appl. Phys. Lett.* **25**, 718 (1974).

54. D. Milam and M. J. Weber: "Measurement of Nonlinear Refractive Index Coefficients using Time-Resolved Interferometry: Application to Optical Materials for High-Power Neodymium Lasers," *J. Appl. Phys.* **47**, 2497 (1976).

55. D. Milam and M. J. Weber: "Nonlinear Refractive Index Coefficients for Nd Phosphate Laser Glasses," *IEEE J. Quantum Electron.* **QE-13**, 512 (1976).

56. D. Milam, M. J. Weber, and A. J. Glass: "Nonlinear Refractive Index of Fluoride Crystals," *Appl. Phys. Lett.* **31**, 822 (1977).

57. M. J. Weber, C. F. Cline, W. L. Smith, D. Milam, D. Heiman, and R. W. Hellwarth: "Measurements of the Electronic and Nuclear Contributions to the Nonlinear Refractive Index of Beryllium Fluoride glasses," *Appl. Phys. Lett.* **32**, 403 (1978).

58. M. J. Weber, D. Milam, and W. L. Smith: "Nonlinear Refractive Index of Glasses and Crystals," *Opt. Eng.* **17**, 463 (1978).

59. N. L. Boling, A. J. Glass, and A. Owyoung: "Empirical Relationships for Predicting Nonlinear Refractive-Index Changes in Optical Solids," *IEEE J. Quantum Electron.* **QE-14**, 601 (1978).

60. A. J. Glass: *Laser Program Annual Report 74*, Lawrence Livermore National Laboratory, Livermore, CA, UCRL-50021-84 (1975), p. 260.

61. J. L. Emmett, W. F. Krupke, and W. R. Sooy: "Future Development of High-Power Solid State Laser Systems," Lawrence Livermore National Laboratory, Livermore, CA, UCRL-53344 (1982): *Sov. J. Quantum Electron.* **13**, 1 (1983).

62. R. W. Hopper and D. R. Uhlmann: "Mechanism of Inclusion Damage in Laser Glass," *J. Appl. Phys.* **41**, 4023 (1970).

63. J. H. Pitts: "Modeling Laser Damage Caused by Platinum Inclusions in Laser Glass," Lawrence Livermore National Laboratory, Livermore, CA, UCRL-93249 (1985): to be published in *Proceedings of the 17th Annual Symposium—Optical Materials for High Power Lasers, Boulder, Colorado*, National Bureau of Standards, Washington, DC, NBS Special Publication.

64. C. Brecher, L. A. Riseberg, and M. J. Weber: "Line-Narrowed Fluorescence Spectra and Site-Dependent Transition Probabilities of Nd^{3+} in Oxide and Fluoride Glasses," *Phys. Rev. B* **18**, 5799 (1978).

65. C. Brecher, L. A. Riseberg, and M. J. Weber: "Site-Dependent Variation of Spectroscopic Relaxation Parameters in Nd Glasses," *J. Luminescence* **18/19**, 651 (1979).

66. V. I. Nikitin, M. S. Soskin, and A. I. Khizhnyak: "Influence of Uncorrelated Inhomogeneous Broadening of the 1.06 μm Band of the Nd^{3+} Ions on Laser Properties of Neodymium Glasses," *Sov. J. Quantum Electron.* **8**, 788 (1978): *Kvantovaya Elektron.* **5**, 1375 (1978).

67. V. I. Nikitin, M. S. Soskin, and A. I. Khizhnyak: "New Data About Internal 1.06 μm Luminescence Band Structure of Nd^{3+} in Silicate Glass," *Sov. Tech. Phys. Lett.* **2**, 64 (1976); *Pis'ma Zh. Tekh. Fiz.* **2**, 172 (1976).

68. V. I. Nikitin, M. S. Soskin, and A. I. Khizhnyak: "Uncorrelated Non-uniform Spreading—A Basic Reason for Narrow-Band Generation in Phosphate Glass with Nd^{3+}," *Sov. Tech. Phys. Lett.* **3**, 5 (1977): *Pis'ma Zh. Tekh. Fiz.* **3**, 14 (1977).

69. W. E. Martin and D. Milam: "Gain Saturation in Nd:Doped Laser Materials," *IEEE J. Quantum Electron.* **QE-18**, 1155 (1982).

70. S. M. Yarema and D. Milam: "Gain Saturation in Phosphate Laser Glasses," *IEEE J. Quantum Electron.* **QE-18**, 1941 (1982).

71. B. R. Judd: "Optical Absorption Intensities of Rare-Earth Ions," *Phys. Rev.* **127**, 750 (1962).

72. G. S. Ofelt: "Intensities of Crystal Spectra of Rare-Earth Ions," *J. Chem. Phys.* **37**, 511 (1962).

73. R. D. Peacock: "The Intensities of Lanthanide f↔f Transitions," *Struct. Bonding* **22**, 83 (1975).

74. W. F. Krupke: "Induced-Emission Cross Sections in Neodymium Laser Glasses," *IEEE J. Quantum Electron.* **QE-10**, 450 (1974).

75. T. S. Lomheim and L. G. DeShazer: "New Procedure of Determining Neodymium Fluorescence Branching Ratios as Applied to 25 Crystal and Glass Hosts," *Opt. Comm.* **24**, 89 (1978).

76. S. E. Stokowski: Lawrence Livermore National Laboratory, Livermore, CA: measurements made in 1978.

77. H. T. Powell, J. E. Murray, and K. S. Jancaitis: Lawrence Livermore National Laboratory, Livermore, CA, private communication (1987).

78. C. B. Layne, W. H. Lowdermilk, and M. J. Weber: "Multiphonon Relaxation of Rare-Earth Ions in Oxide Glasses," *Phys. Rev. B* **16**, 10 (1977).

79. C. B. Layne and M. J. Weber: "Multiphonon Relaxation of Rare-Earth Ions in Beryllium-Fluoride Glass," *Phys. Rev. B* **16**, 3259 (1977).

80. N. E. Alekseev, V. P. Gapontsev, M. E. Zhabotinskii, V. B. Kravchenko, and Yu. P. Rudnitskii: *Laser Phosphate Glasses* (Nauka, Moscow, 1980): Lawrence Livermore National Laboratory, Livermore, CA, UCRL-TRANS-11817 (1983), p. 3-97.

81. T. Förster: *Ann. Phys.* **2**, 55 (1948).

82. D. L. Dexter: "A Theory of Sensitized Luminescence in Solids," *J. Chem. Phys.* **21**, 836 (1953).

83. D. L. Dexter and J. H. Schulman: "Theory of Concentration Quenching in Inorganic Phosphors," *J. Chem. Phys.* **22**, 1063 (1954).

84. S. E. Stokowski: "Glass Lasers" in *CRC Handbook of Laser Science and Technology*, Vol. 1, *Lasers and Masers*, ed. by M. J. Weber (CRC Press, Boca Raton, FL, 1982), p. 215.

85. S. E. Stokowski and D. Krashkevich: "Transition-Metal Ions in Nd-Doped Glasses: Spectra and Effects on Nd Fluorescence," *Mat. Res. Soc. Symp. Proc.* **61**, 273 (1986).

86. V. F. Egorova, V. S. Zubkova, G. O. Karapetyan, A. A. Mak, D. S. Prilezhaev, and A. L. Reichakhrit: "Influence of Glass Composition on the Luminescence Characteristics of Nd^{3+} Ions," *Opt. Spectrosc. USSR* **23**, 148 (1967); *Opt. Spectrosk.* **23**, 275 (1967).

87. P. H. Sarkies, J. N. Sandoe, and S. Parke: "Variation of Nd^{3+} Cross Section for Stimulated Emission with Glass Composition," *J. Phys. D: Appl. Phys.* **4**, 1642 (1971).

88. R. R. Jacobs and M. J. Weber: "Dependence of the $^4F_{3/2} \rightarrow {}^4I_{11/2}$ Induced-Emission Cross Section for Nd^{3+} on Glass Composition," *IEEE J. Quantum Electron.* **QE-12**, 102 (1976).

89. H. G. Lipson, J. R. Buckmelter, and C. O. Dugger: "Neodymium Ion Environment in Germanate Crystals and Glasses," *J. Non-Cryst. Solids* **17**, 27 (1975).

90. N. B. Brachkovskaya, A. A. Grubin, S. G. Lunter, A. K. Przhevuskii, E. L. Raaben, and M. N. Tolstoi: "Intensities of Optical Transitions in Absorption and Luminescence Spectra of Neodymium in Glasses," *Sov. J. Quantum Electron.* **6**, 534 (1976).

91. G. O. Brachkovskaya, G. O. Karapetyan, A. L. Reishakhrit, and M. N. Tolstoi: "Luminescence of Neodymium in Alkali Silicate Glasses," *Opt. Spectrosc.* **29**, 173 (1970).

92. K. Hauptmanova, J. Pantoflicek, and K. Patek: "Absorption and Fluorescence of Nd^{3+} Ion in Silicate Glass," *Phys. Status Solidi* **9**, 525 (1965).

93. C. Hirayama: "Nd Fluorescence in Alkali Borate Glasses," *Phys. Chem Glasses* **7**, 52 (1966).

94. C. Hirayama, F. E. Camp, N. T. Melamed, and K. B. Steinbruegge: "Nd^{3+} in Germanate Glasses: Spectral and Laser Properties," *J. Non-Cryst. Solids* **6**, 342 (1971).

95. N. E. Alekseev, A. A. Izyneev, Yu. L. Kopylov, V. B. Kravchenko, Yu. P. Rudnitskii, and N. F. Udovenko: "Activated Nd^{3+} Laser Glasses Based on the Metaphosphates of Divalent Metals," *J. Appl. Spectrosc.* **24**, 691 (1976): *Zh. Prikl. Spektrosk.* **24**, 976 (1976).

96. N. E. Alekseev, A. A. Izyneev, Yu. L. Kopylov, V. B. Kravchenko, and Yu. P. Rudnitskii: "A Study of Neodymium Glasses Based on Alkali Metal Metaphosphates," *J. Appl. Spectrosc.* **26**, 87 (1977): *Zh. Prikl. Spektrosk.* **26**, 116 (1977).

97. S. E. Stokowski, R. A. Saroyan, and M. J. Weber: " Nd Doped Laser Glass Spectroscopic and Physical Properties," Lawrence Livermore National Laboratory, Livermore, CA, M-095 Rev. 2 (1981).

98. Yu. K. Voronko, B. I. Denker, A. A. Zlenko, et al.: "Spectral Lasing Properties of Li–Nd Phosphate Glass," *Opt. Commun.* **18**, 88 (1976).

99. L. M. Cook, A. J. Marker III, and S. E. Stokowski: in *Proc. SPIE*, Vol. 505 (Soc. Photo-Optical Inst. Engineers, Bellingham, Wash., 1984). pp. 102–111.

100. J. E. Marion: "Strengthened Solid-State Laser Materials," *Appl. Phys. Lett.* **47**(7), 694–696 (1985).

101. J. E. Marion: "Development of High Strength Solid State Laser Materials," *Proc. Amer. Inst. Physics* **146**, 234 (1986).

102. Owens-Illinois, Inc., Optical Products Division, Product Information on ED-2S Strengthened Glass (1978).

103. S. D. Stookey: *High Strength Materials*, ed. by V. F. Zakey (Wiley, New York, 1964), p. 669.

104. K. A. Cerqua, S. D. Jacobs, B. L. McIntyre, and W. Zhong: to be published in *Proceedings of the Boulder Damage Conference, 1985*, National Bureau of Standards, Washington, DC, NBS Special Publication.

One Is Not Enough: Intra-Cavity Spectroscopy with Multi-Mode Lasers

P.E. Toschek[1]* and V.M. Baev[2]†

[1]Joint Institute for Laboratory Astrophysics, University of Colorado and National Bureau of Standards, Boulder, CO 80309, USA
[2]1. Institut für Experimentalphysik, Universität Hamburg, D-2000 Hamburg 36, Fed. Rep. of Germany

1. Introduction

When the laser was discovered more than 25 years ago as the ultimate outgrowth of spectroscopic work, it kept the physics community busy for more than a decade observing, surveying, and classifying all those surprising features of the new type of light and its interactions with matter. The early lasers, lacking technical sophistication, usually oscillated in a vast multitude of radiative modes. These systems were considered overly complex, and accordingly theoretical models and explanations were constructed for ideal single-mode lasers, which, however, barely could be demonstrated in experiments. It is perhaps not so surprising that it took ten years before emission spectra of a real multi-mode laser were subjected to detailed inspection. A close look revealed that the light flux of individual modes in their broad spectral comb did not vary smoothly -- as was anticipated -- across the full emission band. Instead, crests and crevices, rifts and ridges showed up in the recorded spectra. Soon these observed features were traced to either very weak absorption of matter or spurious interference inside the laser resonator [1]. This interpretation suggested that this ultra-sensitive response to weak extinction could be used for a novel spectroscopic technique.

An early example of an "intra-cavity" absorption (ICA) spectrum -- of HN_3 vapor [2] -- is compared in Fig. 1 with the analogous spectrum recorded by conventional differential absorption [3]. The new technique turned out to be 4×10^7 times more sensitive with the exposure time reduced from hours to 1 ms -- a considerable advancement!

The superior sensitivity of a multi-mode laser to intra-cavity frequency-selective extinction was noticed independently in some laboratories in the U.S. [4-6]. At Stanford, an experiment was devised that combined the demonstration of ultra-sensitive I_2 absorption inside a cw dye laser with a specific narrow-band technique of light detection [6]. The schematics is shown in Fig. 2. The rhodamine-6G laser with a Z-shaped cavity

---

*1986-87 JILA Visiting Fellow. On leave from 1. Institut für Experimentalphysik, Universität Hamburg, D-2 Hamburg 36, F. R. Germany.

†On leave from Lebedev Physical Institute, Academy of Sciences of the USSR, Moscow, USSR.

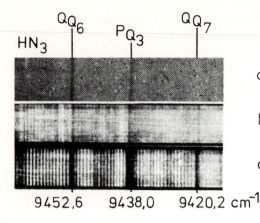

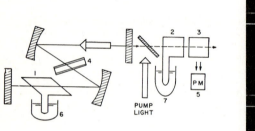

Fig. 1. (a) HN₃ absorption spectrum; 4 m absorber lengths; 0.4 bar [3]. ICA spectra of HN₃ in a cell of 0.4-m length, at 10^{-4} mbar (b), and at 10^{-1} mbar (c) (from Ref. 2).

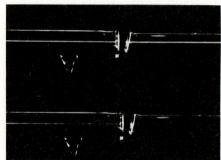

Fig. 2. Left: cw Rh 6G laser with ICA cell (I_2^{127}, 1) and fluorescence cells (I_2^{129}, 2; I_2^{127}, 3). Dye cell (4), photomultiplier (5), liq. N_2 (6,7). Right: Side view of fluorescence cells with I_2^{127} in ICA cell (1) frozen out (upper portion), and at room temperature (lower portion) (from Ref. 6).

contains a flowing dye cell and an absorption cell filled with isotopically pure I_2^{127}, which can be frozen out in a side arm. The laser emission, covering the spectral range of some 3 nm, could have been analyzed by a spectrograph in order to detect the quenching of the light at the absorption lines. However, a simpler approach makes use of a fluorescence cell external to the laser: If the cell's content matches the absorber, little or no fluorescence is excited, since the laser modes required for excitation are quenched (see Fig. 2). With another species (a different isotope, for example) in the external cell, however, the fluorescence appears due to the resonance mismatch of the absorber and fluorescent. The enhancement in sensitivity over that of extra-cavity absorption was estimated to be 10^5.

Such dramatic effects deserve further investigation. During the one-and-a-half decades elapsed since then, many aspects of laser intra-cavity spectroscopy (ICS) have been studied in quantitative detail and their results have been condensed in more than 500 papers. Today we understand rather well the fundamental dynamics of multi-mode lasers with spectrally selective perturbations, and also their intrinsic benefits and restrictions when used as analytic tools. Although some features are still being clarified, ICS today deserves application on a broader scale.

In Sec. 2, we outline the dynamics of multi-mode lasers with the help of a rate-equation model, and we describe the time evolution of the mean values of the mode amplitudes and the effects of fluctuations. The energy accumulated in the spectral bins corresponding to the modes defines line shapes for both the emission and the perturbation. The latter is dealt with in Sec. 3. In Sec. 4, examples are discussed for the application of high-sensitivity ICS to the detection of weak linear and nonlinear absorption, of gain, and of weak light. The limitations on spectral and time resolution of ICS are described in Sec. 5.

2. Dynamics of a Multi-Mode Laser

We use rate equations for the modeling of a multi-mode laser. Possible effects of coherence are briefly discussed later.

2.1 Rate equations for a multi-mode laser

The system of rate equations for the photon number in the field modes and for the inversion of the laser medium with homogeneously broadened gain is [7,8]

$$\dot{M}_q = -\gamma M_q + B_q N(M_q + 1) - \kappa_q c M_q \quad , \tag{1}$$

$$\dot{N} = P - N/\tau - N \sum_{q=1}^{n} B_q M_q \quad , \tag{2}$$

where M_q is the photon number of mode q, N is the density of the inversion, $\gamma = T_{ph}^{-1}$ is the broadband cavity loss rate, $\kappa_q c$ is the narrow-band absorption rate, P is the pump rate, and τ is the decay time of the inversion. The mode number q extends up to the total number of modes, n. The first term on the right-hand side of Eq. (1) describes the cavity loss, the second one the gain, and the third one the narrow-band loss by intra-cavity absorbers. In Eq. (2), the second and third terms on the rhs describe spontaneous and stimulated decay of the inversion, respectively.

So far, Eqs. (1) and (2) give the mean values of the parameters that characterize the laser dynamics. Fluctuations of the light field due to its quantum nature require us to amend Eq. (1) by a Langevin term $F_q(t)$, normalized as

$$\langle F_q(t) \rangle = 0 \quad ; \quad \langle F_q(t) F_p(t') \rangle = \gamma \langle M_q \rangle \delta_{qp} \delta(t-t') \quad .$$

Interaction of field modes can play a significant role at elevated power levels. This interaction is taken into account by an additional term $\sum_i \Phi(M_q M_{q+i})$ on the rhs of Eq. (1), where $\Phi(M_q M_{q+i})$ gives the transfer of light to mode q from the i-th neighboring mode by nonlinear mode coupling, e.g., by Brillouin scattering.

91

2.2 Solutions of the rate equations

For some simple but important cases, Eqs. (1) and (2) are easily solved analytically:

(i) With lack of intra-cavity absorption ($\kappa_q = 0$), negligible fluctuations due to time averaging [$F_q(t) = 0$], same gain for all modes ($B_q \equiv B$), and with the inversion adiabatically following the field ($\gamma < \tau^{-1}$), the stationary solution of Eqs. (1) and (2) is

$$N^o = \gamma/B \quad , \tag{3}$$

$$M_q^o = (\eta-1) \, P_{th}/\gamma n \quad , \tag{4}$$

where $P_{th} = \gamma/B\tau$ is the pump threshold, and $\eta = P/P_{th}$ is the normalized pump rate.

(ii) If we add loss κ in one particular mode q, i.e., $\kappa_q = \kappa$, threshold and inversion are determined by the remaining modes with small and non-specific loss. After N and M_q have reached stationary values N^o and M_q^o for $t > \gamma^{-1}$, the light flux in mode q_o decays according to

$$M_{q_o} = M_{q_o}^o \, e^{-\kappa ct} \quad . \tag{5}$$

The minimum detectable absorptivity is defined as

$$\kappa_{min} = (ct_\ell)^{-1} \equiv \ell_p^{-1} , \tag{6}$$

where t_ℓ is the laser pulse duration. Thus, an ICA line is equivalent to a line of conventional absorption with the optical path length in the absorber $\ell = \ell_p$. In Fig. 1b, e.g., this length is $\ell_p = 300$ km.

The minimum detectable IC absorption κ_{min} is estimated to be equal to spontaneous emission $\kappa c M_q^o = BN^o = \gamma$ [see Eq. (1)]; then

$$\kappa_{min} = \gamma/cM_q^o \quad . \tag{7}$$

With parameters of a typical laser,

$$\gamma = 3 \times 10^7 \, s^{-1} \quad , \quad M_q^o = 3 \times 10^7 \quad ,$$

we have

$$\kappa_{min} = 3 \times 10^{-11} \, cm^{-1} \quad .$$

The laser pulse duration which is required if we want to exploit this sensitivity is

$$t_\ell = M_q/\gamma = 1 \, s$$

with the equivalent pulse length 3×10^5 km.

(iii) With homogeneously broadened gain of bandwidth Γ, the laser modes differ in their amplification. Close to its center, the gain profile may be approximated by a parabola [9]:

92

$$B_q = B_o \{1 - |2(q-q_o) \, \nu/\Gamma|^2\} \quad , \tag{8}$$

where ν is the mode separation, and B_o is the gain of the central mode q_o, which controls threshold and inversion in the stationary state:

$$N_o = \gamma/B_o \quad , \quad P_{th,o} = \gamma/B_o\tau \quad .$$

For high enough pump rate, maintained over the time γ^{-1}, the laser starts oscillating all over the gain profile. Soon, the inversion stabilizes itself at N_o, and the emission narrows [9,10] according to Eq. (1):

$$M_q = M_q^o \, (\nu/\Gamma) \, (\gamma t/\pi)^{-1/2} \, \exp[-\gamma t \, (\frac{(q-q_o)\nu}{\Gamma/2})^2] \quad . \tag{9}$$

The number of oscillating modes decreases in time as

$$n \approx \Gamma/\nu(\gamma t)^{1/2} \quad . \tag{10}$$

This "spectral condensation" is shown in Fig. 3 [10]. The spectrochronogram represents the spectrum of a multimode cw dye laser along the ordinate direction, which is time-scanned by a streak camera in the abscissa direction. After an interruption of its oscillation, the laser starts anew with a broad emission band whose width decreases in time, according to Eq. (9). In addition, the figure shows the decay of the light at wavelengths coinciding with atmospheric water absorption [compare Eq. (5)]. Similar phenomena have been observed under various conditions [9,11,12]. In a recent experiment [12], the three-mirror dye laser could be bandwidth-limited (<20 cm^{-1}) by a glass etalon. The pump light was switched on for 1 ms, and the laser spectrum was recorded with 10 μs gate time and variable decay, controlled by an acousto-optical deflector. Up to 500 μs delay, the validity of Eqs. (5) and (9) was confirmed with and without intra-cavity etalon, i.e., for different Γ.

In contrast with spectral condensation, which shows up with a homogeneously broadened gain medium, the emission band of a multi-mode laser widens in time with an inhomogeneously broadened gain profile: Spectral subgroups of molecules farther in the wings reach the threshold at a later time. The decay of the light flux at absorption lines has been proved valid, however, for at least 300 μs after start of the oscillation [13].

(iv) If the emission bandwidth of the laser is small compared with the absorption linewidth, we attribute the same absorptivity to all modes,

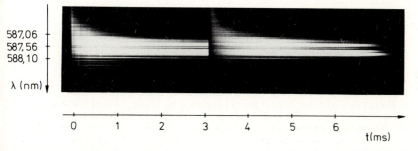

587,06
587,56
588,10

λ (nm)

0   1   2   3   4   5   6

t(ms)

Fig. 3.   Spectrochronogram of a multi-mode cw Rh-6G laser with water vapor ICA (from Ref. 10).

$\kappa_q = \kappa$. Now, the extra absorption is simply compensated by enhanced inversion,

$$N = (\gamma + \kappa c)/B \quad , \tag{11}$$

and the total photon number

$$M = P_{th} \frac{\eta - 1 + \kappa c/\gamma}{\gamma(1 + \kappa c/\gamma)} \tag{12}$$

is equally distributed among all modes. Then, the absorption coefficient is

$$\kappa = (1 - M/M_o) \, (\eta-1) \, \gamma/c \quad , \tag{13}$$

where M_o is the photon number when no absorber is placed inside the laser resonator. We notice that the sensitivity of narrow-band lasers to ICA is high for small loss, and when operating close to the threshold ($\eta-1 \ll 1$). With the typical parameter values $\gamma = 3 \times 10^7$ s^{-1}, $\eta = 1.5$, and $M/M_o = 0.5$, we have $\kappa \simeq 10^{-4}$ cm^{-1}. The smallest detectable absorption could perhaps be pushed down by an order of magnitude closer to threshold. However, further improvement of sensitivity is hardly feasible in a narrow-band laser: Increasing the threshold by extra loss raises the steady-state gain, and that loss in turn has a less significant effect on the output power. In contrast, a broad-band laser acts differently upon narrow-band absorption: the pump threshold is controlled by the modes of minimum loss, whereas all other modes suffer exponential decay of their output. This decay is faster for modes subject to extra absorption. Only the total inversion couples the modes in the model used so far, and accordingly this contrast becomes more and more entrenched until the stationary state is reached after $t \simeq 1$ s. On the other hand, this is the time the spectrum would require to recover its initial shape if the extra absorption were suddenly removed [14].

2.3 Quantum fluctuations

So far we have dealt with a model that describes the dynamics of a multi-mode laser in terms of mean values of the light flux in individual modes q. However, the weakness of mode coupling and the concomitant sensitivity to perturbation leaves this system susceptible to large variation of the light flux M_q with mode number q. The values of M_q, initiated by spontaneous emission, also vary in time about their mean values. These "quantum fluctuations" are suitably modeled by an additional term on the rhs of Eq. (1) which describes a Langevin random force [15,16]. The analysis of the model thus modified shows that the values of M_q are exponentially distributed, and the fluctuations are large: The probability that mode q accumulates the photon number M_q is

$$P(M_q) = \langle M_q \rangle^{-1} \exp(-M_q/\langle M_q \rangle) \quad . \tag{14}$$

The mean absolute deviation of the photon number in mode q from its average is $\langle |\Delta M_q| \rangle \propto \langle M_q \rangle$, which is, above threshold, much larger than the corresponding deviation of the total photon number M, i.e., $\langle |\Delta M| \rangle \propto \sqrt{\langle M \rangle}$, or the fluctuations in a single-mode laser. Therefore those quantum fluctuations of individual modes are easily detectable [17]. A LiF:F$_2^+$ color center laser of 300 ns pulse length, pumped by a Xe ion laser, has been used for this purpose. Its spectral emission range was 899 nm through 907 nm, its cavity length 25 cm. With a 1-m spectrograph, individual modes have been resolved in its output. Figure 4 shows densitograms of

94

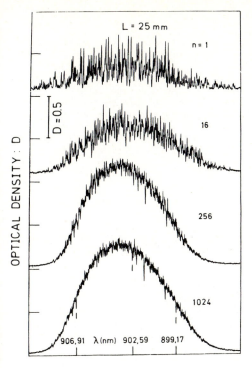

Fig. 4.
Densitograms of spectra of the output of a LiF:F_2^+ color center laser (n superimposed light pulses) (from Ref. 17).

recorded spectra with the number of superimposed light pulses varied. The top trace demonstrates a spectrum whose height distribution approximately obeys an exponential law. The deviations of the spectral flux from its mean value in <u>multi-pulse spectra</u> — or in spectra averaged by insufficient resolution — have a Gaussian distribution:

$$P(\Delta M_q) = (\sigma \sqrt{2\pi})^{-1} \exp\left|-(\Delta M_q)^2/2\sigma^2\right| \quad , \tag{15}$$

where $\sigma^2 = \sigma_0^2 \langle M_q \rangle^2 /jn$ is the dispersion, n is the number of superimposed pulses, and j the number of unresolved modes. Examples of spectra that are characterized by this distribution are shown in the lower traces of Fig. 4. A long laser cavity with its narrowly spaced modes suppresses the observation of the large quantum fluctuations in the frequency spectrum. In time domain, their characteristic duration is:

$$t_q \simeq \langle M_q \rangle \gamma^{-1} \quad . \tag{16}$$

For typical laser parameters [see evaluation of Eq. (7)], $t_q \simeq 1$ s. However, fluctuations on the order of milliseconds have been observed; their duration decreases with increasing laser output power. Thus, another mechanism is required to which the observed fluctuations can be attributed. This mechanism is a nonlinear coupling of the laser modes.

2.4 Mode coupling

There is much evidence for the existence of fluctuations on a millisecond time scale, whose mean period <u>decreases</u> upon increasing laser power [7,8, 10,18-25]. An example is shown in Fig. 5. The experimental setup for this recording was similar to the one used for the spectrochronograms of Fig. 3, but with spectral resolution being 10^6, in order to resolve individual modes. Satisfactory explanation of this phenomenon has been achieved recently on the basis of mode coupling by stimulated Brillouin scattering (SBS) [7,8]. To model this nonlinear interaction of modes, we assume that only neighboring modes interact significantly. Accordingly, Eq. (1) has to be amended by an extra term on the rhs which is given by $DM_q(M_{q+1}-M_{q-1})$.

When a realistic value on the order of 10^{-3} s^{-1} [7] is inserted for the coupling constant D, numerical solutions of Eq. (1), extended by this term, show strong fluctuations of mode power with characteristic times that are between tens of microseconds and a few milliseconds, in fair agreement with the observations. These time periods are much shorter than those inferred from quantum fluctuations in the absence of other dynamic perturbations.

An analytic estimate of the characteristic time of deterministic fluctuations is derived as follows [7]: Under stationary conditions, the first two terms on the rhs of the modified Eq. (1) cancel, and we set κ_q = 0, and $M_{q+1} - M_{q-1} \simeq \langle M_q \rangle$, as the amplitude of spectral fluctuations, according to Sec. 2.3, is on the order of the mean spectral power density. Since $F_q \simeq \sqrt{\langle M \rangle} \ll \langle M \rangle$ for large photon number M, Eq. (1) reduces to

$$\dot{M}_q = D \langle M_q \rangle M_q \quad , \tag{17}$$

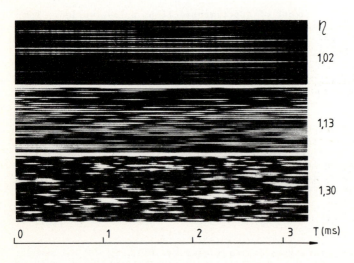

Fig. 5. Spectrochronograms of the output of a cw Rh-6G laser at high spectral resolution, for various values of the relative pump rate η. The horizontal lines are individual modes separated by $\Delta\nu$ = 0.033 cm^{-1} (from Ref. 25).

which gives, for the characteristic time of fluctuations in the output power of individual modes,

$$t_n \simeq (D \langle M_q \rangle)^{-1} \quad . \tag{18}$$

With $\langle M_q \rangle \simeq 10^6$, this expression gives the realistic value $t_n \simeq 1$ ms. Being the mean time for unperturbed oscillation in long enough dye laser pulses, t_n is the width of the intensity autocorrelation of individual modes and determines the sensitivity of ICAS. In general,

$$\kappa_{min} = (ct_m)^{-1} \quad ; \quad t_m = \min\{t_\ell, t_q, t_n\} \quad , \tag{19}$$

where t_ℓ and t_q are the laser pulse duration and the mean length of quantum fluctuations, respectively. Figure 6 shows how the sensitivity to absorption varies upon pump power and pulse length, when t_ℓ is either smaller or larger than t_n.

Numerical solutions of Eq. (1) for 100 modes and with different approximations discussed above are shown in Fig. 7 [7]. The spectrum on the top represents the simple case of an ICA spectrum described in Sec. 2.2(ii). The second spectrum includes quantum fluctuations (2.3), which are reduced by averaging over 10 laser pulses in the third spectrum (compare with the data of Fig. 4). The bottom spectrum includes strong mode coupling. An interesting feature of this last spectrum is the conspicuous asymmetry as a result of the nonlinear mode interaction. We present more detail on this problem in Sec. 3.

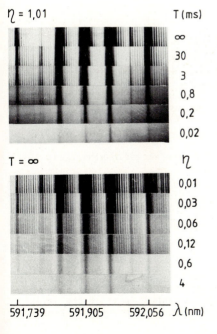

Fig. 6. ICA spectra of water vapor obtained with various pulse lengths of Rh–6G laser pulses at the relative pump rate $\eta = 1.01$ (top). Corresponding spectra at various pump rates, for cw operation (bottom) (from Ref. 25).

97

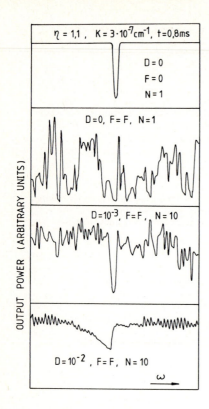

Fig. 7. Numerical solutions of rate equations for a 100-mode laser with T = 0.8 ms pulse duration, single-line ICA with $\kappa = 3 \times 10^{-7}$ cm^{-1} in center of emission spectrum, and $\eta = 1.1$, for various values of mode coupling constant D, Langevin random force F, and number of superimposed laser pulses, N (from Ref. 7).

2.5 Spatial hole burning

For some time in the history of ICAS, the role of spatial hole burning was controversial. It now appears that it does not severely limit the sensitivity of the method under most conditions met in practice. For quantitative evaluation, M_q and N must be written spatially dependent, and a Fourier decomposition is useful for the former quantity:

$$M_q \rightarrow M_q(z) = M_q^o \sin^2 q\nu z \quad ; \quad N \rightarrow N(z) \quad . \tag{20}$$

With this modification, Eqs. (1) and (2) have been solved analytically for some limiting cases [6,9,26,27].

(i) The gain medium is thought to fill the entire cavity. In the limit of a large mode number, n, the decrement of the output power of an attenuated mode is [6]

$$\frac{dM_q}{M_q} = -d\zeta \cdot 2n \cdot \frac{\alpha/\zeta}{\alpha-\zeta} \quad , \tag{21}$$

where α and $\zeta = \kappa L$ are the gain and extinction, respectively, taken to be equal for all modes unaffected by the absorber; L is the length of the resonator. The second and third factor show the enhancement of the attenuation due to ICA over the corresponding single-pass interaction. Enhance-

98

ment according to the third factor, $(\alpha/\zeta)(\alpha-\zeta)^{-1}$, is expected for a single-mode laser with homogeneously broadened gain, if its output adjusts such that the saturated gain equals the loss. The second factor, $2n$, gives the genuine enhancement due to multi-mode oscillation. We estimate $\kappa_{min} = d_\zeta/L \simeq 10^{-7}$ cm^{-1} from this analysis. Necessarily, this situation is approximated, at best, since the absorber also requires resonator space.

(ii) The gain medium is concentrated in a thin slab of thickness $l \ll L$. An analytical solution of Eqs. (1) and (2) including the spatial inhomogeneity shows [9] that the modification of the system is slight, and still absorptivity values as small as $\kappa \simeq 10^{-14}$ cm^{-1} can be detected. Under this condition, the previously discussed limitations have more severe effects on the sensitivity of the method.

If the gain medium fills only part of the laser cavity ($l < L$), spatial inhomogeneities dominate the sensitivity of ICA only if the location of that medium is close to one of the cavity mirrors in a standing-wave configuration [26,27].

2.6 Sensitivity limits of ICA

We are now in a position to evaluate and compare the physical factors that can limit the sensitivity of ICA. With long enough laser pulses, or with cw operation, spontaneous emission and nonlinear mode coupling by SBS (providing the laser fluctuations) may effectively limit that sensitivity. The former effect fades with increasing laser power, the latter one grows. For best sensitivity the laser power is optimum when both mechanisms equally contribute. Equating t_ℓ and t_n [Eqs. (5a) and (17)] we find

$$\langle M_q \rangle_o = \sqrt{\gamma/D} \qquad (22)$$

with $\gamma = 3 \times 10^7$ s^{-1} and $D = 10^{-3}$ s^{-1}, $\langle M_q \rangle_o = 2 \times 10^5$. With 50 oscillating modes, this value corresponds to 0.1 mW of laser output power, which requires us to keep the laser closely above threshold, $\eta = 1.004$. Practically, we may achieve $P/P_{th} \gtrsim 1.01$ when the power stability of the pump laser is on the order of 1%. Thus, in most cw dye lasers, mode coupling by SBS determines the ultimate sensitivity.

From Eqs. (6), (18), and (22), we find the minimum absorptivity, which indicates the absolute ICA sensitivity of a cw dye laser:

$$\kappa_{min} = \sqrt{D\gamma}/c \quad . \qquad (23)$$

This value is $\kappa_{min} = 6 \times 10^{-9}$ cm^{-1} for the example given above, and represents the maximum sensitivity reported so far with cw dye lasers [7,9,10,28,29]. Of course, this sensitivity can be accomplished only if fluctuations of technical origin are avoided, such as gas bubbles in the dye liquid, and, for a broader range of lasers, mechanical vibrations, dust in the optical path of the resonator, instability of the pump light level, and etalon effects due to reflection and/or scattering off resonator components.

In particular situations, other physical mechanisms not discussed so far can affect ICA sensitivity. In relevant studies, the saturation of the absorber [21], partial mode locking [30,31], self-focusing [32,33], and coherent contributions to the interaction of the light with the absorber [34] have been considered. Their effects are important under certain particular experimental conditions only.

3. Intracavity Line Shapes

When the multimode laser emission is analyzed by a spectrally selective detector with integration time τ_d, the light flux in individual modes, according to Sec. 2, is supposed to vary in time and frequency. The detection unavoidably includes time averaging over τ_d or the laser pulse length, t_ℓ, whichever is shorter. If the accumulated energy of the modes — or mode groups, with lower spectral resolution — varies slowly enough with frequency, we may define its envelope as a line shape. With the general condition $\gamma_0 \ll \Gamma$, this line shape is a superposition of the broad emission band and the narrow ICA line shape due to the perturber, of width γ_0.

Experimentally, symmetric and inconspicuous ICA line shapes have been observed [1,2,9,11-13,28]. Quantitative evaluation of ICA line shapes has been achieved with the recording of rotational overtones of O_2 in the visible [35]. Fitting of the lines using Voigt profiles with suitable contributions of Doppler and pressure broadening has been satisfactory, and no systematic line distortion was noticed. On the other hand, several observations have revealed complex asymmetric lines [7,8,23,30-34,36-41], sometimes with spectral light condensation on one or both wings of the line. These line distortions, if intrinsic to ICAS, would severely limit the applicability of the method. Fortunately, the distortions can be traced to two types of effects which can be avoided under most experimental conditions.

3.1 Mode stability of the cavity shifted by extra dispersion

The anomalous dispersion of the intra-cavity absorber modifies the diffraction loss of the cavity, κ_0, such that a frequency-dependent contribution develops [37]. For simplicity we assume that the cavity is formed by two spherical mirrors. We expand that contribution up to the first order in terms of the stability parameter $G(n) = g_1(n) \cdot g_2(n)$ of the cavity,

$$\kappa_d(\omega) = \kappa_0 + (d\kappa/dG)(dG/dn)\Delta n.$$

Here, $n-1$ is the dispersion of the absorber, $g_i(n) = 1 - L/r_i - \ell/nr_i$, and $L + \ell$, r_i, and ℓ are the distance and radii of curvature of the mirrors, and the absorber length, respectively. We combine this spectrally varying loss with the line absorption

$$\kappa(\omega) = k \,\mathscr{L}(\omega) \tag{24}$$

to form the overall extinction $\varepsilon(\omega)$. Here, $k = 2\pi\gamma_0 \cdot N\, c^2/\omega_0^2$, $\mathscr{L}(\omega) = \Gamma_0[\Gamma_0{}^2 + (\omega-\omega_0)^2]^{-1}$, N is the absorber density, γ_0 is the radiative absorption linewidth (HWHM), and ω_0 is the resonance frequency. This extinction is

$$\varepsilon(\omega) = k\,\mathscr{L}(\omega) + (d\kappa_d/dG) \times$$

$$[c_1 k\,\mathscr{L}(\omega) \cdot (\omega-\omega_0)/\Gamma_0 + c_2 k^2\,\mathscr{L}^2(\omega) \cdot (\omega-\omega_0)^2/\Gamma_0^2] \,, \tag{25}$$

where $c_1 = (g_1(1) - \ell/r_1)\ell/r_2 + (g_2(1) - \ell/r_2)\ell/r_1$, and $c_2 = 2\ell^2/r_1 r_2$.

In $\varepsilon(\omega)$, the first term in the brackets represents a line asymmetry, the second one a symmetric distortion. Extra contributions to ε by lensing may exist because of spatial inhomogeneity of the absorber (or of the light field) if nonlinear interaction is appreciable. To test this model, ICA spectra have been recorded with a pulsed LiF:F_2^+ color center laser, using Cs vapor in an intracavity heat pipe as absorber [37]. The observed spectral features are satisfactorily explained by the outlined model.

The analyzed type of line distortion is avoided when a resonator configuration is chosen which corresponds to minimum diffraction loss, i.e., $dk_d/dG = 0$. Then, all the spectrally varying loss is genuine intracavity absorption.

3.2 Nonlinear interactions of the light

In certain experiments including cw dye lasers a characteristic line shape with enhanced red wing has been observed [7,8,23,33,36]. This distortion affects all ICA lines, and all narrow spectral structures, irrespective of their strength. Since it is related to the presence of light tightly focused into the dye jet, it is traced to the same nonlinear interaction of the light with the gain medium, which is responsible for the coupling of individual modes, i.e., SBS (see Fig. 7). To remedy this line distortion the laser pump power must be reduced, and the mode density increased by choosing a long laser cavity.

Enhancement of the light flux on those modes, which lie on one wing, or on both wings, of an ICA line, has been reported many times [30-34,38-41]. This phenomenon seems to develop upon modulation of the (inverted) population in the gain medium. If the period of modulation slightly exceeds that of a round-trip of the light in the cavity, mode coupling is maximum and loss is minimum in the wings of the absorption line, where the mode separation is smaller due to anomalous dispersion and matches the modulation. Stable mode locking somewhat off the transferred modulation frequency has been observed with a bichromatic helium-neon laser [42].

The above phenomenon has been demonstrated with external modulation of the laser gain [40,41]. This parametric interaction of the light can also show up, however, upon (partial) mode locking of the laser to a saturated IC absorber line [30,31,34,38,39]. Due to time-dependent phase shifts between neighboring spectral polarization components, the absorber imposes time-varying gain, rather than loss, upon light modes, which may time-average as net gain at a suitable frequency distance. Thus, this spectrally selective amplification gives rise to enhanced light flux in the wings of the line. The existence of parametric gain, which is similar in its orgin to this gain, has been pointed out in the context of saturation spectroscopy in simple systems that include only two light modes [43,44].

These nonlinear phenomena usually occur at elevated IC laser power levels, and with strong IC absorbers. They can be avoided in many situations of experimental interest.

4. Applications of ICS

In the following we describe the preconditions and several experiments for sensitive and time-resolved detection of linear absorption, of nonlinear absorption, of gain, and of light pulses.

4.1 Detection of linear absorption

From the very beginning, laser interactivity spectroscopy has been considered a technique for the detection of very weak optical extinction. In Sec. 2 we have shown that the spectral output, in mode q, of the broadband laser including loss obeys the Lambert-Beer law $J_q(\omega, t) = J_q(t) \times \exp(-\kappa_q(\omega)ct)$, where the optical path ct corresponds to the length of the uninterrupted wave of a particular laser mode. From the temporal decay of the light flux, the absorption can be determined from at least two measurements of the instantaneous spectral flux at the absorption line, which are separated in time by the interval θ, and corresponding normalization measurements outside the line:

$$\kappa_q = (c\theta)^{-1} \ln \left[\frac{J_q(t)}{J_{q+\delta q}(t)} \Big/ \frac{J_q(t+\theta)}{J_{q+\delta q}(t+\theta)} \right] \quad . \tag{26}$$

In this way, e.g., the absorption of rotational lines of the $000 \rightarrow 043$ transition of the CO_2 molecule was measured with a streak camera for time-resolved spectral recording [13]. With the laser light gated by an electro-optical switch for 10 μs, ro-vibrational overtone spectra of atmospheric H_2O [12] and O_2 [35] were measured. On the other hand, κ_q may be determined by a measurement of the time-integrated light flux, normalized to the flux outside the absorption line:

$$R \simeq \frac{\int_0^\theta J_q(t) \exp(-\kappa_q ct)dt}{\int_0^\theta J_{q+\delta q}(t)dt} \quad . \tag{27}$$

The relationship between κ_q and $\ln R$ is, in general, nonlinear. With constant flux during the pulse time t_ℓ, e.g., we have [13]

$$R = (\kappa_q ct_\ell)^{-1} [1 - \exp(-\kappa_q ct_\ell)] \quad . \tag{28}$$

As pointed out before, the light flux of a homogeneously broadened laser is not constant in time, but rather varies as $\sqrt{t}$ for modes in the central part of the emission profile. For these modes, the analytic result of Eq. (27) is [45]

$$R = \exp[-\kappa_q ct_\ell] \sum_{n=0}^{\infty} \frac{[\kappa_q ct_\ell]^n}{(\frac{3}{2} + n)!} \quad . \tag{29}$$

If $\kappa_q ct_\ell < 1$, we may restrict the sum to the leading term such that κ_q varies linearly with $\ln R$. In this approximation, R is underestimated by less than 10%. Although accuracy and sensitivity of this version of IC measurements is slightly lower than that of time-resolved observation, it is much more convenient and has been used in most experiments so far.

Intracavity spectra are usually recorded photographically or photoelectrically with a spectrograph. Spectral resolution in these measurements is usually limited by the spectrograph. Since a laser beam is analyzed, resolving power need not be traded off for satisfactory transmission. In fact, resolution as high as 10^6 has been achieved with special designs [7].

The photographic recording of time-resolved spectra, or spectrochronograms, includes a streak camera for real-time observations, or some kind
of gating of the laser light, for sampling techniques.

Photoelectric recording makes use of a vidicon, of photodiode arrays,
or of a complete optical multi-channel analyzer. These photoelectric devices allow direct processing of the acquired data.

In various ways, the application of ICAS has extended the spectroscopic
sensitivity limits of conventional absorption measurements. With a Nd-
glass laser for example, extinction values as small as 10^{-9} cm^{-1} have
been measured routinely. Consequently, extremely weak overtones of polyatomic molecules have been detected, e.g., HN_3 ($00000 \to 30000$) (Fig. 1),
CO_2 ($00°0 \to 04°3$), CH_4 ($000 \to 123$), NH_3, C_2H_2, and C_2HD [2].

With the use of cw dye lasers and color center lasers, the number of
accessible molecules and radicals is immense. The sensitivity of a cw dye
laser to very weak absorption has been demonstrated by detecting the 3–0
band of the $b^1\Sigma_g^+ \leftarrow X^3\Sigma_g^-$ magnetic dipole transition of O_2 [15,36]. With
0.016 cm^{-1} resolution, absorption coefficients as low as 10^{-8} cm^{-1} have
been measured, and line shapes have been recorded.

Other experiments were devoted to systematically recording atmospheric
absorption [9,28,29,46]. The duration of undisturbed emission in individual modes in these experiments was several milliseconds, giving rise to
3×10^{-9} cm^{-1} minimum detectable absorptivity. Within the spectral range
accessible with lasing of rhodamine 6G (584 nm through 603 nm), 717 lines
have been recorded, 340 of which were not previously observed in the atmospheric absorption of the solar spectrum. Twenty-eight of these lines
were identified as being caused by weak NO_2 pollution of the air in the
laboratory [28].

4.2 Detection of two-photon absorption

The extension of ICAS to studies of nonlinear absorption is straightforward. A two-photon absorption line, e.g., shows up in ICAS, if the absorber is irradiated simultaneously by additional strong narrow-band
light. A photon from the broad-band laser is chosen, which makes up for
the energy defect between a photon of the narrow-band light and a two-
photon resonance transition. Excitation of this line is characterized by
a cross section that is resonantly enhanced when a real level, whose
parity is opposite to the parity of the ground state, is close to the
intermediate virtual level.

This situation can be met conveniently in alkali vapors. Two-photon
absorption of sodium in a flame is easily detectable this way [47]. A
measurement of the cross section for two-photon absorption in potassium
vapor included a narrow-band ruby laser and a broad-band DOTS dye laser
excited by the same ruby laser [48]. In fact, two states, the fine structure levels $^2P_{1/2}$ and $^2P_{3/2}$, enhance the absorption probability but can
give rise to a destructive interference [49]. This interference of indistinguishable transitions has been demonstrated in two-photon absorption
by monitoring the rate of excitation with the subsequent fluorescence
[50]. Direct two-photon absorption observed recently in an experiment on
sodium vapor is shown in Fig. 8 [51]. A pulsed broad-band dye laser was
tuned to the frequency range of Na 3P-5S absorption, and a pulsed narrow-
band laser close to the 3S $\to$ 3P resonance transitions, say, to the
wavelength λ_{12}. One observes, in the spectrum of the broad-band laser, an

103

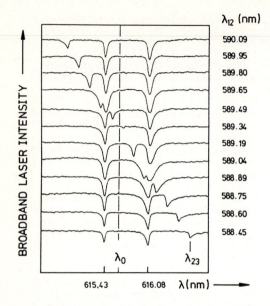

λ_{12} (nm)

590.09
589.95
589.80
589.65
589.49
589.34
589.19
589.04
588.89
588.75
588.60
588.45

Fig. 8.
Two-photon absorption in Na, observed with broad-band dye laser, on the transition 3S-5S. λ_{12}, λ_{23} correspond to first and second photons, respectively. The lines at 615.43 nm and 616.08 nm represent single-photon absorption at the 3P-5S transitions (from Ref. 51).

additional absorption line of wavelength λ_{23}, which varies with λ_{12}, such that $\nu_{12} + \nu_{23} = \nu_{3s-5s}$. This absorption line corresponds to the second photon of the two-photon transition. For $\lambda_{12} = 589,40$; $\lambda_{23} = \lambda_U = 615,64$ the amplitude of the two-photon line vanishes due to destructive interference of the two excitation channels.

4.3 Detection of amplification

Certain groups of modes of a multi-mode laser with a broad emission band may be enhanced in amplitude rather than quenched if frequency-selective gain is made available inside its resonator. In an ICAS experiment using a LiF:F_2^+ color center laser, which was pumped by the green emission of a pulsed xenon ion laser, the absorption of I_2 was to be studied [52]. Surprisingly, gain was observed on numerous ro-vibrational lines (see Fig. 9), since left-over pump light inverted the corresponding transition. In this way, pumping conditions and available inversible transitions can be studied systematically.

4.4 Detection of injected light

So far we have considered applications of ICAS where the gain of the broad-band laser was decremented by spectrally selective absorber loss or incremented by extra gain. There is a third way to exploit the stunning sensitivity of that system to frequency-selective perturbation: the injection of narrowband light pulses. In contrast to the previous variants, the light field is manipulated by placing an excess photon number in a particular mode, or group of modes. In a single-mode laser, the response of the inversion, characterized by time τ, is strong, and it quickly reestablishes the light amplitude. In multi-mode lasers, on the other hand, the instantaneously enhanced light amplitude of a particular mode (or mode group) does not much affect the total inversion. Stimulated emission funnels light into that mode whose amplitude is preserved over a

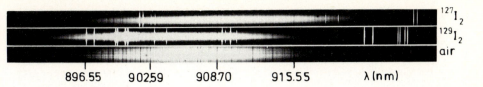

$$^{127}I_2$$

$$^{129}I_2$$

air

896.55   902.59   908.70   915.55   λ(nm)

Fig. 9. IC gain spectra of I_2 isotopes, and atmospheric absorption (from Ref. 52).

much longer time. In practical situations, this time duration is determined by the time of uninterrupted oscillation of the modes in the broad-band laser. In a recent experiment [53] (see Fig. 10), 2-ns pulses of a N_2-pumped coumarine-153 laser of 0.8 cm^{-1} spectral width were injected into a broad-band (30 cm^{-1}) jet-stream Rh-6G laser of 1,5-μs pulse length, excited by a xenon ion laser. The near concentric cavity of the broad-band laser consisted of mirrors with 5 cm and 8 cm radius of curvature. The multi-mode light was analyzed by a 1-m spectrograph and a 1728-channel diode array. Simultaneously, the total instantaneous light flux was temporally resolved by a fast transient digitizer. Figure 11 shows spectra of the broad-band laser emission, and pulse shapes of the total light flux, both for various values of the time delay of the injection. Upon earlier arrival, the injected light pulses redistribute the light generation in the multi-mode laser more efficiently into the spectral channel subject to the injection. This light flux extension enables one to detect weak spectrally selective light pulses, since the time-integrated power of these pulses is amplified at the expense of the multi-mode laser gain. Although the sensitivity has not been pushed to its limits in this exploratory experiment, we stress the point that the prodigious sensitivity of multi-mode lasers to spectrally selective perturbation offers considerable prospects for the detection of light.

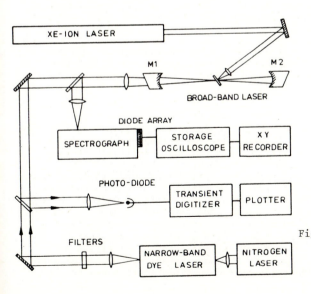

Fig. 10. Experiment for IC detection of narrow-band light emission (from Ref. 53).

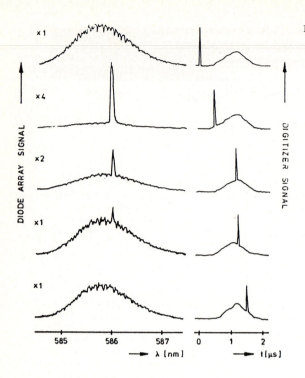

Fig. 11. Emission spectra of broad-band dye laser (left), and time-resolved superposition signal of injected light and spectrally integrated light of the broad-band laser. The injection delay increases from top to bottom (from Ref. 53).

5. The Ultimate Resolution of ICS

The high sensitivity of ICS is complemented by a considerable potential with respect to time and frequency resolution.

5.1 Time resolution

The sensitivity of the ICA technique with multi-mode lasers is determined, as shown above, by the laser pulse duration t_ℓ, if the operating conditions of the laser can be set so as to extend the mean fluctuation time beyond t_ℓ. Then, we have

$$\kappa_{min} \cdot t_\ell = 1/c \quad , \tag{30}$$

and the time resolution, determined by t_ℓ, is limited only by the required sensitivity of the detection. Indeed, time resolution can be traded for sensitivity within wide limits, with 10 ns being the shortest practical resolution time. With the minimum detectable absorptivity 10^{-5} cm^{-1}, e.g., the resolvable time is on the order of a microsecond [45,52]. Thus, ICS is applicable, if we want to track the variation of the macroscopic parameters of an absorber, as an afterglow, and/or the microscopic kinetics of its constituents. As an example, time-resolved absorption spectra of He_2 in the afterglow of a pulsed electric discharge are shown in Fig. 12 [45]. The multi-mode $LiF:F_2^+$ laser, with wavelength range between 900 and 930 nm and with 200 ns pulse duration, probed a pulsed helium discharge with variable delay. Each of the laser pulses was marked by spectral modulation from the absorption by two ro-vibrational bands of the

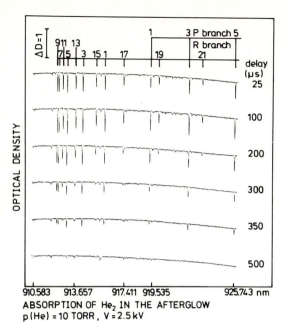

Fig. 12. ICA by He_2 in the afterglow of a He discharge. Excitation voltage 2.5 kV, He pressure 13.3 mbar (from Ref. 45).

electronic transition $c\Sigma_g^+ - a^3\Sigma_u^+$ starting in the lowest metastable state. Hoenl-London plots derived from these measurements enable one to determine the rotational temperature of the He_2 molecules, its deviation from thermal equilibrium, and the variation of temperature and metastable population as functions of the time delay in the afterglow.

High sensitivity along with time resolution is a prerequisite for studies of the kinetics of radicals [54,55]. In fact, ICAS has been shown to be the only analytic technique for measurements of the kinetic constants of reactions involving radicals such as NH_2, HCO, HNO, or PH_2 [55].

5.2 Spectral resolution

The spectral distance of modes in a multi-mode laser determines the spectral resolution in the IC spectrometers used to date. For 0.5 m resonator length, this spectral distance is 0.01 cm^{-1} or 300 MHz. The corresponding resolution is satisfactory for many Doppler-broadened lines and for molecular absorption at atmospheric pressure.

High spectral resolution, corresponding to the emission bandwidth of an individual mode, could be obtained if a variant of the detection technique described above is used. Instead of broadband-detecting the entire spectrum of the laser with a spectrograph, one would use a monochromator as a narrow-band filter for a particular mode. The frequency of this mode would be scanned across the laser spectrum in the conventional way, e.g., by piezo-electric drive of a cavity mirror. A measurement of the light flux in this mode would combine ultra-high sensitivity with high resolution. The latter is, in principle, limited by the Schawlow-Townes condi-

107

tion [56] for the ultimate laser bandwidth only, which is 30 mHz, or smaller. Of course, at present, technical noise keeps the limit at a level several orders of magnitude less favorable, on the order of 1 MHz for commercial laser systems. However, by use of intricate control circuits, cw dye lasers with a bandwidth of some hundred Hz, and He-Ne lasers with a bandwidth of 100 mHz have been demonstrated in the laboratory [57].

6. Conclusions

The scope of this presentation is necessarily limited: we have not aspired to offer a complete account of work on, or with, ICS. In particular, attempting to cover all the pertinent work that has been done in molecular spectroscopy and chemical kinetics would far exceed the limits of this article. Instead, we have tried to point out two facts that are, it seems, not fully appreciated by the broader spectroscopists' community: First, the considerable measure of understanding we now have of the fundamentals that underlie ICS, and second, the level of sophistication and maturity this field has acquired recently.

In adaptation of, but also — with due respect — challenging Arthur Schawlow's popular definition of a diatomic molecule, we claim that there indeed exist niches in physics where ONE is not enough! Eventually, what is lost in precious simplicity of the system under scrutiny is, perhaps, more than regained with inherent features, that allow the spectroscopist to sometimes extend spectacularly the limits previously put on his techniques.

P.E.T. is indebted to the JILA Visiting Fellows Program. V.M.B. appreciates support by the Alexander von Humboldt Foundation.

References

1. L. A. Pakhomycheva, E. A. Sviridenkov, A. F. Suchkov, L. V. Titova, and S. S. Churilov, ZhETP Pisma Red. 12, 60 (1970) [JETP Lett. 12, 43 (1970)].

2. T. P. Belikova, E. A. Sviridenkov, and A. F. Suchkov, Opt. Spektrosk. 37, 654 (1974) [Opt. Spectrosc. 37, 372 (1974)].

3. G. Herzberg, Molecular Spectra and Molecular Structure (Van Nostrand, New York, 1960), p. 427.

4. N. C. Peterson, M. J. Kurylo, W. Braun, A. M. Bass, and R. A. Keller, J.O.S.A. 61, 746 (1971).

5. R. J. Thrash, H. von Weyssenhof, and T. S. Shirk, J. Chem. Phys. 55, 4559 (1971).

6. T. W. Hänsch, A. L. Schawlow, and P. E. Toschek, J. Quant. Electron. QE-8, 802 (1972).

7. Yu. M. Ajvasjan, V. M. Baev, V. V. Ivanov, S. A. Kovalenko, and E. A. Sviridenkov, Kvantovaja Elektron. 14, No. 2 (1987).

8. H. Atmanspacher, H. Scheingraber, and V. M. Baev, Phys. Rev. A 35, 142 (1987).

9. V. M. Baev, T. P. Belikova, E. A. Sviridenkov, and A. F. Suchkov, Zh. Eksp. Teor. Fiz. 74, 43 (1978) [Sov. Phys.-JETP 74, 21 (1978)].

10. V. M. Baev, T. P. Belikova, S. A. Kovalenko, E. A. Sviridenkov, and A. F. Suchkov, Kvantovaja Elektron. 7, 903 (1980) [Sov. J. Quantum Electron. 10, 517 (1980)].

11. E. N. Antonov, V. G. Koloshnikov, and V. R Mironenko, Opt. Comm. 15, 99 (1975).

12. F. Stoeckel, M. A. Melieres, and M. Chenevier, J. Chem. Phys. 76, 2191 (1982).

13. T. P. Belikova, B. K. Dorofeev, E. A. Sviridenkov, and A. F. Suchkov, Kvantovaja Elektron. 2, 1325 (1975) [Sov. J. Quantum Electron. 5, 722 (1975)].

14. N. A. Raspopov, A. N. Savchenko, and E. A. Sviridenkov, Kvantovaja Elektron. 4, 736 (1977) [Sov. J. Quantum Electron. 7, 409 (1977)].

15. S. A. Kovalenko, Kvantovaja Elektron. 8, 1271 (1981) [Sov. J. Quantum Electron. 11, 759 (1981)].

16. V. R. Mironenko and V. J. Yudson, Opt. Comm. 34, 397 (1980).

17. V. M. Baev, G. Gaida, H. Schröder, and P. E. Toschek, Opt. Comm. 38, 309 (1981).

18. S. J. Harris, J. Chem. Phys. 71, 4001 (1979).

19. E. N. Antonov, A. A. Kachanov, V. R. Mironenko, and T. V. Plakhotnik, Opt. Comm. 46, 126 (1983).

20. S. J. Harris, Appl. Opt. 23, 1311 (1984).

21. H. Atmanspacher, H. Scheingraber, and C. R. Vidal, Phys. Rev. A 32, 254 (1985).

22. H. Atmanspacher, H. Scheingraber, and C. R. Vidal, Phys. Rev. A 33, 1052 (1986).

23. F. Stoeckel, G. H. Atkinson, Appl. Opt. 29, 3591 (1985).

24. Yu. M. Ajvasjan, V. M. Baev, A. A. Kachanov, and S. A. Kovalenko, Kvantovaja Elektron. 13, 1723 (1986).

25. Yu. M. Ajvasjan, V. M. Baev, T. P. Belikova, S. A. Kovalenko, E. A. Sviridenkov, and O. I. Yushchuk, Kvantovaja Elektron. 13, 612 (1986) [Sov. J. Quantum Electron. 16, 397 (1986)].

26. V. R. Mironenko, Kvantovaja Elektron. 7, 2069 (1980).

27. W. Brunner and H. Paul, Opt. Quantum Electron. 12, 393 (1980).

28. V. M. Baev, T. P. Belikova, M. B. Ippolitov, E. A. Sviridenkov, and A. F. Suchkov, Opt. Spektrosk. 45, 58 (1978) [Opt. Spectrosc. 45, 31 (1978)].

29. A. A. Kachanov and T. V. Plakhotnik, Opt. Comm. 47, 257 (1983).

30. Y. H. Meyer and M. N. Nenchev, Opt. Comm. 41, 292 (1982).

31. V. M. Baev, T. P. Belikova, O. P. Varnavskij, V. F. Gamalij, S. A. Kovalenko, and E. A. Sviridenkov, ZhETP Pisma Red. 42, 416 (1985) [JETP Lett. 42, 514 (1985)].

32. T. S. Zeilikovich, S. A. Pulkin and L. S. Gaida, Zh. Eksp. Teor. Fiz. 87, 125 (1984) [Sov. Phys.-JETP 60, 72 (1984)].

33. W. T. Hill III, T. W. Hänsch, and A. L. Schawlow, Appl. Opt. 24, 3718 (1985).

34. V. S. Egorov and I. A. Chekhonin, Opt. Spektrosk. 52, 591 (1982) [Opt. Spectrosc. 52, 355 (1982)].

35. M. Chenevier, M. A. Melieres, and F. Stoeckel, Opt. Comm. 45, 385 (1983).

36. W. T. Hill, R. A. Abreu, T. W. Hänsch, and A. L. Schawlow, Opt. Comm. 32, 96 (1980).

37. H. Schröder, K. Schultz, and P. E. Toschek, Opt. Comm. 60, 159 (1986).

38. Ya. I. Khanin, A. G. Kagan, V. P. Novikov, M. A. Novikov, I. N. Polushkin, and A. I. Shcherbakov, Opt. Comm. 32, 456 (1980).

39. Y. H. Meyer, Opt. Comm. 30, 75 (1979).

40. A. N. Rubinov, M. V. Belobon, and A. V. Adamushko, Kvantovaja Elektron. 6, 723 (1979) [Sov. J. Quantum Electron. 9, 433 (1979)].

41. V. M. Baev, V. F. Gamalij, E. A. Sviridenkov, and D. D. Toptygin, Kratk. Soob. po Fiz. No. 8, 6 (1986) [Sov. Phys. Lebedev Institute Report].

42. W. Neuhauser and P. E. Toschek, Opt. Comm. 11, 331 (1974).

43. S. Haroche and F. Hartmann, Phys. Rev. A 6, 1280 (1972).

44. M. Sargent III and P. E. Toschek, Appl. Phys. 11, 107 (1976).

45. B. Stahlberg, V. M. Baev, G. Gaida, H. Schröder, and P. E. Toschek, J. Chem. Soc. Faraday Trans. 81, 207 (1985).

46. V. M. Baev, S. A. Kovalenko, E. A. Sviridenkov, A. F. Suchkov, and D. D. Toptygin, Kvantovaja Elektron. 7, 1112 (1980) [Sov. J. Quantum Electron. 10, 638 (1980)].

47. J. P. Reilly and J. H. Clark, Proc. Int. Conf. on Advances in Laser Chemistry, California Institute of Technology, Pasadena, Calif., 1978, ed. by A. W. Zewail (Springer-Verlag, Berlin, 1978), p. 355.

48. V. M. Baev, T. P. Belikova, V. F. Gamalij, E. A. Sviridenkov, and
A. F. Suchkov, Kvantovaja Elektron. 11, 2413 (1984) [Sov. J. Quantum
Electron. 14, 1596 (1984)].

49. V. M. Baev, V. F. Gamalij, E. A. Sviridenkov, and D. D. Toptygin,
Kratk. Soob. po Fiz. No. 8, 3 (1986) [Sov. Phys. Lebedev Institute
Report].

50. J. E. Bjorkholm and P. F. Liao, Phys. Rev. Lett. 33, 128 (1974).

51. K. Boller, V. M. Baev, and P. E. Toschek, Opt. Comm., in press.

52. V. M. Baev, H. Schröder, and P. E. Toschek, Opt. Comm. 36, 57 (1981).

53. V. M. Baev, K.-J. Boller, A. Weiler, and P. E. Toschek, Opt. Comm., in
press.

54. D. M. Sarkisov, E. A. Sviridenkov, and A. F. Suchkov, Chimicheskaya
Fizika 9, 1155 (1982).

55. F. Stoeckel, M. Schuh, N. Goldstein, and G. H. Atkinson, Chem. Phys.
95, 135 (1985).

56. A. L. Schawlow and C. H. Townes, Phys. Rev. 112, 1940 (1958).

57. J. L. Hall, private communication.

Spectroscopic Applications
of Frequency Modulated Dye Lasers

A.I. Ferguson, S.R. Bramwell, and D.M. Kane

Department of Physics, University of Southampton,
Southampton SO9 5NH, UK

1. INTRODUCTION (AIF)

My interest in frequency modulated lasers was stimulated when I was a visiting scholar at Stanford in the Schawlow/Hänsch group from October 1977 to February 1979. During the first part of my visit to Stanford I worked with Jim Eckstein on the use of a coherent train of mode-locked dye laser pulses as a source for Doppler-free coherent multiple pulse spectroscopy [1]. The idea behind these experiments was that a mode-locked laser could provide a source of coherent radiation with high peak power but yet could be used for high-resolution spectroscopy. At Jim's PhD viva, Tony Siegman asked if he thought that frequency modulated lasers could be used for this kind of spectroscopy. On hearing about this question, most of us smiled knowingly and then began to wonder exactly what is a frequency modulated laser! I can't remember how Jim managed to extricate himself from this question but the next day several of the people involved in this project went off to the library to find out about frequency modulated lasers.

It did not take long to discover that nearly all the important work on frequency modulated lasers had been done at Stanford. Steve Harris and Russel Targ had discovered in 1964 that, when a phase modulator crystal of KDP, inserted into the cavity of a He Ne laser operating at 633 nm, was driven close to the cavity mode spacing, a broadband frequency modulated output could be obtained [2]. In the time domain the amplitude was constant with the carrier frequency sinusoidally sweeping over the Doppler width. In the frequency domain this appeared as a series of equally spaced modes with Bessel function amplitudes. In 1965 Steve Harris and Otis McDuff developed a theory of frequency modulated lasers [3]. In this linear theory for an inhomogeneously broadened laser they assumed that each mode was coupled to its nearest neighbours by an intracavity phase perturbation. Several characteristic features came out of this theory. Firstly, they discovered that the modulation index, Γ of the FM oscillation was given by

$$\Gamma = \left(\frac{\delta}{\pi} \right) \left(\frac{\Omega}{\nu_m - \Omega} \right) , \tag{1}$$

where δ is the single pass phase retardation of the phase modulator, Ω is the cavity mode spacing and ν_m is the modulation frequency. We refer to $\Delta\nu = \nu_m - \Omega$ as the detuning. They also discovered that there were three regions of operation, called the FM, unquenched and phase-locked regions. The FM region corresponds to constant amplitude with Bessel function mode amplitudes and is characterised by a relatively large detuning. As the detuning is reduced the unquenched region is reached where several FM oscillations may occur. At the smallest detunings the mode amplitudes become Gaussian in shape and the laser consists of a periodic train of pulses. This is sometimes called FM mode-locking. Many of these predictions were confirmed by the work of Amman et al [4] using a He Ne laser.

In 1970 Dirk Kuizenga and Tony Siegman investigated the FM operation of a Nd:YAG laser [5]. They had already worked on the theory and experiment of FM mode-locking of a Nd:YAG laser [6,7]. They discovered that the FM bandwidth was much greater in Nd:YAG than the mode-locked bandwidth. They also emphasised that despite the large bandwidth, FM lasers are coherent and can be viewed in a generalised sense as 'single mode'.

As far as we could see in the Schawlow/Hänsch group, not much more had been done on frequency modulated lasers. The idea that larger bandwidths could be obtained in the FM region than in the mode-locked region was tantalising. Jim Eckstein and I had managed to generate bandwidths of up to 500 GHz coherently, with a synchronously pumped mode-locked dye laser [8]. Could an FM dye laser give bandwidth well in excess of a THz?

The spectroscopic aspects of FM dye lasers looked interesting. For example, lasers separated by a large frequency interval could be compared by heterodyning the reference and unknown laser frequency together with the FM laser. The low frequency beat between the reference laser and one of the modes of the FM laser and a similar beat with the unknown laser, together with a knowledge of the equally spaced FM laser mode frequency could be used to measure the laser frequency differences modulo the FM mode spacing. Less precise methods could establish the integer mode separation between the lasers unambiguously. The difference frequency between the lasers could be as large as the bandwidth but only low frequency beats (~ 100 MHz) would have to be measured.

Other intriguing possibilities came up. For example Heinz Weber and Ernst Mathieu [9] had predicted in 1967 that an FM beam could be divided into two halves, one beam being delayed by half the modulation period and the two beams being combined in a nonlinear crystal in such a way that only the sum frequency was generated. The resulting output would be a single frequency at twice the carrier frequency of the FM oscillation. In a similar vein, we at Stanford worked out that if an FM beam was used for Doppler-free two-photon spectroscopy, with the counter-propagating beam delayed by half the modulation period, a spectrum identical to the single frequency spectrum could be obtained as the carrier frequency was swept through the two-photon resonance. The theory of this process was worked out by Ted Hänsch and N.C. Wong in 1980 [10].

Our ideas about FM dye lasers went untested until I moved to Southampton in 1984 and was fortunate to obtain a Metrology Award and research contract from the National Physical Laboratory. What is described in the rest of this paper is the development of an FM dye laser together with some of the performance characteristics, experiments on single frequency UV generation using an FM laser and experiments on Doppler-free two-photon spectroscopy using an FM laser.

2. FM DYE LASER

When we started developing FM dye lasers it was not clear which would be the best configuration to pursue. Dramatic differences between the performance of single frequency lasers had been observed for standing-wave and ring dye lasers. We decided to purchase a commercial dye laser which could be operated in a ring configuration or standing-wave configuration and was versatile enough to enable a variety of tuning elements to be tested (Coherent 699-21). The dye was chosen to be Rhodamine 6G since it was thought to be typical and because it was efficient and easy to use. Detailed performance of ring and standing-wave lasers have been reported in previous publications [11, 12, 13]. The main difference between ring and standing wave was that the bandwidth of the standing wave laser was twice that for a ring and that the phase modulator had to be placed close to one of the end mirrors of the standing-wave laser. It was found that the bandwidth of the FM laser was well described by eqn. (1). In Fig. 1 we show the spectrum of a standing-wave FM dye laser restricted to run over a narrow range of frequencies by two intracavity etalons and a 3-plate birefringent filter as a function of detuning for a fixed single pass phase retardation. We compare this spectrum with the spectrum predicted for a pure FM oscillation. The observed and predicted spectra show excellent agreement.

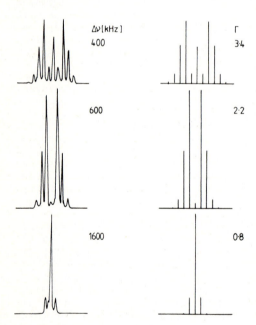

Fig.1 FM optical spectra obtained by modulating a single frequency dye laser as a function of detuning. The line spectra on the right are computer generated Bessel function spectra with the modulation index indicated. The mode spacing is 127 MHz

In order to attempt to develop a laser with as large a bandwidth as possible, we have studied dye lasers without etalons. In the standing-wave and bidirectional ring we found that it is difficult to obtain a single pure FM oscillation. We have now settled on the use of a unidirectional ring laser design. When such a laser is free-running it tends to operate on a single mode which is rather free to wander. The application of phase modulation tends to stabilise this mode and gives rise to a single FM oscillation.

A schematic diagram of this laser is shown in Fig. 2. The basic laser was unchanged from the commercial laser (Coherent 699-21) consisting of four mirrors, one of which was mounted on a piezoceramic. This could be used to make rapid adjustments to the cavity length when frequency stabilising. Slow changes in the cavity length were provided by a fused silica plate, mounted at Brewster's angle on a galvanometer, and intersecting two beams. The laser was made to operate in one direction by use of a device based on the Faraday effect. Tuning was accomplished with a 3-plate birefringent filter. An etalon of 225 GHz free spectral range and 25% reflectivity could be incorporated for further spectral narrowing and control. Several different phase modulators have been tested and are located in the laser where etalons are usually placed.

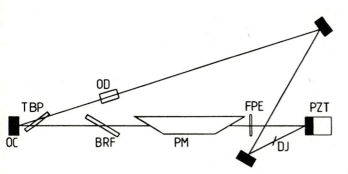

Fig.2 Schematic diagram of the FM dye laser. The symbols represent the following: OC, output coupler; OD, optical diode; BRF, birefringent filter; PM, phase modulator; DJ, dye jet, PZT, piezoelectric tweeter mirror; FPE, etalon; TBP, tipping Brewster plate.

Most of the phase modulators that have been tested consist of 45° y-cut ADP crystals. ADP is available in large sizes of excellent optical quality but has a rather small electro-optic coefficient. Several schemes for driving these crystals have been studied. These include inductively coupling to a coil which forms part of a resonant circuit with the capacitance of the crystal. One of the more successful phase modulators is a commercial unit (Gsänger) consisting of two Brewster angled ADP crystals configured to give no deflection or displacement of the laser beam. This modulator has rather a large capacitance and was normally directly driven. The usual source for driving the systems was a frequency synthesiser and 5W amplifier. For the ring laser the modulation frequency was in the region of 181 MHz. In Fig. 3 we show some typical spectra taken using this modulator at a single pass phase retardation of 1 mrad for different detunings. Bandwidths of greater than 30 GHz could easily be obtained.

In more recent tests we have obtained $MgO:LiNbO_3$ modulators. These tend to be smaller and of poorer optical quality than ADP but have much larger

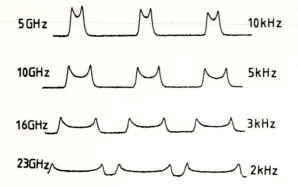

5GHz ... 10kHz

10GHz ... 5kHz

16GHz ... 3kHz

23GHz ... 2kHz

Fig.3   Spectra of the FM dye laser as a function of detuning observed with 30 GHz free spectral range interferometer.   The detuning $\Delta\nu = \nu_m - \Omega$ is given on the right and the measured bandwidth is given on the left.

electro-optic coefficients.   We have obtained a single pass phase retardation of δ = 3 radians at 181 MHz using this material.   This gives a bandwidth of up to 6 THz, although this appears to be strongly dependent on the bandpass of the tuning element.

At these exceptionally large bandwidths it is difficult to analyse the FM spectrum to see how closely the output approximates to a pure FM oscillation.   We have used the techniques of single frequency UV generation and Doppler-free two-photon spectroscopy to help us analyse the spectrum.   In both of these methods all modes contribute to the signal.   Experiment and analysis show that amplitude distortions of the FM output appear as sidebands around the UV or two-photon signal whereas frequency distortions give rise to broadened signals.   We will now describe these techniques in more detail.

3. SINGLE FREQUENCY ULTRAVIOLET GENERATION

The simplest way to understand the principle of single frequency UV generation using an FM laser is to consider the instantaneous electric field $E(t)$ of an FM oscillation at carrier frequency ω_0 given by

$$E(t) = \frac{1}{2}\ E_o\ \exp\ [-i\ \{\omega_0 t + \Gamma \sin \Omega t\}] + c.c. ,$$

where we have arbitrarily set the phase of the FM oscillation to zero. If we now take this same FM oscillation but introduce a phase delay of θ radians, the field is given by

$$E'(t) = \frac{1}{2}\ E_o\ \exp\ [-i\ \{\omega_0 t + \Gamma \sin (\Omega t + \theta)\}] + c.c.$$

If these two beams interact by a second-order nonlinear process we can pick out the sum-frequency oscillation given by

$$E''(t) \propto E(t) \; E'(t) + c.c$$

$$\propto E_o^2 \; \exp \; [i\{2\omega_o \; t_o + 2 \; \Gamma(\cos(\theta/2)) \; \sin(\Omega t + \theta/2)\}] + c.c.$$

This represents an FM oscillation centred on a carrier frequency of $2\omega_o$ with modulation frequency Ω and modulation index $2\Gamma\cos(\theta/2)$. If we set the phase delay to be $\theta = \pi$ the modulation index becomes zero and we are left with a single frequency oscillation at twice the original carrier frequency. We can investigate the spectrum we would expect to see when we detune from $\theta = \pi$ by introducing a detuning phase given by $\Delta\theta = \pi - \theta$. The modulation index of the FM oscillation will then become $2\Gamma\sin(\Delta\theta/2)$.

The experimental arrangement for demonstrating its effect consists of our standard FM ring dye laser which is split into two components. One component is delayed and recombined, with the aid of a lens, at a crystal of ADP, phase-matched for sum frequency mixing at the laser wavelength. The sum frequency is generated at the bisector of the two beams and then enters a UV confocal interferometer for spectral analysis. A typical pair of traces is shown in Fig. 4. This shows the spectrum of the dye laser observed with a 30 GHz free spectral range confocal interferometer at the bottom together with the spectrum of the sum frequency UV observed with a 2 GHz free spectral range interferometer. The bandwidth of the dye laser in this case was 7.7 GHz whereas the UV linewidth was less than

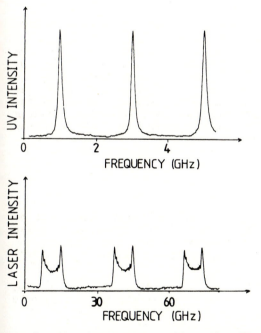

Fig.4 The lower trace shows the FM laser bandwidth of 7.7 GHz as observed with a 30 GHz free spectral range interferometer. The upper trace shows the UV spectrum observed on a 2 GHz free spectral range interferometer.

the resolution of the interferometer. The effect of changing the delay from the θ = π condition is shown in Fig. 5. These curves were taken with the laser operating in the same conditions as those applying to Fig. 4. We have indicated on this figure the measured phase delay Δθ and compared the generated UV spectra with ideal FM spectra to deduce the effective modulation index Γ'. We have found excellent agreement between the predicted and measured modulation index.

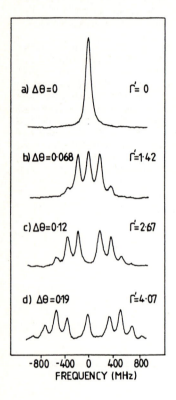

Fig.5 Sum-frequency spectrum as a function of phase delay Δθ. The effective modulation index Γ' is also indicated.

The single frequency UV radiation can be used to perform atomic or molecular spectroscopy. Alternatively, the generated UV could be locked to a stabilised interferometer. Either method would enable the complete FM spectrum to be stabilised to provide a comb of reference frequency over a large bandwidth.

4. DOPPLER-FREE TWO-PHOTON SPECTROSCOPY

Doppler-free two-photon spectroscopy using an FM laser can be thought of in a similar way to single frequency UV generation. However, in this case counter-propagating beams are required and instead of considering sum-frequency generation we consider two-photon excitation.

The experimental arrangement consists of the FM dye laser focussed into a cell containing sodium, at a temperature of about 180°C and tuned to the 3S - 4D transition. This beam is then recollimated and propagates to a mirror after which it returns to the cell [15]. The distance between the cell and mirror is adjusted to be half of the modulation period, corresponding to π phase delay. The two-photon excitation is detected by UV fluorescence from the sodium cell. The dye laser frequency is stabilised by locking to a 2 GHz free spectral range confocal interferometer. This means that the dye laser bandwidth is restricted to at most 2 GHz. The laser carrier frequency is swept by slaving the FM laser to this cavity.

A typical Doppler-free two-photon spectrum is shown in Fig. 6b at π phase delay. Also shown on the same figure is a Doppler-free two-photon spectrum obtained with the same laser system but operated in a single mode. The two spectra are almost identical. Note however, that the Doppler-broadened background is slightly higher in the single frequency case. This is predicted by the Hänsch and Wong theory [10] and is due to the fact that in the FM case the laser bandwidth exceeds the Doppler width and hence the laser power within the Doppler broadened line is smaller than in the single frequency case. The strength and width (8MHz) of the Doppler-free signals are the same in both cases as is expected [10].

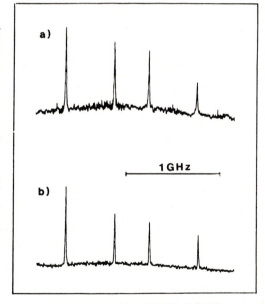

LASER FREQUENCY SCAN

Fig.6a Doppler-free two-photon spectrum of the 3S-4D transition in sodium using single frequency dye laser.
6b The same transition observed using an FM laser of 1.6 GHz bandwidth.

5. CONCLUSIONS

We have developed an FM dye laser capable of producing an FM oscillation over a region in excess of 1THz. We have devised ways in which the spectrum of the FM oscillation can be analysed and compared with a pure FM oscillation. In these techniques of sum frequency UV generation and two-photon Doppler-free spectroscopy, amplitude distortions appear as sidebands and frequency distortions appear as spectral broadening. These methods can also be used to stabilise the FM dye laser to an absolute reference.

As phase modulators are improved we expect to be able to generate even wider bandwidths. This opens up many possibilities for optical metrology and spectroscopy. A further intriguing possibility is that of using dispersion, perhaps in a fibre or grating, to compress this broad spectrum into a train of ultrashort pulses. The bandwidths that we have already achieved suggest that sub-picosecond pulses are possible.

We see no reason why the techniques of frequency modulation that we have developed should not be used in other dyes, and indeed other homogeneously broadened lasers. This will open up possibilities of investigations of atoms throughout the visible region of the spectrum.

REFERENCES

1. J.N. Eckstein, A.I. Ferguson, T.W. Hänsch: Phys. Rev. Lett. 40, 847 (1978).

2. S.E. Harris, R. Targ: Appl. Phys. Lett. 5, 202 (1964).

3. S.E. Harris, Otis McDuff: IEEE J. QE-1, 245 (1965).

4. E.O. Ammann, B.J. McMurtry, M.K. Oshman: IEEE J. QE-1, 263 (1965).

5. D.J. Kuizenga, A.E. Siegman: IEEE J. QE-6, 673 (1970).

6. D.J. Kuizenga, A.E. Siegman: IEEE J. QE-6, 694 (1970).

7. D.J. Kuizenga: IEEE J. QE-6, 709 (1970)

8. A.I. Ferguson, J.N. Eckstein, T.W. Hänsch: J. Appl. Phys. 49, 5389 (1978).

9. H.P. Weber, E. Mathieu: IEEE J.QE-3, 376 (1967).

10. T.W. Hänsch, N.C. Wong: Metrologia 16, 101 (1980).

11. D.M. Kane, S.R. Bramwell, A.I. Ferguson: Appl. Phys. B39, 171 (1986).

12. D.M. Kane, S.R. Bramwell, A.I. Ferguson: Appl. Phys. B40, 147 (1986).

13. D.M. Kane, S.R. Bramwell, A.I. Ferguson: In Laser Spectroscopy VII ed. by T.W. Hänsch, Y.R. Shen, Springer Ser. Opt. Sci. 49 (Springer Verlag, Berlin 1985) pp.362-365.

14. S.R. Bramwell, A.I. Ferguson, D.M. Kane: Opt. Commun. 61, 87 (1987).

15. S.R. Bramwell, A.I. Ferguson, D.M. Kane: Opt. Commun. (submitted for publication).

Reminiscence of Schawlow at the First Conference on Lasers

D.F. Nelson

AT&T Bell Laboratories, Murray Hill, NJ 07974, USA

Art Schawlow's legendary good naturedness, humor, and showmanship evidenced themselves early in the development of lasers, in fact, at the first session of a conference ever to be concerned with lasers. That conference was the Ann Arbor Conference on Optical Pumping organized by Peter Franken and Richard Sands of the Physics Department of the University of Michigan in June 1959.

I was just completing a post-doctoral appointment at Michigan and had accepted a research position in optical physics at Bell Telephone Laboratories in Murray Hill, New Jersey starting in July of that year. I was interested in pursuing both efficient electroluminescence and the attainment of an "optical maser", as the laser was then generally called. Thus, I attended the Optical Pumping Conference with considerable interest. Little did I realize then that within a little over a year I would be coauthoring a paper on lasers with Art.

Art, having coauthored the now famous Schawlow-Townes paper published in the Physical Review in December 1958, was chairman of Miscellaneous Session I which included a number of papers on proposed optical masers. There were papers by Gordon Gould on pumping processes and optical cavities, J. H. Sanders on possible optical maser action in helium, Ali Javan on attaining a "negative temperature", as a population inversion was then generally called, in neon by resonant collisional excitation with helium, and Irwin Wieder on optical pumping of ruby. The session concluded with remarks by Art Schawlow on the "neighbor lines" in dark ruby arising from chromium ion pairs and their possible use as optical maser transitions, a use that Art was successful in attaining a year and half later.

The paper by Gordon Gould is well remembered by me because of three interesting uses of language, two by Gould and one by Schawlow in a comment. Gould in his paper entitled "The LASER, Light Amplification by Stimulated Emission of Radiation" first introduced the acronym LASER in place of the then current term optical MASER. Many of us would resist that new name for a time, fearing the need to call other devices IRASERs, UVASERs, etc. depending on their wavelength of emission. But the attainment of the laser a year later led to so many dramatically new results that the device demanded a name of its own, not a name derived from the original, but less useful, maser.

The second use of language memorable to me arose when Gordon Gould reported he had applied for $50,000 of federal funding to pursue the laser idea but had been awarded $250,000 (as best I can recall the numbers)! Thus, he referred to funding from Uncle Sam as coming from "Uncle Cornucopia". I have found many occasions in the years since to refer to the largesse of federal funding in many programs as coming from Uncle Cornucopia.

As Gould's paper concluded, Session Chairman Schawlow could not resist - as always - a note of humor. His humor correctly foresaw the laser's main use as an oscillator, not an amplifier, and also expressed the great uncertainty, even improbability, felt by everyone at the time about the possible attainment of a laser. Beginning in mock solemnity and ending in belly-shaking laughter, Art opined that the LASER was likely to be most used as an oscillator and so should be named "Light Oscillation by Stimulated Emission of Radiation", or the "LOSER". Art was certainly right that the LASER is usually a LOSER, but as time has amply shown certainly not a loser!

Part II

Atomic and Molecular Spectroscopy

*A diatomic molecule is
a molecule with one atom
too many*

Laser and Fourier Transform Techniques for the Measurement of Atomic Transition Probabilities

J.E. Lawler

Department of Physics, University of Wisconsin, Madison, WI 53706, USA

ABSTRACT

Radiative lifetimes for many atoms and ions are determined using time-resolved laser-induced fluorescence on slow atomic and ionic beams. These lifetimes are combined with precise branching ratios measured using a Fourier transform spectrometer to determine accurate absolute atomic transition probabilities for the elements in low stages of ionization.

1. INTRODUCTION

A powerful combination of techniques from laser spectroscopy and Fourier transform spectroscopy is playing a central role in determining accurate atomic transition probabilities for the elements in low stages of ionization. This has been one of the most persistent problems of atomic spectroscopy. Classical spectroscopists measured the wavelengths of the major spectral lines of most elements to part per million accuracy. Techniques for measuring Einstein A coefficients for these spectral lines lagged far behind techniques for wavelength measurements. For some elements not a single transition probability was known to within a factor of two before the recent application of laser techniques. The lack of reliable transition probability data has been a serious problem for many fields. The problem has been severe for astronomy, which is in some sense the parent field of spectroscopy. Accurate absolute atomic transition probabilities are essential for determining the abundances of chemical elements in the sun, other stars, and interstellar clouds. These abundance studies provide critical information on the distribution and production of chemical elements throughout the observable universe.

Many techniques have been used to measure atomic transition probabilities including: the Hanle effect, the Hook method, absolute emission measurements on thermal arcs, beam-foil time-of-flight techniques, time-correlated photon counting with pulsed electron beam excitation, and others. Each technique has its own strengths and limitations. None of these older techniques provided the accuracy, convenience, and extremely broad applicability of laser techniques that are now available. Radiative lifetime measurements to ~5% absolute accuracy are now routine using laser techniques. The 5% figure is typical of the systematic plus random uncertainties on many lifetime measurements, and is typical of the level of agreement between measurements by independent groups.[1,2] Accuracy to better than 1% is possible. Measurements on highly refractory species such as Nb, Ta or even Nb$^+$ or Ta$^+$ using lasers and hollow cathode atomic/ionic beam sources are now as routine as measurements on the alkalies.[3-7]

The radiative lifetimes do not by themselves determine individual Einstein A coefficients unless one branch dominates decay from the level. The lifetimes provide the essential absolute normalization needed to convert relative intensity measurements or branching ratios into absolute transition probabilities. Fortunately, the vast improvement in radiative lifetime measurements coincided with very important progress on branching ratio measurements. Fourier transform spectrometers, such as the 1.0 meter FTS at the National Solar Observatory on Kitt Peak, are now used to measure branching ratios.[8,9] The 1.0m FTS has extraordinary advantages for spectrophotometry on complex atoms and ions. The 1.0m FTS offers Doppler limited resolution, wavenumber accuracy to one part in 10^8, a very high data collection rate, and has the intrinsic advantage of simultaneous measurement of all spectral elements. An elegant technique for producing a relative intensity calibration of an FTS is available.[9] The potent combination of radiative lifetimes measured using time-resolved laser-induced fluorescence and of branching ratios measured using the 1.0m FTS is rapidly producing an enormous improvement in knowledge of atomic transition probabilities.

This article is a review of recent work with laser and FTS techniques on only part of the periodic table, specifically the transition metals with open 3d, 4d and 5d shells. The transition metals include the Fe group which is quite important in astrophysics, and include other highly refractory metals which have significant technological applications. The very high melting points of many of these metals had previously provided a challenge to experimentalists. The complexity of the spectra and atomic structure makes theoretical calculations of the transition probabilities for these elements very difficult.

2. RADIATIVE LIFETIMES

A low-pressure gas phase sample or an atomic/ionic beam is needed in order to measure radiative lifetimes using time-resolved laser-induced fluorescence. Some of the laser measurements of transition metal lifetimes involved thermal atomic beam sources, including early work on FeI, CoI, and NiI.[10-14] It has become increasingly apparent that discharge (sputter) sources are superior for this work. Hannaford and Lowe at C.S.I.R.O. developed a sputtering cell used for radiative lifetime measurements.[15] Duquette, Salih, and Lawler developed a hollow cathode atomic/ionic beam source.[5,16] Both devices have extremely broad applicability and have been used with substantial success. The hollow cathode beam source has an advantage in that there are no effects due to collisional quenching in a beam environment. Although collisional quenching of longer lived levels is observed at the 0.1 to 5 Torr pressures used in the cell experiments, the effect of collisional quenching can be eliminated by extrapolation to zero pressure. The sputtering cell technique has an advantage in that it limits contamination during work on radioactive elements such as Tc. Figure 1 is a schematic of the sputtering cell experiment.

The hollow cathode atomic/ionic beam experiment which is shown in Fig. 2, will be described in some detail. Many of the ideas which led to the hollow cathode beam source were developed during the author's two year stay at Stanford with the Schawlow-Hänsch group. The hollow cathode atomic/ionic beam source is based on a low pressure, large bore hollow cathode discharge tube. The hollow cathode is used as a beam source by sealing one end of the cathode except for a 1.0 mm diameter opening. The opening is flared outward at 45° to serve as a nozzle for forming an

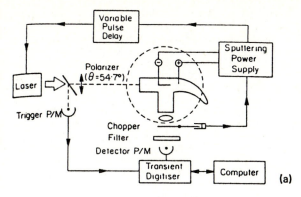

Fig. 1(a) Schematic diagram of the sputtering cell experiment developed by P. Hannaford and R. M. Lowe at C.S.I.R.O. in Melbourne, Australia. (b) The timing sequence for time-resolved laser-induced fluorescence measurement of atomic and ionic lifetimes.

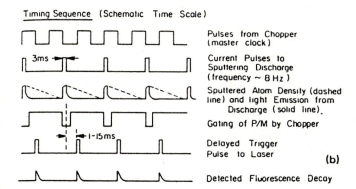

Timing Sequence (Schematic Time Scale)

Pulses from Chopper (master clock)

Current Pulses to Sputtering Discharge (frequency ~ 8 Hz)

Sputtered Atom Density (dashed line) and light Emission from Discharge (solid line).

Gating of P/M by Chopper

Delayed Trigger Pulse to Laser

(b)

Detected Fluorescence Decay

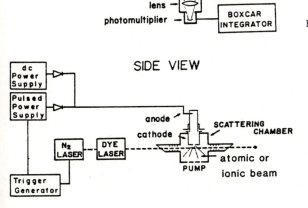

TOP VIEW

SIDE VIEW

Fig. 2 Schematic diagram of the hollow cathode atomic/ionic beam experiment developed by D. W. Duquette, S. Salih and J. E. Lawler at the Univ. of Wisconsin. The experiment is used for time-resolved laser-induced fluorescence measurements of atom and ion lifetimes in a beam environment.

uncollimated atomic or ionic beam. The 29 mm inside diameter, 50 mm long stainless steel cathode is lined with a 0.25 mm thick foil of the metal to be studied. The hollow cathode and the scattering chamber are at ground potential. Argon, the carrier gas for the discharge flows continuously into the hollow cathode discharge. A 10 cm diffusion pump is used to evacuate the scattering chamber. The scattering chamber is sealed from the hollow cathode discharge, except for the nozzle, and is maintained at a much lower pressure. The argon pressure in the discharge is usually 0.3 Torr, and the resulting scattering chamber pressure is approximately 10^{-4} Torr of argon. A direct current of 50 mA to 200 mA in the hollow cathode discharge is typically used to produce a beam of neutral atoms. A pulsed power supply which delivers up to 25A for 5 μsec is used to produce an intense burst of slow ($\leq$1eV) ions in the beam. Figure 3 is a scale drawing which shows materials and dimensions for the source.

The source has been used to produce atomic or ionic beams of Mo, Zr, W, Nb, Hf, Rh, Ta, Re, Al, Be, Fe, Co, Ni, Ti, V, Mg, Pt, Ru, and other species. It is believed that the source will work well with all metallic elements. The absolute intensity of the metal atom beam was measured by depositing and weighing a Nb film. The Nb atom beam intensity on axis is 1.8×10^{14} atoms/(sec sr) at 180 mA of discharge current.[3] It is assumed that all Nb atoms stuck to the glass substrate, thus the intensity should be considered a lower limit.

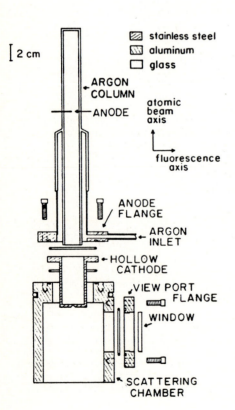

2 cm

▨ stainless steel
▨ aluminum
☐ glass

ARGON COLUMN

ANODE

atomic beam axis

fluorescence axis

ANODE FLANGE

ARGON INLET

HOLLOW CATHODE

VIEW PORT FLANGE

WINDOW

SCATTERING CHAMBER

Fig. 3 Scale drawing of the hollow cathode atomic/ionic beam source developed by D. W. Duquette, S. Salih, and J. E. Lawler.

128

The hollow cathode atomic/ionic beam source and the discharge cell approach share other advantages besides convenience and very broad applicability throughout the periodic table. The free atoms and ions they produce are distributed among many metastable levels. It is routine to study levels which are not directly connected to the ground level, or which are connected to the ground level by VUV transitions. Metastable levels as high as 30,000 cm^{-1} above the ground level of MoI were found to have useful populations when the discharge is pulsed.[17] Metastable levels as high as 14,000 cm^{-1} above the ground level of the ion in NiII were found to have useful populations.[18] Metastable levels in inert gases, including the metastable levels of He at 166,000 cm^{-1} above the ground state, have useful populations. Certain levels in He, which are accessible from the metastable levels, have lifetimes known to better than 1% from theoretical calculations. These levels are very useful for testing the overall performance of the lifetime measuring apparatus.[19] The hollow cathode atomic/ionic beam source is gaining increased acceptance, even among groups which previously preferred thermal sources.[6,7]

The pulsed dye lasers used in the time-resolved laser-induced fluorescence studies are in most cases not too different from the dye laser developed by Hänsch.[20] The intracavity telescopic beam expander introduced by Hänsch has in many cases been replaced by a prismatic beam expander[21], or a grazing incidence grating[22], but much of the dye laser cavity is unchanged. These pulsed dye lasers are now pumped by N$_2$, Nd-YAG, or eximer lasers. The dye lasers typically produce a pulse with a duration of 3 to 8 nsec and a bandwidth of 3 to 8 GHz. The transverse pumping of the dye cell produces an exceeding high (~10^6) unsaturated gain. The high unsaturated gain makes the high losses of the cavity tolerable. The 20 cm long cavity used for lifetime experiments at the Univ. of Wisconsin is typical. It has a round trip loss of 99.5%. Only 0.5% of the light survives a round trip because of the 4% reflection of output mirror, the 25% transmission of the double passed prismatic beam expander, and the estimated losses of 50% due to the dye cell and grating. The short, very lossy cavity has an important advantage for lifetime work: it has negligible capacity to store light. Once the pump laser power drops below threshold the dye laser output decreases very abruptly. The abruptness of the dye laser pulse termination is far more important than the pulse duration when measuring short lifetimes. It is not necessary to deconvolute the excitation pulse from the fluorescence data, if the first five nanoseconds of the fluorescence decay curve is discarded. Clean, single exponential decays are observed for lifetimes as short as 3 nsec.

These pulsed dye laser oscillators are often followed by one or more amplifiers in order to produce sufficient power for nonlinear generation of UV light. Frequency doubling crystals of KDP and KB5 provide tunability to 217 nm in the UV. Dye lasers pumped by high power Nd:YAG or eximer lasers will provide tunability into the VUV. Raman shifting of the dye laser frequency in H$_2$ provides continuous tunability to 138 nm.[23,24] The shorter wavelengths will result in excitation of very short lived levels. Fortunately each nonlinear step, whether frequency summing or doubling in a crystal, Raman shifting in H$_2$, or four wave sum frequency mixing in metal vapor, will further sharpen the temporal distribution of the dye laser pulse.

The simplicity and very broad tunability of the pulsed dye lasers is an enormous advantage. Spatially resolved laser-induced fluorescence on fast ionic beams is a good technique[25], but it has not been used as widely as time-resolved laser-induced fluorescence on slow atomic and ionic beams.

Spatially-resolved laser-induced fluorescence requires an accelerator and is limited to wavelength ranges where c.w. dye lasers are available. The requirement for a c.w. laser is not fundamental, but is dictated by signal-to-noise considerations.

Several different approaches are used to filter, detect, and log fluorescence signals in time-resolved laser-induced fluorescence experiments.[1,3,10] The use of a grating monochromator to filter the fluorescence is not necessary because of the selective laser excitation. A grating monochromator offers certain advantages: it can provide additional confidence that the laser excitation line is not blended, has been correctly classified, and is correctly identified in the experiment. The substantial loss of fluorescence signal intensity due to the grating monochromator is not a disadvantage if time-correlated photon counting is used in detecting and logging the fluorescence. Typically ~0.1 photon per laser pulse is detected in time-correlated photon counting experiments. The low counting rate is essential to prevent severe distortion of the fluorescence decay curve.

It is desirable to collect as much fluorescence as possible when using a boxcar integrator or transient digitizer to log the fluorescence signals. The very fast f/1 collection system shown in Fig. 2 is used with dye or interference filters.[3] Minimal spectral filtering provided by dye or interference filters can sometimes be used to block all laser light scattered from windows etc., if the branching ratios are favorable. It remains necessary to detect fluorescence at the laser wavelength in many cases, thus it is important to minimize the amount of scattered laser light. The use of very high quality Brewster windows and extensive light baffling is essential to minimize scattered laser light. Background light from the discharge in cell experiments and from hollow cathode beam source in beam experiments is not a problem. The discharge is pulsed in the cell experiments and the laser is fired after an appropriate delay as indicated in Fig. 1.[1] The several millisecond delay is chosen such that light emission from the discharge has died away, but sputtered metal atoms are still present in the cell. Background light from the hollow cathode atomic beam source shown in Fig. 2 is often 10^4 times weaker than the peak fluorescence signals.[3] This high signal-to-background ratio is achieved without spectral filtering.

The use of a boxcar integrator provides a substantial (10^2 or 10^3) advantage over time-correlated photon counting in data collection rate.[3] A transient digitizer, though, provides a 10^4 to 10^5 advantage over time-correlated photon counting in data collection rate.[1] Thousands of photons can be logged per laser shot using a transient digitizer. It is often possible to collect an excellent decay curve in seconds using a transient digitizer. The Tektronix 7912AD digitizer provides mV sensitivity, an analogue bandwidth of 0.5 GHz, and a sampling rate up to 100 GHz.

The primary advantage of a fast boxcar or a time-correlated photon counting system is that they have the potential to measure subnanosecond lifetimes. This advantage has not yet been exploited in work on transition metals, but it should be important for VUV studies.

Many good photomultipliers are available which are well suited for time-resolved laser-induced fluorescence studies. Several groups use the 1P28A.[15,16] It is inexpensive and rugged. It provides high gain (~10^6), linearity, and freedom from ringing when used in a properly designed base.

The base should be designed for very low inductance, and it should include bypass capacitors and small damping resistors.[26] The 1P28A provides excellent linearity to peak anode currents of 10 mA. Although single photon spikes are 2 nsec wide, they have very steeply rising and falling edges. The 1P28A has adequate electronic bandwidth for lifetimes as short as 2 or 3 nsec. Faster photomultipliers will be necessary in the VUV. Several faster, new types of photomultipliers are available.[27]

The large signals achieved in time-resolved laser-induced fluorescence experiments are important in reducing statistical errors, but the greatest strengths of the experiments are their nearly complete freedom from systematic errors. The selective excitation provided by tunable dye lasers eliminates the radiative cascading problem which plagued beam-foil time-of-flight experiments. Minimal spectral filtering of the fluorescence is occasionally necessary to block cascade fluorescence from lower levels in laser-induced fluorescence experiments. Collisional quenching is not observed in experiments involving laser excitation of atoms or ions in a beam. Systematic studies to test for collisional quenching of long-lived levels are routinely performed by varying the beam intensity, but no collision quenching has been observed. Collisional quenching of long-lived levels is observed at the 0.1 to 5 Torr pressures used in the cell experiments, but it is not difficult to extrapolate the observed lifetime to zero pressure.[1] The corrections are usually comparable to the 3 to 8% uncertainties on the lifetimes. Radiation trapping on strong resonance lines of metals is not a problem. Although routine tests are performed by varying the beam intensity when studying strong resonance lines, radiation trapping on resonance lines of metal atoms has not been observed in beam experiments using a hollow cathode source. Radiation trapping can also be easily avoided in the cell experiments. Distortion of fluorescence decay curves by Zeeman quantum beats is avoided by making zero field (<20 mG) measurements on short lifetimes (<300 nsec), and high field (30 G) measurements on long lifetimes.[3] Any possible distortion of the fluorescence decay curve due to zero-field hyperfine quantum beats can be avoided by using "magic" angle polarization.[1]

Time-resolved laser-induced fluorescence is used routinely on lifetimes from a few nanoseconds to a few microseconds. New detection systems and new lasers are making it possible to measure lifetimes substantially less than a nanosecond.[28] The accuracy of time-resolved laser-induced fluorescence measurements of lifetimes over a microsecond has been limited by error due to atoms leaving the observation region before radiating. Recent experiments demonstrate that this effect can be largely eliminated by using time-of-flight selection in the atomic beam experiments.[29] Lifetimes greater than a microsecond are now being measured to accuracies of 5%. Time-of-flight selection is achieved by pulsing the hollow cathode atomic beam source. The laser excitation occurs some distance (~1 cm) from the source nozzle. The delay between the current pulse to the hollow cathode discharge source and the laser pulse is used to select atoms of a particular velocity. Error due to atoms leaving the observing region before radiating is greatly reduced by extrapolating to infinite delay (zero atomic velocity). Some recent measurements on the $4s^2 4p\ ^2P_{1/2}$ level of ScI are shown in Fig. 4.

The various precautions and systematic studies are important in establishing the accuracy of the lifetime measurements. The high level of agreement routinely achieved by independent groups using time-resolved laser-induced fluorescence gives a great deal of credibility to the 3 to 8% accuracy claimed for the lifetime measurements. Three independent groups published lifetime measurements in NbI during 1982.[3,30,31] All three

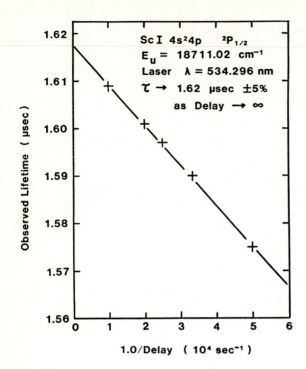

Fig. 4 Lifetime measurements on a ScI velocity selected beam using time-resolved laser-induced fluorescence. The horizontal axis which is labeled 1.0/Delay in units of 10^4 sec^{-1} may also be read as atomic velocity in units of 10^4 cm/sec. Each data point is an average lifetime determined from thousands of fluorescence decay curves recorded using a transient digitizer. The ±5% total uncertainty includes systematic and random effects.

groups used time-resolved laser-induced fluorescence on Nb atomic beams. To the best of this author's knowledge the groups were unaware of each others work prior to publication. Duquette and Lawler measured lifetimes for 50 NbI levels.[3] Rudolph and Helbig measured lifetimes for 6 NbI levels.[30] Kwiatkowski, Zimmerman, Biemont, and Grevesse measured lifetimes for 11 NbI levels.[31] The superior performance of the hollow cathode atomic beam source used by Duquette and Lawler is evident from the larger number of measurements they made. All lifetimes measured by more than one of the groups are compared in Table I. The excellent agreement indicated by the comparison in Table I is now considered routine for lifetime measurements made using time-resolved laser-induced fluorescence.

Table II is a list of some recent research papers reporting transition metal radiative lifetimes measured using laser-induced fluorescence. Accuracy of 3 to 8% is achieved in most of the measurements reported in Table II. Measurements accurate to better than 1% are possible with extraordinary care using time-resolved laser-induced fluorescence.

3. BRANCHING RATIOS

A radiative lifetime by itself determines an Einstein A only if a single transition dominates the decay of the level. Branching ratios for many levels in metal atoms and some levels in ions can be derived from the Corliss and Bozmann (CB) monograph.[80] The extensive CB monograph is from absolute emission measurements on intense arcs. Absolute transition

Table I. A comparison of NbI radiative lifetimes measured using time-resolved laser-induced fluorescence. Uncertainties of approximately 5% were claimed for all measurements.

	Level	Energy (cm^{-1})	Lifetime (nsec)		
			Ref[3]	Ref[30]	Ref[31]
$4d^35s(a^3F)5p$	$z^4F_{5/2}$	23574	44.5		42.7
	$z^4F_{7/2}$	24015	38.6		37.3
	$z^4F_{9/2}$	24507	35.9		33.2
$4d^4(a^5D)5p$	$y^6F_{1/2}$	23985	8.5		8.0
	$y^6F_{3/2}$	24165	8.6		8.3
	$y^6F_{7/2}$	24770	8.5		8.1
	$y^6F_{9/2}$	25200	8.2	8.6	7.9
	$y^6F_{11/2}$	25680	7.8	7.5	
$4d^35s(a^5P)5p$	$y^6D_{7/2}$	26832	25.8		25.3
	$y^6D_{9/2}$	27420	15.1	14.7	
$4d^4(a^5D)5p$	$x^6D_{1/2}$	26552	8.0		8.3
	$x^6D_{3/2}$	26713	7.3	7.1	
	$x^6D_{5/2}$	26983	7.0	7.3	6.7
	$x^6D_{7/2}$	27427	7.7	7.9	7.9

Table II. Recent radiative lifetime measurements on 3d, 4d, and 5d transition metals using laser-induced fluorescence.

Atom or Ion	Number of Levels	Reference	Atom or Ion	Number of Levels	Reference
ScI	13	[32]	CrI	3	[37]
	59	[29]		1	[38]
				28	[39]
ScII	8	[25]			
	15	[29]	MnI	2	[40]
TiI	17	[33]	MnII	5	[41]
TiII	18	[34]	FeI	10	[10]
				1	[11]
VI	9	[33]		6	[12]
	12	[35]		25	[14]
	39	[36]	FeII	13	[42]

Table II (continued)

Atom or Ion	Number of Levels	Reference	Atom or Ion	Number of Levels	Reference
CoI	8	[12]	ZrII	1	[15]
	10	[43]		20	[51]
				5	[1]
CoII	14	[19]			
			NbI	50	[3]
NiI	8	[13]		6	[30]
	11	[44]		11	[31]
NiII	12	[18]	NbII	7	[5]
				27	[54]
CuI	4	[45]			
	2	[1]	MoI	6	[16]
				9	[55]
AgI	2	[1]		6	[30]
	1	[61]		56	[56]
				14	[17]
CdI	14	[62]			
	16	[63]	MoII	15	[57]
LaI	4	[64]	RuI	12	[58]
				50	[59]
LaII	7	[65]			
			RhI	13	[31]
HfI	25	[66]		22	[4]
TaI	35	[4]	PdI	9	[60]
TaII	6	[6]	WII	3	[6]
	10	[7]			
			ReI	9	[70]
WI	15	[67]			
	13	[68]	OsI	6	[71]
	39	[69]			
			IrI	25	[72]
ZnI	11	[46]			
	5	[47]	PtI	15	[73]
YI	9	[48]	AuI	2	[74]
	10	[49]			
	34	[50]	HgI	3	[75]
				1	[76]
YII	14	[50]		1	[77]
				1	[78]
ZrI	11	[15]			
	34	[51]	HgII	1	[79]
	11	[52]			
	15	[53]			
	36	[1]			

probabilities from this source are known to have many large errors, but branching ratios derived from it are reasonably accurate for branches greater than 0.1.[50] Errors in CB absolute transition probabilities due to lack of local thermodynamic equilibrium, and to inaccurate knowledge of the electron temperature and metal atom density, do not affect the accuracy of branching ratios derived from the CB monograph. Intensity and wavelength-dependent errors in the photographic calibration do affect the branching ratios. The CB monograph is nevertheless a useful source of branching ratios for stronger ($\geq$0.1) branches.

Remeasurement of the branching ratios is desirable because of the enormous advantage of photoelectric detectors over the photographic system used by CB. Some branching ratios have been remeasured using grating monochromators, but it has become increasingly clear that Fourier transform spectrometers (FTS) are very well suited to this work. The 1.0m FTS at the National Solar Observatory is probably the best instrument in the world for spectrophotometry on complex atoms and ions.[8] High spectral resolution and good absolute wave number accuracy are important to reduce line blending and to insure correct line identification. The 1.0m FTS provides Doppler limited resolution and absolute wave number accuracy to one part in 10^8. A high data collection rate is also desirable for spectrophotometry on complex atoms or ions. The 1.0m FTS records a million point interferogram covering a spectral region from the edge of the VUV to the IR in approximately one hour. The spectra are produced in a convenient computer format for analysis, because the interferograms are recorded and transformed digitally. The 1.0m FTS has a five decade dynamic range with very good linearity. Fourier transform spectrometers also have an intrinsic advantage over grating monochromators for spectrophotometry. A grating monochromator which sequentially scans each spectral element will map any small drift in source intensity into a branching ratio error. The interferogram recorded by a FTS represents simultaneous measurements on all spectral elements. A FTS is therefore less sensitive to small drifts in source intensity.

Conventional hollow cathode discharge lamps are often used in spectrophotometry with the 1.0m FTS. These lamps are simple, very reliable, and broadly applicable. The lamps usually contain a few Torr of Ne or Ar. The metal emission lines are strong and are broadened primarily by the Doppler effect. The translational temperatures are often only slightly above room temperature. The low collision rate in the lamps implies severe departures from local thermodynamic equilibrium, but this is not a concern in branching ratio measurements. It is important to verify that the lamps are optically thin. A test for radiation trapping is performed by recording spectra at several discharge currents, and verifying that the branching ratios are unchanged. Radiation trapping is sometimes observed. Accurate branching ratios are then determined by extrapolating to zero discharge current.

The relative intensity calibration as a function of wave number is the primary concern when using the 1.0m FTS for spectrophotometry. Whaling and collaborators, who were the first to use the 1.0 FTS for spectrophotometry in the visible and UV, devised an elegant method for determining the relative efficiency versus wave number of the FTS.[9] Overlapping sets of Ar branching ratios are used to construct the efficiency versus wave number curve for the FTS. The selected ArI and ArII branching ratios are for transitions between excited levels which are not prone to radiation trapping. The selected branching ratios cover the region from 4300 cm^{-1} to 34000 cm^{-1}, and are reasonably accurate ($\sim$4%).[9,81] The use of ArI and

ArII branching ratios to calibrate the FTS has important advantages. Argon is often the carrier gas in the hollow cathode discharge. Thus a calibration spectrum is recorded with each metal spectrum. Furthermore the calibration automatically includes any effects due to variations in the transmission of the source window or due to variations in the reflectance of the back of the hollow cathode. The 1.0m FTS has recently been used to measure branching ratios in FeI [82,83], CoI [84], YI and YII [50], MoI [17,56], CoII [19], VI [36], CrI [85], RhI [86], NbI [87], HfI [87], Ta [69], and WI [69].

4. OTHER DEVELOPMENTS

It is essential to include all major branches when combining radiative lifetimes and emission branching ratios to determine transition probabilities for individual spectral lines. Spectra from the 1.0m FTS are usually analyzed to determine branching ratios using an interactive computer. All known energy levels are entered in the computer and the spectra are searched for every possible transition between known energy levels. The tables of Atomic Energy Levels by C. E. Moore are often used in the search.[88]

Although the 1.0m FTS is well suited to spectrophotometry in the IR, the possibility of strong IR branches beyond the long wavelength limit of the spectra is occasionally a cause for concern. If the level of interest has a relatively short lifetime (~10 nsec) and strong visible or UV branches, then one can be quite confident that the visible and UV branches are dominant. An IR branch from such a level would need an extraordinarily large oscillator strength to have a significant branching ratio.

Short-lived levels with strong visible or UV branches serve as reference levels in an experiment by Duquette, Den Hartog and Lawler.[87] This "missing branch experiment" is used to screen emission branching ratios for strong unidentified IR branches. The magnitudes of the laser-induced fluorescence signals produced by driving two transitions sharing a common lower level are compared in this experiment. One transition is to the reference level and the second is to the level to be tested for missing IR branches. If the test level has strong unidentified IR branches with a total branching ratio R, then the fluorescence from the test level will be suppressed by a factor of $(1-R)^2$. The technique is sensitive to the square of the missing branches because the failure to include IR branches in predicting the visible or UV fluorescence affects both laser excitation and subsequent fluorescence. The use of a common lower level eliminates any need to determine an atomic density. The spectral fluence produced by each dye laser as it is scanned across the spectral lines must be measured, and the relative efficiency versus wavelength of the fluorescence collection system must be known. The missing branch experiment is performed on atoms in a beam using much of the same apparatus used in the lifetime experiments. The beam environment eliminates any problems from collisional quenching of the fluorescence. The hollow cathode atomic beam source makes the missing branch experiment very broadly applicable. It can be applied to any metallic element, and many metastable levels have sufficient populations to serve as lower levels for laser excitation using the hollow cathode beam source. The missing branch experiment has been applied to long-lived levels of Hf, Nb, W, and Ta.[69,87] Only the decays of the z^5F levels of HfI were found to be dominated by IR branches. The strong IR branches were subsequently identified and measured on FTS spectra.

The missing branch experiment described here is used to verify the completeness of emission branching ratio. Relative magnitudes of laser-induced fluorescence signals from transitions sharing a common lower level could also be used in a fashion analogous to the powerful "bowtie" method of Cardon et. al.[89] The bowtie method combines emission oscillator strengths and relative absorption measurements in a least square adjustment. The use of relative laser-induced fluorescence signals instead of relative absorption signals will make the bowtie method far more broadly applicable.

5. APPLICATIONS IN ASTRONOMY

Although accurate atomic transition probabilities are needed in many fields, the need is most acute in astronomy. The wealth of new data on atomic transition probabilities from laser spectroscopy and Fourier transform spectroscopy is already having an impact in astronomy. The new transition probability data have resulted in greatly improved solar abundance determinations for La [65], Zr [51], Co [84], Pd [60], Re [70], Y [50], Rh [4,31], Mo [90], Ru [58], Os [71], V [36], and Nb [54].

These very accurate spectroscopically determined solar abundances are in excellent agreement with elemental abundances determined from chondritic meteorites. The discrepancy between a Nb solar abundance determined using NbI lines and the Nb abundance in chrondritic meteorites has been resolved.[31,54] Line misidentification and/or blending in the solar spectrum was the source of the problem. The NbI lifetimes and oscillator strengths are accurate. Recent work on NbII transitions has resolved the apparent disagreement between the spectroscopic solar abundance and that determined from chrondritic meteorites.[54] The high level of agreement between spectroscopically determined solar abundances and abundances from chrondritic meteorites is not accidental. The meteorites are believed to be a representative sample of matter in the solar system. The elemental abundances should agree for most non-volatile elements.

The precision solar abundance studies are important for many reasons. In some sense the solar abundances work paves the way for precision spectroscopic studies on other stars with elemental abundances much different than the sun. An exciting example is recent work on χ Cygni and o Ceti by Dominy and Wallerstein.[91] Certain red giants with enhanced heavy element abundances contain the unstable element Tc. The longest lived Tc isotope has a half-life of only 2×10^5 years which is short on stellar time scales. The enhanced heavy element abundances and the existence of Tc is believed to be due to neutron capture during shell flashes. Spectroscopic elemental abundance studies of the solar and stellar atmospheres, and of interstellar clouds,will ultimately produce a more quantitative understanding of the production and distribution of chemical elements.

6. SUMMARY

A combination of developments in laser spectroscopy and in Fourier transform spectroscopy is leading to enormous progress in determining atomic transition probabilities for the elements in low stages of ionization. Time-resolved laser-induced fluorescence is widely used to determine radiative lifetimes with small (~5%) systematic and random uncertainties. The development of sputter techniques such as the hollow cathode atomic/ionic beam source makes it possible to apply laser-induced fluorescence to neutral atoms and ions of all metallic elements. The application of powerful Fourier transform spectrometers in the visible and UV is providing branching ratios which complement the lifetime measurements. It is hoped that in a few years it will be possible to assemble a compendium of atomic transition probabilities, as extensive as the CB monograph, but with a vastly improved accuracy of 5 to 10%. The recent progress in measuring atomic transition probabilities is resulting in precision elemental abundance determinations in the sun, other stars, and in interstellar clouds. The abundance studies are essential for developing an understanding of the production and distribution of elements in the observable universe.

ACKNOWLEDGMENTS

The preparation of this manuscript was supported by NSF Grant AST85-20413. I wish to acknowledge the profound influence of Art Schawlow on my life and work.

LITERATURE

1. P. Hannaford, R. M. Lowe: Opt. Engineering 22, 532 (1983)
2. J. Richter: Physica Scripta T8, 70 (1984)
3. D. W. Duquette, J. E. Lawler: Phys. Rev. A 26, 330 (1982)
4. S. Salih, D. W. Duquette, J. E. Lawler: Phys. Rev. 27, 1193 (1983)
5. S. Salih, J. E. Lawler: Phys. Rev. 28, 3653 (1983)
6. M. Kwiatkowski, F. Naumann, K. Werner, P. Zimmermann: Phys. Lett. A103, 49 (1984)
7. W. Schade, V. Helbig: Phys. Lett. A115, 39 (1986)
8. J. W. Brault: J. Opt. Soc. Am. 66, 1081 (1976)
9. D. L. Adams, W. Whaling: J. Opt. Soc. Am. 71, 1036 (1981)
10. H. Figger, K. Siomos, H. Walther: Z. Physik 270, 371 (1974)
11. K. Siomos, H. Figger, H. Walther: Z. Physik A272, 355 (1975)
12. H. Figger, J. Heldt, K. Siomos, H. Walther: Astron. and Astrophys. 43, 389 (1975)
13. J. Heldt, H. Figger, K. Siomos, H. Walther, Astron. and Astrophys. 39, 371 (1975)
14. J. Marek, J. Richter, H. J. Stahnke: Physica Scripta 19, 325 (1979)
15. P. Hannaford, R. M. Lowe: J. Phys. B14, L5 (1981)
16. D. W. Duquette, S. Salih, J. E. Lawler: Phys. Lett. A83, 214 (1981)
17. W. Whaling, P. Chevako, J. E. Lawler: J. Quant. Spectrosc. Radiat. Transfer 36, 491 (1986)
18. J. E. Lawler, S. Salih: Phys. Rev. A (in press).
19. S. Salih, J. E. Lawler, W. Whaling: Phys. Rev. A31, 744 (1985)
20. T. W. Hänsch: Appl. Optics 11, 895 (1972)
21. G. K. Klauminzer: U. S. Patent 4, 127, 828, Nov. 28 (1978)
22. M. G. Littman, H. J. Metcalf: Appl. Optics 17, 2224 (1978)
23. H. S. Schomburg, H. F. Dobele, B. Ruckle: Appl. Phys. B30, 131 (1983)
24. K. G. H. Baldwin, J. P. Marangos, D. D. Burgess, M. C. Gower: Optics Comm. 52, 351 (1985)
25. O. Vogel, L. Ward, A. Arnesen, R. Hallin, C. Nordling, A. Wännstrom: Physica Scripta 31, 166 (1985)
26. J. M. Harris, F. E. Lytle, T. C. McCain: Analytical Chem. 48, 2095 (1976)
27. B. Leskovar, Laser Focus/Electro-Optics 20, 73 (1984)
28. I. Yamazaki, N. Tamai, H. Kume, H. Tsuchiya, K. Oba: Rev. Sci. Instrum. 56, 1187 (1985)
29. E. A. Den Hartog, G. Marsden, J. E. Lawler, J. T. Dakin, V. Roberts: Bull. Am. Phys. Soc. (in press) Presented at the 39th G. E. C. (1986)
30. J. Rudolph, V. Helbig: Phys. Lett. A89, 339 (1982)

31. M. Kwiatkowski, P. Zimmermann, E. Biemont, N. Grevesse: Astron. and Astrophys. $\underline{112}$, 337 (1982)
32. J. Rudolph, V. Helbig: J. Phys. B$\underline{15}$, L1 (1982)
33. J. Rudolph, V. Helbig: J. Phys. B$\underline{15}$, L599 (1982)
34. M. Kwiatkowski, K. Werner, P. Zimmerman: Phys. Rev. A$\underline{31}$, 2695 (1985)
35. A. Doerr, M. Kock, M. Kwiatkowski, K. Werner, P. Zimmerman: J. Quant. Spectrosc. Radiat. Transfer $\underline{33}$, 55 (1985)
36. W. Whaling, P. Hannaford, R. M. Lowe, E. Biemont, N. Grevesse: Astron. Astrophys. $\underline{153}$, 109 (1985)
37. R . M. Measures, N. Drewell, H. S. Kwong: Phys. Rev. A$\underline{16}$, 1093 (1977)
38. H. S. Kwong, R. M. Measures: Appl. Optics $\underline{19}$, 1025 (1980)
39. M. Kwiatkowski, G. Micali, K. Werner, P. Zimmermann: Astron. Astrophys. $\underline{103}$, 108 (1981)
40. H. D. Kronfeldt, J. R. Kropp, A. Subaric, R. Winkler: Z. Physik A$\underline{322}$, 349 (1985)
41. M. Kwiatkowski, G. Micali, K. Werner, P. Zimmermann: J. Phys. B$\underline{15}$, 4357 (1982)
42. P. Hannaford, R. M. Lowe: J. Phys. B$\underline{16}$, L43 (1983)
43. J. Marek, K. Vogt: Z. Physik A$\underline{280}$, 235 (1977)
44. U. Becker, H. Kerkhoff, M. Schmidt, P. Zimmermann: J. Quant. Spectrosc. Radiat. Transfer $\underline{25}$, 339 (1981)
45. H. Kerkhoff, G. Micali, K. Werner, A. Wolf, P. Zimmermann, Z. Physik-A$\underline{300}$, 115 (1981)
46. H. Kerkhoff, M. Schmidt, P. Zimmermann: Z. Physik A$\underline{298}$, 249 (1980)
47. M. Chantepie, J. L. Cojan, J. Landais, B. Laniepce, A. Moudden, M. Aymar: Opt. Comm. $\underline{51}$, 391 (1984)
48. P. Hannaford, R. M. Lowe: J. Phys. B$\underline{15}$, 65 (1982)
49. J. Rudolph, V. Helbig: J. Phys. B$\underline{15}$, L1 (1982)
50. P. Hannaford, R. M. Lowe, N. Grevesse, E. Biemont, W. Whaling: Astrophys. J. $\underline{261}$, 736 (1982)
51. E. Biemont, N. Grevesse, P. Hannaford, R. M. Lowe: Astrophys. J. $\underline{248}$, 867 (1981)
52. D. W. Duquette, S. Salih, J. E. Lawler: Phys. Rev. A$\underline{25}$, 3382 (1982)
53. J. Rudolph, V. Helbig: Z. Physik A$\underline{306}$, 93 (1982)
54. P. Hannaford, R. M. Lowe, E. Biemont, N. Grevesse: Astron. Astrophys. $\underline{143}$, 447 (1985)
55. M. Kwiatkowski, G. Micali, K. Werner, P. Zimmermann: Phys. Lett. A$\underline{85}$, 273 (1981)
56. W. Whaling, P. Hannaford, R. M. Lowe, E. Biemont, N. Grevesse: J. Quant. Spectrosc. Radiat. Transfer $\underline{32}$, 69 (1984)
57. P. Hannaford, R. M. Lowe: J. Phys. B$\underline{16}$, 4539 (1983)
58. E. Biemont, N. Grevesse, M. Kwiatkowski, P. Zimmermann: Astron. Astrophys. $\underline{131}$, 364 (1984)
59. S. Salih, J. E. Lawler: J. Opt. Soc. Am. B$\underline{2}$, 422 (1985)
60. E. Biemont, N. Grevesse, M. Kwiatkowski, P. Zimmermann: Astron. Astrophys. $\underline{108}$, 127 (1982)
61. K. P. Selter, H. J. Kunze, Astrophys. J. $\underline{221}$, 713 (1978)
62. H. Kerkhoff, M. Schmidt, U. Teppner, P. Zimmermann, J. Phys. B$\underline{13}$, 3969 (1980)
63. M. Chantepie, J. L. Cojan, J. Landais, B. Laniepce, A. Moudden, M. Aymar: Optics Comm. $\underline{46}$, 93 (1983)
64. B. R. Bulos, A. J. Glassman, R. Gupta, G. W. Moe: J. Opt. Soc. Am. $\underline{68}$, 842 (1978)
65. A. Arneson, A. Bengtsson, R. Hallin, J. Lindskog, C. Nordling, T. Noreland: Physica Scripta $\underline{16}$, 31 (1977)
66. D. W. Duquette, S. Salih, J. E. Lawler: Phys. Rev. A$\underline{26}$, 2623 (1982)
67. D. W. Duquette, S. Salih, J. E. Lawler, Phys. Rev. A$\underline{24}$, 2847 (1981)

68. M. Kwiatkowski, G. Micali, K. Werner, M. Schmidt, P. Zimmerman: Z. Physik A304, 197 (1982)
69. E. A. Den Hartog, D. W. Duquette, J. E. Lawler: J. Opt. Soc. Am. B 4, 48 (1987)
70. D. W. Duquette, S. Salih, J. E. Lawler: J. Phys. B15, L897 (1982)
71. M. Kwiatkowski, P. Zimmermann, E. Biemont, N. Grevesse, Astron. Astrophys. 135, 59 (1984)
72. D. S. Gough, P. Hannaford, R. M. Lowe: J. Phys. B16, 785 (1983)
73. D. S. Gough, P. Hannaford, R. M. Lowe, J. Phys. B15, L431 (1982)
74. P. Hannaford, P. L. Larkins, R. M. Lowe: J. Phys. B14, 2321 (1981)
75. E. N. Borisov, A. L. Osherovich, V. N. Yakovlev, Opt. Spectrosc. 47, 109 (1979)
76. E. N. Borisov, A. L. Osherovich: Opt. Spectrosc. 50, 346 (1981)
77. J. A. Halstead, R. R. Reeves: J. Quant. Spectrosc. Radiat. Transfer 28, 289 (1982)
78. P. van de Weijer, R. M. M. Cremers: J. Appl. Phys. 57, 672 (1985)
79. J. C. Bergquist, D. J. Wineland, W. M. Itano, H. Hemmati, H. U. Daniel, G. Leuchs: Phys. Rev. Lett. 55, 1567 (1985)
80. C. H. Corliss, W. R. Bozman: Experimental Transitional Probabilities for Spectral Lines of Seventy Elements, U. S. Natl. Bur. of Stand. Monograph 53, (U. S. G. P. O. Washington, DC 1962)
81. K. Danzmann, M. Kock: J. Opt. Soc. Am. 72, 1556 (1982)
82. D. L. Adams, W. Whaling: J. Quant. Spectrosc. Radiat. Transfer 25, 233 (1981)
83. M. Kock, S. Kroll, S. Schnehage: Physica Scripta T8, 84 (1984)
84. B. L. Cardon, P. L. Smith, J. M. Scalo, L. Testerman, W. Whaling: Astrophys. J. 260, 395 (1982)
85. G. P. Tozzi, A. J. Brunner, M. C. E. Huber: Mon. Not. R. Ast. Soc. 217, 423 (1985)
86. D. W. Duquette, J. E. Lawler: J. Opt. Soc. Am. B2, 1948 (1985)
87. D. W. Duquette, E. A. Den Hartog, J. E. Lawler: J. Quant. Spectrosc. Radiat. Transfer 35, 281 (1986)
88. C. E. Moore, Atomic Energy Levels, U. S. Natl. Bur. Stand. Natl. Stand. Ref. Data Ser. 35, (U. S. G. P. O. Washington DC 1971)
89. B. L. Cardon, P. L. Smith, W. Whaling: Phys. Rev. A20, 2411 (1979)
90. E. Biemont, N. Grevesse, P. Hannaford, R. M. Lowe, W. Whaling: Astrophys. J. 275, 889 (1983)
91. J. F. Dominy, G. Wallerstein: Astrophys. J. (in press) (1987)

Atomic Engineering of Highly Excited Atoms

M.H. Nayfeh

Department of Physics, University of Illinois at Urbana-Champaign, 1110 W. Green Street, Urbana, IL 61801, USA

1 Introduction

Recent astronomical observations in the direction of the object Cassiopeia A have detected interstellar radio emissions at the lowest frequencies observed to date, in the band 7 - 16 MHz [1]. These probably originate from atoms of enormous size, about a million times larger than a normal hydrogen atom and larger than the size of a bacterium, which would be the largest atoms yet encountered in any environment. The idea of gigantic atoms is fascinating in itself. However, its scientific appeal is not due just to its *Guinness Book of Records* aspect. In fact work in progress in laboratories is not competing with this record, but is rather aimed at the deliberate manipulation of atoms. This concept has long been sought but it has only become possible using highly excited (large) atoms. Although the production and detection of such large atoms involve some of the most advanced techniques of laser and atomic physics, their study represents in some sense a dramatic return to the earliest modern ideas of atomic structure: the model of Niels Bohr, the centenary of whose birth has just been celebrated [2].

2 Bohr's Atomic Theory

In 1913 Niels Bohr developed the first quantum theory of hydrogen [2]. It retained major elements of a classical theory in that it represented the hydrogen atom in terms of an electron in orbit about a proton, just like a miniature model of a planet orbiting the sun. However, it introduced a new assumption which was not based on any principle of classical mechanics: Nature allows only those orbits for which the angular momentum is an integral multiple of Planck's constant h. For a circular orbit of speed v and radius r, this condition is expressed as

$mvr = nh$, where m is the mass of the electron and n takes the integer values 1, 2, 3, ... for the different allowed orbits. Classical mechanics requires that the centripetal force which holds the electron in orbit be provided by the inverse-square-law electrical force between electron and proton: $(mv^2)/r = q^2/(4\pi\epsilon_o)r^2$ (using MKS electrical units). These two relations together yield values for the radii, speeds, periods $(T = 2\pi r/v)$, and binding energies E of the allowed orbits, which are summarized in Table 1.

The main prediction of the Bohr theory concerns the wavelengths of light that will be emitted or absorbed when the electron makes a transition from one orbit to another. The predicted values of wavelength agree very well with those actually observed: to within a few parts per million, which was at the time virtually the limit of experimental accuracy. On this basis alone the Bohr model can be judged a spectacular success. However, there is no obvious way to extend this model to the other atoms in the periodic table. For example, the neon atom has 10 electrons, and in its normal state all of these are contained within a volume comparable to that of the most tightly bound orbit of hydrogen (n=1). Despite many ingenious attempts, it was never found possible to describe many-electron atoms in terms of classical mechanics with a subsidary quantization assumption. About a decade after Bohr's theory, quantum mechanics was developed by Schrödinger, Heisenberg, and others.

Table 1: Quantum mechanical results for hydrogen

Quantity	Dependence on n	Value for $n = 1$	and for $n = 100$
Orbital radius, r	n^2	5.3×10^{-11} m	5.3×10^{-7} m
Speed, v	n^{-1}	2.2×10^6 m/s	2.2×10^4 m/s
Period of Revolution, T	n^3	1.5×10^{-16} m/s	1.5×10^{-10} m/s
Binding energy, E	n^{-2}	13.6 eV	1.36×10^{-3} eV
Radiative Lifetime, τ	$n^{4.5}$	—	~ 40m/s
Cross section for excitation from the ground state σ	n^{-3}	—	$\sim 10^{-26}$ m^2

The new theory confirmed many of the predictions of Bohr's theory, put them on a more solid basis, and solved all of its shortcomings. One of these is the inability of Bohr's theory to give a method for calculating excitation cross sections σ, lifetimes τ, and the intensities of the spectral lines, the former two of which are summarized in Table 1. Electronic orbits or states in the new theory are described not only by Bohr's original quantum number n, but also by three additional quantum numbers: the orbital angular momentum quantum number l which gives the angular momentum of the electron about the nucleus; the azimuthal quantum number m_l, which gives the direction of l (projection of l along a specified direction); and m_s, the projection of the electron's spin angular momentum along a specified direction. For each n, l can take the values 0, 1, 2, ... up to $n-1$, and m_l takes the values - l, $-l+1$, ...$l-1$, and l, and m_s takes two values: $+1/2$, and $-1/2$. According to our present understanding, the Bohr energy levels of hydrogen are split into fine structure and hyperfine structure, which are due respectively to interaction of electron and proton spins with the magnetic fields associated with their orbits. The fine and hyperfine splittings are 10^{-5} and 10^{-7} of the energy difference between Bohr levels in hydrogen. Much larger fine and hyperfine splittings are present in complex atoms as a result of the interaction of the active electron with the other electrons. In some cases the "fine" structure splittings are larger than the energies given by the Bohr model. Bohr's theory has thus come to be regarded as an important, but merely transitional, state between classical and quantum mechanics.

3   Emergence of Interest in Highly Excited States

In the past ten years there has been a great revival of interest in the Bohr picture and its consequences [3,4]. Surprisingly, the initial stimulus for this renewal came not from studies of atomic hydrogen, but rather from those of complex atoms – the very species that contributed to the downfall of the model! How can this be?

The simple technical answer is that any atom becomes hydrogenic when it is in a highly excited state. From Table 1 we see that the radius of a stable Bohr orbit increases as the square of n (now called the principal quantum number). As n increases, the electron is placed ever further from the proton; and if the electron is sufficiently far out it will be unable to discern the difference between a proton and any other object with one unit of positive electric charge (this is analogous to saying that the orbit of a planet about the sun depends only on the total mass of the sun, and not how that mass is distributed within the sun). In particular, the proton can be replaced by a complex atom with one electron removed. Thus the real question to be answered is how and why highly excited states have become a topic of interest. We believe it is originally due to three key developments in different areas of physics.

The first of these developments that brought about this field took place in 1965 following the remarkable astrophysical discovery of radio recombination lines by Höglund and Metzger [5]. These lines are due to transitions between adjacent highly excited states (Rydberg states) of atomic hydrogen, helium, and carbon that are populated primarily by radiative recombination, in which a free electron emits a photon and is captured into a high-n orbit about an ion. The rates at which such recombination processes occur are very low compared to those of ordinary (three-body) chemical reactions, in addition to the fact that it is not selective, and so it is very difficult to study or to utilize radiative recombination in terrestrial laboratories. (However, another recombination process — dielectronic recombination — has been the focus of much recent experimental work[6], and it too takes place via highly excited states - see article in [4].)

Since electron-ion recombination is a weak and nonselective process, and hence cannot be used to produce highly excited states in the laboratory, alternative means involving excitation of the atomic ground state must be employed. Two major problems then arise: first, the cross section for excitation decreases rapidly with n (see Table 1), so that a very high spatial density of excitation is required; second, the difference in energy between successive Rydberg states also decreases with n, so that a highly monochromatic source of excitation is required for selective population of a given state. Both of these problems were solved by another major development, in an area completely distinct from radio astronomy: the invention of the narrow-band tunable dye laser by Hänsch [8], which advanced laser technology initiated by Schawlow and Townes [7] by a major step. Although this laser design has since found applications in diverse problems in physics, chemistry and biology, the initial motivation for its development was the study of the fine structure of optical transtions of hydrogen - a structure that was not predicted

144

by Bohr's theory, and whose accurate measurement is necessary for the determination of the cornerstone of the fundamental constants, the Rydberg constant, which is the binding energy of the ground state of atomic hydrogen [9].

Finally, during the 1960s evidence was accumulating for the validity of the quark model of elementary particles. The great predictive success of this model was shadowed by a fact that was difficult to explain: no isolated quark had ever been observed. These particles are postulated to have distinctive electrical properties (i.e. a charge of -e the charge of the electron), and would give rise to characteristic spectra if they were bound to an atomic nucleus. It became apparent that the new laser technology offered the possibility of detecting a single atom of a desired species in a sample containing up to 1020 other atoms and molecules, and so it provided a technique for searching for very rare elementary particles. This was the primary stimulus behind the effort [10] to achieve single atom detection by selective excitation and ionization, a process which necessarily involved the production of excited states. It should be noted that the further development of single-atom detection techniques has not yet resulted in any major discovery in particle physics, although work in this area continues, but it has had a significant impact on analytical chemistry and materials science [11].

4   The Preparation of Large Atoms

With this initial motivation for Rydberg state excitation, derived from interest in astrophysics and in the possibilites for single-atom detection schemes, basic spectroscopy of Rydberg states soon blossomed into a major subfield of atomic physics.

The first generation of experiments involved complex atoms because their excitation is within the reach of the new visible tunable dye lasers. This is certainly true for most alkalis and some rare earth atoms whose ionization potentials are less than 5 to 6 eV and whose low lying excited states typically lie at about 2 eV. Double or triple photon excitation by one or two lasers at visible wavelengths of 2 - 2.5 eV is sufficient to excite high Rydberg states of these atoms. Because of the narrow band and high intensity available from these lasers, excitation efficiencies of near 10 % can be achieved.

Hydrogen, on the other hand, has an ionization potential of 13.6 eV, and its first low lying excited state occurs at 10.2 eV, thus making it out of reach of the dye lasers. Thus, hydrogen, the simplest of all atoms continues to be the most popular atom to atomic theorists, while sodium became the most popular atom in laser spectroscopy experiments. Since sodium has one electron outside a closed shell, it does in many respects behave as a true one-electron atom.

Nevertheless, highly excited states of the hydrogen atom possess unique properties, most notably the ways in which they respond to external electric and magnetic fields. Thus there has been constant interest in applying tunable laser technology to the excitation of hydrogen, and in the past few years several approaches have been successful. We review some of these in order to indicate the diversity of experimental effort on this problem.

One method, developed largely at Yale University, utilizes a process in which a beam of protons of a few keV kinetic energy is shot through a gaseous target [12]. The protons pick up an electron from the gas (charge exchange) to form hydrogen atoms. In this charge exchange process, the atomic hydrogen beam exciting the target contains atoms in a large number of excited states, with relative populations determined by the various charge exchange cross sections and the state lifetimes. The basic process therefore suffers from a lack of selectivity. It is possible, however, to employ an elaborate scheme of electric field quenching of many of the excited states and further excitation of a remaining state with an infrared laser to produce a well-defined population of a given highly excited state.

Another method that bypasses the use of tunable dye lasers relies on excitation with a fixed frequency laser beam and an appropriate utilization of the Doppler effect. This approach has been developed at Los Alamos Laboratory [13]. A large particle accelerator, originally designed to generate mesons for nuclear and particle physics, produces a beam of H^{-1} ions (this is a hydrogen atom with an extra electron attached to it) travelling at 84% of the speed of light; these are passed through a thin foil, resulting in the stripping of the extra electron to yield a relativistic beam of atomic hydrogen. This beam is crossed with a laser beam of fixed wavelength. By varying the angle at which the two beams cross, the energy of the photons in the rest frame of the atoms can be adjusted into resonance with transitions between the ground state and various high-lying states. However, due to the energy spread of available atomic hydrogen beams, the resolution attainable with this technique is fairly limited, and to date only states with $n < 14$ have been selectively excited.

The complexity and unavailablity of accelerators has kept highly excited hydrogen out of reach of small laboratory experiments. Recent advances, however, in nonlinear optical processes in certain types of crystals and gases have allowed conversion of tunable visible dye laser radiation into tunable ultraviolet radiation with reasonable efficiency. These methods have been applied to excitation of high-n states of atomic hydrogen at the University of Bielefeld [14], in West Germany, and at our laboratory at the University of Illinois [15]. In this first scheme, hydrogen is excited to the 2p state by absorption of a single photon from a 1215 Å pulsed tunable beam followed by an absorption of a single photon from a 3660 Å pulsed tunable beam that excites a high n state. In our scheme (shown in Fig. 1) two photons from a 2430 Å tunable beam excite the 2s state from the ground state, and subsequent absorption of a single photon from a 3660 Å tunable beam excites a high n state. Although single-photon excitation with 1215 Å is more efficient than the two photon excitation with 2430 Å, it is in some respects more convenient to utilize the two-photon process. This is due to the fact that it is much easier and more efficient to produce the 2430 Å radiation than the 1215 Å radiation and it does not suffer from resonance trapping (emission and reabsorption of radiation) which prevents a well-defined geometry of excitation.

In our scheme the exciting radiation (2430 Å) is capable of directly photoionizing the $n = 2$ states, which constitutes a loss of the efficiency

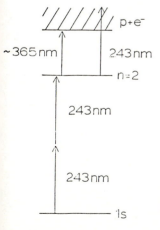

Fig. 1. Scheme for exciting a high n state.

of the excitation of the highly excited states. Such competition is usually taken into consideration in the design of the experiment, and excitation efficiency of 10^{-7} has been achieved from the ground state with a loss of 10^{-8} to direct photoionization, under atomic beam conditions. A simplified experimental setup is shown in Fig. 2. Atomic hydrogen produced by a discharge is pumped out of the discharge, collimated into an atomic beam, and enters a region between two electrodes that supply a dc electric field. The atomic beam is crossed by the two collinear laser beams in the field region. The produced charge is pushed by the electric field and travels about 1m before it gets detected by a single ion detector. The time of flight provides mass analysis of the detected charge. By this means we are able to excite states with n up to about 60, as shown in Fig. 3. The loss of resolution beyond $n = 60$ is due to the 1 cm^{-1} bandwidth of the 3660 Å pulsed laser [15,17].

When continuous wave lasers (CW) rather than pulsed lasers are used it is possible to selectively excite states of higher n because one can attain much narrower bandwidth. Laser technology has just been extended to allow CW excitation of hydrogen [18]. The highest n reached to date is that of $n = 280$ in barium taken at the Free University of Berlin [16] using CW lasers. The excitation of a hydrogen n state was monitored by ionizing it by an external pulsed electric field of a few milliseconds which is turned on after the excitation has taken place in

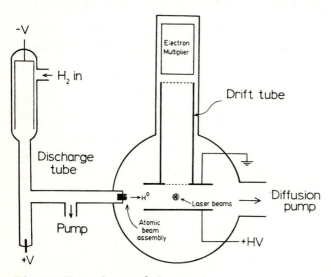

Fig. 2. Experimental Apparatus.

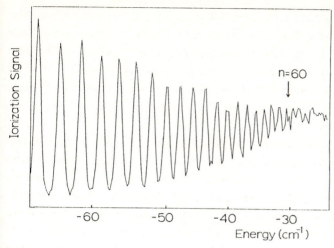

Fig. 3. Rydberg Spectrum of hydrogen.

order not to interfere with the excitation process itself. This method is very sensitive, selective (by appropriate choice of field amplitude one can ionize a specific $n > n_o$ without affecting the $n < n_o - 1$ state, and so on), and is actually at the heart of strong field effects that will be discussed in the latter sections. The barium Rydberg states on the other hand were monitored by colliding them with a beam of other atoms, thus utilizing another property of excited states: their sensitivity to collisional interactions. However this technique does not have the n selectivity of the field ionization method.

5 Sensitive Probes

We have seen that highly excited states are gigantic in size (geometrical cross section πr^2 is very large), thus making them very fragile, i.e. the highly excited electron can be easily removed by environmental effects such as collision with atoms of the same kind or with foreign atoms and molecules, ions or electrons if present, walls of the container, etc. Moreover highly excited atoms are also very sensitive to external electric or electromagnetic fields. As such, these atoms promise to be the most sensitive detectors of weak external fields whether they are of collisional or of electromagnetic origin. The collsional sensitivity has been illustrated most dramatically by the recent astrophysical observation alluded to above [1]. The observation of the 16 MHz emission point to the existence of very large atoms in the interstellar gas clouds. These are large enough such that collisional effects by particles in the medium are becoming observable in spite of the extremely low density (1 atom/cc) via the broadening they cause to the radio lines characteristic of their transitions. This has provided new diagnostic methods for the density and temperature of the clouds whose study provides information about the process of star formation.

One promising application of highly excited states is in the development of very sensitive detectors of low-energy radiation, especially microwaves. It is relatively easy to build instruments operating in the visible wavelength range which are capable of detecting a single photon. This is because a visible photon has enough energy (a few eV) to eject an electron from a solid, and the free electron can be accelerated to arbitrarily high energy by an electric field. Thus unlimited amplification can be achieved if the photon can produce a free electron. The energies of microwave photons lie in the range $10^{-3} - 10^{-4}$ eV, which is far too small to liberate an electron from ordinary matter. However, this range coincides with the energies of transitions between experimentally accessible Rydberg states: for instance, the transition between $n = 30$ and $n = 31$ requires 10^{-3} eV. As noted above, we can use a pulsed electric field to monitor the excitation of n to $n + 1$. Thus it is often possible in principle to adjust the electric field strength so that the state n is long-lived while state $n + 1$ is ionized fairly quickly. A group of Rydberg atoms prepared in the state n could then serve to detect photons of the appropriate energy. Unfortunately, it has not yet been possible to implement this scheme in a practical fashion, due mostly to the difficulty of maintaining a large population of highly excited states.

6   Atomic Engineering

As the strength of the external field increases, so will the atomic response. It is interesting to consider sufficiently high fields such that the electronic interactions with the external and the Coulomb fields are comparable, that is when none of them dominates. In this case the interaction is highly nonlinear in the external field amplitude. However, it is almost impossible to create fields in the laboratory which are strong enough to disrupt atoms in their normal states: for instance, the electric field on an electron in the ground state of the hydrogen atom has a strength of 5×10^9 V/cm. A similar value is obtained for the valence electrons in all atoms, and it is about 1000 times greater than fields which can be maintained steadily in the laboratory. On the other hand, for the $n = 30$ state in hydrogen, the Coulomb field of the nucleus can be overcome by an external electric field of only 5 kV/cm. Thus entry of atomic physics into the strong-field regime has been accomplished by dealing with highly excited states, rather than by generating enormous laboratory fields. The same statement also applies to the interaction of atoms with external magnetic fields.

These points are clearly illustrated by examining the potential energy of the electron of a hydrogen atom placed in an external electric field $\mathbf{F}$ (along the z axis), and simultaneously in a magnetic field $\mathbf{B}$.

$$V = -\frac{e^2}{r} - e\mathbf{F} \cdot \mathbf{r} + \frac{1}{8}\frac{e^2}{c^2}(\mathbf{r} \times \mathbf{B})^2 + \frac{e}{2c}\mathbf{l} \cdot \mathbf{B},$$

where e is the elementary unit of charge and $\mathbf{l}$ is the orbital angular momentum. The Coulomb term dominates at small r (normal atomic size), whereas at very large r (highly excited atoms), the external fields dominate. At some intermediate distances the three potentials become comparable and none of them dominate. It is in this intermediate regime that the concept of atomic engineering has been realized.

6.1   Electric Fields

To explain the concept of atomic engineering we consider the case where hydrogen is immersed in a static electric field only. We utilize Fig. 4a which shows the potential of a hydrogen atom for a cut along the z axis in the presence of an external electric field along the z axis. Several interesting features of the combined Coulomb and Stark potential should be noted. First, it is evident that the potential is not spherically symmetric; the electricfield leads to a lowering of the ionization potential of the atom and creation of a potential barrier in the $z > 0$ half space. Thus the state of energy higher than the top of the barrier E_c can classically escape, that is ionize; in other words the motion for this state in this half space is *unbounded*. On the other hand in the $z < 0$ half space the electric field leads to a rise in the potential towards infinity; consequently the motion of the electron is *bounded* for all energies including positive energies in this half spac, something that does not occur in the isolated atom case.

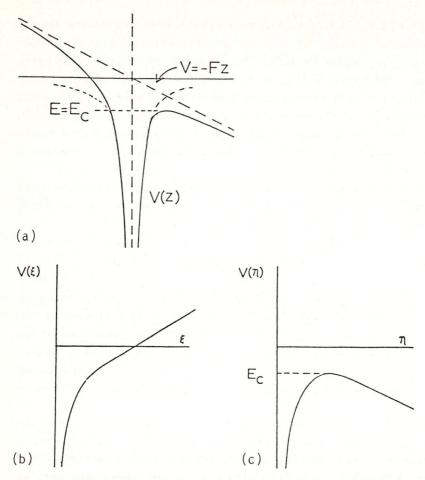

Fig. 4. Potential of hydrogen in a strong external electric field.

Thus the atom in the presence of the field will have quantized energy levels for all energies including positive energies that will *spontaneously* ionize for energies above E_c. The positions of these levels are governed by the binding well in the $z < 0$ region, while their lifetime against ionization is governed by the degree of coupling to the free motion in the $z > 0$ region.

Although the potentials in the $z = 0$ plane are very useful in bringing out some features of the interaction, they are not very useful for quantitative calculation. This is because the nonspherical symmetry of the potential makes the interaction non- separable: that is, it cannot be separated into three independent one dimensional motions in spherical coordinates. The interaction, however, is separable in parabolic coordinates $\xi = r + z$, $\eta = r - z$, and ϕ the azimuthal angle, with quantum numbers n_1, n_2, and m respectively. The effective potentials for the ξ and η motions shown in Figs. 4 b,c in fact, have good resemblance to that of the z cut in the $z < 0$ and $z > 0$ regions respectively, and hence govern the energy of the system (location of the energy) and the ionization lifetimes of these levels, respectively. The quantum number m_l is common to both parabolic and spherical descriptions, and the principle quantum number $n = n_1 + n_2 + |m_l| + 1$. The spherical l and parabolic n_1, n_2 quantum numbers do not have a one-to-one correspondence: a state with definite values of n_1 and n_2 is composed of many different values of l.

One important property of the atom that comes out of this procedure is the fact that only a fraction of the nuclear charge $Z_1 < 1$ drives the ξ motion and hence dictates the energy of the system, while the rest of the charge $Z_2 = 1 - Z_1$ drives the free η motion and hence dictates its lifetime. Thus the presence of an external electric field provides us with a situation where the nuclear charge that drives the bounded motion can be varied, in a near continuous fashion. Considering the fact that the physical and chemical identity of isolated atoms is defined by the nuclear charge, then it is clear that we have at our hand a means for creating new "types" of atoms.

We will now discuss the preparation and nature of the new types of atoms by discussing their spectroscopic properties such as ionization lifetimes, charge distributions (or excitation dipole moments), and branching ratios (or excitation strengths). To do so we will consider the positive and negative energy regimes separately, starting with the former.

Let us assume that atomic hydrogen is immersed in laser radiation of energy just larger than 13.6 eV, the ionization potential of hydrogen, and whose polarization is along the external dc electric field in which the atom is immersed. Because of this choice of polarization, the electron gets an initial kick along the dc field, and the energy of the system is raised by 13.6 eV, thus rising to zero total energy. The electron can now execute bound motion even for this positive energy. The motion of the electron is nearly a one-dimensional motion with the orbit resembling a cigar whose axis is along the external field, the nucleus being located inside it near its lower tip (Fig. 5a). This specifically tailored atom lives on the order of 5×10^{-13} s (giving very broad widths), and the electron

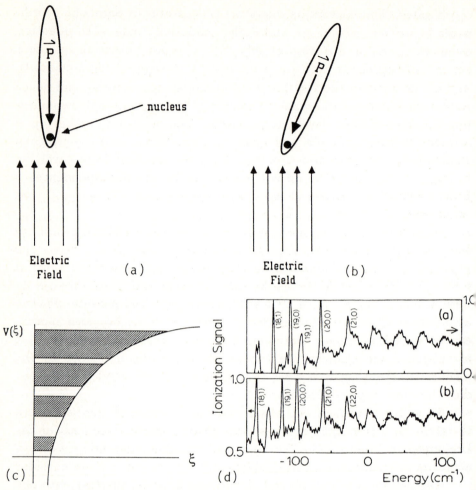

Fig. 5(a). Schematic orbit of a giant dipole aligned along the field.
Fig. 5(b). Schematic orbit of a giant dipole at an angle to the field.
Fig. 5(c). Schematic energy levels above $E = 0$.
Fig. 5(d). Spectrum of hydrogen in a field of 5 kV/cm (top) and 3.5 kV/cm (bottom).

executes on the average about 5 rounds before it breaks away from the proton on its own, and it is found to spend most of its time away from the nucleus, near the upper tip of the cigar. If the electron were initially kicked perpendicular to the field (laser polarization perpendicular to the external field), the cigar would have been created at an angle with the field. (See Fig. 5b.)

154

The extraordinary thing about this cigar atom is that such a "separated charge" distribution gives a dipole moment **P** which points *opposite* to the external field. Moreover, the dipole is very large since the separation of the charge (length of the cigar) is about 1600 Å, hence giving dipole moments that are 3000 times larger than those of normal atoms. For this reason we call these atoms "giant dipole" atoms. However, in general, one cannot exclusively prepare these types of atoms without preparing the highly excited normal atom since, first of all, the excitation has to start from the ground state of the normal atom which is only weakly affected by electric fields and secondly both Coulomb and Stark fields will have to compete. Therefore, after the excitation process we always have a superposition of these two types with the branching ratio depending on a number of parameters including the total energy of the system, field strength, and the properties of the exciting laser radiation [19-23]. For example, the "visibility" of the giant dipoles, which is a measure of how much they rise above the accumulated smooth continuum, tends to be very small (4% at 5 kV/cm). This visibility gets worse at higher energy because these states get closer to each other in energy as shown in Fig. 5c. Those states were first seen in complex atoms such as rubidium, sodium, barium, krypton, and yttrium during 1978-1983 [25-29], but were found to have strengths that are smaller than is predicted for hydrogen. Theories that included the effect of core electrons explained the reduction of the strength [21]. The first observation of the giant dipoles in hydrogen was made in 1984 in our laboratory at the University of Illinois [30]. (See Fig. 5d.) Similar observation was also achieved at the University of Bielefeld [14].

Considering the shortness of their lifetimes, and the low efficiency of excitation, it is clear that experimentation with these "new atoms" will not be easy unless these two properties are enhanced. Recently we have been able to improve the efficiency [31] and to produce giant dipoles that live much longer than 10^{-12} s [17] in atomic hydrogen. We will discuss the efficiency first. The scheme we devised for this purpose relies on a process we call multistage shaping or charge shape tuning of the charge of the atom. In one-photon excitation from the ground state one effectively starts from a spherically symmetric charge distribution (zero dipole moment), and tries to mold it by a single operation into a giant dipole whose charge is highly focussed along the field. On the other hand in multistage shaping one uses one photon to create from a ground state a not too large dipole of charge distribution that is focussed along the field at an intermediate state followed by another photon absorption from this intermediate state that produces larger dipole whose charge is even more focussed along and so on till one excites the giant dipole in a highly focussed distribution along the field.

The ability to create moderately large focussed dipoles as intermediates is the key to the success of the multistage shaping operation [17,32]. This is explained in Fig. 6 for a two-stage process using as an intermediate $n = 2$ of hydrogen. Because the level splittings in $n = 2$ of hydrogen are small enough (0.3 cm^{-1}) such that an electric field imposed on the atom which is larger than 5 kV/cm will be able to mix all of these sublevels and hence their charge distributions (each has a zero dipole) to produce distinct dipole distributions needed for the shaping process.

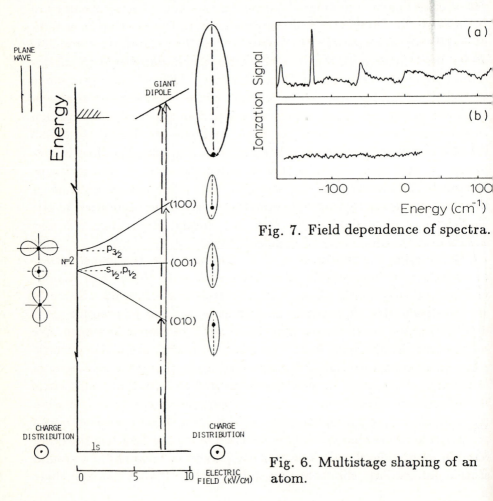

Fig. 7. Field dependence of spectra.

Fig. 6. Multistage shaping of an atom.

156

Our calculations show that by utilizing the up-field extended dipole of $n = 2$ as an intermediate, the efficiency can be increased from 10 to 30 %, whereas by utilizing the down-field extended dipole the efficiency is reduced to 1 %. These were confirmed in our hydrogen experiment as shown in Fig. 7. Our further calculations using higher n states whose charge can be focussed along the field more easily, as shown in Fig. 8, and hence can be matched or tuned more closely to the charge of the giant dipoles, showed dramatic effects on the efficiencies [33]. The use of, for example, $n = 9$ as the intermediate step in the process, rejects almost completely the excitation of the spectrum of the normal atom in favor of the one-dimensional atom. Results for $n = 1$, $n = 2$, and $n = 9$ are shown in Fig. 9, along with a schematic of the focussing effect on the intermediates.

The enhanced efficiency is very nice, but it is found that it is practically not possible to increase the lifetime of these giant dipoles in this positive energy region by too much. Such inability is related to the fact

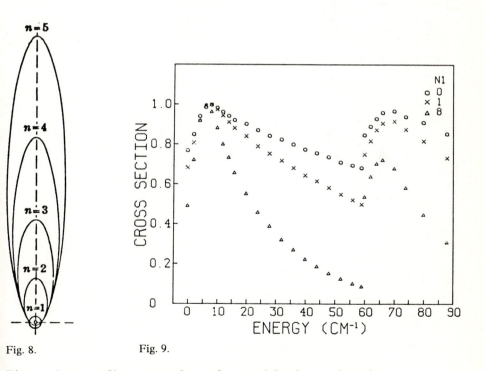

Fig. 8.   Fig. 9.

Fig. 8. Intermediate state dependence of final wavefunction.
Fig. 9. Cross sections for different intermediate states.

that the bound motion of all these giant dipoles in this region are driven by nearly the same charge, most of the nuclear charge $Z_1 \sim 1$, which also dictates very similar orbits where the nucleus is located at the lower tip of the cigar. However, it is found that such enhanced efficiency can be extended to the negative energy region where it is also possible to

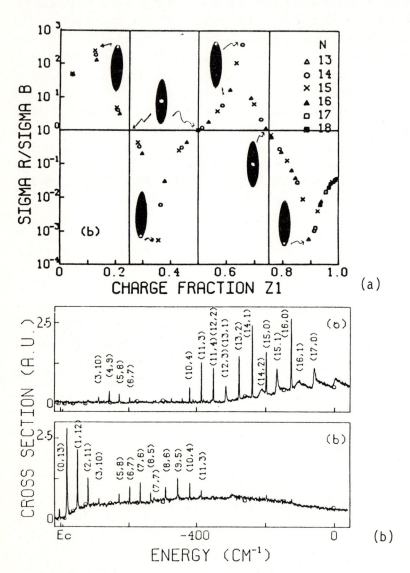

Fig. 10(a). Position of nucleus as a function of charge fraction.
Fig. 10(b). The giant dipole spectrum.

produce giant dipoles that live quite long. Therefore we will now discuss such promising negative energy regions between $E = 0$ and $E = E_c$.

In this region the giant dipole atoms take on different properties than the one in the positive energy region. Firstly, the fraction of the charge that drives the bound motion can be varied from 0 to 1 by varying the energy of the system, and consequently the position of the nucleus inside the cigar can also be controlled. In Fig. 10a we plotted an indicator of the location of the nucleus inside the cigar as a function of Z_1, along with sketches of some possible orbits. This indicator is related to the ratio of the area of the orbit below the nucleus to that above the nucleus. The figure shows a remarkable property: for the fractional charges $Z_1 = 1/4$, $1/2$, and $3/4$ the nucleus is located at the center of the cigar (the atom has zero dipole moment). These fractional charges thus constitute lines across which the direction of the giant dipole reverses. In the first and third quarters the dipole is along the imposed field whereas in the other two quarters it is opposite to it. Given the size of the orbit (which can be calculated), one can use the above indicator to determine the magnitude of the giant dipole. Moreover, we have recipes to cook up giant dipoles of given Z_1 values. These features and others have been recently confirmed by our experiments. The giant dipole spectrum is shown in Fig. 10b.

Examination of the spectrum indeed shows a variety of widths (lifetimes) that range from quite short to quite long. In fact there are giant dipoles that do not show up in our spectrum because they live longer than the time of measurement, which is 100 ns, or because they radiatively decay before they ionize and hence do not get detected. Again there are systematics to the ionization lifetime as a function of Z_1 and hence as a function of energy that makes the selection of a giant dipole of given specification possible.

6.2 Magnetic Fields

Atomic engineering can also be achieved using external magnetic fields or simultanous electric and magnetic fields. The effect of a magnetic field on a highly excited atom is very different from that of an electric field. As can be seen from Fig. 11a, a magnetic field is described by a potential which increases in all directions away from the nucleus. Thus it does not polarize the atom in a given direction, but rather confines the motion of the electron in the plane perpendicular to the field thus

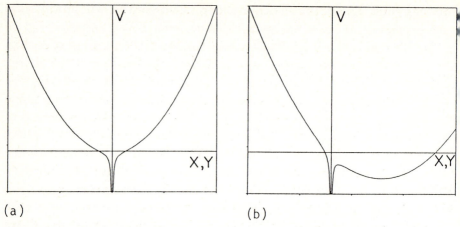

Fig. 11. Hydrogen potential in a magnetic field (a) and in crossed electric and magnetic fields (b).

giving a near two-dimensional atom. This confinement makes it possible to produce localized states with energies in excess of the ionization energy. Such states were first seen [34] in photoabsorption by barium in 1969. They are called "quasi-Landau resonances" because of an analogy between their theoretical description and Landau's theory of a free electron in a magnetic field. The quasi-Landau resonances were observed in hydrogen just two years ago [14,35]. Recently, new experiments with higher resolution have suggested the existance of many long lived states with energies above $E = 0$. These findings have stimulated further theoretical efforts. Isolated classical orbit calculations show good agreement with the experimental results, however, the problem is not yet fully understood.

As indicated by our previous discussion of the dynamics of hydrogen in an electric field, the key to understanding is the identification of a coordinate system in which the motion along each coordinate can be treated independently. For hydrogen in an electric field, the parabolic coordinates ξ, η, and ϕ provide such a system; in the case of a magnetic field, on the other hand, it is almost certain no such coordinate system exists. Many such nonseparable systems are encountered in other branches of physics and no general method has been developed for their theoretical description. At present some initial progress on the theoretical understanding of highly excited hydrogen in a magnetic field has been achieved with the aid of large scale numerical calculations [36].

6.3 Simultaneous Electric and Magnetic Fields

The simultaneous application of electric and magnetic fields to a Rydberg atom represents the ultimate in atomic engineering. It is an area which at present is not as well explored as that of pure electric or magnetic field studies, but it has yielded some interesting results. First of all, one can investigate in greater detail the dynamical symmetry of the hydrogen atom, which is responsible for the l-degeneracy of states in zero field and the stability of the blue giant dipole states. This dynamical symmetry, which can be represented in terms of the four-dimensional rotation group $O(4)$, yields a remarkably orderly spectrum for the hydrogen atom in combined electric and magnetic fields of moderate strength: remarkable in the sense that its simplicity cannot be attributed to any geometrical symmetry of the field configuration. This effect has recently been observed in high angular momentum states of alkali atoms [37].

The regime in which both electric and magnetic fields are strong compared to the atomic field, as shown in Fig. 11b, has not yet been studied closely. However, there are theoretical predictions of a new type of motion in a highly excited atom in perpendicular electric and magnetic fields. In this case the electron experiences an electric force that tends to stretch it in a cigar shape, and a magnetic force that tends to squeeze it in the perpendicular plane. The balance of the two forces results in forcing the electron to orbit around two centers: the nucleus and the saddle point [38]. Experiments to search for such states are in progress.

6.4 Circular Atoms

There is a case of atomic engineering of a Rydberg state n which actually utilizes weak dc electric field but with additional help from absorption of $n - 1$ microwave photons [39]. With each photon imparting to the atom, in an ordered fashion, a unit of angular momentum, the atom ends up in the state of the highest possible angular momentum, $l = n - 1$. The high angular momentum prevents the electron from coming close to the nucleus, and in fact the electron executes a circular orbit around the nucleus which is in fact the closest approximation to the original Bohr model of the atom. Such states are the counterpart to the giant dipole atoms; they have zero electric dipole moments but quite large magnetic dipole moment. The production of such states in sodium atoms has been achieved by researchers at the Massachusettes Institute of Technology (MIT).

6.5 Atoms Between Large Plates: Squeezed States

Finally a property of large atoms that makes another type of atomic engineering possible is the fact that radiation emitted or absorbed between the adjacent highly excited levels is in the microwave region of the electromagnetic spectrum. When such a highly excited atom is injected into a microwave cavity or waveguide of a size that is comparable to the atom's characteristic transition wavelength λ_o, then a number of interesting effects may occur [40,41]. For example, if the cavity size is such that the wave spontaneously emitted by the atom cannot propagate in the guide because λ_o is larger than the so called cutoff wavelength of the guide λ_c, then the wave will not be emitted, thus preventing the spontaneous emission of the atom and consequently lengthening its lifetime. Alternatively this process can be looked at as an engineering process of the properties of the generated electromagnetic radiation. Electromagnetic radiation can propagate in free space with any wavelength and direction, but the introduction of a guide restricts the propagation to only certain wavelengths and directions. Such propagation effects are well known in electrical engineering, but it is only recently that their implications for the generation of radiation from a quantum system such as highly excited atoms placed inside cavities have been realized. One exciting possibility is the engineering of what is called "squeezed states" of electromagnetic waves. In a normal wave, the levels of noise associated with its amplitude and its phase are balanced; however, a squeezed state is a state where some of the noise in either one of them is channeled to the other. Since some detectors can be designed to utilize either the phase or the amplitude separately, then such process promises extreme improvement in detection sensitivity [42].

6.6 Complex Atoms

It is relevant to examine the possibility of atomic engineering in complex atoms. In complex atoms, the highly excited electron interacts with the rest of the electrons in the system (electronic core) in addition to its interaction with the Coulomb field of the nucleus and with the external field. This additional interaction is short range, i.e. it is only important for distances on the order or less than the size of the ground state atom. If the highly excited electron stayed away from the electronic core, the effect of this interaction would be negligible and as a result the excited atoms would be nearly hydrogenic. In atomic engineering, however, the orbit of the highly excited electron does penetrate the core, thus making the system not quite hydrogenic. The effect of such penetration has the following implications. First of all it mixes and smears the Coulomb-Stark interaction to the degree of preventing selective excitation of the giant dipoles, resulting in the diminishing of their dipole moment. Also it causes mixing among states of short and long lifetimes, resulting in shortening of the lifetimes of the long-lived ones. Another serious effect is the prevention of the multistage shaping concept that we use for the enhancement of excitation strength. This is because the interaction with the electronic core destroys the near degeneracy in l, that is, it causes very big splittings in the l substates of low lying states, thus rendering the external field ineffective in partially molding their charge distributions into distinct dipole distributions, operations that are the backbone of the concept of the enhancement.

7 One Dimensional Atoms in Microwave Fields

There has been a much interest in the question of existence of chaotic behavior in quantum mechanical systems whose classical analogs are known to be nonintegrable and exhibit chaotic behavior [50]. The interest stems from the fact that the allowance for quantum effects in the systems gives rise to substantial anomalies in the manifestation of stochasticity even when the initial population of the system is quasiclassical. The periodically kicked classical and quantum rotors have been analyzed extensively by many researchers [51-53]. In some studies it was found that the quantum correlation functions attenuate much more slowly than in the classical limit, ultimately reducing the rate of quantum diffusion. In another study [53] it was noted that one cannot tell with certainty whether a quantum state of the rotor describes a chaotic or a regular state beyond the quantum resolution $\sqrt{\hbar}$. Other studies of quantum systems, however, have shown that under certain conditions and for evolution over finite times, quantum dynamics can be brought closer to the classical stochastic approach [50].

Experiments to study chaotic behavior in the response of atoms and molecules to radiation fields are underway in a number of laboratories including ours [54-57]. One experiment deals with the collisionless dissociation of polyatomic molecules interacting with coherent IR laser radiation. Another system of interest is that of highly excited atomic hydrogen ionized by a microwave field [55-57].

Because an analysis of the hydrogen system involves many degrees of freedom (the electron moves in a 6-dimensional phase space), the detailed numerical and analytical investigation of this process is time consuming, and at this date remains incomplete [58]. A system that is amenable to theoretical analysis is the one-dimensional hydrogen atom (involving two dimensions in phase space) that we have been experimenting with recently [59]. These atoms are prepared by laser excitation of hydrogen in the presence of strong dc electric fields. They offer not only the nearest approximation to the ideal one-dimensional problem, but they offer systems of low quantum numbers as well as high ones so that a wide range of quantum numbers can be studied.

We have theoretically analyzed stochasticity in highly excited atomic hydrogen in the presence of a microwave field and a dc electric field using a classical one-dimensional model similar to that of an electron system over a helium surface. We have determined in detail the effect of the dc field on the threshold of global stochasticity and on the number of states trapped in the nonlinear resonances [60]. Our results indicate that the number of trapped states may not always increase, and may decrease with increasing dc field contrary to the case of a system of surface electrons.

8 Detection by Electric Field Ionization

The concept of detection of neutral atoms by electric field ionization is a consequence of atomic engineering, although it has been known since the mid 1970s [43]. In fact this technique is one among many that have been developed during the past decade for the detection of ever smaller concentrations of neutral atoms, including the ultimate detection of single atoms [10].

Traditional optical spectroscopy has been widely used to identify and characterize various substances through the measurement and interpretation of the spectra arising from the absorption or emission of radiant energy by samples under investigation. In techniques measuring absoprtion, the test material is illuminated by a continuous light source, such as an incandescent lamp, and the transmission of light is monitored. The spectral lines of the sample appear as dark lines in the transmitted light. Although this method can be very selective, its sensitivity is only moderate ($10^9/\text{cm}^3$).

In traditional emission methods, samples are subjected to flames or electrical discharges to excite the electrons of the sample to higher orbits, and the radiant energy (fluorescence) they emit during relaxation to the ground state is monitored. This method is a more sensitive measurement than the absorption method; however, it is less selective because many excited states can be populated from all the species present, with the result that the spectral lines of the predominant materials swamp the weaker emission of the dilute constituents.

Spectroscopic technique was improved in both sensitivity and selectivity with lasers. Atoms are excited from the ground state by a resonant intense laser beam, with no ions produced and with many fewer competing resonances, and the emission light can be observed directly, making the method simultaneously very sensitive and very selective. It was possible with this laser-based method to measure sodium densities down to $100/\text{cm}^3$, an improvement over the sensitivities of existing analytical methods of eight orders of magnitude. Further studies using the resonance-fluorescence method resulted in measurements of $3 \times 10^3 /\text{cm}^3$ sodium atoms and $10^3/\text{cm}^3$ uranium atoms.

There is a scheme that matches the selectivity of resonance fluorescence and surpasses its sensitivity which was developed by Hurst, Nayfeh and Young (then at Oak Ridge National Laboratory) [10]. Atoms are stimulated to a sequential series of excited states by causing them to absorb a photon of frequency ν_1, then a photon of frequency ν_2, and so on. The number of allowable absorptions is arbitrary depending on the atom being investigated. In all cases, however, the final step involves the photoionization of the atom: the absorption of a photon results in breaking the atom into an ion pair – an electron and a positive ion, or the application of a pulsed dc electric field that also results in the ionization of the atom. The produced charge – either the electron, the ion, or both – is detected in the method, which is called resonance ionization spectroscopy (RIS). Resonance ionization differs from resonance fluorescence in a very basic way. With ionization, the detection scheme involves the measurement of massive particles–electrons or ions that can be easily controlled through their diffusion times (this time can be on the order of a few microseconds) — whereas with fluorescence, measurements are made of photons, which are difficult to control. The ability to control electrons makes elimination of wall events possible, resulting in considerable reduction of background emission interference. In the resonance-fluorescence scheme, the detector must be focused at less than 4π solid angle to avoid the main unscattered beam; in the ionization scheme, 100% collection efficiency is achieved. In addition, the measurement of small numbers of electrons or ions is much less complicated than the measurement of small numbers of photons — another advantage of the resonance ionization method over the fluorescence method. In fact, with this improved sensitivity even detection of single atoms has been achieved. Later efforts and variations on the resonance fluorescence have also pushed its sensitivity to the single atom limit.

The concept of electric field ionization of excited atoms can be easily explained using Fig. 4. One can see that the presence of the electric field lowers the ionization potential of the atom in half of the space by an amount given by $E = 2\sqrt{F}$.

Thus all excited states of the atom whose energies occur higher than the new potential spontaneously ionize with different rates. For states lying below this threshold (the top of the barrier), the ionization rates are very slow for the atom to be detected as a pair of charges. In atomic engineering one excites the atom from the ground state in the presence of the electric field, however, when the field is used as a detector only, the electric field is switched on after the excitation has taken place. What is very attractive about this method is that it has some energy selectivity. If we have an ensemble of atoms in which a number of excited states are populated, then by increasing the strength of the pulsed electric field continuously from zero, we can map out the population of all of these excited states.

Although in principle one can use field ionization or photoionization in the final step of the detection, for low lying excited states (2-3 eV binding) photoionization is more appropriate since the electric field needed is too large, while the required radiation is visible radiation that is readily available. On the other hand, when one is dealing with highly excited states one finds it is more convenient to use field ionization, since photoionization with visible laser is not efficient because the binding energy (a few meV) is much too small relative to the photon energy. Moreover, the required electric field is no more than a few kV/cm which can be generated easily and cheaply. In most cases, ionization efficiencies of near unity can be achieved.

9 Large Atoms in Physical Environments

One question which arises immediately is the extent to which a highly excited state can survive in a typical gaseous medium. For sufficiently large n, the electron's orbit will enclose a number of atoms of the gas, and it might be thought that these would disrupt the stability of the Rydberg state. However, as was first pointed out by Fermi [44], when many neutral gas atoms intervene between the Rydberg electron and the ion, they can be represented by a continuous dielectric medium. This effectively reduces the electric charge on the ion, but it does not qualitatively alter the basic Bohr picture of the dynamics. Thus it is possible for a Rydberg atom to enclose tens of thousands of neutral atoms of a background gas and still remain stable - a feat which, in a famous passage, Breene [45] has compared to the ability of a nanny goat to pick its kid from a large herd. What limits that stability of a Rydberg state is the effect of hard collisions between the slowly moving, nearly free electron and individual atoms of the gas: the distance the atom travels between collisions must be greater than the de Broglie wavelength of the electron. Analysis of this effect has made it possible to deduce from Rydberg spectra the cross sections for electron scattering by atoms and molecules at very low energies, which are extremely difficult to measure directly.

In the presence of strong external fields, giant dipoles in complex atoms can be produced with picosecond lifetimes. Giant dipoles in hydrogen on the other hand can be produced with lifetimes on the order of nanoseconds and even microseconds [46-48]. How do collisions affect a giant dipole or a disk-like atom? How does the lifetime depend on the direction of the relative velocity with respect to the quantization axis of the giant atom? What is the role of the geometrical cross section of the atom? Our recent work has shown that large distortions of the electronic charge distribution, in the case of giant dipoles, make these atoms extremely active and render the geometrical cross section (area of the orbit) irrelevant as an indicator of the collisional activity [46]. We have recently investigated the simultaneous interactions of highly excited atoms with external dc electric fields and depolarizing collisional interactions with electrons, ions, or neutral atoms [60]. We find for the first time that the electric field enhances by many orders of magnitude the depolarization cross section. The interaction is so long range that the excitation duration (pulse width) governs the time over which the interaction takes place. Although this long range dipole lives less than 10^{-11} seconds and the excitation pulses are less than 10 ns, the interaction is strong enough to cause appreciable depolarization even at number densities of ions as low as $10^8/cm^3$. In other words, the electric field renders the geometrical cross section irrelevant as an indicator of the collisional activity [60,45]. An alternative indicator of the collisional activity is the distance from the nucleus to the classical turning point.

Another aspect to consider is whether the "size" of Rydberg atoms could ever be perceived in the familiar mechanical sense. Atoms of $n = 280$ have been prepared where the orbital radius is of a few micrometers. This is a distance scale just at the limits of common optical microscopy, and one on which machining of objects can be carried out e.g. by electroforming techniques. A recent experiment [49] at the Ecole Normale Supérieure in Paris has taken a step towards direct comparison of the size of a Rydberg atom with that of a classical object. A gold foil was perforated with an array of rectangular slits several micrometers wide. A beam of Rydberg atoms was directed towards the slit, and the fraction of atoms passing through was measured as a function of principal quantum number. The results were found to be consistent with a simple "hard sphere" model, in which a Rydberg atom passes through a slit if its "edge" doesn't hit the side. The apparent size of the Rydberg atoms as determined from these results does indeed vary as n^2, though with a constant of proportionality larger than that given by the Bohr model. Of course, the specific mechanism by which the Rydberg beam is attenuated must be understood in terms of electromagnetic interaction between the atoms and the foil. Nevertheless this experiment does provide some sort of classical measurement of atomic dimensions, as well as a demonstration of an effect of a macroscopic object upon a highly excited atom.

10   Acknowledgements

I would like to acknowledge the input of Dr. Charles Clark of NBS Gaithersburg, MD, to this article. Some parts of it were prepared by both of us for different purposes. Finally, I would like register deep gratitude to my graduate student, Tom Sherlock, for his patience and technical expertise in typesetting this paper.

References

[1] K. R. Anantharamaiah, W. C. Erickson, and V.Radhakrishnan: Nature 315, 647 (1985); W. D. Watson, ibid, p. 630 - 631.

[2] Physics Today 38, 23-72, (1985).

[3] R. F. Stebbings: Science 193, 537 (1976); T. F. Gallagher: In *Advances in Atomic and Molecular Physics*, 14, ed. by B. Bederson, (Academic, New York 1978) p. 365; D. Kleppner: *Progress in Atomic Spectroscopy*, ed. by W. Hanle and H. Kleinpopper, (Plenum, New York 1979) p. 713.

[4] See articles in *Atomic Excitation and Recombination in External Fields*, ed. by M. H. Nayfeh and C. W. Clark (Gordon and Breach, New York 1985).

[5] B. Höglund and P. G. Metzger: Science 150, 339 (1965).

[6] J. B. A. Mitchell, C. T. Ng, J. L. Foraud, D. P. Levac, R. E. Mitchell, A. Seu, D. B. Miko, and J. Wm. McGowan: Phys. Rev. Lett. 50, 335 (1983); D. S. Belic, G. H. Dunn, T. J. Morgan, D. W. Mueller, and C. Timmer: ibid 50, 339 (1983); P. F. Dittner, S. Datz, P. D. Miller, C. D. Heath, P. H. Stelson, C. Bottcher, W. B. Dress, G. D. Alton, and N. Neskovic: ibid 51, 31 (1983).

[7] A. L. Schawlow and C. Townes: Phys. Rev. A 112, 1940 (1958).

[8] T. W. Hänsch: Appl. Optics 11, 895 (1972).

[9] T. W. Hänsch, M. H. Nayfeh, S. A. Lee, S. M. Curry, and I. S. Shahin: Phys. Rev. Lett. 32, 1336 (1974); B. P. Kibble, W. R. C. Rowley, R. E. Shawyer and G. S. Series: Proc. Phys. Soc. of London 6, 1079 (1973).

[10] M. H. Nayfeh: Am. Sci. 67, 204 (1979); G. S. Hurst, M. G. Payne, S. D. Kramer, and J. P. Young: Rev. Mod. Phys. 51, 767 (1979).

[11] See articles in *Resonance Ionization 1984*, ed. by G. S. Hurst, and M. G. Payne (The Institute of Physics, Bristol 1984).

[12] J. E. Bayfield, L. D. Gardner, and P. M. Koch: Phys. Rev. Lett. 39, 76 (1977);
P. M. Koch: Phys. Rev. Lett. 41, 99 (1978).

[13] H. C. Bryant : Phys. Rev. A 27, 2889, 2912 (1983);
W. W. Smith, C. Harvey, J. E. Stewart, H. C. Bryant, K. B. Butterfield, D. A. Clark, J. B. Donahue, P. A. M. Gram, D. Macarthur, G. Comtet, T. Bergman, in *Atomic Excitation and Recombination in External Fields*, ed. by M. H. Nayfeh, and C. Clark (Gordon and Breach, New York 1985).

[14] K. H. Welge and H. Rottke: In *Laser Techniques in the Extreme Ultraviolet-OSA*, Boulder, Colorado, 1984, ed. by S. E. Harris and T. B. Lucatorto, AIP Conf. Proc. No. 119 (AIP, New York 1984), pp. 213-219;
H. Rottke and K. H. Welge: Phys. Rev. A 33, 301 (1986).

[15] W. L. Glab and M. H. Nayfeh: Opt. Lett. 8, 30 (1983);
W. Glab: Ph.D Thesis, University of Illinois, 1984 (unpublished).

[16] H. Rinneberg, J. Neukammer, G. Jonsson, H. Hieronymous, A. Konig, and K. Vietzke: Phys. Rev. Lett. 55, 382 (1985).

[17] M. H. Nayfeh, K. Ng, and D. Yao: In *Atomic Excitation and Recombination in External Fields*, ed. by M. H. Nayfeh and C. W. Clark, (Gordon and Breach, New York 1985).

[18] T. W. Hänsch: In *Proceedings of the International Laser Science Conference*, ed. by W.C. Stwalley and M. Lapp (Am. Inst. of Phys. 1986).

[19] E. Luc-Koenig and A. Bachelier: Phys. Rev. Lett. 43, 921 (1979).

[20] A. R. P. Rau and K. T. Lu: Phys. Rev. A 21, 1057 (1980).

[21] D. A. Harmin: Phys. Rev. A 24, 2491 (1981); Phys. Rev. Lett. 49, 128 (1982); Phys. Rev. A 26, 2656 (1982).

[22] U. Fano: Phys. Rev. A 24, 619 (1981).

[23] W. D. Kondratovich and V. N. Ostrovsky: Zh. Eksp. Teor. Fiz. 4, 1256 (1982).

[24] C. W. Clark, K. T. Lu, and A. F. Starce: In *Progress in Atomic Spectroscopy C* ed. by H. Beyer and H. Kleinpoppen (Plenum, New York 1984) p. 247

[25] R. R. Freeman, N. P. Economou, G. C. Bjorklund, and K. T. Lu: Phys. Rev. Lett. 41, 1463 (1978).

[26] T. S. Luk, L. DiMauro, T. Bergeman, and H. Metcalf: Phys. Rev. Lett. 47, 83 (1981).

[27] S. Feneuille, S. Liberman, E. Luc-Koenig, J. Pinard, and A. Taleb: Phys. Rev. A 25, 2853 (1982).

[28] W. Sandner, K. A. Safinya, and T. F. Gallagher: Phys. Rev. A 23, 2448 (1981).

[29] W. Glab, G. B. Hillard, and M. H. Nayfeh: Phys. Rev. A 28, 3682 (1983).

[30] W. L. Glab and M. H. Nayfeh, Phys. Rev. A 31, 530 (1985).

[31] W. L. Glab, K. Ng, D. Yao, and M. H. Nayfeh: Phys. Rev. A 31, 3677 (1985).

[32] M. H. Nayfeh, K. Ng and D. Yao In *Laser Spectroscopy VII*, T. Hänsch and R. Shen, eds. Springer Ser. Opt. Sci., Vol. 49 (Springer, Berlin Heidelberg 1985) p. 71
K. Ng, D. Yao and M. H. Nayfeh, Phys. Rev. A (april) (1987).

[33] Y. P. Ying and M. H. Nayfeh: Phys. Rev. A 35, (1985).

[34] W. R. S. Garton and F. S. Tomkins: Astrophys. J. 158, 839 (1969).

[35] A. Holle and K. Welge: In *Laser Spectroscopy VII*, ed. by T. Hänsch and R. Shen, Springer Ser. in Opt. Sci., Vol. 49 (Springer, Berlin, Heidelberg 1985) p. 81

[36] C. W. Clark and K. T. Taylor: Nature 292, 437 (1981); J. Phys. B 15, 1175 (1982);
G. Wunner: Phys. Rev. A (in press 1986);
P. O'Mahony and K. T. Taylor: J. Phys. B (in press 1986).

[37] F. Peuent, D. Delande, F. Biraben, and J. C. Gay: Opt. Comm. 49, 184 (1984).

[38] C. W. Clark, E. Korevaar, and M. G. Littman: Phys. Rev. Lett. 54, 320 (1985).

[39] R. G. Hulet and D. Kleppner: Phys. Rev. Lett. 51, 1430 (1983).

[40] P. Dobiasch, G. Rempe and H. Walther: In *Laser Spectroscopy*, ed. by T. Hänsch and R. Shen, Springer Ser. Opt. Sci., Vol. 49 (Springer, Berlin, Heidelberg 1985) pg. 62

[41] S. Haroche, C. Fabre, P. Goy, M. Gross, J. M. Raimond, A. Heidmann and S. Reynaud: In *Laser Spectroscopy VII*, ed. by T. Hänsch and R. Shen, Springer Ser. Opt. Sci., Vol. 49 (Springer, Berlin, Heidelberg 1985) pg. 62

[42] M. D. Levenson and R. M. Shelby: In *Laser Spectroscopy VII*, ed. by T. Hänsch and R. Shen, Springer Ser. Opt. Sci., Vol. 49 (Springer, Berlin, Heidelberg 1985) pg. 250

[43] T. W. Ducas, M. G. Littman, R. R. Freeman and D. Kleppner: Phys. Rev. Lett. 35, 36 (1975);
V. S. Letokhov, In *Tunable Lasers and Applications*, ed. by A. Mooradian, T. Jaeger, and P. Stokseth, Springer Ser. Opt. Sci., Vol. 3 (Springer, Berlin, Heidelberg 1976) p. 122

[44] F. Fermi: Nuovo Cimento 11, 157 (1934).

[45] R. G. Breene, Jr. : *The Shift and Shape of Spectral Lines*, (Pergamon, Oxford 1961), p. 301.

[46] T. Yoshizawa and M. Matsuzawa: J. Phys. B17, L 485 (1984);
R. L. Becker and A. D. MacKellar: Bull. Am. Phys. Soc. 29 (1984).

[47] M. H. Nayfeh, G. B. Hillard, and W. Glab: Phys. Rev. A 32, (1985).

[48] K. T. Lu: In *Atomic Excitation and Recombination in External Fields*, ed. by M. H. Nayfeh, C. W. Clark, (Gordon and Breach, New York 1985) p.507

[49] C. Fabre, M. Gross, J. M. Raimond and S. Haroche: J. Phys. B16, L 671 (1983).

[50] G. Casati, B. V. Chirikov, F. M. Israelev, J. Ford: In *Stochastic Behavior in Classical and Quantum Systems*, ed. by G. Casati and J. Ford, **Lecture Notes in Physics**, Vol. 93, (Springer, New York 1979).

[51] D. R. Grempel, S. Fishman, and R. E. Prange: Phys. Rev. Lett. 49, 833 (1982).

[52] M. V. Berry: Physica (Amsterdam) 10D, 369 (1984).

[53] S. J. Chang and K. J. Shi: Phys. Rev. Lett. 55, 269 (1985).

[54] E. V. Shuryak: Zh. Eksp. Teor. Fiz. 71, 2939 (1975); Sov. Phys. JETP 44, 1070 (1976).
P. I. Belobrov, G. P. Berman, G. M. Zaslavski, and A. P. Slivinskii: ibid. 76, 1960 (1979); 49, 993 (1979).

[55] J. E. Bayfield and L. A. Pinnaduwage: Phys. Rev. Lett. 54, 313 (1985).

[56] P. M. Koch: J. Phys. (Paris), Colloq. 43, C2-187 (1982).

[57] D. Humm and M. N. Nayfeh: to be published.

[58] R. V. Jensen: Phys. Rev. A 30, 386 (1984); in *Chaotic Behavior in Quantum Systems*, ed. by G. Casati, (Plenum, London 1985).

[59] M. H. Nayfeh, D. Yao, Y. Ying, D. Humm, K. Ng and T. Sherlock: In *Advances in Laser Science - 1*, AIP Conference Proceedings 146, 370 (1986);
See also [32a]

[60] M. H. Nayfeh and D. Humm in *Proceedings of the International Workshop on Photons and Continuum States*, ed. by N. Rahman, (Cortona, Italy 1986), (Springer, Berlin, Heidelberg 1987)

Study of Small but Complete Molecules - Na$_2$

Hui-Rong Xia and Zu-Geng Wang

Department of Physics, East China Normal University,
Shanghai 200062, People's Republic of China

1. INTRODUCTION

Several years ago we felt somewhat puzzled to hear that "A diatomic molecule is a molecule with one atom too many." This was said by Professor Arthur L. Schawlow during one lecture in a series on High Resolution Laser Spectroscopy at East China Normal University in Shanghai. We have no doubt of this now. In fact, a diatomic molecule as simple as a neutral sodium dimer, with only two outer-orbital electrons was not well understood. Schawlow's comment:

> "To find something new, you never need to know
> everything about a subject. You only need to
> recognize one thing that is unknown."

This impressed us the most! Therefore, which "unknown" would be a good one for Mrs. Xia to study when she began research in Schawlow's group during April of 1980?

At that time, KENNETH C. HARVEY [1] had observed two-photon lines in Na$_2$, and J. P. WOERDMAN [2] had reported the two-photon line at 16601.88 cm^{-1} in Na$_2$ and had identified the rotational quantum number, but the upper electronic state was not identified. Success in determining the energy level constants of the unknown high-lying gerade states in Na$_2$ was achieved by Professor Schawlow and his talented students through the development of two-step polarization labeling spectroscopy. As many as 20 new states were found.[3] One was the level identified by WOERDMAN, however it was uncertain as to which one it was.

Professor Schawlow proposed an experiment for Mrs. Xia with his graduate student Gerard P. Morgan (Fig. 1), and it made a good beginning for our study on sodium dimers. Our paper summarized the research on this subject during the time we were Visiting Scholars at Stanford in 1980 and 1982, respectively, to the present time at East China Normal University. We combined various nonlinear laser spectroscopic techniques for the study of Na$_2$. This was done on near-resonant enhanced two-photon transitions to populate high-lying singlet (earlier) and triplet states (recently) and on various mechanisms of optically pumped stimulated emission of radiation distributed from infrared (earlier) to violet or ultraviolet regions (recently), see Fig. 2.

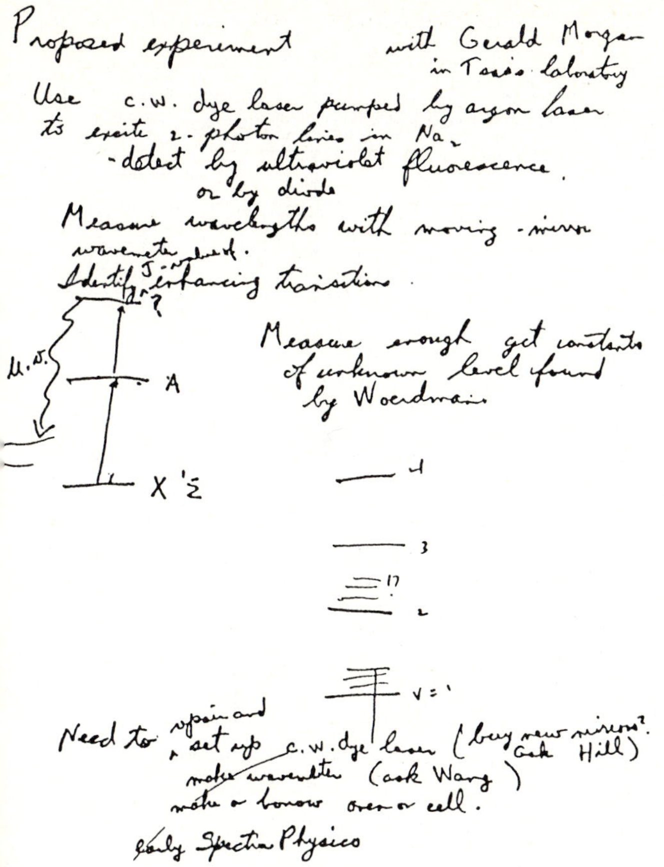

Fig. 1. A proposed experiment by Professor Schawlow for Xia in June, 1980

2. COMPREHENSIVE IDENTIFICATION OF EQUAL-FREQUENCY MOLECULAR TWO-PHOTON TRANSITIONS

It was surprising to us initially, that with careful scanning of the single mode cw dye laser, covering 1000 cm^{-1} to around 6000A, we observed only 79 two-photon lines with very different intervals in Na$_2$. The total number of observed two-photon transitions was much smaller than that which could occur between the numerous rovibrational levels of the ground state and excited states. The conventional one-photon molecular spectral structure was distorted beyond recognition. A concrete analysis of the transition rates for a branch of lines of a particular two-photon band revealed the truth. In fact, each observed transition was mainly enhanced by a near coincidence of the laser frequency with a suitable one-photon transition. Only those transitions with intermediate levels less than about 1 cm^{-1} from two-photon resonances were strong enough to be measured. It

175

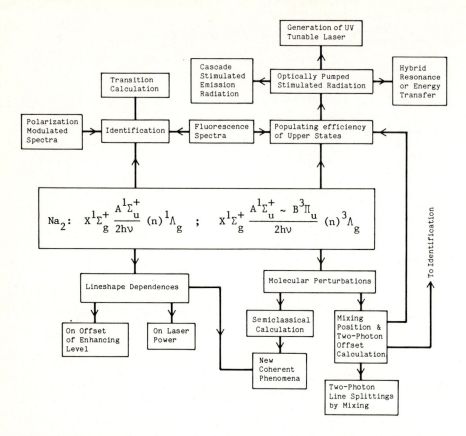

Fig. 2. The general scheme of study for Na$_2$ from the time at Stanford University to the present time at East China Normal University

stimulated our curiosity to calculate the two-photon absorption frequencies and their respective enhancing levels, as described in Ref. 4. We then plotted curves of optical frequency as a function of rotational energy for the one- and two-photon transitions, seeking the rotational quantum number at which the two curves cross. This level provides the strongest enhancement of the two-photon transition. With this figure we then could identify the two-photon transitions and analyze the dramatically altered band structures if the constants of the related states were known. Or we could fit the constants for an unknown high-lying state to the observed two-photon lines.

Indeed there are two types of excitation for two-photon transitions with equal frequencies, as illustrated in Fig. 3. The first is shown in Fig. 3b, this occurred between the singlet states in Na$_2$. The observed two-photon lines were calculated and 57 of 79 lines were recognized as having frequencies very close to those of transitions between the ground state $X^1\Sigma_g^+$ and either of two excited g-parity states, $^1\Sigma_g^+$ or $^1\Pi_g$, at 33000 cm^{-1}, observed by SCHAWLOW, et al.[3] The identifications were associated with

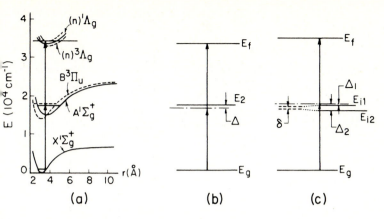

Fig. 3. Schemes of the near-resonant molecular two-photon transitions in Na₂: (a) respective electronic states,
(b) two-photon transition between singlet states,
(c) a mixing-level enhancing two-photon transition

two complementary experiments: taking the fluorescence spectrogram around the exciting wavelength and recording cw Doppler-free two-step excitation spectra, for part of the lines.[4]

Recently, the above-mentioned methods have been used to study the second type of molecular two-photon transitions, efficiently populating the excited triplet states as shown in Fig. 3c. We observed dozens of lines between 6200 and 6800A distinguished from the mentioned spectral structures. It was interesting to note that the spectral density for the second type of transition was in some wavelength regions much greater than that for the first type of transition. In fact, the ground state in Na₂ is singlet, so this kind of two-photon transition is allowed and strong enough to be noticed only if there was a near resonant enhancing level to be singlet-triplet mixed. Due to their wavefunction sharing, there are two branches of one-photon transitions enhancing one branch of the two-photon band, thus corresponding to two different regions providing the strong enhancement of the two-photon transitions (see framed in Fig. 4b). Furthermore, a new coherent effect, predicted by the semiclassical calculation for the four-level-system two-photon processes may make the spectral density even heavier.[5] We have undertaken to determine the energy constants of the triplet states.

While working on triplet states we met new problems. Where the levels mix, there are level shifts of the zero-approximation energy values. The suggested polarization modulated two-photon spectroscopy was demonstrated to be helpful and to distinguish molecular two-photon branches of lines in the whole scanning region.[6] However, for examining the specifications of the upper states, the complementary experiment of recording fluorescence spectra was extended over a much wider range from UV (3000A) to near infrared (8000A), as shown in Fig. 5.

The comparison of the fluorescence spectra for different excitation approaches, including three energy transfer processes in addition to molecular two-photon excitation transition, helps us to determine the population efficiencies for the high-lying states.[7]

177

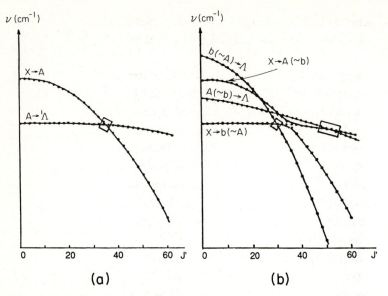

Fig. 4. Frequency of light for one- and two-photon transitions:
(a) for the first approach shown in Fig. 3a, $X:X^1\Sigma_g^+$; $A=A^1\Sigma_u^+$; $^1\Lambda:(n)^1\Lambda_g$,
(b) for the second approach shown in Fig. 3b. Dots represent
unperturbed frequencies; Frames indicate near-resonant enhancing
positions for equal-frequency two-photon transitions; $X:X^1\Sigma_g^+$;
$^3\Lambda:(n)^3\Lambda_g$; $A(\sim b)$ or $b(\sim A)$: $A^1\Sigma_u^+ \sim b^3\Pi_u$ mixing

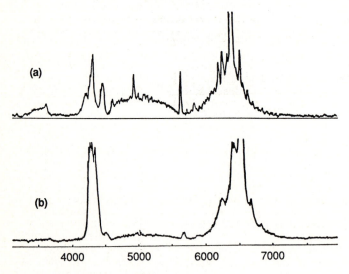

Fig. 5. Fluorescence spectra of two-photon absorption, corresponding to the
upper states:
(a) singlet-triplet coupled level,
(b) quite pure triplet state

178

3. LINESHAPE DEPENDENCES AND SPECTRAL STRUCTURES OF MOLECULAR TWO-PHOTON TRANSITIONS

Every step of our studies was encouraged by Professor Schawlow. His frequent questions always began with "Anything new?... ." During the measurements of the Doppler-free two-photon lines we found that the strongest lines eventually occurred with wide linewidths. When Professor Schawlow heard this, he suggested that they may be two-photon transitions and we should check this by changing the laser power. Thus it was confirmed that the wide linewidth was the Doppler-broadened pedestal formed by absorbing two photons from one beam. Compared to the Doppler-free peak above it, the Doppler-broadened background increased so fast for some of the lines that the high ratio of DF/DB was as small as 1/4! (Fig. 6a). To find the lineshape dependences, Mrs. Xia and Mr. Yan observed one- and two-photon transitions simultaneously to determine the offset of the enhancing level by measuring the frequency interval between the Lamb dip of the one-photon transition center and the DF peak of the related two-photon transition.[8] The measured offset ranged from about 3 GHz down to as little as 34 MHz. The Doppler-free peak became weaker as the offset decreased.

The source of the violently reduced peak ratio was proved to be due to the real population of the intermediate level with an offset smaller than the Doppler width. The effect was aggravated by increasing laser power. The background can be eliminated by intermodulation or polarization modulation spectroscopic approaches.[9]

Recently we have observed more complicated spectral structures for the two-photon transitions linked with triplet states. As mentioned above, we observed, for example, over 10 lines within a few wavenumbers with different lineshapes, except one peak line similar to the appearance as shown in Fig. 6a. Also, there were symmetric multi-peak lines, shown in Fig. 6b, and different kinds of asymmetric shapes. The lineshapes help us to recognize the mixing positions and the structures help us to determine

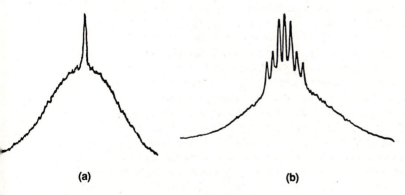

(a)   (b)

Fig. 6. Molecular two-photon line shapes:
    (a) a typical trace for the near-resonant two-photon transitions between singlet states (Ref. 8),
    (b) one kind of the line structure for the two-photon transitions enhanced by the singlet-triplet mixing levels up to high-lying triplet states.

the total number spin of the molecule, as well as the total angular
momentum of the electrons projected on the internuclear axis and the total
angular momentum of the molecules inclusive of the nuclear spin.[10]

4. GENERATION OF THE OPTICALLY PUMPED STIMULATED EMISSION OF RADIATION BASED ON MOLECULAR ELECTRONIC TRANSITIONS IN Na_2

Generating stimulated emission of radiation by optical pumping is an
important part of the studies concerning Na_2, as pointed out by Professor
Schawlow. At that time, we demonstrated stimulated emission of radiation
in the infrared region near 0.91 μm with vibrational and rotational quantum
numbers identified, and observed violet diffuse band stimulated emission at
4300A (Fig. 7c).[11] Stimulated radiation in the UV region was not observed,
although we did try to avoid the parity limitation of homonuclear molecules
by adding potassium to the heat-pipe oven to form NaK molecules to increase
energy transfer efficiency.

In addition, we have performed a number of stimulated emission processes
between the excited bound states in Na_2 providing stimulated emission
spectral lines distributed at different wavelength regions. We have
observed the generation of UV signals by different mechanisms, including (1)
molecular two-photon pumping,[12] (2) atomic two-photon pumping with energy
transfer to Na_2,[13] and (3) molecular and atomic hybrid resonance
processes.[14] The stimulated emission spectra were different, as shown
in Fig. 7.

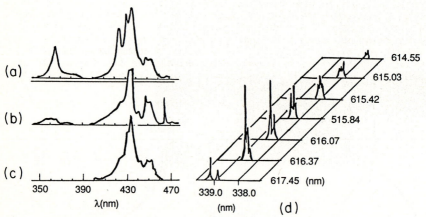

Fig. 7. Optically pumped UV stimulated tunable radiation:
(a) by molecular two-photon pumping (Ref. 12),
(b) by energy transfer from atoms to molecules (Ref. 13),
(c) without UV stimulated radiation by UV one-photon pumping,
(d) by molecular and atomic hybrid resonance pumping

5. CONCLUSION

Our time working with Professor Schawlow at Stanford University was
very joyful and fruitful, and even today there still remains a great
indelible influence on our research. We will always remember the words

e wrote for us during his second visit to East China Normal University
n 1984:

*There are a lot of simple and beautiful
things left for us to find.*

Arthur L. Schawlow
October 30, 1984

Professor Schawlow has shown us that **"OPTICS IS LIGHT WORK,"** and we
elieve this to be true!

. REFERENCES

1. K. C. Harvey, Thesis: M.L. Report No. 2442,
   Stanford University (1975).
2. J. P. Woerdman, Chem. Phys. Lett., 43, 279 (1976).
3. N. W. Carlson, A. J. Taylor, K. M. Jones, and A. L. Schawlow,
   Phys. Rev. A24, 822 (1981).
4. G. P. Morgan, H.-R. Xia, and A. L. Schawlow,
   J. Opt. Soc. Am. 72, 315 (1982).
5. H.-R. Xia, J.-W. Xu, and I.-S. Cheng, to be published.
6. J.-G. Cai, H.-R. Xia, and I.-S. Cheng,
   ACTA Optica Sinica 6, 212 (1986).

7. H.-R. Xia, J.-W. Xu, J.-G. Cai, and I.-S. Cheng, in <u>Proceedings of the National Conference on Laser Spectroscopy</u> (China, 1986), p. 16-19.
8. H.-R. Xia, G.-Y. Yan, and A. L. Schawlow, Opt. Comm. <u>39</u>, 153 (1981).
9. G.-Y. Yan and H.-R. Xia, Scientia Sinica, <u>28</u> 505 (1985).
10. H.-R. Xia, L.-S. Ma, J.-W. Xu, and I.-S. Cheng, to be published.
11. Z.-G. Wang, Y.-J. Wang, G. P. Morgan, and A. L. Schawlow, Opt. Comm. <u>48</u>, 398 (1984).
12. Z.-G. Wang, K.-C. Zhang, X.-L. Tan, and I.-S. Cheng, Acta Optica Sinica, <u>6</u>, 1081 (1986).
13. Z.-G. Wang, L.-S. Ma, H.-R. Xia, K.-C. Zhang, and I.-S. Cheng, Opt. Comm. <u>58</u>, 315 (1986).
14. Z.-G. Wang, X.-L. Tan, K.-C. Zhang, and I.-S. Cheng, to be published.

Laser-Driven Ionization and Photoabsorption Spectroscopy of Atomic Ions

W.T. Hill III[1],[a] *and C.L. Cromer*[2]

[1]University of Maryland, IPST, College Park, MD 20742, USA
[2]National Bureau of Standards, Gaithersburg, MD 20899, USA

Abstract

The application of laser-driven ionization techniques to photoabsorption spectroscopy of atomic ions will be discussed. A summary of the experimental results which confirm that a collisional mechanism is responsible for the nearly complete ionization following laser irradiation is given along with a bibliography of the photoabsorption measurements reported to date. The importance of these investigations, demonstrated in two studies involving the Ba nuclear sequence (i.e., Ba, Ba^+, and Ba^{++}) and the Xe isoelectronic sequence (i.e., Xe, Cs^+ and Ba^{++}), will be discussed. Finally, a tabulation of quantum defect parameters for Xe, Cs^+ and Ba^{++} based on a re-analysis of the Xe-sequence spectra via a shifted R-matrix quantum defect approach, will be presented.

1. Introduction

Uniform, nearly complete ionization of an atomic vapor following resonant laser irradiation has proven to be an extremely useful way to prepare atomic ions for photoabsorption studies. To date, this resonant laser-driven ionization (RLDI) technique has been successful in producing quantifiable densities of Li^+ [1], Na^+ [2,3], Cs^+ [4], Sr^+ [5], Ca^+ [5,6], Ba^+ [5,7-13], Ba^{++} [12-15] and Mn^+ [16] ions. The technique empowers one with an ability to strip away a prespecified number of electrons from a given nucleus and, conversely, to select the strength of the Coulomb field (i.e., nuclear charge) for a specific valence configuration. Thus, systematic changes along sequences of ions, which have either the same nuclear charge or valence structure, can be investigated. When coupled with recent advances in theoretical and empirical approaches to analyze complicated spectra, RLDI provides the atomic physicist with a unique opportunity to study the details of many electron effects, such as autoionization, heretofore not possible.

The focus of this article is the application of laser-driven ionization techniques to study atomic physics problems. To that end, the paper is divided into three parts. To familiarize the reader with RLDI, the first section will present a brief review of: (1) the mechanism responsible for the uniform ionization; (2) the key experimental results which validate the belief in the mechanism; and (3) the conditions and parameters necessary to achieve nearly complete ionization. The second section will be devoted to photoabsorption studies in the vacuum ultraviolet (VUV) region of the spectrum which have been made with the aid of RLDI. The discussion will center primarily on two important experiments involving Cs^+, Ba^+ and Ba^{++} ions but this section will include a bibliography of all the VUV measurements employing RLDI reported to date. The paper will conclude with a short discussion of future experiments that laser-driven ionization techniques could make possible.

2. Resonant Laser-Driven Ionization Mechanism

The first observation of nearly 100% ionization of a Na atomic vapor following resonant excitation by a pulsed dye laser was reported by LUCATORTO and MCILRATH [2] more than ten years ago. Several processes contribute to the ionization, but the universality of the phenomenon, independent of atomic species, corroborates the belief that the primary mechanism is based on superelastic collisions between free electrons and laser-excited atoms followed by collisional ionization by the free electrons, as first suggested by MEASURES [17]. It has also been demonstrated that doubly charged ionic vapors can be produced, with nearly the same efficiency, by resonantly exciting the singly charged ions [12-15]. In principle, the technique can be applied repeatedly to achieve any desired stage of ionization provided that an appropriate laser and resonant transition can be found. (At the writing of this article there have been no reports of ionization beyond the second stage.)

Experimental observations show that RLDI is capable of producing complete ionization of a 10 - 20 cm column of an atomic vapor on a microsecond time scale when: (1) the atomic densities fall in the range of 10^{13} - 10^{16} cm^{-3}; (2) the laser energy is $\geq$ 0.15 J; and (3) the laser pulse length is of the order of 500 ns. (These laser requirements are met by flashlamp pumped dye lasers.)

The mechanism can be summarized as follows. The energy necessary to drive the process is extracted from the pulsed laser field via resonant absorption. Upon irradiation, an excited state population roughly equal to the ground state population is created while, at the same time, a few free electrons are produced. The electrons are generated through a variety of processes, such as multiphoton ionization, associative ionization, or laser-assisted Penning ionization [3,7]. These processes, whose rates vary with atomic species, are generally too inefficient to be solely responsible for the total ionization observed; however, they do provide a source of initial electrons to seed the primary ionization mechanism [3, 7].

The primary mechanism can be qualitatively understood by considering the tendency for the electrons to be in equilibrium with the atomic vapor. In the case of a Na vapor, for example, with the laser saturating the $3\ ^2S_{1/2} \rightarrow 3\ ^2P_{1/2}$ transition (i.e., creating nearly equal population in the two states) the "apparent temperature" of the vapor (treating Na as a two-level system) will be almost infinite. Thus, the electrons will undergo superelastic collisions with the laser-excited atoms (a process in which the electrons gain energy while de-exciting the $3\ ^2P$ atoms) in an attempt to bring the electron-atom system into equilibrium [3]. In order for equilibrium to exist between the electrons and the atoms, however, collisional processes must dominate radiative processes. As a consequence, equilibrium will generally exist between the electrons and the closely spaced high-lying levels (which have long radiative lifetimes) but not with the lower levels [11,18]. Nevertheless, the electrons will gain energy through superelastic collisions and heat up the entire electron-vapor system--the energy gained by the electron will be collisionally transmitted back to the vapor resulting in the high-lying levels becoming

184

populated. A fraction of the atoms in high-lying states will be photo-ionized during the laser pulse which adds to the pool of free electrons while the remainder of the atoms will be ionized by the free electrons. The percentage of ionization will depend on the temperature of the electron-vapor system through the Saha-Boltzmann equation [19].

Complete ionization can only be achieved if the laser pulse is long enough to maintain the excited population so that the electron-vapor system can gain sufficient energy, otherwise only partial ionization will occur. Under long pulse excitation the mechanism is self-terminating because once the neutral population is depleted the energy link to the laser field is broken and, as a consequence, the vapor comes to equilibrium at the next higher stage of ionization.

A few investigators have suggested that the ionization is principally due to radiative effects [9,20]. Some of these suggestions have been quite controversial [21]. Although radiative processes are very important in some systems and contribute to some degree to the production of seed electrons in other systems, experimental results do not support a generic radiative process. On the other hand, there have been several important experiments which provide convincing evidence in support of a collisional mechanism. A few key results are summarized in Table 1.

Table 1. Evidence for Collisional Ionization Mechanism

Experimental Result	Reference
1. Direct observation of superelastic collisions between electrons and excited Ba and Na atoms.	22, 23
2. Ion yields depend on the laser pulse length; when pulse length is too short ($\leq$ 50 ns) electron-atomic vapor does not have time to heat up.	24
3. Ion yields depend on the atomic density; when densities are too low ($\leq 10^{13}$ cm^{-3}) collision rates are too slow.	5, 25
4. The onset of ionization lags behind the beginning of the laser pulse and the peak ion densities occur near the end of the pulse or when the pulse has ended (see Fig. 1).	1, 5-9
5. When several atomic vapors are contained in the same cell and only one is resonantly excited all species show appreciable ionization.	5
6. Ionization can be quenched by momentum-changing elastic collisions with He atoms which cool the electrons.	26

It should be emphasized that complete ionization is reached with nearly the same laser fluence ($\sim$ 5 J/cm^2) for all atomic systems even though the number of photons necessary to ionize the laser-excited atoms varies from one in Cs [4] to three in Ba$^+$ [7]. At the same time, the intermediate near-resonance structure, enhancing multiphoton ionization out of the excited states is vastly different for the different atomic systems. Typically, 100% ionization is reached when the oscillator strength (f) of the transition to which the laser is tuned is of the order of 1. In contrast, even for Ca [6] and Mn [16] in which f is of the order of 10^{-5}, 50% ionization or better was still achieved. These differences in the atomic structure and the transition strengths would produce a strong intensity dependence in the ionization if radiative processes were primarily responsible. One also observes that the ionization time development is of the same order of magnitude for all systems [1,5-9,11]. Figure 1 shows this for Li (f ~ 1) and Ca (f ~ 10^{-5}, intercombination line). In summary, the experimental results overwhelmingly support a primary mechanism based on collisional processes and not on multiphoton ionization processes.

The ionization process has also been studied numerically in an attempt to prove or disprove the collisional or radiative mechanisms [3,7,9,11,17, 27-29]. The results of these model calculations have been mixed. This is largely due to the complexity of the calculation, which requires an accurate knowledge of level populations, radiative rates and collision cross sections. Unfortunately, many of these parameters are not well known, making the numerical results less decisive than the experimental results.

Figure 1 shows that in addition to ions, appreciable densities of excited neutrals are present in the vapor at early times. Less transient excited neutrals are also a by-product of quenching of RLDI by over pressuring the vapor with He gas. (See entry 6 in Table 1 and [22].) Thus, RLDI can be used to prepare excited neutrals for photoabsorption as well.

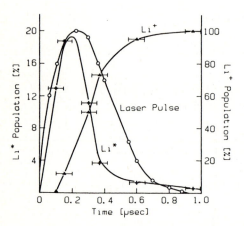

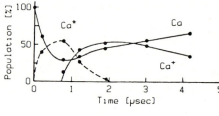

Fig. 1. Evolution of excited neutral and ion populations for Li (from [1]) and Ca (from [6])

Finally, it is interesting to note that it has been recently shown that appreciable ionization occurs on a similar time scale after non-resonant excitation of atomic vapors in the presence of large Ar atmospheres [30]. The mechanism responsible here is based on the laser-fragmentation of metal clusters catalyzed by the Ar atoms. (Cluster formation in He atmospheres is much less efficient.) Cluster fragmentation generates many excited state atoms in addition to ions, so that free electrons can gain energy as they do after resonant excitation and thus create significant ionization, albeit less than 100%.

3. Photoabsorption Spectroscopy of Atomic Ions and Excited Neutral Atoms

Attention will now be turned toward photoabsorption experiments. The studies summarized in Table 2 were obtained with an apparatus similar to that schematically shown in Fig. 2. When doubly charged ions are desired, a second flashlamp pumped dye laser must be added. The studies include photoabsorption from both inner and valence shell electrons and from ions as well as excited states of neutral atoms.

The background radiation in these experiments was generated by either a BRV electrical spark [32] or a laser-produced plasma light source [33] both of which emit radiation in the VUV in the 5 - 200 nm region. These light sources provide time resolution on the order of 10 - 50 ns. The vapors were contained in oven/heatpipe cells and the photoabsorption spectrum was dispersed by a grazing-incidence spectrograph with either photographic film [7] or a photoelectric detector [34] placed in its image plane. Since this radiation does not propagate through the air, the optical paths connecting the light source, the absorption region and the detector were enclosed and evacuated. For the two experiments discussed below, the dye lasers were tuned to 459.3 nm in Cs^+, 553.5 nm in Ba^+ while a second laser tuned to the 493.4 nm resonance line of Ba^+ was used to generate Ba^{++}. More details about the experiments can be found in the references (see [4,12-15]).

Table 2. Photoabsorption Spectra of Ions and Excited Neutrals

System	Lower State Configuration		Electron Shell(s) Excited	Reference
Li^*	$(1s^2 2p)$	^2P	1s	1
Li^+	$(1s^2)$	^1S	1s	1
Na^*	$(2p^6 3p)$	^2P	2p	31
Na^+	$(2p^6)$	^1S	2p	2
Cs^+	$(5p^6)$	^1S	5p	4
Ca^*	$(3p^6 4s4p)$	^3P	3p	6
Ca^+	$(3p^6 4s)$	^2S	3p	6
Mn^+	$(3p^6 3d^5 4s)$	^7S	3p	16
Ba^+	$(4d^{10} 5p^6 6s)$	^2S	4d	12,13
	$(4d^{10} 5p^6 5d)$	^2D	4d	12,13
Ba^{++}	$(4d^{10} 5p^6)$	^1S	4d, 5p	12-19

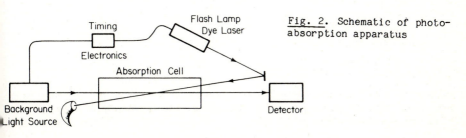

Fig. 2. Schematic of photoabsorption apparatus

3.1. Ba, Ba+, Ba++ Sequence

Figure 3a shows the photoabsorption spectra between 14.6 and 8.3 nm (85 – 150 eV) in Ba, Ba^+ and Ba^{++} involving inner shell $4d^{10} \to 4d^9$ nf,εf transitions (taken from [12]). In Ba and Ba^+ the spectrum consists of a broad resonance feature (i.e., excitation of an εf continuum state) while in Ba^{++} there are several sharp resonances (i.e., excitation of nf autoionizing states) prior to the onset of the broad feature. The main structural difference in the 4d → f photoabsorption spectra between Ba (and Ba^+) and Ba^{++} is due to the reduction in the screening associated with the removal of the 6s valence electrons. As a consequence, the f potential, which has a double well shape in Ba and Ba^+, is flattened out in Ba^{++}. In response, there is more overlap between the 4d wavefunction and the nf wavefunctions (autoionizing states) because the nf wavefunctions are contracted closer to the nucleus in Ba^{++}.

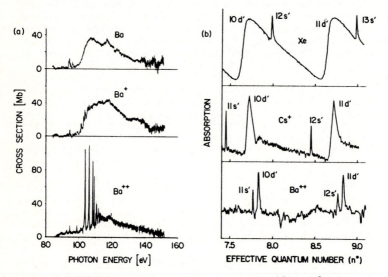

Fig. 3. Photoabsorption spectra of: (a) $4d^{10} \to 4d^9$ nf, εf for Ba, Ba^+ and Ba^{++} (from [12]) and (b) $5p^6 \to 5p^5$ $^2P_{1/2}$ $ns_{1/2}$, $nd_{3/2}$ for Xe (from [35]), Cs^+ (from [4]) and Ba^{++} (from [14])

The significance of this observation extends beyond gas-phase atomic physics when the Ba and Ba^{++} spectra are compared with 4d → f photoabsorption spectra of divalent Ba compounds. In particular, Fig. 14 of [13] shows that the spectrum of $BaBr_2$ is similar to that of atomic Ba while that of BaF_2 begins to resemble Ba^{++}. This progression suggests that the Ba 6s electrons start to become de-localized as the electron affinity increases from Br to F.

3.2. Xe, Cs+, Ba++ Sequence

Figure 3b shows several densitometer traces of autoionizing Rydberg resonances of Xe, Cs^+ and Ba^{++} involving transitions from the ground $5p^6$ $^1S_0$ state to states designated as $5p^5$ $^2P_{1/2}$ $ns_{1/2}$, $nd_{3/2}$. In the figure, the

$ns_{1/2}$ and $nd_{3/2}$ resonances are given the labels ns' and nd', respectively. These states lie above the $^2P_{3/2}$ ionization limit and autoionize into the concomitant εs and εd continua. The s and d continua are the above threshold continuation of the $5p^5\ ^2P_{3/2}\ ns_{1/2}$, $nd_{3/2}$, $nd_{5/2}$ Rydberg series which converge to the $^2P_{3/2}$ ionization limit. These five interacting Rydberg series and their associated continua comprise the five channels of excitation observed in the Xe-like ions.

The spectra in Fig. 3b are plotted vs. the effective quantum number n^* which is defined through:

$$E = I \frac{Z_{eff}^2 R}{(n^*)^2} \tag{1}$$

with E being the energy of the transition (in cm^{-1}), I the ionization limit to which the autoionizing resonances are converging (i.e., $^2P_{1/2}$), Z_{eff} the effective charge on the nucleus and R the Rydberg constant. On an n^* scale, the ns' and nd' resonances appear with a period of one. Displaying the spectra in this way eliminates wavelength dependence and allows the resonances with the same (ns' or nd') classification to be compared on an equal footing.

Figure 3b shows that, as the Coulomb field strength is increased (i.e., in going from Xe to Ba^{++}), the resonances respond in the following manner: (1) the positions of the s' and d' resonances are shifted to higher n^* and (2) the widths of the d' resonances (which are measures of the probabilities for autoionization) are reduced. This behavior is associated with the change in the effective potential and the tendency of the system to approach the hydrogenic limit as Z_{eff} increases [14].

This striking demonstration of the influence Z_{eff} has on the autoionizing widths also shows the intimate connection between autoionization widths and the strength of perturbations in the bound portion of the spectrum. This can be seen by comparing Figs. 3b and 4. Figure 4 displays the energy positions of the bound states (i.e., all Rydberg states below the $^2P_{3/2}$ ionization limit) for Xe, Cs^+, and Ba^{++} in a Lu-Fano diagram [37]. The energy positions are taken from [36,4,15] respectively for Xe, Cs^+ and Ba^{++}. Each experimental point is plotted with the abscissa being the effective quantum number, $\nu_{1/2}$ ($\equiv n^*$ from (1)), relative to the upper $^2P_{1/2}$ ionization limit and the ordinate being the effective quantum number modulo 1, $\nu_{3/2}\{mod\ 1\}$, relative to the lower $^2P_{3/2}$ ionization limit. (An equation similar to (1) with I given by the $^2P_{3/2}$ threshold defines $\nu_{3/2}$.) The solid curves connecting the points are derived from the parameters given in Table 3 which were determined from a numerical fit to the experimental data using a shifted R-matrix formulation of the Multichannel Quantum Defect Theory (MQDT) equations [38,39]. Theoretical MQDT parameters obtained from a relativistic random phase approximation calculation by CHENG [40] were used as starting values for the fit. Some of the parameters could not be determined from the experimental data and the theoretical values were retained; these values are indicated by an asterisk (*) in Table 3.

The Lu-Fano plot is composed of three curves running across the unit cell for the three series converging to the lower limit and two curves running up the cell for the series converging to the upper limit; these curves are repeated for each unit cell. The parameters denoted by δ_i in

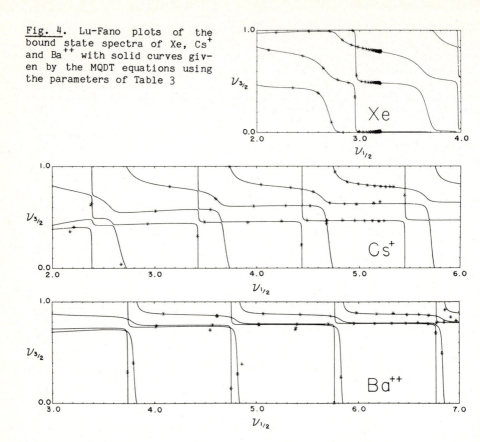

Fig. 4. Lu-Fano plots of the bound state spectra of Xe, Cs$^+$ and Ba^{++} with solid curves given by the MQDT equations using the parameters of Table 3

Table 3 correspond to the quantum defect of each channel in the absence of interchannel perturbations. If there were no perturbations, the curves running across the cell would be straight and horizontal and would pass through the points $(1 - \delta_1)$, $(1 - \delta_2)$, and $(1 - \delta_3)$ while the other two curves would be straight and vertical and pass through the points $(1 - \delta_4)$ and $(1 - \delta_5)$. Because of interchannel interaction, the curves are not straight and avoid crossing each other. The size of the avoided crossings is given by the R_{ij} parameters in Table 3 which are a measure of the strength of the perturbations among the bound resonances below the $^2P_{3/2}$ threshold. For example, R_{24} for Xe describes the avoided crossing in the center of the Xe Lu-Fano plot. (The separation of the parameters into those responsible for intrachannel (δ_i) and interchannel (R_{ij}) interactions is the primary reason for using the shifted R-matrix formulation of MQDT to fit the experimental levels. In addition, the fitting procedure using this parameterization is more transparent because each parameter is associated with a particular, localized section of the Lu-Fano curve [38, 39].) One can immediately see, as pointed out by HILL et al. [14], that (1) the large avoided crossings are linked to large autoionization widths above the $^2P_{3/2}$ threshold (compare Figs. 3b and 4) and (2) the perturbations are reduced as Z_{eff} increases (see R_{24} and R_{25} values in Table 3).

Table 3. Shifted R-matrix MQDT parameters for Xe, Cs^+ and Ba^{++}

Parameters[†]	Xe	Cs^+	Ba^{++}
δ_1^0	0.9938(20)	0.5289(25)	0.2028(51)
δ_2^0	0.5131(16)	0.3617(26)	0.2053(55)
δ_3^0	0.2042(19)	0.1940(31)	0.1201(51)
δ_4^0	1.0374(17)	0.5487(26)	0.2292(33)
δ_5^0	0.3260(19)	0.2917(37)	0.1635(36)
R_{14}^0	0.0442(100)	0.0537(100)	0.0435[*]
R_{24}^0	-0.0729(47)	-0.0072[*]	0.0056[*]
R_{34}^0	-0.0113(296)	0.0085[*]	-0.0018[*]
R_{15}^0	0.0635(59)	0.0122[*]	0.0367[*]
R_{25}^0	0.2397(35)	0.1004(95)	0.1045[*]
R_{35}^0	-0.4209(35)	-0.2814(68)	-0.1255(292)
δ_1^1	-0.5324(201)	-0.2029(74)	-0.1722(169)
δ_2^1	-0.8990(112)	-0.4980(82)	-0.2301(186)
δ_3^1	-0.0759(169)	-0.0467(114)	-0.0357(195)
δ_4^1	-0.4592[*]	-0.2285(106)	-0.1379(85)
δ_5^1	1.1516[*]	-0.2939(220)	-0.1688(94)
R_{14}^1	-0.0430[*]	-0.0418[*]	-0.0270[*]
R_{24}^1	0.6324[*]	0.0799[*]	0.1982[*]
R_{34}^1	-0.2913[*]	-0.0720[*]	-0.0260[*]
R_{15}^1	-0.2516[*]	0.0546[*]	-0.2712[*]
R_{25}^1	-1.1251[*]	-0.1262[*]	0.0955[*]
R_{35}^1	-0.6291[*]	0.6828[*]	0.2144[*]

[*]These values could not be determined from experimental data [36,4,15] and are equal to theoretical values [40] which were used as starting values for the reduction.

[†]The channels:

$$5p^5\ ^2P_{3/2}\ ns_{1/2},\ nd_{3/2},\ nd_{5/2};\ 5p^5\ ^2P_{1/2}\ ns_{1/2},\ nd_{3/2},$$

are numbered consecutively. The energy dependence in the parameters is defined by:

$$\delta_i = \delta_i^0 + \epsilon\delta_i^1 \text{ and } R_{ij} = R_{ij}^0 + \epsilon R_{ij}^1,$$

with ϵ being the energy in atomic units. Values in parentheses indicate the uncertainty in the last digits.

4. Future Experiments

There are several new spectroscopic experiments yet to be performed in conjunction with RLDI; a few examples will now be given. As the flashlamp pumped dye laser technology improves, allowing more intense radiation to be generated in the blue and near UV regions of the spectrum, higher stages of ionization can be achieved. The rare earth elements would provide good candidates for creating the third state of ionization, for example. Furthermore, the cluster fragmentation ionization process (in Ar atmospheres [30]) can be employed to prepare moderate ion densities of samples which do not have suitable transitions for RLDI to be used. At the same time, the fragmentation process probably generates ionized clusters which could be the subject of investigations. The ability to select the percentage of ionization by adding an appropriate amount of He buffer gas [26] will permit one to investigate partially ionized vapors. For instance, line broadening of neutral atomic states in the presence of their own ions might be studied. These mixed vapors might also produce the right environment to create molecular ions. In addition to the spectroscopic studies, pure or partially ionized vapors might be employed as new nonlinear optical media for frequency conversion. Although the suggestions cited are related to atomic physics, laser-driven ionization also has application to plasma physics problems [17,27].

Acknowledgments

The authors would like to thank T. B. Lucatorto, T. J. McIlrath and K. Yoshino for kindly permitting their data to be reproduced; K. T. Cheng for making available the unpublished results of his RRPA calculation for Xe, Cs^+ and Ba^{++}; T. B. Lucatorto for helpful discussions; and B. P. Turner, D. J. Davis and M. G. Spell for technical assistance in preparing this manuscript. This work is sponsored in part by the National Science Foundation under grant PHY-84-51284, the Research Corporation and the National Bureau of Standards.

[a] National Science Foundation Presidential Young Investigator

References

1. T. J. McIlrath, T. B. Lucatorto: Phys. Rev. Lett. 38, 1390 (1977).
2.. T. B. Lucatorto, T. J. McIlrath: Phys. Rev. Lett. 37, 428 (1976).
3. B. Carré, F. Roussel, P. Breger, G. Spiess: J. Phys. B 14, 4289 (1981).
4. T. J. McIlrath, J. Sugar, V. Kaufman, D. Cooper, W. T. Hill, III: J. Opt. Soc. Am. B3, 398 (1986).
5. C. H. Skinner: J. Phys. B 13, 55 (1980).
6. B. F. Sonntag, C. L. Cromer, J. M. Bridgeis, T. J. McIlrath, T. B. Lucatorto: In Proc. of the Third Topical Meeting of Short Wavelength, Coherent Radiation, Monterey, CA (March, 1986).
7. T. B. Lucatorto, T. J. McIlrath: Appl. Opt. 19, 3948 (1980).
8. H. A. Bachor, M. Koch: J. Phys. B 13, L369 (1980).
9. H. A. Bachor, M. Kock: J. Phys. B 14, 2793 (1981).
10. R Künnemeyer, M. Koch: J. Phys. B 16, L607 (1983).
11. L. Jahreiss, M. C. E. Huber: Phys. Rev. A28, 3382 (1983).
12. T. B. Lucatorto, T. J. McIlrath, J. Sugar, S. M. Younger: Phys. Rev. Lett. 47, 1124 (1981).

13. T. B. Lucatorto, T. J. McIlrath, W. T. Hill, III, C. W. Clark: In Inter. Conf. X-Ray and Atomic Inner-Shell Physics, ed. by B. Crassman, AIP Conf. Proc. 94, 584, (1982).
14. W. T. Hill, III, K. T. Cheng, W. R. Johnson, T. B. Lucatorto, T. J. McIlrath, J. Sugar: Phys. Rev. Lett. 49, 1631 (1982).
15. W. T. Hill, III, T. B. Lucatorto, J. Sugar, K. T. Cheng: to be submitted to Phys. Rev. A. (1987).
16. J. W. Cooper, C. W. Clark, C. L. Cromer, T. B. Lucatorto, B. F. Sonntag, F. S. Tomkins: to be submitted to Phys. Rev. A Rapid Communications (1987).
17. R. M. Measures: J. Quant. Spectrosc. Radiat. Transfer 10, 107 (1970).
18. M. Mitchner, C. H. Kruger: In Partially Ionized Gases (John Wiley and Sons, NY, 1972).
19. K. R. Lang: In Astrophysical Formulae, (Springer-Verlag, NY, 1974), p. 244.
20. J. M. Salter: J. Phys. B12, L763 (1979).
21. C. H. Skinner: J. Phys. B13, L637 (1980); T. J. McIlrath, T. B. Lucatorto, J. Phys. B 13, L641 (1980).
22. D. F. Register, S. Trajmar, G. Csanak, S. W. Jensen, M. A. Fineman, R. T. Poe: Phys. Rev. A28, 151 (1983); J. M. Bizau, B. Carré, P. Dhez, D. L. Ederer, P. Gerard, J. C. Keller, P. Koch, J. C. LeGouët, J. L. Picqué, F. Roussel, G. Spiess, F. Wuilleumier: In Laser Spectroscopy VI, Proc. 6th Intern. Conf. Interlaken, Springer Series in Optical Sciences, ed. by H. P. Weber, W. Lüthy (Springer-Verlag, NY, 1983).
23. I. V. Hertel, W. Stell, Adv. At. Mol. Phys. 13, 113 (1977); J. L. LeGouët, J. L. Picqué, F. Wuilleumier, J. M. Bizau, P. Dhez, P. Koch, D. L. Ederer: Phys. Rev. Lett. 48, 600 (1982).
24. T. Stacewicz: Opt. Commun. 35, 239 (1980); C. Bréchignae, P. H. Cahuzac: Opt. Commun 43, 270 (1982); J. L. Bowen, A. P. Thorne: J. Phys. B 18, 35 (1985); T. J. McIlrath, J. L. Carlsten: J. Phys. B6, 697 (1973).
25. B. Carré, F. Roussel, P. Breger, G. Spiess: J. Phys. B14, 4271 (1981).
26. W. T. Hill, III: J. Phys. B 19,359 (1986).
27. R. M. Measures, P. G. Cardinal: Phys. Rev. A23, 804 (1981); R. M. Measures, P. G. Cardinal, G. W. Shinn: J. Appl. Phys. 52, 1269 (1981); R. M. Measures, N. Drewell, P. Cardinal: J. Appl. Phys. 50, 2662 (1979).
28. W. L. Morgan: Appl. Phys. Lett. 42, 790 (1983).
29. P. G. Cardinal: Ph.D. Thesis, University of Toronto (1985).
30. W. T. Hill, III: Opt. Commun 54, 283 (1985).
31. J. Sugar, T. B. Lucatorto, T. J. McIlrath, A. W. Weiss: Opt. Lett 4, 109 (1979).
32. T. B. Lucatorto, T. J. McIlrath, G. Mehlman: Appl. Opt. 18, 2916 (1979).
33. J. M. Bridges, C. L. Cromer, T. J. McIlrath: Appl. Opt. 25, 2205 (1986).
34. C. L. Cromer, J. M. Bridges, J. R. Roberts, T. B. Lucatorto: Appl. Opt. 24, 2996 (1985).
35. K. Yoshino: private communication based on data of [36] (1986/7).
36. K. Yoshino, D. E. Freeman: JOSA B2, 1268 (1985).
37. K. T. Lu, U. Fano: Phys. Rev A 2, 81 (1970); K. T. Lu: Phys. Rev. A 4, 579 (1971).
38. W. E. Cooke, C. L. Cromer: Phys. Rev. A 32, 2725 (1985); A. Giusti-Suzor, U. Fano: J. Phys. B 17, 215 (1984).

39. C. L. Cromer: to be published (1987).
40. K. T. Cheng: private communication based on the procedure of [41] (1986/7).
41. W. R. Johnson, K. T. Cheng, K.-N. Huang, M. Le Dourneuf: Phys. Rev. A <u>22</u>, 989 (1980).

Two-Photon Resonant Parametric and Wave-Mixing Processes in Atomic Sodium

Pei-Lin Zhang and *Shuo-Yan Zhao*

Department of Physics, Tsinghua University, Beijing, China

The generation of coherent radiation in the UV and VUV region through four wave mixing processes has been the subject of several recent reports. In the first, the Na 4d D_J level is populated by two-photon excitation and a cascade of IR stimulated emission ensues. Photons from these cascades of frequency ω_{IR} can then mix with two laser photons with total energy related to $2\omega_L$, yielding UV photons at a frequency $\omega_{UV} = \omega_{IR} + 2\omega_L$. Using various members of the cascade, HARTWIG [1] generated radiation at 330 nm and 333nm, while Wu and Chen [2] have produced radiation at 383 nm and 388nm using this procedure.

Coherent UV radiation may also be generated using parametric oscillations [1],[3]. In this process the signal and the idler wave grow together following two-photon excitation and the coherent radiation produced undergoes a frequency shift because of the presence of atomic transitions. Thus in parametric generation more UV frequencies may be attained.

In this paper, we report results obtained in a study of UV generation using sum and difference-frequency four-wave mixing processes. We also report on the UV outputs obtained in parametric oscillation experiments using the Na 3P and 4P levels as near resonances. The wavelength and the intensity of the parametric oscillations are reported; the dependence of these parameters on pump-laser wavelength and intensity, on the oven temperature and other factors have been investigated. A theoretical analysis based on a wave equation and on atomic polarization is given for these two methods of coherent generation of UV radiation.

2. Theoretical Background

Theoretical analysis of four-wave mixing process and parametric oscillation process requires an analytical expression for the atomic nonlinear polarization. For this purpose we first use a density matrix method to describe the interaction between atom and electric field [4][5], then we use Maxwell equations to relate electric field to electric polarization. The details are described elsewhere.

For four-wave mixing processes, we obtain

$$I_{IR} = I_{IR}^0 \exp(gz) , \tag{1}$$

$$I_{UV} = 256\pi^4\omega_{UV}^2|\chi^{(3)}(\omega_{UV})|^2 I_L^2 I_{IR}/c^4 n_{IR}n_L^2 n_{UV}[g^2+4(\Delta k)^2],$$

where $g = -4\pi\omega_{IR}\mathrm{Im}\chi^{(1)}(\omega_{IR})/cn_{IR}$ is the gain coefficient; $\Delta k = 2k_L - k_{IR} - k_{UV}$ is the phase mismatch of the processes; $\chi^{(1)}$, $\chi^{(3)}$ are the first- and third-order electric susceptibilities,

$$\chi^{(1)}(\omega_{IR}) = -N|(j|\mu|i)|^2(\rho_{jj}-\rho_{ii})/\hbar D_{ji}^*,$$

$$\chi^{(3)}(\omega_{UV}) = \frac{N(0|\mu|k)(k|\mu|n)(n|\mu|m)(m|\mu|0)}{\hbar^3(\omega_m-\omega_L)}\left[\frac{\rho_{00}-\rho_{nn}}{D_{k0}^*D_{n0}^*} + \frac{\rho_{nn}-\rho_{kk}}{D_{k0}^*D_{nk}}\right],$$

for difference-frequency mixing process, and

$$\chi^{(3)}(\omega_{UV}) = \frac{N(0|\mu|k)(k|\mu|n)(n|\mu|m)(m|\mu|0)}{\hbar^3(\omega_m-\omega_L)}\left[\frac{\rho_{00}-\rho_{nn}}{D_{k0}^*D_{n0}^*} - \frac{\rho_{nn}}{D_{k0}^*D_{kn}^*}\right],$$

for sum-frequency mixing process; with $|0)$, $|n)$ denoting ground level and two-photon excited level respectively, N density of Na atom, $D_{ji} = \omega_j - \omega_i - \omega_{IR} + i\Gamma_{ji}$, $D_{nk} = \omega_n - \omega_k - \omega_{IR} + i\Gamma_{nk}$, $D_{k0} = \omega_k - \omega_0 - \omega_{UV} + i\Gamma_{k0}$ and $D_{n0} = \omega_n - \omega_0 - 2\omega_L + i\Gamma_{n0}$.

For parametric oscillation process, the intensities of the id wave and the signal wave are given by

$$I_I = I_I^0\exp\left\{[-\alpha+g+\mathrm{Re}\sqrt{(\alpha+g-2i\Delta k)^2+B^2}]z/2\right\},$$

$$I_S = I_S^0\exp\left\{[-\alpha+g+\mathrm{Re}\sqrt{(\alpha+g-2i\Delta k)^2+B^2}]z/2\right\}$$

with α absorption coefficient of the signal wave,

$$B^2 = (8\pi/c)^2(\omega_I\omega_S/n_In_S)\chi^{(3)}(\omega_I)\chi^{(3)*}(\omega_S)|E_L|^4,$$

$$\chi^{(3)}(\omega_I) = \frac{N(0|\mu|k)(k|\mu|n)(n|\mu|m)(m|\mu|0)}{\hbar^3(\omega_m-\omega_L)}\left[\frac{\rho_{00}-\rho_{kk}}{D_{nk}^*D_{k0}} - \frac{\rho_{00}-\rho_{nn}}{D_{nk}^*D_{n0}^*}\right]$$

and $\chi^{(3)}(\omega_S)$ similar to (4).

3. Experimental Setup

The experimental apparatus is similar to that described previous [6]. A pulsed dye laser pumped by a frequency-doubled Nd:YAG las is used to reach $4d^2D_J$ level of sodium by two-photon resonance excitation. Mixed R590 and R610 dyes in methanol are used for pr ducing laser wavelength 578.73nm. The laser beam is focused into center of a heat-pipe oven through a lens (f=300mm). The dimensi of the stainless steel heat-pipe oven are 22mm in diameter and

460mm in length. Coherent radiation is detected by a SPEX
0.75m spectrometer followed by a EMI9656QB photomultiplier and a
boxcar. The wavelengths of the laser and the generated coherent
radiation are carefully calibrated by Hg, Na, and Ne spectral
lamps with precision of 0.01nm near the atomic transition lines.

4. Results and Discussions

We divide the coherent radiation lines into three groups according to the generation mechanism.

4.1 Coherent lines through 4D–3P–3S parametric oscillation

We report on lines obtained through parametric oscillation with 3P
as a near resonance level. Two-photon excitation produces a population inversion between 4D and 3P levels, which causes stimulated emissions at the same wavelengths as atomic transitions
4 D $3/2$ -3P $1/2$, 4D $5/2$ -3P $3/2$. (Energy level shift due to optical
Stark effect is small as the pumped laser energy is less than
1mJ/pulse). At the oven temperature 280-450°C besides the stimulated emissions we observed two coherent radiation lines with
a little broader width. Their wavelengths are shorter than those
of the corresponding stimulated emissions. The wavelength
shifts from atomic transitions increase with increasing oven temperature as shown in Table 1. The parametric oscillation lines
near atomic 3P-3S transition are also shown in this table. The
wavelengths of the latter are longer than those of atomic transitions and the wavelength shifts increase with increasing oven
temperature too. As the laser wavelength is detuned from two-
photon resonance to the longer wavelength side by about 0.05nm,
the intensities of stimulated emissions decrease rapidly while
the intensities of parametric oscillation vary relatively slowly.
The wavelength shifts of parametric oscillation lines (to
shorter side for 4D-3P transition, and to longer side for 3P-3S
transition) increase as the laser wavelength increases.

Table 1. Temperature dependence of 4D-3P-3S parametric
         oscillation wavelength (with 4 Torr of Ar)

Temperature [°C]	Wavelength [nm]			
332	568.263	568.805	589.030	589.620
356	568.240	568.790	589.047	589.635
380	568.225	568.770	589.072	589.650
Atomic transition	568.263	568.820	588.995	589.592

We have also investigated the intensities versus the oven
temperature. As an example, the thermal dependence of parametric
oscillations near $3P_{3/2}$-$3S_{1/2}$ transition is shown in Fig. 1.
There exists an optimum temperature which yields maximum output
intensities. This can be explained if we consider the parametric
oscillations as noncollinear phase-matched lines. By utilizing
Sellmeier equation and the oscillator strength of sodium atom
[7][8],

$$n-1=\frac{Nr_e}{2}\sum_{ij}\frac{g_{ii}f_{ij}}{(y_{ij}^2-y^2)}$$ (10)

with the classical electron radius $r_e = 2.818 \times 10^{-13}$ cm, γ being the energy in cm^{-1}, and f_{ij} the oscillator strength of the transition from level i to j; we have calculated phase-matched angles θ and found they are proportional to $\sqrt{N}$. Then we calculate I_S by substituting the intensity distribution of the laser beam

$$I_L = I_L^0 \exp(-\theta^2/\theta_0^2) \tag{11}$$

in (7), where θ_0 is the beam divergence. The calculated intensity versus oven temperature curve is also shown in Fig. 1.

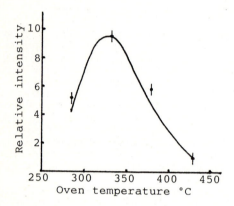

——— Theoretical

╪ Experimental

Fig. 1. Intensity of 589nm parametric oscillation line versus oven temperature

When the oven temperature increases to 400°C or higher, the foregoing parametric oscillation decreases rapidly, at the same time another pair of parametric oscillation lines begins to appear. Their wavelengths are 568.446nm and 589.407nm. The main characteristic is that their wavelengths do not depend upon oven temperature, and their intensities increase as oven temperature increases. Therefore it is clear that these two lines are a pair of collinear phase-matched parametric oscillations. The calculated wavelengths are 568.445nm and 598.404nm which are in agreement with the experimental results (pumped laser wavenumber γ_L =17274 cm^{-1}, being 0.1cm^{-1} less than the resonance wavenumber). As the wavelength shift is as large as 0.18nm, they take place only if noncollinear phase-matched lines are in unfavourable condition due to their large phase-match angles at higher oven temperature.

4.2 Coherent lines through 4P-3S parametric oscillation

For parametric oscillation resonant enhanced by 4P level we have only observed UV coherent radiation near 4P-3S transition. Using a spectrometer of 0.01nm resolution and calibrating the output wavelengths by the same atomic transition 4P$_{3/2}$ –3S$_{1/2}$ 330.237nm and 4P$_{1/2}$-3S$_{1/2}$ 330.298nm, we have studied the generated coherent radiation carefully. It is shown in Table 2 that the output coherent radiation has four components. Two of them correspond to collinear phase-match condition. Their wavelengths do not depend upon the oven temperature and their intensities

increase with increasing oven temperature. The other two components correspond to noncollinear phase-match condition. Their
wavelengths increase as oven temperature increases. From the
relation of the refractive index versus the wavenumber it can be
shown   theoretically that the coherent radiation line should
split into four components and that the wavelengths of noncol-   •
linear phase-matched components shift to the longer side as
temperature increases[9].

Table 2. Temperature dependence of 4P-3S parametric
oscillation wavelength (with 4 Torr of Ar)

Temperature [°C]	Wavelength [nm]			
284		330.237		330.298
428	330.229	330.249		330.309
452	330.228	330.252	330.298	330.314
476	330.227	330.260	330.299	330.320
Theoretical λ of collinear phase- matched lines	330.220		330.293	

4.3 Coherent lines through four-wave mixing processes

The observed coherent lines through four wave-mixing processes
are tabulated in Table 3. In order to obtain stronger signals
the pumped laser energy is increased to about 25mJ/pulse at
578.73nm with linewidth 0.02nm. The heat-pipe oven is heated to
temperature 405°C at sodium density 5.9×10^{15} cm^{-3}. We have observed five lines, 255.78, 257.51, 257.54, 280.47, and 280.51nm,
through processes of the type $\omega_{UV} = 2\omega_L + \omega_{IR}$. For difference-
frequency mixing processes, we have observed two lines, 298.83nm
and 298.87nm, through processes $\omega_{UV} = 2\omega_L - \omega(4P_J - 3D_{J'})$, as well as
two lines, 330.04nm and 333.11nm, through $\omega_{UV} = 2\omega_L - \omega(4P_J - 4S_{1/2})$.

Table 3. Wavelength, relative intensity, and coupling scheme
of four-wave mixing lines.

λ [nm] Exptl.	log I	λ [nm] Theor.	coupling scheme	θ [mrad]	L_C [nm]
255.78	0.5	255.80	$2\omega_L + \omega(4P_{3/2} - 4S_{1/2})$		5.94
		255.84	$2\omega_L + \omega(4P_{1/2} - 4S_{1/2})$		5.94
257.51	0.5	257.45	$2\omega_L + \omega(4D_{3/2} - 4P_{1/2})$		5.94
257.54	0.5	257.49	$2\omega_L + \omega(4D_{5/2} - 4P_{3/2})$		5.94
280.47	1	280.44	$2\omega_L + \omega(4P_{3/2} - 3D_{5/2})$		5.93
280.51	1	280.48	$2\omega_L + \omega(4P_{1/2} - 3D_{3/2})$		5.93
298.83	1	298.82	$2\omega_L - \omega(4P_{1/2} - 3D_{3/2})$	1.27	
298.87	1.5	298.87	$2\omega_L - \omega(4P_{3/2} - 3D_{5/2})$	1.27	
333.04	2	333.00	$2\omega_L - \omega(4P_{1/2} - 4S_{1/2})$	2.73	
333.11	2.5	333.06	$2\omega_L - \omega(4P_{3/2} - 4S_{1/2})$	2.73	

Except the group of the shortest wavelength, observed lines of
all other groups are doublets, because cascade stimulated emission of either D-P or P-S transition is doublet. We can qualitatively explain the intensities of generated UV lines by the
gain coefficient of stimulated emission, nonlinear susceptibility,

and phase-match condition. For example, lines through sum-frequency mixing processes are weak in general because they ca not satisfy the phase-match condition. We have calculated the coherence lengths

$$L_c = \pi / |\Delta k|$$ (12

instead of phase-matched angles in Table 3.

This project has been supported by Science Fund of the Chine Academy of Sciences.

References

1. W.Hartig: Appl. Phys. 15, 427 (1978)
2. C.Y.R.Wu and J.K.Chen: Opt. Commun. 50, 317 (1984)
3. A.V.Smith and J.F.Ward: IEEE J. Quantum Electron. 17, 525 (1981)
4. Y.R.Shen: The Principles of Nonlinear Optics (John Wiley and Sons, New York, 1984)
5. Yu.Malakyan: Sov. J. Quantum Electron. 15, 905 (1985)
6. P.-L.Zhang, Y.C.Wang, and A.L.Schawlow: J. Opt. Soc. Am. B1, 9 (1984)
7. R.B.Miles and S.E.Harris: IEEE J. Quantum Electron. 9, 470 (1973)
8. E.M.Anderson and V.A.Zilitis: Opt. Specktrosk. 16, 99 (1964)
9. P.-L.Zhang and A.L.Schawlow: Canadian J. Phys. 62, 1187 (1984

On the Nature of Hochheim Alloy

G.W. Series

Clarendon Laboratory, Parks Road, Oxford, OX1 3PU, UK

It is not, I think, widely known that Art has access to sources of information that are denied to most of us. Since his earliest days in research - and possibly earlier than that, but I have no knowledge of anything earlier - the cerebrations of Art have transcended those of your ordinary mortal.

The story came from Fred Kelly who was a research student with Art at Toronto in the early nineteen fifties. (As this note was being revised the sad news came of Fred's death on 29 July 1986). Fred and Art were part of a team under the direction of M. F. Crawford applying to the study of hyperfine structure and isotope shifts in the spectra of magnesium the latest techniques in high resolution spectroscopy - emission from an atomic beam combined with Fabry-Perot interferometry. And because one of the lines they wished to study was in the u-v, the high reflectivity available by coating the interferometer plates with silver was useless. What should they use to maximize the finesse at 2,796 Å?

This was long before the days of multi-layer dielectric films. Remember, too, that one generally used interferometer plates of 6 or 7 cm diameter in those days, not the tiny little buttons you mount in laser beams nowadays. Silver was by far the best material to use for coating over most of the visible spectrum, though its absorption increases towards the blue. There the alternative was aluminum which, though definitely superior in the u-v is, even at its best, nothing like as good as is silver in the red.

But was that all the choice there was? Copper had been used as a mirror coating, and gold, and there was something else which turned up occasionally in the literature of the late nineteen-thirties: Hochheim alloy. You find it referred to in some of the articles and text-books on high resolution spectroscopy. It certainly deserved looking into - but there was a problem: what was Hochheim alloy, and what quantitative information was there about its alleged superiority to aluminum for use in the u-v? I quote from a well-known text-book: 'Unfortunately no data appear to be available concerning the numerical performance of the Hochheim alloy, in fact the nature of the alloy does not seem to be generally known. These reflecting alloys are prepared personally by Hochheim, apparently by an evaporation method. According to Murakawa (review of hyperfine structure, 1940, in Japanese) this alloy consists of aluminum and silver. The films made by Hochheim for work with Fabry-Perot interferometers are very thick and almost opaque in the visible region but behave excellently in the u-v. In the region between 2,000 Å and 4,000 Å they are much superior to aluminum and are capable of yielding fringes which are quite fine and indeed comparable with those given by silver in the longer wavelength regions.

A challenge indeed, and one which had not been overlooked in other laboratories. There was widespread activity in the nineteen-fifties on the best methods of depositing reflecting films for use in interferometry, and on their treatment <u>after</u> deposition, and - pending the re-discovery of Hochheim alloy (for Professor Hochheim was no longer on the scene after World War II) - one had to do the best one could with aluminum, for the u-v. In the Clarendon Heini Kuhn and his colleagues - Bradley, Wilson, Pery (now Thorne), and Burridge had carrie out systematic studies and had come to the conclusion (for aluminum) that the ke to obtaining the best films was to trap as little gas as possible in the film as is was being laid down. Thus, one aimed to have the best possible vacuum in the evaporation tank, especially while the pellet of aluminum was being evaporated, an to take the shortest possible time in securing a film of the desired density. If the pressure went up a bit, you had to speed up the evaporation to get a film of comparable quality.

At about this time Fred Kelly joined us in the Clarendon as a post-doc. He s about studying hyperfine structure in the resonance lines of gold. They, too, are i the u-v, and Fred brought to our research group in Oxford the wisdom of North America. Hochheim alloy: yes, of course we in Toronto were puzzled. Yes, indee we managed to sort it out; things like that happen when Art's on the scene. The literature - hopeless: Art has private ways and means. He came into the lab one morning and he said, 'What d'you think? - I know about Hochheim alloy.'

'You're crazy.'

'Maybe I'm crazy, but I met Professor Hochheim last night.'

'You're crazy.'

'Maybe I'm crazy, but it was him all right. It must have been. I spoke to him I said, 'Professor Hochheim, I should like to ask you a question.'

And Professor Hochheim replied, 'Go ahead, my boy, what is it?'

'Professor Hochheim, what <u>is</u> Hochheim alloy?'

And Professor Hochheim replied, 'It's aluminum chum, it's aluminum - but pu it on <u>faaast</u> !'

Part III

Solid State Spectroscopy

Anything worth doing
is worth doing twice —
the first time quick and dirty,
and the second time the
best way you can

Optical Spectral Linewidths in Solids

R.M. Macfarlane

IBM Almaden Research Center, 650 Harry Road,
San Jose, CA 95120, USA

1. INTRODUCTION

Physicists and chemists have long realized that the information about material systems which can be obtained by spectroscopic studies is strongly dependent on the resolution which can be achieved. This may be limited by the physical system itself or by the measurement process. A number of early papers studied sharp-line spectra in solids. One of the pioneers was Jean Becquerel [1], who worked extensively with rare earth materials in Paris in the early 1900's. He recognized that at low temperatures, spectral lines could become quite sharp, and in 1908 took his samples and spectrograph to Leiden and collaborated with Kammerlingh Onnes [2] on a measurement of the absorption spectra of rare earth ions in naturally occurring crystals such as tysonite (LaF_3) and xenotime (YPO_4) at the temperatures of liquid and solid hydrogen. The resulting spectra, recorded with a theoretical resolving power of $\sim 10^4$, were extremely rich but certainly contained many lines which were instrumentally broadened. In the 1930's, Otto Deutschbein working in Marburg [3] carried out extensive measurements of the spectra of chromium ions in natural crystals of many materials including ruby, spinel, alexandrite and tourmaline, as well as synthetic materials. He found that at liquid nitrogen temperatures, a number of spectral lines were ~ 3 cm^{-1} wide, but again, since he was working with a prism spectrograph with a linear dispersion of 10Å/mm, many of these lines were broadened by the measurement process. In much of the work on the line spectra of ions in solids, even a half century or more after the work of Becquerel and Deutschbein, the sharpest lines are still broadened by the resolution of the spectrometer, which even in very favorable cases is ~ 3 GHz.

Although sharp spectral lines are more or less universal in gas phase systems, their occurrence in solids is much less widespread. At the second Quantum Electronics Conference in Berkeley in 1961 [4], Schawlow noted that "It seems possible that under some conditions, very sharp lines may be obtained (in solids), perhaps even rivalling atomic spectra in sharpness." This has certainly been borne out in recent years in solid-state rare-earth spectroscopy and to some extent also for transition metal ions. However as we will see below, the measurement of these narrow linewidths had to await the development of single-mode tunable lasers.

The study of the narrow-line spectra of solids containing Cr^{3+}, specifically the R-lines arising from the $^4A_2 \leftrightarrow {}^2E$ transition and the B-lines from the $^4A_2 \leftrightarrow {}^2T_2$

transition, received great stimulus from the invention of the laser by Schawlow and Townes [5] and the operation of the first laser using the R-lines of ruby by Maiman [6]. Schawlow and his collaborators in the 1960's contributed substantially to this field with elegant studies of the effects of external perturbations such as magnetic fields [7] and stress [8] and isotope shifts [4] in materials containing trivalent chromium, using the narrow Cr^{3+} lines as a probe of the local environment. The sensitivity of such probes depends strongly on their spectral linewidths. This focussed attention on ways of measuring these widths and the factors which controlled them. The advent of single frequency tunable cw dye lasers has had an enormous impact on our knowledge and understanding of the sources of optical line broadening in solids. Here, I will discuss some examples which illustrate this progress, without attempting a comprehensive review of the subject which is given elsewhere [9]. These examples use rare earth ion impurities in single crystals, but the techniques used, and the mechanisms for line broadening which have been demonstrated, are much more generally applicable. Art Schawlow was always fascinated by the occurrence of sharp spectral lines and the origin of their width, so it is fitting that the brain-child of Schawlow and Townes should have enabled this field to progress so rapidly.

Before proceeding to individual examples, a few general remarks on inhomogeneous and homogeneous broadening are in order. Inhomogeneous broadening, in solid-state spectroscopy, is the spread of resonance frequencies produced by the range of static micro-environments in which atoms, ions or molecules are found. It is inhomogeneous in the sense that different optical centers have different resonant frequencies. The sensitivity of an optical transition frequency of an ion in a crystal to strain fields tells us something about the likely magnitude of the inhomogeneous broadening [4]. The ability to specify the resonance frequency for a given center depends on its homogeneous width. This is the width exhibited by all ions (ideally independent of where their resonance falls in the inhomogeneous profile), due to dynamical perturbations from phonons or spin fluctuations or lifetime effects. The homogeneous linewidth (Γ_h) can be expressed in terms of the optical dephasing time T_2:

$$\Gamma_h = (1/\pi T_2) = (1/2\pi T_1) + \left(1/\pi T_2'\right)$$

where T_1 is the ("longitudinal") population decay time and T_2' is the ("transverse") pure dephasing term due essentially to frequency modulation of the optical transition. Figure 1 illustrates schematically the case where inhomogeneous broadening dominates homogeneous broadening. This is often the case for transitions to metastable levels at low temperatures. In this context, metastable levels are those with lifetimes of approximately 1 μsec-1 msec. These have relatively large energy gaps below them which reduce spontaneous phonon emission processes and the associated lifetime broadening (Fig. 2). For a group of levels such as illustrated in Fig. 2, optical transitions to the upper levels are usually homogeneously broadened by spontaneous phonon emission, and for transitions to the lowest one, inhomogeneous broadening dominates the small homogeneous part. This will be illustrated with an example below. It has been implied that this

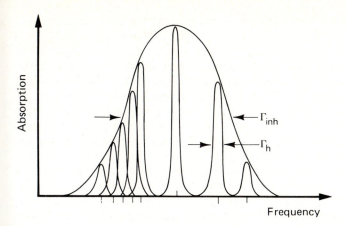

Figure 1. Schematic illustration of a spectral line profile in which inhomogeneous broadening (Γ_{inh}) dominates homogeneous broadening (Γ_h).

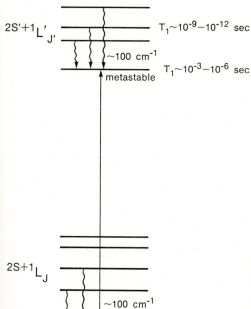

Figure 2. Schematic energy level diagram for trivalent rare earth ions in crystals showing ground and excited state J-manifolds split by the crystal field.

distinction between inhomogeneous and homogeneous broadening is clean and simple. Unfortunately, this is not always the case. For example, the "homogeneous" width may depend on the position in the inhomogeneous line, or what is considered homogeneous and what inhomogeneous may depend on the timescale of the measurement. Generally, however, the separation is a clear and useful one.

2. INHOMOGENEOUS BROADENING

There are several reasons to be interested in inhomogeneous broadening apart from finding ways to eliminate its effect on spectral resolution. In the first place, it can provide a measure of single crystal perfection and strains. It can also be used to monitor the damage produced by ion implantation, the strains at interfaces and in superlattices, or the reduction of strain by thermal annealing for example. There may even be geological applications, using naturally occurring probe ions in crystals of known geological origin. It has been proposed [10] that information can be stored in the frequency domain using a laser addressed optical memory in which permanent but reversible spectral holes are bleached in an inhomogeneous line of a material held at low temperatures. This frequency domain multiplexing has the potential to increase storage densities by the ratio Γ_{inh}/Γ_h which can be a factor of 10^4 or greater [11]. Finally, of course, an understanding of the origins of inhomogeneous broadening may make it possible to control it and significantly influence the optical properties of materials especially at low temperatures.

Very little is known about the microscopic origins of inhomogeneous broadening in solids which, in general, are exceedingly complex. This is in contrast to the case of gases where the distribution of atomic or ionic velocities determines inhomogeneous broadening, and a single parameter − the velocity − describes the position in the line. In solids, it is usually assumed that a Gaussian distribution of resonance frequencies results from the random strain fields due to dislocations and point defects. Many variables control the inhomogeneous width and it is generally assumed that Γ_{inh} is all that we need to specify, indeed is all that we can specify. In some cases, such a simple description fails. Figure 3, for example, shows three traces of the inhomogeneous profile of the $^7F_0 \leftrightarrow {}^5D_0$ transition of EuP_5O_{14} at 2K taken at different parts of the crystal separated by less than 1 mm. Not only are the profiles not Gaussian, but they vary with position in the crystal and also the spot size of the probe laser. These are macroscopic inhomogeneities. Non-Gaussian lineshapes also arise from ions in special environments, for example, distant pairs of ions that are spatially correlated and have an interaction energy that shifts their resonance by a small amount. Vial and Buisson [12] used an elegant method to separate this contribution: they measured an excitation spectrum of up-conversion fluorescence which occurs only for pairs and not for single ions. This showed, in the case of LaF_3:Pr^{3+}, considerable structure in the wing of the line. As the pairs become more distant, their resonance frequency approaches that of the single ion.

Non-Gaussian distributions are also expected when the number of ions being probed is very low, such as in small volumes at low concentrations. In this situation, fluctuation in the number of centers in different homogeneous packets would occur, giving structure in the inhomogeneous profile.

The lack of a simple parametric dependence of the optical transition frequency on strain makes the specification of a unique position in the inhomogeneous line difficult. There is no equivalent of the "zero-velocity packet" which can be determined by an analog of Lamb-dip saturation spectroscopy [13]. There are,

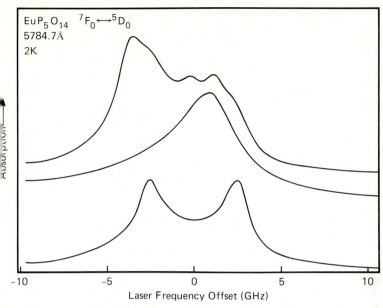

Figure 3. Inhomogeneous line profile for the $^7F_0 \leftrightarrow {}^5D_0$ transition of EuP_5O_4 measured at three different positions in the crystal showing the effect of macroscopic inhomogeneities.

however, some cases where another quantity, such as an rf-optical double resonance frequency, varies smoothly as a function of the optical resonance frequency [14] and could provide an additional piece of information to fix the optical transition frequency.

What are some of the factors which control inhomogeneous broadening? In addition to crystal defects, the introduction of the dopant ions themselves produces lattice strains, so that inhomogeneous linewidths can be quite sensitive to defect concentration. This is illustrated in Fig. 4 for Pr^{3+} in F^- compensated sites of C_{4v} symmetry in CaF_2. Three crystals were used, all with quite low concentrations of Pr^{3+} ($\approx 0.05\%$). The relative concentrations measured from absorption vary by more than a factor of 8. For the lowest doping, the hyperfine structure associated with the ^3H_4E level is clearly resolved and the inhomogeneous linewidth of the transition to the metastable $^1D_2$ level is only 700 MHz [15]. This, in itself, is quite remarkable since the ground state is a non-Kramers' doublet whose degeneracy could be lifted by strain, and this might be expected to produce large inhomogeneous broadening. The observation of this hyperfine structure also provides an example of the information which can be extracted simply by probing an inhomogeneous line profile with a narrow-band laser. This spectrum had been studied many times by conventional spectroscopy and only a single broad line observed. There are numerous cases of this. The high resolution laser can be thought of as a frequency domain "microscope" which enables new interactions and

209

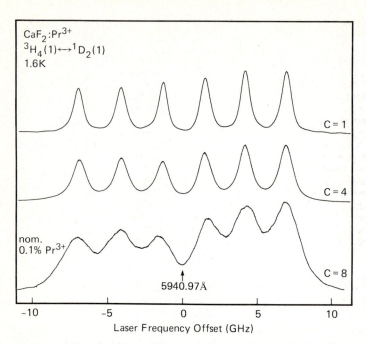

Figure 4. Resolved hyperfine structure on the absorption from the ground state to the lowest component of $^1D_2$ of CaF_2:Pr^{3+} at 5941Å showing the concentration dependence of the inhomogeneous broadening. Relative concentrations are shown on the right and one nominal concentration, i.e., dopant added to the melt, is shown on the left. This is the F^- compensated C_{4v} site.

new dynamics to be discovered when its resolution is increased. Referring to Fig. 4, we see that for a higher doping of Pr^{3+}, significantly increased broadening is observed and at a nominal doping level of 0.1%, the hyperfine structure is barely resolved. This is rather qualitative but that is the nature of the subject at this time. The growth and annealing of the individual crystals (all supplied by Optovac Inc.) is believed to be rather uniform but may contribute somewhat to the variations observed. There are many examples known where there is a strong dependence of Γ_{inh} on the doping level. Very little has been written about the absolute magnitude of Γ_{inh}, however, which is appropriate because little is yet understood. At high doping levels, pair structure will become evident in the inhomogeneous profile leading both to a broadening and the appearance of complex structure [12,16]. Materials in which the optical centers are present in stoichiometric quantities can again exhibit narrow lines (see below) so as a function of concentration the inhomogeneous linewidth at first increases, and then decreases.

Two other factors controlling inhomogeneous broadening should be noted. The first is that different optical transitions may show very different strain sensitivity and, hence, inhomogeneous broadening. A study of the effect of external stress on

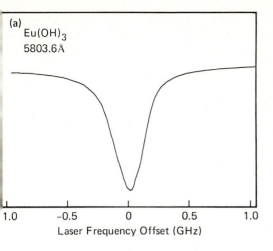

(a) Eu(OH)$_3$
5803.6Å

1.0	−0.5	0	0.5	1.0

Laser Frequency Offset (GHz)

Figure 5. Inhomogeneous line profiles of the $^7F_0 \leftrightarrow {}^5D_0$ transition of Eu^{3+} in three materials.
(a) Eu(OH)$_3$ where the narrowest width exhibited by a single crystal is 170 MHz. For the sample shown here, Γ_{inh} is 260 MHz; (b) SrF$_2$:Eu^{3+} showing a representative linewidth for a good quality doped single crystal of $\Gamma_{inh} = 1.2$ GHz; (c) a Eu^{3+} doped silicate glass showing the large inhomogeneous broadening typical of disordered materials. Here, $\Gamma_{inh} = 100$ cm^{-1} or 3000 GHz.

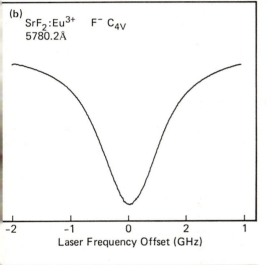

(b) SrF$_2$:Eu^{3+} F$^-$ C$_{4V}$
5780.2Å

−2	−1	0	2	1

Laser Frequency Offset (GHz)

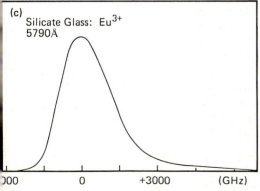

(c) Silicate Glass: Eu^{3+}
5790Å

)00	0	+3000	(GHz)

the optical transition frequencies of rare earth ions, similar to the studies made by Schawlow and co-workers on ruby [4] and $MgO:Cr^{3+}$ [8] would be very worthwhile in this context.

Another factor which strongly influences inhomogeneous broadening of defect absorption lines is the specific host material. For substitutional doping, the ease with which a dopant can be incorporated into the lattice without introducing local deformations of the structure, clearly influences Γ_{inh}. It also determines the distribution coefficient K, which is the ratio of dopant concentration in the crystal to that in the melt. For rare-earth and transition-metal doping, it is sometimes found that a wide range of solid solutions can be formed, e.g., $Pr_xLa_{1-x}F_3$ for $0 < x < 1$. In other cases, such as $YAG:Nd^{3+}$, K has values between 0.1 and 0.3 depending on the method of crystal growth [17]. An inverse correlation between the distribution coefficient and Γ_{inh} might be expected, and some examples support this. For example, for the lowest $^3H_4 \leftrightarrow {}^1D_2$ transition of $LaF_3:Pr^{3+}$ at 5925.2Å, the linewidth is ~5 GHz for $x = 5 \times 10^{-4}$, and in YAG, where K is small (~0.1), it is 50-60 GHz for the same concentration. This host dependence of Γ_{inh} is illustrated in Fig. 5 for the $^7F_0 \leftrightarrow {}^5D_0$ transition of some different Eu^{3+} systems. For this transition, the free-ion levels are nondegenerate so the effect of the crystalline environment is to shift the $^5D_0$-$^7F_0$ separation, which typically varies between $17200\ cm^{-1}$ and $17350\ cm^{-1}$ in different ionic solids. Inhomogeneous broadening results from the shift of this energy from site to site within a given material. Perhaps the most remarkable of the cases illustrated in Fig. 5 is that of $Eu(OH)_3$. For this material, the inhomogeneous linewidth is extremely small, varying between 170 MHz and 280 MHz in different samples [18]. This appears to be the narrowest inhomogeneous linewidth yet reported in a solid. It is not a doped system so questions of lattice matching of the Eu^{3+} ion do not arise. As we have seen in Fig. 4 for the case of EuP_5O_{14}, such narrow lines are not always found in stoichiometric europium compounds so it reflects a high degree of crystal perfection and the low strain of the hydrothermally grown $Eu(OH)_3$ crystals which were produced by Dr. Stanley Mroczkowski of Yale University. In doped single crystal systems, Γ_{inh} is typically 1-10 GHz (e.g., Fig. 5b) and in a silicate glass, for example, where disorder produces large inhomogeneities, Γ_{inh} is $\sim 100\ cm^{-1}$ [19]. Thus, the extreme range of linewidths observed for the Eu^{3+} $^7F_0 \leftrightarrow {}^5D_0$ transition in these different hosts is 2×10^4.

2.1 Inhomogeneous broadening and ground state hyperfine structure in $LaF_3:Ho^{3+}$

Before leaving the subject of inhomogeneous broadening, the example of $LaF_3:Ho^{3+}$ is introduced, and it will be further developed below. The optical spectrum associated with transitions from the $^5I_8$ ground state to the $^5F_5$ manifold very effectively illustrates the situation shown in Fig. 2.

Figure 6a shows fluorescence excitation traces made at 2K with a cw laser having a frequency width of ~1 MHz. The six absorption lines originate from transitions to the six lowest crystal field components of $^5F_5$ separated by Δ_n from the lowest one (Fig. 6b). Because of the C_2 site symmetry for Ho^{3+}, all electronic states are nondegenerate. The lowest level is metastable with a lifetime of

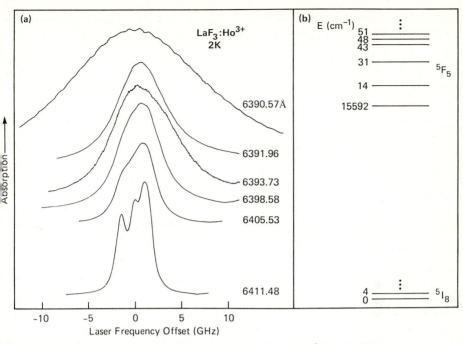

Figure 6. (a) Fluorescence excitation spectra of $LaF_3:Ho^{3+}$ at 1.6K from the ground state ($^5I_8$) to the six lowest crystal field components of $^5F_5$ at the wavelengths given. The lowest transition is inhomogeneously broadened, and for higher transitions, homogeneous broadening due to spontaneous phonon emission becomes increasingly important. Spectral resolution of the dye laser used in those scans is ~1 MHz. (b) The energy level scheme for the transitions shown on the left. At 1.6K only the ground state has appreciable population.

0.55 msec which contributes a negligible 290 Hz to the homogeneous linewidth, and at 2K, thermally induced broadening is absent. The linewidth of ~1 GHz is dominated by inhomogeneous broadening and shows structure which is assigned to the four components of the singlet ground state, coupled to the holmium nuclear spin I = 7/2. These hyperfine splittings are a combination of a pure quadrupole contribution and a pseudoquadrupole or second-order hyperfine contribution [20]. The latter dominates because of the close proximity of the neighboring electronic states (Fig. 6b). The pseudoquadrupole Hamiltonian has the form

$$\mathcal{H}_{pq} = D_{pq}\left[I_z^2 - I(I+1)/3\right] + E_{pq}\left[I_x^2 - I_y^2\right]$$

with

$$D_{pq} = A_J^2[(\Lambda_{xx} + \Lambda_{yy})/2 - \Lambda_{zz}], \quad E_{pq} = A_J^2(\Lambda_{yy} - \Lambda_{xx})/2 \qquad (1)$$

213

where the Λ coefficients express the magnetic coupling between the singlet electronic states $|0>$, $|n>$ of the J-manifold separated by Δ_n, with

$$\Lambda_{\alpha\beta} = \sum_{n=1}^{2J+1} \frac{<0|J_\alpha|n><n|J_\beta|0>}{\Delta_n} . \tag{2}$$

For the ground state, Δ_1 is only 4 cm^{-1}. The matrix elements of J can, in principle, be obtained from the nonlinear Zeeman effect, but this has not yet been done. They will, however, be large because of the high angular momentum of the ground state. The pseudoquadrupole splittings are therefore extremely large (~ 1 GHz, see Table 1) and this appears to be the only case where they can be resolved outside the inhomogeneous width. In Table 1, the three splittings are labelled δ_1, δ_2 and δ_3 in order of increasing energy.

For the transition to the next level of the $^5F_5$ manifold at 6405.5A, the resolution of the hyperfine structure is reduced because of homogeneous broadening due to spontaneous emission of 14.5 cm^{-1} phonons. This emission rate is proportional to the phonon density of states at the frequency differences Δ_n, and also to the coupling strength between pairs of electronic levels and the phonons. This coupling may be significantly different for acoustic and optic modes for example. As Δ_n increases, the lines get broader and there is a dramatic increase for the level at 51 cm^{-1} where the lowest optic phonon mode contributes a peak in the density of phonon states. The linewidth of 15 GHz corresponds to a lifetime due to spontaneous phonon emission of 10 psec. Spectral holeburning, or time resolved fluorescence measurements [21] are required to obtain more precise values of the relaxation rates for each level. For values of Δ_n greater than the highest frequency lattice phonons, two-phonon processes are required and, typically, lines become narrow again. This behavior is seen quite generally in solid-state systems and, in addition to providing important information on relaxation processes, it can also be a useful probe of the effective phonon density of states. Figure 6 illustrates the transition from inhomogeneous broadening to homogeneous broadening with increasing excitation energy above the metastable level.

3. HOMOGENEOUS BROADENING

When homogeneous broadening dominates the total linewidth, its measurement presents little problem for conventional spectroscopy. This is usually the case at high temperatures ($\lesssim 50$K) where phonon absorption and scattering processes [22,23] are responsible for Γ_h. Figure 7 illustrates schematically the temperature dependence of the linewidth observed for metastable levels. On cooling from room temperature, the linewidth narrows by one or two orders of magnitude and becomes inhomogeneous at the lowest temperatures. This is the regime of interest to us here, i.e., where $\Gamma_{inh} > > \Gamma_h$. A number of techniques of laser spectroscopy have been developed to measure Γ_h in the presence of inhomogeneous broadening which is often many orders of magnitude larger. As indicated in Fig. 7, this makes it possible to study new mechanisms for dephasing such as coupling to nuclear-spin fluctuations. A catalog of these techniques includes fluorescence line narrowing

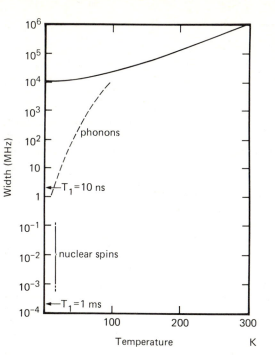

Figure 7. Schematic diagram of the temperature dependence of the homogeneous linewidth of a rare earth impurity ion. At the lowest temperatures, the measured width (solid curve) is dominated by inhomogeneous broadening. By using a variety of techniques of nonlinear laser spectroscopy, the inhomogeneous contribution can be eliminated and this enables the homogeneous width to be followed over many decades to the limit set by the population decay time T_1. For sufficiently long T_1, nuclear-spin fluctuations determine the homogeneous linewidth typically in the range of 1 kHz to 1 MHz.

[24] and spectral holeburning [25,26] in the frequency domain and photon echoes [27], optical free induction decay [28] and optical phase switched transients [29] in the time domain. For a detailed exposition of these techniques, see the original references, or the review by Macfarlane and Shelby [9].

On the subject of eliminating the effects of inhomogeneous broadening, it is interesting to quote another section from Schawlow's 1961 paper at the second Quantum Electronics Conference [4], i.e., "For example, if the lines are inhomogeneously broadened due to strain, then a sufficiently strong source can saturate the absorption of a part of the line. This technique of 'eating a hole in the line' is widely used with microwaves but could not be applied optically before the existence of maser sources." This form of holeburning was observed 14 years later by Szabo in ruby [25]. Since that time, with the availability of tunable lasers, spectral holeburning activity has mushroomed and now includes many mechanisms other than the two-level saturation envisaged by Schawlow, for example, optical pumping of hyperfine, superhyperfine or Zeeman split electronic levels and selective photochemistry of many kinds which can lead to essentially permanent holes. Recent reviews have been given by Macfarlane and Shelby [9,26]

Before the application of these techniques of laser spectroscopy, very little was known about the magnitude of Γ_h in the low temperature regime (<4K) or of the mechanisms responsible for it.

3.1 Homogeneous broadening in $LaF_3:Pr^{3+}$

A large amount of work was devoted to the study of the $^3H_4(Z_1) \rightarrow {}^1D_2(D_1)$ transition of $LaF_3:Pr^{3+}$. Out of this, a much greater understanding of the low temperature dephasing mechanisms evolved. This, and related work, has been reviewed by Macfarlane and Shelby [9], but some important results will be mentioned here. It was shown that fast, time domain, photon echo [30] and optical free induction decay [31] techniques could reliably measure Γ_h at low temperatures and the values were lower than at first expected, i.e., 56 kHz at zero magnetic field and 12 kHz in a field of 80G. Spectral holeburning by optical pumping of hyperfine levels is a much longer timescale experiment (~sec) which puts more serious demands on laser frequency stability for measurements requiring resolution ~10's of kHz. In addition, it was shown [32] that there are slowly varying local fields due to spin flips on ^{19}F nuclei which are close to the Pr^{3+} ion and hence strongly perturbed. These can broaden the hole. The ^{19}F spin flips which contribute on a faster timescale to the photon echo and FID decays are further from the praseodymium ion and outside the "frozen core" of perturbed fluorine nuclei. The role of nuclear-spin flips as the dominant source of low temperature homogeneous broadening was clearly demonstrated by nuclear-spin decoupling experiments [31,33]. Here, the ^{19}F nuclear spins were coherently driven by strong rf fields to average out their local field fluctuations, or their flip rates were slowed down by "magic-angle" decoupling. It is now believed that nuclear-spin coupling is a universally important source of optical coherence loss in solids at temperatures where the phonon contributions have been frozen out. Figure 8 summarizes the progressive steps used in determining the homogeneous linewidth of the 5925Å transition of $LaF_3:Pr^{3+}$. It can be thought of as an example of the frequency domain "microscope" concept. By cooling the sample to 1.6K, homogeneous broadening due to phonons was eliminated and the line profile became dominated by inhomogeneous broadening. Spectral holeburning yielded hyperfine structure [34] but the hole width was limited by laser frequency jitter. Time domain techniques were therefore used, and photon echo measurements gave the homogeneous linewidth of 54 kHz which was narrowed to 4 kHz by decoupling the ^{19}F nuclear spins and essentially removing their contribution from the linewidth. This example demonstrates many of the factors which control coherence loss in solids at temperatures where the phonon contributions have been frozen out.

3.2 Homogeneous broadening and excited state hyperfine structure in $LaF_3:Ho^{3+}$

We return now to the illustrative example of $LaF_3:Ho^{3+}$. A spectral holeburning experiment was carried out on the 6411.5Å line using two single frequency cw dye lasers — one to "saturate" the absorption and the other to probe the resulting hole. The result is shown in Fig. 9. In addition to a hole at the pump laser frequency, a pattern of side holes was observed which measures the excited state hyperfine splittings. Again, these are due to second-order interactions, but in this case, the splittings are smaller than in the ground state. This is mainly because the separation between the two closest lying excited states is now 14.5 cm^{-1}, almost four times as large as in the ground state, but matrix elements of J are also smaller.

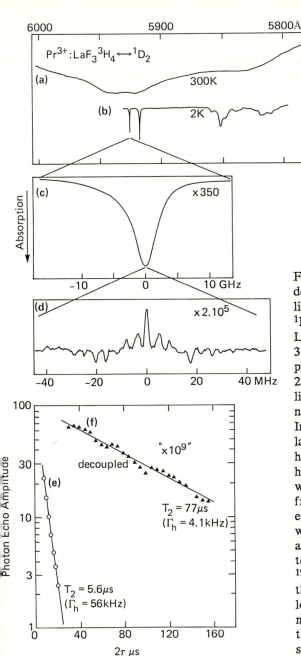

Figure 8. Progressive steps in determining the homogeneous linewidth of the $^3H_4(Z_1)$ to $^1D_2(D_1)$ transition of LaF$_3$:Pr^{3+} at 5925.0Å. (a) The 300K spectrum broadened by phonon interactions. (b) At 2K, the linewidth of 6 GHz is limited by static inhomogeneous strain as shown in (c). In (d), this broadening is largely eliminated by holeburning, showing hyperfine structure. The hole width is limited by laser frequency jitter. Using photon echoes, the true homogeneous width $\Gamma_h = 56$ kHz is revealed, and this can be further reduced to 4.1 kHz by decoupling the ^{19}F nuclear spins, thus showing that they dominate Γ_h at very low temperatures. The magnification factors refer to the effective increase in spectral resolution.

217

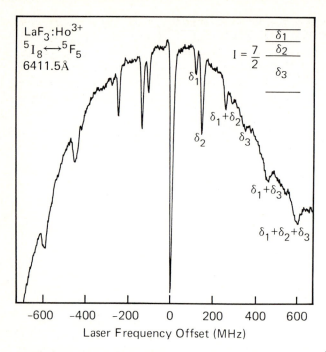

Figure 9. Holeburning spectrum measured in the 6411.5Å transition of LaF$_3$:Ho^{3+} using two cw dye lasers of resolution ~1 MHz. The side holes are due to excited state hyperfine structure. It is thought that holeburning occurs due to population storage in the $^5I_7$ level.

The mechanism for holeburning in this case appears to be population storage in the metastable $^5I_7$ level which has a lifetime of 10's of msec. This allows depletion of the ground state population for those ions whose environment puts them into resonance with the laser. From the fluorescence excitation spectrum and the holeburning, we have obtained approximate but reasonably good values for ground and excited state hyperfine splittings (see Table 1). The pseudoquadrupole splittings are very approximately in the axial ratio of 1:2:3, which is consistent with a dominant contribution coming from hyperfine interactions with a single low-lying level.

Table 1. Pseudoquadrupole Splittings in LaF$_3$:Ho^{3+}

	δ_1(MHz)	δ_2(MHz)	δ_3(MHz)
$^5I_8(Z_1)$	400	800	1500
$^5F_5(D_1)$	110	150	350

If we expand the frequency scale and look at the width of the central hole, we find that it is W = 18 MHz (Fig. 10). This is broader than the 2-3 MHz contribution from laser frequency jitter, but does it give the homogeneous linewidth (i.e., Γ_h = 9 MHz)? To further investigate this question, we measured the optical dephasing time using optical-free induction decay (Fig. 11) and found that T_2 = 380 nsec [9], i.e., Γ_h from this measurement is 0.83 MHz. The resolution

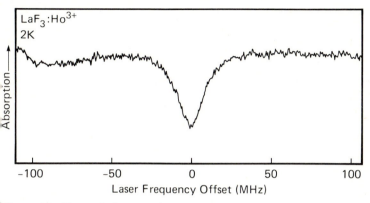

Figure 10. Expanded trace of the central hole of Fig. 7 showing a hole width, W = 18 MHz.

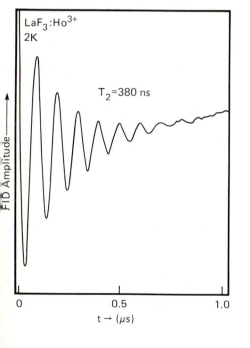

Figure 11. Optical free induction decay measurement of the homogeneous linewidth of the $^5I_8 \longleftrightarrow {}^5F_5$ transition at 6411.5Å. From the dephasing time of 380 nsec, Γ_h = 0.83 MHz. The measurement shown here was made in collaboration with Dr. Richard Meltzer.

219

of this dilemma, as discussed already, for LaF_3:Pr^{3+}, almost certainly arises from the timescale on which the two measurements were carried out. For the FID, it is ~1 μsec, and for holeburning, it is at least as long as the storage time of $^5I_7$ which is a factor of 10^4 longer. The homogeneous width and the hole width are broader here than was the case for LaF_3:Pr^{3+} because of the much larger nuclear magnetic moment of the Ho^{3+} hyperfine levels. The second-order hyperfine interaction which gave rise to such large pseudoquadrupole splittings also induces an enhanced nuclear moment [35], which for the ground state is ~1 MHz/G or almost as large as a typical electronic moment.

3.3 Y_2O_3:Eu^{3+} – The narrowest optical linewidth in a solid

We have already seen that the $^7F_0 \leftrightarrow {}^5D_0$ transition of Eu^{3+} provides the narrowest inhomogeneous linewidth in a solid (170 MHz) due in part to the relative insensitivity of the optical transition frequency to crystal strains. This transition is also noteworthy in that the ground and excited states have no electronic magnetic moment, and the ground state nuclear moment is typically quenched by hyperfine coupling to $^7F_1$ [36]. For this reason, homogeneous broadening due to nuclear-spin fluctuations is generally very small. This depends, of course, on the nature of the host material. In Y_2O_3, only the yttrium ions have nuclear spin and the magnetic moment of yttrium is small (−0.137 nuclear magnetons), contributing approximately 200 Hz to the optical linewidth. The homogeneous linewidth of the

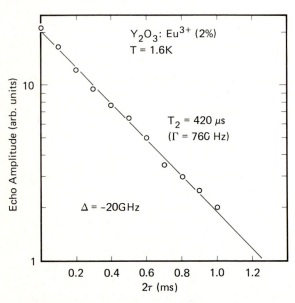

Figure 12. Decay of the photon echo amplitude measured for the $^7F_0 \leftrightarrow {}^5D_0$ transition of Y_2O_3:Eu^{3+} at 1.6K. The homogeneous linewidth measured in this experiment has the very small value of Γ_h = 760 Hz.

$^7F_0 \leftrightarrow {}^5D_0$ transition of Y_2O_3:Eu^{3+} was measured by Macfarlane and Shelby [37] from the photon echo decay. They found a concentration dependent linewidth, and also a dependence of Γ_h on position in the inhomogeneous line. Because the inhomogeneous line profile is essentially a plot of concentration versus frequency, a lower effective concentration of Eu^{3+} ions can be found in the wing of the line. A width of 760 Hz was found at 2K, in a 2% Y_2O_3:Eu^{2+} sample at a frequency 20 GHz off the peak of the inhomogeneous line (Fig. 12). This shows that very narrow homogeneously broadened resonances can be observed in solids at low temperatures, and many more examples await study.

4. CONCLUSION

As techniques for measuring very narrow linewidths become refined and more understanding of the sources of optical homogeneous broadening is obtained, it is natural to ask how narrow these linewidths can become and how to measure them. The simple answer to the ultimate limit on the linewidth is just that it is $(1/2\pi T_1)$, where T_1 is the population decay time. For metastable levels, this may only contribute several Hz to hundreds of Hz. Eliminating the effect of nuclear-spin coupling is clearly necessary, and as linewidths become narrower, interaction between the optical centers themselves may lead to spectral diffusion and line broadening. These ideas have been illustrated principally by the examples of LaF_3:Pr^{3+}, LaF_3:Ho^{3+} and Y_2O_3:Eu^{3+}. The case of the trivalent europium ion has the distinction of providing both the narrowest inhomogeneous linewidth (170 MHz in $Eu(OH)_3$) and the narrowest homogeneous linewidth (760 Hz in Y_2O_3:Eu^{3+}) in solid-state spectroscopy. There is every reason to believe that narrower lines will be observed, providing more and more detailed probes of the nature of crystalline materials and of the complex magnetic interactions between nuclear and electron spins.

When Art Schawlow presented his paper at the Quantum Electronics Conference 26 years ago, and projected that the "optical maser" would enable us to measure narrow spectral lines in solids, did he envisage just how far this field would develop? Probably he did.

ACKNOWLEDGEMENTS

I wish to acknowledge the invaluable contributions of my collaborators in previously published work, particularly Robert Shelby, and also Art Schawlow for his encouragement and guidance during the "impressionable years."

REFERENCES

1. J. Becquerel: Compt. Rend. Acad. Sci. (Paris) 142, 775 (1906); ibid. 145, 413 (1907); ibid. 145, 1150 (1907); Le Radium 4, 49 (1907); ibid. 4, 328 (1907); ibid. 5, 5 (1908).
2. J. Becquerel and H. Kammerlingh Onnes: Le Radium 5, 227 (1908).
3. O. Deutschbein: Ann. Phys. 14, 712 (1932).

4. A.L. Schawlow: In Advances in Quantum Electronics, ed. by J.R. Singer (Columbia University Press 1961) p.50.
5. A.L. Schawlow and C.H. Townes: Phys. Rev. 112, 1940 (1958).
6. T.H. Maiman: Nature 187, 493 (1960).
7. S. Sugano, A.L. Schawlow and F. Varsanyi: Phys. Rev. 120, 2045 (1960).
8. A.L. Schawlow, A.H. Piksis and S. Sugano: Phys. Rev. 122, 1469 (1961).
9. R.M. Macfarlane and R.M. Shelby: In "Spectroscopy of Crystals Containing Rare Earth Ions," Modern Problems in Condensed Matter Sciences, ed. by A.A. Kaplyanskii and R.M. Macfarlane (North-Holland, Amsterdam 1987) in press.
10. G. Castro, D. Haarer, R.M. Macfarlane and H.P. Trommsdorff, "Frequency Selective Optical Data Storage System," U.S. Patent No. 4,101,976 (1978).
11. W.E. Moerner, W. Lenth and G.C. Bjorklund: In "Persistent Spectral Holeburning: Science and Applications," Topics in Current Physics, ed. by W.E. Moerner (Springer-Verlag, Heidelberg 1987) in press.
12. J.C. Vial and R. Buisson: J. Physique Lett. 43, L339 (1982).
13. L.S. Vasilenko, V.P. Chebotayev, A.V. Shishaev: JETP Lett. 12, 113 (1970); B. Cagnac, G. Grynberg and F. Biraben: Phys. Rev. Lett. 32, 643 (1974); M.D. Levenson and N. Bloembergen: Phys. Rev. Lett. 32, 645 (1974).
14. L.E. Erickson and K.K. Sharma: Phys. Rev. B24, 3697 (1981).
15. R.M. Macfarlane, R.M. Shelby and D.P. Burum: Opt. Lett. 6, 593 (1981).
16. P. Kisliuk, N.C. Chang, P.L. Scott and M.H.L. Pryce: Phys. Rev. 184, 367 (1969).
17. R.F. Belt, R.C. Puttbach and D.A. Lepore: L. Pryce: J. Cryst. Growth 13/14, 268 (1972); A.G. Petrosyan, Kh.S. Bagdasarov, T.I. Butaeva, A.M. Kevorkov and A.A. Shakhnazaryan: Kristallogr. 20, 1089 (1975) [Sov. Phys. Crystallog. 20, 665 (1975).
18. M.S. Otteson, R.L. Cone, R.M. Macfarlane and R.M. Shelby: J. Opt. Soc. Am. 73, P1391 (1983).
19. C. Brecher and L.A. Riseberg: Phys. Rev. B 13, 81 (1976).
20. J.M. Baker and B. Bleaney: Proc. Roy. Soc. (Lond.) A 245, 156 (1958).
21. R. Bayerer, W. Schneider, J. Heber and D. Mateika: Z. Phys. B 64, 195 (1986).
22. D.E. McCumber and M.D. Sturge: J. Appl. Phys. 34, 1682 (1963).
23. W.M. Yen, W.C. Scott and A.L. Schawlow: Phys. Rev. 136, A271 (1964).
24. Yu. V. Denisov and V.A. Kizel: Opt. i Spektr. 23, 472 (1967) [Opt. Spectr. 23, 251 (1967)]; A. Szabo: Phys. Rev. Lett. 25, 924 (1970).
25. A. Szabo: Phys. Rev. B11, 4512 (1975).
26. R.M. Macfarlane and R.M. Shelby: In "Persistent Spectral Holeburning: Science and Applications" Topics in Current Physics, W.E. Moerner (Springer-Verlag, Heidelberg), in press.
27. N.A. Kurnit, I.D. Abella and S.R. Hartmann: Phys. Rev. Lett. 13, 567 (1964).
28. R.G. Brewer and R.L. Shoemaker: Phys. Rev. A 6, 2001 (1972).
29. A.Z. Genack, D.A. Weitz, R.M. Macfarlane, R.M. Shelby and A. Schenzle: Phys. Rev. Lett. 45, 438 (1980).
30. R.M. Macfarlane, R.M. Shelby and R.L. Shoemaker: Phys. Rev. Lett. 43, 1726 (1979).

31. S.C. Rand, A. Wokaun, R.G. DeVoe and R.G. Brewer: Phys. Rev. Lett. $\underline{43}$, 1868 (1979).
32. R.M. Shelby, C.S. Yannoni and R.M. Macfarlane: Phys. Rev. Lett. $\underline{41}$, 1739 (1978).
33. R.M. Macfarlane, C.S. Yannoni and R.M. Shelby: Opt. Commun. $\underline{32}$, 101 (1980).
34. L.E. Erickson: Phys. Rev. $\underline{B16}$, 4731 (1977).
35. B. Bleaney: Physica $\underline{69}$, 317 (1973).
36. R.J. Elliott: Proc. Phys. Soc. (Lond.) Sec. B $\underline{70}$, 119 (1957); R.M. Shelby and R.M. Macfarlane: Phys. Rev. Lett. $\underline{47}$, 1172 (1981); K.K. Sharma and L.E. Erickson: Phys. Rev. Lett. $\underline{B23}$, 69 (1981); R.M. Macfarlane and R.M. Shelby: Opt. Commun. $\underline{39}$, 169 (1981).
37. R.M. Macfarlane and R.M. Shelby: Opt. Commun. $\underline{39}$, 169 (1981).

Spectroscopy of Solid-State Laser Materials

S. Sugano

The Institute for Solid State Physics, The University of Tokyo, Roppongi, Tokyo 106, Japan

1. Before the Dawn of Laser History

After finishing the work "On the Absorption Spectra of Complex Ions"[1], Y. Tanabe and myself were looking for clear-cut experimental evidence to justify our energy level diagrams for d^N (N=2, 3, $\cdots$, 8) electron configurations in a cubic field. As is well known, the diagram predicts the co-existence of gaslike narrow lines and broad bands in the absorption spectra of transition-metal ions in cubic environments. We thought that detailed studies of the gaslike lines could justify the diagram as they could provide us with much information such as fine structure, Zeeman effects, and so on. We noticed that the gaslike lines were first reported in 1893 by Lapraik[2] and that many detailed spectroscopic studies had been done on the narrow lines arising from the $t_{2g}^3\ {}^4A_{2g} \to t_{2g}^3\ {}^2E_g$, ${}^2T_{1g}$, and ${}^2T_{2g}$ transitions of Cr^{3+} ions in complex salts[3] and oxide crystals such as spinel, alexandrite, ruby etc.[4]

Shortly after examining these experimental data, we arrived at the conclusion that detailed studies of the ruby spectrum would be the most appropriate for our purpose. The reasons are as follows; (1) The crystal structure is simple with uniaxial trigonal symmetry, and the crystal is stable undergoing no phase transition at low temperatures as in the case of complex salts. (2) The spectral intensity is relatively high because of the absence of inversion symmetry at the Cr^{3+} site, and the spectral widths of the gaslike lines are so narrow at low temperatures that Zeeman experiments may be performed with a readily available magnetic field.

Soon we finished the calculation of the fine structure of the R, R', and B absorption lines arising, respectively, from the $t_{2g}^3\ {}^4A_{2g} \to t_{2g}^3\ {}^2E_g$, ${}^2T_{1g}$, and ${}^2T_{2g}$ transitions by using the even-parity trigonal field and the spin orbit interaction. We also calculated the optical anisotropy, or the optical polarization, of the absorption lines and the broad U and Y absorption bands arising, respectively, from the $t_{2g}^3\ {}^4A_{2g} \to t_{2g}^2 e_g\ {}^4T_{2g}$ and ${}^4T_{1g}$ by using the odd-parity trigonal field and the spin-orbit interaction. We created symbols U and Y for the broad bands to form RUBY in the order of increasing wavenumbers, where symbols R and B had already been used for the narrow lines. The results seemed to explain all the observed features of the spectrum in the absence of a magnetic field.[5]

When we proceeded to the study of Zeeman patterns of the absorption lines, however, we encountered a serious difficulty. We could not find any Zeeman splitting of g $\sim$2 of the ground state, which had been observed by the paramagnetic resonance[6], in the Zeeman patterns of the R_1 and R_2 absorption lines observed by H. Lehmann[7] with a magnetic field (27,000 Gauss) parallel to the trigonal axis and the polarization perpendicular to it: the Zeeman patterns consisted of three split components and the apparent g-values were reported to be ~1.47. As far as I remember, we had no difficulty in explaining the other features of the Zeeman patterns. The difficulty just described, however, was so serious that we were eager to find some experimentalist who could repeat the Zeeman experiment to examine the extraordinarily small g-value reported by Lehmann.

Meanwhile we met I. Tsujikawa, who had just finished Zeeman experiments on chrome alums in collaboration with Madame L. Couture at Bellevue and was constructing a new grating spectrograph in Tohoku University. Examining, with Tsujikawa, very carefully the propriety of performing the Zeeman experiment of ruby by using his new experimental facility, we finally decided that Tsujikawa and myself should start the Zeeman experiment of ruby as soon as his spectrograph was completed. At that time I was a research associate in an experimental laboratory of optics.

Tsujikawa's spectrograph in Eagle mounting with dispersion 2.5 A/mm in the first order was located on the premier étage and a large magnet on rez-de-chaussée. In spite of this separation, adjustment of the optical path was not so difficult. The magnet was used by several groups, so that our experiment was mostly done at night. Liquid Helium and liquid hydrogen were supplied once a week on different days, and I had enough time to enjoy exploring an attractive old city, Sendai, in the north-east (Tohoku) part of Japan. Furthermore, since it took several hours to take a photograph of the emission spectrum, my ability in playing flute showed good progress. In some preliminary experiment, we placed an unpolished pink ruby in a Dewar vessel. We were very excited to see the R_1 emission and the R_2 absorption lines on a photographic plate, as shown in Fig.1.

The experiment with Tsujikawa was successful. We obtained the Zeeman patterns of the R lines at 20 K with a magnetic field parallel to the trigonal axis and the polarization perpendicular to it as shown at the top of Fig.2. The qualitative features of the patterns are quite similar to those observed at T=-190°C by Lehmann: they consist of three components, being asymmetric for R_1 and symmetric for R_2 in shape. Only a difference may be found in the magnitude of the apparent g-value. In our patterns, the averaged apparent g-values for R_1 and R_2 are 1.78 (±0.04) and 1.74 (±0.04), respectively, while they are 1.47 in Lehmann's experiment. Comparison showed that the effective magnetic field was 20% smaller than the value reported in his experiment. The existence of two unresolved lines at the central component was confirmed by observing the relative shift of the cen-

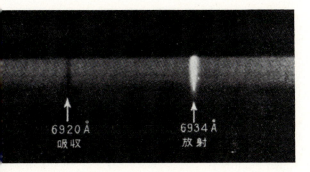

Fig.1. Photograph showing simultaneous observation of the R1 emission line of 6934 Å and the R2 absorption line of 6920 Å of an unpolished ruby at 20 K. The observation is made in the same direction as the incident light.

6920 Å
吸収

6934 Å
放射

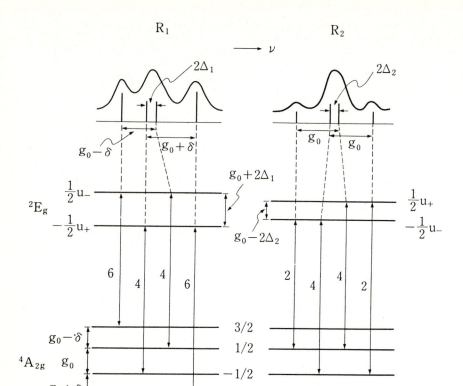

Fig.2. The observed Zeeman patterns of the R_1 and R_2 absorption lines at 20K and the corresponding transition diagrams for a magnetic field parallel to the trigonal axis and the polarization perpendicular to it. (g_0= 1.98, δ=0.34, Δ_1=0.22, Δ_2=0.26)./5/

tral component at 4K. As shown in Fig.2, we can find in our patterns the Zeeman splitting of the ground state observed by paramagnetic resonance. Analyzing our Zeeman patterns, we obtained the g-shifts in the R_1 and R_2 excited Kramers doublets as Δ_1=0.22 and Δ_2=0.26.

Shortly after publishing the papers on ruby in August, 1958/5/, we received a preprint of short communication on Maser Action in Ruby by G. Makhov, C. Kikuchi, J. Lambe and R. W. Terhune from C. Kikuchi of the University of Michigan in October, 1958./8/ This preprint brought the following idea to my notice. As easily seen in Fig.2, the transitions from the R_2 excited state populate the excited Zeeman levels of the ground state. Therefore, if one could find a trigonal Cr^{3+} system in which the R_2 excited state is lower than the R_1 excited state contrary to the case of ruby, one may expect to induce maser action by optical pumping as long as non-radiative transitions could be ignored. I wrote about this idea to Andy Liehr of Bell Telephone Laboratories who was trying to invite me to Bell Labs. At the end of that year, I read the epoch-making paper of Schawlow and Townes on "Infrared and Optical Masers"/9/ with strong excitement.

On February 19 of 1959, I received a letter from Andy, in which the following statement was made: [I have taken the liberty of showing your letter to Al Clogston, Art Schawlow, Martin Peter, Stan Geschwind, and Derek Scovil, all of whom have had ideas concerning optical pumping in Ruby. They have asked me to include the following paragraph concerning the connection between your ideas and theirs: "Since reading your very interesting paper on the ruby spectrum, we have been aware that there are several different kinds of optical pumping experiments which might be done on ruby. Several of our people are thinking about this and some of these experiments will probably be tried fairly soon. However, it would be interesting to know more about the one you propose."] Later Art Schawlow told me that a trigonal crystal is available in which the sign of the trigonal field is opposite to that in ruby. However, I am unaware if the optical pumping experiment has ever been performed on that crystal.

When I arrived at Bell Telephone Laboratories at the end of summer in 1959, Art Schawlow showed me many big ruby crystals of different shapes. It took some time for me to understand the purpose these crystals were grown for. On some day soon after my arrival, Art pushed me into his car. When we finally got off the car, I realized that we were attending the First Quantum Electronics Conference held at Shawanga Lodge. Listening to several talks, I could easily understand the purpose of this conference. Speakers seemed to be insisting that all the variety of the optical transitions they were studying could be used for possible optical masers to be realized in the near future.

Many interesting experiments on ruby were conducted at Bell Labs. in 1959. Among them, the experiment on "Self-Absorption and Trapping of Sharp-Line Resonance Radiation in Ruby" by Frank Varsanyi, Dar Wood and Art Schawlow /10/ reminded me of the spectrum in Fig.1, where the R_1 emission and the R_2 absorption lines are seen on a photographic plate. I suspected that this might be due to the relaxation, within the ^2E state, of the trapped photons at the rough surface. However, I did not pursue this problem further.

I would like to comment upon the experiment performed in this period by the group of Stan Geschwind on "Optical Detection of Paramagnetic Resonance in an Excited State of Cr^{3+} in Al_2O_3" /11/. They detected paramagnetic resonance in the R_1 excited levels ($\bar{E}:\pm1/2u_\mp$) by observing change of the emission intensities from these levels. The presence of the resonance signal with a static magnetic field parallel to the trigonal field indicated the existence of non-zero linear Zeeman splitting with a magnetic field perpendicular to the trigonal axis. Later, we studied the Zeeman

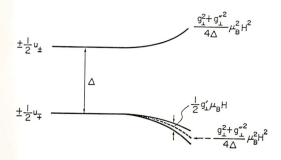

Fig.3. Zeeman splitting of the t_{2g}^3 ^2E excited states of ruby in a magnetic field perpendicular to the trigonal axis /12/.

227

splittings of the R_1 and R_2 excited states by using the method of effective Hamiltonian giving the exact results /12/. The result for a magnetic field perpendicular to the trigonal field is shown in Fig.3, which indicates that the linear Zeeman splitting exists in the R_1 excited state but not in the R_2. This means that, if the sign of the trigonal field was reversed making the R_1 excited state higher than the R_2 in energy, we could not observe the paramagnetic resonance at low temperatures where only the lower excited state is populated. This conclusion is very interesting when compared with that given before for the optical pumping of the excited Zeeman levels of the ground state.

2. Spectroscopic Work at Bell Labs

In 1960, I collaborated in a series of the spectroscopic work on the cubic-field line of Cr^{3+} in a MgO crystal developed in Art's group. The line has to arise from either an electric dipole transition slightly released by the asymmetric distortion of the system due to lattice vibration, or by a magnetic dipole one. Zeeman studies can decide which transition is responsible for the line. As an example, the experimental and the calculated Zeeman patterns for the electric dipole and magnetic dipole transitions with a magnetic field parallel to the [001] axis are shown in Fig.4/13/, which makes one conclude that the transition is of a magnetic dipole nature. As far as I know, no Zeeman effect has been reported on crystalline-field lines allowed by the coupling with odd-parity vibrations. This would be due to the broadening of these lines by the coupling of the local modes of vibration.

At the beginning of this work, we used a Bausch and Lomb Dual Grating Spectrograph. The photographic plates used were calibrated and photo-metered for intensities. During the course of the work, however, we got a Jarrell-Ash high-resolution photoelectric spectrometer. It was very nice to find that the results obtained with this new instrument agreed well with those obtained photographically in the cases where comparisons were made. In this period Frank Varsanyi carried me between Bell Labs and my apartment in his car every day, so that we had enough time to exchange new ideas during the ride.

The study of the longitudinal Zeeman effect of the same emission line was more interesting. As an example, we show the experimental and theoretical longitudinal Zeeman patterns with a magnetic field parallel to the [111] axis for circular polarization. In Fig. 5 the central component is linearly polarized as it is given by superposition of the opposite circular polarizations with a fixed relative phase. However, this phase difference would vary from train to train of the emitted light, so that the linear polarization would take any direction in the plane perpendicular to the [111] axis. Unfortunately we did not confirm this experimentally.

The degenerate cubic-field line may be split by applying uniaxial pressure. The experiment on such a splitting was performed by Art Schawlow and A. H. Piksis./14/ I joined the analysis of the experiment. Piksis was also enthusiastic in making some calculations by himself. The stress, simple pressure P, was applied normal to the (001), (110), and (111) planes. The strain-induced change of the crystal field V linear to the strain can be written in the form,

$$V = \sum_{\Gamma\gamma} C_{\Gamma\gamma} V_{\Gamma\gamma}(\vec{r}) e_{\Gamma\gamma},$$

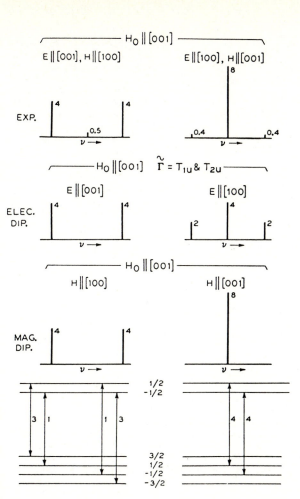

Fig.4. The experimental and the calculated Zeeman patterns of the electric dipole and magnetic dipole transitions for a cubic-field emission line $(t_{2g}^3 \, ^2E_g \rightarrow t_{2g}^3 \, ^4A_{2g})$ of Cr^{3+} in MgO with a magnetic field H_0 parallel to the [001] axis. E and H stand for the oscillating electric and magnetic fields of the emitted light, and $\tilde{\Gamma}$ for the vibrational mode coupled with the electric dipole transitions. The intensities are relative and not normalized /13/.

where $\Gamma\gamma$ runs over the irreducible representation Γ ($\Gamma=A_1$, E, T_2) and their components γ of the cubic symmetry group. $V_{\Gamma\gamma}(\vec{r})$ is a function of electron coordinates $\vec{r}$ transforming as the γ basis of the Γ representation. $e_{\Gamma\gamma}$ is the strain component transforming in the same way as $V_{\Gamma\gamma}$;

$$e_{A_1} = e_{xx} + e_{yy} + e_{zz}, \qquad e_{T_2\xi} = e_{yz},$$

$$e_{Eu} = 2e_{zz} - e_{xx} - e_{yy}, \qquad e_{T_2\eta} = e_{zx},$$

$$e_{Ev} = e_{xx} - e_{yy}, \qquad e_{T_2\zeta} = e_{xy},$$

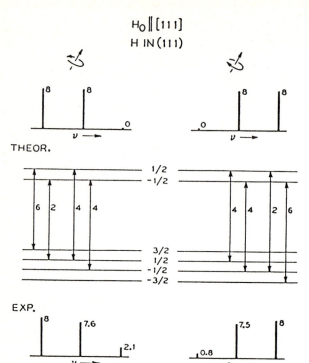

Fig.5. The experimental and the calculated longitudinal Zeeman patterns of the R cubic-field emission line of Cr^{3+} in MgO with a magnetic field H_0 parallel to the [111] axis for circular polarization. The straight arrow shows the direction of the applied magnetic field, and the circulating one around the straight arrow indicates the direction of the rotating magnetic vector of the radiation /13/.

where e_{ij}'s are the strain components ordinarily used. Further, $V_{\Gamma\gamma}(\vec{r})$ is expanded in a power series of the electron coordinates, and the expansion coefficients are calculated by using the point-charge model and summing up appropriately the electrostatic field coming from the positive and negative charges at all the lattice points. The results of the calculation are given in Table 1 together with those of the experiment. We used Watson's analytical Hartree-Fock wavefunction of a free Cr^{3+} ion/25/ to calculate the average values;

$$\langle r^2/R^2 \rangle = 0.0918, \qquad \langle r^4/R^4 \rangle = 0.0179,$$

$$R = 3.97 \ (a.u.),$$

where R is the lattice constant.

In Table 1, the sign of the observed splitting was determined to give agreement between the calculated and observed polarization. The agreement between the calculated and observed splitting is surprisingly good. One might suspect that the agreement is accidental, but the fact that it is uniformly good in all the cases seems to suggest something significant. It should be noticed that, if only the point charges at the nearest neighbor are taken into account in the calculation of the strain-induced change of the crystal field, the sign of the calculated splitting turns out to be wrong in some case. It is well known that the calculation of the cubic-field strength by the use of the point-charge model and the Hartree-Fock wavefunction gives a too small value; here we have to take into account the

Table 1. The strain-induced splitting, shifts, and the polarization of the split components of the R cubic-field emission line of Cr^{3+} in MgO. The polarization is defined as the ratio of the magnetic dipole strength of the shorter-wavelength split component to that of the longer-wavelength one when the splitting is positive. Δ and λ are given in cm^{-1} and P in dynes/cm^2 (P < 0) /14/.

Case	Calc. split. $\Delta/(-P)$	Obs. split. $\Delta/(-P)$	Calc. pol.	Obs. pol.	Obs. shift $-\lambda/(-P)$
	$\times 10^{-10}$	$\times 10^{-10}$			$\times 10^{-10}$
$P_\perp(001)$	6.8	6.3±0.2	H$\perp$P 3	2.4	2.9
			H$\parallel$P 0	0.4	
$P_\perp(110)$	-3.8	-(3.6±0.1)	H$\parallel$[1$\bar{1}$0] 1.7	1.2	2.9
			H$\parallel$[00$\bar{1}$] 0.05	0.01	
			H$\parallel$[110] 4.6	1.5	
$P_\perp(111)$	-2.0	-(2.8±0.1)	H$\perp$P 0.6	0.61	2.6
			H$\parallel$P 3	1.6	

non-orthogonality and the covalency effects in addition to the point-charge crystal field./12/ However, the contribution of these effects to the low-symmetry crystal field may be expected to be small, as the contribution of the crystal field from distant point-charges is appreciable. This would be a reason why the agreement is so good. In Table 1, we also see that the observed shifts λ of the center of the split components are almost the same in the three cases. This is reasonable, as the shifts are caused by the strain-induced change of the spherically symmetric and cubic crystal fields which should be the same in all the three cases. Later, such a shift will be discussed again more quantitatively.

In 1960, T. H. Maiman reported the first observation of laser action of the R lines of ruby /16/, and Art was busy in observing the laser action of the satellite lines of concentrated ruby./17/ On January 31, 1961, a continuously-operating He-Ne gas laser built by A. Javan, W. R. Bennett Jr., and D. R. Herriott/18/ was demonstrated at a press conference of Bell Labs. In the 1961 spring meeting of the Optical Society of America, the red spot of a ruby laser was projected on a screen in Ballroom of Penn-Sheraton Hotel when Charles Townes counted down to "zero". In this meeting Art gave an invited talk on Solid-State Optical Masers, and I gave an invited talk on Spectroscopy for Solid Optical Masers./19/ The title of the present article comes from the title of my talk: at that time, as far as I remember, Bell people used the term "Optical Masers" instead of "Lasers". When my talk ended, A. Kastler, chairman, asked me the question "Do you think that a new type of laser could be designed by use of chemical reaction?" My answer "May be" induced an explosion of laughter. Later, Kastler asked me to spend one year at Ecole Normale Supérieure.

In 1961, which was the last year of my stay at Bell Labs., I was very busy in studying the electronic structure of KNiF$_3$ with Bob Shulman. One day in the spring, W. Kaiser and Dar Wood came to my office and showed me the preliminary results of their experiment on the electric field

effect on the R emission line of ruby. The result seemed to show clearly
the splitting of the R line, but it was not clear whether the splitting was
quadratic or linear to the field strength. Since I had been working on
ruby so long, I knew the existence of two sites of Cr^{3+} in ruby at which
the directions of the odd-parity crystal fields are opposite. It did not
take much time for me to arrive at the conclusion that the apparent splitting is due to the superposition of the linear spectral shifts of the opposite signs. Since the odd-parity crystal field E_{cryst} at the Cr^{3+} site
with C_3 symmetry may transform like an electric field along the trigonal
axis, the application of an external electric field E_0 along the trigonal
axis induces the spectral shift proportional to E_0 as shown by

$$(E_{cryst} + E_0)^2 \sim E_{cryst}^2 + 2E_{xryst}E_0. \qquad (E_{cryst} \gg E_0)$$

In Fig.6, we show the final result of the experiment/20/, which ver-
ifies our reasoning. It is easy to show that such a linear shift does not
exist if the site symmetry is D_3./12/ The apparent splitting thus observed
is called "Pseudo-Stark Splitting". Later, an apparent linear splitting
with the same origin was also observed in the paramagnetic resonance of ruby
in Bloembergen's group.

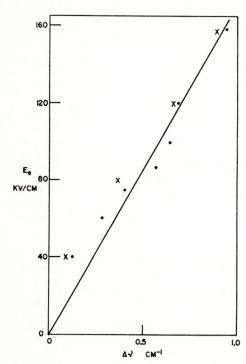

Fig.6. Pseudo-Stark splitting $\Delta\nu$ of the R emission
lines of ruby vs applied
electric field E_0. Data
are obtained from the $R_1(\cdot)$
and $R_2(\times)$ lines /20/.

3. Spectroscopy with a Strong Pulsed Magnetic Field

When I came to the Institute for Solid State Physics (ISSP), I found that
some experimental facilities for producing a strong pulsed magnetic field
were available in ISSP. I asked K. Aoyagi and A. Misu, who were working

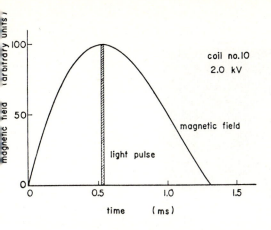

coil no.10
2.0 kV

magnetic field

light pulse

time   (ms)

Fig.7. The principle of spectroscopic experiment by using a strong pulsed magnetic field./21/

in the experimental laboratory of optics and spectroscopy where I worked on ruby before my stay at Bell Labs., if they were interested in the research project of solid state spectroscopy by using a strong pulsed magnetic field. Our project started in the summer of 1962.

The principle of our experiment is simply illustrated in Fig.7. Let us assume that the field is a simple sine curve such as $H_0 \sin \omega t$, and the pulsed light is of a rectangular shape having the width 2Δ. Then, the fluctuation of the magnetic field, H_0, during an exposure is given by $\delta H_0/H_0 \sim (\omega\Delta)^2/2$, where $\omega\Delta \ll 1$ is assumed. For example, by assuming $T \equiv \pi/\omega$ =1 ms and Δ=1 μs, we obtain $\delta H_0/H_0 \sim 0.05\%$ which is negligible compared with the inhomogeneity of the field inside a sample. The maximum field of 230 kOe was produced by a copper wire coil surrounded by a cylindrical ring of berryllium-copper alloy immersed in liquid nitrogen.

Our experiment was very useful in determining the g-values of the B_1, B_2 excited states arising from the $t_{2g}^3 \, ^2T_{2g}$ state. The values obtained are given in Table 2. In our strong field, we could observe the onset of Paschen-Back effect in the Zeeman patterns of the R lines with a magnetic field perpendicular to the trigonal axis. Instead of showing this Zeeman pattern, however, I would like to show in Fig.8 the change of the Zeeman pattern when the field is increased up to 140 T (1.4 MOe). This data was recently obtained by N. Miura's group in ISSP./23/ This change shows the perfect Paschen-Back effect as plotted in Fig. 9.

Table 2. g-values of the excited states of the B lines of ruby

	Exp.	Theor./22/
$g_\perp (B_1)$	< 0.5	0.01
$g_\perp (B_2)$	< 0.6	0.13
$g_{//} (B_1)$	0.69 ± 0.09	0.96
$g_{//} (B_2)$	-2.97 ± 0.15	-2.97

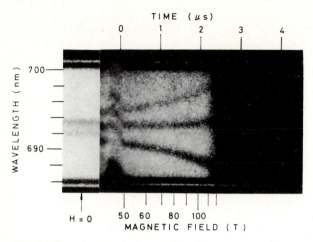

Fig.8. The observed change of the Zeeman splitting of the R lines of Ruby at T=150 K when a pulse magnetic field is applied perpendicular to the trigonal axis./23/

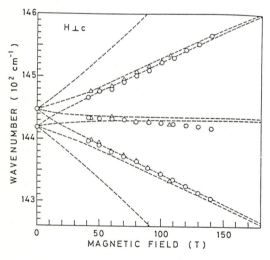

Fig.9. Plot of the Zeeman splitting of the R lines of Ruby with a magnetic field perpendicular to the trigonal axis. The broken curves are theoretical curves taking into account the Paschen-Back effect. The circles and triangles represent data for two different samples. T=150-160 K./23/

Besides observing Zeeman patterns, we were successful in observing optically the process of electron spin relaxation in the ground state of ruby in the magnetic field of ~100 kOe at low temperatures, by using a pulsed magnetic field and a light flash./24/ The principle of the experiment is illustrated in Fig.10. The Zeeman pattern is observed at time t measured from the time when the field was applied. We vary t in the repeated experiments making the field strength at t to be constant. Then, we found that the relative intensities of the Zeeman components vary as t varies. This is shown in Fig.11, where the Zeeman pattern of the B_1 and B_2 lines is observed at t=0.27, 0.67, and 1.13 ms in a magnetic field, $H_0(t)$=110 kOe, as shown in Fig.10. The transition diagram giving the Zeeman pattern of Fig.11 is shown in Fig.12.

234

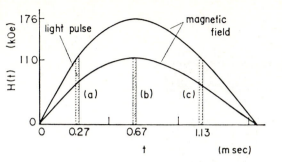

Fig.10. The combinations of a pulsed magnetic field and a light flash. /24/

H(t) ∥ C₃ ,
E ⊥ C₃ .

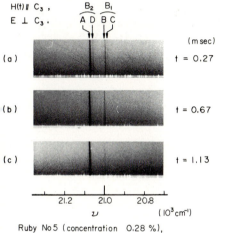

(a) t = 0.27

(b) t = 0.67

(c) t = 1.13

21.2   21.0   20.8

ν (10^3 cm⁻¹)

Ruby No 5 (concentration 0.28 %),
H(t) = 110 kOe ,   4.2°K .

Fig.11. The Zeeman patterns taken by light flash (a), (b), and (c) shown in Fig. 10. /24/

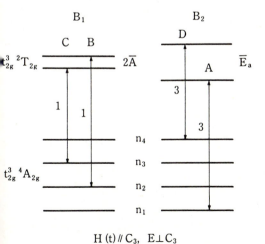

$H(t) \parallel C_3$, $E \perp C_3$

Fig.12. The transition diagram giving the Zeeman pattern of Fig.11. n_i (i=1, 2, 3, 4) are populations of the Zeeman levels of the ground state.

235

Since we know the relative transition probabilities of A, B, C, and D components as indicated in Fig.12, we can determine from the observed relative intensities, the population n_i ($i=1, 2, 3, 4$) of the Zeeman levels of the ground state. The populations thus determined are plotted in Fig.13 against t. Since the energy separations between the neighboring Zeeman levels are almost equal, we may expect that n_i's are equally spaced at a fixed value of t if an effective temperature of the spin system can be defined. Fig.13 shows that it is not the case.

Another spectroscopic experiment by using the pulse property of the magnetic field was done on rare earth compounds. In this experiment, we applied a light flash at delay time t and took the photograph of the spectrum in a magnetic field H(t) in the same way as was done in the spin relaxation experiment of ruby. At first we observed the Zeeman spectrum of $Dy_3Al_5O_{12}$(DAG) at 4.2K as shown in Fig.14. It is surprising to see that the d_{21} and d_{22} lines, which arise from the transitions from the Z_2 excited state 70.3cm^{-1} above the ground state as shown in Fig.15, begin to appear when the magnetic fields are above 30 kOe even at 4.2K. If a pulse field is replaced by a static field, no d_{21} and d_{22} lines appear at 4.2K as shown in Fig.16, since the population of the Z_2 excited state is negligibly small.

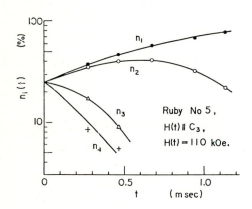

Ruby No 5,

$H(t) \parallel C_3$,

$H(t) = 110$ kOe.

Fig.13. The population of the Zeeman levels of the ground state versus t./24/

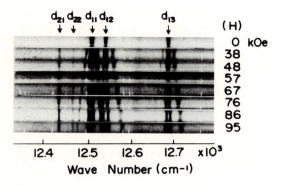

Fig.14. The Zeeman spectra of DAG in the axial polarization at 4.2 K. The pulsed magnetic fields are applied along the (001) cubic axis. d_{ij} indicate the transitions shown in Fig.15. /25/

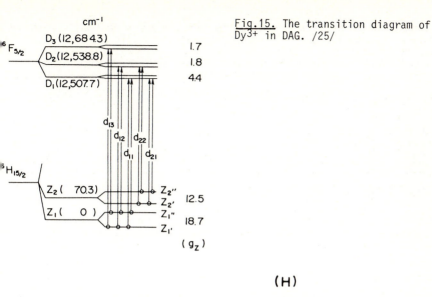

Fig.15. The transition diagram of Dy^{3+} in DAG. /25/

Wave Number (cm⁻¹)

Fig.16. The Zeeman spectra of DAG in the axial polarization with the static magnetic field along the <001> cubic axis. /25/

Using the relative transition probabilities determined from the absorption measurement at 77K, we can determine population N_i (i=1", 2' and 2") of the Z_i level. Then, assuming a Boltzmann distribution, we can determine the effective temperature T_{eff} to satisfy $N_i/N_{1'} = \exp(-E_i/kT)$, where E_i is the energy difference between Z_i and $Z_{1'}$. The effective temperatures thus determined from $N_{2'}$ and $N_{2''}$ are almost the same, but T_{eff} from $N_{1''}$ is a little higher as shown in Fig.17. The important result is the observation that T_{eff} does not change when t is changed as long as H(t) is fixed. This suggests that there is no energy flow from the electron system to the heat path surrounding the crystal within the duration of the pulsed magnetic field.

Summarizing the experimental results, we may conclude that (i) the electron system is thermally isolated from the heat reservoir in the duration of the pulse field and (ii) electrons in the $Z_{2'}$ and $Z_{2''}$ levels are in thermal equilibrium allowing us to define an effective electron temperature. These facts seem to show that the anomalous effect we observed may be discussed from a purely thermodynamic point of view, i.e. an effec-

237

tive temperature higher than that of the heat reservoir may be interpreted as due to heating by adiabatic magnetization.

To calculate the entropy of the electron system, we assume that the electron system has only two Kramers doublets Z_1 and Z_2 with g-values, $g_1 = 10.8$ and $g_2 = 7.2$, and energy separation $E_2 = 70.3$ cm^{-1}. The calculated temperature dependence of the entropy is given in Fig.18 for magnetic fields up to 100 kOe. A broken line in the figure shows the path taken by the system when the field is increased from zero to H. The crossing points of this path with the entropy curves determine the T_{eff} versus H curve which is shown in Fig.17. Fig.17 shows that the calculated T_{eff} are in fair agreement with the experimental ones.

Thermal isolation of the electron system from the heat bath may be due to either a long lattice-bath relaxation time τ_ℓ or a long spin-lattice

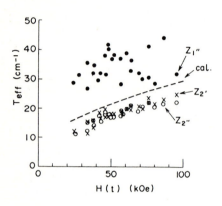

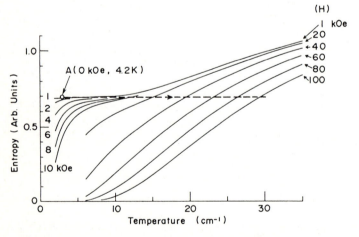

Fig.17. Magnetic field dependence of T_{eff}. The broken line is T_{eff} numerically calculated by assuming adiabatic magnetization. /25/

Fig.18. The calculated temperature dependence of the entropy of the electron system. The broken line shows the path taken by the system when the magnetic field is increased from zero to H. Point A indicates the starting point of the path. /25/

238

relaxation time $\tau_{s\ell}$ as compared with the delay time of the light flash. The former possibility, however, may be excluded by considering the fact that T_{eff} suddenly decreases when the temperature of the heat bath decreases below the Néel temperature of DAG, $T_N=2.5$ K, as shown in Fig.19: there is no reason why τ_ℓ should change abruptly at T_N. Thus, our conclusion may be summarized as

$$\tau_s, \ \tau_\ell < t < \tau_{s\ell} \qquad \text{at} \qquad T > T_N,$$

$$\tau_s, \ \tau_\ell, \ \tau_{s\ell} < t \qquad \text{at} \qquad T < T_N,$$

where τ_s is the cross relaxation time of the effective spins and t the delay time of the light flash. The abrupt change of $\tau_{s\ell}$ at T_N might be due to the crossing of the dispersion curves of effective-spins in the magnetic field with those of phonons below T_N. The adiabatic magnetization effect mentioned here was also observed in the optical spectra of $DyAlO_3$, where $E_2=54$ cm^{-1} and the maximum g-value of the ground state is 13.2, and in $Er_3Ga_5O_{12}$, where $E_2=46$ cm^{-1} and $g_{max}=11.3$ in the ground Kramers doublet.

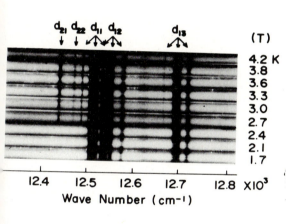

Wave Number (cm^{-1})

Fig.19. Temperature variation of the Zeeman spectra of DAG in the axial polarization with the magnetic field parallel to <001>. H_{max} and t are 76 kOe and 0.7 ms, respectively. The d_{21} and d_{22} lines suddenly disappear at $T < T_N=2.5$ K. /25/

4. Absorption Spectrum of Optically Pumped Ruby/26/

In 1965, T. Kushida of Central Research Laboratory of Tokyo Shibaura Co. showed me beautiful absorption spectra of ruby optically pumped by a xenon flash lamp. M. Shinada, my research associate, and myself immediately joined Kushida to help in the analysis of his data.

The energy level diagram of ruby is shown in Fig.20, in which the locations of the spin doublets of the $t_{2g}^2 e_g$ electron configuration had not been experimentally confirmed. Kushida's experiment measuring the absorption transitions from the $t_{2g}^3 \, {}^2E_g$ and ${}^2T_{1g}$ states could determine these locations, as the transitions are spin-allowed: for the conventional absorption in the ground state, they are spin-forbidden and difficult to be observed being masked by the spin-allowed transitions. In the experiment, it is reasonable to assume that thermal equilibrium is established among the five Kramers doublets of $t_{2g}^3 \, {}^2E_g$ and ${}^2T_{1g}$, as the relaxation time between

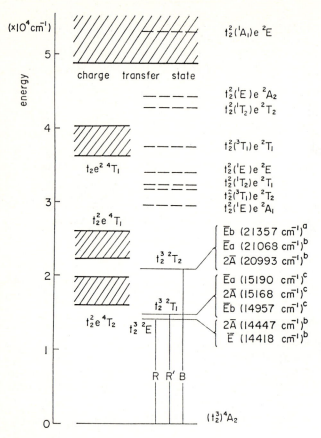

Fig.20. The energy level diagram of ruby. /26/
a. O. Deutschbein/4/, b. S. Sugano and I. Tsujikawa/5/, c. J. Margerie/27/.

these doublets is shorter than a fraction of 1 μs while the lifetime of these levels is as long as ∿3.4 ms at room temperature. Thus, the transition probabilities from these levels to the $t_{2g}^3\ ^2T_{2g}$ Kramers doublets can also be determined in Kushida's experiment.

Let us first discuss the transitions to the $t_{2g}^3\ ^2T_{2g}$ Kramers doublets in the optically pumped ruby. The transitions lie in the infrared region. The experimental spectrum due to these transitions is compared with the theoretical one in Fig.21, where we use the locations of the relevant Kramers doublets indicated in Fig.20. The agreement of the relative intensities is surprisingly good between theory and experiment. The important points are as follows; (1) The theory predicts no absorption due to the $\bar{E}_b(^2T_{1g}) \rightarrow \bar{E}_b(^2T_{2g})$ transition expected at ∿6,400 cm^{-1} and the experimental absorption curve shows a minimum: (2) The theory predicts very weak π intensities of the $2\bar{A}(^2T_{1g}) \rightarrow 2\bar{A}(^2T_{2g})$ and $\bar{E}_a(^2T_{1g}) \rightarrow \bar{E}_a(^2T_{2g})$ transitions expected at ∿5,900 cm^{-1} and no π-absorption is observed. The un-

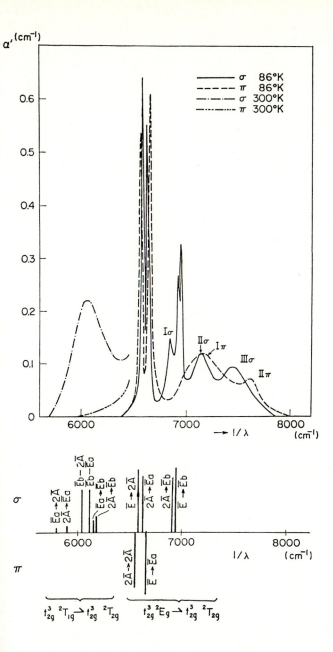

Fig.21. Comparison of the calculated absorption spectrum at 300 K with the experimental one of the optically pumped ruby in the infrared region./26/

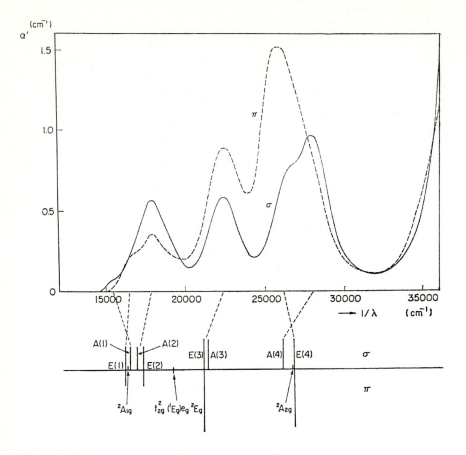

Fig.22. Comparison of the predicted absorption spectrum with the experimental one in the visible region./26/

identified peaks, I_σ, II_σ, III_σ, I_π, and II_π, are tentatively assigned to the electric dipole transitions exciting one phonon of the odd-parity mode.

Now, we discuss the transitions from the t_{2g}^3 ^2E_g state to the spin doublets of the $t_{2g}^2 e_g$ electron configuration. The calculated energies/22/ of these spin doublets measured from the t_{2g}^3 ^2E_g state are given in Table 3. Comparison is made of the calculated absorption spectrum with the observed one in the visible region in Fig.22. For A(i) and E(i) (i=1, 2, 3, 4), see Table 3, the agreement between theory and experiment is fair. In this way, we confirmed the applicability of the ligand field theory to such highly excited states as those lying 40,000 cm^{-1} above the ground state.

Table 3. The calculated energies of the $t_{2g}^2 e_g \, ^2\Gamma_g$ states measured from the $t_{2g}^3 \, ^2E_g$ excited state./22/ Here, the spin-orbit splitting is averaged and the centers of gravity of the spin-orbit split components are indicated. The indicated splittings are due to the trigonal field: A's are mainly M=0, and E's are mainly M=±1 components.

Excited states		Calculated energies
$t_{2g}^3 \, ^2E_g$		0 cm^{-1}
$t_{2g}^2(^1E_g)e_g \, ^2A_{1g}$		16,300
$t_{2g}^2(^3T_{1g})e_g \, ^2T_{2g}$	E(1)	16,200
	A(1)	16,500
$t_{2g}^2(^1T_{2g})e_g \, ^2T_{1g}$	A(2)	16,900
	E(2)	17,300
$t_{2g}^2(^1E_g)e_g \, ^2E_g$		19,200
$t_{2g}^2(^3T_{1g})e_g \, ^2T_{1g}$	E(3)	21,100
	A(3)	21,400
$t_{2g}^2(^1T_{2g})e_g \, ^2T_{2g}$	A(4)	26,100
	E(4)	26,800
$t_{2g}^2(^1E_g)e_g \, ^2A_{2g}$		26,700
$t_{2g}^2(^1A_{1g})e_g \, ^2E_g$		38,600

5. High-Pressure Effects on the Optical Spectrum of Ruby/28/

In 1978, S. Ohnishi of the Geophysics Department joined our group. He was naturally interested in high-pressure effects on solids. We started studies of high-pressure effects on the optical spectrum of ruby.

In the ligand field theory, the most important physical parameters are cubic field splitting parameter 10Dq and Racah parameter B: parameter C is approximately related to B. The dependence of these physical parameters upon the metal-ligand distance, R, has been one of the main subjects of the optical studies of transition-metal compounds under high-pressure. It was embarrassing that the pressure experiments show/29/ the dependence of 10Dq upon R to be fairly close to that given by the point charge model. However, it is now believed that this coincidence is rather accidental and more elaborate quantum mechanical treatments including the covalency in the metal-ligand bond are necessary to explain this observed pressure dependence. Moreover, the pressure experiments show a decrease of parameter B with pressure increase as observed in the spectral red shift of the R emission line of ruby on the pressure application. The same red shift was also observed in the R cubic field line of Cr^{3+} in MgO as seen in Table 1. The red shift may be interpreted to be due to an increase of the covalency with a decrease of the metal-ligand distance.

To calculate the high-pressure effects, we used a cluster model, $[CrO_6]^{9}$ in a flat potential well. The use of this cluster model seems to be a fair approximation/30/ as long as we are concerned with the localized d-electrons. Further, we assumed that the cluster has O_h symmetry. It was carefully determined that the pressure-induced change of the trigonal field gave negligible spectral shifts of the U band and the R lines as compared with those due to the change of 10Dq and B./28/

For the calculation, we applied the discrete-variational $X\alpha$ method/30/ taking into account the spin-polarization in the transition-state scheme. Spectral positions of the U band peak, E_U, and the R line, E_R, were calculated for transition-states $(t_{2g\uparrow})^{2.5}(e_{g\uparrow})^{0.5}$ and $(t_{2g\uparrow})^{2.5}(t_{2g\downarrow})^{0.5}$, respectively. Nine cases of the Cr-O distance from 2.000 Å through 1.754 Å were examined: they are enough to cover the pressure range of more than 500 kbar for ruby. The distance in the absence of pressure is assumed to be 1.9 Å. The calculation was made self-consistent and the degree of convergence was within 0.1 % of the Mulliken charge.

The peak energy of the U band is given by

$$E_U = \varepsilon(e_{g\uparrow}) - \varepsilon(t_{2g\uparrow}),$$

where $\varepsilon(e_{g\uparrow})$ and $\varepsilon(t_{2g\uparrow})$ are the spin-polarized $e_{g\uparrow}$ and $t_{2g\uparrow}$ eigenvalues of the transiton-state $(t_{2g\uparrow})^{2.5}(e_{g\uparrow})^{0.5}$. The energy of the R line is given by

$$E_R = \frac{4}{5}\{\varepsilon(t_{2g\downarrow}) - \varepsilon(t_{2g\uparrow})\},$$

where $\varepsilon(t_{2g\downarrow})$ and $\varepsilon(t_{2g\uparrow})$ are the spin-polarized $t_{2g\downarrow}$ and $t_{2g\uparrow}$ eigenvalues of the transition-state $(t_{2g\uparrow})^{2.5}(t_{2g\downarrow})^{0.5}$. The numerical factor 4/5 comes from the fact that the $\{\varepsilon(t_{2g\downarrow}) - \varepsilon(t_{2g\uparrow})\}$ is supposed to give the excitation energy, $15(3B+4)/4$, of the center of gravity of the multiplets, t_{2g}^3 ^2E_g, $^2T_{1g}$, and $^2T_{2g}$, while the excitation energy of the R line in the strong crystal field limit is $3(3B+C)$./1/

The R dependence of E_U and E_R in our calculation turns out to be

$$\delta(\ln E_U)/\delta(\ln R) = -4.5,$$

$$\delta(\ln E_R)/\delta(\ln R) = 0.375.$$

The value, -4.5, in the first equation should be compared with -5.0 of the point-charge model. Sato and Akimoto/31/ have given two sets of values of isothermal bulk modulus K_T and its pressure derivative K_T' for α-Al_2O_3: (A) $K_T = 2.26$, $K_T' = 4.0$, (B) $K_T = 2.39$, $K_T' = 0.9$. If one assumes the incompressibilities of the $[CrO_6]^{9-}$ cluster in ruby to be that of α-Al_2O_3, one obtains from the above-given second equation the wavelength (Å) derivative of the pressure (kbar) at p=0 as follows;

$$dp/d\lambda\big|_{p=0} = 2.60 \quad \text{for set (A),}$$
$$= 2.75 \quad \text{for set (B).}$$

those values should be compared with the linear relationship /32/ between pressure p(kbar) and the wavelength shift $\Delta\lambda(\text{Å})$ from $\lambda_0 = 6942$ Å for the pressure calibration;

$$p = 2.74 \ \Delta\lambda.$$

The relation between λ/λ_0 and p was calculated as shown in Fig.23.

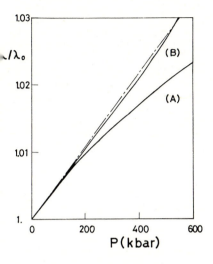

Fig.23. The calculated pressure dependence of wavelength λ of the R line of ruby for $K_T = 2.26$, $K_T' = 4.0$ (A), and $K_T = 2.39$, $K_T' = 0.9$ (B). The chain line shows the linear relationship used for the pressure calibration. /28/

. Further Developments

In the present article, I have not mentioned the spectroscopy of magnetically ordered materials. The research group of Art Schawlow at Stanford University is one of the pioneers of this field. When R. L. Greene, D. D. Sell, W. M. Yen, A. L. Schawlow, and R. M. White published a famous paper on the magnon sideband of MnF_2 /32/. We, Y. Tanabe, T. Moriya and myself, immediately responded by publishing a theory on the magnon-induced electric dipole transition moment. /33/ We have many interesting stories of research developments before and after the dawn of the history of this field. It is my great regret, however, that the space left for me in this volume is too small to accommodate the stories. I would like just to mention that I enjoyed the collaboration /34/ with K. Tsushima, K. Aoyagi and many others to perform a research project on the spectroscopy of antiferromagnetic rare-earth orthochromites at Broadcasting Science Research laboratories of NHK where I worked on the absorption spectra of complex ions /1/ in 1952-56. In rare-earth orthochromites, the R lines of Cr^{3+} played as usual an important role in guiding us through the maze of data.

In concluding this article, I would like to point out an additional theoretical development of the spectroscopy of transition-metal compounds including ruby, which has not been confirmed by experiments. This is on a light-induced change in multiplet satellites of 3p-photoelectron spectra of transition-metal compounds. In this work, I collaborated with Y. Miwa and T. Yamaguchi who preformed careful algebraic manipulation as well as tedious numerical calculations. /35/ The idea is as follows: The energy

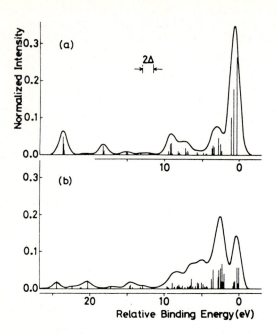

Fig.24. The calculated 3p-photoelectron spectra of a Cr^{3+} compound for the two cases of the t_{2g}^3 $^4A_{2g}$ initial state (a), and the 100% excited initial state of the t_{2g}^3 ^2E_g (b)./35/

spectrum of photoelectrons emitted by exciting 3p electrons of a transition-metal compound consists of a number of multiplet satellites due to the interaction between the incomplete d-shell and the 3p-hole. This spectrum depends upon the initial state of the compound. Therefore, if the initial state is changed by optical pumping, the photoelectron spectrum would be changed. As an example, the calculated 3p-photoelectron spectra of a Cr^{3+} compound are shown in Fig.24(a) and (b) for the two cases of the t_{2g}^3 $^2A_{2g}$ initial state and the 100% excited initial state of the t_{2g}^3 ^2E_g, respectively. The change of the spectum can be seen clearly.

Finally the author would like to express his hearty thanks to the following friends and collaborators for allowing him to use the figures and quote the essence of the joint papers; Professor A. L. Schawlow, Professor Y. Tanabe, Professor I. Tsujikawa, Professor H. Kamimura, Dr. F. Varsanyi, Dr. A. H. Piksis, Dr. D. L. Wood, Professor W. Kaiser, Professor G. Kuwabara, Professor K. Aoyagi, Professor A. Misu, Professor T. Kushida, Professor M. Shinada, Dr. S. Ohnishi, Dr. K. Tsushima, and Professor T. Yamaguchi. The author would also like to express his sincere thanks to Professor N. Miura for providing him with the beautiful photograph (Fig.8) and the plot of the Zeeman splitting (Fig.9) of the R lines of ruby in ultra-high magnetic fields showing the perfect Paschen-Back effect. The author is also indebted to Miss T. Oto for her careful preparation of this camera-ready manuscript.

References

1. Y. Tanabe and S. Sugano: J. Phys. Soc. Japan 9, 753 (1954); ibid. 9, 766 (1954)
2. Lapraik: J. prakt. Chem. 47, 305 (1893)
3. F. H. Spedding and G. C. Nutting: J. Chem. Phys. 3, 369 (1935)

4. O. Deutschbein: Ann. d. Phys. [5] 14, 712 (1932); ibid. 14, 729 (1932); ibid. 20, 828 (1934)
   B. V. Thosar: Phys. Rev. 60, 616 (1941); J. Chem. Phys. 10, 246 (1942)
5. S. Sugano and Y. Tanabe: J. Phys. Soc. Japan 13, 880 (1958)
   S. Sugano and I. Tsujikawa: J. Phys. Soc. Japan 13, 899 (1958)
6. J. E. Geusic: Phys. Rev. 102, 1252 (1956)
7. H. Lehmann: Ann. d. Phys. [5] 19, 99 (1934)
8. G. Makhov, C. Kikuchi, J. Lambe and R. W. Terhune: Maser Action in Ruby, The University of Michigan 2616-1-T, June 1958
9. A. L. Schawlow and C. H. Townes: Phys. Rev. 112, 1940 (1958)
0. F. Varsanyi, D. L. Wood, and A. L. Schawlow: Phys. Rev. Lett. 3, 544 (1959)
1. S. Geschwind, R. J. Collins, and A. L. Schawlow: Phys. Rev. Lett. 3, 545 (1959)
2. S. Sugano, Y. Tanabe, and H. Kamimura: Multiplets of Transition-Metal Ions in Crystals (Academic Press, 1970) p.187
3. S. Sugano, A. L. Schawlow, and F. Varsanyi: Phys. Rev. 120, 2045 (1960)
4. A. L. Schawlow, A. H. Piksis, and S. Sugano: Phys. Rev. 122, 1469 (1961)
5. R. E. Watson, Mass. Inst. Technol., Solid-State and Molecular Theory Group, Tech. Rept. No.12 (June 15, 1959)
6. T. H. Maiman: Nature 187, 493 (1960)
   R. J. Collins, D. F. Nelson, A. L. Schawlow, W. Bond, C. G. B. Garrett, and W. Kaiser: Phys. Rev. Lett. 5, 303 (1960)
7. A. L. Schawlow: Quantum Electronics edit. by C. H. Townes (Columbia University Press, New York, 1960)
8. A. Javan, W. R. Bennett Jr., and D. R. Herriott: Phys. Rev. Lett. 6, 106 (1961)
9. S. Sugano: Applied Optics 1, 295 (1962)
0. W. Kaiser, S. Sugano, and D. L. Wood: Phys. Rev. Lett. 6, 605 (1961)
1. K. Aoyagi, A. Misu, and S. Sugano: J. Phys. Soc. Japan 18, 1448 (1963)
2. S. Sugano and M. Peter: Phys. Rev. 122, 381 (1961)
3. N. Miura and F. Herlach edit: Strong and Ultrastrong Magnetic Fields and their Applications edit. by F. Herlach (Springer Verlag, Berlin, Heidelberg, 1985) p.330
4. K. Aoyagi, A. Misu, G. Kuwabara and S. Sugano: J. Phys. Soc. Japan 19, 412 (1964)
5. K. Aoyagi and S. Sugano: J. Phys. Soc. Japan 45, 837 (1978)
6. T. Kushida: J. Phys. Soc. Japan 21, 1331 (1966)
   M. Shinada, S. Sugano, and T. Kushida: J. Phys. Soc. Japan 21, 1342 (1966)
7. J. Margerie: CR Acad. Sci. (Paris) 255, 1598 (1962)
8. S. Ohnishi and S. Sugano: Japanese J. Appl. Phys. 21, L309 (1982)
9. H. G. Drickamer: Solid State Physics edit. by F. Seitz and D. Turnbull (Academic Press, New York, 1965) vol.17, p.1
0. H. Adachi, S. Shiokawa, M. Tsukada, C. Satoko and S. Sugano: J. Phys. Soc. Japan 47, 1528 (1979)
1. Y. Sato and S. Akimoto: J. Appl. Phys. 50 (8), 5285 (1979)
2. R. L. Greene, D. D. Sell, W. M. Yen, A. L. Schawlow, and R. M. White: Phys. Rev. Lett. 15, 656 (1965)
3. Y. Tanabe, T. Moriya, and S. Sugano: Phys. Rev. Lett. 15, 1023 (1965)
4. S. Sugano, K. Aoyagi, and K. Tsushima: J. Phys. Soc. Japan 31, 706 (1971)
5. S. Sugano, Y. Miwa, and T. Yamaguchi: Phys. Rev. Lett. 44, 1527 (1980)

Ruby – Solid State Spectroscopy's Serendipitous Servant

G.F. Imbusch[1] *and W.M. Yen*[2]

[1]Department of Physics, University College, Galway, Ireland
[2]Department of Physics and Astronomy, The University of Georgia,
Athens, GA 30602, USA

1. Introduction

Ruby is a handsome gemstone whose color can vary from pale pink to a deep purple-red, depending on the percentage of chromium present in the crystal. Large naturally-occurring good quality specimens are rare and highly-prized. The hardness and high refractive index of the material gives the cut gemstone a particular brilliance. In addition, the "cold fire" of ruby - its deep red luminescence - adds to the mystique and aura of this beautiful crystal.

Ruby is also a most interesting scientific material. It is crystalline Al_2O_3 (sapphire) with a dilute doping of Cr^{3+} ions (usually much less than 1%) replacing some Al^{3+} ions. Its spectroscopic properties have been studied for over one hundred years; the early work has been described brief by MOLLENAUER /1/. In a treatise on light (*La Lumière, Ses Causes et Ses Effets*) published in 1867 Edmond BECQUEREL /2/ devoted nearly a full chapter to *Alumine et ses Combinaisons*, whose luminescence he studied with the aid of his "phosphoroscope" (Fig. 1). In this device the sample was excited by a beam of sunlight and its luminescence viewed through a prism spectrometer a controllable time after the exciting sunlight beam had been shut off. With this apparatus Becquerel was able to observe the two intense sharp luminescence lines of ruby and to correctly estimate the lifetime at around 4 ms. Because the luminescence could be observed from apparently "pure" samples of alumina Becquerel claimed that the emission was the result of some intrinsic property of the Al_2O_3 crystal and that chromium merely played the role of activator. This view was challenged by BOISBAUDRAN /3/, and a protracted debate ensued which is well chronicled in the literature /1/.

The first high-resolution study of the ruby spectrum was that of DU BOIS and ELIAS /4/ in 1908 who measured the absorption and emission spectra. It was they who first used the label R ("rot") to designate the sharp luminescence lines. These workers, as well as J. BECQUEREL /5/ and MENDENHALL and WOOD /6/, observed and attempted to analyze the Zeeman splitting of the R lines. In addition, du Bois and Elias reported the presence of neighboring sharp lines (N lines) in the vicinity of the R lines. In a series of papers beginning in 1932 DEUTSCHBEIN /7/ reported extensive studies of chromium luminescence in a number of crystalline host materials.

The proper theoretical analysis of the ruby spectrum awaited a number of major steps. First the method of quantum mechanics had to be developed to permit a proper description of the interaction between radiation and atoms. The method of determining the splitting of atomic energy levels by the electrostatic fields of neighboring ions in a solid was explained by BETHE /

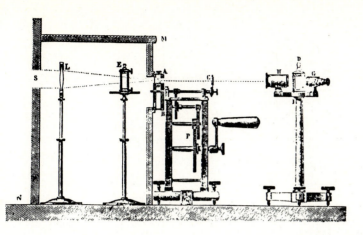

Fig. 1. Becquerel's phosphoroscope. Sunlight enters at S and is focussed onto the sample through a filter at E. By turning the handle a pair of segmented disks, one on either side of the sample, are rotated, which alternatively block the sunlight and the path of the luminescence to the spectrometer.

in 1929. Finally, the systematic analysis of the spectroscopic properties of the Cr^{3+} ion in ruby was included in a series of papers, beginning in 1954, by SUGANO, TANABE, and KAMIMURA /9/ which have provided a solid basis for modern studies of the spectroscopy of transition metal ions in solids. As a check on the theory in the case of ruby SUGANO and TSUJIKAWA /10/ made precise measurements on the anisotropy and Zeeman effect in ruby. These experiments, incidentally, were carried out under difficult conditions, as Sugano relates in his review paper ("Spectroscopy of Solid-State Laser Materials") in this volume.

Electron paramagnetic resonance was observed in the ground state of Cr^{3+} in ruby in 1955 by MANENKOV and PROKHOROV /11/ and maser operation was successfully achieved in ruby by MAKHOV et al./12/. In thinking about materials which might exhibit optical maser action Art Schawlow was attracted to ruby by virtue of its sharp intense luminescence lines, to which his attention had been drawn by the experiments of Sugano and Tsujikawa. The credit for producing the first operational laser goes to MAIMAN /13/; the active medium was ruby.

Mainly because of its importance as a laser material methods have been perfected for growing large single-crystal specimens of ruby of excellent quality, and the availability of these crystals has greatly assisted research into its spectroscopic properties. We both began our training in solid state optical spectroscopy in Art Schawlow's laboratory at Stanford University where we were introduced by him to this unique material which was then, and which continues to be, the testing ground for many of the ideas of solid state spectroscopy. We will describe some of these studies in this paper.

2. The Spectroscopy of the Cr^{3+} Ion in Ruby

Part of the $\alpha-Al_2O_3$ crystal structure is shown in Fig. 2(a), the smaller dark circles represent Al^{3+} ions while the larger open circles represent O^{2-} ions.

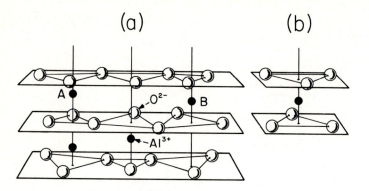

(a)   (b)

Fig. 2(a).   Part of the α-Al₂O₃ crystal showing the A and B sites of the Al
ions.   The Al site has trigonal symmetry about the optic axis (vertical
line through the Al ion).
Fig. 2(b).   Al^{3+} ion in a site of octahedral symmetry.

There is a similar arrangement of oxygen ions about each Al^{3+} ion - a large
triangle and a small triangle of oxygen ions, one above and one below the Al
ion - but these arrangements are rotated relative to each other on the
different Al^{3+} sites, as Fig. 2(a) shows.   The vertical lines through the
Al^{3+} ions in Fig. 2(a) give the direction of the optic axis of the crystal.

In ruby a small fraction of the Al^{3+} ions are substituted by Cr^{3+} ions.
When present in very dilute amounts these ions can be considered too far apart
to interact with each other.   We then have the equivalent of a dilute "gas"
Cr^{3+} ions in an inert sapphire host matrix.   The interesting spectroscopic
properties of the ruby crystal derive from these Cr^{3+} ions.   In contrast to
the case of a normal gas, whose particles are constantly in translational
motion, the constituents of the Cr^{3+} gas are confined to specific Al^{3+} sites
in the sapphire lattice.   All of these ions are, in principle at least, in
identical sites and are subject to the same kind of electrostatic field (the
crystal field) of neighboring O^{2-} and Al^{3+} ions.   Because they are identical
and isolated from each other in dilute ruby we need only confine our attention
to a single representative Cr^{3+} ion and calculate its spectroscopic properties
Then, using suitable statistical considerations, we can predict the behaviour
of a macroscopic sample of dilute ruby.   Although each chromium ion is
confined to a specific site in the sapphire host it is not rigidly fixed in
space since it participates in the vibrational motion of the host lattice.
However, it is convenient, initially at least, to ignore this vibrational
motion and to imagine that the representative Cr^{3+} ion and its neighboring
Al^{3+} and O^{2-} ions are at rest.

In this paper we are concerned with optical and, to a lesser extent,
paramagnetic resonance studies of ruby.   The properties of interest derive
from the interaction of the electromagnetic radiation with the outer unpaired
3d electrons of the Cr^{3+} ion.   The inner core electrons do not actively
participate in the interaction and so can be ignored.   This greatly
simplifies the analysis of the behaviour of the Cr^{3+} ions.   The filled inner
core of electrons can be considered to form a time-average spherically-
symmetric charge distribution.   This and the nucleus can be regarded as
creating an electrostatic *central field* potential $V_{central}(\vec{r})$, in which the
outer electrons move.   In such a potential orbital and spin angular momenta

f the individual outer electrons are good quantum numbers, and these one-electron states are classified by their n, ℓ values. The outer three electrons of Cr^{3+} are in n = 3, ℓ = 2 states, so Cr^{3+} has the configuration $3d^3$.

Next we must take into account the interaction among these outer 3d electrons. This strong Coulomb interaction is described by the energy term $H_{Coulomb} = \sum e^2/(4\pi\varepsilon_0 r_{ij})$, where $r_{ij} = |\vec{r}_i - \vec{r}_j|$, i and j being indices labelling the outer electrons. This Coulomb interaction and the requirement of the Pauli principle split the energy of the $3d^3$ configuration into a number of distinct states (called *terms*) classified by the total orbital angular momentum L and total spin angular momentum S. For the three outer electrons S can have values 1/2 or 3/2, so the terms are divided into spin doublets (S = 1/2) and spin quarters (S = 3/2). The lowest energy term is ^4F, in conformity with Hund's rule. The wavefunctions of these free-ion LS terms are of even parity and, to very good approximation, are derived solely from single-electron 3d states.

When the Cr^{3+} ion is substituted for an Al^{3+} ion in the Al_2O_3 crystal the outer 3d electrons of the chromium ion are affected by the electrostatic crystal field of the neighboring O^{2-} and Al^{3+} ions, which for the present we regard as rigidly fixed in position in the Al_2O_3 lattice. As Fig. 2(a) shows the arrangement of O^{2-} ions about each Al^{3+} site is seen to have three-fold symmetry around the optic axis through the Al^{3+} ion. This would be much more symmetrical if the two parallel triangles of O^{2-} ions (above and below the Cr^{3+} ion) were identical and symmetrically arranged, as in Fig. 2(b). In this symmetrical arrangement the perpendicular separation between the two triangles is $\sqrt{2}/\sqrt{3}$ times the length of the side of the oxygen triangle, and the Al^{3+} ion is midway between the triangles. This arrangement has octahedral (O_h) symmetry. The crystal field potential in this case is written $V_{O_h}^{(g)}(\vec{r})$, where the (g) superscript indicates that this potential function has even parity. We find that the actual arrangement of oxygen ions in ruby has a distorted octahedral arrangement; the crystal field potential is predominantly octahedral, but two weaker potential terms with lower symmetry (trigonal) also occur. We write these as $V_{trig}^{(g)}(\vec{r})$ and $V_{trig}^{(u)}(\vec{r})$. The first is of even parity, the second is of odd parity, as denoted by the superscript (u) and is a reflection of the fact that the Al^{3+} site in ruby lacks inversion symmetry. (Dopant Cr^{3+} ions in all insulating materials occupy sites with six negatively-charged anions nearby. The crystal field potential sometimes has full octahedral symmetry, but more often additional smaller terms of lower symmetry also occur.)

To solve for the electronic energy levels of the Cr^{3+} ion in ruby we can start by neglecting the small terms in the Hamiltonian, such as $V_{trig}^{(g)}$, $V_{trig}^{(u)}$, and spin-orbit coupling, and calculate the effect of $V_{O_h}^{(g)}$ on the LS terms. This octahedral crystal field term splits the free ion terms into crystal field *multiplets*. The energy levels of the crystal field multiplets for all $3d^n$ configurations have been calculated by Sugano et al./9/, and the separations between these multiplet levels are given in terms of three parameters (i) two Racah parameters, B and C, which are a measure of the strength of the Coulomb interaction among the outer electrons, and (ii) a single parameter, Dq, which describes the strength of the octahedral crystal field. Since C $\simeq$ 4B for all $3d^n$ systems the separations between the energy levels can be approximately described by two parameters, B and Dq. The splittings of LS terms for Cr^{3+} in an octahedral crystal field are shown in Fig. 3, where E/B is plotted against Dq/B. All of these wavefunctions have even parity, as they are formed from even-parity free-ion LS terms by the even-parity octahedral crystal field term. Odd-parity crystal field states,

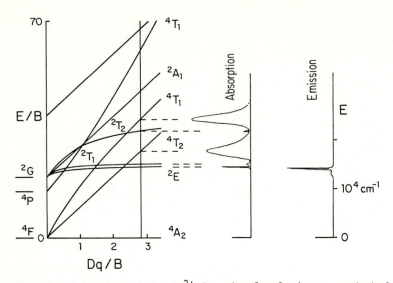

Fig. 3. Splitting of the Cr^{3+} free-ion levels in an octahedral crystal fiel
Low-temperature absorption and emission spectra are shown on the right.

formed from $3d^2 4p$ orbitals, occur some 50,000 cm^{-1} higher in energy.   The
solid vertical line drawn at $Dq/B = 2.8$ shows the value of Dq/B for the
average octahedral crystal field in ruby.

Since we are neglecting spin-orbit coupling the total spin S is a good
quantum number and the states are still labelled by their spin multiplicity
(2S+1).   Now, however, the orbital states are labelled according to the
irreducible representations of the octahedral symmetry group.   These
representations are A_1, A_2, E, T_1, T_2, and this labelling scheme is used to
describe the orbital levels in Fig. 3.   On the right-hand side of the figur
the low-temperature absorption and luminescence spectra of ruby are drawn.
There is good agreement between the predicted energy levels and the absorpt
peaks.   One observes that the spin-allowed transitions, $^4A_2 \rightarrow {}^4T_2$, $^4T_1$, a
much more intense than the spin-forbidden transitions, $^4A_2 \rightarrow {}^2E$, $^2T_1$, $^2T_2$
as expected.

Next, the smaller terms, $V_{trig}^{(g)}$ and spin-orbit coupling, must be taken int
account.   These cause a splitting of all the octahedral levels, and we are
particularly interested in the 29 cm^{-1} splitting of the ^2E level and the 0.
cm^{-1} splitting of $^4A_2$.   Since much of the subsequent discussion will invol
the optical transitions between these two states, we show in Fig. 4 how the
octahedral levels are split by the lower-symmetry crystal field and spin-or
coupling, and how these levels are further split by a magnetic field applied
parallel to the optic axis.

Our analysis so far predicts $sharp$ electronic energy levels and, hence,
sharp optical transitions between the levels.   However, as the absorption
spectrum in Fig. 3 shows, some of the observed transitions are quite broad.
This broadening is a consequence of the vibrational motion of the Al_2O_3 lat
including the dopant Cr^{3+} ions.   These vibrations are quite large; even a
the absolute zero of temperature the root-mean-square lattice displacement
an ion is around 5% of the inter-ion separation.   As a result the Cr^{3+} io

252

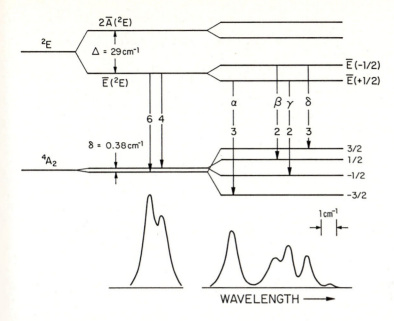

Fig. 4. Under the trigonal crystal field and spin-orbit interaction the ^2E and $^4A_2$ levels exhibit splittings of 29 cm^{-1} and 0.38 cm^{-1}, respectively. These levels are further split by a magnetic field, B, along the optic axis. The R_1 line ($\bar{E}(^2E) \rightarrow {}^4A_2$) in emission at 2 K is shown for B = 0 and B $\simeq$ 1.2 T.

experience a *dynamic* crystal field as well as the average *static* crystal field we have discussed earlier. Since the crystal field is modulated by the lattice vibrations the optical transitions are *frequency modulated* by the lattice vibrations. Consequently, the optical transitions appear as sharp zero-phonon lines accompanied by FM sidebands. As is clear from Fig. 3, a variation of around ±5% in Dq/B significantly changes the energies of the $^4T_1$ and $^4T_2$ levels relative to the $^4A_2$ ground state. Hence the $^4A_2 \rightarrow {}^4T_1, {}^4T_2$ transitions are strongly modulated and are dominated by broad FM sidebands. On the other hand, the $^4A_2 \leftrightarrow {}^2E$ transition, which appears in absorption and in emission, is seen to be affected to a much smaller extent by the variation in crystal field, and the zero-phonon lines are the dominant feature of this transition.

Finally we take account of the odd-parity crystal field term, $V_{trig}^{(u)}$. This has a negligible effect on the energy and splitting of levels; its most important consequence is a mixing of some odd-parity wavefunctions with the even-parity $3d^n$ wavefunctions. As a result the $^4A_2 \leftrightarrow {}^2E$ zero-phonon lines in ruby occur by a weak electric dipole process, which is an order of magnitude stronger than the analogous magnetic dipole process.

Ions raised to the $^4T_2$ and $^4T_1$ levels by optical absorption lose some energy to the lattice (as lattice vibrations) and quickly drop to the ^2E level, which is the lowest excited state and is about 14,000 cm^{-1} above the ground state. The radiative transition between the ^2E and $^4A_2$ levels is spin-forbidden and has a relatively long lifetime ($\approx$ 3.6 ms at low temperatures). Despite its small transition probability the radiative emission process is the most effective decay process out of the ^2E level, the luminescence process at room temperature and lower temperatures being almost 100% efficient.   Two sharp zero-phonon lines dominate this transition: the R_2 line between the upper ^2E level ($2\bar{A}(^2E)$) and the $^4A_2$ ground state, and the R_1 line between the lower ^2E level ($\bar{E}(^2E)$) and the ground state.   The strong broad absorption bands of ruby in the green and blue permit efficient optical pumping of the ^2E state.   The long lifetime of this state, then, allows a sizeable population to be maintained there by continuous optical pumping. These properties were put to good use when the first laser was constructed using ruby as the active medium, with laser action taking place in the R_1 line.

3. The Zero-phonon R_1 and R_2 Luminescence Lines

A matter of some importance in the early days of the ruby laser was to gain an understanding of the origin of the broadening of the R_1 line.   That this broadening could be due in part to strains in the crystal was suspected by Schawlow who noticed that the more perfectly-grown specimens exhibited the narrowest lines.   This was confirmed by him and others when the R lines were found to shift in wavelength if the material was subjected to static stresses in controlled laboratory experiments.   Random strain fields occur in all ruby crystals which lead to variations in the time-average crystal field among the Cr^{3+} sites and, consequently, to a distribution of R_1 frequencies. The situation is illustrated schematically in Fig. 5. Cr^{3+} ions in sites of identical strain have a small homogeneous broadening at low temperatures and emit a very sharp R_1 line.   The observed R_1 line from a ruby crystal at low temperatures is a composite of such emissions from Cr^{3+} ions in a range of different crystal field sites;   the observed line exhibits *inhomogeneous* broadening.   The fluorescence line-narrowing (FLN) experiment of SZABO /14/ illustrates this inhomogeneous broadening very clearly.   In this experiment a narrow band laser, tuned to a frequency within the R_1 line, excites a subset of the Cr^{3+} ions - those in resonance with the laser - and these excited Cr^{3+} ions emit their luminescence in a sharp line whose width is determined by the width of the laser beam and by the *homogeneous* width of the transition, not by the width of the inhomogeneous line.   In his initial FLN experiments Szabo achieved a linewidth of around 30 MHz, a factor of 100 smaller than the normal inhomogeneous linewidth.   Szabo's experiments in rub were the first demonstration of fluorescence line narrowing in a solid. Subsequent experiments employing FLN and hole-burning techniques by MURAMOTO et al./15/, and JESSOP, MURAMOTO and SZABO /16/ have succeeded in reducing th observed low-temperature homogeneous linewidth of the R_1 line to a few MHz.

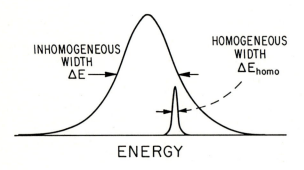

Fig. 5.   The inhomogeneously broadened emission line is a composite of narrower emission lines from ions in different strain environments.

As the temperature is raised the homogeneous broadening of the R lines increases significantly, as SCHAWLOW's early measurements showed /17/. The origin of the broadening is a second-order effect due to interaction with lattice vibrations, in which two lattice vibrations of the same frequency interact with the Cr^{3+} ion to cause a broadening of the zero-phonon line. In a quantum description this broadening is ascribed to a Raman scattering of phonons, and a theoretical formula for the broadening was first developed by McCUMBER and STURGE /18/. They also made accurate measurements of the temperature-dependent widths of the R_1 and R_2 lines of ruby and found excellent agreement with the theoretical formula. There is also a variation with temperature of the central frequency of the zero-phonon line which comes about because the mean crystal field energy varies slightly as the lattice vibrational energy increases. The calculations of McCumber and Sturge show that the frequency of the zero-phonon line varies as the mean-square lattice displacement. Fig. 6 shows the good agreement between their measurements (open circle) of the shift of the R_1 line of ruby and the temperature-dependence predicted theoretically (solid curve). McCumber and Sturge included some measurements of the variation of the R_1 line with temperature (the triangles in Fig. 6) made by GIBSON /19/ in 1916. The temperature-dependent shift and broadening of sharp optical transitions, first elucidated in ruby, are general effects in the spectroscopy of doped transition metal and rare earth ions.

The occurrence of an energy shift proportional to the mean-square lattice displacement has an interesting consequence. Chromium occurs in four

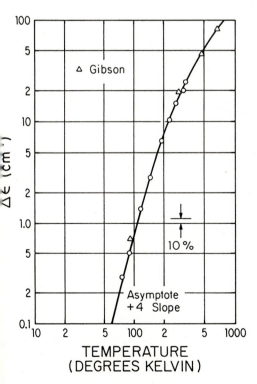

Fig. 6. Shift of the R_1 line as a function of temperature (from McCumber and Sturge /18/).

255

reasonably abundant isotopes: $Cr^{50}(4.3\%)$, $Cr^{52}(83.6\%)$, $Cr^{53}(9.6\%)$, and $Cr^{54}(2.4\%)$. Now the lighter isotopes are expected to vibrate with larger amplitude than the heavier isotopes, so the mean-square lattice displacemen and consequently the energy shift, is greater for the lighter isotopes. Th should lead to a slight variation of the position of the zero-phonon line w isotopic mass. Such an isotope shift was originally found in the R_1 line ruby by SCHAWLOW /20/ and his spectrum was seen in Fig. 7(a). This spectru was taken at helium temperatures, where the homogeneous broadening is a minimum but where there is still a sizeable mean-square lattice displacemen due to zero-point vibrations. Figure 7(a) shows the two components of the R_1 line due to the 0.38 cm^{-1} ground state splitting, and the isotope shift seen as barely resolvable structure on these components. JESSOP and SZABO /21/ utilized FLN techniques to resolve the isotope shift with much greater clarity. Their spectrum is shown in Fig. 7(b) and illustrates the great experimental advantages of laser excitation spectroscopy.

Even since the development of the first laser spectroscopists have employed this unique tool in diverse ways in their laboratory. In the beginning, when only the ruby laser was available, it was natural to study resonant interaction of the ruby laser beam with a target ruby sample. Th phenomena of *photon echoes* analogous to the concept of spin echoes original demonstrated by HAHN /22/, was observed in ruby by ABELLA, HARTMANN, and KURNIT /23/. This is but one example of a class of phenomena which can be classified as *coherent radiation spectroscopy.*

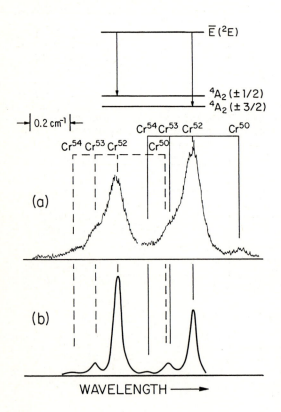

Fig. 7. Structure on the R_1 line due to ground state splitting and isotope shift: (a) Schawlow's original spectrum, (b) the spectrum of Jessop and Szabo obtained by laser excitation.

256

Because of the narrow width of the R_1 and R_2 lines at low temperatures and the high density of Cr^{3+} ions present in a typical ruby sample, the peak absorption in the R lines can be quite strong. Thus when an excited Cr^{3+} ion decays with emission of R_1 or R_2 radiation the emitted photon may be reabsorbed ("trapped") by an unexcited Cr^{3+} ion before it leaves the crystal. That this phenomenon was occurring in ruby was first recognized by VARSANYI, WOOD, and SCHAWLOW /24/ and was the first example of this effect, well known to occur in gaseous spectroscopy, to be reported in a solid. Trapping acts as a mechanism for the radiative transfer of resonant optical excitation throughout the full ruby crystal. Its most noticeable consequence is a lengthening of the observed radiative lifetime, from its intrinsic value of around 3.6 ms found in dilute powdered samples, to over 10 ms in large heavily-doped samples (NELSON and STURGE /25/).

Under intense optical pumping into the broad absorption bands at low temperatures the final step of the nonradiative relaxation process brings the Cr^{3+} ion across the 29 cm^{-1} gap from $2\bar{A}(^2E)$ to $\bar{E}(^2E)$ and can result in the generation of a high density of 29 cm^{-1} phonons /26/. Interesting experiments can be carried out to study the dynamics of these nonequilibrium phonons /27,28,29,30/.

4. Magnetic Resonance Studies in Ruby

Electron paramagnetic resonance (EPR or ESR) is a high-resolution technique for probing the fine structure of electronic levels. The material is irradiated with microwave power of a specific frequency, the electronic splitting can be tuned by applying a variable magnetic field, and the occurrence of microwave absorption is detected by use of a balanced bridge network. The microwave photon is a finer probe than the optical photon, but the detection of microwave photons is a much more difficult task than the detection of optical phonons, and as a result EPR is only possible when a large number of ions interacts with the microwave radiation. EPR studies, then, are restricted to ground electronic states. (Under constant optical pumping a large population can be maintained in some *excited* triplet states in some organic solids, and conventional EPR studies can be made in these states. Such situations do not occur in doped inorganic materials.) EPR in the ground state of ruby was originally achieved in 1955 /11/; a very comprehensive study of ruby ground state EPR was published by SCHULZ-DU BOIS /31/.

In order to carry out an EPR study in an excited electronic state where, even with intense optical pumping, the density of ions will be small, a much more sensitive detection technique than that employed in conventional EPR must be used. Such a technique was proposed by GESCHWIND, COLLINS, and SCHAWLOW /32/ in the case where the excited electronic state emits luminescence. The scheme is based on the fact that when one microwave photon is absorbed in the excited state the luminescence pattern is changed by one optical photon; the occurrence of microwave absorption can then be detected with much greater sensitivity by monitoring the luminescence.

We have seen how the ^2E state of Cr^{3+} in ruby is efficiently pumped by broadband optical sources and has a relatively long lifetime, so a sizeable population can be maintained in this state. Geschwind proposed to carry out his experiment for the optical detection of magnetic resonance (ODMR) in this state. A number of detection schemes were outlined by him, one of which we can illustrate with the aid of Fig.4, which shows the Zeeman splitting of the $\bar{E}(^2E)$ level, from which the R_1 line originates. At helium temperatures the

lower Zeeman level ($\bar{E}(\frac{1}{2})$) has the greater population and, as a consequence, t
α line, which originates on this level, is the most intense optical Zeeman
component.   If resonant microwave power is strongly absorbed in the $\bar{E}(^2E)$
state the populations in the two Zeeman levels become equal and the α line i
reduced in intensity.   Thus the onset of EPR in the $\bar{E}(^2E)$ state is detected
monitoring the intensity of the α component.   The pioneering ODMR experimen
of Geschwind et al./32/ were carried out in ruby in 1959.   Since then the
technique of ODMR has been developed into a very useful and sensitive probe,
particularly in studies of doped semiconductor materials.

When the resonant microwave power is switched off the intensity returns t
its equilibrium value with a time constant which is determined in part by th
spin-lattice relaxation between the two Zeeman levels, $\bar{E}(1/2)$ and $\bar{E}(-1/2)$.
ODMR measurements of this spin-lattice relaxation time (T_1) were made by
GESCHWIND et al./33/ and these are shown in Fig. 8.   The temperature
dependence shows that this is clearly a two-phonon Orbach process, proceedin
via the $2\bar{A}(^2E)$ level 29 cm^{-1} higher in energy, the Orbach process predicting

$$T_1 = C \exp(\Delta/kT), \qquad\qquad\qquad (1)$$

where Δ is the energy gap (29 cm^{-1}) between $2\bar{A}(^2E)$ and $\bar{E}(^2E)$.

At 3 K the spin lattice relaxation time in the $\bar{E}(^2E)$ state has the same
value as the radiative decay time of the $\bar{E}(^2E)$ state;   below this temperatur
the spin-lattice relaxation rate is slower than the radiative decay rate.
This fact could have a bearing on the optical detection of magnetic resonanc
in ruby.   In order for magnetic resonance to occur there must be a greater
population in the lower $\bar{E}(1/2)$ level than in $\bar{E}(-1/2)$ so that microwave power
will be absorbed.   How is this population difference achieved?   The optica
pumping process involves first raising the ions to the $^4T_2$ or $^4T_1$ levels,
after which the ions relax by phonon emission until the $\bar{E}(^2E)$ state is

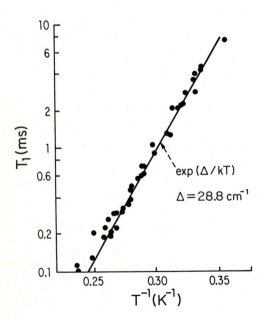

Fig. 8.   Spin-lattice
relaxation time in the
$\bar{E}(^2E)$ state of Cr^{3+} in   ruby
(after Geschwind et al./33/)

reached.   At the conclusion of this indirect pumping process it might be assumed that both the $\bar{E}(1/2)$ and $\bar{E}(-1/2)$ levels are equally populated.   Next spin-lattice relaxation in the $\bar{E}$ state causes an adjustment of population ("thermalization"), giving a larger population in the lower level.   However, thermalization can only occur if the spin-lattice relaxation rate is faster than the radiative decay rate (which, as Fig. 8 shows, occurs above 3 K in ruby); if there is not enough time for thermalization to occur the populations stay equal, no microwave absorption occurs, and no ODMR signal should be found.   Such, however, was not found experimentally; the ODMR signal became stronger as the temperature was reduced below 3 K.   Clearly the optical pumping process itself leads *directly* to a population difference between the $\bar{E}(-1/2)$ and $\bar{E}(1/2)$ levels.   This would occur if there were spin selection rules in the broadband ($^4A_2 \rightarrow {}^4T_2$, $^4T_1$) optical pumping of the ^2E state.   At low temperatures the Cr^{3+} ions in the ground $^4A_2$ state are preferentially in the lower spin states, under broadband optical pumping they retain ground state *spin memory*, and at the end of the pumping cycle they preferentially populate the lower spin level of the $\bar{E}(^2E)$ state.   That such a spin memory occurs in the optical pumping of ruby was experimentally verified by changing the population in the *ground* spin states during optical pumping and detecting the resultant change in population in the ^2E spin states (IMBUSCH and GESCHWIND /34/).   Although it was unambiguously demonstrated for the first time in ruby, the phenomenon of spin memory is of very general occurrence and has been demonstrated in many other luminescence systems.

5. Interaction between Cr^{3+} Ions in Ruby - Excitation Transfer

The picture of ruby so far presented is of isolated Cr^{3+} ions randomly distributed in Al^{3+} sites in the sapphire crystal.   If the ions are close enough for a weak interaction to occur between them new effects will manifest themselves.   For example, an excited Cr^{3+} ion may transfer its excitation *nonradiatively* to a nearby Cr^{3+} ion with which it interacts, and there may be a number of such transfers before the excitation is released as a photon. We can try to distinguish *resonant* nonradiative transfer - in which excitation is transferred between ions whose excited levels coincide to within the homogeneous linewidth - from *nonresonant* nonradiative transfer between nearby ions whose excited levels are sufficiently different that a phonon must be included in the process to compensate for the energy mismatch.

The occurrence of nonresonant transfer among the Cr^{3+} ions in ruby was clearly demonstrated in the laser excitation studies of SELZER et al. /35/. Using a pulsed narrowband laser tuned to a frequency within the R_1 inhomogeneous line they observed the usual FLN signals (the sharp lines in Fig. 9) immediately after the excitation pulse.   As the narrow lines decreased in intensity with time a broad background, identical with the R_1 inhomogeneous lineshape, grew, as Fig. 9 shows.   This *spectral diffusion* of excitation from the FLN lines to the full inhomogeneous line is a manifestation of excitation transfer from the Cr^{3+} ions excited by the laser to the main body of Cr^{3+} ions.   This nonresonant transfer process was analyzed by HOLSTEIN, LYO, and ORBACH /36/.   The question of resonant transfer in ruby will be discussed in a later section.

6. Exchange-coupled Pairs of Cr^{3+} Ions

As the chromium concentration is increased the probability also increases that two Cr^{3+} ions in ruby may be situated sufficiently close to each other for a strong exchange interaction to occur between them.   Such an exchange-coupled

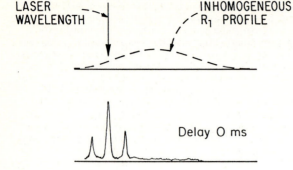

LASER WAVELENGTH

INHOMOGENEOUS R_1 PROFILE

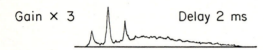

Delay 0 ms

Gain × 3   Delay 2 ms

Gain × 5   Delay 5 ms

Fig. 9. FLN signals from a 0.9 at.% ruby at 10 K. The laser excitation is on the high energy side of the line center. The delay indicates the time after the pulsed laser excitation (from Selzer et al./35/).

ion pair is a distinct spectroscopic center with distinct optical transitions. Some additional sharp luminescence lines are found in the vicinity of the R lines in ruby of medium to heavy doping, these were labelled N lines ("Nebenlinien") by Deutschbein /7/, and were correctly interpreted by SCHAWLOW, WOOD, and CLOGSTON /37/ as originating on exchange-coupled pairs of Cr^{3+} ions. In the Al_2O_3 lattice first-, second-, third-, and fourth-nearest Al^{3+} neighbors are sufficiently close for a strong exchange interaction to occur between Cr^{3+} ions occupying such near-neighbor sites. Thus there is a plethora of sharp lines due to exchange-coupled Cr^{3+} pairs in heavily-doped ruby. The process of identifying these lines with specific pair types was a formidable task which was addressed by a number of workers. The technique employed by MOLLENAUER and SCHAWLOW /38/ and by KAPLYANSKII and PRZHEVUSKII /39/ was to apply uniaxial stress along various crystallographic directions in ruby and observe the effects on the pair lines. The principle of these experiments is that when the stress is along the line joining the two ions of a specific pair the optical transitions on this pair are most strongly affected. In this way particular N lines can be ascribed to specific pair types.

When both Cr^{3+} ions of the pair are in the $^4A_2$ ground state the interaction between the ions can be written in the form

$$H_{EX} = -J \vec{S}_1 \cdot \vec{S}_2 + j(\vec{S}_1 \cdot \vec{S}_2)^2 . \qquad (2)$$

The first term is the bilinear exchange term, where J is the exchange integral whose value is determined by experiment. The second term

260

(biquadratic exchange) is small and arises from exchange striction effects. We will only concern ourselves with the values of J.

The eigenstates of this Hamiltonian can be classified according to the values of the total spin, $\vec{S} = \vec{S}_1 + \vec{S}_2$ of the two-ion system, and in the case of two-spin 3/2 ions S can have values 0, 1, 2, 3. If J is negative (positive) the exchange is antiferromagnetic (ferromagnetic). Fig. 10 shows the luminescence transitions on a fourth-nearest neighbor pair at 1.6 K. The pattern of lines corresponds to ferromagnetic coupling with J = 7.0 cm^{-1}.

Table 1 summarizes the findings for the first four types of exchange-coupled Cr^{3+} ion pairs in ruby. The values of dJ/dr for first- and second-nearest neighbors were obtained by HEBER and PLATZ /40/ from stress experiments. The data were used by them to compute values for J for the first- and second-nearest neighbors in the antiferromagnetic material Cr_2O_3. This has the same crystal structure as ruby, but the separations between ions are larger than in ruby. For the first-nearest neighbors the separation is 2.55 A in ruby and 2.65 A in Cr_2O_3. Taking the values of J and dJ/dr for ruby and using a linear extrapolation one estimates J = -78 cm^{-1} for the first-nearest neighbor pair in Cr_2O_3. The measured value is -86±6 cm^{-1}. One similarly calculates J = -30 cm^{-1} for the second-nearest neighbor pair in Cr_2O_3; the measured value is -38±6 cm^{-1}.

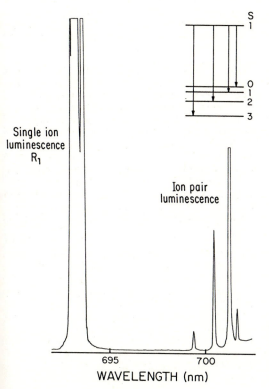

Fig. 10. Luminescence from a ruby of medium concentration at 2 K. The four weak lines originate on fourth-nearest neighbor ion pairs. The strongest pair line above is the N_2 line.

261

Table 1. Values of the exchange parameter for near-neighbor pairs in ruby

Neighbor type	Separation /A/	J /cm^{-1}/	dJ/dr /cm^{-1}A^{-1}/
1	2.55	-115	3.7×10^{-2}
2	2.78	-109	8.0×10^{-2}
3	3.18	-11.7	
4	3.50	+ 7.0	

7. More on Excitation Transfer

In their paper in which they identified the N lines as luminescence transitions on exchange-coupled pairs Varsanyi et al. /24/ pointed out that the intensity of the N lines was larger than one would expect from statistical considerations of the probability of the occurrence of pairs. They suggested that the pair lines might derive part of their intensity by energy transfer from single ions. That such a transfer occurs is clear from a study of the decay patterns of R and N lines after broadband pulsed excitation (IMBUSCH /41/); the R line decay is single exponential while the N line pattern has an initial fast decay rate (the intrinsic decay rate of the pairs) followed by a slow decay indistinguishable from the R line decay. The slowly-decaying component was attributed to excitation transfer from the single ions. A more direct demonstration of this transfer was provided by SELZER et al./42/ who used a pulse of narrowband laser excitation at the R_1 line frequency to excite single Cr^{3+} ions directly into the $\bar{E}(^2E)$ level in a ruby of 0.51 at.% chromium. After the laser pulse was extinguished R_1 line and N line luminescence was observed; the N_2 luminescence pattern had an initial rise followed by a decay whose rate strongly resembled the R_1 decay rate, precisely what is expected if the pairs derive excitation only by transfer from the single Cr^{3+} ions excited by the laser pulse. Figure 11, taken from a later study by SELZER et al./35/, shows these decay patterns.

Because of the strong resemblance between the R_1 decay pattern and the slowly-decaying part of the N_2 line decay pattern, the view was advanced that in heavily-concentrated ruby samples the pairs derived excitation not only from nearby excited single ions but from the main body of excited single ions and this came about through a strong *resonant* nonradiative transfer through the single ions. Not all research workers were agreed that such a strong resonant energy transfer occurs in ruby.

That a rapid nonradiative resonant transfer might be occurring among the Cr^{3+} ions prompted intense theoretical interest in the resonant transfer process in ruby since it might permit the observation of an *Anderson localization* phenomenon. This would show up when the chromium concentration was gradually decreased, as an *abrupt* change from a condition of excitation migration over a macroscopic portion of the ruby sample to a condition where the excitation was localized in microscopic sections of the sample. Indirect experiments to demonstrate the existence of an Anderson localization gave conflicting results and underlined the need for a more direct method of measuring the spatial extent of the nonradiative migration of chromium excitation.

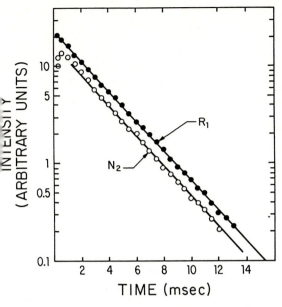

Fig. 11. Luminescence decay patterns of the R_1 and N_2 lines after pulsed laser excitation by a narrowband laser tuned to the R_1 transition (after Selzer et al./35/).

The first attempts to directly measure nonradiative spatial transfer employed the techniques of degenerate four-wave mixing (EICHLER et al./43/, LIAO et al./44/, HAMILTON et al./45/). In one such experiment two counter-propagating laser beams, $\vec{k}_1$ and $\vec{k}_2$, and a third beam, $\vec{k}_3$, making a small angle with $\vec{k}_1$, are directed at the ruby sample (Fig. 12). All the beams are derived from the same laser. The interference between beams $\vec{k}_1$ and $\vec{k}_3$ produces a spatial periodic variation in the density of excited Cr^{3+} ions which leads to a spatial periodic modulation in the refractive index in the ruby sample. This acts like a diffraction grating (Fig. 12) whose period is

$$d = \frac{\lambda}{2 \sin \frac{\theta}{2}} \cdot \tag{3}$$

where λ is the laser wavelength. Beam $\vec{k}_2$ is diffracted by this grating and by Bragg's law a diffracted beam ($\vec{k}_4$) will be observed along the $-\vec{k}_3$ direction. If beams $\vec{k}_1$ and $\vec{k}_3$ are turned off the gradual fading of the grating causes the intensity of the diffracted beam $\vec{k}_4$ to decrease in time as

$$I_4(t) = I_4(0) \exp(-t/\tau), \tag{4}$$

where

$$\frac{1}{\tau} = \frac{2}{\tau_r} + 2D.\left(\frac{4\pi}{\lambda} \sin \frac{\theta}{2}\right)^2, \tag{5}$$

τ_r is the radiative decay time of the Cr^{3+} ions, and D is the diffusion coefficient describing the transfer of excitation among the Cr^{3+} ions. All

263

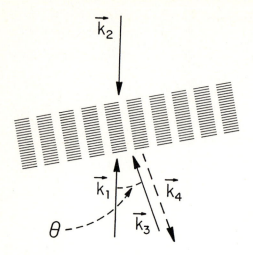

Fig. 12. Representation of the grating formed in a sample of ruby by laser beams $\vec{k}_1$ and $\vec{k}_3$. Laser beam $\vec{k}_2$ is diffracted by the grating, and a coherent diffracted beam is produced in the $\vec{k}_4$ direction.

the experiments concluded that there is no spatial energy diffusion within the limits of resolution; the upper limit for the diffusion was put at around 30 nm. This is still a large distance. What was needed was a much finer probe, and this was provided by the very ingenious experiment of CHU et al. /46/.

This experiment takes advantage of the inequivalence of the two types of Al^{3+} sites, labelled A and B in Fig. 2, of the α-Al_2O_3 lattice. Because of the existence of the odd-parity $V_{trig}^{(u)}$ crystal field term there is an internal electric field, E_0, (pointing along the optic axis), acting at each Al^{3+} site and this field acts in opposite directions on the Al^{3+} (or Cr^{3+}) ions in A and B sites. If an external electric field, E, is applied along the optic axis it adds to the internal field at one site but subtracts from the internal field at the other site. Because of the different resultant electric fields acting on them the Cr^{3+} ions in the A and B sites have their energies shifted relative to each other. The resultant linear *pseudo-Stark splitting*, which was originally discovered by KAISER, SUGANO, and WOOD /47/, is quite large. Thus, by applying an external electric field the R_1 lines from the A and B sites can be separated by about 1 cm^{-1}, enough to move the ions out of resonance with each other.

We can visualize the experiment of Chu et al. with the aid of Fig. 13. (a) With no external field the R_1 lineshapes of the A and B ions are identical. (b): With an external field applied the R_1 lines of the A and B ions are separated from each other. A fast pulse from a narrowband dye laser excites a subset of the A ions. An FLN signal is seen immediately after the pulse. This is indicated as a single line (shaded) in the figure, although in practice the FLN signal may consist of a number of sharp lines (because of ground state splitting). (c): The external field is switched off for a specified time; this brings the A and B ions back into resonance so that the excited A ions may transfer excitation to those B ions which are in resonance with them - if such a resonance transfer is possible. (d): The external field is switched on, again separating the A and B ions. One looks for a FLN signal from the B ions which is indicated by the second shaded component in the figure. The strength of this second component is a measure of the resonant transfer probability between the ions. The conclusion of Chu et al

264

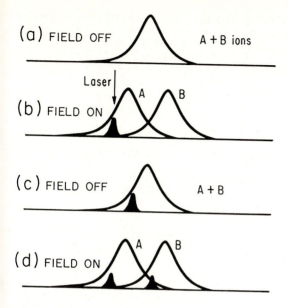

(a) FIELD OFF   A + B ions

Laser

(b) FIELD ON   A   B

(c) FIELD OFF   A + B

(d) FIELD ON   A   B

Fig. 13. Schematic representation of the separation of A and B ions by an electric field in the experiment of Chu et al. /46/

is that there is nonradiative resonant transfer but it is very slow, much slower than had been postulated. A similar study by JESSOP and SZABO /48/ gave the same result. Although these findings rule against the existence of a rapid resonance transfer between A and B sublattice ions do they also rule out rapid resonance transfer between Cr^{3+} ions within the A or B sublattices? MONTEIL and DUVAL /49/ claim that intrasublattice transfer can be rapid. However, the electric field results cited above, along with the theoretical analysis of GIBBS et al. /50/ are generally accepted as convincing evidence against a rapid resonance transfer among single Cr^{3+} ions in ruby.

Although in this instance ruby appears not to fulfil our expectations and exhibit rapid resonant excitation transfer and the much-sought Anderson transition, the theoretical and experimental investigations made in search of these elusive phenomena have greatly deepened our understanding of the general excitation transfer process. As a bonus, however, it was found in the course of the laser excitation and four-wave mixing experiments that when concentrated ruby was irradiated with an intense laser beam an internal electric field of high strength (MV/cm) was generated in ruby. This field manifests itself by causing a pseudo-Start splitting of the R lines /51,52/. Furthermore, this internal electric field persisted after the irradiation had ceased.

8. Conclusion

The impressive body of accumulated knowledge of the crystallographic and spectroscopic properties of ruby, the availability of large good-quality specimens, and the strong narrow easily excited luminescence lines have all contributed to the acceptance of ruby as an excellent material with which to test new spectroscopic techniques and to look for new spectroscopic phenomena. It has served us well in the past and continues to be an object of our investigations in new and interesting areas. For example, at the present time much attention is being focussed on the spectroscopic properties of amorphous and disordered materials. A recent interesting study in this area is the laser excitation experiments and detailed analysis reported by WASIELA, BLOCK, and MERLE D'AUBIGNE /53/ on $\alpha-Al_{(2-x)}Ga_{2x}O_3:Cr^{3+}$, which can be regarded as disordered ruby in that some of the Al^{3+} sites are randomly occupied by Ga^{3+} ions.

References

1. L.F. Mollenauer: Microwave Laboratory Report No. 1325, Stanford University (1964)
2. E. Becquerel: La Lumière - ses causes et ses effets (Librairie de Firmin Didot Freres, Fils et Cie, Imprimeurs de L'Institut, Paris, 1867)
3. L. de Boisbaudran: Comptes Rendus 103, 1107 (1886)
4. H. du Bois and G.J. Elias: Annalen der Physik 27, 233 (1908)
5. J. Becquerel: Phys. Zeit. 8, 932 (1907)
6. C.E. Mendenhall and R.W. Wood: Phil. Mag. 30, 316 (1915)
7. O. Deutschbein: Annalen der Physik 14, 712 (1932); 20, 828 (1934); Zeitschrift für Physik 77, 489 (1934)
8. H. Bethe: Annalen der Physik 3, 133 (1929)
9. Y. Tanabe and S. Sugano: J. Phys. Soc. Japan 9, 753 (1954); J. Phys. Soc. Japan 11, 864 (1956); Y. Tanabe and H. Kamimura: J. Phys. Soc. Japan 13, 394 (1958); S. Sugano and Y. Tanabe: J. Phys. Soc. Japan 13, 880 (1958); also S. Sugano, Y. Tanabe, and H. Kamimura: in Multiplets of Transition-Metal Ions in Crystals (Academic Press, New York 1976)
10. S. Sugano and I. Tsujikawa: J. Phys. Soc. Japan 13, 899 (1858)
11. A.A. Manenkov and A.M. Prokhorov: Sov. Phys. - JETP 1, 611 (1955)
12. G. Makhov, C. Kikuchi, J. Lambe, R.W. Terhune: Phys. Rev. 109, 1399 (1958
13. T.H. Maiman: Nature 187, 493 (1960)
14. A. Szabo: Phys. Rev. Lett. 25, 924 (1970)
15. T. Muramoto, S. Nakamishi, T. Hashi: Opt. Comm. 21, 139 (1977)
16. P.E. Jessop, T. Muramoto, A. Szabo: Phys. Rev. B21, 926 (1980)
17. A.L. Schawlow: In Advances in Quantum Electronics, ed. by J.R. Singer (Columbia University Press, New York and London 1961) p.50
18. D.E. McCumber and M.D. Sturge: J. Appl. Phys. 34, 1682 (1963)
19. K.S. Gibson: Phys. Rev. 8, 38 (1916)
20. A.L. Schawlow: J. Appl. Phys. Suppl. 33, 395 (1962)
21. P.E. Jessop and A. Szabo: Opt. Comm. 33, 301 (1980)
22. E. Hahn: Phys. Rev. 80, 580 (1950)
23. N.A. Kurnit, I.D. Abella and S.R. Hartmann, Phys. Rev. Lett. 13, 567 (1964
24. F. Varsanyi, D.L. Wood, and A.L. Schawlow: Phys. Rev. Lett. 3, 544 (1959)
25. D.E. Nelson and M.D. Sturge: Phys. Rev. 137A, 1117 (1965)
26. R. Adde, S. Geschwind, L.R. Walker: in Proceedings of the Fifteenth Colloque Ampere, ed. by P. Averback (North Holland, Amsterdam 1969) p.460
27. K.F. Renk and J. Peckenzell: J. Phys. (Paris) 33, C4 (1972)
28. J.I. Dijkhuis, A. van der Pol, H.W. de Wijn: Phys. Rev. Lett. 37, 1554 (1976)
29. R.S. Meltzer and J.E. Rives: Phys. Rev. Lett. 38, 421 (1977)
30. A.A. Kaplyanskii, S.A. Basun, V.A. Rachin, R.A. Titov: JETP Lett. 21, 200 (1975)
31. E.O. Schulz-du Bois: Bell System Technical Journal 38, 271 (1959)
32. S. Geschwind, R.J. Collins, A.L. Schawlow: Phys. Rev. Lett. 3, 545 (1959
33. S. Geschwind, G.E. Devlin, R.L. Cohen, S.R. Chinn: Phys. Rev. A137, 1087 (1965)
34. G.F. Imbusch and S. Geschwind: Phys. Rev. Lett. 17, 238 (1966)

35. P.M. Selzer, D.L. Huber, B.B. Barnett, W.M. Yen: Phys. Rev. B17, 4979 (1978)
36. T. Holstein, S.K. Lyo, R. Orbach: Phys. Rev. Lett. 36, 891 (1976)
37. A.L. Schawlow, D.L. Wood, A.M. Clogston: Phys. Rev. Lett. 3, 271 (1959)
38. L.F. Mollenauer and A.L. Schawlow: Phys. Rev. 168, 309 (1968)
39. A.A. Kaplyanskii and A.K. Przhevuskii: Sov. Phys. - Solid State 9, 190 (1967)
40. J. Heber and W. Platz: J. Luminescence 18/19, 170 (1979)
41. G.F. Imbusch: Phys. Rev. 153, 326 (1967)
42. P.L. Selzer, D.S. Hamilton, W.M. Yen: Phys. Rev. Lett. 38, 858 (1977)
43. H.J. Eichler, J. Eichler, J. Knof, C.H. Noak: Phys. Status. Solidi 52, 481 (1979)
44. P.F. Liao, L.M. Humphrey, D.M. Bloom, S. Geschwind: Phys. Rev. B20, 4145 (1979)
45. D.S. Hamilton, D. Heiman, J. Feinberg, R.W. Hellwarth: Opt. Lett. 4, 124 (1979)
46. S. Chu, H.M. Gibbs, S.L. McCall, A. Passner: Phys. Rev. Lett. 45, 1715 (1980)
47. W. Kaiser, S. Sugano, D.L. Wood: Phys. Rev. Lett. 6, 605 (1961)
48. P.E. Jessop and A. Szabo: Phys. Rev. Lett. 45, 1712 (1980)
49. A. Monteil and B. Duval: In Energy Transfer Processes in Condensed Matter, ed. by B. Di Bartolo (Plenum Press, New York and London 1984) p. 643
50. H.M. Gibbs, S. Chu, S.L. McCall, A. Passner: in Coherence and Energy Transfer in Glasses, ed. by P.A. Fleury and B. Golding (Plenum Press, New York and London 1984) p. 373.
51. P.F. Liao, A.M. Glass, L.M. Humphrey: Phys. Rev. B22, 2276 (1980)
52. S.A. Basum, A.A. Kaplyanskii, S.P. Feofilov: Z.E.T.F. 87, 2047 (1984)
53. A. Wasiela, Y. Merle d'Aubigne, D. Block: J. Luminescence 36, 11 (1986);
    A. Wasiela, D. Block, Y. Merle d'Aubigne: J. Luminescence 36, 24 (1986)

Four-Wave Mixing Spectroscopy
of Metastable Defect States in Solids

S.C. Rand

Hughes Research Laboratories, 3011 Malibu Canyon Road,
Malibu, CA 90265, USA

1. Introduction

Early four-wave mixing experiments[1] relied on the high power
available from pulsed lasers to produce third harmonic generation, field-induced second harmonic generation and a variety of
other weak processes which combined three optical waves to
yield a fourth. It was immediately recognized however that
electronic resonances of various kinds could enhance nonlinear
mixing of light waves. This was confirmed by a plasma
experiment[2] in which input beams of frequencies ω_1, ω_2 and ω_3
produced an enhanced intensity of the four-wave mixing output
wave at $\omega_4 = \omega_1 + \omega_2 - \omega_3$ when $\omega_2 - \omega_3$ and $k_2 - k_3$ were adjusted to match
the plasma resonance frequency and wave vector. Experiments
followed which showed that intermediate electronic states[3],
as well as Raman[4] resonances enhanced third order mixing.
Resonant enhancement soon permitted narrowband, continuous-wave
lasers to be used successfully for degenerate four-wave
mixing[5] ($\omega_4 = \omega_3 = \omega_2 = \omega_1$) and coherent anti-Stokes Raman (CARS)
generation in liquids[6] and gases[7], which demonstrated the
high spectral resolution capability and sensitivity of these
techniques. Also, important applications of four-wave mixing
which were non-spectroscopic in nature were found, including
phase conjugation and amplified reflection[8,9,10].

From the beginning, applications of four-wave mixing in
solids were equally varied. Numerous investigations were
carried out to obtain spectroscopic information on Raman[11],
libron[12], phonon[13], vibron[14], and polariton[15]
excitations. Continuous-wave degenerate four-wave mixing due
to saturated absorption was discovered in ruby[16]. In
dielectrics, transient grating techniques were developed to
measure the rate of spatial migration of electronic excitation
over distances comparable to the wavelength of light[17], as
well as mechanisms of energy transfer between impurities[18].
Other workers determined fast excited state relaxation[19] and
coherence dephasing[14] times for localized and delocalized[20]
excitations. In semiconductors, similar studies led to time-
resolved studies of picosecond carrier dynamics[21], carrier
concentration[22] and nonlinear spectroscopy of excitons[23].
Also, Landau levels in InSb were observed to produce Raman-like
resonances in four-wave mixing processes[24].

In gases, spectroscopic studies based on four-wave mixing
rapidly developed into high resolution, Doppler-free

echniques[25]. In condensed matter however, initial studies
ere not able to exploit the high spectral resolution
apabilities of the new methods, even after continuous-wave,
ackward-wave generation was observed[16]. The narrowest
features encountered in solids, for example the zero phonon
ransitions of rare earth ions in crystals cooled to liquid
elium temperatures, were typically ~5 GHz wide. It therefore
eemed pointless to develop and apply methods in solids with
esolution in excess of ~1 GHz. However, when a method was
ound to tune the relative frequency of input beams derived
rom a single laser source while preserving the correlation of
heir frequency fluctuations, extremely narrow resonances were
nexpectedly discovered which did justify high resolution
ethodology[27]. Features as narrow as a few Hertz were
bserved in a variety of crystals and were shown to provide
ccurate measurements of very long decay times from metastable
tates of impurities[28] and color centers[29] in solids.

In this article the theoretical and experimental basis for
tudying relaxation processes of defect metastable states in
olids by this nearly degenerate four-wave mixing (NDFWM)
echnique are reviewed. Results are presented for defects
omprising transition metal ions, rare earth ions and color
enters in a variety of media. Although the method works
qually well in photorefractive media, results of experiments
e have performed in $LiNbO_3$ and $BaTiO_3$ are not discussed here.
he basic correspondence between theory and experiment is
erified for the simple three-level systems formed by dilute Cr
ons in $YAlO_3$ and Al_2O_3, and for F_2 color centers in LiF.
inewidth and saturation measurements in $Nd:\beta''$-Na-Alumina
eveal a transition from closed to open quantum system behavior
s Nd density increases. For N3 color centers in diamond,
DFWM spectroscopy provides the first evidence for an unknown
eep level of this structure. The results as a whole show that
his method provides information about metastable states which
an be quite different from that provided by direct
luorescence or phosphorescence measurements. Temporal decay
xperiments measure depopulation of metastable excited states
hereas high resolution NDFWM spectroscopy measures
epopulation kinetics of the ground state. NDFWM is therefore
ensitive to dark processes and system saturation. As shown in
his paper, an understanding of ground state behavior
omplements the picture furnished by radiative decay methods,
iving either additional or completely new information on the
nteraction of light with defects in condensed media.

. Theory

early degenerate four-wave mixing resonances arise from the
nteraction of two pump beams at frequency ω with an
ndependently tunable probe beam at frequency $\omega+\delta$ through the
hird order susceptibility $\chi^{(3)}$ of a medium. The geometry of
he light beams is shown in Fig. 1. The pump beams are chosen
o be counter-propagating so as to guarantee phase-matching,
ndependent of incident probe direction. They create a
tanding wave excitation with which the forward-going probe
nteracts to generate an output wave radiated in the backward
irection, as dictated by momentum conservation. By energy

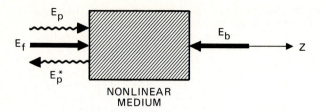

Fig.1. Geometry of the light beams assumed in nearly degenerate
four-wave mixing calculations. Forward and backward pump waves
E_f and E_b are counter-propagating along z so that a collinear
probe wave E_p gives rise to a collinear, conjugate signal E_p^*.

conservation the frequency of the fourth or signal wave is $\omega-\delta$
and it is phase conjugate to the probe.

Although there are 48 possible permutations of the light
fields, conjugate waves and collision events in the most
general formulation of four-wave mixing[30], we assume here
that only one or two terms dominate the mixing. Other
nonlinear processes in the sample are ignored. By virtue of
the geometry of the experiment and the near degeneracy of the
input frequencies, this simplification is well justified in the
study of many systems because only one or two terms are phase-
matched and therefore strong. Accidental occurrence of states
with energies near 2ω or 3ω above the ground state or the
generation of magnetic sub-level coherences can naturally
complicate the simple picture presented here.

In Sections 2.1 and 2.2, perturbation theory for NDFWM in 3-
and 4-level systems illuminated continuously by weak,
monochromatic beams is developed. The frequency of the pump
waves is assumed to be in resonance with an allowed one-photon
transition to an excited state which is connected by fast
relaxation processes to some metastable state of interest.
This state creates a population bottleneck, leading to
saturation of ground state absorption, generation of a large
third order susceptibility and efficient four-wave mixing at
zero pump-probe detuning. The NDFWM spectrum, obtained by
tuning the probe frequency relative to that of the pumps, is
shown to consist of a single Lorentzian-shaped resonance with a
width numerically related to the inverse lifetime of the
metastable state.

The proper interpretation of the NDFWM spectrum is, however,
that its width reflects the decay time of ground state spatial
hole-burning, since the resonance first appears in the second
order perturbation expression for the ground state density
matrix element, $\rho_{11}^{(2)}$. It does not appear in any other matrix
elements in second order. The NDFWM spectrum therefore
measures a repopulation rate, something quite different from
the customary radiative relaxation or population decay times of
excited states. Direct measurements of relaxation are
therefore possible on states which emit no light. The
technique should also be sensitive to the onset of non-

radiative processes which do not affect the population decay time of radiating metastable states, a point discussed further in Sections 3 and 4.

In Section 2.3, a non-perturbative approach to the theory is presented which extends previous work[31] on saturation effects due to pump waves of arbitrary intensity to 3-level systems. Power-broadened NDFWM spectra are shown to furnish useful information for the accurate determination of saturation intensity in homogeneous or inhomogeneously broadened systems. Results presented in Section 3 illustrate this application in an inhomogeneously broadened material.

2.1 Perturbation Theory: 3-Level System

In Fig.2 we show the 3-level model to be considered in this section. Absorption on the $|1\rangle$-$|2\rangle$ transition at frequency ω is followed by decay either to the ground state or to metastable state $|3\rangle$. The associated relaxation rates are given in the figure and lead to the following starting equations for the density matrix.

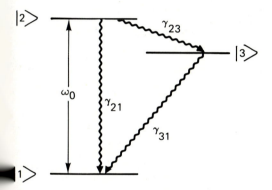

Fig.2. Schematic diagram of a 3-level system with an allowed one-photon absorption at ω_0. The γ_{ij} refer to relaxation rates from state i to state j.

$$i\hbar\frac{d}{dt}\rho = [H,\rho] + [V,\rho] + i\hbar\frac{d}{dt}\rho\Big|_{decay} \tag{2.1.1}$$

$$i\hbar\frac{d}{dt}\rho_{11} = (V_{12}\rho_{21} - \rho_{12}V_{21}) + i\hbar\gamma_{31}\rho_{33} + i\hbar\gamma_{21}\rho_{22} \tag{2.1.2}$$

$$i\hbar\frac{d}{dt}\rho_{22} = (V_{12}\rho_{21}\ \rho_{12}V_{21}) - i\hbar(\gamma_{21}+\gamma_{23})\rho_{22} \tag{2.1.3}$$

$$i\hbar\frac{d}{dt}\rho_{33} = i\hbar\gamma_{23}\rho_{22} - i\hbar\gamma_{31}\rho_{33} \tag{2.1.4}$$

$$i\hbar\frac{d}{dt}\rho_{12} = -\hbar\omega_0\rho_{12} + (V_{12}\rho_{22} - \rho_{11}V_{12}) + i\hbar\Gamma_{21}\rho_{12} \tag{2.1.5}$$

$$\rho_{21} = \rho_{12}^{*} \qquad (2.1.6)$$

Here H is the atomic Hamiltonian in the absence of light, $\Gamma_{21} = (\gamma_{21} + \gamma_{23})/2 + \gamma_{ph}$ and V is the optical perturbation given by

$$(V_{12})_j = -\frac{1}{2}\mu_{12}(E_j e^{i(\omega_j t - k_j z)} + c.c.), \qquad (2.1.7)$$

where the subscript $j = f, b, p$ denotes the forward pump, backward pump and probe waves respectively. Let

$$\Omega_j = \mu_{12}E_j/2h, \qquad (2.1.8)$$

$$\phi_j = \omega_j t - k_j z, \qquad (2.1.9)$$

and

$$\Delta_j = \omega_j - \omega_\emptyset. \qquad (2.1.10)$$

Because there are three distinct incident fields, we proceed to third order in the perturbation sequence. According to (2.1.2) -(2.1.5) a third-order off-diagonal matrix element $\rho_{12}^{(3)}$ can only be obtained from second-order elements $\rho_{11}^{(2)}$ and $\rho_{22}^{(2)}$ which are in turn derived from a first order $\rho_{12}^{(1)}$. From (2.1.5) we find directly that

$$(\rho_{12}^{(1)})_j = \frac{\Omega_j}{-\Delta_j + i\Gamma_{21}} e^{i\phi_j} \rho_{11}^{(0)}. \qquad (2.1.11)$$

Using (2.1.11) and (2.1.3) we obtain

$$i\hbar\frac{d}{dt}(\rho_{22}^{(2)})_{jk} = -\hbar \sum_{j,k} \Omega_j^* e^{-i\phi_j} e^{i\phi_k} \Omega_k (\frac{\Omega_k}{-\Delta_k + i\Gamma_{21}})\rho_{11}^{(0)}$$

$$+ \hbar \sum_{j,k} \Omega_j e^{-i\phi_k} e^{i\phi_j} \frac{\Omega_k^*}{(-\Delta_k - i\Gamma_{21})}\rho_{11}^{(0)} - i\hbar(\gamma_{21} + \gamma_{23})\rho_{22}^{(2)}. \qquad (2.1.12)$$

Setting $\gamma_2 = \gamma_{21} + \gamma_{23}$ and retaining only phase-matched terms which contain conjugate wave Ω_p^* we find

$$\rho_{22}^{(2)} = \frac{\rho_{11}^{(0)}}{\delta + i\gamma_2} [\Omega_p^* \Omega_f e^{-i(\phi_p - \phi_f)}(\frac{1}{-\Delta_p - i\Gamma_{21}} - \frac{1}{-\Delta_f + i\Gamma_{21}})$$

$$+ \Omega_p^* \Omega_b e^{-i(\phi_p - \phi_b)}(\frac{1}{-\Delta_p - i\Gamma_{21}} - \frac{1}{-\Delta_b + i\Gamma_{21}})]. \qquad (2.1.13)$$

Substituting (2.1.13) and the solution for $\rho_{33}^{(2)}$ in terms of $\rho_{22}^{(2)}$ from (2.1.4) into (2.1.2) we also find

272

$$\rho_{11}^{(2)} = \frac{\rho_{11}^{(0)}}{\gamma_2 - \gamma_{31}} \left(\frac{\gamma_{23}}{\delta + i\gamma_{31}} + \frac{\gamma_{21} - \gamma_{31}}{\delta + i\gamma_2} \right)$$

$$\cdot \left[-\Omega_p^* \Omega_f e^{-i(\phi_p - \phi_f)} \left(\frac{1}{-\Delta_p - i\Gamma_{21}} - \frac{1}{-\Delta_f + i\Gamma_{21}} \right) \right.$$

$$\left. -\Omega_p^* \Omega_b e^{-i(\phi_p - \phi_b)} \left(\frac{1}{-\Delta_p - i\Gamma_{21}} - \frac{1}{-\Delta_b + i\Gamma_{21}} \right) \right]. \qquad (2.1.14)$$

Notice that resonant denominator $(\delta + i\gamma_{13})^{-1}$ appears in the ground state element $\rho_{11}^{(2)}$, but not in $\rho_{22}^{(2)}$. Finally, we use (2.1.13) and (2.1.14) in (2.1.5) to obtain

$$\rho_{12}^{(3)} = \rho_{11}^{(0)} \left[\left(1 + \frac{\gamma_{21} - \gamma_{31}}{\gamma_2 - \gamma_{31}} \right) \left(\frac{1}{\delta + i\gamma_2} \right) + \left(\frac{1}{\gamma_2 - \gamma_{31}} \right) \left(\frac{1}{\delta + i\gamma_{31}} \right) \right]$$

$$\cdot \left[\frac{\Omega_p^* \Omega_f \Omega_b e^{i(\omega - \delta)t + ikz}}{(-\Delta_b + \delta + i\Gamma_{21})} \left(\frac{1}{-\Delta_p - i\Gamma_{21}} - \frac{1}{-\Delta_f + i\Gamma_{21}} \right) \right.$$

$$\left. + \frac{\Omega_p^* \Omega_b \Omega_f e^{i(\omega - \delta)t + ikz}}{(-\Delta_f + \delta + i\Gamma_{21})} \left(\frac{1}{-\Delta_p - i\Gamma_{21}} - \frac{1}{-\Delta_b + i\Gamma_{21}} \right) \right]. \qquad (2.1.15)$$

In this derivation we implicitly assume an optically thin sample, since field amplitudes throughout the sample are taken to be constant. From (2.1.15) it is clear that for excitation on resonance (Δ-0), a system with large homogeneous broadening ($\Gamma_{21} \gg \gamma_{31}, \delta$) and a slow decay rate from $|3\rangle$ to ground ($\gamma_{31} \ll \gamma_{21}, \gamma_{23}$) exhibits an NDFWM spectrum which is just a single Lorentzian peak with a full width at half maximum (FWHM) equal to γ_{31}.

2.2 Perturbation Theory: 4-Level System

If a fourth level is considered, as shown in Fig.3, and it is assumed that $\gamma_{31} = \gamma_{24} = 0$, we start with the equations below.

$$i\hbar \frac{d}{dt}\rho_{11} = (V_{12}\rho_{21} - \rho_{12}V_{21}) + i\hbar\gamma_{21}\rho_{22} + i\hbar\gamma_{41}\rho_{44} \qquad (2.2.1)$$

$$i\hbar \frac{d}{dt}\rho_{22} = -(V_{12}\rho_{21} - \rho_{12}V_{21}) - i\hbar(\gamma_{21} + \gamma_{23})\rho_{22} \qquad (2.2.2)$$

$$i\hbar \frac{d}{dt}\rho_{33} = i\hbar\gamma_{23}\rho_{22} - i\hbar\gamma_{34}\rho_{33} \qquad (2.2.3)$$

$$i\hbar \frac{d}{dt}\rho_{44} = i\hbar\gamma_{34}\rho_{33} - i\hbar\gamma_{41}\rho_{44} \qquad (2.2.4)$$

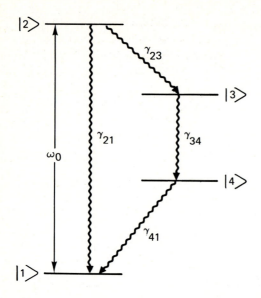

Fig.3. Schematic diagram of a 4-level system. The decay pathways are restricted by the assumption that $\gamma_{31}=\gamma_{24}=0$.

$$i\hbar\frac{d}{dt}\rho_{12}=-\hbar\omega_0\rho_{12}+V_{12}(\rho_{22}-\rho_{11})-i\hbar\Gamma_{21}\rho_{12} \qquad (2.2.5)$$

(2.2.3) and (2.2.4) may be solved for ρ_{44} in terms of ρ_{22} whereupon the solution for $\rho_{12}{}^{(3)}$ proceeds exactly as in the previous section with the result that

$$\rho_{12}^{(3)}=\rho_{11}^{(0)}\{(\frac{1}{\delta+i\gamma_2})(1+[\gamma_{34}\gamma_{41}(\gamma_{34}-\gamma_{41})+(\gamma_2\gamma_{41}-\gamma_{41}\gamma_{23})(\gamma_{41}-\gamma_2)$$

$$+(\gamma_2\gamma_{34}-\gamma_{34}\gamma_{23})(\gamma_2-\gamma_{34})]/D)$$

$$+(\frac{1}{\delta+i\gamma_{34}})\gamma_{23}\gamma_{41}(\gamma_{41}-\gamma_2)/D$$

$$+(\frac{1}{\delta+i\gamma_{41}})\gamma_{23}\gamma_{34}(\gamma_2-\gamma_{34})/D\}$$

$$\cdot[\frac{\Omega_p^*\Omega_f\Omega_b e^{i(\omega-\delta)t+ikz}}{(-\Delta_b+\delta+i\Gamma_{21})}(\frac{1}{-\Delta_p-i\Gamma_{21}}-\frac{1}{-\Delta_f+i\Gamma_{21}})$$

$$+\frac{\Omega_p^*\Omega_b\Omega_f e^{i(\omega-\delta)t+ikz}}{(-\Delta_f+\delta+i\Gamma_{21})}(\frac{1}{-\Delta_p-i\Gamma_{21}}-\frac{1}{-\Delta_b+i\Gamma_{21}})]\,, \qquad (2.2.6)$$

where $D=\gamma_{34}\gamma_{41}(\gamma_{34}-\gamma_{41})+\gamma_2\gamma_{41}(\gamma_{41}-\gamma_2)+\gamma_2\gamma_{34}(\gamma_2-\gamma_{34})\,.$

Notice that the presence of the fourth level introduces an additional resonance with a width γ_{41}, which corresponds to the decay rate from the new level. However, if only state $|3\rangle$ is metastable, a single resonance again dominates. That is, if $\gamma_{34} \ll \gamma_{21}, \gamma_{23}, \gamma_{41}$ the NDFWM spectrum consists of a single Lorentzian with a width (FWHM) of γ_{34}. However, anticipating results for the N3 color center given in Section 3.2, we remark that if $|4\rangle$ is also metastable and $\gamma_{41} < \gamma_{34} \ll \gamma_{21}, \gamma_{23}$ the resonance due to $|4\rangle$ will dominate and the NDFWM linewidth will be narrower than γ_{34}.

2.3 NDFWM Theory Including Saturation Effects

In this section we extend earlier calculations of NDFWM[31] to show that for low intensities power-broadening in a 3-level system is linear in intensity for the homogeneous case but varies with the square root of incident intensity in inhomogeneous systems. This broadening occurs at intensities far below those necessary to cause AC Stark shifting of the levels and is therefore not accompanied by AC Stark sidebands, although the formalism incorporates the AC Stark effect. Even in the absence of AC Stark effects, when pump waves approach the 3-level saturation intensity, the perturbation theory presented in Sections 2.1 and 2.2 understandably fails. It is then necessary to take a different approach, one in which the pump waves are accounted for exactly in the first step of the calculation. The probe and signal waves are subsequently added in as perturbations in the second step.

The zero-order optical interaction Hamiltonian with equal amplitude pump waves is taken to be

$$V^{(0)} = -\frac{1}{2} \mu_{12} E_0 (e^{ikz} + e^{-ikz}) e^{-i\omega t} + c.c. \tag{2.3.1}$$

while the first order term contains only probe and signal waves

$$V^{(1)} = -[\frac{1}{2} \mu_{12} E_1 (e^{ik_1 z - i\omega_1 t}) + \frac{1}{2} \mu_{12} E_2 (e^{-ik_2 z - i\omega_2 t})] + c.c. \tag{2.3.2}$$

The probe wave is assumed to have frequency $\omega_1 = \omega + \delta$ as before, and the signal wave frequency is $\omega_2 = \omega - \delta$. To develop the starting equations for each order, we set

$$\rho = \rho^{(0)} + \lambda \rho^{(1)}, \tag{2.3.3}$$

$$V = V^{(0)} + \lambda V^{(1)}, \tag{2.3.4}$$

where λ is a perturbation series parameter. We use (2.3.3) and (2.3.4) in (2.1.1) to find

$$i\hbar \frac{d}{dt} \rho^{(0)} = [H, \rho^{(0)}] + [V^{(0)}, \rho^{(0)}] + i\hbar \frac{d}{dt} \rho^{(0)} \Big|_{decay}, \tag{2.3.5}$$

$$i\hbar \frac{d}{dt} \rho^{(1)} = [H, \rho^{(1)}] + [V^{(1)}, \rho^{(0)}] + [V^{(0)}, \rho^{(1)}] + i\hbar \frac{d}{dt} \rho^{(1)} \Big|_{decay} \tag{2.3.6}$$

275

with

$$V_{21}^{(0)} = -\hbar\Omega_{21}(e^{ikz}+e^{-ikz})e^{i\omega t} + \text{c.c.} , \qquad (2.3.7)$$

$$V_{21}^{(1)} = -\hbar\Omega_{21}(\epsilon_1 e^{i(k_1 z-\omega_1 t)} + \epsilon_2 e^{-i(k_2 z+\omega_2 t)}) + \text{c.c.} , \quad (2.3.8)$$

where $\epsilon_1 = E_1/E_\emptyset$, $\epsilon_2 = E_2/E_\emptyset$ and $\Omega_{21} = \mu_{21}E_o/2\hbar$. Making use of the rotating-wave approximation (RWA), we drop complex conjugate terms in $(2.3.7)-(2.3.8)$. The zero-order starting equations for a 3-level system are

$$i\hbar\frac{d}{dt}\rho_{11}^{(0)} = (V_{12}^{(0)}\rho_{21}^{(0)} - \rho_{12}^{(0)}V_{21}^{(0)}) + i\hbar\gamma_{21}\rho_{22}^{(0)} + i\hbar\gamma_{31}\rho_{33}^{(0)} , \quad (2.3.9)$$

$$i\hbar\frac{d}{dt}\rho_{22}^{(0)} = -(V_{12}^{(0)}\rho_{21}^{(0)} - \rho_{12}^{(0)}V_{21}^{(0)}) - i\hbar(\gamma_{21}+\gamma_{23})\rho_{22}^{(0)} , \quad (2.3.10)$$

$$i\hbar\frac{d}{dt}\rho_{33}^{(0)} = i\hbar\gamma_{23}\rho_{22}^{(0)} - i\hbar\gamma_{31}\rho_{33}^{(0)} , \quad (2.3.11)$$

$$i\hbar\frac{d}{dt}\rho_{21}^{(0)} = \hbar\omega_0\rho_{21}^{(0)} + V_{21}^{(0)}(\rho_{11}^{(0)} - \rho_{22}^{(0)}) - i\hbar\Gamma_{12}\rho_{21}^{(0)} , \quad (2.3.12)$$

$$\rho_{11}^{(0)} + \rho_{22}^{(0)} + \rho_{33}^{(0)} = N . \quad (2.3.13)$$

Proceeding as in Section 2.1, we use $(2.3.9)-(2.3.13)$ to obtain

$$\rho_{21}^{(0)} = \frac{-iN(V_{21}^{(0)}/\hbar)(+i\Delta+\Gamma_{12})}{\Delta^2 + \Gamma_{12}^2[1+(1+[\gamma_{23}/2\gamma_{31}])|\Omega_{21}|^2)/\Gamma_{12}\gamma_2]} , \quad (2.3.14)$$

$$\rho_{11}^{(0)} - \rho_{22}^{(0)} = \frac{N(\Delta^2 + \Gamma_{12}^2)}{\Delta^2 + \Gamma_{12}^2 + |\Omega_{21}|^2(1+[\gamma_{23}/2\gamma_{31}])\Gamma_{12}/\gamma_2} . \quad (2.3.15)$$

The first-order starting equations are

$$i\hbar\frac{d}{dt}\rho_{11}^{(1)} = (V_{12}^{(0)}\rho_{21}^{(1)} - \rho_{12}^{(1)}V_{21}^{(0)}) + (V_{12}^{(1)}\rho_{21}^{(0)} - \rho_{12}^{(0)}V_{21}^{(1)})$$

$$+ i\hbar\gamma_{21}\rho_{22}^{(1)} + i\hbar\gamma_{31}\rho_{33}^{(1)} , \quad (2.3.16)$$

$$i\hbar\frac{d}{dt}\rho_{22}^{(1)} = -(V_{12}^{(0)}\rho_{21}^{(1)} - \rho_{12}^{(1)}V_{21}^{(0)}) - (V_{12}^{(1)}\rho_{21}^{(0)} - \rho_{12}^{(0)}V_{21}^{(1)})$$

$$- i\hbar\gamma_{21}\rho_{22}^{(1)} - i\hbar\gamma_{23}\rho_{22}^{(1)} , \quad (2.3.17)$$

$$i\hbar\frac{d}{dt}\rho_{33}^{(1)} = i\hbar\gamma_{23}\rho_{22}^{(1)} - i\hbar\gamma_{31}\rho_{33}^{(1)} , \quad (2.3.18)$$

$$i\hbar\frac{d}{dt}\rho_{21}^{(1)} = \hbar\omega_0\rho_{21}^{(1)} + V_{21}^{(0)}(\rho_{11}^{(1)} - \rho_{22}^{(1)}) + V_{21}^{(1)}(\rho_{11}^{(0)} - \rho_{22}^{(0)}) - i\hbar\Gamma_{12}\rho_{21}^{(1)}$$

$$\tag{2.3.19}$$

$$\rho_{11}^{(1)} + \rho_{22}^{(1)} + \rho_{33}^{(1)} = 0. \tag{2.3.20}$$

Because frequencies are no longer degenerate in first order, complex conjugates must be retained in expressions for purely real, diagonal elements of the density matrix.

$$\rho_{11}^{(1)} = \frac{1}{2}(\tilde{\rho}_{11}^{(1)}e^{i\delta t} + \text{c.c.}) \tag{2.3.21}$$

$$\rho_{22}^{(1)} = \frac{1}{2}(\tilde{\rho}_{22}^{(1)}e^{i\delta t} + \text{c.c.}) \tag{2.3.22}$$

$$\rho_{33}^{(1)} = \frac{1}{2}(\tilde{\rho}_{33}^{(1)}e^{i\delta t} + \text{c.c.}) \tag{2.3.23}$$

The off-diagonal matrix element is

$$\rho_{21}^{(1)} = \frac{-iN\Omega_{21}|\Omega_{21}|^2}{1+I'}(e^{ikz} + e^{-ikz})^2\{L_2L_3e^{-i\omega_2 t}(2+\frac{\gamma_{23}}{\gamma_{31}-i\delta})$$

$$\cdot \ [(L+L_1^*)\epsilon_1^*e^{-ik_1 z} + \text{c.c} + (L+L_2)\epsilon_2 e^{-ik_2 z} + \text{c.c}] + L_1L_3^*e^{-i\omega_1 t}$$

$$\cdot \ (2+\frac{\gamma_{23}}{\gamma_{31}+i\delta})[(L^*+L_1)\epsilon_1 e^{ik_1 z} + \text{c.c.} + (L^*+L_2^*)\epsilon_2 e^{ik_2 z} + \text{c.c.}]\}$$

$$+ \frac{iN\Omega_{21}}{1+I'}\{\epsilon_1 e^{i(k_1 z - \omega_1 t)}L_1 + L_2\epsilon_2 e^{-i(k_2 z + \omega_2 t)}\}, \tag{2.3.24}$$

where the Lorentzian resonant factors are

$$L = (\Gamma_{21} - i\Delta)^{-1}, \tag{2.3.25}$$

$$L_1 = (\Gamma_{21} - i\Delta - i\delta)^{-1}, \tag{2.3.26}$$

$$L_2 = (\Gamma_{21} - i\Delta + i\delta)^{-1}, \tag{2.3.27}$$

$$L_3 = (\gamma_2 + \frac{\gamma_2\gamma_{31}}{2\gamma_{31} + \gamma_{23}}(2+\frac{\gamma_{23}}{\gamma_{31} - i\delta})I' + i\delta)^{-1}, \tag{2.3.28}$$

and $I' = \dfrac{2|\Omega_{21}|^2(2\gamma_{31} + \gamma_{23})}{\gamma_2\gamma_{31}}(1+\cos 2kz)(L+L_1^*) \doteq 2(1+\cos 2kz)I/I_{\text{sat}}.$

Dropping all but the phase-matched term, we get

$$\rho_{21}^{(1)} = \frac{-iN\Omega_{21}|\Omega_{21}|^2}{1 + I'} (e^{ikz} + e^{-ikz})^2 L_2 L_3$$

$$\cdot \; (2 + \frac{\gamma_{23}}{\gamma_{31} - i\delta}) \epsilon_1^* e^{-i(k_1 z + \omega_2 t)} \quad (L + L_1^*) \qquad (2.3.29)$$

The saturation behavior of the microscopic polarization is contained in L_3. Its magnitude squared varies as

$$|L_3|^2 = (\frac{1}{\gamma_{31}^2 + \delta^2}) \; \gamma_2^2 \; (\gamma_{31}^2 \; [1+I']^2 + \delta^2), \qquad (2.3.30)$$

where we have assumed $\gamma_2 \gg \gamma_{31}, \delta$.

For optically thin samples, we calculate signal intensity by using (2.3.29) as the nonlinear polarization source term in Maxwell's equations. This procedure[32] performs the appropriate spatial average of the sinusoidal distribution of pump light intensity in (2.3.30). This merely changes the intensity-dependent lineshape in (2.3.30) for the homogeneous case from $|\rho_{21}^{(1)}|^2 \; \alpha \; (\gamma_{31}^2[1+I/I_{sat}]^2 + \delta^2)^{-1}$ to

$$I_{sig} \; \alpha \; (\gamma_{31}^2[1+4I/I_{sat}]^2 + \delta^2)^{-1}. \qquad (2.3.31)$$

For inhomogeneous 3-level systems, an additional integration of $\rho_{21}^{(1)}$ over the frequency distribution of the absorption line is necessary. The result for a Gaussian distribution is that $|\rho_{21}^{(1)}|^2 \; \alpha \; (\gamma_{31}^2[1+I/I_{sat}] + \delta^2)^{-1}$. After spatial averaging this gives

$$I_{sig} \; \alpha \; (\gamma_{31}^2[1+4I/I_{sat}] + \delta^2)^{-1}, \qquad (2.3.32)$$

at low intensities. These results are illustrated in Fig. 4.

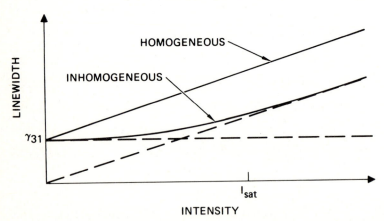

Fig.4. Broadening of the 3-level NDFWM Lorentzian linewidth as a function of incident intensity for homogeneous and inhomogeneous systems.

3. Experiments

3.1 Rare Earth and Transition Metal Dopants

Two sets of results are presented in this section. First, data on Cr ions in $YAlO_3$ and Al_2O_3 is used to verify the basic theory of Section 2.1, establishing the foundation of the high resolution NDFWM technique. Second, the interesting case of trivalent Nd ions in an unusual crystalline phase of alumina is discussed. β''-Na-Alumina accepts higher densities of rare earth ions than other materials without experiencing fluorescence quenching of the dopant. Curiously, the saturation intensity of Nd:β''-Na-Alumina is found to depend strongly on Nd density. NDFWM experiments reveal that this is due to the onset of non-radiative pair interactions between Nd neighbors which cause little change in upper state lifetime but increase the ground state decay time substantially.

The experimental configuration is depicted in Fig.5. Measurements on Cr:$YAlO_3$ were performed using a dye laser to irradiate a 3.9 mm-thick sample doped with approximately 0.05% Cr and exhibiting an optical density of 0.29 at 570 nm. The crystal was positioned with the c axis parallel to the counter-propagating pump beams. The probe beam was nearly collinear with the forward pump beam and intersected the pump beams in the sample at an angle of 0.2°. As shown in the figure, two acousto-optic modulators were used to synthesize the appropriate wavelengths for the pump and probe beams and to permit tuning through zero offset frequency. The two synthesizers were phase-locked together and stable to better than 1 Hz throughout the duration of each experiment.

Typically, ω_1 was held fixed at 40 MHz offset, and ω_2 was scanned over a range of a few hundred Hertz centered at 40 MHz offset frequency. The fundamental laser frequency was tuned near 570 nm for peak signal from the sample. The probe beam

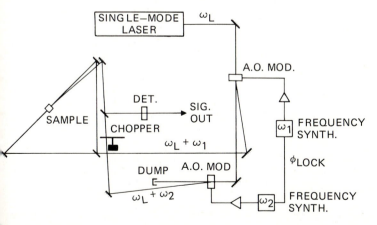

Fig.5. Experimental configuration for NDFWM spectroscopy using synthesized frequencies for the three input beams.

was chopped at a low frequency and the signal was detected phase-sensitively with a lock-in amplifier. The signal was recorded on a signal averager and then sent to a VAX 11-780 computer for analysis. The measurements were confirmed to be independent of intensity in the range 1-20 W/cm^2.

Figure 6 shows the NDFWM signal in dilute $Cr:YAlO_3$ versus pump-probe detuning δ together with the best fit theoretical curve and the ^2E fluorescence decay curve. When the incident lasers are tuned to the $^4A_2$-$^4T_2$ absorption, Cr^{3+} forms a simple 3-level system with $|1>=^4A_2$, $|2>=^4T_2$ and $|3>=^2E$. It is not surprising then to find that the observed Lorentzian linewidth is 9.8 Hz, in excellent agreement with the fluorescence decay data which yields γ_{31}^{-1}= 33±1 ms or a linewidth of 9.7 Hz. There is a single pathway open for relaxation from the metastable state to ground, so agreement with the theory of Section 2.1 is expected. Similarly, for $Cr:Al_2O_3$ in which the fluorescence decay is somewhat shorter, the observed NDFWM spectral width of 138 Hz agrees well with the fluorescence measurements (τ_{31}=2.6 ms or $\Delta\nu$=123 Hz).

Now consider the 4-level system formed by $Nd:\beta"$-Na-Alumina and shown in Fig.7. This material has the unusual property that Nd fluorescence is not quenched even at densities as high as 10^{21} Nd/cm^3. Because $\beta"$-Na-Alumina has significant potential for new applications in nonlinear optics and laser technology it is important to understand why the fluorescence lifetime is constant and what limitations might be imposed on uses of the material due to saturation effects.

NDFWM spectra were recorded as a function of incident intensity for five samples, together with fluorescence decay

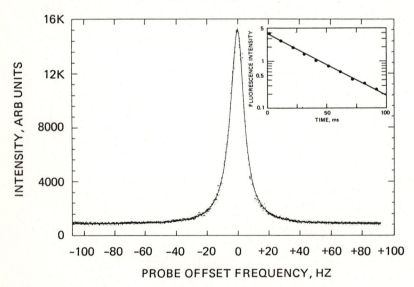

Fig.6. The NDFWM spectrum in $Cr:YAlO_3$ and (inset) the corresponding ^2E fluorescence decay curve.

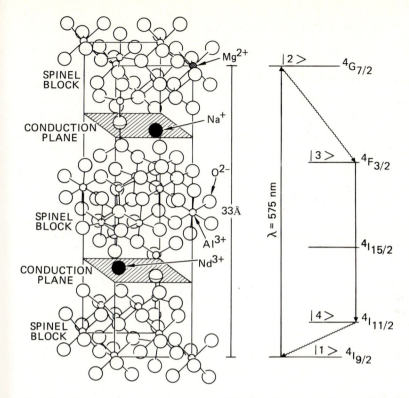

Fig.7. Crystal structure of β''-Na-Alumina ($Na_{1+x}Mg_xAl_{11-x}O_{17}$) with Nd^{3+} dopant ions exchanged for Na^+ via the conduction plane. Energy levels and the excitation/relaxation pathway of a single Nd^{3+} ion are shown on the right.

curves measured at a wavelength of 1.06 μm. In Fig.8a, several spectra recorded at different intensities are shown for a sample containing 6.4×10^{20} Nd/cm^3. Power broadening is clearly evident at the higher intensities and in Fig.8b the Lorentzian linewidth of these traces is seen to scale with the square root of the incident intensity, in excellent agreement with (2.3.32).

A lineshape of the form (2.3.32) was fitted to data similar to that in Fig.8 to determine the unsaturated linewidth and saturation intensity of each sample. The results are shown in Fig.9. The signal here originates from spatial hole-burning in the ground state and exhibits saturation behavior identical to that of the gain coefficient in an inhomogeneous system [33].

The NDFWM linewidth fluctuates below 5×10^{20} Nd/cm^3 in response to crystal field variations[28] and exhibits a small overall decline as density increases. The saturation intensities show a much larger decrease as a function of

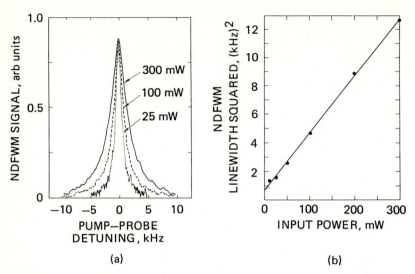

(a)   (b)

Fig.8. (a) NDFWM signal versus pump-probe detuning for total incident powers of 25, 100 and 300 mW, showing power-broadening. Nd^{3+} density=6.4×10^{20} cm^{-3}. (b) The square of the NDFWM linewidth (FWHM) versus total input power. The size of the data points is indicative of the experimental uncertainties and the solid curve is a least squares fit to the data.

density, diminishing to a value of 12 kW/cm^2 at $8.4 \times 10^{20} Nd/cm^3$, in excellent agreement with Boyd[34]. This corroborates the trend in the linewidth data and confirms a genuine <u>increase</u> in the ground state grating lifetime, albeit small, over a density range in which the excited state decay time actually <u>decreases</u> slightly[28,35]. In other words, the relaxation of the excited state is qualitatively different from that of the ground state above a density of $\sim 6 \times 10^{20} cm^{-3}$.

These NDFWM results are surprising. Ordinarily the onset of the process of cross-relaxation, familiar as the mechanism of fluorescence quenching in virtually all other highly doped Nd materials[36], would cause more rapid decay of the emitting state $|3>$. In a system following a closed relaxation path, the ground state grating would then wash out faster at higher densities. The NDFWM linewidth and saturation intensity would therefore increase with increasing Nd density. Instead, just the opposite is observed in Nd:β''-Na-Alumina. The reason for this is that as density increases, Nd-Nd pairs consisting of one optically excited $^4F_{3/2}$ ion and one $^4I_{9/2}$ neighbor interact more readily in a nearly resonant, mutual decay process yielding two $^4I_{15/2}$ ions which relax more <u>slowly</u> to the <u>ground</u> state than the $^4I_{11/2}$ ions usually produced by spontaneous emission. At high enough densities, the ground state grating which gives rise to four-wave mixing can therefore persist longer due to pair interactions. Processes other than this pair interaction would enhance the ground state relaxation

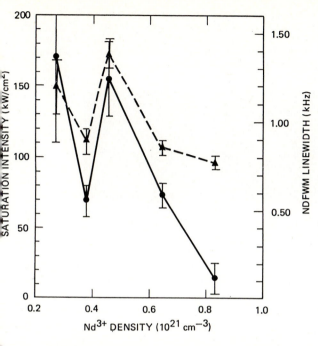

Fig.9. NDFWM saturation intensity (solid) and linewidth
(dashed) measurements versus Nd^{3+} dopant density. The
linewidth data give the best fit values of Lorentzian full
widths at half maximum intensity.

state, in disagreement with our measurements. The NDFWM
results provide clear evidence of a transition at high Nd
density from a closed to an open quantum system, mediated by
pair interactions which permit the ground and excited states to
evolve separately.

3.2 Color Center Studies

Two sets of data are again used in this section to illustrate
the simplicity and unique capabilities of NDFWM spectroscopy in
solids. NDFWM techniques are applied to F$_2$ color centers to
establish intersystem crossing as the mechanism for efficient,
continuous-wave phase conjugation with these structures.
Similarly, signals from N3 color centers in diamond arise due
to intersystem crossing, but the NDFWM spectrum is narrower
than expected from temporal decay measurements. This indicates
the presence of a hidden deep level for the N3 defect.

When pure LiF receives a dose of 100 MRad ^{60}Co γ-rays, high
densities of F-aggregate color centers are formed, particularly
F$_2$ centers. These radiation products consist of vacancy pairs
occupied by two electrons, and are stable for years in the

presence of low intensity light, although high intensities typically ionize the centers, forming short-lived F_2^+ defects. Cw phase conjugation is relatively efficient at the first resonance energy of the F_2 defects[29] and corrects optical aberrations in the usual manner (Fig.10).

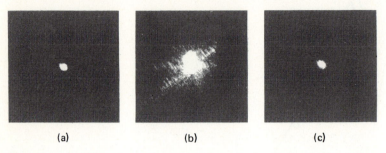

(a)   (b)   (c)

Fig.10. (a)Phase-conjugate signal from F_2:LiF without aberrator. (b)Aberrated probe beam with mirror replacing the sample. (c)Restored phase-conjugate signal with aberration as in (b).

The two electrons form singlet and triplet manifolds with the three lowest states constituting a 3-level system similar to Cr^{3+}[37]. In this case however, intersystem crossing can populate the metastable triplet level and phosphorescence accounts for relaxation to the ground state. Observations of the NDFWM spectrum[29] have shown that the linewidth agrees very well with the inverse of the phosphorescence decay time measured at the same temperature, confirming that this picture of saturated absorption due to intersystem crossing is correct. In Fig.11, the linewidth also narrows as temperature is decreased, in accord with previous studies of F_2 phosphorescence decay in KCl[37].

In diamond, color centers can be formed from aggregates of substitutional nitrogen atoms and are more stable to light and heat than the F-centers in alkali halides. For example the N3 center consists of three nitrogen atoms bonded to a common carbon or vacancy[38] with the absorption spectrum shown in Fig.12. The energy levels of this center are not fully known but it has been proposed[39] that the small features around 450-500 nm are due to partially allowed absorption to a state with the same symmetry as the ground state. Light absorption out of this state has been reported and the lifetime measured to be $0.73\pm.03$ ms by a double resonance technique[40], although no emission from this state has been reported.

Figure 13a presents the energy level diagram determined by these earlier measurements. By analogy with the Cr and F_2 systems, it is apparent that given the one known metastable level (and perhaps others) below the main resonance, the ground state absorption of the N3 center should be easily saturable under continuous excitation. With a single metastable state, NDFWM spectroscopy should reveal a linewidth corresponding to

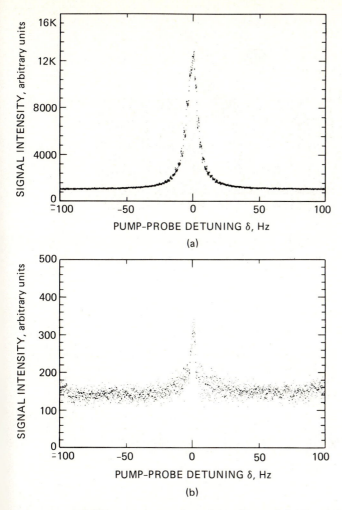

Fig.11. NDFWM spectra of γ-irradiated LiF. (a)T=20°C. The least squares fit Lorentzian linewidth is 4.70 Hz(FWHM). (b)T=-25°C. The spectral width has narrowed to 2.1 Hz (FWHM).

0.73 ms. With an additional deep level however, the linewidth could be narrower, as explained in Section 2.2.

Indeed this is the case for the N3 center in diamond which is accessible to study using several lines of the Kr$^+$ laser. As indicated by the preliminary data in Fig.14, the observed width of 265 Hz is roughly a factor of two narrower than expected on the basis of the energy levels of Fig.13a and the earlier decay measurement[40]. It therefore provides the first evidence of additional metastable energy levels for the N3 center below the $^2A_2(^2A_1)$ excited state.

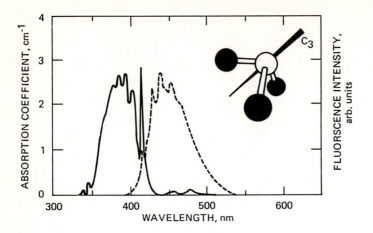

Fig.12. Absorption (solid) and emission (dashed) spectra of the
N3 center with structural model inset. The solid circles are
substitutional nitrogen impurities.

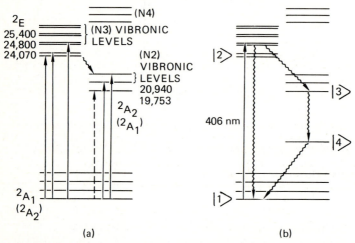

Fig.13. Energy levels of the N3 center in diamond, (a) as in
Ref.[39] and (b) with an additional, metastable level as
indicated by the present NDFWM results. Solid (dashed) lines
indicate allowed (forbidden) transitions.

Although a complete interpretation must await a full analysis
of the data, it can be concluded that the energy level diagram
should be modified as shown in Fig.13b. On the basis of the
NDFWM measurement, an additional deep level (or levels) with a
decay rate of $(265\pi)^{-1}$ must be present and in keeping with the
current ground state assignment[39] it should transform as a
singlet representation of the C_{3v} (3m) group.

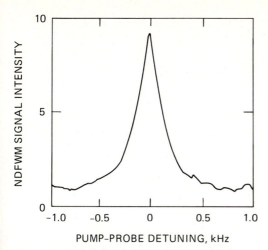

Fig.14. NDFWM signal versus pump-probe detuning for the N3 center. The sample was 1.75 mm thick and exhibited a peak optical density in the N3 absorption band of 0.23. Excitation wavelength was 406 nm.

4. Discussion

The excellent agreement between theory and the measured NDFWM spectrum of Cr establish the equivalence of fluorescence decay experiments and ground state relaxation measurements in simple "closed" systems. This correspondence can elucidate four-wave mixing mechanisms, as illustrated by the results on the 3-level system formed by F_2 color centers in LiF. However, in more complex systems with several metastable states which may emit no radiation at all, NDFWM spectroscopy provides unique information, inaccessible to conventional optical techniques. This is illustrated by detecting the onset of weak non-radiative pair interactions in Nd:β"-Na-Alumina over a dopant density range in which the fluorescence decay rate is constant. Also, the existence and relaxation rate of a hidden level of the N3 color center in diamond are established despite the fact that no emission from the metastable manifold has ever been reported.

5. Acknowledgements

The author wishes to thank D.G. Steel and J. Lam for many useful discussions regarding the theory of four-wave mixing. R.S. Turley and R.A. McFarlane contributed invaluable assistance with data analysis. R.A. McFarlane also gave a critical reading of the manuscript which was prepared for publication by J. McNulty. β"-Na-Alumina samples were grown by D. Dunn of UCLA and O.M. Stafsudd kindly furnished unpublished data on this material. Natural diamonds were loaned by V. Manson of the Gemological Institute of America. J. Brown and R. Cronkite provided superb technical assistance. The author also wishes to thank A. Au for the generous loan of a Kr ion laser. This research was funded under AFOSR contract F49620-85-C-0058.

References

1. P.D. Maker and R.W. Terhune, Phys.Rev. A137, 801(1965); R.L. Carman, R.Y. Chiao and P.L. Kelley, Phys.Rev.Lett. 17, 1281(1966); R.Y. Chiao, P.L. Kelley and E. Garmire, Phys. Rev.Lett. 17, 1158(1966).
2. B.L. Stansfield, R. Nodwell and J. Meyer, Phys.Rev.Lett. 26, 1219(1971).
3. G.C. Bjorklund, J.E. Bjorkholm, P.F. Liao, and R.H. Storz, Appl.Phys.Lett. 29, 729(1976).
4. F. DeMartini, F. Simoni, E. Santamato, Opt.Comm.9, 176(1973).
5. P.F. Liao, D.M. Bloom, and N.P. Economou, Appl.Phys Lett. 32, 813(1978).
6. S.A. Akhmanov, A.F. Bunkin, S.G. Ivanov, N.I. Koroteev, A.I. Kovrigin and I.L. Shumay, in Tunable Lasers and Applications, eds. A. Mooradian, T. Jaeger, P. Stoketh (Springer, Berlin, Heidelberg, New York, 1976), p.389, and references therein; S.A. Akhmanov, F.N. Gadjiev, N.I. Koroteev, R. Yu Orlov, I.L. Shumay, JETP Lett. 27, 243(1978).
7. J.J. Barrett, R.F. Begley, Appl.Phys.Lett. 27, 129(1975); M. Henesian, L. Kulevskii, R.L. Byer, J. Chem. Phys. 65, 5530(1976).
8. D.M. Bloom and G.C. Bjorklund, Appl.Phys.Lett. 31,592(1977.
9. D.M. Bloom, P.F. Liao, N.P. Economou, Opt.Lett.2, 58(1978).
10. A. Yariv and D.M. Pepper, Opt.Lett. 1, 16(1977).
11. M.D. Levenson and N. Bloembergen, Phys.Rev. B10,4470(1974).
12. K.Duppen, B.M.Hesp, D.A.Wiersma, Chem Phys.Lett.79, (1981).
13. M.D. Levenson, C.Flytzanis, N. Bloembergen, Phys.Rev. B6,3962(1972).
14. D.D. Dlott, C.S. Schosser and E.L. Chronister, Chem.Phys. Lett. 90, 386(1982).
15. J.J. Wynne, Phys.Rev.Lett.29, 650(1972); F. DeMartini, G.Giuliani, P. Mataloni, E. Palange, Y.R. Shen, Phys.Rev.Lett. 37, 440(1976).
16. P.F. Liao and D.M. Bloom, Opt.Lett. 3, 4(1978).
17. J.R. Salcedo, A.E. Siegman, D.D. Dlott and M.D. Fayer, Phys.Rev.Lett. 41, 131(1978); D.S. Hamilton, D. Heiman, J. Feinberg and R.W. Hellwarth, Opt.Lett 4, 124(1979).
18. J.K. Tyminski, R.C. Powell and W.K. Zwicker, Phys.Rev. B29, 6074(1984).
19. D. W. Phillion, D.J. Kuizenga, and A.E. Siegman, Appl. Phys. Lett. 27, 85(1975).
20. R.W. Olson, F.G. Patterson, H.W. Lee, and M.D. Fayer, Chem. Phys.Lett. 79, 403(1981).
21. R.K. Jain, Opt.Eng. 21, 199(1982); K. Jarasiunas and J. Vaitkus, Phys.Stat.Solidi a44, 793(1977).
22. P. Kupacek, M. Comte, and D.S. Chemla, Appl.Phys.Lett. 38, 44(1981).
23. A. Maruani and D.S. Chemla, Phys.Rev. B23, 841(1981).
24. E. Yablonovich, N. Bloembergen, and J.J. Wynne, Phys.Rev.B3, 2060(1971).

25. R.K. Raj, D. Bloch, J.J. Snyder, G. Camy and M. Ducloy, Phys. Rev. Lett. 44, 1251(1980).
26. P.F. Liao and D.M. Bloom, Opt.Lett. 3, 4(1978).
27. D.G. Steel and S.C. Rand, Phys.Rev.Lett. 55, 2285(1985).
28. S.C. Rand, J. Lam, R.S. Turley, R.A. McFarlane and O.M. Stafsudd, to be published.
29. S.C. Rand, Opt. Lett. 11, 135(1986) and to be published.
30. N. Bloembergen, in Quantum Electronics, Proc. 3rd Quantum Electronics Conference, Paris, 1963, ed. by N. Bloembergen,P. Grivet (Dunod, Paris 1964)pp.1501-1512; also Proc.IEEE 51, 124-131(1963).
31. R.L.Abrams and R.C. Lind, Opt.Lett. 2, 94(1978); D. Harter and R. W. Boyd, IEEE J.Q.E. QE-16, 1126(1980).
32. Y.R. Shen, The Principles of Nonlinear Optics, J. Wiley (New York, 1984), p.48.
33. See for example A. Yariv, Quantum Electronics, 2nd Edition Wiley (New York, 1975), p.170.
34. R.W. Boyd, M.T. Gruneisen, P. Narum, D.J. Simkin, B. Dunn and D.L. Yang, Opt.Lett. 11, 162(1986).
35. M. Jansen, A. Alfrey, O.M. Stafsudd, B. Dunn, D.L. Yang and G.C. Farrington, Opt.Lett. 10, 119(1984).
36. A.A. Kaminskii, Laser Crystals, Springer-Verlag (Berlin,1981), p.327.
37. Y. Farge, J.M. Ortega and R.H. Silsbee, J.Chem.Phys 69, . 3972(1978) and references therein.
38. G. Davies, Diamond Res.15-24(1977);M.D. Crossfield, G.Davies, A.T. Collins and E.C. Lightowlers, J.Phys.C7, 1909(1974); J.A.Van Wyk, J.Phys.C15, L981(1982).
39. G. Davies, C.M. Welbourn and J.H.N. Loubser, Diamond Research (1977), pp. 23-30; M.F. Thomaz and G. Davies, Proc.Roy.Soc. London A362, 405(1978).
40. S.C. Rand and L.G. DeShazer, Opt.Lett.10,481(1985).

The Turvy Topsy Contest

In 1975, when Schawlow was President of the Optical Society of America, he proposed a contest for the best "Turvy Topsy." A "Turvy Topsy" is, of course, the inverse of a Topsy Turvy, whereas a Topsy Turvy is a picture which looks like some recognizable object when right side up or upside down. A "Turvy Topsy", therefore, must show an object which is always obviously upside down.

The purpose of the contest was to advance the art of slidesmanship, by demonstrating how to produce a slide that can never be presented the right way up. Four prizes were offered: First Prize, $10, Second Prize, $5, Third Prize, a copy of Schawlow's latest paper, and Fourth prize, copies of Schawlow's two most recent papers.

The First Prize was won by the figure below, submitted by Bruce D. Hansche of the University of Michigan, which shows a mathematical integral, that cannot be made to end in "dx".

$$xd^8\left[\frac{\delta H}{\Delta \phi}\right] \int \left[\frac{\phi \nabla}{H\S}\right]_8 px$$

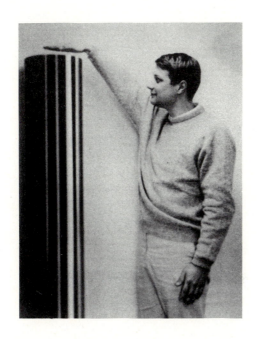

*Linn F. Mollenauer with a large
ruby rod used at Stanford in 1965*

Part IV

Miscellaneous Ideas

A scientist typically does his best work on the back of an envelope. To enhance productivity, provide envelopes with two backs. Such envelopes were available from "Doublethink, Inc." a division of "Nocturnal Aviation", A.L. Schawlow, Proprietor

Using a Tokamak to Study Atomic Physics: Brightness Ratios of Transitions Within the $n=2$ Levels of Be *I*-like Ions (C *III* to Cr *XXI*)

H.W. Moos

Department of Physics and Astronomy, Johns Hopkins University, Baltimore, MD 21218, USA

Abstract

Although the primary motivation for constructing devices which produce magnetically confined high-temperature plasmas is controlled thermonuclear fusion, these devices can also be used to study the physics of highly ionized atoms. This paper reviews an example of such a study in which the n=2 to n=2 transition lines in the Be I isoelectronic sequence were measured with photometrically calibrated extreme ultraviolet instrumentation. The brightness ratios of the 2s2p ^{3}P-2p^2 ^{3}P (R) and the 2s^2 ^{1}S$_0$-2s2p ^{3}P$_1$ (R*) lines to the resonance 2s^2 ^{1}S$_0$-2s2p ^{1}P$_1$ line were determined and compared with theoretical predictions. At low nuclear charge, Z, the experimental values agree with those computed by the R-matrix method which includes the effect of resonances. At higher Z, the data agrees very well with calculations based on the distorted wave approximation. In addition to confirming the theoretical methods, this agreement lends confidence to the use of line ratios for diagnostic studies of high-temperature plasmas.

1. Introduction

A. L. Schawlow frequently has noted the mutually beneficial interaction between science and technology. Advances in scientific knowledge are used to produce improved and new technology. In turn, these technological improvements are utilized by the scientific community for further advances, etc. Similar relationships also exist between less and more applied branches of science. Although this observation was based on a different field of endeavor, it is also true of controlled thermonuclear fusion research. A reliable understanding of the excitation physics of atoms and ions in a high–temperature plasma is necessary both to minimize the radiative losses from a plasma and to utilize spectroscopic data for plasma diagnostics. In turn, the existence of these devices and improvements in the electron temperature make possible new experimental checks of the theoretical techniques used to compute the atomic parameters for the highly ionized species present in these plasmas. Indeed, it can be argued that measurements performed in plasmas (as against beam experiments) are extremely important in that they check both the accuracy of the atomic physics calculations and for previously unsuspected excitation mechanisms such as inner shell ionization [1].

Over the past decade, a new generation of spectroscopic instruments has come into use for diagnostic studies of the magnetically confined high temperature plasmas produced for fusion research by devices such as tokamaks. These spectrographs use microchannel plate image converters optically coupled to photo-diode or CCD arrays to provide simultaneous time resolved measurements of a large number of spectral emission lines. In addition, these instruments can be calibrated against absolute photometric

295

standards such as synchrotron radiation. Although designed for diagnostics, these instruments have provided the opportunity to check the theoretical calculations of excitation rates. It is expected that the experimental checks on the theory will lead to a more confident application of the spectroscopic data for density and temperature measurements in both laboratory and astrophysical plasmas. In magnetic fusion plasmas, these diagnostics will be particularly important for complex magnetic geometries where electron density and temperature measurements are not always available at the point where the spectroscopic measurements are made.

Stimulated by the requirements of both astrophysics and fusion, a number of workers have computed relative level populations in ionized atoms as a function of both density and temperature. The relative population of any pair of levels can be determined by measuring the brightness ratio of a pair of emission lines. Thus, if the calculations are correct, it is possible to use these ratios to measure the electron density or temperature (depending on the line pair selected) in the plasma. This laboratory has used diagnostic instrumentation to measure these ratios in laboratory fusion plasmas where the electron temperature and density have been well determined by other means as a check on the theoretical calculations. Experiments of this type have been performed on the Princeton Large Torus at the Princeton Plasma Physics Laboratory [2,3], the Tandem Mirror Experiment at Lawrence Livermore National Laboratory [4,5] and the TEXT Tokamak at the University of Texas [4-8].

This article will review the work performed on the Be I-like ions ranging from C III to Cr XXI using the TEXT Tokamak [6,8]. This work has shown that although the more complex R-matrix calculations which include resonances are necessary at low nuclear charge, at higher Z very good agreement is found with the distorted wave approach.

Figure 1 shows a simplified Grotian diagram of the $n=2$ levels of Be I-like ions. It was possible to measure two line brightness ratios with some precision along the isoelectronic sequence:

$$R = B(2s2p\ ^3P_2-2p^2\ ^3P_2)/B(2s^2\ ^1S_0-2s2p\ ^1P_1) \tag{1}$$

and

$$R^* = B(2s^2\ ^1S_0-2s2p\ ^3P_1)/B(2s^2\ ^1S_0-2s2p\ ^1P_1). \tag{2}$$

(For low Z, all of the lines in the multiplet $^3P - ^3P$ were used to compute R.) These ratios are relatively insensitive to the electron temperature and those of low Z ions have been used to determine solar plasma densities [9]. At high Z, these ratios are not sensitive to changes in the density at the values near the center of a tokamak. Hence, the high Z ratios are not useful tokamak diagnostics, although the low Z ratios can be used at the plasma edge (a region where spectroscopic density diagnostics may be especially valuable). However, note that for both ratios the denominator is proportional to the spin-allowed excitation rate, whereas at tokamak densities the numerator for R^* depends on the intercombination $^1S_0 - ^3P$ excitation rate and the numerator for R depends on both the $^1S_0 - ^3P$ and the $^3P - ^3P$ excitation rates. Consequently, the numerator and denominator brightnesses depend on very different types of electron excitations (spin allowed vs. intercombination). Thus, the experimentally determined values of the ratios over a wide range of Z serve as an important test of the theoretical calculations. The good agreement with theory shown by the experimental measurements lends confidence in using the theory both for edge diagnostics and to seek other line ratios more appropriate for tokamak density diagnostics near the center.

296

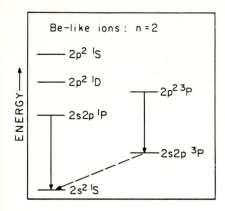

Fig. 1. Simplified Grotian diagram
of the n=2 levels of Be I-like ions
showing the transitions used.

2. Experimental Method

The plasma of the TEXT Tokamak [10] plasma is confined in a toroidal vacuum vessel. The major radius of the torus is 100 cm; the minor radius,is 27 cm. The toroidal magnetic field is 15-30 kG, and the plasma current is 150-400 kA (Fig. 2a). A TEXT discharge lasts ~ 400-600 ms, with a 200-400 ms steady state phase, during which the plasma current, the electron density, and the temperature are essentially constant (Fig. 2b). There is a slight rise in the electron density during injection of an impurity. In the standard TEXT discharge, the toroidal field is 28 kG, the plasma current is 300 kA, the average density along a chord through the center of the plasma is $3.5 \times 10^{13} cm^{-3}$ and the electron temperature at the toroidal axis is ~ 1 keV. The electron temperature and density across the minor radius of the torus are displayed in Figure 2c. In order to study the density dependence of the line ratios of intrinsic oxygen, plasmas were produced at three different chord-averaged densities, 1.5×10^{13}, 3.5×10^{13}, and $8 \times 13 cm^{-3}$ (most of the data was obtained at the middle density), with central electron temperatures varying between 700 and 1100 eV. The electron density is measured by far infrared interferometry for six different chords across the plasma diameter and then inverted to obtain a radial profile. The electron temperature is obtained by Thomson scattering. Since some of the lines of interest were emitted near the plasma edge, electron temperatures and densities were measured in this region by Langmuir probes.

The spectra were recorded in the 50-2000 Å range by means of two time-resolving spectrometers: a grazing incidence instrument and a normal incidence instrument. Both are equipped with microchannel plate image converter detectors. The instruments, the detectors and their properties have been described in detail elsewhere; for the normal-incidence instrument see [11] and for the grazing-incidence instrument see [12]. The two instruments were located at the same toroidal position. Radial profiles of the ion emissivities were obtained by scanning the plasma on a shot by shot basis with the grazing incidence instrument and Abel inverting the brightness profiles of emission lines in this spectral region (Δn=1 transitions for low Z and Δn=0 for high Z).

Absolute brightness calibrations of the two instruments over their entire wavelength range were obtained by using the SURF synchrotron radiation source at the National Bureau of Standards [13,14]. The spectral resolution of the normal-incidence instrument varies from 0.7 to 4 Å

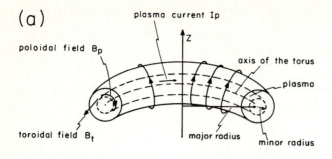

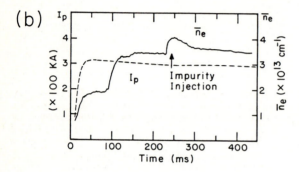

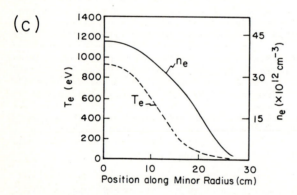

Fig. 2. TEXT Tokamak characteristics (a) Configuration of TEXT Tokamak;
(b) time histories of plasma current I_p (dashed line) and line average
density $\bar{n}_e$ (solid line) of a nominal discharge with impurity injection and
(c) radial profiles of electron temperature (dashed line) and electron
density (solid line) of a standard TEXT discharge. Reference [6].

depending on the breadth of the spectral range covered; several gratings
having the same radius of curvature but different ruling densities permit
the recording of either large spectral ranges with low resolution or smaller
spectral domains with higher resolution. The grazing-incidence spectrograph
has a resolution of 0.7 Å, a range of 15 to 360 Å and the detector
simultaneously covers ~70 Å. The time resolutions of the two instruments

are slightly different, 4.1 ms in the case of the normal-incidence instrument, 5.4 ms for the grazing-incidence one.

A number of intrinsic impurities such as carbon and oxygen were always present in the discharge, introduced by plasma-wall interactions. Elements which were not intrinsic could be introduced into the plasma either by a laser-ablation technique [15] or by a gas puff through a fast valve.

As the temperature decreases from the center to the edge, the ions are located in shells at a plasma radius with an electron temperature near that of ionization equilibrium. Thus the emission could be identified by wavelength, radius of the emission shell and time history. (Lower Z ions peak earlier if intrinsic and decay away more rapidly if laser injected.) Some of the spectral features were partially blended, but the linear properties of the detector permitted separation of the blends in many cases. In the case of injected impurities, the spectrum before the injection was subtracted from that after the injection to isolate the spectrum due to the injected elements. Fig. 3 presents an example for the case of titanium.

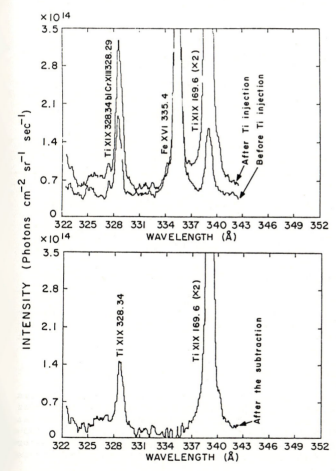

Fig. 3. Example of removal of blends by subtraction of a spectrum before the injection. The Ti XIX 328.34 Å line is blended with the Cr XIX 328.29 Å line. The signal due to the intrinsic Cr XIII 328.29 Å was eliminated after the subtraction as was that due to the intrinsic Fe XVI 335.4 Å. Reference [8].

3. Results

The wavelength range of the emissions divided the study into two parts; the low Z n=2 to n=2 emissions [6] were measured with the normal incidence instrument and the high Z emissions [8] with the grazing incidence instrument.

3.1. Low Z

Table 1 compares the measured values of R with R-matrix calculations which include the effect of resonances. The agreement between theory and experiment is quite good. It appears that the R-matrix calculations provide reliable predictions of the intensity ratios.

A widely used theoretical technique is the distorted wave method which is simpler to use than the R-matrix method but does not include the effects of resonances. This technique is expected to be unreliable at low Z but to improve with increasing charge of the Be I-like ion. As a comparison, a distorted wave calculation for C III (at T_e =5eV) gave a value of 1.2 for R, far outside the experimental uncertainty [6].

Table 1. Experimental and R-matrix Calculated Ratios

	Wavelengths of transitions (Å)		Line ratio $R = \dfrac{I(2s2p^3P-2p^2\ ^3P)}{I(2s^2\ ^1S-2s2p^1P)}$			
Ion			Experiment	Theory		
	$2s^2\ ^1S-2s2p^1P$	$2s2p^2P-2p^2\ ^3P$		Value at	T_e(eV)	n_e(cm^{-3})
C III	977.0	1174.9–1176.4	0.56	0.54[a]	3	5×10^{11}
O V	629.7	758.7–762.0	0.82	0.81[a]	13	2×10^{12}
F VI	535.2	644.0–648.5	0.80	0.78[b]	22	4×10^{12}
Ne VII	465.2	558.6–564.5	0.80	0.75[c]	54	2×10^{13}

a) Reference [16]
b) Reference [17]
c) Reference [18]

Figure 4 displays the results for O V, F VI and Ne VII. Measurements of the O V emissions were obtained in discharges with three different densities, permitting a study of the density dependence. The agreement between the experimental values and the R-matrix calculations is quite good. Distorted wave calculations for O V [6] gave values (0.88 and 1.06 at a electron temperature of 15 eV and electron densities of 1.0×10^{12} and 1.0×10^{13} respectively) which were above those of the R-matrix method, but the differences were not as large as the case of C III.

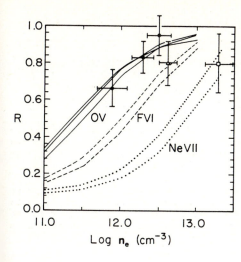

Fig. 4. Comparison of low Z experimental values with theoretical computations using the R-matrix method (1) O V – solid circles (experiment); solid curves (theory) for three electron temperatures, 14 eV, 22 eV, and 34 eV. (At the low-density limit the curves are ordered so that increasing R corresponds to decreasing temperature; at the high-density limit increasing R corresponds to 14 eV, 34 eV, and then 22 eV.) (2) F VI-empty triangle (experiment); dashed curves (theory) for two temperatures, 22 eV (upper curve) and 43 eV (lower curve). (3) Ne VII – empty square (experiment); dotted curves (theory) for two temperatures, 27 eV (upper curve) and 54 eV (lower curve). Reference [6].

3.2 High Z

The discrepancies between the R-matrix and the distorted wave methods are expected to decrease with increasing Z. A comparison of the relative brightnesses of fifteen lines in Ca XVII by [8] indicated agreement between R-matrix [19,20] and distorted wave [21] calculations of better than 25%. Of course, even though the predictions of two techniques agreed, this did not guarantee that they were correct, and it was extremely important to compare these predictions with experimental measurements. However, the convergence with Z indicated that a comparison with R-matrix calculations was necessary only at low Z and the appropriate comparison for high Z was with the more widely available distorted wave calculations.

For this range of Z, the line intensity ratios, R and R*, are expected to be insensitive to either the electron temperature or the electron density over the range of the tokamak plasma conditions [8]. A comparison of the experimental and theoretical values of R and R* for Sc XVII and Ti XIX at different electron densities showed that there was good agreement except for the case of R* for Sc XVIII, where two of the experimental values are significantly below the theoretical values. In no case was there evidence for a strong electron density dependence.

The measured and computed line ratios for Cl XIV to Cr XXI at an average line electron density near 3.5×10^{13} cm^{-3} are listed in Table 2 and presented as a function of Z in Fig. 5. Calculated line ratios were available for Ar XV, Ca XVII, Ti XIX and Cr XXI [21] and were extrapolated or interpolated for the other elements. Fig. 5 shows that the measured R line ratio points are scattered around a straight line connecting the points predicted by distorted wave calculations. The calculated R* ratios were scattered about a horizontal line at 0.034. The agreement between experiment and theory is quite good except for the case of R* for Sc XVIII. Table 2 lists the ratios of measured to calculated values, R_m/R_c and R^*_m/R^*_c. The two arithmetic means for the ratios of the experimental to calculated values are 1.01 ± 0.19 and 0.84 ± 0.17, respectively. Thus, it appears that the agreement between experiment and theory is quite good, well within the experimental uncertainties.

Table 2. Relative brightnesses of Be I-like ion lines at $\langle n_e \rangle$ = 3.5E13/cm³

	Cl XIV	Ar XV	K XVI	Ca XVII	ScXVIII	Ti XIX	V XX	Cr XXI	average
R*(×10⁻²):									
m					2.05	3.12	2.87	3.53	
c	3.21e	3.26	3.39i	3.57	3.33i	3.11	3.49i	3.81	
R_m^*/R_c^*:					0.62	1.00	0.82	0.93	0.84±.17
R(×10⁻²):									
m	3.70		2.51	1.66	1.54	1.76	1.17	0.92	
c	2.78e	2.53	2.28i	2.02	1.77i	1.52	1.28i	1.05	
R_m/R_c:	1.33		1.10	0.82	0.87	1.16	0.91	0.88	1.01±.19

R and R* are defined in Eq. 1 and Eq. 2. m = measured, c = calculated by reference [21], and interpolated between electron densities i = interpolated along Z, e = extrapolated along Z

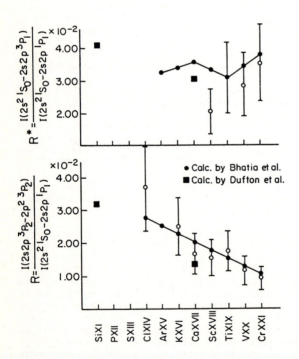

Fig. 5. Comparison of the measured line ratios, R and R*, at the average electron density 3.5 × 10¹³cm⁻³ with calculations based on the distorted wave approximation [21] and the R-matrix method [17,19,20]. Reference [8].

R-matrix method calculations are available for only two cases in this range, Si XI [17] and Ca XVII [19,20]. The values are close to the distorted wave approximation results for Ca XVII and to an extrapolation of the distorted wave values for Si XI. A direct comparison with experimental data is available only in the case of Ca XVII for which the R line ratio was measured and is within 25% of both theoretical calculations.

Which relative collision strengths does an experiment such as this check and to what extent are the results of these experiments valid at other densities? Since one is normally interested in the percentage change in R for a percentage change in the collision strength, Ω_{ij}, the sensitivity of the ratio R to a given collision strength is best expressed by

$$S = abs[(\Omega_{ij}/R)(\partial R/\partial \Omega_{ij})] . \tag{3}$$

A similar equation holds for R*. The quantity S was estimated for Ca XVII using the collision strengths of [22]. For R*, the values of S are about 1.0, 0.44 and 0.40 for the $2s^2\ ^1S_0-2s2p\ ^1P_1$, $2s^2\ ^1S_0-2s2p\ ^3P_1$ and $2s^2\ ^1S_0-2s2p\ ^3P_2$ transitions, almost independent of density. Other collision strengths contribute in negligible amounts. For R, the situation is slightly more complex. Fig. 6 shows the dependence of the most important

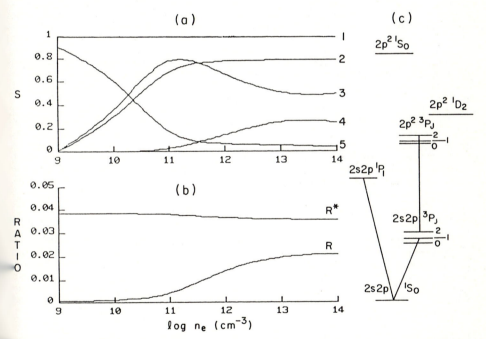

Fig. 6. S, the sensitivity of the value of R to uncertainties in the collision strengths. The values for the most important collision strengths in Ca XVII are plotted as functions of electron density. The curves are numbered with respect to the electron collisional transitions marked in part c. (b) The brightness ratios, R and R* defined in Eqs. 1 and 2, are plotted as functions of electron density for Ca XVII. (c) The Grotrian diagram of λ=2 levels of Ca XVII. The important collisional transitions are numbered according to part a. Modified from reference [8].

sensitivities as a function of electron density along with the density dependence of R and R*. At a density of $3 \times 10^{13} cm^{-3}$, where the tokamak measurements were made, the $2s^2\ ^1S_0-2s2p\ ^1P_1$, $2s^2\ ^1S_0-2s2p\ ^3P_2$ and $2s2p\ ^3P_2-2p^2\ ^3P_2$ transitions dominate followed by the $2s2p\ ^3P_2-2p^2\ ^3P_1$ transition. These transitions are important down to densities of about $3 \times 10^{10} cm^{-3}$ where the $2s^2\ ^1S_0-2p^2\ ^3P_2$ transition becomes significant. The tokamak data provides no data on this collision strength, which involves a spin change and the promotion of two electrons from $2s^2$ to $2p^2$. Note, however, that at densities below this point, the value of R is near 10^{-3} and accurate spectroscopic measurements become more difficult. Thus, it appears that the tokamak experiments have checked the collision strengths for the most important transitions and it is likely that theoretical values of R will be accurate in the electron density range from 10^{14} down to $10^{11} cm^{-3}$ for Ca XVII. For other ions, the lower limit will shift with Z.

4. Conclusions

Brightness ratios for emission lines ratios emitted between the n=2 levels of Be I-like ions ranging from C III to Cr XXI have been measured in a tokamak plasma in which the electron temperatures and densities were measured independently by non-spectroscopic methods. At low Z, the ratios agreed with those computed by the R-matrix method. These results at low Z confirm the importance of resonances by experimentally verifying the electron impact excitation rates computed by the R-matrix approach. For high Z (Cl XIV to Cr XXI), the experimentally determined line ratios, R and R*, agree extremely well with theoretical calculations based on the distorted wave approximation; the ratios of the measured to calculated values of R and R* were 1.01 ± 0.19 and 0.84 ± 0.17 respectively. For the one case where an R-matrix method value was available, Ca XVII, good agreement also exists.

Finally, note that this agreement also gives assurance that the theoretical calculations can be relied on for plasma diagnostics and predictions of radiative loss from plasmas. The relationship between science and technology discussed in the introduction has now come full circle. Indeed, the next step is to seek out additional ratios which are sensitive to electron density and temperature, and determine which are the most practical for high-temperature plasma diagnostics.

Acknowledgements

This article is a review of an extensive study with a large number of collaborators. (See [6] and [8].) I wish to acknowledge each of them and in particular the ideas and efforts of M. Finkenthal. The TEXT Tokamak research group produced and characterized the large number of similar discharges necessary for this study. This work was supported by the Department of Energy under grant DE-FG02-85ER53214 to the Johns Hopkins University.

References

1. M. Finkenthal, B. C. Stratton, H. W. Moos, A. Bar Shalom, and M. Klapisch: Physics Letters 108A, 71, (1985).
2. B. C. Stratton, H. W. Moos and M. Finkenthal: Ap.J. Letters 279, L3 (1984).
3. B. C. Stratton, H. W. Moos, S. Suckewer, U. Feldman, J. F. Seely and A K. Bhatia: Phys. Rev. A 31, 2534 (1985).
4. T. L. Yu, M. Finkenthal and H. W. Moos: Ap.J. 305, 880 (1986).

5. M. Finkenthal, T. L. Yu, S. L. Allen, L. K. Huang, S. Lippmann, H. W. Moos, B. C. Stratton, P. L. Dufton and A. E. Kingston: Astron.Astrophys, submitted (1987).
6. M. Finkenthal, T. L. Yu, S. Lippmann, L. K. Huang, H. W. Moos, B. C. Stratton, A. K. Bhatia, R. D. Bengtson, W. L. Hodge, P. E. Phillips, J. L. Porter, T. R. Price, T. L. Rhodes, B. Richards, C. P. Ritz, and W. L. Rowan: Ap.J. 313, 920 (1987).
7. S. Lippmann, M. Finkenthal, L. K. Huang, H. W. Moos, B. C. Stratton, T. L. Yu, A. K. Bhatia and W. L. Hodge, Ap.J., in press 1987.
8. L. K. Huang, S. Lippmann, T. L. Yu, B. C. Stratton, H. W. Moos, M. Finkenthal, W. L. Hodge, W. L. Rowan, B. Richards, P. E. Phillips, A. K. Bhatia: Phys.Rev., in press, (1987).
9. P. L. Dufton and A. E. Kingston, Adv.Atom.Molec.Phys. 17, 355 (1981).
10. K. W. Gentle et al.: Plasma Physics and Contr.Fusion 26, 1407 (1984).
11. R. E. Bell, M. Finkenthal and H. W. Moos: Rev.Sci.Instrum. 52, 1806 (1981).
12. W. L. Hodge, B. C. Stratton and H. W. Moos: Rev.Sci.Instrum. 55, 16 (1984).
13. D. L. Ederer, E. B. Saloman, S. C. Ebner and R. P. Madden: Journal of Research of NBS-A.Physics and Chemistry 79A, 761 (1975).
14. D. L. Ederer and S. C. Ebner: A Users Guide to Surf, National Bureau of Standards, Washington, D.C. 20234 (1985).
15. D. R. Terry, W. Rowan, W. C. Connolly and W. K. Leung: Proc. 10th Symp. on Fusion Energy, 14, 959 (1983).
16. P. L. Dufton, K. A. Berrington, P. G. Burke and A. E. Kingston: Astron.Astrophys. 62, 111 (1978).
17. F. P. Keenan, D. A. Berrington, P. G. Burke, A. E. Kingston and P. L. Dufton: Mon.Not.R.Astron.Soc. 207, 459 (1984).
18. P. L. Dufton, J. G. Doyle and A. E. Kingston: Astron.Astrophys. 78, 318 (1979).
19. P. L. Dufton, A. E. Kingston, J. G. Doyle and K. G. Widing: Mon.Not.R.Astron.Soc. 205, 81 (1983).
20. P. L. Dufton, A. E. Kingston and N. S. Scott: J.Phys.B 16, 3053 (1983).
21. A. K. Bhatia, U. Feldman and J. F. Seely: Atomic Data and Nuclear Data Tables, in press.
22. A. K. Bhatia and H. E. Mason: Astron. Astrophys. Suppl. Ser. 52, 115 (1983).

How to Squeeze the Vacuum, Or, What to Do When Even No Quantum Is Half a Quantum Too Many

M.D. Levenson

IBM Research, Almaden Research Center, 650 Harry Road, San Jose, CA 95120, USA

The vacuum is conventionally defined as the absence of matter and energy. One might imagine that such a vacuum would be simple and without interest to physicists. Quantum mechanics, however, provides a detailed and complex model of the vacuum, pregnant with possibilities and well worth careful study [1]. The quantum mechanical vacuum is defined as the ground state of all fields. The electromagnetic field is the field most familiar to spectroscopists, and is the model used to understand more complex forces. The electromagnetic field is conventionally modelled as an assembly of harmonic oscillator modes. It is well known that the ground state of a harmonic oscillator does not have zero energy; instead, it contains one half quantum. This vacuum energy allows an oscillator in the ground state to have a slightly fluctuating position and momentum and thus fulfil the uncertainty principle [2].

The analogous zero point motion of the electromagnetic field corresponds to fluctuations of the electric and magnetic fields around their zero average value [3]. The mean square of the fields are, however, nonzero. The uncertainty principle similarly allows vacuum fluctuations of all the other force and matter fields of quantum mechanics.

Vacuum fluctuations affect optical processes in a variety of ways. Students of laser physics are familiar with the explanation of spontaneous emission as stimulated emission due to the vacuum fluctuation input [4]. Spontaneous emission into a laser mode in turn causes fluctuations which broaden the laser linewidth according to the famous Schawlow-Townes formula [5]. One explanation of the Lamb shift is that it results from perturbations of an electron's motion due to fluctuating vacuum fields [6].

The modern theory of quantum optics also implies that the familiar "shot noise" from a high efficiency photon detector is a direct result of the zero point motion of the vacuum field [7,8]. This viewpoint is in direct contradiction to the conventional model where the "shot noise" appears in a light detector as the result of the quantization of the electron charge. That model would predict an irreducible noise floor for the detection of any optical field, even a perfectly constant classical wave. This noise floor is sometimes called the "standard quantum limit" or the "vacuum noise level." This vacuum noise limits the sensitivity of all optical measurements; in particular, absorption spectroscopy is especially sensitive to this noise and thus, has a much lower sensitivity than emission spectroscopy. The newer theory allows states of the electromagnetic field with less noise, a clear indication of nonclassical behavior [9]. Some of these low noise states are states where some of the vacuum fluctuations appear to be smaller than normal. Such states correspond to a "squeezed vacuum."

To understand why a squeezed vacuum is possible and and how it might be produced and detected, one must begin by explaining the modern theory of quantum noise. Light is an electromagnetic oscillation, and the electric field amplitude can be expressed as $E(t) = E_1 \cos \omega t + E_2 \sin \omega t$. The quantities E_1 and E_2 are termed "quadrature amplitudes" or, simply, "quadratures." In the quantum mechanical theory, these quadratures are conjugate operators like position and momentum. The Heisenberg uncertainty principle requires that the product of the uncertainty in the two quadrature amplitudes be greater than a dimensional quantity that is proportional to Planck's constant. The root mean square average of the zero point motion is the uncertainty in the quadrature amplitude.

The vacuum state must be independent of phase, and thus the uncertainties in the two quadrature amplitudes must be equal. The coherent states familiar from laser theory are constructed by adding a coherent amplitude to the fluctuating vacuum field [3]. If that coherent amplitude is in the cosine quadrature, $E(t) = (E_1 + \delta E_1) \cos \omega t + \delta E_2 \sin \omega t$. Where δE_1 and δE_2 are understood to represent the quantum mechanical fluctuations. On the average, the electric field is $<E(t)> = E_1 \cos \omega t$. The instantaneous intensity is $I = (c/8\pi)[(E_1 + \delta E_1)^2 + (\delta E_2)^2]$, but the average intensity is just $< I > = (c/8\pi)|E_1|^2$. The term containing δE_1 has zero average, and the terms quadratic in the vacuum fluctuations reflect only the vacuum energy. A quantum detector produces an instantaneous current proportional to the instantaneous power $P = IA$, where A is the detector area and the vacuum energy has been omitted. The proportionality constant for perfect quantum efficiency is $e/\hbar\omega$ where e is the electron charge. The number of electrons produced in time t is just $N = Pt/\hbar\omega$, the number of photons incident in time t.

The noise is best parameterized as the mean square deviation of the optical power (or current or number of detected electrons) from the corresponding average value. In terms of the field fluctuations, the mean square deviations are, respectively:

$$< \delta P^2 > \, = \, < P^2 > - < P >^2 \, = \, \frac{cA}{2\pi} < P > \ < \delta E_1^2 > ; \tag{1a}$$

$$< \delta i^2 > \, = \, \left(\frac{e}{\hbar\omega} \right)^2 < \delta P^2 > \, = \, \frac{ecA}{2\pi\hbar\omega} < i > \ < \delta E_1^2 > ; \tag{1b}$$

$$< \delta N^2 > \, = \, < N^2 > - < N >^2 \, = \, \frac{Act}{2\pi\hbar\omega} < N > \ < \delta E_1^2 > . \tag{1c}$$

These mean square deviations are proportional to the nonzero mean square vacuum fluctuations in the E_1 quadrature. For the vacuum state and coherent states, the mean square fluctuations in E_1 and E_2 are equal and constant [9]. In cgs units:

$$< \delta E_1^2 > \, = \, < \delta E_2^2 > \, = \, < \delta E_c^2 > \, = \, \frac{2\pi\hbar\omega}{Act} . \tag{2}$$

The mean square fluctuations depend only on the sum of the squares of the average quadrature amplitudes, that is, the average intensity.

$$< \delta P^2 > \, = \, \frac{\hbar\omega}{t} < P > ; \tag{3a}$$

$$< \delta i^2 > \, = \, \frac{e}{t} < i > ; \tag{3b}$$

$$< \delta N^2 > \, = \, N. \tag{3c}$$

The mean square fluctuation in the number of electrons produced in period t is equal to N, the average number produced. This is one signature of Poisson statistics, and is just the noise level predicted by the conventional "shot noise" model [9]. The time t is inversely proportional to the bandwidth of the measurement. That model correctly predicts the properties of coherent states and incoherent sums of coherent states. With a few exceptions, all experimental results up until 1985 could be explained in terms of such states.

A squeezed vacuum would have mean square fluctuations in one quadrature less than those characteristic of an unsqueezed vacuum. For example:

$$< \delta E_1^2 > = < \delta E_c^2 > e^{2r}; \tag{4a}$$

$$< \delta E_2^2 > = < \delta E_c^2 > e^{-2r}; \tag{4b}$$

$$< \delta E_1 \delta E_2 > = 0. \tag{4c}$$

The parameter r describes the degree of squeezing [8]. The product of the uncertainties remains equal or greater than the limit required by the uncertainty principle. If the average quadrature amplitudes are

$$< E_1 > = E_0 \cos \theta \quad \text{and} \quad < E_2 > = E_0 \sin \theta. \tag{5}$$

The quadrature fluctuations at phase angle θ ($\delta E_\theta = \delta E_1 \cos \theta + \delta E_2 \sin \theta$), are responsible for the mean square fluctuations in the power, current and number of detected electrons. For the quadrature fluctuations in (4), the relevant mean square quadrature fluctuation is $< \delta E_\theta^2 > = < \delta E_c^2 > \{e^{2r} \cos^2\theta + e^{-2r} \sin^2\theta\}$. The power, current and number fluctuations, then, depend on the phase angle θ as:

$$< \delta P_\theta^2 > = \frac{\hbar\omega}{t} < P > \{e^{2r} \cos^2\theta + e^{-2r} \sin^2\theta\}; \tag{6a}$$

$$< \delta i_\theta^2 > = \frac{e}{t} < i > \{e^{2r} \cos^2\theta + e^{-2r} \sin^2\theta\}; \tag{6b}$$

$$< N_\theta^2 > = N\{e^{2r} \cos^2\theta + e^{-2r} \sin^2\theta\}. \tag{6c}$$

For the correct choice of θ, the mean square fluctuation for a squeezed vacuum is reduced by a factor of $\exp(-2r)$ [8].

The squeezed vacuum has more energy than the vacuum state, but substantial reductions in the detected noise occur for minimal increase in field energy. The average value of the field remains zero. However, even one quantum radiated or scattered randomly into the detector can obliterate the squeezing effect. One of the major experimental challenges of squeezed state generation and detection is avoiding all additional noise sources.

Squeezed light is generated by nonlinear interactions in various kinds of materials. Strong pump waves are required for these interactions, and even very small scattering cross sections can scatter enough light to wipe out the noise reduction. Moreover, lasers are never perfect. Laser-produced noise above the quantum level must somehow be suppressed. Experimentally overcoming these difficulties can be just as challenging as squeezing the vacuum.

In attempts to avoid problems with unstable dye lasers or strongly scattering sodium vapor, we chose to squeeze the vacuum fluctuations using the nonlinear index of refraction of an optical fiber [10]. In an optical fiber, the index of refraction depends on the intensity

308

of the light propagating through the core [11]. In terms of the electric field amplitude, the index is

$$n(E) = n_0 + n_2 |E|^2. \tag{7}$$

After propagating through a length ℓ of optical fiber, the phase of a light wave has shifted by $\phi = 2\pi n(E)\ell/\lambda$. Fluctuations in the amplitude of the pump wave cause correlated fluctuations in the phase of the wave:

$$\delta\phi(\ell) = \frac{4\pi n_0 n_2 \ell}{\lambda} <E_1> \delta E_1 + \delta\phi(0) \tag{8}$$

where $\delta\phi(0) = \delta E_2(0)/E_1$ is the phase fluctuation at the input to the fiber. This equation applies even for pump amplitude fluctuations caused by the vacuum fields [12].

After propagating through a length ℓ of optical fiber, part of the phase fluctuations of a light beam are correlated with the amplitude fluctuations. The amplitude fluctuations cannot be changed by a fluctuating index of refraction and are thus equal at input and output.

The result is illustrated in Fig. 1. At the input to the fiber the fluctuations around the average optical amplitude can be represented as a fuzzy circle, without phase dependence. After propagating through the fiber, correlated amplitude and phase fluctuations arise. These fluctuations can be represented by a tilted ellipse as in Fig. 1b. The minor axis of the ellipse is smaller than the radius of the initial circle. The fluctuations along that axis can be made to appear as intensity modulation of a wave by phase shifting the average amplitude to be parallel to the minor axis. The result is a light beam with lower noise than the noise level at the input to the fiber, even when the input noise is at the vacuum level [10].

This analysis can be made more formal by introducing operators for the creation and destruction of photons, Hamiltonians and similar paraphernalia. The physics of the interaction, however, is captured in the treatment here [13].

Propagating the light through a length ℓ of optical fiber causes fluctuations δE_1 and δE_2 to become correlated according to:

$$\delta E_2(\ell) = \frac{4\pi n_0 n_2 \ell}{\lambda} <E_1>^2 \delta E_1(\ell) + \delta E_2(0); \tag{9a}$$

$$\delta E_1(\ell) = \delta E_1(0). \tag{9b}$$

The fluctuations are detected at the output of the fiber by beating them against a phase-shifted local oscillator wave derived from the transmitted average amplitude. If the local oscillator is in phase with the transmitted amplitude, only δE_1 is detected. If it is phase shifted by $90°$, δE_2 appears. In general, the detected quadrature is a linear superposition of these two:

$$\delta E_\theta = \delta E_1 \cos\theta + \delta E_2 \sin\theta \tag{10}$$

where θ is the phase shift of the local oscillator wave. The mean square averages for the quadrature operators are equal for vacuum states, but not for the squeezed states produced by the fiber. The ratio of the mean square fluctuations $- <\delta E_\theta(\ell)^2>$ for the squeezed states and $<\delta E_c^2>$ for coherent and vacuum states is:

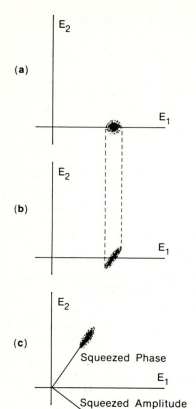

(a)

(b)

(c)

Squeezed Phase

Squeezed Amplitude

Figure 1. Quadrature amplitude plots for coherent and squeezed states. A coherent state with $< E_2 > = 0$ appears in Fig. 1a. The average amplitude is at the center of the fuzzy circle which represents the quantum fluctuations. The kind of squeezed state made by an optical fiber appears in Fig. 1b. The projection of the quantum fluctuations along the E_1 axis remains as in the coherent state in Fig. 1a, but fluctuations correlated to δE_1 are added to E_2. Figure 1c shows phase – shifted average amplitudes which produce squeezed states with phase or amplitude fluctuations less than a coherent state. Only the squeezed amplitude fluctuations reduce the noise as measured on a simple light detector.

$$\frac{< \delta E_\theta^2(\ell) >}{< \delta E_c^2 >} = V(\theta) = 1 + 2r \sin 2\theta + 4r^2 \sin^2\theta \qquad (11)$$

where $r = 2\pi n_0 n_2 \ell < E_1 >^2/\lambda$, is analogous to the squeeze parameter introduced in (4).

The mean square fluctuations in the optical power, detector current and photon number all vary with phase in this way. At $\theta = 0$, the mean square fluctuations for the fiber-produced squeezed state are equal to those for the coherent states in (3). The projection of the ellipse in Fig. 1 on a line at angle θ is $\sqrt{V}$.

Unfortunately, all of this can be disrupted by inelastic light scattering, fluorescence, absorption and loss. All of these processes add fluctuations. In optical fibers, the light scattering processes are most troublesome. It happens that the forward light scattering at low frequencies consists of correlated Stokes and anti-Stokes emission which together add to δE_2, but not δE_1 [14].

Figure 2 shows a spectrum of this light scattering for a bare optical fiber. The narrow peaks are resonances corresponding to vibrational modes of the cylindrical optical fiber.

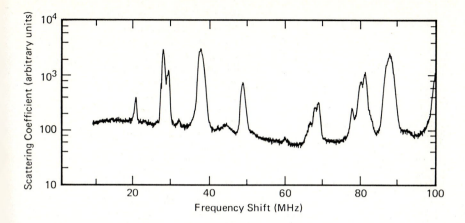

Figure 2. Spectrum of light scattering from a bare optical fiber at 60K. The narrow peaks correspond to Brillouin scattering by phonons guided by the cylindrical fiber structure (so-called Guided Acoustic Wave Brillouin Scattering or GAWBS). The noise floor between the peaks is due to another light scattering mechanism, probably related to structural relaxation.

Between those peaks is a background of light scattering due to structural relaxation. These two processes have different dependences on temperature. The phonon scattering varies linearly with temperature as shown in Fig. 3a; the background scattering is very different, as shown in Fig. 3b. At present, the details of this background light scattering are not well understood.

In the context of squeezing the vacuum, light scattering is a source of noise to be eliminated. Even at 2K, the noise from light scattering almost eliminates the noise reduction. At such a low temperature, the threshold for stimulated Brillouin oscillation is very low. Strong phase modulation is necessary to suppress the oscillation and allow a strong enough pump field [10].

The experimental noise level normalized to the vacuum noise level is plotted as a function of local oscillator phase in Fig. 4. For a narrow range of angles around –22°, the measured level is 12% below that of the vacuum. Other experiments with different technology have shown noise levels a factor of 3 below the vacuum [15].

A remarkable feature of these squeezed vacuum states is that attenuation increases the measured noise level. This is illustrated by Fig. 5 where the normalized noise level for a squeezed vacuum state and incandescent light (which is a sum of coherent states) are plotted as a function of the transmission of a neutral density filter [13]. When the transmission is 100%, the squeezed vacuum gives a noise level 10% below the vacuum level. As the transmission decreases, the normalized noise level for the squeezed vacuum rises. The normalized noise of the incandescent light remains constant at the vacuum level. At optical frequencies and ambient temperatures, black body radiation and vacuum fluctuations are synonymous. The partly absorbing filter radiates vacuum fluctuations ("darkness waves") and thus enforces the standard quantum limit.

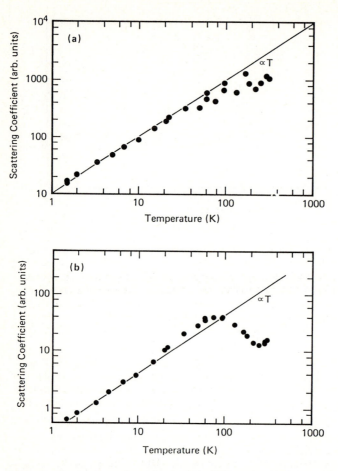

Figure 3. Temperature dependence of the light scattering in an optical fiber. The strong (GAWBS) phonon peak at 49.6 MHz in Fig. 2 shows linear temperature dependence as in Fig. 3a. The baseline noise level between the peaks (measured at 55 MHz) varies as shown in Fig. 3b.

The nonlinear interactions in an optical fiber can produce correlations among more than the two quadratures of (9a) and (9b). When there are two average amplitudes (E_x and E_y) at two frequencies in an optical fiber, the phase fluctuations analogous to those in (8) correlate with the amplitude fluctuations of both waves:

$$\delta\phi_x(\ell) = \delta\phi_x(0) + \frac{4\pi n_0 \ell}{\lambda} \left[n_2(\omega_x, \omega_x) <E_{1x}> \delta E_{1x} + n_2(\omega_x, \omega_y) <E_{1y}> \delta E_{1y} \right]. \quad (12)$$

In this equation, $n_2(\omega_x, \omega_y) = (24\pi/n_0)\chi^{(3)}(-\omega_x, \omega_x, \omega_y, -\omega_y)$ is a complex and spectroscopically interesting quantity which can be enhanced by Raman resonances that do not contribute to $n_2(\omega_x, \omega_x)$. If $|\omega_x - \omega_y|$ approaches a Raman mode frequency, the Raman resonance dominates all other effects. Figure 6 is a plot of the variation of

312

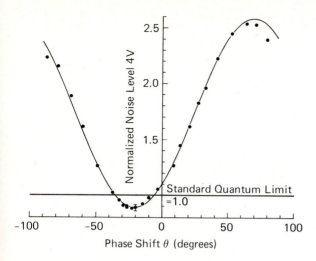

Figure 4. The noise-to-current ratio for the fiber produced squeezed state compared to the noise from an incandescent source (Standard Quantum Limit). The phase dependence of the noise with a net noise level below the standard quantum limit shows that this is a squeezed state. The radius of the solid circles is three times the estimated uncertainty of all of the measurements used at a particular phase while the error bracket shows the uncertainty of a single measurement.

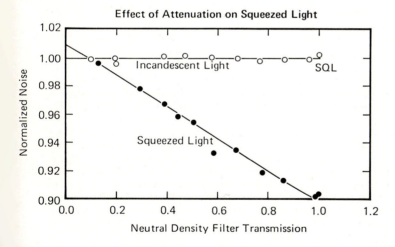

Figure 5. The effect of attenuation on squeezed light. The noise-to-current ratio is plotted here as a function of the transmission of a neutral density filter. The ratio is independent of attenuation for light at the standard quantum limit as produced by our incandescent source. However, the noise-to-current ratio of squeezed light rises toward the standard quantum limit as the transmission of the filter is decreased

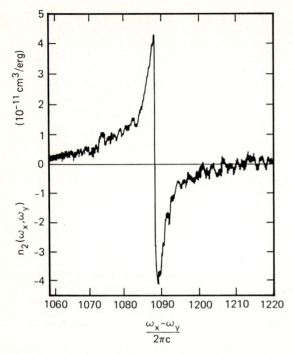

Figure 6. A plot of the Raman enhanced Kerr nonlinearity of calcite. The nonlinearity n_2 responsible for the ordinary squeezing interaction would be 4×10^{-13} esu. The resonance can be used to enhance the sensitivity of quantum nondemolition detection when the interacting waves are shifted by 1087 cm^{-1} from one another. For fused silica, $n_2 = 1.3 \times 10^{-13}$ esu.

$n_2(\omega_x, \omega_y)$ for a material with a strong Raman resonance. There is nearly a 100 fold enhancement of the nonlinearity over the value obtained when $\omega_x = \omega_y$. In many such materials, even $n_2(\omega_x, \omega_x)$ is larger than fused silica. The potential of this Raman enhanced quantum correlation has yet to be exploited, but the interactions observed in fused silica optical fibers show what might be expected. The fluctuations in $\delta\phi_x$ are correlated with the amplitude fluctuations of the other wave (*i.e.*, δE_{1y}). This effect allows one to measure the vacuum fluctuations - without perturbing them [16]. This is a form of back-action evading measurement [17] or quantum nondemolition detection (QND). Subsequent measurements of the amplitude fluctuation would result ideally in identical values. The uncertainty necessarily added by a quantum measurement appears in the phase of the measured wave, not the amplitude. Thus, not only can the vacuum be squeezed, it can be measured – repeatedly.

Squeezing and measuring the vacuum can enhance the sensitivity and precision of the techniques of laser spectroscopy. While remarkable, the techniques required are not unduly complex. The difficult thing is believing that an apparently intrinsic quantum limit can be transcended. The experimental proof is now at hand, the future will disclose the applications.

314

One spectroscopic application deserves to be mentioned in any volume honoring Arthur Schawlow: it is possible to build a dual beam spectrometer with this technology which makes absorption spectroscopy as sensitive as fluorescence spectroscopy. The basic idea is shown in Fig. 7. Light waves with two different frequencies are coupled in an optical fiber. The two wavelengths are separated at the output, and the phase fluctuations of one are measured as in squeezed state experiments. That phase fluctuation signal measures the amplitude fluctuations at the other wavelength. The second beam passes through a weakly absorbing sample. The absorption may be modulated by dithering the laser frequency, *etc.*,

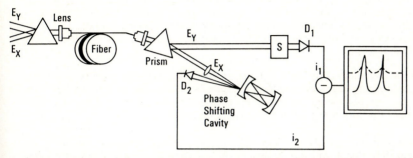

Figure 7. A dual beam "spectrometer" employing QND in an optical fiber. The phase fluctuations of the beam E_x become correlated with the amplitude fluctuations of the beam E_y in the optical fiber. The current produced by the detector D_2 is a QND measurement of the amplitude fluctuations. A weakly absorbing sample at S modifies the amplitude of the wave reaching detector D_1. The quantum fluctuations (which had been previously measured at D_2) are subtracted off allowing the detection of small changes caused by the sample.

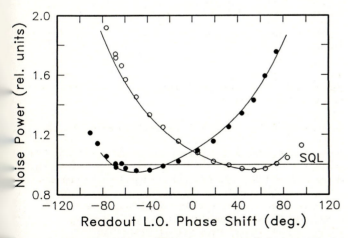

Figure 8. Noise level for the "QND Spectrometer" shown in Fig. 7 normalized to the standard quantum limit noise level on detector D_1, plotted as a function of the cavity phase shift. The noise level is reduced by 5% below the standard quantum limit for the appropriate choice of optical phase. The open circles correspond to subtracting the electrical currents from the two detectors as shown in Fig. 7, while the solid circles correspond to the sum of the currents.

and the transmitted intensity is detected. The signals from the detectors which monitor the transmitted intensity and the phase of the other beam are then subtracted. Since the quantum noise for the beam passing through the sample causes phase fluctuations on the other beam which can be accurately measured, the quantum noise of the beam passing through the sample can be subtracted from the detector signal. All that remains are the effects of the sample on the light that passed through.

Fluctuations from the light source have long been subtracted from spectroscopy signals using dual beam techniques. The two beams are conventionally split optically using a partially reflecting mirror. Such a technique can eliminate classical fluctuations, but the quantum noise would still constitute a noise floor – the standard quantum limit. The back-action evading spectrometer proposed here would get below that noise floor, enhancing sensitivity even further.

Figure 8 shows a very early result of this sort, the noise level in a system such as shown in Fig. 7. The ordinary quantum noise level is shown as a horizontal line labelled SQL. The curves and data points show the noise level in a back-action evading system. Even in this early case, the noise level is 5% below the ordinary limit [12].

Lasers have already revolutionized spectroscopy by providing sources of strong coherent light. The next great breakthrough might be light with much less noise than even fully coherent light. The future remains incredibly bright.

REFERENCES

1. E. G. Harris: In A Pedestrian Approach to Quantum Field Theory (Wiley-Interscience, 1972), Chapter 4.
2. L. I. Schiff: In Quantum Mechanics (McGraw-Hill, 1955), p.60.
3. B. R. Mollow: Phys. Rev. 188, 1969 (1969).
4. C. Cohen-Tannoudji, B. Diu, F. Laloe: In Quantum Mechanics (Wiley-Interscience/Hermann, 1977), p. 618.
5. A. L. Schawlow, C. H. Townes: Phys. Rev. 112, 1940 (1958).
6. T. A. Welton: Phys. Rev. 125, 804 (1962).
7. R. Bondurant, J. H. Shapiro: Phys. Rev. D30, 2548 (1984).
8. C. M. Caves: Phys. Rev. D23, 1693 (1981).
9. D. F. Walls: In Nature (London) 306, 141 (1983).
10. R. M. Shelby, M. D. Levenson, S. H. Perlmutter, R. G. DeVoe, D. F. Walls: Phys. Rev. Lett. 57, 691 (1986).
11. Y. R. Shen: In The Principles of Nonlinear Optics (John-Wiley and Sons, 1984), 303 ff.
12. M. D. Levenson, R. M. Shelby: In Four Mode Squeezing and Applications, Optica Acta (to be published).
13. G. J. Milburn, M. D. Levenson, R. M. Shelby, D. F. Wall, S. H. Perlmutter, R. G. DeVoe: J. Opt. Soc. Am. B (to be published).
14. R. M. Shelby, M. D. Levenson, P. W. Bayer: Phys. Rev. 31, 5244 (1985).
15. L. A. Wu, H. J. Kimble, H. Wu, J. L. Hall: Phys. Rev. Lett. 57, 2520 (1986).
16. M. D. Levenson, R. M. Shelby, M. Reid, D. F. Walls: Phys. Rev. Lett. 57, 2473 (1986).
17. C. M. Caves: In Quantum Optics, Experimental Gravitation and Measurement Theory, ed. by P. Meystre, M. O. Scully (Plenum Press, New York, 1983), p. 567; C. M. Caves, K. S. Thorne, R. W. P. Drever, V. D. Sandberg, M. Zimmerman: Rev. Mod. Phys. 57, 341 (1980).

Raman Spectroscopy of Biomolecules

T.J. O'Leary

Department of Cellular Pathology, Armed Forces Institute of Pathology, Washington, DC 20306, USA

1. INTRODUCTION

Raman spectroscopy is now so highly dependent on excitation by laser light sources that it hardly seems possible that C.V. Raman could have first observed this faint scattering phenomenon using filtered sunlight [1-4]. First predicted in 1923 by Adolf Smekel [5], the Raman effect is the scattering of an incident photon shifted in frequency by an allowed vibrational frequency of the scattering molecule. Thus, the Raman spectrum, like the infrared spectrum, yields information about the vibrational frequencies of molecules. Unlike infrared absorption, which results from an interaction of the incident radiation with the electric dipole of the molecule, Raman scattering results from an interaction of this radiation with the polarizability tensor [6]. The practical result is that the Raman spectrum does not yield exactly the same vibrational information as does the infrared spectrum, since the selection rules are different. A second result is that the Raman effect results in a very weak scattering spectrum. In classical Raman spectroscopy, only about one photon in 10^8 which reaches a sample is scattered (hence the need for sources with high spectral brightness). In spite of this severe limitation, investigations of the Raman effect and its use in chemical spectroscopy proceeded relatively rapidly even prior to the availability of laser sources, and hundreds, if not thousands, of papers on Raman spectroscopy had been published by the end of the 1950's. After the laser replaced the Toronto arc as the standard light source [7] an even wider range of investigations became possible.

In this short review I summarize a small subset of the published work on Raman spectroscopy applied to polypeptides, proteins, and lipids which relates to my own work on the structure of biological membranes. I attempt to identify some of "firsts" in this field, to delineate some types of biological problems which can now be tackled using Raman spectroscopy, and to point out current developments in theory and instrumentation which may well bring about even greater utility. I will ignore a great many important contributions, particularly if they have not had a fairly direct impact on my own research. Readers who wish a more extensive exploration of these studies are referred to the many excellent review articles and monographs which have appeared in the last few years [8-12].

2. THEORY

When a collection of molecules is illuminated by monochromatic light with frequency ν_0, the exciting radiation induces oscillation of the electrons in the sample. The induced polarizability P is equal to the product of the molecular polarizability tensor α and the electric field E arising from the incident light:

$$P = \alpha E. \tag{1}$$

The intensities I_{nm} of the vibrational Raman bands are proportional to the square of the induced transition moment matrix elements P_{nm}

$$I_{nm} = [64\pi^2/(3c^2)](\nu_o + \nu_{nm})^4 P_{nm}^2, \tag{2}$$

where ν_{nm} is the frequency of the transition between state n and state m, and c is the speed of light. The P_{nm} in turn are given by

$$P_{nm} = (1/h) \sum_r \{[M_{nr}M_{rm}/(\nu_{rn} - \nu_o)] + [M_{nr}M_{rm}/(\nu_{rm} + \nu_o)]\}E, \tag{3}$$

where the various M are transition moments, and the subscript r refers to an infinite number of virtual states. The matrix elements P_{nm} may also be obtained by taking the appropriate quantum mechanical integral

$$P_{nm} = (\int \phi_m *\alpha\phi_n dQ)E, \tag{4}$$

where ν_m and ν_n are the vibrational wave functions of the final and initial states respectively of the scattering molecule, and α is a function only of the nuclear normal vibrational coordinates Q. If the force constants and geometry of the scattering molecule are known accurately, then the vibrational frequencies may be calculated relatively easily using the Wilson GF matrix method [13]; from a knowledge of the molecular symmetry one can also usually determine which vibrations will be observed in the infrared spectrum and which will be seen in the Raman; the intensities are much more difficult to estimate on theoretical grounds.

If the scattering molecule has an absorption band that overlaps or is near the frequency of the exciting light, then the efficiency of the Raman scattering process may be increased by many orders of magnitude, as predicted by (3). The theory behind this process, which is known as resonance Raman scattering, will not be given here because it is both complex and incomplete. I have included several references for anyone who wishes to delve further [14.

3. INSTRUMENTATION

The basic Raman spectrometer system is illustrated in Fig. 1. The light source is almost always a laser; most commonly either the 488 or 514 nm argon ion laser line is employed. Nearby plasma lines are filtered out using a grating or a Claassen filter. The sample is placed in a capillary tube, and scattered light is typically analyzed using a double or triple grating spectrograph; since shortly after the inauguration of laser excitation a photomultiplier tube has been used to record the scattered photons.

Several difficulties plague the Raman experiment in comparison with the infrared absorption measurement. The extreme inefficiency of the scattering process often necessitates the use of relatively wide spectrometer slits to record a spectrum in a reasonable amount of time, with a necessary reduction of spectral resolution. In addition, even small amounts of fluorescence, common in biological materials, can swamp the small Raman signal. The use o quenching agents, such as iodine, is only sometimes successful in eliminatin this interference, and the use of infrared excitation sources has been hampered by the lower sensitivity of detectors for this wavelength region an the much lower scattering efficiency which results from the fourth power dependence of scattering on the frequency of the incident light, as seen in

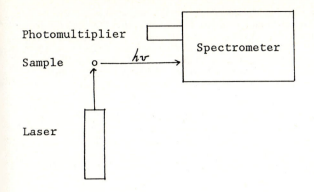

Fig. 1. Schematic illustration of typical Raman spectrometer system

Photomultiplier

Sample

Laser

4. AMINO ACIDS, POLYPEPTIDES AND PROTEINS

Edsall [19] credits Wright and Lee [20] with first investigating the Raman scattering properties of the amino acids glycine, alanine, tyrosine, and cystine. Nevertheless, the first systematic Raman spectroscopic investigations of these compounds really started with Edsall himself, who carried this work on for a number of years. His early studies [19,21,22] were concerned primarily with the effects of amino acid ionization on the Raman spectrum, but during his later studies he also analyzed the first spectra of polypeptides and proteins [23]. In the course of these studies, Edsall correctly deduced that bands centered at approximately 1650, 1526, 1400, 1250 and 915 cm^{-1} represented vibrational modes of the Amide bonds. Subsequent work by a number of people [24-30] demonstrated the sensitivity of Amide I (1630 - 1690 cm^{-1}) and Amide III (1220-1300 cm^{-1}) band positions and shapes to the secondary structure of the proteins (Table I). Lippert [31] developed a method for utilizing the spectra of proteins dissolved both in H_2O and D_2O to quantitate the components of the secondary structure. Recently, Williams has developed "composite" spectra typical of the various types of protein secondary structure (α-helix, β-sheet, β-turn and disordered structures) and has demonstrated that they may be used to accurately determine the fractional contributions of these secondary structures to the spectra of proteins and polypeptides [32-35]. Empirical use of the Raman spectrum to deduce secondary structure has become an important tool for studying membrane systems, since circular dichroism spectroscopy, the most commonly used method for protein secondary structure determination, cannot easily be applied to turbid or absorbing solutions and dispersions.

Table 1. Typical Frequencies for Amide I and III Bands of Various Protein Structures

	Frequency (cm^{-1})	
	Amide I	Amide III
α-helix	1645-1657	1260-1300
β-sheet	1660-1680	1225-1245
β-turn	1660-1690	1230-1290
Random coil	1660-1670	1240-1250

Figure 2 illustrates the use of the Amide I region Raman spectrum in studying systems which are not amenable to analysis by circular dichroism spectroscopy. The spectrum of the crystalline polypeptide immunosuppressant Cyclosporine A (spectrum A), for which a detailed molecular structure is known from X-ray diffraction experiments is very similar to that for the membrane associated polypeptide (spectrum B), demonstrating that no significant conformational changes occur when the polypeptide is incorporated into the membrane [36].

Although normal coordinate calculations of amino acids and polypeptides contributed greatly to our understanding and interpretation of amide region spectra [37-39], they have not until recently provided information which is particularly useful in analysing the secondary structure of particular molecules. This is changing, largely because of the extensive and careful

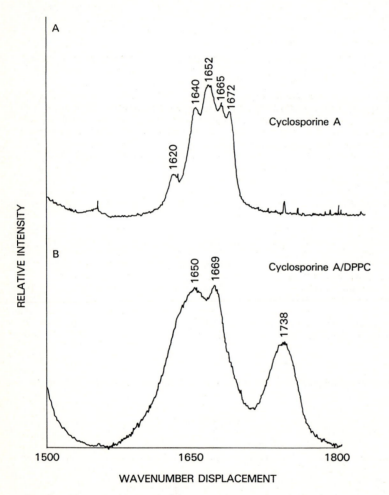

Fig. 2. Raman spectra of (A) polycrystalline cyclosporine A and (B) cyclo-sporine A/DPPC dispersion in the Amide I spectral region. The peak at 1738 c is due to a DPPC C=O stretching mode

studies of Krimm and his coworkers [40-43], who have recently applied the calculations to two very similar putative structures for a gramicidin A - ion complex, and have been able to unambiguously differentiate between the two [44]. This work strongly suggests that much more detailed structure determinations will be possible as the force fields are refined and measurements improved.

5. LIPIDS AND LIPID MEMBRANES

Lippert and Peticolas [45] first published the Raman spectrum of a lipid multilayer assembly, and demonstrated that the spectroscopic features of the $1000 - 1200$ cm^{-1} C-C stretching mode region are characteristic of the state (gel or liquid crystalline) of the lipid assembly and can be used to follow the thermally induced phase transitions, such as the gel to liquid crystalline phase transition, and the perturbation of these transitions by non-phospholipid membrane constituents, such as cholesterol. Although Raman spectroscopy is most often used to study the structure of the hydrocarbon chains of lipid molecules, distinct spectral features are available which reflect the structure of all three regions of a lipid membrane - headgroup, interface, and acyl chain (Fig 3, Table 2).

Figure 5 illustrates schematically the major changes which occur during the phospholipid pretransition and gel to liquid crystalline (main) transition. The pretransition is characterized by a reduction of acyl chain tilt with respect to the plane of the bilayer, the appearance of "ripples", and the introduction of a few *gauche* isomers into the chain termini. The gel to liquid crystalline phase transition results in a loss of these ripples,

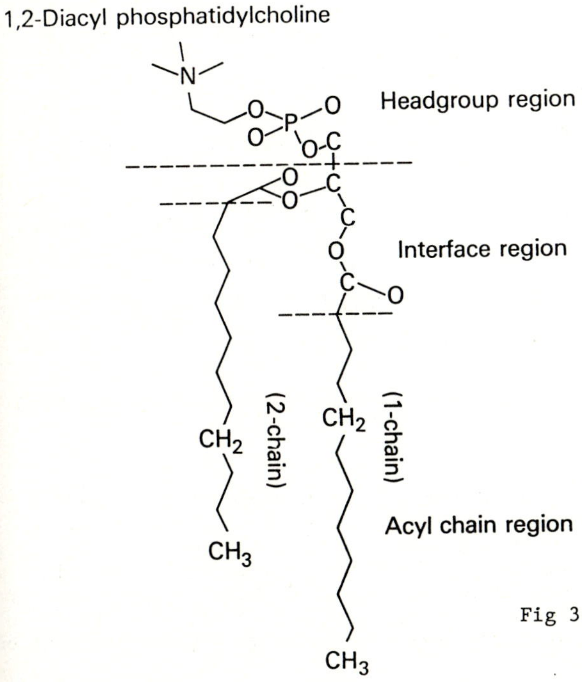

1,2-Diacyl phosphatidylcholine

Fig 3. Chemical structure of diacylphosphatidylcholine molecule

Table 2. Spectral regions particularly useful for the investigation of membrane structure [from 46]

Frequency range (cm^{-1})	Chemical Moiety	Assignment	
710- 720	Headgroup	C-N	symmetric stretch
750- 760	Headgroup	O-P-O	symmetric stretch
1700-1800	Interface	C=O	stretch
1000-1200	Acyl chain	C-C	stretch
1400-1500	Acyl chain	CH$_2$	deformation
2000-2300	Acyl chain	C-D	stretch
2800-3100	Acyl chain	C-H	stretch

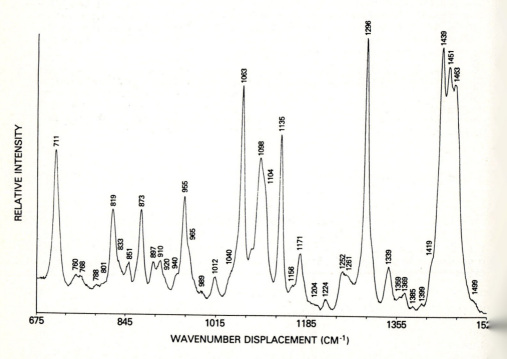

Fig 4. Raman spectrum of polycrystalline diisopalmitoylphosphatidylcholine

lateral expansion of the bilayer, and a much larger amount of *trans-gauche* isomerization. The CH stretching mode region has proven to be especially valuable for following the phase transitions because the dramatic change (Fig 6) in the Raman spectrum, as quantitated by the I_{2935}/I_{2880} cm^{-1} intensity ratio is linearly related to the entropy of transition [47], suggesting that this ratio is a direct probe of the number of gauche conformers in the lipid acyl chain, as well as providing confirmation of calorimetric measurements or lipid phase transitions.

322

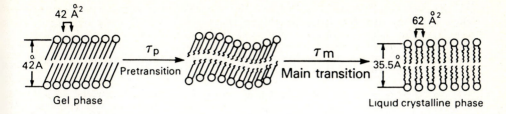

Fig 5. Schematic illustration of changes which occur at lipid bilayer phase
transitions

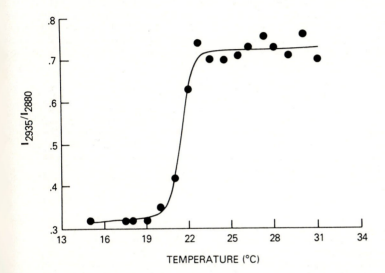

Fig. 6. Temperature profile of diisopalmitoylphosphatidylcholine CH stretching
mode region band intensity ratios, showing dramatic change at the gel to liquid
crystalline phase transition temperature

My own work on lipid spectroscopy has focused on understanding the role of
acyl chain and headgroup chemical structure in determining the physical state
and phase transitions of membrane assemblies, and in understanding how
incorporation of nonlipid membrane constituents influences the details of
membrane structure and phase transitions. For example, in work with John
Silvius of McGill, we were able to show that phosphatidylcholines with acyl
chains which are branched at preterminal methylene group have low temperature
gel phases in which the acyl chains are packed in an orthorhombic or
monoclinic subcell, rather than the quasihexagonal subcell typically formed by
nonbranched phosphatidylcholines under similar conditions. This conclusion can
be drawn by the appearance of a single spectroscopic feature, a shoulder at
approximately 1418 cm^{-1}, in spectra of the isobranched species which is not
present in spectra of the straight chain phosphatidylcholines [48]. By
determining a gel phase I_{2935}/I_{2880} ratio which is substantially lower for gel
phase diisopalmitoylphosphatidylcholine than for dipalmitoylphosphatidylcholine,
we could show that fewer *gauche* isomers were present in the acyl chains of the
former than of the latter. When used together with scanning calorimetry, to

323

more precisely determine phase transition temperatures, Raman spectroscopy allows a rather complete mapping of the phase diagram of lipids and lipid mixtures. For this reason, we have used Raman measurements to explore not only the effects of changes in acyl chain structure, as above, but also effects due to changes in headgroup structure [49], solvation [50-52], and addition of general anesthetics [53,54].

By using molecules in which isotopic labels have been introduced into lipid molecules at specific lipid sites, it is possible to increase the precision of the structural information obtained by Raman spectroscopy. For example, Bansil and coworkers [55] have shown that selective deuteration of lipid acyl chains at specific methylene sites can be used to probe the structure of the acyl chain region with great detail. Together with Ira Levin I was able to use these lipids to study the differences in how cholesterol alters the structure of phosphatidylcholine acyl chains near the headgroup and those near the terminal methyl group [56]. In earlier studies, Mendelsohn found that he could independently study the structure of two different types of lipid molecu within a lipid bilayer by completely deuterating the acyl chains of one while leaving unaltered those of the other species [57].

A second type of specific probe is provided by the availability of "reporter molecules" which demonstrate resonance Raman scattering. The carotenoids and similar molecules may, under some circumstances, be used to monitor the phase of lipid membranes. Mendelsohn [58] has shown that either the absolute intensity or the frequency of the 1525 cm^{-1} zeaxanthin feature may be used to monitor the physical state of phosphatidylcholine membranes. Similarly, the amphotericin B resonance Raman spectrum reflects the state of phosphatidylcholine-cholesterol membranes [59].

6. VIRUSES, CELLS AND TISSUES

Although there has not been as much work done on the Raman spectroscopy of intact organisms as there has on biomolecules, that which has been done provides encouragement that further studies may be worthwhile. Thomas [12] has shown that Raman spectroscopy is an extremely effective tool for studying the structure of both viral proteins and nucleic acids, both isolate and in intact virions. Raman spectra have been obtained on intact erythrocyt membrane [60], mitochondria [61], bacteria [62], and lymphocytes [63,64]. Although these studies have not yet demonstrated great utility for Raman spectroscopy in the study of living systems more complicated than viruses, they have clearly demonstrated that interpretable spectra may be obtained. It's a lot of fun to see the Raman spectrum of a bunch of lymphocytes come o the spectrometer! As reporter molecules are developed which reflect specific structural or metabolic characteristics of these organisms and their constituent molecules, it seems likely that Raman spectroscopy will provide another useful tool for cell biologists.

7. SUMMARY

Raman spectroscopy is a very useful tool for studying the structure of a number of biological structures, particularly proteins and membranes. Advances in the understanding and interpretation of these spectra make Raman spectroscopy second only to X-Ray crystallography in elucidating the structures of these compounds. A number of advances in instrumentation and technique are underway which may significantly enhance the usefulness of Ram spectroscopy in real biological problems.

Use of photodiode array detectors has greatly increased the speed with which spectra may be acquired, making possible ever more detailed studies of protein and membrane dynamics. By using resonance Raman active "reporter molecules" or probes, which are sensitive to the structural state of a biological molecule, or to such things as membrane potential and ion concentration, it is likely that Raman spectroscopy will find increasing use in studying kinetics of not only isolated biomolecules, but also of intact cells and tissues. In addition, resonance Raman studies of peptide bonds performed using ultraviolet excitation may improve both the speed and accuracy of protein secondary structure determination, making possible sensitivity superior to that currently afforded by circular dichroism techniques.

Although use of photodiode array detectors is limited at present to low resolution studies, the use of Fourier transform spectrometers to acquire Raman spectra promises to increase the obtainable resolution of Raman systems by several orders of magnitude, and, because of the "multiplex advantage", make possible the increasing use of infrared radiation for acquisition of Raman spectra. This will allow ready acquisition of conventional Raman spectra in preparations which fluoresce when excited with visible radiation, thus providing an alternative to coherent anti-Stokes Raman scattering (CARS), which has often been the only practical alternative for measuring Raman spectra on such compounds.

When I was a graduate student, Dr. Schawlow reminded me that bromine, a molecule on which I was working, was "a molecule with one atom too many." The molecules that I've been studying lately have a few more atoms than one atom too many. Through the laser, Dr. Schawlow has made it possible to learn a bit more even about messy molecules like these, and even perhaps, about how people are put together. I suspect that is pretty far beyond what he and Townes were thinking about when they first proposed "optical masers."

8. REFERENCES

1. C.V. Raman: Ind. J. Physics 2, 387 (1928)
2. C.V. Raman, K.S. Krishnan: Ind. J. Physics 2, 399 (1928)
3. C.V. Raman, K.S. Krishnan: Nature 121, 501 (1928)
4. C.V. Raman, K.S Krishnan: Proc. Royal Soc. (London) 122A, 23 (1928)
5. A. Smekal: Naturwiss. 11, 873 (1923)
6. G. Placzek: In Raleigh and Raman Scattering, UCRL Trans. 526L from Handbuch der Radiology, Vol 2, ed. by E. Marx, (Akademische Verlagsgesellshaft, Leipzig 1934), p 209
7. S.P.S. Porto, D.L. Wood: J. Opt. Soc. Amer. 52, 251 (1962)
8. A.T. Tu: Raman Spectroscopy in Biology: Principles and Applications (John Wiley and Sons, New York 1982)
9. F.S. Parker: Applications of Infrared, Raman, and Resonance Raman Spectroscopy in Biochemistry, (Plenum Press, New York 1983)
10. P.R. Carey: Biochemical Applications of Raman and Resonance Raman Spectroscopies (Academic Press, New York, 1982)
11. T.G. Spiro, B.P. Gaber: Annu. Rev. Biochem. 46, 553 (1977)
12. G.B. Thomas: In Infrared and Raman Spectroscopy, Ed. by E.G. Brahme and J.G. Grasselli, Vol 1 (Dekker, New York 1977), p717
13. E.B. Wilson, J.C. Decius, P.C. Cross: Molecular Vibrations, The Theory of Infrared and Raman Vibrational Spectra (McGraw-Hill, New York 1955)
14. A.C. Albrecht: J. Chem. Phys. 34, 1476 (1956)
15. A.C. Albrecht, M.L. Hutley: J. Chem. Phys. 55, 4438 (1971)
16. J. Tang and A.C. Albrecht: In Raman Spectroscopy, Ed. by H.A. Szymanski, Vol 2 (Plenum Press, New York 1970) p 33

17. L.A. Nafie, P. Stein and W. Peticolas, Chem. Phys. Lett. 12, 131 (1971)
18. W. Peticolas, L. Nafie, P. Stein and B. Fanconi, J. Chem. Phys. 52, 1576 (1970)
19. J.T. Edsall: J. Chem. Phys. 4, 1 (1936)
20. N. Wright, W.C. Lee: Nature 136, 300 (1935)
21. J.T. Edsall: J. Chem. Phys. 5, 225 (1937)
22. J.T. Edsall: J. Chem. Phys. 5, 508 (1937)
23. D. Garfinkel, J.T. Edsall: J. Amer. Chem. Soc. 80, 3318 (1958)
24. N.T. Yu, C.S. Liu: J. Amer. Chem. Soc. 94, 5127 (1972)
25. G.S. Bailey, J. Lee, A.T. Tu: J. Biol. Chem. 254, 8922 (1979)
26. R.L. Lord, N.T. Yu: J. Mol. Biol. 50, 509 (1970)
27. H. Brunner, M. Holz: Biochim. Biophys. Acta 379, 408 (1975)
28. W.S. Craig, B.P. Gaber: J. Amer. Chem. Soc. 99, 4130 (1977)
29. V.C. Lin, J.L. Koenig: Biopolymers 15, 203 (1978)
30. M. Pezolet, M. Pigeon-Gosselin, L. Coulombe: Biochim. Biophys. Acta 453, 502 (1976)
31. J.L. Lippert, D. Tyminski, P.J. Desmeules: J. Amer. Chem. Soc. 98, 7075 (1976)
32. R.W. Williams: J. Mol. Biol. 166, 581 (1983)
33. R.W. Williams, A.K. Dunker: J. Mol. Biol. 152, 783 (1981)
34. R.W. Williams, M.M. Teeter: Biochem. 23, 6796 (1984)
35. R.W. Williams: J. Biol. Chem. 260, 3937 (1985).
36. T.J. O'Leary, P.D. Ross, M.L. Lieber, I.W. Levin: Biophys. J. 49, 795 (1986)
37. S. Susuki, T. Shimanouchi, M. Tsuboi: Spectrochim. Acta 19, 1195 (1963)
38. K. Itoh, T. Shimanouchi: Biopolymers 9, 383 (1970)
39. K. Itoh, T. Shimanouchi: Biopolymers 10, 1419 (1971)
40. W.H. Moore, S. Krimm: Biopolymers 15, 2439 (1976)
41. W.H. Moore, S. Krimm: Biopolymers 15, 2465 (1976)
42. J. Bandekar, S. Krimm: Proc. Natl. Acad. Sci. USA 76, 774 (1979)
43. J.F. Rabolt, W.H. Moore, S. Krimm: Macromolecules 10, 1065 (1977)
44. V.M. Naik, S. Krimm, Biophys. J. 49, 1147 (1986)
45. J. L. Lippert, W.L. Peticolas: Proc. Natl. Acad. Sci. USA 68, 1572 (1971)
46. I. W. Levin: Adv. Infrared. Raman. Spect. 11, 1 (1984)
47. C.H. Huang, J.R. Lapides, I.W. Levin: J. Amer. Chem. Soc. 104, 5926 (1982)
48. J. R. Silvius, M. Lyons, P.L. Yeagle, T.J. O'Leary: Biochem. 24, 5388 (1985)
49. J.R. Silvius, P.M. Brown, T.J. O'Leary: Biochem. 25, 4249 (1986)
50. T.J. O'Leary, I.W. Levin: J. Phys. Chem. 88, 1790 (1984)
51. T.J. O'Leary, I.W. Levin: J. Phys. Chem. 88, 4074 (1984)
52. T.J. O'Leary, I.W. Levin: Biochim. Biophys. Acta 776, 185 (1984)
53. T.J. O'Leary, P.D. Ross, I.W. Levin: Biochem. 23, 4636 (1984)
54. T.J. O'Leary, P.D. Ross. I.W. Levin: Biophys. J. 50, 1053 (1986)
55. R. Bansil, J. Day, M. Meadows, D. Rice, E. Oldfield: Biochem. 19, 1938 (1980)
56. T.J. O'Leary, I.W. Levin: Biochim. Biophys. Acta 854, 321, (1986)
57. R. Mendelsohn, J. Maisano: Biochim. Biophys. Acta 506, 192 (1978)
58. R. Mendelsohn, R.W. van Holten: Biophys. J. 27, 221 (1979).
59. M.R. Bunow, I.W. Levin: Biochem. Biophys. Acta 464, 202 (1977)
60. J.L. Lippert, L.E. Gorczyca, G. Meiklejohn: Biochim. Biophys. Acta 382, (1975)
61. F. Adar, M. Erecinska: Biochem. 17, 5484 (1978)
62. W.F. Howard, W.H. Nerslon, J.F. Sperry: Appl. Spectrosc. 34, 72 (1980)
63. L.V. DelPrior, A. Lewis, K.A. Shat: Membrane Biochem. 5, 97 (1984)
64. T.J. O'Leary, unpublished data
65. The opinions or assertions contained herein are the private views of the author, and are not to be construed as official or as reflecting the views of the Department of the Army or the Department of Defense

For Arthur Schawlow on His Sixty-fifth Birthday

1)   There once was a man named Johannes Balmer,
       He lived in eighteen eighty-five.
       At the time he looked at the hydrogen spectrum
       None of us were alive.

2)   Now we have a friend named Arthur Schawlow,
       This year he is sixty=five.
       When he last looked at the hydrogen spectrum
       All of us were alive.

3)   When they invented the optical maser,
       And sent off a paper to the Physical Review,
       Then they went back to the lab and redid the calculation --
       They forgot a factor of 2.‡

4)   Now we stimulate the atoms with coherent radiation
       M, A, S E R.
       It's a Money Acquisition Scheme for Expensive Research...
       But we won't use it for war.

5)   Now back in the old days with the graduate students
       The Boss kept us busy and on the run.
       In those days the life was pretty exciting,
       We had some fun.*

6)   We took a crystal of magnesium oxide,
       And doped it up with chromium.
       Lo and behold: there's a beautiful R line at
       Sixty nine eighty one.

7)   Now when the Boss comes down the hall in the morning,
       He always has a story to tell;
       But with such a big laugh when he gets to the punch line,
       We miss the point of it all.

---

‡ Forgive the "poetic license". Of course they did <u>not</u> forget a factor of 2.

* For some of the verses the last line is a beat or so shorter, and the alternate music can be used. Also, of course, feel free to change the note values or the syncopation so as to fit the words in each of the verses.

8)   Now our friend Arthur went up to the city,
     Got a double balloon from the S.F. zoo:
     Mickey Mouse on the inside and clear on the outside,
     Mickey Mouse is blue.*

9)   So then he pulls the trigger on his Buck Rogers Laser,
     Flashes ruby crystal mounted inside.
     While the beam passes through the outer balloon,
     Mickey Mouse is fried.*

10) Now the Boss had the idea of the laser eraser,
     Helpin' out typists who make mistakes.
     Pull the trigger on the laser that is pointed at the character:
     It evaporates.*

11) We can measure the wavelength of light with a ruler
     In the optics class, we can demonstrate:
     From the pattern of diffraction we can calculate the wavelength:
     Sixty three twenty eight.

12) Now just last night I went down to the Safeway,
     And there were the lasers at the check-out stands.
     Reading the prices -- hopefully correctly,
     From the Zebra bands.*

13) Yes we have a friend in Arthur Schawlow,
     This year he is sixty-five,
     When he last looked at the hydrogen spectrum
     We were all alive.*

Peter Scott
U. C. Santa Cruz
June 1986

The Schawlow-Hänsch group when the Nobel Prize was announced, October 9, 1981.
Left to Right: Frans Alkamade, Gerard Morgan, Steven Rand, Philipp
Dabkiewics, Fred-a Jurian, Louis Bloomfield, Arthur Schawlow, Theodor
Hänsch, Edward Hildum, Antoinette Taylor, Kenneth Sherwin, Harald Gehrhardt,
Kevin Jones, and Li-Shing Lee.

The Authors

Valery M. Baev was born in Kamchatka in the USSR in 1948. He received his diploma in physics from the Moscow Physico-Technical Institute in 1972, and Ph.D. from the Lebedev Physical Institute in 1980. Since 1972, he has been a research fellow of the Lebedev Institute in Moscow. The main subject of his research has been intracavity laser spectroscopy, a field in which Arthur Schawlow published one of the pioneering papers in 1972, with Peter Toschek and Theodor Hansch as co-authors. Dr. Baev has visited Prof. Toschek's laboratory to collaborate on intra-cavity laser spectroscopy experiments, both at Heidelberg in 1979-1981 and at Hamburg in 1985-1987, most recently as an Alexander von Humbolt fellow.

Allister I. Ferguson graduated with a B.Sc degree in Physics from the University of St. Andrews, Scotland,in 1974. After spending a year at Imperial College, London, where he first met Art Schawlow, he completed the degree of Ph.D.in 1977 at the University of St. Andrews. His thesis work involved the development of a frequency doubled dye laser for spectroscopy. In 1977 he was awarded a Lindemann Fellowship and moved to Stanford University where he was a visiting scholar until February 1979. While at Stanford he worked on coherent multiple pulse spectroscopy using mode-locked dye lasers and spectroscopy of atomic hydrogen. He collaborated with Art Schawlow and others on Doppler-free optogalvanic spectroscopy. Since that time he has held posts in St. Andrews, Oxford and has lectured at Southampton University since 1984. His current interests include hydrogen spectroscopy, FM dye lasers, fibre lasers, diode pumped Nd:YAG lasers.

Theodor W. Hänsch is Professor of Physics at the University of Munich and Director at the Max-Planck-Institute for Quantum Optics in Garching. He returned to his native Germany in 1986, after spending 16 years at Stanford University. Joining the laboratory of Arthur L. Schawlow as a post-doc in 1970, he was appointed an associate professor at the Department of Physics in 1972 and a full professor in 1975. Working at times in close association with Arthur Schawlow, he has pursued research in laser physics, quantum electronics, nonlinear spectroscopy, and atomic physics. Among his principal research interests have been methods of high resolution Doppler-free laser spectroscopy and their applications for the measurement of fundamental constants and for the test of basic laws of physics. In 1973, he was named "California Scientist of the Year", jointly with Arthur Schawlow. He earned his doctorate from the University of Heidelberg, Germany, in 1969.

Serge Haroche, born in 1944, was a student at Ecole Normale Supérieure in Paris. He got his Ph.D. in Physics under the supervision of Claude Cohen-Tannoudji in 1971. His thesis was a theoretical and experimental study of optically pumped atoms interacting with strong radiofrequency fields. The concept of "dressed atom" was first introduced during this work, before being generalized in quantum optics to atoms interacting with intense laser fields. In 1972, Serge Haroche came as a postdoc to Stanford where, with A. Schawlow and J. Paisner, he performed the first quantum beat spectroscopy experiment using dye lasers.

At Ecole Normale since 1973, Serge Haroche and his co-workers have performed a large number of experiments in the general area of spectroscopy and quantum optics of Rydberg atoms. Recently, they have been interested in the field of cavity quantum electrodynamics, dealing with such topics as the enhancement of the spontaneous emission rate of excited atoms in microwave cavities, superradiance of small samples of Rydberg atoms, microscopic Rydberg masers, etc...

Serge Haroche was a visiting Professor at Harvard in 1981. Since 1984, he has spent part of his time at Yale where he has engaged in experiments studying the properties of atoms confined in small metallic structures (Rydberg atom-surface Van der Waals interactions, inhibition of optical spontaneous emission in micronsized cavities, etc...).

Wendell T. Hill, III is a National Science Foundation Presidential Young Investigator and an Assistant Professor in the Institute for Physical Science and Technology at the University of Maryland. As a Schawlow student from 1976 to 1980, he gained experience with several laser spectroscopic techniques such as two-photon and polarization spectroscopies. His thesis project, "Intracavity Absorption Spectroscopy", involved constructing one of the most sensitive apparatuses at the time for detecting very weak absorption and characterizing the peculiar lineshapes associated with this technique; the apparatus was used to study the visible absorption bands of molecular oxygen. After completing graduate school, he was a National Research Council postdoc at the National Bureau of Standards where he studied many-electron correlation effects in the vacuum ultraviolet spectra of atomic ions. His current research includes investigating the details of the atomic and molecular dynamics associated with photofragmentation of small molecules and developing techniques for quantitative measurements of population densities.

John F. Holzrichter is Deputy Associate Director for Advanced Lasers at the Lawrence Livermore National Laboratory. His early responsibilities included the design and construction of the SHIVA and NOVA lasers. From 1980 to 1984 he was responsible for the Inertial Fusion Program. Since 1984 he has been responsible for the Advanced Laser Development Program. Dr. Holzrichter received his Ph.D from Stanford in 1971 having worked under Professor Arthur Schawlow. During his thesis research period he collaborated closely with Dr. Roger Macfarlane (IBM) who was a research associate in the Schawlow group at that time. He built the first dye laser at Stanford (flashlamp pumped) to photo-induce magnetic signals in the anti-ferromagnetic MnFe . Other activities at Stanford included flashlamp studies and materials studies. In an interesting aside, he removed tatoos from a young lady and from a young pig (both living). He joined the Naval Research Laboratory in 1971 and LLNL in 1972. He is the author of over 30 papers and has given numerous invited talks on high power lasers and inertial fusion.

George Francis Imbusch was born in County Limerick, Ireland on October 7, 1935. He obtained his baccalaureate at the Galway campus of the National University of Ireland. He then obtained his Ph.D. degree from Stanford University on 1964. He attended Stanford under the aegis of a Joseph P. Kennedy, Jr. Fellowship (1959-61). He and Dr. Linn Mollenauer were the first two graduate students to join Professor Art Schawlow's group shortly after Art's move from Bell Telephone Laboratories in 1961. Dr. Imbusch is on record as holding the first Ph.D. granted by Professor Schawlow. After a short postdoctoral appointment with Dr.

S. Geschwind at Bell Labs, Dr. Imbusch was promoted to the technical staff and remained there until 1967 when he accepted the Chair of Experimental Physics at his Alma Mater in Galway. He is presently serving as the Chairman of Physics Department of University College, Galway, and he continues to maintain an active research program in the optical properties of solids. Dr. Imbusch has also established many collaborative programs with groups in the United States including those at Stanford, Wisconsin, Oklahoma State and Georgia. He was named to a Brittingham Professorship at the University of Wisconsin-Madison in 1977-78 and has served on numerous Scientific Committees and Boards.

James E. Lawler is currently a faculty member at the University of Wisconsin supervising research programs on optogalvanic spectroscopy and laser techniques for determining atomic transition probabilities. He received his B.S. degree from the University of Missouri at Rolla in 1973 and his M.S. and Ph.D. degrees from the University of Wisconsin in 1978. Between 1978 and 1980 he was a Research Associate in the Schawlow group at Stanford where he developed new techniques for high resolution spectroscopy of plasmas.

Marc D. Levenson is currently "head manager" in Optical Storage Technology at the IBM Almaden Research Center. He received his B.S. degree in physics from MIT in 1967 and his Ph.D. degree from Stanford in 1972. His Ph.D. thesis was conducted under the supervision of Prof. Schawlow and concerned saturated absorption spectroscopy of iodine. He was later a post-doctoral fellow at Harvard University and an Associate Professor of Physics at the University of Southern California.

Roger Macfarlane was born in New Zealand and received his formal education there, graduating with a Ph.D. in Physics from the University of Canterbury in 1964. From 1965-68 he was a Research Associate with Art Schawlow in the Physics Department at Stanford University. During that time he collaborated with John Holzrichter and Art Schawlow on one of the early applications of dye lasers to spectroscopy, measuring photomagnetism in MnF_2. This kindled an enthusiasm for resonant pumping experiments and high-resolution laser spectroscopy of solids which were carried out subsequently at the IBM Research Laboratory in San Jose. The application of frequency selective photochemistry to frequency domain information storage has been his major interest for the past several years. Much of the work at IBM was done in collaboration with Robert Shelby.

Warren Moos joined the fledgling laser spectroscopy group at Stanford as a Research Associate late in 1961, after completing his Ph.D. work at the University of Michigan. At present, he is Professor in the Department of Physics and Astronomy at the Johns Hopkins University. His current research interests include high-temperature plasmas, planetary atmospheres and magnetospheres, and space astronomy. He looks back on his stay at Stanford as a time of unusual excitement and creativity." We were always looking for new things to do with the laser. A person would come in the morning with an idea and within a day or two we would decide if it was within the experimental state of the art. A. L. Schawlow, of course, was most often the source for these ideas. Over the years, it has been fascinating to see how many of the ideas which had to be rejected because the technology was not available then have come to fruition with improvements in laser technology."

Munir Nayfeh is Associate Professor of Physics at the University of Illinois at Urbana-Champaign. He received his BSc from the American University of Beirut, and his PhD from Stanford University. He has been a research physicist at Oak

Ridge National Laboratory and a lecturer and research associate at Yale University. His early work at Stanford with the Schawlow-Hansch group involved a precision measurement of one of the fundamental constants of nature, the Rydberg constant. His research interests include experimental and theoretical studies of atomic collisions utilizing atomic scattering in the presence of laser fields, detection of low levels of atoms--including rare events--and the coherent interaction of light beams with atoms and molecules. More recently, he has been conducting experimental and theoretical work on the structure of highly excited atoms, including hydrogen, in strong external electromagnetic fields (dc electric, magnetic, microwave, and optical) and their interactions with other atoms and surfaces.

Timothy J. O'Leary received a B. S. in Chemistry from Purdue in 1972. Following a summer spent with Lowell Wood at the Lawrence Livermore Laboratory, during which time he became familiar with laser isotope separation, especially the Schawlow-Tiffany work, he came to Stanford as a Fannie and John Hertz Foundation Fellow. Although officially enrolled as a student in Chemistry, he persuaded John Brauman and Dr. Schawlow to "co-direct" his thesis work. During his stay he worked on developing some wideband chelate lasers (successful), hydrogen isotope enrichment by selective photodecomposition in HBr (successful), and enrichment of bromine by selective photofragmentation from a molecular beam (unsuccessful, and scooped by Bell Labs). After receiving his Ph.D. in 1976, he became a medical student at the University of Michigan (where he also served as a research associate in physiology). Since that time, he has served as a resident, fellow, attending pathologist and Chief of the Autopsy Service at the National Institutes of Health, medical officer in the Food and Drug Administration and, currently, as Chairman of the Department of Cellular Pathology, Armed Forces Institute of Pathology.

S. C. Rand received his doctorate in physics from the University of Toronto in 1978 for work with Boris Stoicheff on Brillouin light scattering in rare gas solids and molecular crystals. This was followed by optical coherent transient studies of "magic-angle" spin decoupling as a World Trade Fellow at IBM San Jose with Richard Brewer (1978-80). Later he was a Research Associate with Arthur Schawlow and Theodor Hansch at Stanford, where he performed high resolution spectroscopy of helium and investigated cooperative absorption and emission processes in rare earth solids (1980-82). Then in 1982 he joined Hughes Research Lab and developed the LiF color center Q-switch, the diamond laser, ultrahigh resolution techniques to study metastable defects in solids and also predicted the electric magic angle effect.

Stanley E. Stokowski , a native of Lewiston, Maine, received a Bachelor of Science degree from the Massachusetts Institute of Technology in 1963. While at M.I.T. he did an undergraduate thesis on laser-induced breakdown of air,under the direction of Professor C. H. Townes. In 1964 he received his masters degree from Stanford University after which he was accepted as a Ph.D. candidate in Arthur L. Schawlow's research group. After receiving his Ph.D. from Stanford in 1968, he did post-doctoral work at the National Bureau of Standards from 1968 to 1970. From 1970 to 1972 he was a member of the technical staff at Bell Telephone Laboratories in Murray Hill, N.J. In 1972 he joined the Martin Marietta Research Laboratories near Baltimore, Maryland. After coming to Lawrence Livermore National Laboratory in Livermore, California, in 1977, he has been actively involved in developing new materials for the large laser drivers for inertial confinement fusion. He is the author of about 50 published papers and two articles on Nd-doped laser glass and filter glasses in the CRC Handbook on Laser Science and Technology.

Satoru Sugano is currently Professor of Solid State Physics at the Institute of Solid State Physics, University of Tokyo, Japan. He received his Bachelor and Doctor of Science Degrees in Physics from the University of Tokyo in 1952 and 1959, respectively. From 1959 until 1961, he was a Member of the Technical Staff at Bell Telephone Laboratories, Murray Hill, N.J. where he met and collaborated with Dr. Schawlow. He has been a visiting Professor at the University of Colorado and at the Université Paris VII.

Peter E. Toschek was born in Hindenberg, Germany, and received his Ph.D. from the University of Bonn in 1961. Subsequently, he was research assistant and professor of physics at the applied physics laboratory of Heidelberg University. In 1981 he was appointed to the "I. Institut für Experimentalphysik" of Hamburg University. His topics of research include quantum optics, atomic physics, and laser spectroscopy. He first encountered Arthur Schawlow during a brief visit to Stanford in 1968. More extended visits in 1970 and 1972 led to more extensive interactions with Schawlow and with Toschek's former student, Ted Hänsch, as well as with Tony Siegman.

Zugeng Wong is presently Professor of Physics at East China Normal University, Shanghai, Peoples Republic of China. He was a visiting scholor at the Schawlow group at Stanford 1982-1983.

Hui-Ron Xia is Lecturer of Physics at the East China Normal University, Shanghai, Peoples Republic of China. She also was a visiting scholar at Stanford during 1982-83.

William M. Yen was born in Nanking, China on April 5, 1935. He obtained his B.S. degree from the University of Redlands in 1956 and his Ph.D. in Physics from Washington University (St. Louis) in January 1962. He joined Professor Schawlow's group at Stanford in the Summer of 1962 as a Research Associate, increasing the size of the team to four. He accepted a position as Assistant Professor at the University of Wisconsin, Madison, in the Fall of 1965 and was subsequently promoted to Associate and to Full Professor there. He has been the recipient of a Guggenheim Fellowship (1979) and of a Humboldt Senior US Scientist Award. He has held visiting positions at the University of Tokyo, the University of Paris, the Australian National University, the Goethe University of Frankfurt, Harvard and UCSB. In 1986, he was named to the Graham Perdue Chair of Physics at the University of Georgia, Athens. He has served as Chairman of a number of international scientific conferences including the International Conference on Luminescence held in Madison in 1984. His research interests have centered on the optical properties of ordered and disordered material.

Pei-Lin Zhang was born in Cheffoo, Shandung, China, on June 13, 1933. He received undergraduate degree in Radioelectonics and graduate degree in Physics from Tsinghua University, Beijing, China, in 1953 and 1956 respectively. He is the recipient of National Invention Award in 1981 by National Scientific and Technological Commission of China. From November 1982 to December 1983 he was with Department of Physics, Stanford University, as a visiting scholar, where he did parametric wave-mixing processes research with Prof, A. L. Schawlow. Since 1956, he has been with Tsinghua University, Beijing, People's Republic of China, where he is currently Professor of Physics. His research interests are in the area of lasers, spectroscopy and nonlinear optics.

Shuo-Yan Zhao received undergraduate degree in Engineering Physics from Tsinghua University, Beijing, People's Republic of China, in 1958. Presently she is an Associate Professor of Physics at Tsinghua University. Her research interests are in the area of laser spectroscopy and its applications.

Index of Contributors

PACIFIC AND WORLD STUDIES, NO. 2

Pacific Dynamics:
The International Politics of Industrial Change

Edited by   Stephan Haggard and Chung-in Moon

CIS-Inha University   Westview Press

PACIFIC AND WORLD STUDIES SERIES, NO. 2

Copyright © 1989 by the **Center for International Studies, Inha University**

Published in 1989 in the Republic of Korea by the **Center for International Studies, Inha University**; 253 Yonghyun-Dong, Nam-ku, Inchon 402-751.

Published in 1989 in the United States of America and distributed exclusively throughout the world excluding Korea, by **Westview Press, Inc.**, 5500 Central Avenue, Boulder, Colorado 80301

Library of Congress Cataloging-in-Publication Data
PACIFIC DYNAMICS: THE INTERNATIONAL POLITICS OF INDUSTRIAL CHANGE
(Monographs of the Center for International Studies, Inha University)

    Includes index.
    1. Pacific Area—Industries.
    2. East Asia—Industries.
    3. Pacific Area—Foreign Economic Relations.
    4. East Asia—Foreign Economic Relations.
HC681, P274 1987   338.099—dc19
ISBN 0-8133-0583-7
ISBN 0-8133-0587-X (pbk.)

Printed and bound in the Republic of Korea by Seoul Computer Press

Contents

v

Tables ———————————————————————

Figures

Charts

CHAPTER 1
Introduction: The International Politics of Industrial Change

Stephan Haggard

The shift in the world's economic geography towards East Asia has been heralded as opening the "Pacific Century."[1] Though it is virtually a cliché to note the growing economic interdependence between the United States and East Asia, there is sharp disagreement about the broader economic and political implications of this development. Economists emphasize the advantages of allowing integration to advance through open policies toward trade and investment. Free trade and investment, it is argued, promote an efficient allocation of resources and dynamic gains from accelerated technological change. Liberals have long stressed the link between international commerce and peaceful relations among nations. The realist tradition of international relations, by contrast, notes that the emergence of new centers of economic power is likely to generate conflict. Shifting comparative advantage may reflect rapid economic growth, but it also entails painful adjustments and can even threaten mature industrial powers such as the United States with deindustrialization.

This book examines the politics of industrial change resulting from trade and investment between the United States and the major economies of Northeast Asia: Japan, Korea and Taiwan. To date, the debate about this process has revolved largely around the question of protectionism. This focus is not unwarranted; protection is the most obvious manifestation of the conflicts associated with shifts in the international division of labor. Yet the papers in this volume suggest that the dichotomy between "free trade" and "protection" has broken down. Protection is usually associated with the defense of declining sectors but commercial policy has also been used to create new export-

I would like to thank Nancy Gilson, Chung-in Moon and Greg Noble for comments on this chapter and Marie Park for research assistance.

1. Staffan Burenstam Linder, *The Pacific Century: Economic and Political Consequences of Asian-Pacific Dynamism* (Stanford: Stanford University Press, 1986).

oriented industries or as a strategic weapon to force liberalization abroad. The "managed" trade of the 1970s and 1980s rests on quota restrictions and is quite different than the tariff protection of the past. In addition, the menu of issues being negotiated has lengthened. Trade policy is only one dimension of the international politics of industry. The policy agenda now includes trade in services, the protection of intellectual property, investment, exchange rates and macroeconomic policy, all areas in which international norms are evolving and uncertain.

How are we to analyze the mixed-motive games—partly cooperative, partly conflictual—that result from growing economic interdependence across the Northern Pacific? The answers are not straightforward, since political variables operate at several distinct levels of analysis. The first is the level of the international system; the overall distribution of power and economic capabilities. The "high" politics of the region and the economic pre-eminence of the United States set the context within which regional economic interdependence developed in the postwar period. America's relative economic position has subsequently declined. A central puzzle addressed in this volume is the effects of this change on the management of interdependence. Is hegemonic decline necessarily associated with increased economic conflict, or can there be cooperation "after hegemony"?[2]

Second, attention must be paid to the "process" level of international negotiations, agreements and institutions. A number of multilateral initiatives are of relevance to the economic integration of the Pacific Basin, including the Uruguay Round of GATT negotiations, multilateral discussions of exchange rates and macroeconomic policy coordination, and continuing efforts to institutionalize regional cooperation. Nonetheless, bilateralism remains the dominant form of economic diplomacy in the region. Bilateralism is usually associated with protectionism and with a departure from the norm of non-discrimination. Bilateral agreements have grown up to "manage" trade in semiconductors and autos, and the restrictionist Multifibre Agreement ultimately rests on negotiated bilateral accords. Each of these three sectors is the subject of a case study in Part II. But further liberalization is also being negotiated bilaterally. The effects—and wisdom—of bilateralism constitute a second major theme of this volume.

Since interdependence muddies the boundaries between domestic and foreign economic policies, the national level must be treated as a crucial component of the region's political structure. Macroeconomic policy, protection and industrial policies influence competitiveness and com-

---

2. See Robert Keohane, *After Hegemony* (Princeton: Princeton University Press, 1984).

parative advantage, trade and financial flows. These policies, in turn, are conditioned by domestic politics, including pressures from organized interest groups and the effects of national political institutions. As the essays in Part I demonstrate, the political conflicts between the United States and the East Asian countries reflect in no small measure deeply rooted differences in state-society relations and national policy styles. The remainder of this chapter introduces the studies that follow by examining the logic of these three levels of political analysis in more detail.

High Politics and Regional Integration

The theory of hegemonic stability is the most ambitious attempt to link the distribution of power to international economic outcomes.[3] The theory holds that the existence of a dominant world economic power is conducive to free trade and capital flows and the maintenance of strong systems of international rules of regimes. Being the most technologically advanced and productive economy in the system, hegemons have an *interest* in open trade and investment; being the most powerful state in the system, they have the *capability* to facilitate it. There is some debate about how the hegemon acts to promote an integrated world economy. Those adopting a "benign" view of hegemonic power emphasize the incentives to the hegemon to provide public goods or to perform functions which facilitate the smooth operation of the world economy. These include maintaining an open market, providing long-term and countercyclical lending and developing international institutions that facilitate cooperation. Hegemons also act as providers of security and enforcers of rules and contracts. A more "malign" conception of hegemony sees the hegemon as acting to structure international economic relations in its interests through the exercise of coercive power.

If periods of hegemonic ascendence are ones of stability, liberalization and growth, periods of hegemonic decline or "power transitions" are characterized by increased rivalry, mercantilist policies and even war. The gradual diffusion of economic capabilities to newcomers creates new economic challenges to the hegemon's dominance, and makes its interest in free trade more selective. Rising powers will expand their spheres of economic influence at the expense of the

---

3. For an overview of the theory of hegemonic stability, see Robert Gilpin, *The Political Economy of International Relations* (Princeton: Princeton University Press, 1987), ch. 3.

declining leader, and may seek a rearrangement of the territorial status quo. Hegemons become less willing to bear the costs of leadership and less tolerant of "free-riding." International regimes and cooperative arrangements unravel since the hegemonic power is less willing or able to bear the costs of leadership.

The economic integration of the Pacific Basin suggests two points of theoretical significance that partly confirm and partly modify the hegemonic stability hypothesis. First, economic integration in the region has been shaped by the overall distribution of capabilities and by the strategic as well as economic concerns of the great powers. Power has been exercised in different ways by successive "hegemons," however. Britain's empire, partly formal, partly informal, was different from Japan's tightly integrated colonial system. Both were different than the Pax Americana, with its peculiar emphasis on liberalism, multilateralism and international rules. Despite the emergence of managed trade in some sectors, there is little evidence that the United States has abandoned its liberal objectives. Debates, rather, have centered on how such objectives are best achieved. As Stephen Krasner argues in Chapter Ten, a "diffuse" conception of reciprocity has given way to a more "specific" conception of reciprocity, but within a broadly liberal foreign economic policy.

Second, the development of the Pacific economy confirms the expectation that the emergence of new poles of industrial power generates adjustment problems and conflicts over rules and markets. This rivalry has taken a variety of forms, however. The intra-imperialist rivalries of the late 19th century were the result of the entry of new imperial aspirants from outside the region confronting an established imperial power. The Pacific war of 1941–1945 resulted from the challenge Japan's imperial design for East and Southeast Asia posed to American security. Current conflicts between the U.S., Japan and the East Asian newly industrializing countries (NICs), by contrast, are primarily economic. Jeffrey Hart shows in his study of semiconductors (Chapter Five) that trade can raise security concerns in sensitive sectors, but the emergence of Japan and the NICs does not in itself constitute a security threat to the United States. To the contrary, the reconstruction of East Asia and its integration into the American security network was a primary political objective of U.S. policy in the postwar period.

The difference between the present era and past periods results from variations in the *overall* security structure that theorists of hegemonic stability have ignored. Where economic and geo-political rivalry overlap and reinforce one another, as they did between Japan and the United States in the 1930s, conflict is likely to be more intense than when economic rivalry takes place within the overarching context of an alliance, as it has in the post-war period. Parallels to previous "power transitions" are thus probably misleading. The *overall* political setting remains strongly conducive to continued economic integration in the Pacific, even if declining *economic* hegemony, or more simply, the diffusion of industrial capabilities, has been the source of some conflict.

These modifications of the theory of hegemonic stability can be seen by reviewing the emergence of a regional economy in the Northeast Pacific. The origins of the modern Pacific economy might be traced to the two Opium Wars (1840-42, 1858-1860), when Britain established the political basis for European and American commercial penetration of China through a system of unequal treaties and extraterritoriality. After China's defeat by Japan in 1895, European, American and Japanese investments in shipping, shipbuilding, manufacturing, and mining accelerated as well, concentrated in the treaty ports, Hong Kong, and Manchuria. Western financial consortia lent extensively to the Chinese government to finance railroad construction and to meet war-related indemnities.

Critics of imperialism argue that Japan's subsequent economic success is partly attributable to the good fortune of avoiding imperial domination during the 19th century. As Frances V. Moulder argues, Japan's seclusion facilitated the emergence of an autonomous, centralizing and ultimately developmental state apparatus.[4] China's subjugation to outside political interference and economic penetration, by contrast, prohibited it from developing the kind of state required to foster rapid growth. Moulder does not test the hypothesis against the development of Korea and Taiwan. Both were largely neglected in the first wave of Western imperialism, and both did witness attempts at "self-strengthening" against outside pressures. Neither developed the political capability to stave off Japanese encroachment, however.

The great power competition associated with Britain's relative decline intensified imperial ambitions, and vice versa. Prior to 1894, the Western interest in China was primarily commercial and the imperialist powers tended to act together against the Chinese government. Britain's economically dominant position and the logistical difficulties of direct rule led it to favor a non-discriminatory "open door." By the 1890's, however, Britain's dominance in China was challenged by new entrants into the imperial game.[5] Less secure in their commercial capa-

4. Frances V. Moulder, *Japan, China and the Modern World Economy* (New York: Cambridge University Press, 1977).
5. One of the best accounts of the politics of imperialism remains William Langer, *The*

bilities and harboring different conceptions of empire, the latecomers—Germany, France, Japan and Russia—sought exclusive spheres of interest. Railroads and other concessions were sought for reasons of great power competition and annexation as well as commercial benefit. The new level of inter-imperialist rivalry not only increased the foreign economic penetration of China, but also escalated the political pressures on its periphery in Manchuria, Korea, Taiwan and Southeast Asia.

Japan was to be the most important of the new claimants, but the U.S. also entered the imperial game following the Spanish-American war of 1898.[6] American expansion consisted of "normal" commercial and investment relations, a direct imperial venture in the Philippines and attempts to play a political role in China. The U.S. depended heavily on rubber and tin from Dutch and British colonies, and by the 1930s, had an interest in the Eastern oil market. As Dennis Encarnation notes in Chapter Six, commercial, financial and investment ties with Japan grew rapidly during World War I and over the 1920s. By 1936-38, British Malaya, Japan and the Dutch East Indies were respectively the third, fourth and eighth most important sources of American imports, while Japan was the third largest export market.[7]

America's diplomatic interest in the Open Door and attempts at dollar diplomacy should not be confused with effective economic penetration or political influence. Japan's defeat of China in 1895, the 1902 naval alliance with the British, and the defeat of Russia in 1904 were of much greater significance than America's presence, and signalled a profound shift in the region's power balance.[8] The trend was only accelerated as Europe turned inward against itself in World War I. Japan's emergence as the first non-Western industrial power was accompanied by imperial ambitions that radically reoriented the political economy of Northeast Asia. The Japanese conception of empire involved a particularly close link between economic and military goals, orchestrated by an archetypically "strong" state. Japan colonized neighboring countries and invested heavily in infrastructure, making

Diplomacy of Imperialism 2d edition, (New York: Knopf, 1951). See also Akira Iriye, *Across the Pacific* (New York: Harcourt, Brace and World, 1967), chapters 2-5.

6. The emergence of America as an imperialist power was given new attention by the revisionist historians of the mid-1960s. One of the best works remains Walter Lafeber, *The New Empire* (Ithaca: Cornell University Press, 1963).

7. Norman S. Buchanan and Friedrich Lutz, *Rebuilding the World Economy* (New York: The Twentieth Century Fund, 1947), pp. 328-9.

8. On the changing conception of Japanese empire, see Michael A. Barnhart, *Japan Prepares for Total War* (Ithaca: Cornell University Press, 1987), chapters 1-3.

possible a close integration between colony and metropole.

Initially, Korea and Taiwan were integrated in a typical colonial fashion, supplying inexpensive rice (Korea) and tropical products (Taiwan) while serving as a captive market for manufactured exports.[9] After 1931, Japan's empire entered a new phase, driven by the exigencies of the Great Depression and the political opportunities provided by European weakness and American isolationism. The Greater East Asian Co-Prosperity Sphere rested on a complex regional division of labor, enclosed within a Yen bloc. The southern region would supply oil, processed raw materials and food. Manchuria and North China would provide materials for a heavy industrial complex while the whole would provide a vast protected market for Japanese manufactures. The strategy demanded significant investment and allowed for industrial growth of the colonies, particularly Korea and Manchuria. The most enduring change wrought by the Japanese was social and political as much as economic, however. Bruce Cumings states the case most clearly:

> ... the colonial state stood above and apart from societies that had not yet reached Japan's level of social, political and economic development. Thus a highly articulated, disciplined, penetrating colonial bureaucracy substituted both for the traditional regimes and for indigenous groups and classes that under "normal" conditions would have accomplished development themselves. The colonial state replaced an old weak state, holding society at bay, so to speak: this experience goes a long way toward explaining the subsequent (post–1945) pronounced centralization of Taiwan and both Koreas, and has provided a model for state-directed development in all three.[10]

While Japanese industrialization and imperialism served to integrate the regional political economy, it naturally created international tensions as well. As conflict over China and the Pacific War proved, much more than economic conflict was at stake. Nonetheless, the onset of world depression brought economic issues to the fore. One example associated directly with Japan's manufacturing capabilities concerned textiles.[11] Japan's encroachment into the Chinese market displaced the Americans before World War I and steadily ate into the British share of

9. See Ramon Myers and Mark Peattie, eds., *The Japanese Colonial Empire: 1895-1945* (Princeton: Princeton University Press, 1984).

10. Bruce Cumings, "The Origins and Development of the Northeast Asian Political Economy: Industrial Sectors, Product Cycles and Political Consequences," *International Organization* 38, 1 (Winter 1984), p. 11.

11. See Osamu Ishii, *Cotton Textile Diplomacy: Japan, Great Britain and the United States 1930-1936* (New York: Arno Press, 1981).

the market over the twenties. With cost advantages that gave rise to charges of "social dumping," Japan threatened British textiles in the Dominions and its own colonies as well. Following the breakdown of Anglo-Japanese negotiations for a partition of world markets in 1934, Britain instituted quotas on textiles throughout the empire. Domestic interests in the United States were initially split on how to respond to Japanese imports, with New England textile manufacturers pitted against Southern cotton exporters who supplied Japan. By mid-1936, however, the pressures on Roosevelt were overwhelming, and a series of restrictive agreements were put into place, precursors to postwar agreements. Japan's quest for self-sufficiency in raw materials and oil through the incorporation of Southeast Asia, the heart of the so-called "southern strategy," was of course to have more drastic consequences.[12]

The international economic system constructed by the United States in the postwar period is commonly portrayed as multilateral based on the liberalization of trade and capital movements. As Stephen Krasner argues in Chapter Ten, United States hegemony rested on a posture of "diffuse reciprocity" that ignored relative gains among allies in the pursuit of broader strategic goals. These strategic considerations dictated significant exceptions to the liberal, multilateral norm in East Asia just as they did in Europe. Trade, financial and aid links were self-consciously developed to serve three mutually supportive goals: economic reconstruction, strengthening the internal political position of pro-American political elites, and cementing strategic relations through economic interdependence. These aims provided a new political foundation for a regional economy in which Japan once again came to occupy a central economic position, but within a broader anti-Soviet and anti-Chinese *political* alliance.

Though Japan's reconstruction was seen as integral to the larger geopolitical goal of containing the spread of Communism, its re-emergence as a major industrial power created a new generation of "low" political problems. These conflicts took three forms. The first centered on the adjustment problems created by a low-wage, but technologically adept newcomer. Did the U.S. and Europe have the political capacity to absorb Japan's highly competitive manufactured exports? A second issue concerned access to the *Japanese* market and the control- oriented style of Japanese economic management. The United States was initially indulgent towards Japan's system of economic controls, but as early as 1959 the U.S. began to pressure Japan over currency convertibility, import liberalization and access for American investors.

---

12. See Barnhart, *Japan Prepares...*

Diffuse reciprocity was increasingly supplemented by calls for a more careful equilibration of economic benefits. These two problems—of absorbing exports and securing "reciprocity"—were to be repeated with the East Asian NICs as they began their industrialization. The third set of problems centered on the emergence of a North-South axis between Japan and the developing countries of Asia. As Japan's economy revived, its economic influence on Korea and Taiwan grew. New conflicts appeared over foreign direct investment, aid, trade and more generally, the appropriate division of labor within the region.

Japan was the key to America's postwar strategy in Asia.[13] The reintegration of Japan into the world political economy proved difficult, however, since the "loss" of China and the breakup of the empire severed natural markets. As in Europe, the "dollar gap" emerged as a critical constraint on Japanese recovery. The Korean War made an important contribution to solving the problem by increasing U.S. military expenditures in Japan. Direct economic aid was also important. As a large and already-established industrial power, however, manufactured exports were crucial for Japan's long-term recovery. To Britain, which was experiencing its own post-war economic trauma, a reconstructed Japan posed an even more dire economic threat than it had posed in the 1930s. In 1951, Britain flatly refused U.S. efforts to secure most-favored nation status for Japan, and negotiated an Anglo-Japanese Payments Union in which neither would demand dollars or gold for the accumulated balances of the other. While this agreement permitted an expanded trade with the sterling bloc countries in Asia, it was not until 1955 that Japan was admitted to the GATT. Most European countries immediately invoked Article 35 which freed them from the obligation of extending most-favored nation status. Ironically, that same year, the United States negotiated its first postwar textile restraint with Japan. Nonetheless, the United States played the hegemon's role of absorbing a growing proportion of Japan's exports. In the mid-1930s, Asia took 64 percent of Japan's exports; North America took 16 percent. By 1960, Asia's share stood at 34 percent and North America's just under 30 percent.[14]

The Korean war brought the Cold War to Asia, and guaranteed that Korea and Taiwan, as "frontline" states, would receive massive

---

13. The following is influenced by William S. Borden, *The Pacific Alliance: United States Foreign Economic Policy and Japanese Trade Recovery 1947-1955* (Madison: The University of Wisconsin Press, 1984); and Michael Schaller, *The American Occupation of Japan: The Origins of the Cold War in Asia* (New York: Oxford University Press, 1985).
14. Warren S. Hunsberger, *Japan and the United States in World Trade* (New York: Harper and Row, 1964), p. 185.

American assistance.[15] Aid bolstered conservative regimes in both countries by financing import substitution. As import substitution slowed and aid commitments declined over the late 1950s, American advisors urged a reorientation of economic strategy. Both countries launched policy reforms in the early 1960s that encouraged manufactured exports, foreign direct investment and in Korea's case, commercial borrowing. Historical timing was an important factor in the success of export-led growth. Korea and Taiwan entered the world market during a period of unprecedented trade expansion, rapid liberalization of the American and European markets and a growth of foreign direct investment. Success was not solely a result of changed policies and "getting the prices right," but of a conducive international environment and an open U.S. market. Haggard and Cheng, and Yoffie show in Chapters Eleven and Twelve that success had its political drawbacks. American foreign economic policy toward the export-oriented NICs gradually stiffened, just as it had with Japan. Open markets and diffuse reciprocity gave way to selective protection and increasing pressures for specific reciprocity.

The NICs' turn toward export-oriented policies also redefined their relationship with Japan. Partly to ease its own aid burden, the United States pressed for a rapprochement between Seoul and Tokyo, effected in 1965, which opened the way for a greater economic role for Japan in Korea. Japan's economic relations with Taiwan had returned to normalcy much earlier and survived the diplomatic break in 1972 through a careful separation of the political and economic tracks.[16] Japanese planners and firms initially saw natural economic complementarities with the NICs. Korea and Taiwan became important export markets and sites for foreign direct investment in mature industries. As Moon and Chang show through their analysis of the textile sector (Chapter Seven), economic relations between Korea and Taiwan and Japan were strained over the 1970s by the competitive dimension of their economic relationship. As the NICs developed, and particularly as the yen strengthened after 1985, they forced industrial adjustments on Japan similar to those Japan had forced on the United States decades earlier.

In its broad outlines, the theory of hegemonic decline seems well suited for understanding the conflicts between the United States and its

15. For a general discussion, see Stephan Haggard and Tun-jen Cheng, "State and Foreign Capital in the East Asian NICs," in Fred Deyo, ed., *The Political Economy of the New East Asian Industrialism* (Ithaca: Cornell University Press, 1987).
16. This pattern is nicely documented in Walter Arnold, "Japan and Taiwan: Community of Economic Interest Held Together by Paradiplomacy," in Robert Ozaki and Walter Arnold, eds., *Japan's Foreign Relations* (Boulder: Westview Press, 1985).

major East Asian trading partners. In the period of its ascendence, the U.S. promoted the growth of an open regional economy by keeping its market open and providing security, aid and technical assistance. The U.S. tolerated "free-riding" in the form of closed trade policies, restrictive investment rules and vigorous, even mercantilist, export-promotion policies. The relative decline of the United States has been accompanied by major trade imbalances, increased protectionism and a gradual move away from multilateralism toward bilateralism.

However, some important amendments and caveats are required. The "decline" of the United States was simply the flip side of the *successful* economic and political reconstruction of the major economies of East Asia. Despite their economic successes, the trading states of East Asia remain closely linked to the United States. Japan remains dependent on the United States for its security, and is unlikely to fundamentally reorient either its foreign policy or its trading patterns. Though Korea and Taiwan may seek to diversify their economic relations, they too remain dependent on the United States both politically and economically. The postwar event of greatest geo-strategic significance for the region, the Sino-Soviet split, opened the way to China's closer economic integration with the rest of the region. As Hart demonstrates in his discussion of recent conflicts with the Japanese over semiconductors (Chapter Five), security concerns can be the source of economic friction as well as cooperation. Sharp political divisions on the Korean penninsula and across the Taiwan Straits are still obstacles to economic complementarity and also hinder efforts to institutionalize regional cooperation by complicating the issue of membership. Nonetheless, the current security system is favorable to continued regional growth and integration.

There are also a number of troubling anomalies for the theory of hegemonic stability. The protectionist trend predicted by the theory has not wholly materialized; the 1970s and 1980s have not mirrored the 1930s. New forms of protection have emerged, but as Yoffie shows in Chapter Twelve, these have not been as restrictive as once thought. There have been liberalizing trends as well, for example in the negotiation of the Tokyo Round codes and in the promise of the Uruguay Round. Japan is under strong pressure to liberalize and economic reforms aimed at exploiting comparative advantage more fully are on the agenda in both Seoul and Taipei. Nor should bilateral economic diplomacy be equated with protection. As Haggard and Cheng argue in Chapter Eleven, bilateralism marks a departure from the norm of non-discrimination, but it can nevertheless promote liberalization. International networks of trade and investment have deepened over the last fifteen years, the period during which American hegemony was presumably on the wane.

The theory of hegemonic stability draws attention to the importance of the underlying power structure and political context within which international economic relations develop. It does less well explaining variations within a given structure or the persistence of cooperation and liberalizing trends. To explain these, attention must be paid to the international political *processes* governing trade and investment.

The International Politics of Industry

In *Power and Interdependence,* Robert Keohane and Joseph Nye argued that conditions of complex interdependence changed some of the common assumptions that have governed the study of world politics.[17] Interdependence increased the number of domestic actors with a stake in foreign economic policy, widened the range of issues over which states negotiated and increased the number of bilateral and multilateral political channels connecting countries. Political scientists working on the politics of interdependence have struggled to translate these basic insights into a research program. One approach was to focus greater attention on comparative foreign economic policy; the chapters in Part I reflect this approach.[18] This allowed for more detailed understanding of the politics of interdependence, but faced the problem of how to aggregate across national experiences. A second approach that generated significant research in the 1980s was to study the dynamics of international *regimes;* the norms, rules and procedures governing interactions in a particular issue area.[19] In many "issue areas," however, the norms, rules and procedures were far from clear. Regime analysis ran the risk of assuming a greater degree of rule-regulated behavior than was warranted, and underestimated the continuing importance of bilateral diplomacy.

An alternative to comparative political economy and regime analysis is to examine the international politics of particular *industries* or *sectors*; this approach is adopted in the sector studies of semiconductors by Hart (Chapter Five), autos by Dunn (Chapter Six), and

17. Robert Keohane and Joseph Nye, *Power and Interdependence* (Boston: Little Brown, 1977).

18. This line of thinking is pursued in Peter Katzenstein, ed., *Between Power and Plenty* (Madison: University of Wisconsin Press, 1977).

19. See Stephan Haggard and Beth Simmons, "Theories of International Regimes," *International Organization* 41, 3 (Summer 1987).

textiles by Moon and Chang (Chapter Seven).[20] This approach has several advantages. First, it combines domestic and international levels of analysis. National trade and industrial policies are generally formulated with reference to particular sectors and both business and labor tend to be organized by industry. Taking the industry as the unit of analysis thus provides a good entry into the *domestic* politics of interdependence. At the same time, the sectoral approach necessarily places industries in their global setting, a necessity given increasing multinationalization, international competitive pressures and the growing web of bilateral accords. A sectoral approach also provides a good way of examining the interaction between economic and political variables. On the one hand, industry characteristics such as the level of concentration, the degree of surplus capacity, the maturity of technology and the degree of multinationality can be used to explain the politics of trade and investment. On the other hand, sectoral analysis draws attention to the effects of negotiated agreements on the market dynamics of the industry.

The sectoral approach has yielded a number of insights that are of relevance to an understanding of the political economy of the Pacific Basin. The first, and most obvious, is that the pursuit of export-oriented strategies by Japan and the NICs has created surplus capacity in a number of sectors.[21] The textile and apparel industries provide the starkest example of the adjustment dilemma. In Chapter Seven, Moon and Chang call textiles and apparel "start up" industries because the barriers to entry are low. Technology is relatively simple and labor costs have historically been an important component of total costs, providing opportunities for LDC entrants. In the importing countries, however, exit is both economically and politically difficult. The textile and apparel industries in the United States are large employers, geographically dispersed and labor-intensive, factors that contribute to their political clout and to a protectionist outcome.

Vinod Aggarwal has shown how domestic interest group pressures in the textile and apparel industries cannot be understood without reference to the international dynamics of protection itself. Protection tended to "snowball" over time toward wider and wider coverage and eventually, to institutionalization in an international regime.[22] As cotton textiles were protected in the United States, production in Japan and

---

20. This line of research is developed in John Zysman and Laura Tyson, eds., *American Industry in International Competition* (Ithaca: Cornell University Press, 1983).
21. See the introduction to Zysman and Tyson, *American Industry in International Competition* (Ithaca: Cornell University Press, 1983).
22. Vinod K. Aggarwal, *Liberal Protectionism: The International Politics of Organized Textile Trade* (Berkeley: University of California Press, 1985).

the East Asian NICs shifted toward man-made fibers. This generated political pressures to expand the initial agreements covering cotton textiles to other fibers. As access to the American market was restricted by bilateral agreements, exports were diverted to European markets. The European producers came to support a global accord. As established producers such as Hong Kong, Korea and Taiwan came under restraint, production was diverted to non-restricted countries, often through offshore investment from the NICs themselves. Protection gradually spread to cover these countries as well. As Moon and Chang demonstrate, Japan, and even Korea and Taiwan, are now facing their own adjustment problems in the textile and apparel industries. What began as developmental, infant industry protection has now assumed the same rationale as protection in the U.S. and Europe: support for mature industries.

Import-affected industries seeking protection in the United States have not always succeeded in securing it, however, and where they have succeeded, protection has not necessarily persisted or become institutionalized in an international regime. Footwear and televisions provide two examples, and suggest the importance of industry characteristics in determining the demand for protection.[23] The United States signed orderly marketing agreements with Taiwan and Korea in footwear in 1976, but these were dropped in 1981 and industry efforts to secure protection in the mid-1980s failed. Similarly, the television industry secured orderly marketing agreements against Japan in 1977 and Korea and Taiwan in 1979, but they were allowed to lapse in the early 1980s. Due to industry characteristics and firm-level adjustment strategies, the demand for protection in these sectors was not as consistent as in textiles and apparel. While small footwear manufacturers suffered from stiff competition from imports, larger manufacturers integrated into retailing and imported those lines that they could not produce competitively. This had the effect of splitting the footwear coalition and reducing its effectiveness. The television industry was dominated by two relatively diversified multinational firms that had market-based adjustment options, including offshore investment and the licensing of technology.

Industry characteristics are also important in understanding why some industries have shown an interest in protection not as a form of relief from competition, but as a strategic weapon to secure access to the markets of their competitors. Jeffrey Hart's study of the

---

23. See Vinod K. Aggarwal, Robert O. Keohane and David B. Yoffie, "The Dynamics of Negotiated Protectionism," *American Political Science Review* 81, 2 (June 1987).

U.S.-Japan semiconductor dispute offers an example of an industry with such a "strategic" trade policy interest. The American industry's first preference was for open markets in both Japan and the United States. Given the high costs of innovation and the importance of scale economies in moving down the learning curve, an open U.S. market and a restricted Japanese one provided Japanese firms with decisive advantages. The threat of protection was seen as a way of securing the market access required to compete effectively, even if the ultimate agreement did not have the intended effect.

In Chapter Twelve, David Yoffie seeks to generalize about the effects of protectionist accords, suggesting that they frequently have unanticipated and perverse consequences.[24] These stem from the fact that contemporary protection takes the form of orderly marketing agreements or voluntary export restraints that restrict imports by fixing *quantities* rather than raising *prices,* as a tariff does. Surprisingly, such quota arrrangements allow exporting countries greater leeway for export expansion than tariffs. "Managed trade" is, of course, restrictive when compared to a free trade counterfactual, but has not been as restrictive as the redundantly high tariff levels of the 1930s. This stems partly from the incentives quotas provide to exporters, partly from the political strategies permitted by negotiated protection. Quotas yield monopoly rents to exporters that can be used to upgrade and diversify. Since imports are restricted by quantity, quotas also provide incentives for producers to upgrade within a given product category toward the most profitable "high end" items. In Chapter Six, Dunn demonstrates that this was one component of the Japanese response to the auto restraint agreement with the U.S. Yoffie also argues that quotas provide an incentive to diversify away from protected goods altogether and toward non-protected ones, or simply to cheat on agreements.

Yoffie argues that the negotiation process opens additional political strategies for exporters that are unavailable when protection is imposed unilaterally. Negotiation allows exporters opportunities for delay, linkage to other issues and the formation of transnational coalitions with parties in the importing country with an interest in free trade. Haggard and Cheng are less sanguine, however. In Chapter Eleven, they argue that a tougher U.S. trade policy is seeking to close some of the loopholes left open under previous accords, while making greater demands on the exporting countries for reciprocity.

The papers by Encarnation and Chernotsky (Chapters Eight and

---

24. See also David Yoffie, *Power and Protectionism: Strategies of the Newly Industrializing Countries* (New York: Columbia University Press, 1983).

Nine) and the study of automobiles by Dunn (Chapter Six) show that protection provides incentives for exporters to invest directly, both as a means of evading quotas and as a political strategy to ease trade tensions. As an increasing share of world trade moves through multinational corporations, trade and investment have become more closely linked. Dunn outlines how an initial auto restraint agreement was allowed to lapse as American firms entered into a variety of agreements with their East Asian counterparts. He concludes that the effort to manage auto trade had the paradoxical effect of deepening international cooperation between Western and Asian manufacturers. Instead of shipping whole vehicles, exchanges now consist of capital investments in local assembly and manufacturing, transfer of technological and management know-how, shipment of original equipment components and replacement parts and various joint venture, production and marketing agreements. Helen Milner has made the point in a general way by demonstrating that multinational ties affect firms' preferences for protection.[25] Firms with extensive multinational ties are less likely to seek protection, even when facing distress and strong import competition than those firms without such multinational ties. The growth of cross-investment chronicled by Chernotsky and Encarnation thus provides a political counterweight to protectionist forces, and helps explain why protection has not been as pervasive as was once anticipated.

The Domestic Politics of Industrial Change

A recurrent theme of this book is that national policies cannot easily be derived either from international political position nor from particular industrial characteristics. Similarly situated states respond to the international pressures of shifting comparative advantage differently as a result of domestic political structures. The contributions in Part I by Nelson on the United States (Chapter Two), Noble on Japan (Chapter Three) and Gereffi and Wyman on Korea and Taiwan (Chapter Four) each contain subtexts of controversey on how to unravel the determinants of national policy. They also even suggest disagreements on how to best *characterize* national industrial policies.

Nelson's contribution on the United States challenges purely "societal" or "demand-oriented" models of protection, the most

---

25. See Helen Milner, *Resisting Protectionism* (Princeton: Princeton University Press, 1988).

important "industrial policy" employed by the United States. The assumptions of these models are shared by the pluralist, Marxist and rational-choice traditions alike. Policy is seen as an exchange between politicians seeking to enhance their electoral chances, and constituents, usually individual firms, business organizations or labor unions, whose trade policy preferences are determined by their international market positions. Societal actors deploy political resources, such as lobbying or "getting out the vote," for the purpose of securing desired policies. Since policy outcomes are demand driven, the main puzzle is understanding the determinants of the demand for protection. Hypotheses include, among others, the level of sectoral or general economic distress, measured by such proxies as import penetration or the business cycle; the political strength of the industry, measured by size, geographic dispersion or level of organization; and international competitive position, measured by exchange rate fluctuations or other indicators of comparative advantage.[26]

Nelson argues that these models suffer from lack of attention to how institutions filter the demands of social groups; the demand side of the policy process has been emphasized at the expense of the supply side. Trade policy institutions in the United States are particularly fragmented, and have pushed policy in contradictory directions. On the one hand, the Reciprocal Trade Agreements Act of 1934 shifted authority over trade policy toward the executive. This had a number of consequences that were conducive to international leadership in liberalizing world trade and avoiding recourse to protection. First, it protected legislators from immediate and narrow interest group pressure. Second, it introduced broader economic and foreign policy concerns onto the trade policy agenda. Protection for a particular sector had to be weighed against the interest of consumers and the effects it might have for broader foreign policy initiatives. Finally, the process of negotiating tariff reductions exposed the trade-offs between protection at home and liberalization abroad. To gain concessions in trade negotiations demanded making some in return.

At the same time, however, a different set of trade policy institutions generated *protectionist* biases. Nelson argues that these include the mechanisms of administered protection, and particularly the escape clause process. These protectionist institutions help explain the seemingly contradictory trend, noted by Haggard and Cheng in Chapter Eleven, of continued initiatives for multilateral negotiations coupled simultaneously with growing exceptionalism.

---

26. See Douglas R, Nelson, "Endogenous Tariff Theory: A Critical Review," *American Journal of Political Science,* forthcoming.

The contributions by Noble (Chapter Three), Krasner (Chapter Ten) and Encarnation (Chapter Nine) are indicative of the sharp divisions over how to characterize the Japanese economic policy system. Krasner and Noble both note the fundamental divide between those who see Japan's success as explicable in terms of market-oriented policies, and those who stress the importance of the state and industrial policy. Both agree that Japanese industrial policy has changed over time in a more liberal direction. Both note that even the "statists" now focus more attention on the ability of the government to coordinate than on sheer command. Krasner and Encarnation, however, look at patterns of trade and investment, or *outcomes*, for evidence of continued government intervention. Krasner argues that Japan's low level of manufactured imports, low level of intersectoral trade and relatively low level of inward foreign direct investment are not easily explained in terms of comparative advantage or under the assumption of market-oriented policies. He admits, however, that this result may spring from particular patterns of business organization as much as government policy.

Noble, by contrast, looks directly at the policy *process* for evidence concerning the government's role. Noble makes two arguments, one concerning the influence of business, a second concerning the power of the state to achieve its objectives. Noble argues that "statists" underestimate the political influence of business. MITI does not stand aloof from the conflicts of the private sector; rather, the dominant firms in each industry seek to use MITI to enforce consensus. As Noble's case study of adjustment in the steel industry shows, firms can draw on political resources such as their ties with the ruling party to achieve their goals. The state is less insulated than is commonly thought, and can only achieve its goals by eliciting cooperation from business. Second, Noble contends that the power of the state is also overestimated, and that the ability of the bureaucracy to go against the market is severely limited. Efforts to enforce format standardization were undermined by maverick firms with superior products who gained their way against the wishes both of the government and the industry consensus. Noble concludes that closer attention needs to be paid both to the political power of business and to underlying market forces as determinants of industrial policy.

The relative importance of market and government has also been a source of controversy in the analysis of the newly industrializing countries. Neoclassical economists see the export-oriented NICs as vindicating their market-oriented prescriptions, while sceptics have catalogued a range of government interventions that have supported industrial development. Gereffi and Wyman place the East Asian NICs in comparative perspective by contrasting the export-oriented growth strategy of Korea and Taiwan with the import-substitution strategy of Argentina, Mexico and Brazil. While their account sides generally with the "statist" view of NIC growth, their comparison underlines several difficulties of linking state structures with economic outcomes. It has been argued, for example, that the East Asian NICs grew rapidly as a result of authoritarian political systems.[27] Authoritarian rule limited distributionist pressures and permitted the formulation and implementation of coherent policies. Gereffi and Wyman point out, however, that the majority of developing countries have authoritarian regimes, but few have autonomous or strong states, and most have been unable to devise effective development strategies. The precise features of authoritarian rule that were conducive to rapid growth have still not been adequately specified.

They also point to weaknesses in much coalitional analysis of state strategies, however, particularly attempts to extend dependency thinking to the study of the East Asian NICs. Dependency theorists have argued that peripheral industrialization occurs under a political-economic "triple alliance" of multinational and local business, orchestrated by the state, but at the expense and exclusion of the "popular sectors."[28] Those interests favored by a particular strategy— such as multinational corporations' or domestic business—may not necessarily have been politically important in formulating that strategy, however. More detailed research on the transition to export-led growth in Korea and Taiwan suggests that local and foreign business played little role in the debate over economic policy.[29] Gereffi and Wyman imply that political analysis must be supplemented by an examination of the *situational constraints* determining policy choice, including market size, natural resource endowment and the nature of external economic linkages. Small domestic markets, limited resource endowments and a gradual reduction in aid in the late 1950s and early 1960s were important in pushing Korea and Taiwan toward more export-oriented policies. International competitive pressures forced industrial upgrading over the 1970s and political pressures from the United States provided a powerful impetus for more liberal economic policies in the 1980s.

---

27. See Haggard and Cheng, "State Strategies..."
28. See Peter Evans, *Dependent Development: The Alliance of Multinational, State and Foreign Capital in Brazil* (Princeton: Princeton University Press, 1979).
29. Stephan Haggard, Byung-kook Kim and Chung-in Moon, "The Transition to Export-led Growth in Korea, 1954-1966," unpublished ms., Harvard University.

Conclusion

I have argued that the regional political system must be disaggregated into three distinct components: an international systemic level; a "process" level of international negotiations and agreements; and a domestic level of national policy interventions. Despite the alarming structural imbalances in the world economy in the mid-1980s, particularly the large U.S. budget and trade deficits, the *political* system of the Pacific Basin remains surprisingly conducive to continued regional integration.

First, high politics continues to favor expanded trade and investment. U.S. security interests continue to be served by growing interdependence in the region as they have been throughout the postwar period. As a resource-poor country, Japan's security is integrally tied to the free flow of raw materials, goods and capital. The Chinese entry into the world economy was partly motivated by security considerations, particularly the desire to guarantee access to needed technologies. The NICs are so deeply intertwined with the regional economy that any alternatives are foreclosed. For Taiwan in particular, but for Korea as well, commercial, financial and investment relations also serve political purposes.

Second, the international politics of industry has not proven as protectionist as was once thought. This is partly the result of the very internationalization of the world economy. As firms come to have more extensive trade and investment links, their trade preferences necessarily become more liberal. Despite the apparent growth of protectionist forces in Washington, free trade lobbying has increased as well.[30] The fact that trade policy has not been as protectionist as was anticipated in the early 1970s also has to do with the nature and dynamics of the "new protectionism" itself. Attempts to manage trade through international agreement have proven unsuccessful. Restrictionist arrangements have been circumvented, in part through expanded investment ties.

Finally, the domestic politics of the United States, Japan and the NICs are not as adverse to continued integration as is frequently thought. In the United States, liberal forces continue to present a strong counterweight to protectionists. The most important U.S. trade initiatives in recent years involve using market power to open markets abroad. The wisdom of such a bilateral strategy may be debated, since it marks a clear departure from the commitment to multilateralism and

---

30. I.M. Destler and John Odell, *Anti-Protection: Changing Forces in United States Trade Politics* (Washington D.C.: Institute for International Economics, 1987).

non-discrimination. Nonetheless, it should not be confused with an autarchic policy or a replay of Smoot-Hawley. In Japan and the NICs, the trend is definitely in the direction of greater internationalization, in no small part because of heightened international pressures.

This picture is not meant to paper over the profound adjustment difficulties that loom ahead; the chapters that follow stress these problems clearly. At this writing in mid-1988, the United States international asset position had shifted from that of a net creditor to a net debtor and huge macroeconomic imbalances threaten the world economy with recession. Even more daunting policy problems exist at the microeconomic level. Can firms and workers adjust to the rapid shifts in comparative advantage that a more deeply integrated regional economy implies? What role does government have in easing this transition? How will technological changes, the perpetual wildcard, affect the adjustment process? These are the questions for the 1990s. But if expanded interdependence is seen as a good, the politics of trade and investment in the Pacific Region provide room for cautious optimism about the future.

CHAPTER 2
Determinants of Development Strategies in Latin America and East Asia

Gary Gereffi and Donald Wyman

Introduction

In recent decades a relatively small set of developing nations, commonly known as the newly industrializing countries (or NICs), has distinguished itself in a variety of ways. While observers differ somewhat in their definition and identification of the NICs, they agree that these countries have achieved rapid rates of economic growth, relatively high levels of industrialization accompanied by significant increases in per capita income,[1] and prominence as exporters of manufactured products. For purposes of this paper the term NICs shall refer to the eight largest developing country exporters of manufactured goods, which means Argentina, Brazil, and Mexico in Latin America, Hong Kong, Singapore, South Korea, and Taiwan in East Asia, and India in South Asia.[2] These countries also are among Third World leaders in their levels of industrial output, and they are comparable to the advanced industrial nations in the manufacturing sector's share of gross domestic product.[3]

The NICs have followed different paths in becoming industrialized. The two basic approaches are export-oriented industrialization (EOI), initiated by Japan and later adopted by the East Asian countries of

The authors would like to thank Stephan Haggard for his helpful comments on this paper.

1. One well-known World Bank economist has defined the NICs as those countries that had per capita incomes between $1100 and $3500 in 1978, and in which the manufacturing sector accounted for at least 20 percent of gross domestic product in 1977. By these standards, eighteen countries qualified for NIC status. Bela Balassa, *The Newly Industrializing Countries in the World Economy* (New York: Pergamon Press, 1981), p. x.
2. See Colin I. Bradford, Jr., ''The Rise of the NICs as Exporters on a Global Scale,'' in Louis Turner and Neil McMullen (eds.). *The Newly Industrializing Countries: Trade and Adjustment* (London: George Allen & Unwin, 1982), p. 10 and Table 2.1.
3. World Bank, *World Development Report, 1986* (New York: Oxford University Press, 1986), Annex. Table 3, pp. 184-185.

Hong Kong, Singapore, South Korea, and Taiwan, and import-substituting industrialization (ISI), typified by the Latin American countries of Argentina, Brazil, and Mexico.[4] One of the most interesting features of these NICs is that the choice, timing, and sequence of their development patterns have varied by region.[5]

During the 1930s the large Latin American countries were the first in the developing world to experience a significant increase in the breadth and depth of their industrial activities, and they did so on the basis of supplying their sizable internal markets with locally produced goods. By contrast, the NICs of East Asia industrialized later and in a more selective and outward-oriented fashion. Between 1955 and 1974, Hong Kong, Singapore, South Korea, and Taiwan all sustained growth rates of over 7 percent per year in gross domestic product, a record which was superior to the 4.7 percent average experienced by the industrialized economies.[6] During the 1970s, the East Asian NICs not only grew far more rapidly than their Latin American counterparts, but they also had a more equitable distribution of income.[7]

---

4. Bela Balassa. "The Process of Industrial Development and Alternative Development Strategies," in his *The Newly Industrializing Countries in the World Economy,* pp. 1-26. India also has followed an import-substituting path of industrialization, but with much less reliance on direct foreign investment, foreign loans, or external trade than the other NICs. For an excellent review of this important case, see Aditya Mukherjee and Mirdula Mukherjee. "Imperialism and the Growth of Indian Capitalism in the Twentieth Century," *Economic and Political Weekly* (New Delhi, India), March 12, 1988, pp. 531-546.
5. Gustav Ranis, "Challenges and Opportunities Posed by Asia's Super-exporters: Implications for Manufactured Exports from Latin America," in Werner Baer and Malcolm Gillis (eds.). *Export Diversification and the New Protectionism: The Experiences of Latin America* (Champaign: University of Illinois, 1981), p. 205.
6. The star economic performer in the postwar years is Japan, whose economy grew at the rate of 9.5 percent per annum between 1955 and 1974. Edward K. Y. Chen, *Hypergrowth in Asian Economies: A Comparative Study of Hong Kong, Japan, Korea, Singapore and Taiwan* (London: The Macmillan Press, 1979), pp. 9-10.
7. The distribution of income in the East Asian NICs is treated in Shirley W. Y. Kuo, Gustav Ranis, and John C. Fei, *The Taiwan Success Story: Rapid Growth with Improved Distribution in the Republic of China* (Boulder, Co.: Westview Press, 1981), pp. 38-60, and Gustav Ranis. "Equity with Growth in Taiwan: How 'Special' is the 'Special Case'?," *World Development* 6:3 (1978), pp. 397-409. An argument that income inequality is on the rise in South Korea may be found in Hagen Koo, "The Political Economy of Income Distribution in South Korea: The Impact of the State's Industrialization Policies," *World Development* 12:10 (1984), pp. 1029-1037. On Latin America see Joel Bergsman, "Income Distribution and Poverty in Mexico." *World Bank Staff Working Paper* 395 (Washington, D.C.: World Bank, June, 1980); David Denslow, Jr. and William Tyler, "Perspectives on Poverty and Income Distribution in Brazil," unpublished manuscript, 1983: David Felix, "Income Distribution and the Quality of Life in Latin America: Patterns, Trends, and Policy Implications," *Latin American Research Review* 18:2 (1983), pp. 3-34.

Furthermore, the East Asian NICs appear to have adjusted much more successfully than the Latin American ones to the global economic crisis that originated in the 1970s.[8]

East Asia's superior economic performance during recent years has given renewed impetus to a debate about appropriate development policies in both governmental and academic circles. Officials in key multilateral institutions (e.g., the World Bank), the United States government (e.g., the Agency for International Development), and the internationalized segments of the private sector (e.g., leading commercial banks) have advocated export-promoting policies as the new development orthodoxy for Third World nations, thus denigrating the import-substituting approach to industrialization.[9] This interpretation has been challenged by some scholars as well as by those associated with other multilateral institutions (e.g., the United Nations Economic Commission on Latin America and the U.N. Committee on Trade and Development).[10] In addition, the debate surrounding EOI and ISI has generated substantial disagreement among a variety of political and economic actors in developing countries themselves.

A close examination of the evolution of development patterns in the two regions, however, shows that historically the opposition between ISI and EOI poses a false dilemma. The two approaches have been compatible and even mutually reinforcing, as we will see below. It is a fallacy, though, to believe that it is possible to replicate the East Asian model in substantially different circumstances.

The regional contrast in the development experience of the Latin American and East Asian NICs also raises major theoretical issues in the field of international political economy. One of these is the relative importance of internal and external determinants of national

8. This success has not gone unnoticed in the press. See, for example, Robert Gibson, "Asia's little Dragons Spew Economic Fire," *Los Angeles Times,* July 15, 1984.

9. This view is typified by the March, 1984 issue of the influential report, *World Financial Markets,* issued by the Morgan Guaranty Trust Company. Focusing on whether Korea provides an appropriate adjustment model for the 1980's, the report found that Korea's economic achievements "exemplify the benefits that can flow from an outward-looking, market-oriented strategy for economic adjustment and development...," p. 1. See also the discussion of the contrast between the successes of export promotion and the problems associated with import substitution in *World Economic Outlook* (Washington, D.C.: International Monetary Fund, 1985), pp. 182-183.

10. For an academic economist's challenge to EOI as the new orthodoxy see Colin I. Bradford, Jr., "The NICs: Confronting U.S. 'Autonomy'," in Richard E. Feinberg and Valeriana Kallab (eds.), *Adjustment Crisis in the Third World* (New Brunswick, N.J.: Overseas Development Council, 1984), pp. 121-129.

economic policies. Our contribution here lies in demonstrating that the development sequences of the Latin American and East Asian NICs are marked by significant phases of commonality, divergence, and convergence. Within each region, the NICs have altered their development trajectories in remarkably similar ways and at virtually the same times, which lends strong prima facie support to the notion that factors external to these nations have played a major role in shaping their industrial transformations. These external determinants are not uniform in their strength or regional impact, however. While the Latin American and East Asian NICs moved in divergent directions at the end of their initial stage of ISI, there are recent indications of policy convergence by the NICs in both regions, as well as unique national variations. In order to explain these outcomes, one must assign different weights to the NICs' diverse forms of linkage to the international system, and also look closely at the priorities and interests of local state elites and other domestic actors. The cross-regional analysis can help us explore this complex interplay between external and internal determinants of national policies and development trajectories.

The comparison of Latin American and East Asian NICs is especially appropriate in addressing a central controversy in the dependency literature. This perspective, most commonly applied to Latin American nations, has argued that extensive reliance on direct foreign investment, foreign loans, foreign aid, and exports of primary commodities produces growing inequality (internationally and domestically), inadequate economic integration in the local economy, authoritarian political regimes, social marginality, and slow or erratic rates of economic growth.[11] The assumptions of the dependency literature are challenged by the experience of the East Asian NICs, which during the past two decades have industrialized rapidly, with relatively equitable distributions of income, and without extensive reliance on multinational corporations (South Korea) or foreign loans (Taiwan).[12]

---

11. The literature that falls into what is now referred to as the dependency tradition is extensive. A classic statement is Fernando Henrique Cardoso and Enzo Faletto, *Dependencia y desarrollo en América Latina* (Mexico City: Siglo Veintiuno Editores, S. A., 1969), expanded and revised as *Dependency and Development in Latin America* (Berkeley: University of California Press, 1979). The dependency perspective is reviewed and analyzed in Gary Gereffi, *The Pharmaceutical Industry and Dependency in the Third World* (Princeton: Princeton University Press, 1983), Chapter 1.

12. This challenge is explored explicitly in Alice H. Amsden, "Taiwan's Economic History: A Case of Etàtisme and a Challenge to Dependency Theory," *Modern China* 5:3 (July, 1979), pp. 341-380, and Richard E. Barrett and Martin K. Whyte, "Dependency Theory and Taiwan: Analysis of a Deviant Case," *American Journal of Sociology* 87:5

In our view, a fundamental problem of the dependency approach for comparative purposes is that it has focused too heavily on the specific forms of dependence associated with Latin America's relationship to the world economy and on the hegemonic influence of the United States. The East Asian countries are particularly useful in showing that external linkages may be opportunities and not just constraints for developing countries. The cross-regional analysis of Latin America and East Asia enables us to better identify the conditions under which specific situations of dependency have positive, negative, or mixed consequence for national development.

The objectives of this paper are twofold. The first is to trace the contrasting patterns of development followed by the Latin American and East Asian NICs. We believe that the comparison is sharpest if we examine a pair of NICs from each region: Brazil and Mexico from Latin America, and South Korea and Taiwan from East Asia.[13]

Secondly, we wish to identify and analyze the determinants of these development sequences. We compare the two regions in terms of the following key determinants: natural resource endowments and internal market size, geopolitical factors, transnational economic linkages, state structures, elite policy preferences, prevailing development ideologies, and social coalitions.[14] These determinants, we argue, help explain both similarities and differences in the development sequences that characterize the Latin American and East Asian NICs. Using this approach, we also explore the interplay between external and internal determinants, and the relationship between forms of participation in the international economy and domestic political factors.

(1982), pp. 1064-1089. Hill Gates argues that the costs of dependency are merely disguised in Taiwan's case. See "Dependency and the Part-time Proletariat in Taiwan," *Modern China* 5:3 (July, 1979), pp. 381-408.

13. The three other countries from these regions that are commonly referred to as NICs —Argentina. Hong Kong, and Singapore—share many aspects of this regional contrast in development sequences, but they also present distinctive features that cannot be treated in this paper.

14. The distinctive cultural heritages of Latin America and East Asia, although a fascinating and complex topic, do not seem to be primary determinants of the development patterns we are concerned with in this paper. The cultural traditions of each region have existed for centuries, while the development shifts we focus on are of relatively recent origin. Nonetheless, cultural influences do have an indirect effect on development policies, mediated through some of the variable we have identified (e.g., social coalitions, elite preferences, and development ideologies).

15. In terms of economic policies, the inward- and outward-oriented approaches are differentiated from each other by the presence on absence of an anti-export bias. Anne Krueger, "Export-Led Industrial Growth Reconsidered," in Wontack Hong and

Development Patterns and Development Strategies as Comparative Concepts

The development experience of the Latin American and East Asian NICs is complex. One way to conceptualize their trajectories is in terms of development *patterns* that are historically and structurally situated. These development patterns have three dimensions: (1) the types of industries that are most prominent in each phase of a country's economic development; (2) the degree to which these leading industries are inwardly or outwardly oriented (i.e., whether production is destined for the domestic market or for export); and (3) the major economic agents relied on to implement and sustain development.

Based on a broad historical view of industrialization in Mexico, Brazil, South Korea, and Taiwan, one can identify five main phases of industrial development. Two of these are inward-looking: primary ISI and secondary ISI. The other three are outward-looking: a commodity export phase, and primary and secondary EOI.[15] The subtypes within the outward and inward approaches are distinguished by the kinds of products involved. In the *commodity export phase*, the output typically is unrefined or semiprocessed raw materials (agricultural goods, minerals, oil, etc.). In *primary ISI*, the objective is to produce locally some of the basic consumer goods that are being imported, and in almost all countries the key industries during this phase are textiles, clothing, footwear, and food processing. *Secondary ISI* involves using domestic production to substitute for imports of a variety of capital- and technology-intensive manufactures: consumer durables (e.g., automobiles), intermediate goods (e.g., petrochemicals and steel), and capital goods (e.g., heavy machinery). Both phases of EOI involve manufactured exports. In *primary EOI* these tend to be labor-intensive products, while *secondary EOI* includes higher value-added items that are skill-intensive and require a more extensively developed local industrial base. The principal phases of industrial development for the Latin American and East Asian NICs are outlined in Tables 2.1 and 2.2.

Development *strategies*, on the other hand, can be defined as sets of policies that shape a country's relationship to the global economy and that affect the domestic allocation of resources among industries and social groups. The notion of development strategies links sets of policies and particular structures of production in such a way as to shed light on a country's relationship to international markets and

Lawrence B. Krause (eds.), *Trade and Growth of the Advanced Developing Countries in the Pacific Basin* (Seoul: Korea Development Institute, 1981), pp. 8-10.

resources, and on its decisions about domestic economic growth and equity. A variety of policies may be used to establish a particular pattern of inward- or outward-oriented production, but our focus is on the broad strategy itself rather than on policy oscillations or shifts within specific ISI or EOI approaches.

We need to recognize, however, that if we define development strategies as relatively coherent sets of government policies that serve to relate a domestic economy to the global one, these strategies become a variable (rather than a constant) feature of national economic growth. Thus government policies to promote industrial growth have not been of equal importance in each of the phases of economic development shown in Tables 2.1 and 2.2. A careful examination of the leadership or followership role of government policies is a logical outgrowth of our framework, but it is not a task for this paper. Our working hypothesis is that state-led industrialization becomes the norm in all four NICs following World War II, although the relative importance of local private firms, foreign-owned companies, and state enterprises varies in each country. In prior periods local government policies tend to have played a secondary role to the initiatives of foreign or domestic economic elites. In some instances, like the Japanese colonial era, the issue is moot because independent local governments did not even exist.

In the remainder of the paper, we will use the term "development strategies" for convenience to refer to the broad sequences outlined in Tables 2.1 and 2.2. However, one should bear in mind that there is an important analytical distinction between development "patterns" (which are economic outcomes) and the policies that make up government "strategies," which may be instrumental in bringing about import-substituting or export-promoting industrial transformations.

The Trajectories of Development Strategies: Commonalities, Divergence, and Convergence

Over time, each regional pair of NICs has followed a sequence that includes three development strategies that closely approximate the import-substituting and export-oriented ideal types, plus a mixed strategy in the most recent period. An analysis of these sequences, as shown in Table 2.1, suggests the following conclusions. First, the contrast often made between Latin America and East Asia as representing inward- and outward-oriented development strategies respectively is oversimplified. While this distinction is appropriate for some periods, an historical perspective shows that each of these

Table 2.1 Development Strategies in Latin America and East Asia: Commonalities, Divergence, and Convergence

Mexico and Brazil: 1880-1930	Mexico and Brazil: 1930-1955	Mexico: 1955-1970 Brazil : 1955-1968	Mexico: 1970 to present Brazil : 1968 to present
Commodity Exports →	Primary ISI →	Secondary ISI ↗	Diversified Export Promotion and Continued Secondary ISI
		Primary EOI ↗	Secondary ISI (heavy and chemical industrialization) and Secondary EOI
Taiwan: 1895-1945 Korea: 1910-1945	Taiwan: 1950-1959 S. Korea: 1953-1960	Taiwan: 1960-1972 S. Korea: 1961-1972	Taiwan and S. Korea: 1973 to present

ISI = Import-Substituting Industrialization.
EOI = Export-Oriented Industrialization.

Table 2.2 Patterns of Development in Latin America and East Asia

MEXICO AND BRAZIL

Development Strategies	Commodity Exports	Primary ISI	Secondary ISI	Diversified Exports + Secondary ISI
Main Industries	Mexico: Precious metals (silver, gold), minerals (copper, lead, zinc), oil Brazil: Coffee, rubber, cocoa, cotton	Mexico and Brazil: Textiles, food, cement, iron and steel, paper, chemicals, machinery	Mexico and Brazil: Automobiles, electrical & non-electrical machinery, petrochemicals, pharmaceuticals	Mexico: Oil, silver, apparel, transport equipment, non-electrical machinery Brazil: Iron ore and steel, soybeans, apparel, footwear, transport equipment, non-electrical machinery, petrochemicals, plastic materials
Major Economic Agents	Mexico: Foreign investors Brazil: National private firms	Mexico and Brazil: National private firms	Mexico and Brazil: State-owned enterprises, transnational corporations, & national private firms	Mexico and Brazil: State-owned enterprises, transnational banks, transnational corporations, and national private firms
Orientation of Economy	External Markets	Internal Market	Internal Market	External & Internal Markets

Table continues on the following page.

Table 2.2 Continued

TAIWAN AND SOUTH KOREA

Development Strategies	Commodity Exports	Primary ISI	Primary EOI	Secondary ISI + Secondary EOI
Main Industries	Taiwan: Sugar, rice; Korea: Rice, beans	Taiwan and S. Korea: Food, beverages, tobacco, textiles, clothing, footwear, cement, light manufactures (wood, leather, rubber, and paper products)	Taiwan and S. Korea: Textiles and apparel, electronics, plywood, plastics (Taiwan), wigs (S. Korea), intermediate goods (chemicals, petroleum, paper, and steel products)	Taiwan: Steel, petrochemicals, computers, telecommunications, textiles & apparel; S. Korea: Automobiles, shipbuilding, steel and metal products, petrochemicals, textiles and apparel, electronics, video-cassette recorders, machinery
Major Economic Agents	Taiwan and Korea: Local producers (colonial rule by Japan)	Taiwan and S. Korea: National private firms	Taiwan and S. Korea: National private firms, transnational corporations	Taiwan and S. Korea: National private firms, transnational corporations, state-owned enterprises, transnational banks (S. Korea)
Orientation of Economy	External Markets	Internal Market	External Markets	Internal & External Markets

ISI = Import-Substituting Industrialization.
EOI = Export-Oriented Industrialization.

regional pairs of NICs has pursued *both* inward- and outward-oriented approaches.[16]

Second, the initial phases of industrialization—commodity exports and primary ISI—were common to all four of the Latin American and East Asian NICs. The subsequent divergence in these regional sequences stems from the way in which each country responded to the basic problems associated with the continuation of primary ISI; these problems included balance of payments pressures, rapidly rising inflation, high levels of dependence on intermediate and capital goods imports, and low levels of manufactured exports.[17]

Third, the timing and duration of these development patterns vary by region. Timing helps explain these sequences because the opportunities and constraints that shape development choices are constantly shifting. The East Asian NICs began to emphasize the export of manufactured products at a time of extraordinary dynamism in the world economy. The two decades that preceded the global economic crisis of the 1970s saw unprecedented annual growth rates of world industrial production (approximately 5.6 percent) and world trade (around 7.3 percent), relatively low inflation and high employment rates in the industrialized countries, and stable international monetary arrangements.[18] The expansion of world trade was fastest between 1960 and 1973, when the average annual growth rate of exports reached almost 9 percent. Starting in 1973, however, the international economy began to enter a troublesome phase. From 1973 to the end of the decade, the annual growth in world trade fell to 4.5 percent[19] as manufactured exports from the developing countries began to encounter stiffer protectionist measures in the industrialized markets.

---

16. Indeed, a previous period of import-substitution may be a prerequisite for successful adoption of EOI based on national entrepreneurs. See Stephan Haggard and Tun-jen Cheng, "State and Foreign Capital in the East Asian NICs," in Frederic C. Deyo (ed.), *The Political Economy of the New Asian Industrialism* (Ithaca, N.Y.: Cornell University Press, 1987), pp. 84-135.

17. The problems associated with ISI in Latin America are dealt with by Werner Baer, "Import Substitution and Industrialization in Latin America: Experiences and Interpretations," *Latin American Research Review* 7:1 (Spring, 1972), pp. 95-122, and Albert O. Hirschman, "The Political Economy of Import-Substituting Industrialization in Latin America," *The Quarterly Journal of Economics,* 82 (February, 1968), pp. 2-32. East Asian countries experienced similar problems with primary ISI. See, for example, Ching-yuan Lin, *Industrialization in Taiwan, 1946-72: Trade and Import-Substitution Policies for Developing Countries* (New York: Praeger Publishers, 1973), pp. 68-74.

18. W. W. Rostow, *The World Economy: History and Prospect* (Austin: University of Texas Press, 1978), p. 247.

19. A. G. Kenwood and A. L. Lougheed, *The Growth of the International Economy, 1820-1980* (London: George Allen & Unwin, 1983), p. 299.

Fourth, the development strategies of the Latin American and East Asian NICs show some signs of convergence, yet they also are maintaining their distinctiveness. To support this convergence thesis, we need to distinguish two subphases during the most recent period. In the 1970s Mexico and Brazil began to expand both their commodity (oil, soybeans, minerals, etc.) and manufactured exports, as well as to accelerate their foreign borrowing, in order to acquire enough foreign exchange to finance the imports necessary for furthering secondary ISI. One prominent economist has labeled this approach "export-adequate" to differentiate it from the East Asian "export-led" strategy.[20] South Korea and Taiwan, on the other hand, emphasized heavy and chemical industrialization from 1973 to 1979, with a focus on steel, automobiles, shipbuilding, and petrochemicals. The objective of heavy and chemical industrialization in East Asia was twofold: to develop national production capability in these sectors, justified by national security as well as import substitution considerations; and to lay the groundwork for more diversified exports in the future. Thus, the Latin American and East Asian NICs felt the need to couple their previous strategies from the 1960s (secondary ISI and primary EOI, respectively) with elements of the other strategy in order to derive the complementary benefits of simultaneously pursuing inward- and outward-oriented approaches.

There were further pressures in the 1980s toward convergence. The oil price shock of 1979–1980, rising international interest rates, and growing protectionism in the advanced industrial countries combined to push all four of the Latin American and East Asian NICs to adopt similar adjustments in their development strategies. These shifts may be described as efforts to promote economic stabilization, *privatization,* and internationalization. *Stabilization* measures sought to fiscal restraint and monetary control, together with restrictions on wage increases. *Privatization* meant a turn in all four countries to a more market-oriented style of economic management. This involved a move away from discretionary, sector-specific interventions and toward indirect, non-discretionary supports (such as incentives for research and development, and manpower training), deregulating foreign exchange controls, liberalizing imports, limiting the role of state-owned enterprises, and lessening government influence over

20. Albert Fishlow, "State and Economy in Latin America: New Models for the 1980s," unpublished paper presented at the Workshop on the Impact of the Current Economic Crisis on the Social and Political Structure of the Newly Industrialized Nations, São Paulo, Brazil, February, 1985.

banks and credit. *Internationalization* refers to measures taken in all four NICs to open up their domestic markets by removing restrictions on direct foreign investment, especially in the service sector (including banking, insurance, hotels, and retail stores).

Notwithstanding these notable forms of convergence, there also were significant national variations among the NICs during this period. Heavy and chemical industrialization in the East Asian nations had similar goals, but in Taiwan the principal domestic agents were state enterprises, while in South Korea this phase of development was primarily carried out by large private industrial conglomerates (*chaebols*). This also had implications for the kind of industrial restructuring that was emphasized in each country in the 1980s. There were common export promotion efforts focusing on a new range of "strategic" high-technology industries, such as semiconductors, computers, telecommunications, computerized numerical-control machine tools, etc. In this phase of secondary EOI, there was less of a concern with export volume from labor-intensive industries, and much more emphasis on value-added in capital- and technology-intensive industries. In Taiwan, however, industrial restructuring also implied a rationalization in which mergers and joint ventures were encouraged in sectors like automobiles, advanced electronics, and heavy machinery in order to overcome the limitations of the predominantly small-scale manufacturing firms on the island. In South Korea, industrial restructuring had an opposite meaning: namely, reducing the level of concentration among the *chaebols* by giving greater attention to small and medium-sized firms which are central to the diversified export of light manufactures.

Whereas exchange rate liberalization in East Asia led to the appreciation of local currencies in Japan, Taiwan, and South Korea, in the Latin American NICs there was a sharp devaluation of their currencies. In Mexico, this led to a spectacular increase of labor-intensive manufactured exports from the *maquiladora* (bonded-processing) industries located along the U.S. border, and also to a renewed inflow of direct foreign investment in the mid-1980s. Brazil has had a more diversified profile in terms of its manufactured exports, especially in consumer durables (like automobiles and auto parts), steel, capital goods, and armaments. This reflects Brazil's more successful and sustained secondary ISI investments in the late 1970s. Thus the East Asian and Latin American NICs still show differences in emphasis with regard to their production structures, but all four NICs have moved to a more advanced stage of industrialization in which secondary ISI and secondary EOI are combined.

Determinants of Development Strategies

Having provided a broad outline of the development patterns that characterize the Latin American and East Asian NICs, we now turn to a discussion of some of the key determinants of the strategies that make up these patterns. Each of the factors considered below has its own explanatory logic that can help account for the economic achievements of the Latin American and East Asian NICs, as well as some of the main differences between them. In practice, however, scholars frequently join two or more of these determinants of national policy choice, and assign them different weights. Our intent in treating these determinants individually is merely to highlight some of the causal arguments associated with each. We do not favor single-factor explanations of development trajectories or outcomes; in fact, some of these "determinants" constitute contending hypotheses and approaches.

It also needs to be pointed out that these factors may be consequences, as well as causes, of certain strategic choices. Whether a factor is cause or consequence is not intrinsic to the variable itself, but depends on the circumstances. Even something as apparently immutable as natural resource endowments can be affected by policy choices. A country's comparative advantage in natural resources, for example, can be thought of in "dynamic" rather than "static" terms, since the East Asian nations have done remarkably well in creating a comparative advantage for themselves in the production of steel and petrochemicals despite lacking indigenous supplies of iron ore and oil.

This section of the paper is merely a survey of what we believe is a fruitful research agenda. We raise as many questions as we answer. Future historical and statistical work undoubtedly will be required in examining the individual determinants we consider here, and in addition more detailed inquiry will be needed to look at how these diverse causal factors interact.

Natural Resource Endowments and Internal Market Size

When a particular development strategy is no longer viable, what factors influence a country's choice of its subsequent approach? In particular, why did Latin American countries respond to a crisis in primary ISI by adopting a secondary ISI strategy, while the East Asian countries responded to a similar crisis by adopting an export-oriented development strategy? Part of the answer is found in certain country-specific characteristics, especially natural resource endowments and internal market size. Economists have long argued that foreign trade is much more important for small economies than for large ones.[21] With few natural resources and relatively small markets, this argument goes, the East Asian countries had no choice but to pursue EOI. The Latin American countries, in contrast, had large enough potential markets, coupled with their diverse array of export commodities (e.g., minerals, petrochemicals, agricultural goods), to make secondary ISI feasible.

It would be foolish to argue that factors such as size of the internal market or natural resource endowments do not play a major role in the determination of national development strategies. Confinement to these factors alone, however, asserts an overly deterministic explanation for the outcomes of policy deliberations. The correlation between a country's size and its economic policies, for example, is far from perfect. Furthermore, as Balassa has pointed out, the East Asian NICs are not really small in market terms relative to many other developing countries.[22] A cross-regional comparison of the causes and consequences of development strategies must consider choice as a real phenomenon to be explained. As such, the analysis must go beyond given structural features of these countries, such as resource endowments and internal market size, to consider factors such as geopolitical situation, linkages with the global economy, state structures, elite policy preferences, prevailing development ideologies, and social coalitions.

Geopolitical Factors

For purposes of this paper, we distinguish between two dimensions of the international system: the interstate system and the global economy. The former is discussed in terms of geopolitical factors, in particular the notion of contending regional hegemonies; the latter is analyzed in terms of four types of transnational economic linkages.

Hegemonic powers affect both the development patterns and strategies of countries within their sphere of influence in several ways. Hegemonic powers have the political, military, and economic means to control or significantly alter the flow of goods and resources to subordinate nations. The hegemon is in a position to benefit either

---

21. Simon Kuznets, "Quantitative Aspects of the Economic Growth of Nations, IX. Level and Structure of Foreign Trade: Comparisons for Recent Years," *Economic Development and Cultural Change* 13, part 2 (October, 1964), pp. 1-106.
22. Balassa, "The Process of Industrial Development and Alternative Development Strategies," pp. 2-3. In 1984, for example, South Korea's population of 40 million was larger than Argentina's 30 million. Brazil and Mexico, with populations of 133 million and 77 million respectively, are much larger than Korea or Taiwan (19 million), but the

through market forces or various mechanisms of control, including influence over transnational networks of actors and agencies (aid donors, multinational corporations, banks). In a situation of declining or disputed hegemony, the impact of geopolitical factors is much less certain.[23]

Latin America and East Asia each were subject to two clear periods of external domination, but the successive hegemonies for each region differed in their nature and timing. Latin America experienced British hegemony from the late nineteenth century to the beginning of World War I, and United States hegemony from the end of World War II to the present. In East Asia, Korea and Taiwan were subject to Japanese colonial rule from the late nineteenth century until the end of World War II. Following the defeat of the Japanese empire, the United States assumed a dominant role vis-à-vis newly independent South Korea and Taiwan. Since the early 1970s the predominance of the United States in South Korea and Taiwan has been challenged by the emergence of Japan as a global economic power.[24]

The two periods of external domination and influence that have obtained in East Asia each have significantly affected the nature of political, social, and economic development in South Korea and Taiwan.[25] They help to account for the origins and nature of the "developmental state" in these two countries and for its degree of autonomy relative to their respective civil societies. The Japanese imperium, for example, left a legacy of strong state control. The relatively unscathed condition of the state bureaucracy at the end of the Second World War. gave the state in the East Asian countries substantial autonomy relative to internal groups. United States hegemony contributed to this domestic autonomy as U.S. forces and

---

extensive poverty in these two Latin American countries reduces their effective market size for selected consumer goods to levels nearer those of the East Asian NICs than population figures alone would indicate. World Bank, *World Development Report, 1986*, Annex, Table 1, pp. 180-181.

23. Robert O. Keohane. *After Hegemony: Cooperation and Discord in the World Political Economy* (Princeton: Princeton University Press, 1984), pp. 32-46.

24. Giovanni Arrighi, "A Crisis of Hegemony," in Samir Amin et al. (eds.), *Dynamics of Global Crisis* (New York: Monthly Review Press, 1982), pp. 55-108; Robert O. Keohane, "Hegemonic Leadership and U.S. Foreign Economic Policy in the 'Long Decade' of the 1950s," in William P. Avery and David P. Rapkin (eds.), *America in a Changing World Economy* (New York: Longman, 1982), pp. 49-76.

25. For a discussion of the subsequent consequences of the Japanese colonial rule in Korea and Taiwan, see Bruce Cumings, "The Origins and Development of the Northeast Asian Political Economy: Industrial Sectors, Product Cycles, and Political Consequences," *International Organization* 38:1 (Winter, 1984), pp. 1-40. The economic aspects of colonialism for Taiwan and Korea respectively are treated in Samuel P. S. Ho, *Economic Development of Taiwan, 1860-1970* (New Haven: Yale University Press, 1978),

policies helped to weaken social groups that might have challenged the state, such as the agrarian elite.

United States hegemony over Korea and Taiwan stimulated the adoption of liberal trade and investment policies in those two countries in the 1960s. Moreover, U.S. occupation forces in Korea contributed to the weakening of leftist influences within unions, of urban political parties, and of demands from the rural sector, at the same time as expropriated Japanese properties and U.S. aid provided substantial material resources to the state.[26] Because U.S. hegemonic interests in East Asia and the ideological perspectives and policy preferences of the political elites often coincied, U.S. hegemony in that part of the world tended to be supportive regarding the pursuit of an export-oriented development policy. However, the large aid relationship gave the United States leverage which it did not hesitate to use whenever local elites vacillated in adopting EOI policy reforms.

Development in Latin America, too, has been conditioned by the hegemonies that existed in the region. British hegemony, for example, reinforced the notion of free trade as applied to the export of primary products and the import of manufactured goods.[27] The relative absence of hegemony in Latin America during the 1930s was one of the reasons that the region was able to pursue independent political and economic experiments.[28] Postwar U.S. political pressures had a great deal to do with the establishment of a favorable investment climate for multinational corporations in the region and with the high costs associated with Latin American commercial and financial diversification away from the United States. In contrast to the situation in East Asia, the policy preferences of domestic elites in Latin America clashed frequently with U.S. business interests in the 1960s and 1970s, so U.S. hegemony was more antagonistic to the locally preferred development approach.[29]

pp. 25-102, and Paul Kuznets, *Economic Growth and Structure in the Republic of Korea* (New Haven: Yale University Press, 1977), pp. 8-28.

26. Bruce Cumings, *The Origins of the Korean War: Liberation and the Emergence of Separate Regimes* (Princeton: Princeton University Press, 1981).

27. On economic relations between Great Britain and Latin America see, for example, D.C.M. Platt, *Latin America and British Trade, 1806-1914* (London: Adams & Black,1972) and J. Fred Rippy, *British Investments in Latin America, 1822-1949* (Minneapolis: University of Minnesota Press, 1959). A broad threatment of British influence in one Latin American country is Richard Graham's *Britain and the Onset of Modernization in Brazil, 1850-1914* (Cambridge: Cambridge University Press, 1972).

28. These experiments are described in Rosemary Thorp (ed.), *Latin America in the 1930s: The Role of the Periphery in World Crisis* (New York: St. Martin's Press, 1984).

29. Several essays that explore the instruments and effects of U.S. hegemony in Latin America are contained in Julio Cotler and Richard R. Fagen (eds.), *Latin America and*

Despite Japan's growing prominence as a world economic power, the United States has maintained a considerable degree of direct influence over the adjustment possibilities in the Latin American and East Asian NICs. The United States has tended to link its debt relief and trade negotiations with Brazil and Mexico, for example, to a variety of issues that go well beyond macroeconomic stabilization concerns. At the behest of well organized lobbies, U.S. negotiators have demanded that these Latin American nations free multinational corporations from assorted ownership and performance requirements (e.g., joint venture restrictions, minimum local content levels, and export promotion schemes that make incentives to firms conditional on their export sales). The United States also has successfully urged the repeal of restrictive legislation affecting certain U.S. industries (such as pharmaceuticals and computers), and supported efforts to reduce the overall role of government in these economies.

Whereas foreign debt and direct foreign investment are the primary vehicles for U.S. pressures in Latin America, disputes over aid and trade have been the focal point in the relationship between the United States and East Asia. The aid story in the 1950s and early 1960s is well known. American influence has far from disappeared in the 1980s, however. The United States has used the threat of reducing Generalized System of Preference benefits and curtailing access to the U.S. market in order to successfully pressure Taiwan and South Korea to slash tariffs and lift bans on American imports, remove foreign exchange controls (leading in both cases to sharp currency appreciations), "voluntarily" restrict their exports to the U.S. market, and open their domestic economies to U.S. investments in the previously closed service sector. Thus the Latin American and East Asian NICs have been susceptible to continuing U.S. influence, with distinct regional variations. The United States, despite its relative decline as a hegemonic power, is still in a position to shape the development strategies of the NICs.

Transnational Economic Linkages

Transnational economic linkages or TNELs (economic aid, export trade, direct foreign investment, and foreign loans) affect development strategies in several ways. First, they represent economic

---

the United States: The Changing Political Realities (Stanford: Stanford University Press, 1974). An excellent treatment of country experiences and regional themes is Tulio Halperin Donghi. Historia Contemporanea de América Latina (Madrid: Alianza Editorial, 1969).

resources that may be used, singly or in diverse combinations and sequences, to finance development. The availability of TNELs facilitates the pursuit of particular development strategies. Conversely, the absence of these resources constrains the choices of national policy makers. For example, direct foreign investment sustained secondary ISI in Latin America, much as massive foreign aid flows made primary ISI possible in East Asia.

Second, the availability of these resources is conditioned by factors beyond as well as within the control of nation states. Factors beyond the control of individual countries include global economic conditions (e.g., trends in world trade) as well as geopolitical pressures that help channel these resources toward some countries and away from others. National policies regarding foreign investment, domestic wage levels, and the degree of political stability in a country, on the other hand, can also shape the performance of TNELs.

Third, as Table 2-3 shows, each of these TNELs is associated with different institutional carriers or agents. Once present in a developing country, these actors become part of the constellation of interests that seek to influence policy choice. How this influence is expressed depends on regime type, social coalitions, and prevailing development ideologies.

Finally, the destination and use of TNELs in a country affect not only development outcomes (according to how productively the resources are employed), but also strategy choice (through their influence on the economic power of domestic actors). It matters, for example, whether these resources are used to finance fur coats and fine china or irrigation systems and highways, just as it matters whether the presence of these resources strengthens agrarian elites, urban middle classes, or the industrial working class.

Our analysis of TNELs across regions suggests the following conclusions, which derive from the importance assigned to these linkages in Table 2-4. The "high," "medium," and "low" weights in Table 2-4 are based on estimates of the relative significance of the TNELs in each economy, compared with other developing countries at similar phases in their industrialization process.

Cross-regional variation in the role played by TNELs in the Latin American and East Asian NICs is considerable. First, the salience of TNELs varies markedly within each region according to the strategy pursued, since each phase of industrialization is associated with a different mix of external resources used to finance development. In East Asia, for example, primary ISI relied on a great deal of foreign aid and little in the way of export trade, whereas EOI was defined by

Table 2.3 The Origins, Destination, and Institutional Bases of Transnational Economic Linkages

Transnational Economic Linkages	Institutional Carriers	Origins or Content	Host Country Destination
Foreign Aid	Bilateral agencies Multilateral agencies Private voluntary organizations	U.S. Agency for International Development World Bank Group European or Japanese development agencies	Central government Specific public agencies Private organizations
Foreign Trade	Transnational corporations Domestic manufacturing firms Trading companies	Commodity exports Manufactured exports	United States Western Europe Japan Socialist country markets Developing country markets
Direct Foreign Investment	Transnational corporations	American European Japanese Other	Mining Agriculture Manufacturing Service sector
Foreign Borrowing	Commercial banks Official lending institutions	Transnational banks (private) World Bank Group International Monetary Fund Interamerican Development Bank Asian Development Bank	To: Central government State-owned enterprises Private sector For: Project loans Program loans Structural adjustment loans Undefined

Table 2.4 The Structure of Dependent Development in Latin America and East Asia

Development Strategies	Brazil and Mexico			South Korea and Taiwan			
	Commodity Primary ISI	Secondary ISI	Diversified Exports + Secondary ISI	Commodity Exports	Commodity Primary ISI	Primary EOI	Secondary ISI + EOI
Trans-National Economic Linkages							
Foreign Aid	Low	Medium	Low	Low	High	Medium	Medium
Foreign Trade	Low	Low [Exports] Medium [Imports]	Medium [Imports] High [Exports]	High	Low [Exports] Medium [Imports]	High	High
Direct Foreign Investment	Medium [Brazil] High [Mexico]	High	Medium	Medium	Low	Low [Korea] Medium [Taiwan]	Medium [Korea] High [Taiwan]
Foreign Borrowing	Medium	Medium	High	Low	Low	Low [Taiwan] Medium [Korea]	Medium [Taiwan] High [Korea]

ISI = Import-substituting industrialization
EOI = Export-oriented industrialization

extensive exports and virtually no foreign aid.

Second, the salience of TNELs also varies within similar strategies across the two regions. For example, both regions went through a period of primary ISI, but the dynamics were different. Import-substituting industrialization in Latin America came earlier than in East Asia, lasted longer, and was associated with populism; in East Asia, labor and peasants were excluded from politics in the 1950s and had little direct influence over strategy choices. Moreover, the linkages through which each region related to the international economy while experiencing primary ISI were quite different. In East Asia, import substitution was financed by massive amounts of foreign economic assistance,[30] whereas in Latin America primary ISI tended to be carried out by local industrialists with the support of the state and with limited participation by multinational corporations.

Third, the contrast with regard to TNELs is sharpest when we compare Latin America's secondary ISI with East Asia's EOI. The former phase relied primarily on foreign direct investments and external loans, but was oriented toward supplying local markets; the latter depended on access to overseas markets, but was implemented in large part by domestic entrepreneurs who drew heavily on local financial resources.

Fourth, Latin America and East Asia also differ in terms of the overall weight that TNELs have had in the two regions. Historically direct foreign investment and foreign loans represented the most important external economic resources for the Latin American NICs; in contrast, export trade and foreign aid have been the key forms of East Asian linkage to the international economy. Despite these contrasts, however, a fundamental similarity between these two regions should not be overlooked. The NICs in both Latin America and East Asia have relied extensively on linkages with the international economy in their industrialization efforts. They embody, therefore, variations on the theme of dependent development.

State Structures and Policy Preferences

Development strategies are state-centered policies—i.e., they are policies designed and executed in large part by governments. Different

---

30. Ho, *Economic Development of Taiwan, 1860-1970*, pp. 110-111: Little, "Economic Reconnaissance," p. 457: Edward Mason et al., *The Economic and Social Modernization of the Republic of Korea* (Cambridge, Ma.: Council on East Asian Studies, Harvard University, 1980), p. 181 fn. 19.

groups in society may have strong policy preferences on certain issues, but the formulation of a development strategy implies at least some degree of state leadership. Since the state is the agent that formulates a development strategy, an interest in the sequence and choice of these strategies requires us to examine the ways in which political structures buffer the state from the pressures of civil society, as well as the content of the policy preferences of governmental elites.

A standard reference point for current research on the nature of the state and its role in development has been Guillermo O'Donnell's analysis of the emergence and dynamics of "bureaucratic-authoritarian" (BA) regimes in the southern cone of Latin America.[31] More recently, Bruce Cumings introduced the term "bureaucratic-authoritarian industrializing regimes" (BAIRs) to refer to the strong states in Taiwan and South Korea.[32] It seems useful for researchers of both Latin America and East Asia to push toward some integrating or overarching conceptualization of the state in the context of late, dependent, capitalist development. In both regions, the negative effects of authoritarianism in areas such as the suppression of human rights is putatively counterbalanced by extensive, rapid economic growth. The coexistence of authoritarianism and capitalism appears to be taken for granted in this setting.

The similarities among the states in the newly industrializing countries of the two regions are quite apparent: they tend to be strong, centralized, authoritarian (often under military control), and actively and extensively involved in economic affairs.[33] The differences are equally notable. The origins of the BA regimes and the BAIRs contrast sharply. Exclusionary BA regimes in Latin America (this includes Brazil, Argentina, Uruguay, and Chile, but not Mexico) emerged from the crises produced by periods of populist rule, when organized labor had been one of the important bases of social support of the state. BAIRs, on the other hand, inherited the centralized state apparatus from the Japanese colonial period and enjoyed a significant measure of autonomy from local social groups and classes, including those most likely to be affected by rapid industrialization, such as

---

31. Guillermo O'Donnell, *Modernization and Bureaucratic-Authoritarianism: Studies in South American Politics* (Berkeley: Institute of International Studies, University of California, 1973).
32. Cumings, "The Origins and Development of the Northeast Asian Political Economy."
33. For the East Asian NICs the economic role of the state is analyzed in several essays in Robert Wade and Gordon White (eds.), *Developmental States in East Asia: Capitalist and Socialist,* IDS Bulletin 15:2 (1984).

large landowners and workers. Whereas BA regimes had to repress previously mobilized popular sector organizations, such as free trade unions, the BAIRs did not confront an activated popular sector and were exclusionary from the outset.

A second difference relates to the role of the military. Although the military is prominent in both BA regimes and BAIRs, in the Latin American context the nature of the threat to which the military is responding is largely internal insurgency, while in South Korea and Taiwan there has been concern over the possibility of external invasion from neighboring communist countries.[34] Thus, even though the military penetrates far more deeply into civil society in East Asia than in Latin America, it may be that at some level the military is granted a greater degree of legitimacy in South Korea and Taiwan than in the BA regimes because of national security considerations.

A third contrast has to do with the social alliances on which the two kinds of regimes are based. In most Latin American BA regimes, international capital is an important part of the dominant coalition that also includes the military, select industrial groups, and civilian technocrats. In the East Asian BAIRs, international capital is a relatively minor actor, while the role of domestic economic groups integrated across industrial, trade, and financial sectors is central. Although labor is excluded in both BA regimes and BAIRs, labor until recently has been a more influential actor in Latin America than in East Asia.[35]

If we are to fruitfully compare and contrast the nature and role of interventionist states in a cross-regional setting, we need to go beyond relatively simplistic notions like "strong" versus "weak" states. For example, both the Brazilian and South Korean states over the past two decades can be characterized as "strong" since they have been led by military factions with an authoritarian, centralized approach to economic decision-making and considerable insulation from societal pressures. Korea's export-led economic success, two observers have

---

34. Since 1949, Taiwan has supported on a per capita basis one of the world's largest armies (estimated to be between 500,000 and 600,000 men), far larger than the military force that would be necessary to maintain internal political order. Ho, *Economic Development of Taiwan, 1860-1970*, pp. 107-108.

35. Labor in the East Asian cases, with comparative reference to Latin America, is the subject of Frececic C. Deyo. "Industrialization and the Structuring of Asian Labor Movements: The 'Gang of Four'," in Michael Hanagan and Charles Stephenson (eds.), *Confrontation, Class Consciousness, and the Labor Process: Studies in Proletarian Class Formation* (Westport, Ct.: Greenwood Press, 1986), and "State and Labor: Modes of Political Exclusion in East Asian Development," in Deyo (ed.), *The Political Economy of the New Asian Industrialism*, pp. 182-202.

argued, was dependent upon the change from the "soft" state under Rhee, whose administration was ineffectual in enforcing economic regulations, to the "hard" state under Park. The liberalization of economic policies, in this view, required a government that was able to impose compulsory norms and to direct bureaucratic discretion towards what it identified as economically desirable ends.[36]

The similarities notwithstanding, the Brazilian and Korean regimes also show some notable differences. By the late 1970s, around 600 Brazilian public enterprises existed in a wide variety of economic activities, with the public sector in Brazil accounting for over 40 percent of gross capital formation since the 1940s. South Korea, in contrast, has far fewer public enterprises in a relatively narrow range of industries, with the state sector representing less than 13 percent of total gross national product. Furthermore, the military has permeated society and perhaps legitimated itself to a much greater degree in South Korea than in Brazil (probably reflecting the overt military threat to national security represented by the North Korean forces above the 38th parallel), while the Brazilian state has been more responsive to organized societal pressures and democratic overtures than its South Korean counterpart. Similar kinds of parallels and differences emerge from a comparison of Mexico and Taiwan.

State-led industrialization in the NICs, with the exception of limited democratic interludes, has been characterized by authoritarianism. Yet authoritarianism per se does not guarantee either state strength or state autonomy. The majority of developing countries have authoritarian regimes, but few have autonomous or strong states. (The reverse case of a strong state and a nonauthoritarian regime is postwar Japan.) Political leadership, economic ideology, the role of the techno-bureaucracy, and the organizational resources and discretion of decentralized public agencies all need to be taken into consideration to show how state intervention has come about and been made effective in specific historical contexts. One promising approach would be to compare interventionist states in terms of the various economic roles they perform; these include their extractive capacities (e.g., collecting tax revenues), allocative capacities (e.g., distributing credit), planning capacities (e.g., targeting industries domestically and internationally), facilitative capacities (e.g., providing infrastructures),

36. Leroy P. Jones and II Sakong, *Government, Business, and Enterpreneurship in Economic Development: The Korean Case.* Studies in the Modernization of the Republic of Korea: 1945-1975 (Cambridge, Ma.: Council on East Asian Studies, Harvard University, 1980), p. 296.

and productive capacities (e.g., owning firms and producing goods). According to these criteria, Latin American NICs like Brazil and Mexico have emphasized direct productive activities, while state strength in the East Asian NICs is rooted more in their allocative and planning capacities.

An examination of the policy preferences of state elites also directs attention to prevailing development ideologies. In Latin America, for example, the region's experience with the economic problems of the 1930s cleared the way for generalized acceptance of the development strategy espoused by the United Nations' Economic Commission for Latin America (ECLA). Under the intellectual guidance of Raúl Prebisch, ECLA had begun to argue by 1950 that Latin American nations should pursue the path of increased local industrialization, despite the injunction of neoclassical economists who urged that the principal engine of growth ought to be a renewed emphasis on primary product exports.[37] In the East Asian NICs, the adoption of an export-oriented development strategy was encouraged by the U.S. economic missions that accompanied large amounts of American aid. An analysis of the sources (including the international transmission) of policy-relevant ideas and development ideologies is a subject deserving more detailed analysis.

Social Coalitions

The development experiences of the Latin American and East Asian NICs are affected by social coalitions—i.e., constellations of interests that support or oppose particular development strategies. For example, traditional agrarian and mining export elites tended to support and were favored by the primary product export model, while local manufacturing firms and urban labor generally opposed and were adversely affected by that experience. Conversely, industrial workers, national producers of finished consumer goods, and middle- and low-income consumers were often united in their support of primary ISI, even as the agro-export elite opposed this approach which jeopardized their privileged position.

As the initial industrialization phase ends, however, the political parameters of continuing industrial expansion begin to narrow.[38] For

---

37. Albert O. Hirschman, "Ideologies of Economic Development in Latin America," in Albert Hirschman (ed.), *Latin American Issues: Essays and Comments* (New York: The Twentieth Century Fund, 1961), pp. 12-23.

38. For a discussion of this process as it has occurred in Latin America, see Robert R. Kaufman, "Industrial Change and Authoritarian Rule in Latin America: A Concrete

countries like Argentina, Brazil, and Mexico, which promoted investment in consumer durable as well as intermediate goods sectors (secondary ISI), the social basis of this strategy was the so-called "triple alliance" of multinational corporations, affiliated local manufacturers, and the state, along with skilled workers and upper-middle class consumers who can afford to buy items like automobiles and major electrical appliances.[39] Agro-export elites, non-import-substituting firms, and the poor had little to gain from this strategy, although the political climate was such that they did not effectively challenge it. Finally, industrial export promotion both in the East Asian NICs and subsequently in the large Latin American nations favored producers (both domestic and foreign) of exportable manufactured products, while disfavoring traditional export companies, import-substituting firms, and organized labor.

Non-elite groups in the Latin American and East Asian NICs rarely have played a direct role in the determination of strategies in the postwar period, however. The popular sector's main impact tends to revolve around the implementation of development policies, rather than the goal formulation and agenda setting stage of national affairs.

This sketch of the principal social groupings associated with each strategy is generally a post hoc reconstruction of loosely knit social coalitions or alliances. These are the groups, in other words, that provided the social bases of support for a strategy once it had become stabilized. It is more difficult, yet equally important, to determine the social forces that make possible the transition from one development strategy to another (what Guillermo O'Donnell calls, for some phases of the Latin American experience, "coup coalitions").

These shifting constellations of interests are not likely to be the same across regions, nor are the forces that contribute to the downfall of one strategy necessarily those that initiate or help institutionalize a subsequent strategy. The analysis of social coalitions, therefore, must be flexible enough to account for both periods of stability as well as turning points in the development process. In Brazil, for example, the "triple alliance" that consolidated the industrial deepening of secondary ISI was narrower than the "coup coalition" that brought

Review of the Bureaucratic-Authoritarian Model," in David Collier (ed.). *The New Authoritarianism in Latin America* (Princeton: Princeton University Press, 1979), pp. 165-253.

39. This alliance is detailed for the case of Brazil in Peter Evans, *Dependent Development: The Alliance of Multinational, State, and Local Capital in Brazil* (Princeton: Princeton University Press, 1979).

an end to the populist policies of Goulart in the mid-1960s, yet it was also broader than the "coalitional duo" of multinational corporations and the state that held sway at the beginning of the 1970s when the repressiveness and the technocratic orientation of the military government led to the erosion of some of the Brazilian regime's earlier support.[40] Similarly, the importance of external capital (aid, direct investment, and loans), overseas markets, or the influence of foreign advisors has fluctuated dramatically vis-à-vis domestic counterparts in East Asia depending on whether the issue is maintenance or change of a given development strategy.

Conclusions

This paper has traced the alternative sequences of development strategies in Latin America and East Asia, and analyzed some of the determinants of those contrasting sequences. The NICs of Latin America and East Asia have differed markedly in terms of strategy choices. Most notably, the Latin American NICs chose to pursue secondary ISI and the East Asian NICs chose to pursue EOI when each of them found that they could no longer rely on primary ISI as the main stimulus to economic growth. Their choices differed despite the similarity in the nature of the crisis that made continued adoption of primary ISI unfeasible.

The cross-regional comparison has permitted us to deal with the role of external and internal factors in the determination of national policies and with the interplay between TNELs and domestic arrangements. These linkages have varied by region. We have seen, for example, how debt and investment dependency in the Latin American NICs, and the aid and trade dependency of the East Asian NICs, have allowed the United States to exercise a great deal of influence over these nations with regard to macroeconomic stabilization policies, privatization, and the opening up of their domestic markets to new foreign investment in previously restricted sectors. These adjustments have tended to favor American interests. This does not mean, however, that local elites in these countries have not derived benefits and even initiated some of these policies. The inflow of direct foreign investment into Brazil and Mexico in the late 1950s and 1960s, for example, furthered the sectoral development plans and ambitions of

---

40. Guillermo O'Donnell, "Reflections on the Patterns of Change in the Bureaucratic-Authoritarian State," *Latin American Research Review* 13:1 (1978), pp. 3-38.

the presidents in those countries, just as foreign companies helped Taiwan and South Korea launch EOI in the late 1960s and early 1970s. Similarly, heavy foreign borrowing by Brazil and South Korea in the 1970s was used to bring about secondary ISI in support of domestic objectives. Thus, the role of TNELs can be shaped by national development strategies, and vice versa.

Although we have limited ourselves to focusing on four countries in this paper, there are other very interesting cases that raise related development issues. For example, the ongoing experiment of the People's Republic of China with specialized export processing zones allows the world's most populous nation to compete directly with its Asian neighbors in low-cost, high-value-added export activities.[41] The Philippines, on the other hand, is an East Asian nation which thus far has failed in its efforts to successfully adopt an export-oriented strategy of industrialization.[42] Finally, the so-called "second tier" countries or emerging NICs, such as Colombia, Malaysia, Thailand, and Indonesia, are perhaps on the verge of becoming sufficiently industrialized to qualify for NIC status themselves.[43] They raise the question of whether the paths by which the East Asian and Latin American NICs achieved their relatively high levels of economic development could or should be replicated.

The foregoing analysis has various implications for development in the two regions. Industrialization was greatly accelerated in the East Asian NICs through massive amounts of U.S. aid which facilitated the development of a domestic entrepreneurial class, the avoidance of foreign exchange bottlenecks, and the maintenance of political stability. Industrialization in East Asia also has depended upon the production of manufactured goods for foreign markets. This linkage to the international economy made these NICs vulnerable to global recession and

---

41. Xiangming Chen, "China's Special Economic Zones: Origins and Initial Consequences of a New Development Strategy," paper presented at a conference on "Origins and Consequences of National Development Strategies: Latin America and East Asia Compared," Duke University, March 31-April 1, 1986.

42. Walden Bello, David Kinley, and Elaine Elinson, *Development Debacle: The World Bank in the Philippines* (San Francisco: Institute for Food and Development Policy, 1982); Robert T. Snow, "The Bourgeois Opposition to Export-Oriented Industrialization in the Philippines," unpublished paper (October, 1983).

43. "Developing Country Exports of Manufactured Products: The Experience of the 'Second-Tier' Countries," in *Development Co-operation: Efforts and Policies of the Members of the Development Assistance Committee* (Paris: Organization for Economic Co-operation and Development, November, 1982), p. 123; Roy Hofheinz, Jr. and Kent E. Calder, *The Eastasia Edge* (New York: Basic Books, 1982); John Wong, "Export-Oriented Industrialization in Asia: Pattern and Process," unpublished paper, National University of Singapore (October, 1983).

to protectionism in the industrialized markets where they sold most of their goods. On the other hand, this linkage created an imperative for flexibility in state policy and entrepreneurial behavior that proved important when global economic circumstances turned adverse.

In the Latin American NICs, vertically integrated industrialization was financed during the 1950s and 1960s in substantial measure by direct foreign investment, and in the 1970s by foreign loans. This situation had several consequences for these NICs. It stymied the development of a domestic bourgeoisie, weakened the control of national decision-makers over the most dynamic sectors of their economies, and gave significant influence over national macroeconomic policies to external actors (such as the International Monetary Fund). In consequence, it helped reduce the effectiveness with which these countries have been able to adapt to changing international economic circumstances.

The explanation of strategy choice in developing countries frequently gives primacy to the international political economy. However, since the NICs of the two regions experienced their crises of primary ISI at approximately the same time (i.e., the mid-1950s to early 1960s), and so faced roughly similar global economic environments, external factors alone cannot explain the variation in subsequent strategies. Analysis must consider the interplay between linkages with the world economy and domestic political arrangements. In addition, we have found it important to make explicit the entire set of external linkages through which these four countries relate to international forces in order to identify the factors that condition development options, policies, and outcomes. In this view, particular external linkages are not presumed necessarily to have either positive or negative effects on the development process. The regional contrast makes especially clear that external linkages may be opportunities and not just constraints for developing countries.

CHAPTER 3
The Japanese Industrial Policy Debate

Gregory W. Noble

Introduction

The pace of Japan's postwar economic development has been breathtaking. At the end of the war Japan and its industries lay in tatters. Only in 1964 did Japan attain status as a developed country by entering the OECD. Yet by the early 1970s Japanese productivity in steel, autos, and electronics led the world, and by the beginning of the 1980s Japan challenged the United States and Europe for leadership across the entire spectrum of the most technologically advanced and lucrative industries, from semiconductors, computers, and telecommunications to biotechnology and new materials.

This economic advance has touched off an active academic and popular debate on Japanese industrial policy.[1] How did Japan grow so rapidly, and how much difference did industrial policy make? A "developmental state" school stresses the importance of government guidance and intervention, while advocates of the market argue that Japan's success came precisely because government interference was limited and often ineffective, leaving markets to operate efficiently.

A second debate involves contrasting assumptions about the nature of the political system which *produced* Japan's industrial policy. If, as the developmental school asserts, the bureaucracy protected Japanese industry and directed its development, why were its powers judiciously used and economically effective? On the other hand, if growth resulted from fierce competition among private firms responding to price signals, why were market incentives powerful and undistorted by political interference?

---

1. The definition of industrial policy is itself a matter of considerable contention. The focus here is on government policies designed to affect the efficiency and organization of particular firms and industries, as opposed to fiscal, monetary and other macroeconomic policies, or "horizontal" policies designed to affect the supply or quality of whole factors of production, such as labor or basic research.

If the rival perspectives on the role of government during the 1950s and 1960 present a stark contrast, the picture has changed considerably since the early 1970s. Japanese firms are no longer dependent on the West for technology and are far more secure financially than before the oil shock. They are more than capable of competing internationally. Partly as a result, the government has lost many of its most useful policy tools and has adopted a more liberal style of economic management.

This does not necessarily mean that the government is no longer important and the market perspective unchallenged in the contemporary period. Proponents of the developmental state see industrial policy since the oil shock as focused on industries at the two extremes of the product cycle: high technology and depressed industries. In addition, a third "New Japan, Incorporated" school provides a more nuanced understanding of the role of government, and how it helps structure market competition. As a third "New Japan, Incorporated" school has recently emphasized, Japan has dense formal and informal networks for communication and negotiation. The Ministry of International Trade and Industry (MITI) responds to business interests and tries to help industries develop and sustain a consensus approach to common problems. When, after consultation with MITI, a majority opinion ("consensus") forms within an industry as to what its interests are and what the basis of competition should be, MITI can help enforce compliance through provision of administrative guidance and side payments.

This does not imply that the government is merely an instrument in the hands of business. While MITI's ability to articulate and implement an independent vision has declined, it is not open to capture and colonization, particularly by individual firms. Nevertheless, MITI is quite open about trying to help the majority of firms in each industry develop and carry out a consensus. The support of MITI can be an important asset. Externally, MITI helps coordinate relations with other ministries and industries. Within the majority group, MITI oversight helps ensure that firms can trust each other to abide by common agreements and not to take unilateral advantage. MITI can also bring pressure to bear to prevent recalcitrant firms that refuse to adopt the majority approach from undermining the "consensus." It is precisely the semi-autonomous position of MITI which allows it to articulate the interests of business as a whole.

The networks described by the New Japan, Incorporated school clearly exist in Japan. In most cases, however, they do not succeed in restraining market competition. The key question is whether or not the "victims" of the consensual approach, often newer, smaller and more innovative firms, have sufficient resources to fight off attempts at control by MITI and the industry mainstream. Evidence from the consumer electronics industry and from the minimill sector of the steel industry indicates that determined "outside" firms have been quite successful in opposing MITI-backed mainstream coalitions. Competitive pressures from these outsiders have in turn forced the industry mainstreams to compete on criteria they had originally marked off-limits. More than MITI leadership or government-business cooperation, this inadvertant competition—resort to the market by default—may help to explain the continuing dynamism and international competitive strength of Japanese industry.

Explaining the Rapid Growth of the 1950s and 1960s

Japan as a Developmental State

Many scholars stress that the Japanese government, and especially the elite bureaucracies such as MITI, were critical to the success of postwar Japanese growth.[2] This "developmental state" approach (Johnson

2. Major works which argue that industrial policy made an important contribution to rapid growth, particularly in the 1950s and 1960s, include Cyril E. Black et al., *The Modernization of Japan and Russia* (New York: Free Press, 1985); Andrea Boltho, *Japan: An Economic Survey, 1953-1973* (London: Oxford University Press, 1975); chalmers Johnson, *MITI and the Japanese Miracle: The Growth of Industrial Policy, 1925-1975* (Stanford: Stanford University Press, 1982); T.J. Pempel, "Japanese Foreign Economic Policy: The Domestic Bases for International Behavior," in Peter J. Katzenstein, ed., *Between Power and Plenty* (Madison: University of Wisconsin Press, 1978); Pempel, *Policy and Politics in Japan: Creative Conservatism* (Philadelphia: Temple University Press, 1982); Miyohei Shinohara, *Industrial Growth, Trade and Dynamic Patterns in the Japanese Economy* (Tokyo: University of Tokyo Press, 1981); Takafusa Nakamura, *The Postwar Japanese Economy: Its Development and Structure* (Tokyo: University of Tokyo Press, 1981): Yasusuke Muirakami, "The Japanese Model of Political Economy," in Kozo Yamamura and Yasukichi Yasuba, eds., *The Political Economy of Japan, Volume 1: The Domestic Transformation* (Stanford: Stanford University Press, 1987); Eugene Kaplan, *Japan: The Government—Business Relationship* (Washington: U.S. Department of Commerce, 1972); James C. Ablegglen, ed., *Business Strategies for Japan* (Tokyo: Sophia University, 1970); Ezra Vogel, "Guided Free Enterprise in Japan," in *Harvard Business Review,* May-June 1978, pp. 161-70; Vogel, *Japan as Number One: Lessons for America* (New York: Harper and Row, 1979). The assumption that Japan is the classic example of a developmental state is widespread among analysts of American can European policy-making. See John Zysman, *Governments, Markets and Growth: Financial Systems and the Politics of Industrial Change* (Ithaca: Cornell University Press, 1983), pp. 233-251; Zysman and Laura Tyson,

1982) argues that the government aided in the rationalization of firms and industries and hastened the structural transformation of the Japanese economy. MITI encouraged companies to move capital and workers out of declining industries such as coal and textiles, and into those with high potential for growth. These included heavy industries such as steel, petrochemicals and autos, and later high technology industries, including computers, semiconductors and biotechnology.

MITI and the other economic ministries had control over a wide range of policy tools in their efforts to guide Japanese firms. First, in the 1950s and 1960s MITI and the Ministry of Finance (MOF) controlled all aspects of foreign economic relations, including imports of foreign goods and technology, access to foreign exchange, and direct and indirect investment.[3] The Japanese government carefully protected the domestic market, first through quotas and outright import bans, then tariffs after Japan joined GATT, and finally through non-tariff barriers when foreign pressures forced Japan to lower tariffs. MITI used access to the large domestic market in bargaining with foreign firms to procure the most advanced foreign technologies at the most favorable prices. Control over access to imports and foreign exchange also provided the ministry with leverage over domestic firms.

The argument that protecting the domestic market could strengthen international competitiveness has traditionally been scorned by neo-classical economists. Recently, however, a new group of economic theorists has shown that limiting imports or subsidizing exports may increase home country welfare under conditions such as increasing returns to scale, large externalities, steep learning curve effects and oligopolistic competition. According to this "new international economics," trade policy can be used to deter entry and secure strategic market advantages.[4]

The convergence of views between economists modeling strategic trade and champions of Japan's developmental state is by no means complete. Most economists stress that government policies should be limited to correcting evident market failures, that the informational

---

American Industry in International Competition (Ithaca: Cornell University Press, 1983), Chapter One; Peter J. Katzenstein, *Small States in World Markets: Industrial Policy in Europe* (Ithaca: Cornell University Press, 1985), pp. 19-27, 130-131. For further references, see George Eads and Kozo Yamamura, "The Future of Industrial Policy," in Yamamura and Yasuba, eds., *The Political Economy...*

3. Pempel, "Japanese Foreign..."

4. Elhanan Helpman and Paul R. Krugman, *Market Structure and Foreign Trade: Increasing Returns, Imperfect Competition, and the International Economy* (Cambridge: MIT Press, 1985); Paul R. Krugman, ed., *Strategic Trade Policy and the New International Economics* (Cambridge: MIT Press, 1986).

requirements of active policies are extremely high, and that "political failures," including the impulse to save inefficient but politically influential industries, may be costlier than the market failures they purport to remedy.[5] In assessing Japanese industrial policy, Suzumura and Okuno-Fujiwara[6] conclude that MITI's efforts to reduce "excessive competition," previously dismissed by economists as an absurdity, were actually theoretically justifiable, but in practice impossibly difficult or even counter-productive.[7] Despite these reservations, there is no doubt that the new theorizing on international economics has been inspired in part by Japan's success, and that the new line of analysis strengthens the developmental state approach to Japanese industrial policy.[8]

---

5. More precisely, government intervention is justified only where markets are not competitive, or where there are prior distortions, externalities, or insufficient provision of public goods. Even in these cases, policies that deal with particular market failures directly are preferable to more general policies. See Peter Urban, "Theoretical Justifications for Industrial Policy," in F. Gerard Adams and Lawrence R. Klein, eds., *Industrial Policies for Grwoth and Competitiveness* (Lexington, Mass.: D.C. Heath, 1983). Most American economists are openly suspicious of, or even hostile to, actual attempts to implement strategic trade policies. See, for example, the essays by Krugman, Grossman, Cline and Dixit in Krugman, ed., *Strategic Trade Policy*. Even some who view the new trade theories as a decisive break with the past and see the potential for activist industrial policies are quite cautious (James A. Brander, "Rationales for Strategic Trade and Industrial Policy," in Krugman, ed., *Strategic Trade Policy*).

6. Kotaro Suzumura and Masahiro Okuno-Fujiwara, "Industrial Policy in Japan: Overview and Evaluation," in Ryuzo Sato and Paul Wachtel, eds., *Trade Friction and Economic Policy* (New York: Cambridge University Press, 1987), pp. 68-70.

7. A representative dismissal of "excessive competition" appears in Ryutaro Komiya, "Industrial Policy in Japan," in *Japanese Economic Studies* 14:4 (Summer 1986), pp. 51-81. Imai's excellent analysis of the paradoxical results of MITI's attempts to curb "excessive competition" in the steel industry appears in Ken'ichi Imai, "Iron and Steel," in Kazuo Sato, ed., *Industry and Business in Japan* (White Plains, N.Y.: M.E. Sharp, 1980).

8. Some economists are more positive about the role "strategic trade policies," particularly protection for imports and promotion of certain "marginal" exports, played in Japan. See Motoshige Ital et al., "Sangyo Ikusei to Boeki" (Industrial Promotion and Trade), in Koniyo Ryutaro et al., eds., *Nihon no Sangyo Seisaku* (Industrial Policy in Japan) (Tokyo: University of Tokyo Press, 1984) and Itoh and Kazuharu Kiyono, "Welfare-Enhancing Export Subsidies," in *Journal of Political Economy* 95:1 (February 1987), pp. 115-137; though even these authors carefully limit their case and stress the predominance of market factors over governmental ones. In their study of international competition in the semiconductor industry, Michael Borrus et al., "Creating Advantage: How Government Policies Shape International Trade in the Semiconductor Industry," in Krugman, *Strategic Trade Policy*... cite the new work on strategic trade policy as justifying their statist approach.

In addition to protecting the domestic market, the Japanese government directed various subsidies to industries it wished to promote. Direct budgetary allocations were relatively limited, but the ministries provided tax incentives for investment and exports, channelled investment capital at preferential rates through government financial institutions such as the Japan Development Bank, and directed government procurement to domestic firms even before their products were fully competitive with imports. These supportive policies also had an important indirect effect. Businesses invested more heavily and banks were more willing to lend to industries which the government targeted.

Finally, the ministries used regulatory powers over such issues as energy efficiency and land use to convince firms to rationalize their operations. MITI used exemptions it had won from the Anti-Monopoly Law to encourage mergers and cooperative research projects, and to organize recession and modernization cartels. The ministries also had recourse to informal "administrative guidance," which, in conjunction with their more formal powers and backed by their prestige, could be used to influence a firm on virtually any decision.

Within the developmental state school there are, to be sure, significant differences of emphasis. Some writers have a "statist" orientation that emphasizes MITI's ability to create dynamic comparative advantage for the Japanese economy.[9] Other analysts put more emphasis on communication and interaction between business and government. Nevertheless, the "statists" and the "corporatists" largely agree about the rapid growth period. Even the most fervant admirers of the Japanese state note the tremendous vitality of the private sector and the importance of frequent consultation between businessmen and government officials, while the corporatists share the conviction that the bureaucracy played a leading role in Japan's rapid growth.[10]

9. Black et al., *The Modernization...;* Boltho, *Japan...;* Johnson, *MITI and the Japanese Miracle...;* Pempel, "Japanese Foreign...;" Pempel, *Policy and Politics...;* Shinohara, *Industrial Growth...;* Nakamura, *The Postwar Japanese Economy...;* Zysman, *Governments...;* Katzenstein, *Small States...,* p. 131.

10. Kaplan, *Japan...;* Ablegglen, *Business Strategies...;* Vogel, "Guided Free Enterprise...," *Japan as Number One...,* and *Comeback* (New York: Shimon & Schuster, 1985); T.J. Pempel and K. Tsunekawa, "Corporatism Without Labor? The Japanese Anomaly," in P.C. Schmiller and G. Lehmbruch, eds., *Trends Toward Corporatist Intermediation* (Beverly Hills: Sage, 1979); Ronald Dore, *Flexible Rigidities: Industrial Policy and Structural Adjustment in the Japanese Economy 1970-1980* (Stanford: Stanford University Press, 1986). For a recent application of corporatism to Western Europe and the United States which pays particular attention to sectoral-level industrial policy, see Alan Cawson, ed., *Organized Interests and the State: Studies in Meso-Corporatism* (London: Sage, 1985). The division between the "statists" and the "cor-

Of course, many governments exert great influence over the economy. What saved the potentially heavy-handed "developmental state" approach from stifling control, bureaucratic inefficiency, political interference and pervasive corruption? This leads to a second major debate regarding Japanese industrial policy: its political base.

Four political variables are important for the developmental state interpretation. All reflect Japan's position as a late developer. First, the ministries had great power, prestige and internal coherence. The bureaucracy recruited the top university graduates, controlled its own personnel system, and sent many of its alumni into the Diet. Most important, the bureaucracy's erstwhile competitors for power were greatly weakened. After the war the military leaders and politicians were discredited and purged, and the oligopolistic *zaibatsu* were dismantled. The American occupation chose to govern through the bureaucracy, and entrusted it with even more powers than it had enjoyed before and during the war.[11]

Second, the bureaucracy was politically insulated. After 1955 Japan was governed continuously by the conservative Liberal Democratic Party (LDP), which largely left policy making to the bureaucracy and warded off redistributionist pressures from the unions and the socialist and communist parties.

Also important, particularly to those seeing Japan as a corporatist system, was the organization of the business sector. Strong trade associations in each industry, four functionally differentiated peak business organizations, and a network of business leaders who had worked closely with many of the politicians and bureaucrats in the war effort organized private interests and facilitated contacts with the bureaucracy.[12] Business interests were not particularistic but highly

---

poratists" is a matter of degree and emphasis, and the lines of distinction are far from absolute. Some have at various times written works which fall into first one camp (Pempel, "Japanese Foreign..."), then the other (Pempel and Tsunekawa, "Corporatism..."), then both at the same time (Pempel, *Policy and Politics...*). Vogel's approach is essentially "corporatist," but he stresses the great abilities of the elite bureaucrats. Johnson (*MITI and the Japanese Miracle...*) is seen as the archetypical statist, but one of major themes of his book on MITI is the failure of attempts in the prewar period to institute direct bureaucratic control of the economy. For every bold statement about the unchallenged rule of the bureaucracy and the indispensability of MITI to the Japanese economic miracle Johnson gives a dozen rich and nuanced accounts of the conflicts and negotiations between (and within) the bureaucracy and the business sector.

11. Johnson, "Japan: who Governs? A Lesson on Official Bureaucracy," *Journal of Japanese Studies* 2:1 (Autumn 1975), pp. 1-28).

12. Leonard H. Lynn and Timothy J. McKeown, *Organizing Business: Trade Associations in America and Japan* (Washington: American Enterprise Institute, 1988). See

aggregated. The organization of the private sector, including the existence of a number of *zaibatsu*-descended *keiretsu* groups, encouraged the economic ministries to favor competitive oligopolies rather than doting on monopolistic "national champion" firms.

Finally, throughout the society there was a consensus on the priority of rapid growth and the need to export and to "catch up with the West." As in the Meiji period, Japan's status as a "late developer" justified growth-first policies and a strong role for the government as leader of the catch-up effort.[13]

The Market Interpretation

In sharp contrast to the developmental state approach, the market school argues that government intervention was of minor importance. The competition of private firms in domestic and international markets was the engine powering Japanese economic development.[14] The factors underlying Japan's growth did not differ fundamentally from those responsible for growth in other advanced capitalist countries, such as the availability of labor and capital, the level of skills and technology, and rates of savings, investment and taxation.

If the factors explaining Japanese growth were the same as those elsewhere, the problem then becomes to explain why Japan's growth rates were so much higher and more sustained than those of other countries. History, geography and demography all played a part. Japan had a high level of skills and education relative to its level of

also Gerard L. Curtis, "Big Business and Political Influence," in Ezra Vogel, ed., *Modern Japanese Organization and Decision Making* (Berkeley: University of California Press, 1975).

13. The classic discussion of the role of the state in late developing countries is in Alexander Gersohenkron, *Economic Backwardness in Historical Perspective* (Cambridge, Mass.: Harvard University Press, 1962). For an application to Japan with special emphasis on the acquisition of advanced technology, see Terutumo Ozawa, *Japan's Technological Challenge to the West, 1950-1974: Motivation and Accomplishment* (Cambridge, Mass.: MIT Press, 1974).

14. Important representatives of the market school include Hugh Patrick and Henry Rosovsky, eds., *Asia's New Giant: How the Japanese Economy Works* (Washington, DC: Brookings, 1976); Philip H. Trezise, "Industrial Policy is Not the Major Reason for Japan's Success," *Brookings Review* 1 (3), 1983, pp. 13-18; F. Gerard Adams and Shinichi Ichimura, "Industrial Policy...," in F. Gerard Adams and Lawrence R. Klein, eds., *Industrial Policies for Growth and Competitiveness: An Economic Perspective* (Lexington, Mass.: D.C. Heath, 1983); Komiya, "Industrial Policy...;" Edward J. Lincoln, *Japan's Industrial Policies* (Washington, DC: Japan Economic Institute of America, 1984). For further references see Lincoln, *Japan's Industrial Policies,* and Eads and Yamamura, "The Future..."

economic development as early as the Meiji restoration and before, and was able to use those skills to draw effectively on a huge stock of advanced Western technology. In addition, international trade grew extremely rapidly in the 1950s and 1960s, creating room for emerging exporters such as Japan. Japan's domestic market was large and geographically isolated from its major competitors, while the shift of population from countryside to higher productivity jobs in the cities was more rapid than in any other society. The domestic savings and investment rates which supported this growth were unusually high.[15]

Japan's unusual pattern of labor-management relations was especially conducive to the development of human capital. Imperfections in capital and financial markets made managers act as "mediators" between stockholders and workers, rather than as maximizers of shareholder value. Shielded from takeovers, and pressured to increase employees' level of skill and chance of promotion, Japanese executives had even greater incentive to search for opportunities for rapid growth than managers in other countries.[16] Blue-collar workers in large industrial firms became "white-collarized" in the post-war years, partly because both white and blue-collar workers were included in the same unions and seniority rules were weak. Acquisition of a wide range of skills through on-the-job training increased their productivity, flexibility and morale: "promotion from within leads to wider and deeper careers for Japanese blue-collar workers than it does for their Western counterparts. These workers realize that there is a strong connection between their skill development and the business situation of the company."[17]

To be sure, some of these factors were not unrelated to government policies, including educational and macro-economic policies. Japan's extremely high rates of private savings and investment, for example, were a crucial element of the rapid growth pattern, and were probably related to policies that restrained government spending and encouraged

---

15. Kazushi Ohkawa and Henry Rosovsky, *Japanese Economic Growth: Trend Acceleration in the Twentieth Century* (Stanford: Stanford University Press, 1973); Edward F. Denison and William K. Chung, "Economic Growth and its Sources," in Patrick and Rosovsky, eds., *Asia's New Giant.*
16. Masahiko Aoki, *The Cooperative Game Theory of the Firm* (Oxford: Oxford University Press, 1984); cf. Andrew Gordon, *The Evolution of Labor Relations in Japan: Heavy Industry, 1853-1955* (Cambridge, Mass.: Harvard University Press, 1985).
17. Kazuo Koike, "Human Resource Development and Labor-Management Relations," in Yamamura and Yasuba, eds., *The Political Economy,* p. 322.

household savings.[18]

When it comes to micro-level industrial policies, however, the market school is skeptical that the government was an important force for growth. Government financing was limited and mostly not directed to high-growth sectors.[19] Import protection probably had negative as well as positive effects, and was not that effective or important by the 1960s anyway.[20] Control of foreign exchange and technology was often counter-productive—the government's reluctance to allow Sony to license the transistor is often cited as a particularly egregious example.[21]

Market analysts are also more skeptical about the ability of MITI and the other economic bureaucracies to use the policy tools under their control to influence private firms directly. Kitayama[22] argues that strong state approaches to Japan stress the role of bureaucrats in policy formation, where they are indeed omnipresent, but neglect to investigate policy implementation, where the Japanese bureaucracy is actually very weak—in some cases even weaker than that of the United States. In the steel and textile industries, attempts to encourage mergers, limit

---

18. The causes of Japan's high savings rates are a matter of liveley controversy among scholars. Not all agree that government policies in the high growth period, including holding down both taxes and social security benefits, significantly increased the private savings rate. For two of the many views, see Shinohara, *Industrial Growth...*, and Kazuo Sato, "Economic Laws and the Household Economy in Japan: Lags in Policy Response to Economic Changes," in Gary Saxonhouse and Kozo Yamamura, eds., *Law and Trade Issues in the Japanese Economy: American and Japanese Perspectives* (Seattle: University of Washington Press, 1986). An approach which stresses the government's indirect role in strengthening financial intermediation is in Eisuke Sakakibara and Robert A. Feldman, "The Japanese Financial System in Comparative Perspective," in *Journal of Comparative Economics* 7:1 (March 1983), pp. 1-24. For further analysis and citations, see Sato, "Saving and Investment," in Yamamura and Yasuba, eds., *The Political Economy...*
19. Philip H. Trezise with the collaboration of Yukio Suzuiki, "Politics, Government, and Economic growth in Japan," in Patrick and Rosovsky, eds., *Asia's New Giant...*; Lincoln, *Japan's Industrial Policies.*
20. Yutaka Kosai and Yutaka Harada, "Economic Development in Japan: A Reconsideration," in Robert A. Scalapino, et al., eds., *Asian Economic Development—Present and Future* (Berkeley: Institute of East Asian Studies, 1985), pp. 6-7; Gary Saxonhouse, "Evolving Comparative Advantage and Japan's Imports of Manufactures," in Kozo Yamamura, ed., *Policy and Trade Issues of the Japanese Economy* (Seattle: University of Washington Press, 1982).
21. Tresize and Suzuki, "Politics...," p. 798.
22. Toshiya Kitayama, "Nihon in Okeru Sangyo Seisaku no Shikko Katei: Seni Sangyo to Tekogyo," (Implementing Industrial Policy in Japan: The Textile and Steel Industries), in *Hogakuronso* (Kyoto Law Review), 117:5 (August 1985), pp. 53-76 and 118:2 (November 1985), pp. 76-98.

production during recessions or prevent the entry of new firms failed more often than not.

As with the developmental state school, there are different shades of emphasis within the market school as well. Some ardent liberals argue that industrial policy may actually have impeded economic growth.[23] Others concede that it may have been helpful in some cases, but argue that it in many more it was ineffective or irrelevant. Industrial policy was not particularly important to the development of a range of Japan's most competitive export industries, including cameras, watches, consumer electronics, and bicycles.[24]

Just as the members of the market school differ from developmental state theorists in their evaluation of the effectiveness of industrial policy tools they also differ in their assessments of the political determinants of growth. Whereas the developmentalists stress the power of the bureaucracy and the close ties among big business, the bureaucracy and the LDP, market analysts tend to see the political arena as competitive and pluralistic as well. In one of the most detailed and influential examinations of the political context of growth, Trezise and Suzuki argue that "the Japanese political scene is pluralist, competitive, and subject to inherent and effective checks and balances. In essentials, Japanese politics do not differ from politics in other democracies."[25] Japanese politics were at least as sensitive to demands from uncompetitive sectors and interests, such as agriculture, coal, cotton textiles, and small retailers, as to competitive ones. As in other countries, departures from the market were more likely to support losers than winners.

One way Japan did (and still does) differ from most other democracies, however, was in the length and stability of LDP rule and the baisc continuity in policy it provided. As Trezise and Suzuki put it:

> If private business provided much of the motive force for growth, business also had the assurance at virtually every point that government would be safe and sane, partial to profits and dedicated to business growth and social order.[26]

While this argument about political stability and private energy is rarely so explicit presented, it is implicit in most of the market literature.

23. Komiya, "Industrial Policy..."
24. Patrick and Rosovsky, *Asia's New Giant...*
25. Tresize and Suzuki, "Politics...," p. 782.
26. Tresize and Suzuki, "Politics...," p. 808.

At the extreme, a libertarian version of the market approach downgrades the importance of politics altogether. Building on the work of Tullock, Buchanan and Hirschliefer, Kosai and Harada[27] argue that the real contribution of the political system was to move the attention of firms and individuals from "rent-seeking" to "profit-seeking." The early postwar political and economic reforms, like the Meiji reforms before them, constituted a crucial "revolution from the outside" which created a stable and egalitarian society and eliminated many of the discontents which previously had motivated political activity. Once this society was established, the combination of the need to compete in international markets over which Japan had no political influence and the free enterprise orientation of the LDP turned the attention of most Japanese toward business (seeking profits) rather than political activity (seeking rents).[28]

A number of Japanese authors stress that the postwar reforms not only contributed to social and political stability but also unleashed a remarkable burst of entrepreneurship. Many Japanese managers, often relatively young men suddenly propelled to positions of leadership by the Occupation's purge of collaborators, bucked the collective wisdom of the bureaucracy and the leading firms and made daring investments in the immediate postwar period. Famous examples include Honda Soichiro's moves into motorcycles and autos and the decision by Nishiyama Yataro of Kawasaki Steel, against great opposition from top government officials, to enter the ranks of the integrated steel producers.[29] As a former top official in the Economic Planning Agency puts it: "the secret of Japan's economic success... lies in the fact that... human freedom as a basic element of market economy has advanced. And that human freedom originated in the systemic (economic and political) reforms instituted after World War II."[30]

---

27. Kosai and Harada, "Economic Development..."

28. Notice that in essence the libertarian argument requires a strong state as a prime mover to establish the equitable and competitive market system. Following Hirschliefer, Kosai and Harada call this process "the big game." For Kosai and Harada the U.S. occupation authorities played that role in Japan, and once they left, the Japanese government did not act as an intrusive strong state. In the long run, however, they see a danger that pluralistic politics may increasingly intrude into the economic system, presumably necessitating a new "big game."

29. Hideichiro Nakamura, "Japan, Incorporated and Postwar Economic Growth," in *Japanese Economic Studies* 10:3 (Spring 1982), pp. 68-109.

30. Isamu Miyazaki, "The Real Reason for Japan's Success in Economic Growth," in *Japanese Economic Studies* 10:3 (Spring 1982), p. 103.

Finally, some analysts turn the argument of the statists on its head: the private sector was competitive and innovative not because the Japanese government was able to exert such great power on its behalf but precisely because the capacity of the Japanese state was so limited. John Haley argues that the bureaucracy in Japan was forced to turn to informal "administrative guidance" because of the weakness of formal law enforcement in Japan:

> The predominance of administrative guidance as a regulatory form for government intervention in the economy. . . has helped to preserve a competitive market economy by maximizing the freedom of individual firms over economic decisions although behind the veil of pervasive governmental direction. Japanese postwar economic achievement can thus be credited in part to administrative guidance because it ensured the failure of a bureaucratically set agenda.[31]

Efforts to gain more effective powers in the area of industrial policy failed because of opposition from industry and the uneasy position within the LDP of supporters of the bureaucracy. This failure was actually a good thing for Japan, Haley argues:

> Where, however, formal statutory authority and extralegal sanctions have permitted more effective implementation of government policies, either through administrative guidance or mandatory controls, the consequences appear to have been less positive. In agriculture and financial services restrictive government policies have not achieved the benefits claimed.[32]

Evaluating Industrial Policy Since the Oil Shock

Though the debate on the causes of growth in the 1950s and 1960s continues, the Japanese political economy of the 1970s and 1980s has been modified by a host of changes in the international and domestic environment. Starting in the early 1960s, imports, foreign exchange and direct foreign investment were progressively liberalized. By the late 1960s Japan had largely completed the drive to catch up with the West in technology and financial power. At the same time, the costs of the giddy growth of the 1950s and 1960s were increasingly evident,

---

31. John O. Haley, "Administrative Guidance versus Formal Regulation: Resolving the Paradox of Industrial Policy," in Gary R. Saxonhouse and Lozo Yamamura, eds., *Law and Trade Issues of the Japanese Economy* (Seattle: University of Washington Press, 1986), p. 108.
32. Haley, "Administrative Guidance...," p. 122.

including urban sprawl and environmental degradation, and both the LDP and the Japan Socialist Party suffered electoral declines.

Partly as a result, when Tanaka Kakuei became Prime Minister in 1972 he presided over a massive increase in welfare programs and greatly increased political influence over the budget and some of the Ministries, notably the Ministry of Construction. Inflation, scandal, and the oil shock, which brought an end to rapid growth, ended his term. The oil shock also contributed to a polarization of the economy: higher energy prices shifted demand and investment to assembly and high technology industries, while the many energy and labor-intensive industries became mired in recession. Similarly, the balance of financial power shifted from the government toward the private sector. With lower levels of growth and capital investments, strong firms began to increase their equity, while slower increases in tax revenues and heavy welfare expenditures created evermounting government deficits. The burden of financing those deficits, in turn, forced the Ministry of Finance to begin loosening regulation of financial markets.

Thus by the late 1970s and early 1980s, while the LDP was still firmly in power, it could no longer count on a steady electoral base and economic policy-making was increasingly subject to political pressures. Leading Japanese corporations were among the most wealthy and technologically advanced in the world, making them far less dependent on, and amenable to, government initiatives. Industries in "structural depression" needed government help, but lacked the vitality and prospect of future competitiveness that had justified protection and support in the 1950s and 1960s. At the same time, increasing Japanese trade surpluses led to protectionism and foreign monitoring of Japanese economic policy.

Most, though not all, observers believe that the style of economic management in Japan is more liberal than in the 1950s and early 1960s and that business is more independent of government. To liberals such as Komiya, the role of industrial policy was always minor; now it is virtually irrelevant. MITI has shifted from an "industries" ministry to an "issues" ministry, concentrating on regulatory problems such as alleviating pollution and trade friction.[33]

Those who are impressed by the developmental thrust of the Japanese state, on the other hand, deny that the 1970s and 1980s mark a fundamental change. Having succeeded in making most Japanese firms internationally competitive, MITI is concentrating on those industries at the beginning and end of the product cycle,

33. Komiya, "Industrial Policy...," p. 76.

promoting new industries and easing the transfer of resources out of old ones. The government is encouraging energy conservation and oil stockpiling, promoting high technology industries such as VLSI semiconductors, new generation computers and telecommunications, and giving incentives for textile firms to consolidate and increase value-added.[34] The "liberal" 1980 revision of the Foreign Exchange and Foreign Trade Control Law has also left bureaucrats abundant discretion to protect domestic high technology industries from foreign competitors.[35]

A more common assessment than those of either the pure market or pure developmental approaches is that industrial policy made an important contribution in the 1950s and 1960s, but that it is of minor significance today.[36] Most authors argue that the policy tools available to the bureaucracy since the oil shock are much more limited than in the high growth period. Imai[37] points out that the increasing sophistication of the Japanese economy and higher oil prices shifted the center of economic activity to industries without a history of extensive cooperation with MITI. MITI official Wakiyama Takashi goes back even further: "a dilemma which has increasingly bothered MITI since the early 1960s (is that) the phasing out of other policy tools increases the necessity of administrative guidance but decreases its effectiveness."[38]

---

34. Johnson, "The Institutional Foundations of Japanese Industrial Policy," in *California Management Review* 27:4 (Summer 1985), pp. 59-69; Dore, *Flexible Rigidities...*

35. See Encarnation, this volume.

36. Mototada Kikawa, "Shipbuilding, Motor Cars and Semi-conductors: The Diminishing Role of Industrial Policy in Japan," in Shepard, et al., eds., *Europe's Industries: Public and Private Strategies for Change* (London: Frances Pinter, 1983); Thomas Pepper et al., *The Competition: Dealing with Japan* (New York: Praeger, 1985); Adams and Ichimura, "Industrial Policy...;" Takashi Hosomi and Ariyoshi Okumura, "Japanese Industrial Policy," in John Pinter, ed., *National Industrial Strategies and the World Economy* (London: Croom Helm, 1982); and Murakami, "The Japanese Model..." all stress the decline of bureaucratic influence since the early 1970s. Lincoln, in *Japan's Industrial Policies,* emphasizes the liberalization of imports and foreign exchange in the early and mid-1960s. Yukio Nuguchi, "The Government-Business Relationship in Japan: The Changing Role of Fiscal Resources," in Kozo Yamamura, ed., *Policy and Trade Issues of the Japanese Economy: American and Japanese Perspectives* (Seattle: University of Washington Press, 1982) argues that the failure in 1963 of MITI's attempt to gain explicit statutory authority for its policies and the increasingly stringent government financial position after the early 1960s (after 1964 the budget was in deficit every year) marked the end of effective industrial policy.

38. Takashi Wakiyama, "The Implementation and Effectiveness of MITI's Administrative Guidance," in Stephen Wilks and Maurice Wright, eds., *Comparative Government-Industry Relations* (Oxford: Oxford University Press, 1986), p. 221.

In many cases policies which may have been effective in the past became dysfunctional. In the 1950s and 1960s the ability to form cartels provided security and encouraged investment, but by the 1970s it had entangled MITI in an increasing spiral of cartels and regulations.[39] Moreover, with the increased influence of LDP politicians over economic policy-making, government intervention was more likely to be oriented toward protection or pork barrel benefits than economic development.[40] While government intervention in industry has by no means completely ended, it is not necessarily aimed, much less effective, at promotion of economic development.

Many observers, particularly in Japan, stress not only the declining utility of policy tools but also the loss of social consensus on the priority of rapid growth and the privileged position of the bureaucracy.[41] Murakami[42] argues that the end of the consensus on growth has undermined bureaucratic guidance and will eventually erode the electoral base of the LDP; from this perspective the 1970s and 1980s are the interlude in a transition to a new political arrangement and new economic policies.[43]

As external pressures from Japan's trading partners have shifted policy emphasis from promoting industrial exports to alleviating "trade friction," MITI has come to adopt a less protective stance.[44] Foreigners have demanded and won the right to sit on MITI's Industrial Structure Council, and to participate in MITI-sponsored joint research projects. The Japanese press charges that the MITI has turned to foreign firms as a new constituency, favoring them over domestic

39. Kozo Yamamura, "Success that Sourced: Administrative Guidance and Cartels in Japan," in Yamamura, ed., *Policy and Trade Issues...*

40. Nihon Keizai Shinbunsha, ed., *Gekitotsu! tai Matsushita* [Collision! Sony vs. Matsushita] (Tokyo: Nihon Keizai Shinbunsha, 1978); Takashi Inoguchi, "Politicians, Bureaucrats and Interest Groups in the Legislative Process," paper presented at the Workshop on One Party Dominance, Ithaca, N.Y., April 7d the time of the oil shock, international scrap prices soared, leading many to believe that the competitiveness of the minimill sector would be seriously damaged. In fact, scrap prices soon declined, while iron ore price did not, leaving the minimills in a stronger competitive position. See Takanori Tanabe, *Tekko Gyo* [The Steel Industry] (Tokyo: Toyo Keizai, 1981), p. 220.

41. Eads and Yamamura, *The Future...*

42. Murakami, "The Japanese Model..."

43. As with policy tools, some authors trace the origin of these trends even earlier. Michio Muramatsu and Ellis Krauss, "The Conservative Policy Line and the Development of Patterned Pluralism," in Yamamura and Yasuba, eds., *The Political Economy...*, argue that the precise nature of the consensus changed over time as the LDP adjusted its policy approach to maintain electoral dominance.

44. Hiroshi Iyori, "Antitrust and Industrial Policy in Japan: Competition and Cooperation," in Saxonhouse and Yamamura, eds., *Law and Trade Issues...*, p. 58; Komiya, "Industrial Policy...," p. 62.

companies now that many Japanese firms no longer need or want government intervention.[45] In combination with the rise domestically of new issues such as pollution control and consumer rights, foreign pressure has caused even some government officials in Japan to call for a more formal policy process with improved access for third parties, even if it results in a loss of some informality and flexibility.[46]

The New Japan, Incorporated School

For many Western observers, however, informal but intense links between government and business are at the heart of Japanese political economy and are unlikely to melt away in the competitive heat of liberalization and the pressure of foreign scrutiny. Nor are minor fluctuations in the vote of the LDP likely to affect government-business ties significantly. Observing this continuity in the face of economic liberalization, a "New Japan, Incorporated" school has emerged which begins with analysis of private interests, particularly at the industry level, and then examines how government policy is shaped by those interests.[47] Specific policy tools are less important than networks of communication and negotiation. Within these networks, the bureaucracy remains important for structuring conflict and discussion and providing information, even if it is not necessarily the major source of policy innovation or an authoritative decision-maker.

In a study of Japanese policies toward high technology industries Daniel Okimoto proposes that Japan is a "network state" that "has managed to merge market and organization so effectively that its

---

45. See for example the series in *Nihon Keizai Shinbun* entitled "The Disappearance of National Borders and the Dissolution of the Purported Japan, Inc.," especially January 11, 1986, p. 1.
46. Iyori, "Antitrust and Industrial Policy...," p. 71; Wakiyama, "The Implementation...," p. 227.
47. Richard J. Samuels, *The Business of the Japanese State: Energy Markets in Comparative and Historical Perspective* (Ithaca: Cornell University Press, 1987), pp. 282-283, is explicit in his revival and rehabilitation of the term "Japan, Incorporated." The "New Japan, Incorporated" approach can be seen as a descendent of the corporatist wing of the developmental state school, as well as the tradition of writings critical of the power of big business in Japan (Yamamura, *Economic Policy in Postwar Japan* (Berkeley: University of California Press, 1967); Chitoshi Yanaga, *Big Business in Japanese Politics* (New Haven: Yale University Press, 1968). However, in the past even the corporatists tended to assume that bureaucratic direction was crucial, and that the bureaucracy had significant sticks and carrots at its disposal, while those who focused on big business tended to assume a unity of interest which more recent writiers leave as an open, and important, question.

industrial economy can be considered, in some respects, an alternative model to Adam Smith's invisible hands' market economy".[48] Legal scholars such as Michael Young and Frank Upham conclude that informal, bureaucratically-sponsored negotiation is the heart of Japanese economic and social policy.[49]

Other scholars put even more emphasis on how the private sector can get the government to support its interests. Yamamura and Vandenberg,[50] using documents seized by the Japanese Fair Trade Commission (FTC), show how a small group of consumer electronics firms cooperated over many years to maintain high prices for televison sets in Japan while exporting them at or below cost to the United States. MITI did not direct this campaign, indeed it tried, unsuccessfully, to maintain minimum export prices to prevent U.S. accusations of dumping. Nevertheless, government policies were instrumental in creating the kind of industrial structure in which such actions could occur. The government discouraged entry—no new firms entered television production after 1950, despite the industry's high profit levels— and allowed the formation of captive retailing networks: "the policies of the Japanese government substantially influenced the structure, marketing practices, and development of the industry (including) preferential credit allocation via large banks, lax antitrust enforcement, condoning of de facto recession cartels, MITI-guided investment coordination and various forms of NTBs (non-tariff barriers)."[51]

Richard Samuels' study of the energy sector also focuses on private firms and their responses to changes in the market. While the private sector has often been divided on energy policy, it has been able consistently to garner support from the government.[52] Samuels argues that business and the state are linked in a pattern of "reciprocal consent." In exchange for relinquishing ownership and direct control, the Japanese bureaucracy has gained a broad jurisdictional mandate to intervene in energy issues. Underlying this bureaucratically-centered corporatist system is stability of political leadership. Unlike ruling parties and coalitions in most European countries, the ruling party in

---

48. Daniel I. Okimoto, *Between MITI and the Market: Japanese Industrial Policy for HIgh Technology* (Stanford: Stanford University Press, forthcoming).

49. Michael Young, "Judicial Review of Administrative Guidance: Governmentally Encouraged Consensual Dispute Resolution in Japan," *Columbia Law Review* 84:4, pp. 923-983; Frank K. Upham, *Law and Social Change in Postwar Japan* (Cambridge: Harvard University Press, 1987).

50. Kozo Yamamura and Jan Vandenberg, "Japan's Rapid-Growth Policy on Trial: The Television Case," in Saxonhouse and Yamamura, eds., *Law and Trade Issues*.

51. Yamamura and Vandenberg, "Japan's Rapid-Growth Policy...," p. 266.

52. Samuels, *The Business of the Japanese State...*

Japan has not needed to nationalize coal mines or electric utilities to gain political support from coalition partners, from miners or other unions, or from particular regions of the country. Instead, business has managed to extract subsidies and guarantees from the government, albeit on terms negotiated with MITI and the Ministry of Finance. A stable political base has permitted a market-oriented—but not laissez-faire—approach to policy.

The declining utility of particular policy tools has not changed the essentials of the government-business relationship, which is based on communication and the privileged access to policy-makers enjoyed by business: challenges such as environmental protest "come from without hegemonic elite circles: within them, while the terms of the government-industry equation are constantly disputed, the equation itself, like industrial policy, is non-contentious. Industry resists 'control' (tosei) but regards regulation (chosei) as indispensible; free-market purism is confined to the academy."[53]

Underlying this relationship is the rule of the LDP: "Above all it is the dominance of the LDP which has limited and contained challenges to the existing government-industry relationship and to related networks." While there are problems such as U.S. unhappiness over Japanese trade surpluses and pressures to bail out declining industries, "there is as yet no evidence that any of these problems are threatening the LDP's tenure of office, and until such time as they do, the essential condition of the relationship between government and industry remains intact."[54]

Testing Industrial Policy Since the Oil Shock

The three schools reviewed above offer clearly contrasting predictions about the role industrial policy has played in the Japanese economy since the oil shock. The developmental state model would predict that the state has and will continue to use significant incentives to persuade private actors to move in the directions the bureaucracy considers desirable. If the market approach is correct, we would expect that industrial policies are either non-existant or of no importance. Finally, the New Japan, Incoporated school suggests a more interactive pattern. Economic policies and market outcomes

---

53. Richard Boyd, "Government-Industry Relations in Japan: Access, Communication and Competitive Collaboration," in Wilks and Wright, eds., *Comparative Government-Industry Relations,* p. 85.
54. Boyd, "Government-Industry Relations...," p. 86.

follow not just from competition among firms but also from the way in which interests are aggregated and settlements negotiated by the bureaucracy, industry associations and industry leaders.

Figure 3.1 Ideal-Typical Approaches to Japanese Industrial Policy

	Developmental State	Market	New Japan, Inc.
Nature of Industrial Policy	Developmental	Political (eg. protectionism)	Coordinating
Contribution to Growth	Great	Negligible	Moderate
Nature of Political System	Bureaucracy Dominant	Open, Pluralist	Pluralist, but dominated by conservative elite
Importance of Specific Bureaucratic Authority (Policy Tools)	Useful, but not crucial: bur. can use "administrative guidance"	Crucial: otherwise firms will ignore government	Often irrelevant: key is communication, negotiation and consensus building
Intra-Industry Relations	Competitive— within parameters set by Ind. Pol.	Competitive	Competitive— within parameters set by business; bur. brought in as enforcer

By all accounts, industrial policy since the oil shock is largely irrelevant for industries such as autos or cameras, which are in their prime of competitiveness. The crucial cases for industrial policy occur when rising and declining industries face collective problems which are potentially amendable to government direction and coordinated action. I have examined one case of each type. One involves recurrent efforts to standardize product formats in the rapidly expanding consumer video industry, the other the attempt to reduce chronic overcapacity in the steel minimill industry. Standardization and capacity management are important collective problems. Convincing firms to settle on a single format is difficult, but standardization leads to greater economies of scale and more rapid acceptance by consumers. Similarly, each minimill would prefer that some other firm shut down its furnaces and plants, yet only when overall capacity is reduced can product prices stabilize.

In addition to their positions at different ends of the product cycle, these industries differ in a number of crucial ways. Historically, the consumer electronics industry developed without much help from the government. It is dominated by a handful of large firms turning out a variety of differentiated and rapidly changing products for the domestic and world markets. The minimill sector of the steel industry, on the other hand, is comprised of dozens of firms, mostly small independents. The minimills produce undifferentiated commodity products, largely steel bars and shapes for the domestic construction market. Within the limits of a small number of cases, consumer video and minimills provide a good cross-section of Japanese industry.

Despite these differences, the outcomes in the two industries were surprisingly similar. In both cases the leading firms were able, sometimes after a period of struggle, to reach a consensus of opinion about how to standardize or cut capacity. In both cases they were able to gain the support of MITI and, in the case of steel, related industries. Nevertheless, MITI and the mainstream firms were unable to persuade or coerce stubborn "outsider" firms to go along. In consumer video a number of alternative formats reached the market, while minimill firms actually *increased* capacity. Neither mighty MITI nor the network of dominant private interests was able to prevent recalcitrant opponents from taking their cases to the court of market competition.

Consumer Video Format Standardization

In many industries, including consumer electronics, choosing product formats is a crucial issue. From the perspective of the overall industry, standardization of formats offers efficiency, convenience to consumers (and thus a larger market), and the hope of an international competitive advantage, if the format common to Japanese firms can establish itself as the world standard.[55] However, whatever the potential advantages to the industry as a whole, there is a conflict with the interest of individual firms: once companies have poured their time, resources and prestige into new formats, they are reluctant

---

55. On the other hand a possible danger is that standardization may throttle innovation by shutting out technical advances that cannot be incorporated within the parameters of existing formats. Through economist have become increasingly interested in this question, it is not possible to say a priori when the benefits of standardization will outweigh the costs. See Barry Keating, "Industry Standards and Consumer Welfare," *The Journal of Consumer Affairs* 14:2 (Winter 1980), pp. 471-82; and Joseph Farrell and Garth Saloner, "Standardization, Compatibility, and Innovation," *Rand Journal of Economics* 16:1 (Spring 1985), pp. 70-83.

to abandon them in favor of formats developed by their competitors.

The government and industry association can help persuade firms to cooperate in their larger interest. They can lead discussions and negotiations about which format is superior, or create a compromise format. They can help to ensure each firm that if it abandons its format competitors will do the same, so that the industry will in fact be able to enjoy the advantages of unification.

Consumer VCRs: Beta vs. VHS

Before the early 1970s, standardizing video formats was not a major problem.[56] At first Japanese firms built on American formats. Later, the top few firms agreed on video-cassette (VCR) formats, which were then formally promulgated by the Electronics Industry Association of Japan (EIAJ). The most important was the ¼ " "U-matic" standard for professional color VCRs. Work on U-matic was begun by Sony, which invited the two other leading Japanese video firms to join in development and cross-license patents. These were Matsushita, the largest consumer electronics company in the world and producer of such brands as Panasonic and Technics, and Victor (Victor Company of Japan, or JVC), 50.6% of which is owned by Matsushita.

The U-matic format was formally introduced in 1970, just in time for the Osaka Exposition that proudly announced Japan's technological achievements to the world. Sales, marketing problems and considerations of prestige were limited since these early products were all destined for a relatively small number of professional users, such as broadcasting networks and local telelvision stations.

Once advances in technology made it possible for Japanese firms to make products small and inexpensive enough for the vast home market, however, the situation changed drastically. As early as May, 1974, seeing the feverish competition to develop new consumer models, MITI issued an order to the electronics firms to begin work on standardizing formats. The order had little immediate impact. Toshiba and Sanyo began the market battle with their jointly-developed "V-Code" format in September of 1974. It offered sophisticated features but was too expensive, and sales lagged. Sony then showed plans for its projected new format to Matsushita

---

56. The discussion here of video formats and steel minimills is based on Gregory W. Noble, "Between Competition and Cooperation: Collective Action in the Industrial Policy of Japan and Taiwan," unpublished doctoral dissertation, Harvard University, 1988.

and Victor, hoping to repeat the success of the U-matic. This time, however, Matsushita and Victor each had their own independent development plans, and the response was negative. The problem was nominally technical, with Victor and Matsushita arguing that Sony's projected recording time of one hour was inadequate. The real dispute, however, was strategic. They believed that Sony was already tooling up to begin mass production, and was not sincere about cooperating in further development and competing on an even basis.[57]

Figure 3.2 Consumer Video Cassette Recorder (VCR) Formats

```
            1974    1975    1976    1977    1978    1984    1985    1988

V-Code:          1974.9: Sanyo        Beta (1976.2)
                 1974.9: Toshiba
                                              Beta (3.1977)

Philips (Europe only):       VCR-III            V-2000          VHS
Matsushita (VX-2000):             1976.9              VHS (1977.6)
Beta:        1975.5: Sony                     Adds VHS (1988.1)
             1976.2: Sanyo                    VHS (1985.3)
             1977.3: Toshiba                  VHS (1984.12)
             1977.11: NEC                     VHS (1984.3)

VHS:         1976.10: Victor
             1976.12: Hitachi, Sharp
             1977.1: Mitsubishi
             1977.6: Matsushita
                                      1984.3: NEC
                                      1984.12: Toshiba
                                      1985.3: Sanyo
                                      1988.1: Sony
```

By May of 1975 Sony entered the fray with its "Beta" format. Matsushita followed soon after with its bulky but inexpensive "VX" format and Victor worked feverishly on the fifth and final prototype for its "VHS" format. In the meantime, Matsushita and Victor rejected further overtures from Sony.

By June of 1976 MITI began floating trial balloons in the newspapers about the possibility of standardizing formats. The ministry asked Victor to delay its sales schedule, and began collecting more information. Sanyo and Toshiba, recognizing that their product was not viable, were amenable to unification, while Matsushita was

---

57. Interview with Nakamura Makoto, Japan Victor Corporation, May 14, 1986.

willing to abandon its VX in favor of VHS, but not for Beta. With only the Beta and VHS formats left in contention, MITI suggested a meeting between Sony and Victor, but Victor president Matsuno Kokichi refused to abandon VHS. MITI then suggested developing a compromise format that would combine the best points of each design. A committee composed of representatives from eight firms investigated the issue, but the firms began to take sides, and it was soon obvious that no compromise was possible.

In August MITI made a final attempt to achieve unity. Since Beta was already on the market, and was generally considered to have a slight edge technically, MITI decided to back Sony. Victor President Matsuno was called into MITI's Kasumigaseki offices and given an offer: discontinue development of VHS and join in production of Beta in return for reimbursement by Sony of the costs incurred in developing the VHS system, a figure later estimated by Victor at three billion yen. Matsuno walked out of the meeting without making any reply. The next day he convened a meeting of presidents of firms committed to VHS. The presidents agreed that VHS was the joint product of the VHS group, and that MITI's offer was totally unacceptable. On September 9th, 1976, Victor held a news conference to announce that it would begin marketing its new VHS machines in October.[58]

MITI, under pressure from public opinion to create a unified standard, and facing embarrassing Diet interpellation on the issue from the opposition parties, tried to get the industry association to take up the task. The EIAJ reported back that it would do its best in the future, but that unifying the formats already on the market was totally impossible.

By early 1977 the battle lines were clearly drawn, and all the major Japanese consumer electronics firms backed one format or the other. As the two camps, and the individual firms within them, competed feverishly to pack in new features and squeeze costs, the sales of both formats increased rapidly. Nevertheless, the market share of the Beta group began to shrink. The greater marketing power of the VHS group, both in Japan, where Matsushita was dominant, and in the U.S., where RCA marketed VHS format machines, was simply too great. By the mid-1980s Sony's allies abandoned it for VHS and the 10 year war was all but over. Defacto unification was finally achieved,

58. *Nihon Keizai Shinbunsha,* ed., "Gekitotsu!" pp. 60-61. Representatives from Victor directly confirmed this account. Indirectly, MITI officials also confirmed that such and offer was made. Perhaps not surprisingly, Sony representatives declined to comment of the issue.

but despite the efforts of MITI and the industry association, not involving the format favored by the government, and only after ten years of fierce competition.

Videodisks

Standardization was especially important for the next major video product, videodisks. Like the phonograph records they resembled, videodisks offered random access and higher fidelity than cassettes, but could not be recorded in the home. Ensuring the provision of a large number of inexpensive, commercially-recorded disks was critical to the success of the videodisk.

Repeated efforts at unification were mounted by individual firms, ad hoc industry committees, EIAJ and of course MITI. For a time it appeared that the "family building" skills of Victor, this time supported by MITI, would be sufficient to carry the day by sheer force of numbers. In the end, however, Pioneer, a small but determined firm from outside the ranks of the traditional consumer electronics industry, was able to hold out for its product. By the mid-1980s, the result, as with consumer VCRs, was a pattern of cooperation within format groups, but fierce competition between them.

In the mid-1970s several videodisk formats vied for the affections of Japanese consumer electronics companies. A crude but inexpensive mechanical system called TED was developed by Germany's Telefunken. An optical system offering excellent picture clarity and special features was championed by Holland's Philips, with cooperation on the software side from MCA, the parent of Universal Studios. Since the optical systems were complex and expensive, they were seen as particularly appropriate for industrial applications such as corporate training programs and retail sales. CED, a capacitance system developed by RCA, offered a compromise of price and features. All three formats gained Japanese licensees, and some Japanese firms exported optical and CED players to the U.S. and Europe.

Only the optical format, however, had any impact in Japan. In 1977 Hitachi developed an optical videodisk for industrial use and Pioneer followed with an announcement that it would form a joint venture with MCA to produce both industrial and consumer videodisk players. Addressing fears that adoption of this third format for consumer use would lead to a repeat of the recent dispute in VCRs, Pioneer President Ishizuka replied boldly that "The superiority of the MCA-Philips format over the other two formats is a matter of common sense, and there should be no chance of a 'unification of formats' problem occurring. Consumers will not be inconvenienced."[59]

A couple of months later Matsushita announced two videodisks with mechanical pickups (VISC), but it was much less aggressive and dedicated to its models than Pioneer.

Figure 3.3 Consumer Videodisk Formats

Mechanical

| TED: | 1975.6: Telefunken (gained several Japanese licensees) |
| VISC: | 1977.11: Matsushita   VHD (1980.1) |

Capacitance

CED: 1976.9: RCA Announcement
         1981.3: RCA sales (Hitachi and several other
                 Japanese firms exported CED to U.S., but
                 not sold in Japan)
                         1984.4: RCA discontinues

VHD: 1978.9: Victor
         1980.1: Matsushita
         1980.4: Thorn/EMI (UK, US))
             1980.10: Sharp, NEC, Akai, Toshiba, Yamaha
             1980.11: Mitsubishi
                 1980.12: Sansui
                     1983.4: Sales Begin

Optical ("Laserdisk")

    1974.9: Philips/MCA joint venture
        1977.4: Hitachi (commercial use)
            1977.6: Pioneer/MAC joint venture (commercial/consumer)
            1979.6: Pioneer begins US sales
            1981.10: Pioneer begins Japanese sales
            1983.10: Pioneer begins sales of improved,
                     semiconductor-laser based "Laserdisk"
                1983.12: Sony
                    1984.12: Hitachi, Nippon Columbia, Marantz, Sansui

---

Although only Pioneer was committed to using the optical approach for consumer players, a number of firms had already announced that they were planning to use optical technology for "pulse code modulation," or PCM, a version of which eventually became the basis for compact digital audio discs (CDs). Many in the industry hoped that

---

59. Nihon Keizai Shinbun (Hereafter Nikkei), June 14, 1977, p. 8.

once a standard was set for digital audio it could be used as the basis for a videodisk standard, solving the incompatibility problems and allowing both video and audio to be played on the same machines. First ad hoc and then formal EIAJ committees began meeting to consider standardization of PCM audio formats.

Just two days after the formation of an international Digital Audio Disk Council in the fall of 1978, hopes for quick standardization received a rude shock. Victor announced successful development of a new videodisk system with a familiar and auspicious-sounding name: "VHD," for Video High Density. A related audio version called "AHD" was also planned. This new grooveless capacitance approach combined many of the sophisticated features of the optical systems with the low cost of mechanical and capacitance machines. Victor President Matsuno was not bashful about presenting his new kid on the block as a candidate for videodisk standard, and he was soon supported by Matsushita founder Matsushita Konosuke.

With the entry of this attractive and powerful compromise format, standardization efforts entered a new stage. In the spring of 1979 MITI publicly indicated to EIAJ that it "would like to see" videodisk standardization.[60] The most likely forum for unification work was the Digital Audio Disk Council. While MITI characterized its actions as a mere "request" in the midst of informal talks, the ministry did not rule out the use of firmer methods in the future if necessary.[61]

By October, Sony announced that it would adopt the Philips version of optical technology for industrial uses, but that it was not committed to any particular standard for consumer use, and was extending its cross-licensing agreement with Matsushita and JVC to include videodisks. At about the same time it was reported that despite Pioneer's decision to export Philips-format videodisks to the United States, the firm "has no plans for the Japanese market because the government insists on a single, unified standard before the players are introduced commercially."[62] In January of 1980, Pioneer received a serious blow when Matsushita decided to abandon its own video-disk prototypes in favor of Victor's VHD. Matsushita and Victor announced that they planned to approach the other firms in the industry about adopting VHD as the unified Japanese standard.

The announcement of a unification drive by Matsushita and Victor immediately drew an angry blast from Pioneer's charismatic and

---

60. *Electronic News,* May 28, 1979, p. 105.
61. *Electronic News,* May 28, 1979, p. 105.
62. *Electronic News,* October 15, 1979, p. 87.

outspoken President Ishizuka, who drew on his connections with Philips: "From the standpoint of the principle of free competition, standardizing the format used dometically is very strange."[63] To deflect foreign criticism about the closed nature of the Japanese market Matsushita and Victor also tried to approach foreign firms and their Japanese licensees, including Pioneer, about adopting VHD. Still, the thrust of the effort was clearly to unify the Japanese consumer market before foreign technologies became too powerful to dislodge.[64]

Victor and Matsushita made steady progress in their unification drive. First, Britain's leading consumer electronics firm, Thorn-EMI, agreed to market VHD disks and players in the Britain and the U.S. Then in May MITI reemphasized the need to avoid repeating the Beta-VHS fiasco, and urged EIAJ to set up some kind of special video-disk standardization committee.[65] In the meantime Victor finished "AHD" just in time to present it to the digital audio disk council. Even though the delay in introducing AHD had given the optical format an unassailable lead in audio, Victor's introduction of a rival candidate prevented the council from making a formal recommendation of any kind, and thus from exerting any influence on videodisk formats.

The results of Victor's family-building effort became clear at the October Electronics Show as a slew of major manufacturers, including Mitsubishi, Sharp and Toshiba, displayed VHD prototypes. In the face of this "hi no maru" (rising sun flag) standard, Pioneer was completely isolated in its attempt to promote optical videodisks for consumer use.

Pioneer's struggle was even longer and more arduous than Victor's had been. Faced with a longrun decline in sales of its major product, audio equipment, Pioneer decided in the mid-1970s to move into video. Since it lacked the resources to compete in VCRs, the company decided to specialize in videodisks, and after a few false starts settled on the optical approach as most compatible with its "high fidelity" image. In the United States, early sales were disastrous, but continuing heavy investments were necessary and Pioneer was ultimately forced to take over the entire operation itself. In Japan, legal problems regarding film rights dogged Pioneer. Even more vexing was interference from MITI. Once MITI determined that the large majority

---

63. *Nikkei,* January 22, 1980, p. 7.
64. *Nikkei,* January 22, 1980, p. 7.
65. *Nikkei,* May 28, 1980, p. 9.

in the industry was lined up behind VDH, it tried right up until the time of marketing in October 1981 to keep Pioneer from introducing consumer optical disks to the domestic market.[66] Even then, the combination of high prices, Pioneer's limited marketing power and the entrenched position of the VCR limited videodisk sales.

In April, 1983, after overcoming quality problems and internal squabbling and with the specter of an eventual slowdown in VCRs ever-present, the VHD giants finally entered the Japanese marketplace with their "purely domestic system."[67] Victor pressed all VHD disks and subsidized the cost, reversing the usual formula, "give them the razor cheap, then sell the blades dear," in the interests of creating a large and powerful format family.[68] Victor also provided players to most of the twelve other firms for resale, so they could investigate the videodisk market without making an expensive manufacturing commitment. Initial sales were slow, but by summer the power of the mainstream coalition's marketing channels came into play, as exclusive contracts to carry VHD disks were signed with more than 20,000 consumer electronics shops, in addition to specialized stores and record shops.[69]

MITI implicitly supported this dominant new coalition. Ministry officials emphasized that "MITI cannot force the companies or the public to accept one solution or another," and said they simply hoped to help "speed up the process" of forging a consensus on a uniform standard.[70] Despite this modesty, given that thirteen major firms supported VHD while only Pioneer sold consumer-use optical videodisks, it was clear which consensus MITI hoped to promote.

Interestingly enough, while Pioneer bristled at attempts to exclude its format, it welcomed the entry of VHD into the videodisk market. Pioneer was confident that consumers would find the optical format superior, if they could be persuaded to examine videodisks at all, and that the advertising efforts of so many companies, most far larger than Pioneer, could only stimulate public interest in videodisks.[71]

---

66. *Asahi Shinbun,* December 30, 1981.
67. *Nikkei,* March 30, 1983, Evening Edition, p. 3.
68. *Television Digest,* March 23, 1981, p. 11; October 11, 1982, p. 10.
69. *Television Digest,* January 31, 1983, p. 10; February 7, 1983, p. 12; *Nikkei,* May 23, 1983, p. 9.
70. William G. Ouchi, *The M-Form Society: How American Teamwork Can Recapture the Competitive Edge* (Reading, Mass.: Addison-Wesley, 1984), p. 39.
71. *Nikkei,* April 22, 1983, p. 10. This was a consistent belief and not simply an attempt to put up a brave face. Even several years after the introduction of VHD, a Pioneer marketer said in an interview that he personally hoped that the VHD firms would never pull out from the videodisk market, since their presence legitimized the product, and their withdrawal would shake consumer confidence. Interview with Endo Motohiro, Systems

Pioneer's confidence was not misplaced, because soon after the massive VHD coalition finally puts its product on the market, the tide began to turn in Pioneer's favor. Pioneer settled the copyright infringement case and rapidly increased its software library. Most important of all, technical breakthroughs in laser technologies allowed Pioneer to introduce a new generation of smaller, cheaper machines with improved features. The next step in the creation of a viable challenge to VHD was to enlist a major ally. In December of 1983, Pioneer announced that it would join the faltering Beta camp as a marketer of Sony's Beta Hi-Fi model. In return, Sony agreed to market a Pioneer-made consumer optical videodisk player in the spring, with in-house production to follow. This new alliance exerted a powerful impact on the videodisk competition.[72] Pioneer had suddenly transformed itself from an isolated midget jousting at giants into the leader of a family including one of the most powerful firms in the whole electronics industry.

The final breakthrough came at the end of 1984 when Hitachi, one of the largest mass merchandisers in Japan, switched to the Pioneer camp after the demise of CED. Hitachi affiliate Nippon Columbia, Philips subsidiary Marantz and audio maker Sansui, a member of the VHD camp, followed. By the middle of 1985 at least nine firms, many of them from the audio and camera industries, were marketing the optical format and Hitachi was readying its own production. Thanks partly to the boost in publicity and image provided by the CD boom, sales of players finally began to take off. Figures on market shares for the two formats were hard to come by, but seemed to be roughly even overall, with optical players forging to a 70/30 lead in specialty shops.[73] Despite the massive "family" assembled by Victor and MITI's clear and often repeated preference for standardization, consumer videodisk formats became irreversibly fractured.

A search for the success of optical videodisks could uncover a variety of factors, including the technological superiority of the optical approach, the failure of RCA's CED format, the CD boom, and continuing improvements of the laser technology. What is unquestionable is the indispensable role played by Pioneer, a small high fidelity firm from outside of the consumer electronics mainstream, in championing

Products Division 1, Pioneer Electronic Corporation, March 1, 1986. See also the interview with Pioneer President Ishizuka in *Television Digest,* March 29, 1982, p. 9.

72. *Nikkei,* December 17, 1983, p. 6; see also *Nikkei,* march 28, 1984, p. 10 and April 4, 1984, p. 10.

73. *Nikkei,* July 30, 1984, p. 13; December 30, 1985, p. 12; the figure on market share is from *Zaikai,* October 1985, p. 27.

and sustaining the optical technology as a viable consumer format. Even if its perseverence was unusual, however, the emergence of a rebel like Pioneer was not mere happenstance. For Pioneer, competing with the marketing might of the electronics giants on the basis of the same format, particularly one offering only "medium fidelity," would have been a disaster. In a larger sense, the appearance of stubborn firms like Pioneer was not accidental but a natural result of the growing sophistication and diversity of the Japanese economy. Victor was able to build a massive coalition, but MITI, the industry association, and the VHD leadership were helpless in the face of Pioneer's determination.

Camcorders

After the breakdown of formats in VCRs and videodisks, MITI and the Japanese video industry made an attempt to standardizing a final format. At the beginning of this research, in fact, it seemed to be a successful contrast to the breakdown in videodisk formats. At the urging of MITI, the Japanese electronics industry sponsored a huge international conference in 1982 to standardize specifications of 8mm camcorders (portable camera-VCR units). Though the conference was presented as a great success, neither its origins nor its conclusions were what they appeared to be. The main outlines of 8mm were actually worked out before the conference in semi-secret deliberations by the five leading video firms (four Japanese firms plus Philips). These leading firms were primarily interested not in promoting 8mm but in slowing its development and preventing it from becoming a direct threat to existing formats.[74]

Both the movement to constrain 8mm and the attempt by camera and audio firms to establish it as the new generation VCR failed, leaving no less than four contending formats. Camera and audio firms persisted in developing 8mm camcorders and making them more directly competitive with VHS and Beta "table models," and eventually some of the Japanese electronics firms joined them. Sony and Matsushita continued to sell Beta and VHS versions, while Victor developed a compact version of VHS. By 1986 the industry had completely given up on unification. With the exception of Victor,

---

74. At first Philips disagreed with the Japanese firms. It was anxious to develop 8mm because its own V-2000 format was ging badly beated by VHS. However, once defeat was complete and the switch to VHS made, Philips also adopted a go-slow stance on 8mm. The strategy of Sony was sometimes unclear, but contrary to many reports in Japan, Sony also favored its existing format over 8mm. Sony did not actively push 8mm until early 1985, when all of its erstwhile partners had abandoned (or were about to abandon) Beta.

most firms sold two or even three formats. After three straight failures, MITI officials came to deny that they should or could try to help the industry standardize product formats.[75]

Minimills in Japan

A similar pattern of stubborn independent firms undermining attempts at constraint by a formidable alliance of government and leading firms occurred in the slow-growing and chronically depressed steel minimill industry. The quadrupling of oil prices from 1973 to 1974 had a devastating impact on many energy-intensive industries in Japan. Fewer were hit harder than the steel minimills, which had been feverishly expanding just before the oil shock. A number of firms went bankrupt, thousands of employees were dismissed, and almost all firms lost money; many were left with negative net worth. The underlying problem was excess capacity: demand declined drastically just as new plants came on line, and prices plummeted.

In the wake of the oil crisis a powerful coalition formed to deal with the problem of excess capacity. The coalition, which was centered around the minimill industry association, tried to form cartels, coordinate cuts in capacity, and ban the construction of new furnaces and plants. The mainstream coalition comprised several parts. Less competitive small minimills feared that they would be the losers if the market mechanism were allowed to "solve" the excess capacity problem. The labor unions were worried about the layoffs, "voluntary retirements" and discharges resulting from low prices and bankruptcies.

The large integrated steel mills, particularly industry leaders Nippon Steel and Nippon Kokan (NKK) were also concerned, but for somewhat different reasons. In the short run, recession in the construction industry following the oil shock depressed demand for minimill products, and threatened the smallest and weakest mills with bankruptcy. In the long run, however, the sharp increase in the price of energy and raw materials, including iron ore, actually accelerated a slow but persistant tendency for minimills to gain market share at the expense of the larger integrated firms.

Although the minimills and their electric furnaces appeared to be profligate users of energy, after 1974 it became clear that they were actually quite efficient, since they skipped the iron-making stage of

---

75. Interviews with Saeki Hidetaka and Yamuchi Toru, MITI, April 9, 1986.

steel production, and simply melted and re-rolled steel scrap.[76] In the 1960s and early 1970s this trend had been masked by the extraordinary growth of the entire Japanese steel industry, but with the permanent stagnation in demand which followed the oil shock, the minimills came to present a potentially serious threat to the big integrated producers, particularly if they were allowed to diversify into the more lucrative products which had previously been the exclusive preserve of big steel. New competition was especially unwelcome at a time of weak demand in an industry in which maintaining a high level of capacity utilization was the key to success. Thus, for the large integrated firms, the paternalistic rhetoric of helping floundering little minimills reduce "excessive" capacity concealed a more pressing, but less charitable, goal: constraining any of the more efficient minimills from expanding out of the glutted areas and into their own markets.

Fortunately for the integrated steel firms, they already controlled many of the largest minimills directly. Among them, Toshin, a subsidiary of NKK, and Godo, formed by the merger of several Nippon Steel affiliates, dominated the minimill industry association and contacts with the outside world. In addition to the links of capital and technology, dense webs of communication and control tied together the companies in each group. The top management of these affiliated firms was composed almost exclusively of executives who had retired from the parent integrated firm. The diversification plans of these minimill affiliated were always carefully coordinated with those of the parent firm.[77]

On the minimill issue the integrated steel firms and their subsidiaries were joined by the banks and the large general trading companies, or shosha. The banks and shosha were naturally attentive to the interests of some of their largest and most capital intensive customers. More directly, however, the banks and particularly the shosha needed to protect imprudent investments they had made in minimills during the mad spurt of expansion just before the oil crisis. A market solution would have rendered many of their loans and investments uncollectable, while

---

76. Steel scrap prices are notoriously unstable. Around the time of the oil shock, international scrap prices soared, leading many to believe that the competitiveness of the minimill sector would be seriously damaged. In fact, scrap prices soon declined, while iron ore prices did not, leaving the minimills in a stronger competitive position. See Takanori Tanabe, *Tekko Gyo* [The Steel Industry] (Tokyo: Toyo Keizai, 1981), p. 220.

77. Interestingly, younger executives in the affiliated minimill firms who had come up through the ranks and had no pre-existing loyalties to the parent firms reportedly chafed at the "coordination" process and argued that their firms should adopt more agressive policies to compete with the "outsiders." Interview with Omura Kazuo and Hiranuma Makoto, Nomura Research Institute, March 25, 1986.

a negotiated solution held the promise of extra time in which to reduce their exposure to the industry.

This mainstream coalition had tight links to MITI. MITI's main concern was always order and stability. The ministry wished above all to avoid embarrassing bankruptcies, large-scale layoffs and harmful effects on the economy as a whole.[78] In the 1960s MITI had tried, largely without success, to moderate the decline of the medium-sized steel firms which used the outdated open-hearth furnace by enforcing a division of labor between big and little steel;[79] after the oil shock MITI tried to stabilized producers. This effort was facilitated by MITI's long and intimate ties with the major steel companies, particularly Nippon Steel and NKK.

The industry-banking coalition enjoyed support from the politicians as well as the bureaucrats. When the financial plight of the minimills, and a number of other "depressed industries" reached their peak in 1977, Prime Minister Fukuda played an important role in pushing the bureaucracy to devise a plan to revive depressed industries. Fukuda hoped to forestall bankruptcies, and the political ammunition they provided to his opponents, both in the opposition parties and within the LDP, and he wanted to ensure that continuing weakness in depressed industries would not prevent Japan from reaching the 7% growth rate he was promising to American President Carter.

This powerful coalition of weak minimills, integrated firms and their minimill affiliates, banks, traders and retailers came together in a number of organizations. The Conference on Small Steel Bars (which covered the most important and least stable product of the minimill industry) included representatives from the giant integrated firms and the minimills, as well as specialty and general trading firms. It acted as a "pipe" to the bureaucracy, as well as a forum for stabilization efforts. The minimill industry association created special subcommittees to carry out surveys and hammer out specific policies to meet the problems of overcapacity and weak prices. The industry association was headed by the Presidents of Toshin and later Godo. Numerically, weak small firms, many of them dependent on particular trading firms or banks, dominated the association. Administrative

---

78. cf. Komiya, "Industrial Policy...," p. 66.
79. Hirokatsu Ishikawa, *Nihon Tekkogyo no Saihensi* [The Reorganization of the Japanese Steel Industry] (Tokyo: Shin Hyoron, 1974), pp. 217-228; Shunji Ouchi, *Kogata Boko Gairon* [An Outline of Small Steel Bars] (Tokyo: Mainichi Shinbunsha, 1977), pp. 491-492, 736-749; *Nikkei,* January 24, 1970, p. 5; *Nikkei,* December 24, 1972, p. 9.

work was handled by the Japan Iron and Steel Federation, the organization covering the overall steel industry, which was in turn completely dominated by the integrated producers.

Once broad initial positions were established by these industry organizations, more concrete policies were worked out in the minimill subcommittee of the Industrial Structure Council, a MITI advisory organ. The minimill subcommittee was headed by Professor Ueno Hiroya, a well-known analyst of Japanese industrial policy, and included members from the minimill and integrated sectors of the steel industry, the construction industry (major consumers of minimill products), and bankers, traders and retailers, with administrative and technical assistance provided by MITI officials. The main thrust of proposed policies, however, was clearly set by the steel firms.[80]

These committees and Conferences devised a number of policies to raise prices, limit output and control new investments. As early as 1974, the industry association asked MITI to institute controls on new investment, though the Ministry did not respond immediately. In 1975, affiliates of the integrated firms voluntarily cut production in an attempt to boost prices. When that did not work, a majority of the firms in the industry formed a series of production and price cartels. In response to the continued high production levels of such "outsiders" as Tokyo Steel Manufacturing (TSM), as well as cheating by cartel members, the industry mainstream sought permission to form a stronger "small and medium-sized industries" cartel. This type of cartel gave MITI legal authority to compel the "outsiders" to abide by the cartel. In the fall of 1977, after an unusual public "request" from Prime Minister Fukuda, the Fair Trade Commission granted permission.

At the same time the minimill industry made a number of other efforts to ameliorate the problem of excess capacity and weak prices. The industry created marketing cooperatives in each region of the country in an effort to reduce price competition. The industry association persuaded the government to provide financing to firms in the cartel, using steel bars as collateral, to get through the crisis. Together, the industry and MITI persuaded the general trading companies to purchase 40,000 tons of bars from cartel members and hold them until prices rebounded. The industry association and MITI also persuaded the Ministry of Finance to allocate 8.3 billion yen (and

---

80. Interview with sub-committee member Nishikawa Shunsaku, March 5, 1986. Nishikawa, Professor of Economics at Keio University, stressed that policy proposals were determined by the industry itself. Representatives of other industries were present largely to keep abreast of events. Nishikawa repeatedly referred to himself and Professor Ueno as "goyo gakusha" ["kept scholars"].

substantial, though declining, amounts in subsequent years) to purchase steel bars from the cartel for shipment to developing countries as "commodity aid." [81]

In addition to these short-term measures, with MITI's help the industry procured 350 million yen in loan guarantees from the government and a like amount from private financial institutions to establish a Structural Improvement Association. The association provided financing to participating firms to cover the costs involved in shutting down old capacity, such as severance payments and dismantling fees. The direct benefits of these measures were all carefully limited to firms participating in the cartel and the Structural Improvement Association. [82]

Despite these sustained and wide-ranging efforts, cooperation was' difficult to achieve, and the unconstrained production of the "outsider" firms undermined it even further. As a result, frustration mounted and tempers ran high. The head of the cartel denounced Tokyo Steel as "more outlaws than outsiders" and attributed the success of the outsiders to free-riding off the cartel. [83] In order to keep the unruly outsiders from taking indirect advantage of the reduced industry capacity by increasing their production even more, the industry association, along with the integrated makers, traders and retailers, entreated MITI to force the outsiders to curtail their output. In November MITI issued official compliance orders to the outsiders. Tokyo Steel filed an administrative protest, but to no avail. The firm was ordered to cut production 35%.

In December, 1977, at the behest of industry and of Prime Minister Fukuda, MITI also began drafting a structurally depressed industries bill in order to gain control over industries such as aluminum, fertilizers, textiles and minimills which were having trouble eliminating excess capacity. The draft bill would have regularized and extended many of the measures already tried in the minimill industry. It called for cartels to control capacity disposal and to limit or ban construction of new facilities, MITI authority to control outsiders if so requested by 3/4 of the industry, loan guarantee funds, and exemption from Anti-Monopoly Law restraints on mergers and production cartels.

Not surprisingly, this bill aroused immediate and vociferous oppo-

81. *Japan Metal Bulletin,* cited in Bethlehem Steel, *Japanese Government Promotion of the Steel Industry: Three Decades of Industrial Policy* (Washington: Bethlehem Steel, 1983), p. 29.
82. *Tekko Kai,* January 1978, p. 34.
83. *Nikkei,* December 5, 1977, p. 8.

sition from the outsiders and the Fair Trade Commission, as well as from the opposition parties, economists, and consumer groups. More surprisingly, the opposition forced MITI to make major concessions: MITI agreed to seek the agreement of the Fair Trade Commission on exemptions from the Anti-Monopoly Law and, much to the dismay of the mainstream firms of the depressed industries, abandoned attempts to the outsiders.

The end result of all these efforts by the mainstream coalition to "stabilize" the minimill industry and constrain the more efficient outsider firms was almost complete failure. The regional marketing firms were almost always underbid, and failed to stabilize prices. The cartels had but brief, sporadic impact on prices. By early 1978, even as the fight over the depressed industries bill reached a peak, demand in the notoriously cyclical minimill industry began recovering. MITI dropped first the price cartel, then the outside control cartel. The much-ballyhooed Structural Improvement Fund for capacity elimination was barely drawn upon. Despite all the effort to cut capacity, construction of new furnaces and improvements in technology actually led to a net *increase* in capacity.

Though the boom of 1978 was naturally followed by another recession, by the time the depressed industries law came up for renewal in 1983, the minimill industry had largely given up on collective schemes, and made only a token attempt to regain legal control over outsiders. Moreover, cartels and controls on bar products led many minimill firms to switch production to steel shapes, depressing prices for shapes as well, and into large H-beams and other products originally monopolized by the integrated firms—hardly the result the majors had intended.

There were a number of reasons for the failure of the stabilization campaign pushed by the well-organized, MITI-backed coalition of banks, traders, integrated steel firms and their affiliates, small firms and labor. First, it proved impossible to enforce the cooperation of the weaker small firms, even though they supported the stabilization/containment approach in principle. Most were hard-pressed to keep up cash flow, and the temptation to cheat on the cartels and the regional marketing cooperatives was too strong, especially since they knew that their competitors would succumb to those temptations. Even the elaborate checking schemes concocted by the industry association and implemented by MITI proved inadequate to assure compliance. More important, most of the small firms had but one furnace, and for them, proposals to cut capacity were tantamount to an invitation to close up shop and leave the industry.

Second, the majority coalition was unable to constrain the half dozen or so principled outsiders. Almost by definition, these firms were among the most competent in terms of operations and technology. Financing was not a problem, despite the support of the banks and general traders for the mainstream, because they had little debt. The larger firms, such as Tokyo Steel and Yamato Kogyo, also had strong support from the securities firms, which were attracted by their sound management. Politically, the Fair Trade Commission and the opposition parties were important allies because they supported the Anti-Monopoly Law and opposed the power of huge firms.

The outsiders drew support from many in the conservative camp who were committed to market competition and managerial independence (and sensitive to charges of protectionism from Japan's allies). In addition to such predictable allies as economists, law professors and consumer groups, the outsiders also drew support in their opposition to the more extreme sections of the MITI draft from the *Nihon Keizai Shinbun,* the leading business newspaper, from the influential peak business organization Keidanren, and from parts of the LDP. Tokyo Steel, in particular, not only led a publicity campaign, but also enlisted the aid of LDP politicians and hired legal counsel to oppose MITI's orders. Interestingly, the head of the minimill industry association did not call on the LDP, relying instead on links with the bureaucracy.

If the formal attempt to stabilize and constrain the minimill industry through political and legal methods had little effect, the more direct and informal moves of the integrated firms and their affiliates exerted a more powerful impact. First, the integrated firms restrained the diversification of their own affiliates. Toshin, for example, produced only the smaller H-beams, leaving lrger and more profitable sizes for parent NKK. In addition, many of the affiliates merged with one another and absorbed some of the large independents, concentrating the industry and reducing the number of firms capable of mounting a direct threat to the integrated giants.

Second, Nippon Steel resorted to predatory pricing to discourage threatening moves by minimills. Ever since it was created by a merger in 1970, industry leader Nippon Steel was famous for cutting volume in the face of recessions before resorting to discounts. Nevertheless, when Tokyo Steel announced plans in 1982 to build a new mill right next to Nippon Steel to produce large H-beams, the latter began an "H-beam war" with drastic price cuts. Though there was no doubt that Tokyo Steel could produce at lower cost, Nippon Steel's determination and deep pockets forced TSM to cut production of its medium-sized H-beams and accept a smaller share of the market. When the large-size H-beam mill came on stream in 1984, the pattern was repeated, and Tokyo Steel was forced to shift some of its production to other, less lucrative products. The large-sized H-beam mill ran at 1/3 of capacity, and the most efficient firm in the steel industry ran into the red for the year.

Just as important, the other large outsiders were deterred from diversifying into products which threatened the integrated firms. Yamato Kogyo avoided dramatic expansion projects. It established a joint venture in the United States and quietly stuck with its medium-sized beams. Partly as a result, it was consistently more profitable than the larger and more ambitious, or reckless, Tokyo Steel. Ito Seitetsujo, a smaller outsider mill, stayed out of H-beams all together.

With only one minimill bold enough to challenge the dominance of the integrated firms and their proxies, cooption became a viable strategy. President Iketani of Tokyo Steel was offered the chairmanship of the industry association. He accepted a vice-chairmanship, and the new chairman, a former Nippon Steel executive, opened up channels of communication. The message is clear: Tokyo Steel will be recognized as one of the major steel companies of Japan—*if* it accepts the oligopoly and avoids destabilizing investments and price-slashing.

While the response of Tokyo Steel is still unclear, the prospects for cooperation/cooption are certainly much greater than they would have been if the company had to compete with three or four other aggressive minimills which were also willing to challenge the dominance of big steel. At the same time, the pressures on Japan's integrated steel mills to meet the new challenge and adopt some of the cheaper and more flexible technologies of the minimills are also accordingly less.

As with the format standardization cases, the case of the minimills illustrates the difficulty in making firms cooperate in the low-growth era if their basic interests are opposed. It also illustrates that on issues of intra-industry cooperation, market power and private sector connections are more important restraints on firm behavior than MITI and the consensus decision-making system in cases. Japan does not necessarily have a "strong state," but decisions by firms are not just the outcome of "the market," either.

Conclusions

The cases of format unification in consumer video and stabilization of the steel minimill industry reviewed here do not provide much support for the notion that a strong, bureaucratically-led "developmental state" is a major component of the Japanese economy in the period since the oil shock. Aside from an occasional veiled hint, MITI did not resort to any specific policy tools in consumer video. Indirect administrative guidance, as in the attempt to dissuade Pioneer from introducing its Laserdisk into the Japanese market, was ineffective.

MITI resorted to more policy tools in its attempt to stabilize and constrain the minimill industry, but also to little avail. MITI's efforts, including supervision of cartels, control of the outsiders and financing of the loan guarnatee fund, were more a response to requests from the majority coalition than an independent policy. More important, they were insufficient to enforce compliance within the mainstream and failed to stop the outsiders from increasing production, capacity and product range. The most expensive form of intervention, the purchase of steel bars from the cartel association for shipment to underdeveloped countries as "commodity aid," was actually a form of subsidy that tended to retard the exit of weak firms and exess capacity from the industry rather than accelerate it.

The cases provide more support for the contention of the New Japan, Incorporated school that the real significance of industrial policy lies in the way that the government facilitates communication and bargaining among private interests. In consumer video, MITI constantly reminded firms of the importance of standardization and helped persuade several firms to abandon their own formats in favor of more promising alternatives. The ministry was also the prime mover behind the international conference on 8mm camcorders. In the case of minimills, MITI and its Industrial Structure Council provided background research and a forum for discussion among the various firms and industries involved. MITI helped to enforce the cartels and controls created by the mainstream firms, and provided several types of side payments to induce individual firms to cooperate with the majority approach. Finally, the formation of clear "mainstreams" in both industries supports the New Japan, Incorporated emphasis on the need to examine the alignment of interests within the private sector. Patterns of economic competition were not simply the result of atomistic interaction in the market.

Nevertheless, in three separate instances in the consumer video industry and over a period of ten years in the minimill industry, the attempts at cooperation among MITI, industry associations and dominant firms were unable to prevent maverick firms from pursuing their own interests and undermining the policies established by the majority in conjunction with MITI. With minor and temporary exceptions, such as the slowdown on development of 8mm camcorders and brief controls on the outsider minimill firms, MITI and the majority were unable to constrain the mavericks. In both industries maverick firms were able to bring significant resources to the market. In the minimill case, Tokyo Steel and the other outsiders were also able to elicit political support from LDP Dietmen, the Fair Trade Commission and parts of the business establishment.

Moreover, the provocative actions of the mavericks compelled the mainstream firms to react and compete. The movement of Tokyo Steel into large H-beams forced the affiliates of the integrated firms to follow, albeit in a cautious and limited way, even though this economically rational move had been blocked previously by the interests of their parent firms. Similarly, the movement of Pioneer and the camera firms into videodisks and camcorders eventually induced some of the electronics firms to break away from the majority approach, heightening competition across formats, and undermining any tendencies toward collusion. Despite the presence of many of the elements identified by the New Japan, Incorporated school, these results confirm the applicability of the market approach to contemporary Japan.

Consumer video and minimills are by no means the only cases that reveal a pattern of maverick firms undermining majority coalitions backed by MITI, or, especially in industries with a large number of firms, the inability to form a consensus. Within the designated depressed industries, paper linerboard and cotton spinning looked quite similar to minimills. In several other industries MITI policies were essentially irrelevant, and capacity was reduced by the exit of weak firms.

Only in shipbuilding did the consensus approach work relatively smoothly and effectively.[84] Overcapacity was reduced when the top

---

84. In an important article, Merton J. Peck, Richard C. Levin and Akira Goto, "Picking Losers: Public Policy Toward Declining Industries in Japan," *Journal of Japanese Studies* 13:1, pp. 79-123, argue that while the Japanese approach to depressed industries did not have notable impact on the eight unconcentrated industries (including minimills), it was quite effective in cutting capacity in the six relatively concentrated industries. Closer examination of the individual concentrated industries casts doubt on that generalization. The aluminum industry, as Richard J. Samuels, "The Industrial Destructuring of the Japanese Aluminum Industry," *Pacific Affairs* 56:3 (Fall 1983), pp.495-509, shows, was riven by disputes and collapsed before any effective plan could be devised. Urea was protected from imports and 1/3 of the producers were forced to leave the industry. The polyester filament industry, significantly the least concentrated of the six, was only able to cut 5% of capacity—one-third of its modest goal. The policies *were* arguably effective in one case: the three synthetic fibers industries. Even syntheitic fibers, however, witnessed outsider-insider disputes, and the amount of capacity cut was quite limited

few firms accepted considerably larger cuts in capacity than the small firms, and the Japanese industry regained, at least temporarily, much of its competitive edge.[85] Behind this success was the fact that even the most competitive and dynamic Japanese firms had no choice but to cooperate: the drop in demand was far more severe and extended than in the minimills, and Japanese firms already controlled half the world market.

In the area of product formats, a major cooperative research project sponsored by MITI to produce a new generation of computer workstations has been beaten to the market by an incompatible design from Sony, previously a minor player in computers, which combines lower cost and superior performance.[86] Even in the promotion of high technology industries, a major focus of industrial policy in the 1980s and 1990s, enforcing a consensus is extremely difficult.

To be sure, industrial policies in Japan have been effective in some instances. Government help and cooperative approaches have been crucial to the development of the Japanese computer industry, as even skeptics of industrial policy acknowledge,[87] and the joint research project on VLSI semiconductors in the late 1970s allowed Japan to make a major breakthrough in memory chips.[88] Shipbuilding is an example of a relatively successful policy in a depressed industry. Today ambitious efforts are underway to promote satellites and super conductors, though it is still too early to evaluate the impact of such programs.

When Japanese firms believe that they have a common interest, when the industry is stable and composed of a small number of firms, and particularly when no Japanese firm is fully competitive on its own, the basis for concerted action and government support is available. Since that concerted action is easiest to obtain when competing with foreign firms, it is not surprising that foreign observers have tended to attribute strong bureaucratic leadership or collusive tendencies to Japanese industrial policy. The pattern of competition in the vital computer industry provides a clear and important example:

(15-18%). Peck et al. are certainly correct that the Japanese approach was superior to the interminable and costly subsidies common in Europe. With the exception of shipbuilding, however it is less clear even in concentrated industries that the results in Japan were significantly different from a market outcome.

85. Vogel, *Comeback*.

86. *Far Eastern Economic Review,* 19 December 1987, pp. 84-86.

87. Lincoln, *Japan's Industrial Policies*, p. 36.

IBM vs. the Japanese.[89]

As Japanese firms reach the technological frontiers even in computers, semiconductors, and aerospace, however, that pattern is eroding. Japanese firms increasingly find that they share as many interests with certain foreign firms as they do with Japanese rivals. In addition, as more foreign firms—like Philips—enter Japan, form joint ventures and partnerships with Japanese firms, and join industry associations, Keidanren, and Industrial Structure Council committees, and as foreign firms and governments scrutinize Japanese policy with increasing care, industrial policy will be even more difficult to implement.

The convergence of interests is not a magical result of the effective operation of industrial policy but a prerequisite for it. As the consumer video and minimill cases demonstrate, since the oil shock that prerequisite has been increasingly hard to come by in Japan. Whatever its contributions in the 1950s and early 1960s, in the 1980s Japanese industrial policy is characterized less by a protective government or government-inspired cooperation than by fierce and pugnacious competition among skilled, determined, and often aggressively independent firms.

88. Daniel I. Okimoto et al., eds., *Competitive Edge: The Semiconductor Industry in the U.S. and Japan* (Stanford: Stanford University Press, 1984).

89. Marie Anchordogoy, "Mastering the Market: Japanese Government Targetting of the Computer Industry," *International Organization* 42:3 (Summer 1988), pp. 509-543. etary and other macroeconomic policies, or "horizontal" policies designed to affect the supply or quality of whole factors of production, such as labor or basic research.

CHAPTER 4
On the High Track to Protection:
The U.S. Automobile Industry, 1979-1981*

Douglas Nelson

After nearly half a century, protectionism has made a comeback as a political issue in the United States. One central claim of this paper is that any attempt to evaluate the effects of renewed protectionist pressure on U.S policy needs to consider not only interest group pressure and global structural position, but the institutional framework which transforms political pressure into political action. The claim that institutionalized definitions of policy affect the form of politics has been a standard part of political analysis at least since Lowi's fundamental contribution on policy types.[1] Also standard is the notion that

*This paper was written as part of the work program of the World Bank's International Economic Research Division under the direction of J.M. Finger. Many of the methods and ideas in this paper are the result of our collaboration. I would also like to acknowledge valuable input from: Jagdish Bhagwati, I.M. Destler, Stephan Haggard, Paula Holmes, Dermot McAleese and Patrick Messerlin. Since I am, unfortunately, not able to claim that any of the above agree with everything in the paper, it is important to note that I am solely responsible for any errors of fact, judgement or good taste. The World Bank does not accept responsibility for the views expressed herein which are those of the author and should not be attributed to the World Bank or to its affiliated organizations. The findings, interpretations, and designations employed, the presentation of material, and any maps used in this document are solely for the convenience of the reader and do not imply the expression of any opinion whatsoever on the part of the World Bank or its affiliates concerning the legal status of any country, territory, city, area, or of its authorities, or concerning the delimitation of its boundaries, or national affiliation.

1. Lowi developed this typology in a review of Schattschneider's study of the Smoot-Hawley tariff and Bauer, Pool and Dexter's study of the extension of the Reciprocal Trade Agreements Act during the 1950s. See: Theodore Lowi, "American Business, Public Policy: Case Studies and Political Theory," *World Politics* 16:4, pp. 347-382; Schattschneider, *Politics, Pressures and the Traiff* (New York: Prentice Hall, 1935); and Bauer, Pool, and Dexter, *American Business and Public Policy: The Politics of Foreign Trade* (Chicago: Aldine, 1963). Also see Hays' *Lobbyists and Legislators: A Theory of Political Markets* (New Brunswick: Rutgers University Press, 1981) for a review and reconstruction of the literature on policy typologies.

these institutionalized definitions impart systematic biases to policy outcomes through selection and agenda-setting effects, where the first refers to the systematic biases derived from the inclusion or exclusion of either aspects of an issue or interested parties from the agenda; and the latter refers to systematic biases among included aspects or interests derived from the form of the agenda. Where Lowi's concern was primarily with how policy definition effects the organization and disorganization of groups, selection effects operate by inclusion and exclusion. We are, then, concerned with three categories of effect induced by institutional structures: Lowi effects (ie. the triggering of characteristic patterns of group organization); selection effects (inclusion/exclusion of actors in such characteristic conflicts); and agenda-effects.[2]

As a result of the sustained concern by political scientists on the effects of political institutions, the assertion that institutions are important may strike some as trivial. Most of the literature on the political economy of trade policy, however, proceeds under rather different assumptions. With the exception of Lowi's own work on the difference between the politics of trade before and after the Reciprocal Trade Agreement Act (RTAA) of 1934, the great majority of research on the political economy of trade has proceeded under institution-free (or, more accurately, minimally institutional) assumptions. First, there is a substantial body of neo-realist and regime theoretic research which attempts to derive the trade policies of the regime's core countries from structural attributes of the international system.[3] Second, following in

2. Peter Bachrach and Morton Baratz introduced both selection effects ("mobilization bias") and agenda-setting (necessary to "non-decisions") into their analysis. See: "Two Faces of Power," *American Political Science Review* 56:4, pp. 947-952): "Decisions and Nondecisions: An Analytical Framework," *American Political Science Review* 57:3, pp. 632-642. On the importance of agenda-setting, see also Charles Plott and Michael Levine, "A Model of Agenda Influence on Committee Decisions," *American Economic Review* 68:1, pp. 146-160 and Kenneth Shepsle, "Institutional Arrangements and Equilibrium in Multidimensional Voting Models," *American Journal of Political Science* 23:1, pp. 27-59.

3. Standard works in trade policy include: David Calleo and Benjamin Rowland, *America and the World Political Economy: Atlantic Dreams and National Realities* (Bloomington: Indiana University Press,1973); Stephen Krasner, "State Power and the Structure of International Trade," *World Politics* 28:3, pp. 317-347; Krasner, "The Tokyo Round: Particularistic Interests and Prospects for Stability in the Global Trading System," *International Studies Quarterly* 23:4, pp. 491-531; Charles Kindleberger, "Dominance and Leadership in the International Economy," *International Studies Quarterly* 25:2, pp. 242-254; David Lake, "Beneath the Commerce of Nations: A Theory of International Economic Structures," *International Studies Quarterly* 28:2, pp. 143-170; John Conybeare, "Public Goods, Prisoners' Dilemmas and the International Political Economy," *International Studies Quarterly* 28:1, pp. 5-22; Beth Yarbrough and Robert Yarbrough, "Free Trade, Hegemony and the Theory of Agency," *Kyklos* 38:3, pp. 348-364. There is also a growing literature in economics on strategic interactions

the Schattschneider tradition, we find an even larger literature attempting to derive trade policy (and especially the structure of protection) directly from group preferences.[4] Third, there is a literature which emphasizes primarily the interests of state actors caught between fairly general, but contradictory, domestic and international commitments.[5]

among states over trade issues. See John McMillan, *Game Theory in International Economics* (New York: Harwood, 1986). Douglas Nelson, "Structural Theories of the International Political Economy of Trade" (World Bank manuscript, 1987) presents a survey of theoretical issues in neo-realist and regime theoretic research, while Nelson, "The State as a Conceptual Variable" (World Bank manuscript, 1987) examines the validity of the conceptualization of "the state" implied in that literature.

4. On how inter-industry cleavages affect the politics of trade, see Alexander Gerschenkron, *Economic Backwardness in Historical Perspective* (Cambridge: Harvard-Belknap, 1966), see Peter Gourevitch, *Politics in Hard Times: Comparative Responses to International Economic Crises* (Ithaca: Cornell University Press, 1986); James Kurth, "The Political Consequences of the Product Cycle," *International Organization* 33:1, pp. 1-34. Others have developed approaches which emphasize inter-firm/intra-capital cleavages. See Timothy McKeown, "Firms and Tariff Change: Explaining the Demand for Protection," *World Politics* 36:2, pp. 215-233; Thomas Ferguson, "From Normalcy to New Deal: Industrial Structure, Party Competition, and American Policy in the Great Depression," *International Organization* 38:1, pp. 41-94; Thomas Pugel and Ingo Walter, "U.S. Corporate Interests and the Political Economy of Trade Policy," *Review of Economics and Statistics* (1985), pp. 465-473; and others emphasize inter-factor cleavages: Stephen Magee and Leslie Young, "Endogenous Protection int the United States," in R. Stern, *U.S. Trade Policies in a Changing World Economy* (Cambridge, MIT Press, 1987). Seee also Robert Baldwin, "Rent-Seeking and Trade Policy: An Industry Approach," *Weltwirtschaftliches Archiv* 120:4, pp. 662-676; and Helen Hughes, "The Political Economy of Protection in Eleven Industrial Countries," in R. Snape, ed., *Issues in World Trade Policy: GATT at the Crossroads* (London: Macmillan, 1986), pp. 222-237 for surveys of the econometric literature; and Nelson, "Endogenous Tariff Theory: A Survey," *American Journal of Political Science* 32:3, pp. 796-837 for a survey of the formal literature.

5. While the characterizations of the domestic and international commitments varies, they usually amount to domestic welfare state commitments and international commitments to liberal trade relations. See: Jan Tumlir, *National Interest and International Order* (London: Trade Policy Research Centre, 1978); Melvyn Krauss, *The New Protectionism: The Welfare State and International Trade* (New York: NYU Press, 1978); John Ruggie, "International Regimes, Transactions, and Change: Embedded Liberalism in the Postwar Economic Order," *International Organization* 36:2, pp. 379-415; and Nelson, "The Welfare State and Export Optimism," in D. Pirages and C. Sylvester, *The Transformation of the Global Political Economy* (London: Macmillan, forthcoming). Robert Feenstra and Jagdish Bhagwati, "Tariff Seeing and the Efficient Tariff" and Robert Baldwin, "The Political Economy of Protectionism," both in Bhagwati, ed., *Import Competition and Response* (Chicago: University of Chicago Press, 1982) present formal analyses of the choice between domestic welfare objectives and the efficiency objective implied by international commitments; and John Ikenberry, "The State and Strategies of International Adjustment," *World Politics* 39:1, pp. 53-77, presents an analysis of state choice between domestic and international adjustment options.

Thus, a first goal of this paper is simply to demonstrate that institutional structure has an empirically significant effect. This paper, then, is part of the growing body of research on the effects of trade policy institutions. Most of this work, however, follows Bauer, Pool and Dexter in arguing that the primary effect of the post-RTAA domestic political regime for trade policy was to shield politicians and other state actors from group pressures, permitting them to pursue their own ideological preferences. Furthermore, it is also argued that the modal preference among politicians and state actors is for some form of generally liberal trading relations. The result is what Pastor has called the "cry-and-sigh" syndrome, where members of Congress rend their clothes and tear at their hair in public over the depredations of foreign exporters (to pacify their, presumably, protectionist constituents), but under the protection of the post-RTAA institutions, they are able to continue to support their "real" preference for liberalization.[6]

While these analyses are correct in their evaluation of the effect of the post-RTAA institutions on trade policy, their accounts are fundamentally incomplete. One of the major institutional innovations of the RTAA was to make liberalization and protection politically distinct issues. By changing the institutional structure within which protectionist interests are to be accomodated from direct legislation of tariffs to the various administered protection mechanisms, the politics of trade liberalization was decoupled from the politics of protection. In a sense, liberalization and protection were reconstituted as separate issues and, as a result, it has been possible for liberalization and protection to rise at the same time. It is this ambiguity in the trade policy process (and not any shift in attitudes) that accounts for the historically unparalelled Congressional support for the RTAA and its successors. Thus, while agreeing that the change in institutions was liberalizing (via the Lowi effects), this paper argues that the key institutions of the domestic trade policy regime embody protectionist biases operating via selectivity effects and agenda-setting that substantially qualify the liberalizing Lowi-effects of the RTAA.

The next section briefly describes the institutional structure of the

---

6. Robert Pastor is the major proponent of this position. See: *Congress and the Politics of U.S. Foreign Economic Policy: 1929-1976* (Berkeley: University of California Press, 1983); and "The Cry-and-Sigh Syndrome: Congress and Trade Policy," in A. Schick, ed., *Making Economic Policy in Congress* (Washington: AEI, 1983). See also Judith Goldstein, "The Political Economy of Trade: Institutions of Protection," *American Political Science Review* 80:1, pp. 161-184; Stephan Haggard, "The Institutional Foundations of Hegemony: Explaining the Trade Agreements Act of 1934," *International Organization,* 42:1 (Winter 1988), pp. 91-121; and I.M. Destler, *American Trade Politics: System Under Stress* (Washington: Institute for International Economics, 1986).

post-RTAA domestic regime, especially the institutional division between liberalization and protection; while the remainder of the paper examines the particular case of the politics surrounding the 1981 automobile VER with Japan. The case study approach complements the existing aggregate research on the administered protection mechanisms which demonstrates that countervailing and anti-dumping cases are resolved primarily according to technical rules.[7] Escape Clause (Section 201-203 of the Trade Act of 1974) cases, however, were found to operated under much looser statutory constraint. This led to the hypothesis that the Escape Clause would operate as an administrative mechanism for handling politically sensitive protectionist pressures which could threaten the viability of the system as a whole. While there is general statistical support for this conjecture, because of the small number of cases, their large economic size, and the less.constrained terms of administration, a case study approach seems an appropriate complement to the aggregate research. Finally, the automobile case was chosen because (like textiles and steel) it is an industry of some instrinsic interest given its size, for reasons described below it constitutes a strong test of the protectionist-bias hypothesis, and there is good secondary literature on the economic and political issues tangential to the specific interests of this paper.[8]

---

7. Early work on administered protection includes Irving Kravis, "The Trade Agreements Escape Clause," *American Economic Review* 44:3, pp. 319-338; William Kelly, "The Expanded Trade Agreements Escape Clause, 1955-1961," *Journal of Political Economy* 70:1, pp. 37-63; and Kelly, *Studies in U.S. Commercial Policy* (Chapel Hill: North Carolina University Press, 1963). See also Wendy Takacs, "Pressures for Protectionism: An Empirical Analysis," *Economic Inquiry* 19:4, pp. 687-693; Douglas Nelson, "The Political Economy of the New Protectionism," *World Bank Staff Working Paper* 471, 1981; J.M. Finger, "The Industry-Country Incidence of 'Less Than Fair Value' Cases in U.S. Import Trade," *Quarterly Review of Economics and Business* 21:2, pp. 260-279; Finger, H. Keith Hall and Nelson, "The Political Economy of Administered Protection," *American Economic Review* 72:3, pp. 452-466; Susan Feigenbaum and Thomas Willett, "Domestic versus International Influences on Protectionist Pressures in the United States," in S. Arndt et al., *Exchange Rates, Trade and the U.S. Economy* (Cambridge: Ballinger, 1985), pp. 181-190; Robert Baldwin, *The Political Economy of U.S. Import Policy* (Cambridge: MIT Press, 1985), chapters 3 & 4; James Hartingan, Philip Perry and Sreenivas Kamma, "The Value of Administered Protection: A Capital Market Approach," *Review of Economic Studies* 68:4, pp. 610-617; and Judith Goldstein, "The Political Economy of Trade: Institutions of Protection," *American Political Science Review* 80:1, pp. 161-184.

8. This paper can be seen as complementary to recent papers on the international politics of the auto VER, Gilbert Winham and Ikuo Kabashima, "The Politics of U.S.-Japanese Auto Trade," in I.M. Destler and H. Sato, *Coping With U.S.-Japanese Economic Conflict* (Lexington: D.C. Heath, 1982); as well as work on congressional politics, such as Ikuo Kabashima and Hideo Sato, "Local Content and Congressional Politics," *International Studies Quarterly* 30:2, pp.295-324.

The Institutional Structure of the Domestic Regime for International Trade Policy

The RTAA of 1934 created the institutional foundations on which a revolution in the politics of U.S. international trade policy was built.[9] The statutory content of the RTAA was simple: it gave the President the right to negotiate tariff reductions within Congressionally set limits. The RTAA's effect on the domestic politics of trade, however, was far from simple, and almost certainly not generally understood at the time. We can divide these effects into those flowing from the redefinition of the trade issue; and those flowing from the separation of liberalization from protection as political issues. We will consider each briefly.

Prior to the RTAA, tariff levels were generally perceived to be primarily a distributive issue: a policy whose benefits are concentrated in a particular sub-set of the population (generally some politically relevant geographic area); but whose costs are dispersed over the entire population. In Lowi's words:

> ... in the short-run certain kinds of government decisions can be made without regard to limited resources. Policies of this kind are called "distributive"... Distributive policies are characterized by the ease with which they can be disaggregated and dispensed unit by small unit, each unit more or less in isolation from other units and from any general rule... These are policies that are virtually not policies at all but are highly individualized decisions that only by accumulation can be called a policy. They are policies in which the indulged and deprived, t.ie loser and the recipient, need never come into direct confrontation.[10]

The effect of such a definition in Congress, as Lowi conjectured, and recent choice-theoretic research has confirmed, is to induce norms of universalism and reciprocity. Universalism is a norm under which all requests are accommodated ("something for everyone"), yielding what

9. See Stephan Haggard, "The Institutional Foundations of Hegemony...." Nelson argues that while the RTAA was probably necessary to the revolution, it was far from sufficient. The other part of the story was the continuing commitment on the part of the executive branch to promote the trade liberalization program as a necessary part of foreign and national security policy. See Nelson, "The Domestic Political Preconditions of Trade Policy," *Journal of Public Policy,* forthcoming.

10. Lowi, "American Business...," p. 690. For the choice-theoretic research, see Barry Weingast, "A Rational Choice Perspective on Congressional Norms," *American Journal of Political Science* 23:2, p. 245-262; Morris Fiorina, "Universalism, Reciprocity, and Distributive Policy-Making in Majority Role Institutions," *Research in Public Policy Analysis and Management,* Volume 1, pp. 197-201; and Kenneth Shepsle and Barry Weingast, "Political Preferences for the Pork Barrel: A Generalization," *American Journal of Political Science* 25:1, pp. 96-111.

Lowi called "coalitions of uncommon interest." In the case of the tariff, this simply implied that if a member of Congress requested protection, his request should be accommodated. Reciprocity supports the universalism norm through generalized log-rolling in the issue area, or what Schattschneider called "reciprocal non-interference" and demonstrated extensively for the tariff case. A similar pattern of behavior was elicited from the polity at-large, with requests for protection coming primarily from firm and, to a lesser extent, industry-level participants, for whom the tariff was effectively a private good. Opponents of protection were not a major part of the process, and firms expecting injury from protection generally found their only available recourse to be additional protection for themselves.

Under the RTAA and its successors to date, the issue ceased to be the degree of accommodation of discrete, individual interests and became (simplifying only somewhat) the determination of two general rules: one regulating the degree of tariff-cutting authority available to the executive; and the other regulating the ease of access to an administered protection mechanism and the conditions necessary for accommodation within that mechanism. We can identify two major effects of the shift from a distributive to a regulatory definition of trade: change in the organization of group activity; and change in the levels of group activity. The first of these is a direct extension of Lowi's point that issue type determines the form of politics. By contrast to distributive issues (which induce cooperation under norms of universalism and reciprocity), regulatory issues induce conflictive patterns of behavior because of the broader definition of the issue. That is, groups form (in the polity and in the legislature) as a function of their common interests vis-à-vis the general rule: preferences toward greater-or-lesser liberalization; and greater-or-lesser administered protection. The hypothesized effect of organized pressure for liberalization and against protection in general is: first, to reduce a Congressman's preferred level of protection from what it would have been under the distributive regime; and second, as Bauer, Pool and Dexter argue in their study of the politics of RTAA extension in the 1950's, the pressures of liberal groups to some extent offset those of protectionist groups, with the effect that members of Congress had some discretion in the issue. Thus far it has been the case that Congress has used its discretion to accommodate strong requests for negotiating authority from the executive branch, and not for reasons of ideological support of freer trade.

At least as important as its effect on the organization of group interests, however, is the effect of the redefinition of the domestic regime for trade politics on incentives to group action. On the basis of an Olson-type logic of collective action argument, it is usually argued that the major bias toward protection in democratic political systems emerges from an asymmetry in the distribution of benefits from protection and liberalization. That is, the benefits of protection are concentrated on a relatively small, easily identifiable group; while the costs (benefits of liberalization) are considerably more diffused. As a result, the gainers from protection are more likely to organize for collective action, and/or more likely to be effectively organized, than opponents of protection (proponents of liberalization).[11] What is not generally recognized is that this argument rests on the pre-RTAA definition of the domestic political regime: each act of protection is seen as a function of a discrete firm- or industry-specific choice, with easily identifiable and effectively privatized effects at the firm or industry level. Instead, under the RTAA, each act of protection is the outcome of a bureaucratic process administered by the executive branch under a general rule adopted by Congress.[12] Thus, the expected beneficiaries of protection find themselves lobbying for a broad-based public good and not a policy approaching a private good (as in the distributive case). The result of dramatically enlarging the coalition and eliminating the direct link between lobbying and outcome in Congress should be an "underproduction" of protection by the standard collective action logic.[13]

As institutionalists like Pastor suggest, the Lowi-effects should unambiguously lower both group pressure for protection and the propensity of Congress to offer protection for any given level of pressure. Furthermore, they are absolutely correct to point to the dramatic drop in levels of protection and sustained participation in multilateral liberalization efforts as evidence of the operation of these Lowi-effects.

11. The same argument could be used to demonstrate that Congress acts ideologically to support liberalization. A change in Congressional ideology explains why the response to protectionist pressure has been weaker, and support of liberalization stronger since the RTAA. Krasner, "State Power..." and Goldstein, "The Political Economy of Trade" make this argument. I demonstrate in Nelson, "Domestic Politics of Protection," that it is hard to find direct support for this position. Change in the logic of collective action attendant of the change in the domestic politics of the trade issue supports an institutionally-enriched interest group explanation of Congressional action.

12. As will be noted below, this abstracts from the fact that politically sensitive (i.e. large) cases, like the autos case, may be settled on the more political ("high") bureaucratic track, or even through direct recourse to Congress. This complication does not, however, alter the basic argument made here about the effect of issue definition on the central tendency of trade policy in Congress.

13. Hall and Nelson, "Modelling the Market for Protection: Administered versus Legislated Approaches" (World Bank manuscript, 1983) present a simple formal demonstration of this result in a simple, endogenous policy model.

What these analyses miss is the effect of making liberalization and protection effectively distinct issues. The same piece of legislation that involves a major grant of negotiating authority (or ratification of the results of the application of such authority) can also involve a major increase in the protectionist content of the *administered* protection mechanisms. The Trade Act of 1974, for example, involved both the substantial grant of authority to negotiate on tariff and non-tariff barriers necessary to U.S. participation in the Tokyo Round of GATT negotiations, and a major loosening of the conditions regulating access to administered protection.

The preceding analysis leads us to hypothesize a discontinuous downward shift in the level of protection (reflecting the effect of the changed institutional setting and executive pressure), accompanied by a continuing rise in the level of protection (reflecting continuing asymmetrical interest group pressure and Congressional vulnerability to that pressure). However, by the collective action argument, we would expect that rise to be slower and more stable than under the pre-RTAA system. Furthermore, as a result of the procedural division between the protection and liberalization processes, we expect no necessary relationship between increases at the margin in the former and increases at the margin in the latter.

No one seriously disputes the proposition that there has been a discontinuous drop in the overall level of U.S. protection dating from the RTAA program. The matter of steadily increasing protection at the margin is a matter which may command somewhat less support—a fact which in itself provides some indirect support for the third proposition. In assessing the claim that protection has been steadily increasing, it is necessary to recall that marginal changes in the post-RTAA domestic trade regime take the form of changes in the rules regulating the administered protection mechanisms, not direct acts of protection by Congress. On this point there is also virtually no dispute. It is clear that there has been a systematic process of easing access to the administered protection mechanisms and reducing the conditions necessary to affirmative findings as well as evidence that this has resulted in more filings.[14] Furthermore, there is considerable evidence of increasing coverage of trade affected by various forms of administered protection, and evidence that these measures have real trade diversion effects, as

---

14. On the steady easing of access to administered protection, see Joe R. Wilkinson, *Politics and Trade Policy* (Washington: Public Affairs Press, 1960); Nelson, "The Political Economy of the New Protectionism;" Destler, *American Trade Politics,* chapter 6. For evidence that the changes under the 1974 Trade Act elicited increased demands in the system, see Feigenbaum and Willett, "Domestic versus International Influences..."

well as real welfare effects.[15] Finally, however, we must also note that (as suggested above) the growth in protection has been more stable and slower in the 50 year period after 1934 than in the 50 year period preceding 1934, a fact which is often offered as evidence of a fundamental change in political attitudes toward protection.[16] In the absence of any convincing direct evidence of such a collective conversion (and considerable direct evidence to the contrary), the institutional-structural explanation developed here would seem to be a more useful explanation of Congressional action on the protectionist side of trade policy.

Before turning to a more detailed discussion of administered protection, a final note on system dynamics is appropriate. The argument of this paper has assumed: first, that there has been no change in the fundamental asymmetry between the economic interests of protection-preferring citizens and liberalization-preferring citizens; and second, that members of Congress remain generally responsive to effectively organized constituent demands. The Lowi-effects produced a one-time shift in both the level of protection and the rate at which protectionist pressure is accommodated, but that rate remains positive. The underlying *dynamic* remains protectionist.

Our discussion to the point has involved one really major simplification: the administered protection process has been characterized strictly in terms of its rule-based nature for comparison with the legislated protection process of the pre-RTAA domestic regime. While this is a broadly accurate characterization of the anti-dumping and countervailing duty mechanisms (what we call the "technical" or "low" track), it is not an accurate representation of the more politicized parts of administered protection (what we call the "political" or "high" track). Given the central role played by the political track as in accommodating protectionist pressures that could otherwise threaten the

---

15. For evidence on the increase in administered protection, see S. Anjaria, et al., *Ddevelopments in International Trade Policy* (Washington: IMF, 1982); Julio Nogues, Andrezej Olechowski and Alan Winters, "The Extent of Nontariff Barriers to Industrial Countries' Imports," *World Bank Economic Review* 1:1, pp. 181-199; and Gary Hufbauer et al., *Trade Protectionism in the U.S.: 31 Case Studies* (Washington: Institute for International Economics, 1986). On the trade diversion effects of these barriers, see J.M. Finger and Andrzej Oleschowski, "Trade Barriers: Who Does What to Whom" (World Bank manuscript, 1986), and on the welfare effects, David Tarr and Morris Morkre, *Aggregate Costs to the U.S. of Tariffs and Quotas on Imports* (Washington: Federal Trade Commission, 1984); and Robert Baldwin and Anne O. Kreuger, *The Structure and Evolution of Recent U.S. Trade Policy* (Chicago: University of Chicago Press, 1984).

16. Pastor, *Congress and the Politics of U.S. Foreign Economic Policy...* and "The Cry-and-Sigh Syndrome...;" 1983; Goldstein, "The Political Economy of Trade..."

whole structure of the post-RTAA regime, it must be analyzed explicitly.

Protection on the High Track: How It Works

It is important to understand that the "right to protection" noted by Schattschneider did not disappear under the RTAA and its successors, it was merely transformed into what might be called the "no serious injury" norm: Congress has been willing to extend (even substantial) liberalization authority to the executive only under the clear understanding that the executive must be reasonably responsive (defined by Congress in the trade legislation) to requests for protection from economic actors "injured" as a result of international competition. The political purpose of these safeguard mechanisms is to defend the liberalization process against such protectionist pressure. Without some institutionalized mechanism for dealing with these pressures trade issues would be thrown back into Congress. Since the purpose of such a mechanism is to absorb political pressure, not simply redirect it, it must produce the minimum necessary protection with the minimum publicity —ie. without becoming repoliticized. In the U.S. this is done by treating as many of the demands for protection as possible in a strictly technical fashion. Specifically, the majority of cases are processed in the less-than-fair market value (LTFV) mechanism where precisely defined rules regulate access and define the key conditions for protection (ie. "exports at LTFV" and "injury").

Research on administered protection (see footnote 7) indicates that most of LTFV cases are small (relative to Escape Clause Cases) and filed by politically weak industries. The significance of the LTFV mechanisms is that, by providing small amounts of protection for a large number of small protection seekers, potential members of a protectionist coalition are "bought off." Some politically powerful industrial sectors, however, find the rules regulating access and the burdens of proof necessary to an affirmative finding too restrictive. The possibility that such sectors might use their individual influence in Congress to undermine the system of administered protection, thus threatening the overall liberalization effort, necessitates the introduction of some flexibility into the system. In the U.S., the Escape Clause mechanism is the primary instrument providing this flexibility.[17]

17. Nelson, "Political Economy of the New Protectionism" gives a brief comparison between administered protection in the U.S. and in the European Community.

1. A petition for import relief is filed by the domestic industry with the International Trade Commission (ITC);
2. The ITC conducts an investigation to determine if "an article is being imported into the United States in such increasing quantities as to be a substantial cause of serious injury, or the threat thereof, to the domestic industry producing an article like or directly competitive with the imported article." (19 USC 2201) The ITC reports its findings to the President and, if its injury determination is positive, it includes in its report its determination as to the action "necessary to prevent to remedy such injury."
3. The President then determines what action to take. In this decision the President is not bound to adopt the action proposed by the ITC.

The Escape Clause mechanism is considerably more political than the LTFV mechanism in two ways: on the one hand, the decision criteria are looser and more open to interpretation; and the President's freedom of action vis-à-vis the ITC's recommendation makes his decision a political act as well. Most previous research has been more concerned with the technical elements of the administered protection process, their flexibility, and the aggregate characteristics of cases in the high track; producing the major conclusion that in Escape Clause cases the ITC acts as a gate-keeper, allowing through the gate primarily politically significant cases requiring the President's attention. Unlike the low-track, however, political action directed toward the administered protection bureaucracy, the Executive and the Congress may also play a significant role. This makes the institutional structure of the Escape Clause as well as its broader institutional context quite important.

Given the visibility of the industries that are likely to make it through the Escape Clause process and the high degree of autonomy of the President in acting on the ITC decision, we would expect the politics surrounding that decision to be highly public. To the extent that the President and the industry in question differ in their preferences with regard to the final trade action (or non-action) taken, both sides can be expected to take their cases before the public in an attempt to affect the final outcome. It is in this context that the institutionalized bias is important. The effect of the bias imparted to the trade policy-making/implementation process can be seen in the politics surrounding the decision of the President to negotiate a VER on automobiles with the Japanese in 1981. Generalized public concern for trade policy in automobiles could be expected as a result of the central position of the auto industry in the American economy.[18] As a result, we would expect a

---

18. Not only is the auto industry, both directly and indirectly, a major employer (it has been estimated that as many as 1 job in 6 is related directly to the auto industry), but the

wide range of interests to be represented. Even consumers, whose abstention from the trade policy process can usually be expected on grounds of the relative cost of participation versus magnitude of gain, might be expected to participate given the size of costs involved per car.[19] Thus, one might expect that the very factors which make automobile trade policy a high visibility issue will tend to bring the largest possible number of participants (and thus positions) to the debate— including substantial anti-protectionist representation. However, given the logic developed above, we are led to hypothesize instead that the policy will be systematically biased toward protectionist outcomes.

Before presenting and analysing the data, we briefly review the background conditions to the case.[20]

The Auto Case

Though it experienced cyclical ups and downs, the U.S. auto industry was, during the 1950s and 1960s, extremely prosperous. In 1973, the auto industry had record sales and was enjoying decreased import competition because of the decline of the dollar. Under these conditions, the major concern of the U.S. industry was access to foreign markets. As a result, as recently as 1973, the major trade related policy

performance of the economy as a whole has been dominated by the auto industry. For example, see Emma Rothschild, *Paradise Lost: The Decline of the Auto-Industrial Age* (New York: Random House, 1973); and James Kurth, "Political Consequences..." The employment figures are reported in M.S. Salter et al., "U.S. Competitiveness in Global Industries: Lessons from the Auto Industry," in Scott and Lodge, eds., *U.S. Competitiveness in the World Economy* (Cambridge: Harvard Business School, 1985).

19. The traditional explanation for the exclusion of consumer interests from the political process is the relatively small cost to each consumer of protection. In the case of the auto VER, however, with a 3-5 percent increase in costs per car, this argument is of limited applicability. See Robert Feenstra, "Voluntary Export Restraint in U.S. Autos, 1980-81: Quality, Employment and Welfare Effects," in Baldwin and Krueger, eds., *The Structure and Evolution...*, pp. 35-61; Robert Crandall, "Import Quotas and the Auto Industry: The Costs of Protectionism," *The Brookings Review* 2:4, pp. 8-16; Tarr and Morkre, *Aggregate Costs...;* and the Federal Reserve Bank of New York, "The Consumer Costs of U.S. Trade Restraints" *Federal Reserve Bank of New York Quarterly Review* (Summer 1985), pp. 1-12.

20. Our primary concern in this paper is with the effects of the domestic regime on the public politics of protection-seeking; we present no detailed analysis of the economics of the auto industry. See Robert Cohen, "The Prospects for Trade and Protectionism in the Auto Industry," in W.R. Cline, ed., *Trade Policy in the 1980s* (Washington: Institute for International Economics, 1983), pp. 527-563. This section draws heavily on J.M. Finger and Paula Holmes, "The Auto Case: Autopsy of the Public Decision" (World Bank International Economic Research Division paper).

statement from industry sources was that Congress should adopt protectionist measures unless foreign barriers to U.S. auto exports were dropped.

Through its effect on U.S. consumers' tastes in cars as well as on general economic conditions, the oil price increases of 1974 marked a turning point for the U.S. auto industry. Though imports in 1974 were 20 percent below their 1973 level, the UAW, complaining of layoffs of about 50,000 workers, proposed a quota on imported autos "to save U.S. jobs."

Auto imports traditionally had held a 10-15 percent market share, but during the first half of 1975 this share approached 20 percent, sparking new appeals for protection. In mid-1975, the Treasury Department began investigating dumping charges filed by a Pennsylvania congressman and the UAW against auto imports from seven countries. The auto industry declined to take a position on whether it was being injured, but at an ITC hearing, the UAW claimed a 46 percent loss of sales due to imports. The European Economic Community responded to the investigation that any anti-dumping actions taken would adversely affect upcoming GATT talks. Within the U.S. government, the Council on Wage and Price Stability issued a statement that said not only was there no evidence of dumping, but that imports provided a moderating influence on U.S. prices, and the ITC investigation should be ended. The ITC asked the treasury to obtain additional data on the pricing practices of foreign firms.

In 1976, the Treasury Department decision (supported by the UAW) was that even though dumping existed, it would halt its probe and seek an amicable solution with the foreign companies involved rather than assess penalties. Japanese imports continued to increase, and in November the Labor Department ruled that workers in auto plants were eligible for adjustment assistance, even though total auto imports for 1976 were down 5.5 percent.

In early 1977, industry statements predicted that the increased competitiveness of U.S. product lines would hold imports to no more than 15 percent of the domestic market, but by April the import share was up to 20 percent. For the year, import sales were 40 percent higher than 1976. The UAW was, by the end of the year, calling for restrictions on Japanese imports unless Japanese firms made significant investments in U.S. production. This demand continued to figure prominently in the public statements of the industry (and especially labor) in 1978 and was made an explicit part of government policy when the U.S. Trade Representative pressured Toyota and Datsun to manufacture autos in the U.S.

1979 saw another dollar devaluation, another round of oil price rises, and worldwide inflation. Plans to invest billions of dollars overseas were announced by U.S. auto manufacturers (eg. GM $10-13 billion). By the end of the year, imports had a 22 percent market share, industry production had fallen to a million units below pre-oil shock levels, and more than 200,000 people in the industry were unemployed. Also by the end of the year, the industry became increasingly vocal in demanding some form of trade-related action and, as we demonstrate below, they began to focus their statements on the injury concepts necessary to an affirmative finding in the Escape Clause mechanism.

This strategy took concrete form in June and August of 1980 when the UAW and the Ford Motor Company filed Escape Clause cases with the ITC. Following the procedure outlined above, the ITC initiated both internal studies and public hearings to determine whether or not imports of cars and light trucks were a "substantial cause" of the problems being experienced by the industry. To rather general surprise, the ITC (on a 3 to 2 vote) announced in November, 1980 their determination that imports were not a "substantial cause" (understood to mean that no other cause is more important). The majority of the Commission believed that general recession in the U.S. was a more important cause of poor economic performance by the auto industry, and that shifting tastes of American consumers toward smaller, more fuel efficient cars was at least as important a cause of industry distress as Japanese competition. As a result of these findings, the ITC recommended that the Executive take no action against Japanese imports.

Shortly after the ITC announced its decision, the Presidency passed from Carter to Reagan bringing to the White House the closest thing to a doctrinaire economic Liberal likely to be produced by the American political system at the current time. Furthermore, the Reagan government proceeded to interpret their victory as a mandate for a Liberal economic program of de-regulation and reduction of government participation in the economy.[21] Nonetheless, four months after entering

---

21. A clear distinction should be made between economically Liberal and politically Liberal. In the context of American politics, political conservatism and economic Liberalism tend to go hand-in-hand. Thus, the capitalization of "Liberal" denotes the relationship to classical Liberalism of the Smith-Hume variety. Although candidate Reagan promised "to convince the Japanese... that the deluge of cars into the United States must be slowed," there is no reason to credit this as weighing any more heavily on him than other campaign rhetoric. The commitment to deregulation and reducing the non-military involvement of government in the economy is a far stronger hallmark of the Reagan government than is commitment to any specific campaign "promise." The above quotation is cited by Stephen D. Cohen and Ronald Meltzer, *United States International Economic Policy-making in Action* (New York: Praeger, 1982), p. 75.

office, the Reagan administration announced a three year VER program on automobiles.

The debate over trade policy for autos was very public, focussing not only on the Executive but, following the introduction in February by Senators Danforth and Bentsen of legislation to impose auto quotas (S. 396), on the legislature as well.[22] Although the ITC provided both the analysis and the evidence with which to make a strong Liberal case, we demonstrate below that the debate instead was carried out in protectionist terms and in an institutional structure systematically biased in favor of a protectionist outcome.

Legally and institutionally, the deck appeared to be stacked against the auto industry: the ITC found negatively in its injury decision; and the President (Reagan) was very publicly on record against government intervention in the economy. Nonetheless, the President risked a variety of domestic and international repercussions to negotiate a voluntary export restraint agreement (VER) with the Japanese government.

On the basis of a content analysis of New York Times reports on the public politics of the auto case,[23] I find that a key factor in this decision was the state of the public debate on trade policy (especially its strongly protectionist cast) and the institutionalized biases of the high track mechanism.

Results

The results of our content analysis support two major conclusions: first, the terms of the policy discourse are effectively framed by the institutional framework toward which the politics are ultimately directed; and, second, that the institutional framework imparts a protectionist bias to the politics of trade policy. Both of these results may appear to be *prima facie* unsurprising but they are far from standard in the literature on the political economy of trade. Most of the trade policy related analysis is institution free and even the institutional analysis tends to reason backward from the effects of the institution to institutional causation without any direct evidence. Thus, the direct evidence presented here may simply be taken as generally confirmatory.

The protectionist bias claim, however, is far from uncontroversial.

---

22. As Cohen and Metzler point out (*United States International Economic Policy-making...*), the outcome of such legislation was in considerable doubt at the time. On the one hand, the bill had attracted 21 co-sponsors in the Senate by May, but similar legislation had been defeated in the House (315-57) as recently as December 1980.

23. A detailed description of the data can be found in the appendix to this paper.

As I argue above, most of the institutionalist analysis emphasizes continued liberalization as evidence of a liberal bias. While my purpose is not to deny the operation of a liberal bias derived from the Lowi-effects, as I argue above, those effects produced a one-time downward shift in the level of protection and the rate of accomodation of protectionist pressure, but it eliminated neither pressure from nor accommodation of those forces. Change in the system emerges from a protectionist process which is institutionally mediated either through the administered protection mechanism or through attempts to alter the mechanism in Congress. It is this that makes an understanding of the operation of the system, and its biases so fundamental to any overall understanding of the political economy of trade policy.

Institutional Structuring of the Policy Discourse

Social institutions evolve, at least in part, to reduce the costs of social interaction.[24] Over time, societies evolve bodies of substantive and procedural norms for coping with the virtually infinite variety of conflicts that emerge as a result of social interaction. If a new set of rules had to be developed for every conflict, not only would the direct costs be enormous, but the result of the uncertainty imparted to the system would be an even larger cost. As a result, we institutionalize relatively general rules that treat specific conflicts as specific instances of general categories of conflicts. In addition to reducing the social costs of resolving conflicts, however, these institutions fix the terms of the discourse under which a conflict is resolved.

The normative system of a society is made up of a large number of more "local" normative orders.[25] These local orders are either more-or-less overlapping at the same level or they are hierarchically related. When the normative order at one level ceases to resolve the relevant conflict, it is passed to a higher level normative order for resolution. But, because the higher level normative order must handle a wider range of cases, the normative order must be more general.

24. This has become a standard proposition in the international relations literature (eg. the papers in Kenneth Oye, ed., *Cooperation Under Anarchy* (Princeton: Princeton University Press, 1986), but it has equal application to domestic social systems. See Nelson, "Notes on the Application of Social Choice Theory to the Political Economy of Development Policy" (World Bank manuscript, 1986), part I; and Nelson, "The State as a Conceptual Variable."
25. In the language of systems theory we can refer to "global," "regional," and "local" orders where the first refers to the system as a whole, the next to the major subsystems (eg. the economic political, or cultural systems), and the latter to more localized systems within the regional systems.

In these terms, we can characterize the autos case as follows:

1. Prior to the late-1970s conflict/competition in the U.S. auto industry was resolved according to the norms of (oligopolistic) market competition;
2. A sustained recession in the industry combined with increasingly effective Japanese competition in U.S. and third markets led (at least some) firms and labor to reject the market in favor of a political solution—that is, the normative order in the economy broke down and the issue passed to a higher level—an Escape Clause case was filed with the ITC;
3. When the outcome of that process was unacceptable to the industry, it was powerful enough to force the issue to a higher level.

If the above analysis is correct, we would expect the politics in period 1—the period preceding the ITC decision—to be organized in the terms of the relatively specific discourse defined by the Escape Clause legislation. In contrast, we would expect the politics in period 2—between the ITC finding and the announcement of the Japanese VER—to be more general than that leading to the ITC decision.

Our examination of the politics surrounding the auto VER with Japan is entirely consistent with this picture. Consider Tables 4.1 and 4.2 which tabulate the numbers of statements in which advocates and opponents of protection (by the U.S. unilaterally, or through a VER with Japan) explicitly justify their policy preference. The legislation under which the ITC operates in Escape Clause cases defines an industry's access to protection as a function of whether or not imports constitute a "substantial cause of serious injury," where that is taken to mean that no other cause of injury is more important than imports. Thus in the period prior to the ITC decision, both advocates (69 percent) and opponents (58 percent) of protection overwhelmingly stress injury related concepts in their public statements.[26] It is also comforting to note that the statements by each group are consistent with their presumed interest: over half (54 percent) of the statements by protection seekers stress imports as a cause of injury; while a similar proportion (51 percent) of the statements by opponents stress domestic sources of injury.

Once the ITC has made a finding, however, the process does not end. In the case of an affirmative finding by the ITC it is still necessary to ensure that the President acts on the Commission's recommendation for relief. This is a far from guaranteed result since the President has

---

26. The percentages are even larger if a broader sample is included—statements that the auto industry is in trouble that do not specify a specific cause—advocates 77 percent, opponents 71 percent.

tended to ignore the ITC recommendation. In the autos case the situation was even more serious from the point of view of protection advocates since the ITC decided against the petition for import relief. In either case, the political context is different following an ITC decision. Specifically, neither arguments nor actors permitted in the process are explictly defined. The next section explains why this does not result in an expanded field of actors in the process. Here we simply note the

Table 4.1 Causality Statements by Protection Advocates

	Frequencies			Percentage		
	Per 1	Per 2	Per 3	Per 1	Per 2	Per 3
Imports as Cause of Injury	56	15	0	53.85	26.79	0.00
Domestic Measures of Injury	3	1	0	2.88	1.79	0.00
Exogenous Sources of Injury	7	7	0	6.73	12.50	0.00
Endogenous Sources of Injury	6	8	0	5.77	14.29	0.00
Protection for Ind. Recovery	17	14	1	16.35	25.00	33.33
Avoid Worse Protection Later	9	7	2	8.65	12.50	66.67
Fair Trade & Trade Balance	6	4	0	5.77	7.14	0.00
ROW Retaliation	0	0	0	0.00	0.00	0.00
Classic Costs of Protection	0	0	0	0.00	0.00	0.00
Totals	104	56	3			

Table 4.2 Causality Statements by Protection Opponents

	Frequencies			Percentage		
	Per 1	Per 2	Per 3	Per 1	Per 2	Per 3
Imports as Cause of Injury	8	3	4	5.88	4.17	20.00
Domestic Measures of Injury	5	2	0	3.68	2.78	0.00
Exogenous Sources of Injury	33	26	0	24.26	36.11	0.00
Endogenous Sources of Injury	34	9	4	25.00	12.50	20.00
Protection for Ind. Recovery	0	1	0	0.00	1.39	0.00
Avoid Worse Protection Later	13	1	0	9.56	1.39	0.00
Fair Trade & Trade Balance	0	0	0	0.00	0.00	0.00
ROW Retaliation	15	8	1	11.03	11.00	5.00
Classic Costs of Protection	28	22	11	20.59	30.56	55.00
Totals	136	72	20			

effect on the form of the policy discourse. Thus protection seekers (Chart 4.1) in the period between the announcement of the ITC decision and the announcement of the Japanese VER (period 2) shift the emphasis in their public statements from an overwhelming focus on imports as a cause of injury to a broader range of sources of low industry performance, necessitating a period of import relief in which the industry could recover. Statements attributing the cause of poor industry performance to imports fall from 54 to 27 percent, those attributing the cause to domestic sources rise from 15 to 29 percent. These changes were accompanied by an increase from 16 to 25 percent of statements emphasizing the importance of protection to industry recovery.

Chart 4.1 Aggregation of Statements Supporting Policy Positions

---

I.   Arguments Relevant to ITC/Escape Clause Process
    A. Statements that Imports Injure the Industry
    B. Domestic Causes of Injury Exogenous to Industry Social Rel.
        1. Exchange Rate Changes
        2. General Recession
        3. Demand Shift (from large to small cars)
        4. High Interest Rates
    C. Domestic Causes of Injury Endogenous to Industry Social Rel.
        1. Government Auto Policy
        2. Bad Management
        3. Bad Labor Relations/Excessive Wages
    D. Derivative Causes of Injury
        1. Unemployment
        2. Declining Domestic Sales
        3. Domestic Business Losses
        4. Government Revenue Losses from Unemployment Compensation

II. Arguments for Protection
    A. Protection Necessary for Industrial Recovery
    B. Current Protection Will Avoid Worse Protection Later
    C. Fair Trade and Other Trade Balance Related Arguments

III. Arguments Against Protection
    A. Political/Mercantilist Anti-Protection Arguments
        1. Avoid Retaliation by Rest of World (ROW)
        2. Avoid Spread of Protection to Other Industries
        3. Avoid Effects of Protection on LDCs
    B. Classical Economic Costs of Protection
        1. Static Benefits of Price Competition
            -ie, Low Costs & Low Inflation
        2. Dynamic Benefits of Competition
        3. Product Variety and Quality

A similar shift is apparent in the public strategy of protection opponents. On the one hand, they increase their emphasis on the arguments in the ITC decision (ie. the effects of recession and demand shift on the industry) from 24 to 36 percent, arguing that protection will not help the industry. On the other hand, they also increase their emphasis on classical economic arguments against protection (from 21 to 31 percent) which stress the costs of protection to society as a whole. Thus, both groups shift away from positions defined by the Escape Clause mechanism toward more general positions of the sort appropriate to the more general political context.

Institutionalization of Protectionist Bias

Thus far, all that has been demonstrated is the not terribly controversial proposition that institutional structure affects the form of political discourse. We now argue that this structure imparts a systematic bias to those politics. The major mechanisms by which the structure biases the political process are selection-effects operating on who gets to participate; and agenda-setting, operating via the definition of what the politics are about, and the rules and terms under which the political process occurs. Each of these can be seen in the auto case.

Who Gets to Participate. Since the politics are defined as being a response to an industry crisis and not about protection as a general or industry specific phenomenon, prior to a finding by the ITC groups opposed to protection have no effective point of access to the system. Following the ITC finding, when the politics become more open and more general, the initial bureaucratic process will tend to fix the public agenda in such a way that anti-protectionist groups continue to be effectively excluded.

Table 4.3 shows the position of U.S. groups participating in the public debates. While the positions of industry and labor are extensively represented in the public debate, the general interest of consumers in liberal trading conditions is represented by parts of the executive branch (which tended to favor VERs while opposing restraints administered by the U.S.) and importers (whose representations tend to be discounted as those of foreign agents).[27] It is notable that the major intervention by consumer advocate Ralph Nader was to suggest that the auto industry stress to the American consumer that American made cars are safer than foreign cars.

---

27. Schattschneider, *Politics, Pressures and the Tariff,* pp. 159-162, gives several examples of the operation of this attitude toward importers during the hearings leading up to the Smoot-Hav ley tariff.

Table 4.3 Preference for Protection by Group

Statements By:	Favoring Per 1	Per 2	Per 3	Opposing Per 1	Per 2	Per 3
Industry	41	17	0	8	1	0
Labor	40	15	1	0	1	0
Importers	0	0	1	12	8	2
Citizens	8	2	1	3	1	2
Executive Bureaucracy	19	44	3	25	30	4
The President	3	7	0	14	2	0
Congress	22	29	5	1	7	2
Political Candidates	6	0	0	5	0	0
The Press/Analysts	2	0	1	11	11	2
Total	141	114	12	79	61	12

What the Politics Are About. A sustained recession in the auto industry caused a breakdown in the market-based norms that had previously regulated competition in that industry. The response of a large (and politically powerful) segment of the industry was to attribute its problems to decreased competitiveness vis-à-vis Japan and to seek import relief.[28] Even if we accept the general proposition that competitive problems with the Japanese were a source of industry distress, there are a variety of possible ways of perceiving and addressing the problem. Two alternative definitions of the industry's problems were mentioned at the time each of which implies forms of adjustment alternative to trade barriers:

1. Reallocation: parts of the industry had become non-competitive, requiring the reallocation of factors of production from the auto industry into some other industry. To the extent that there is a government role in this case, economic theory suggests that the role is in subsidizing the adjustment process directly; and
2. Restructuring: the social relations in which the industry is embedded (ie. labor relations, regulation, taxation, etc.) have caused problems for an otherwise competitive industry. Given the central role of the government in creating these social relations, there is an obvious role for the government in ameliorating their effects. Even if that role is not one of dismantling the social relations, it is clear that a direct response (via

---

28. Given the ITC's findings, the question of why the industry settled on trade as its first line of response to industrial crisis. The explanation probably has to do with minimizing conflict within the coalition of groups seeking a political response to the crisis.

some form of tax-cum-subsidy policy) is potentially a superior option to trade barriers.

The institutional norms, however, define an international competitiveness problem as being about unfair trade practices, injury, and an industry's *right* to protection. This is a right attached to production in the U.S. very much like other rights in property. Schattschneider's study of the politics surrounding the Smoot-Hawley tariff demonstrates the widely held acceptance of an industry's right to protection and the limiting effect of such acceptance on the debate surrounding the making of a tariff. This right has been redefined (in terms of the "no serious injury" norm) and is more constrained now, in the post-GATT era, by a more specific definition of the terms of access (under the LTFV and Escape Clauses), but it is still treated as a right. Furthermore, there is no parallel institutionalized right of access to international markets or protection from the costs of trade barriers.

The effect of this on the policy debate is clearly reflected in the data. Table 4.4 shows the numbers of statements made in support of each of the policy options. Over all three periods, policies aimed at reducing Japanese competitiveness dominate the discussion with two-thirds of the statements in periods one and two, and well over half in period three (ie. *after* Japan had agreed to limit auto exports voluntarily). Furthermore, direct barriers to trade (administered by either the U.S. or Japan) receive the most supportive statements—48, 64, and 43 percent of total statements, by period. Thus, the institutional structure reduces an industry crisis with multiple contributing causes first to an international trade problem and then to an issue of whether or not the industry has a right to protection under the Escape Clause legislation.

Order of Participation. In politics, as in many games, having the first move constitutes an advantage. This advantage results from knowing the form of the game prior to its start. In the politics of industrial crisis this advantage crops up many times, but the instance that concerns us here involves the politics of protection versus antiprotection. The labels alone tell us something important: the protection seeker is active, the opponent is reactive. Protection seekers are in a position to set and launch their political strategy before anti-protection forces know anything about the specifics of the strategy. There is no part of this mechanism, and no other parallel mechanism, in which opponents of protection/proponents of liberalization are active.

The effect of this structure on the public strategies of the protection seekers is apparent in the data. Prior to the ITC decision, as we have already suggested, the primary emphasis is on protection generally but with significant mentions of VERs, promotion of Japanese investment, adjustment assistance, and regulatory relief. In period 2, responding in part to the large number of statements in both the executive and legislative branches supporting such a policy, protection seekers focus their public position on protection generally and a VER particularly. The opponents of protection only partially adjust to this situation by intensifying their criticism of protection generally, but not of VERs particularly (see Table 4-4). Having achieved their goal vis-a-vis trade (ie. the VER), the protection seekers shift their strategy completely in period 3 to a much increased emphasis on adjustment assistance of the restructuring sort (with a particular emphasis on wage restraint)—increasing from 17 percent of the policy statements in period 1 to 37 percent in period 3. The opponents of protection, instead of adjusting to this shift, finally make the adjustment to the protection seekers' period 2 shift—they intensify their opposition to a VER. Over the three years covered by these data, then, the protection seekers have been able to keep one step ahead of the protection opponents.

It follows from the analysis of the structure of the technical and the political tracks, however, that anti-protection forces should be better organized for a response at the technical level because the terms of discourse are known in advance—reducing the advantage accruing to the protection seekers. The evidence that this is the case has already been presented. Our analysis of the data in Table 4.2 showed that protection opponents placed the emphasis in their public politics on precisely the issues defined by the Escape Clause legislation and, without imputing a directly causal effect to the lobbying effort, the ITC did determine against the protection seekers.[29]

Language of Discourse. The final institutionally borne bias we will consider is that deriving from the language of discourse. Without attributing transcendant importance to language, it is clear the language used in a political process will have an effect on the outcomes of that process. Our analysis of the public positions of the various actors in this system suggest that the language of discourse in the politics of trade policy is mercantilist in nature. Probably the best indicator of this tendency is the institutional evaluation of the effect of trade in terms of

---

29. Without attempting to oversell the effect of lobbying efforts relative to technical response to legislation, it should be noted that our own work (Finger, Hall and Nelson, "The Political Economy of Administered Protection") as well as that of Takacs ("Pressures for Protectionism...") suggests that the ITC is not unresponsive to political pressure.

injury to import-competing industry.[30] Standard trade theory evaluates the effects of protection by weighing the costs to consumers against the gains to industry[31] but, as we suggest above, the political system internalizes only the gains to industry—ignoring the costs to society from inefficient production and consumption.

With regard to opponents of protection, the great majority of their statements seem to accept that protection is potentially a cause of injury but argue that in the specific instance it is not a substantial cause.[32] The only criticisms of protection on general principles were rooted in protectionist logics as well. There were two of these that were made systematically: that domestic protection would lead to retaliation against U.S. exports by other countries (and Japan in particular); and that protection against Japan threatens political solidarity among industrial countries. Both of these rest on the notion that imports are the price a country has to pay to gain a desired objective: access to foreign markets in the first case and political stability (in opposition to Communist internationalism) in the second. It is also notable that in neither case were these arguments made by export industries that might expect to be threatened but, rather, were made either by the members of the executive branch or editorial writers.[33]

Implications for the Future

The substantive conclusions of this research—that institutions affect outcomes, and that the institutions in which the U.S. politics of trade

---

30. A similar bias exists at the international level where the politics of liberalization are carried on in terms or reciprocity. Liberalization (eg. tariff reduction) is concession which must be paid for by a concession of equal value by one's negotiating partner.

31. Max Corden, "The Calculation of the Cost of Protection," *Economic Record* 33, pp. 29-51.

32. Opponents of protection deny that imports are a source of domestic injury in only 10 percent of their policy-related statements in periods 1 and 2. In most of these cases, however, they state that imports are *not now* a cause of injury. They very rarely presume that imports are unlikely *ever* to be a source of injury.

33. One important exception is the opposition of General Motors, prior to 1981, to protection on the grounds that retaliation by other countries would undermine the export market for its cars. Too much should not be made of this point. As I.M. Destler and John Odell demonstrate, protectionist outcomes are still quite possible when export interests are well represented in the porcess. It is the argument of this paper that it is the institutional structure, not who comes into the institution, that biases the outcomes toward protection. See Destler and Odell, *Anti-Protection: Changing Forces in United States Trade Politics* (Washington: Institute for International Economics, 1987).

policy are embedded contain a protectionist bias—contain implications for the future of trade policy in the U.S. Two are of particular interest. First, that the system as currently structured is prone to catastrophic change; and second, that it may be possible to avert or limit such change through institutional reform. We briefly consider each.

The structure of the administered protection mechanism represented a balance between the executive's desire to pursue trade liberalization (often for generally non-economic reasons) and the demands of politically significant groups for protection. For a variety of reasons, considered briefly above, the capacity of the executive to impose that balance has declined.[34] The effect, as a cursory reading of the Washington Post demonstrates, has been to reopen trade policy as a legislative and highly political issue. It is in this context that the development of a strong protectionist orientation in the structuring of the politics of trade policy takes on particularly alarming overtones, because what is at risk is not a single case (even one as large as the auto case), but the entire institutional structure which protected the trade liberalization process even through the strong tests it received in the 1970s.

The point here is not that actual protection has risen. In fact, the evidence is that, for all the talk about the new protectionism, increases in protection (though not economically trivial) have been relatively modest.[35] We have already argued that this fact should be attributed primarily to the Lowi-effects embedded in the post-RTAA domestic regime for trade-related politics. That is, the domestic institutional structure which has evolved in response to U.S. commitments under the GATT has successfully absorbed the increasing protectionist pressures.

---

34. Nelson treats the political and economic causes of this change in the institutional and political structure in "Political Economy of the New Protectionism;" "The Welfare State and Export Optimism;" and "The Domestic Politics of Economic Cooperation."

35. The limited effect of increases in protection stands out even more clearly if we examine import penetration. Helen Hughes and Anne Krueger, "The Political Economy of Protection in Eleven Industrial Countries" in Snape, *Issues in International Trade Policy*... for example, present data of this sort with reference to manufactured exports from developing countries. This sort of evidence can be taken to mean that protection has no significant effect on trade (eg. Susan Strange, "Protectionism and World Politics," *International Organization* 39:2, pp. 233-259). This analysis makes two fundamental errors. The empirical error lies in an underestimation of the trade effects of administered protection. For general evidence on the economic effects of non-tariff barriers, see the research cited in footnote 15; for the specific case of autos, see the research cited in footnote 19. The evaluative error emerges from the failure to consider the total institutional structure of protection (i.e., the combination of Lowi-effects and selection/agenda-setting effects), thus mistaking marginal increases in protection for large increases, and misunderstanding the relationship between continuing protectionist pressure and the prospects for system transformation.

Table 4.4 Statements in Favor of Policy Positions

	Frequencies			Percentages		
U.S. Unilateral Protection	116	61	4	39.73	34.27	14.29
Japanese VER	25	53	8	8.56	29.78	28.57
Japanese Increase Auto Imports	7	3	1	2.40	1.69	3.57
Japanese Increase Auto FDI	33	3	2	11.30	1.69	7.14
Regulatory Relief	33	13	2	11.30	7.30	7.14
Tax Breaks	15	11	2	5.14	6.18	7.14
Wage Concessions	2	17	5	0.68	9.55	17.86
Adjustment Assistance	55	8	0	18.84	4.49	0.00
Free Trade	6	9	4	2.05	5.06	14.29
Totals	292	179	28			

Table 4.5 Statements Opposing Policy Positions

	Frequencies			Percentages		
U.S. Unilateral Protection	73	56	5	72.80	81.16	33.33
Japanese VER	6	5	9	5.94	7.25	60.00
Japanese Increase Auto Imports	1	0	0	0.99	0.00	0.00
Japanese Increase Auto FDI	7	0	0	6.93	0.00	0.00
Regulatory Relief	2	1	0	1.98	1.45	0.00
Tax Breaks	1	2	0	0.99	2.90	0.00
Wage Concessions	2	3	1	1.98	4.35	6.67
Adjustment Assistance	8	2	0	7.92	2.90	0.00
Free Trade	1	0	0	0.99	0.00	0.00
Totals	101	69	15			

The point is that the entire system is at risk. No institutionalized commitment to liberalization exists to counter-balance the rising protectionist pressure. The institutions through which change in the system occurs are so biased toward protection that as the system becomes increasingly politicized and, therefore, increasingly a legislative issue, the risk of a fundamental shift toward a pre-GATT type of system becomes quite large.[36] In simple terms, as protectionist pressure grows

36. Note that this argument is distinct from that presented by Miles Kahler, "European Protectionism in Theory and Practice," *International Organization* 37:3, pp. 475-502. Kahler argues that there is a politically significant "new, new protectionism" characterized by a willingness of governments to use general trade restrictions as an instrument

continuously, the administered protection mechanisms will absorb them up to some point at which the entire system will change discontinuously.

Against this rather bleak backdrop, our analysis does suggest possible grounds for limited optimism. The grounds for optimism lie in the finding that structure affects behavior, the reason for restraining optimism is the low likelihood of the government acting on the preceding grounds. It is a fundamental finding of the research reported in this paper that outcomes in the trade policy process are biased toward protection by the institutional structure of that mechanism. There is no *a priori* reason why the structures cannot be altered so as to shift the bias in another direction. One element of such a shift is to alter the rules of the existing mechanisms so as to explicitly consider the costs of protection in the decision process. This would not only have the effect of making an affirmative finding harder to get in the technical track but, more importantly, it would place the gains from trade explicitly on the agenda of whatever rounds of politics follow an ITC determination. In a sense, such an action institutionalizes the rights of consumers that parallel the rights of producers. A further step in this direction could involve the creation of a mechanism through which to file for relief from protection. In addition to institutionalizing the rights of those who benefit from imports, such a mechanism would allow anti-protectionists to play an active rather than a reactive role in the politics of trade policy.

These two actions alone, however, would be likely to have only a marginal positive effect and could conceivably have a substantial negative effect. The usual political asymmetries between protection seekers and protection opponents would remain. As in the auto case, a technical finding against a politically strong protection seeker does not stop the political process, it merely pushes the conflict to a higher/more general level. To the extent that the institutional changes in fact mobilize additional anti-protectionist forces this could lead to a major political crisis. At least part of the reason the administered protection mechanism has worked so well is its effectiveness at limiting the political action on trade.

Thus, Finger argues that a necessary part of any such attempt to change the terms of the policy discourse is:

---

of macroeconomic policy. The position taken here is that there is a risk of a discontinuous rise in protection as the limitations on an industry's "right" to protection are reduced to their 1920s level. That is, we should be worried about the old protection, as well as the new, new protection.

> "... to begin a process to create greater public awareness of the domestic costs of protection... and thereby to help bring into the public mind a non-mercantilist sense of the gains from trade."[37]

There is no doubt in my mind as to the *capacity* of the state to successfully undertake a program of reformulating the way the public conceptualizes trade policy. The executive branch successfully engineered a change in the widely held isolationist attitudes of the American people into widespread support for anti-communist internationalism in the late-1940s and early-1950s through a conscious policy of public relations. There are, however, good reasons for entertaining both substantive and philosophical doubts about the wisdom of pursuing such a policy. The substantive doubts relate to the magnitude of the public relations program necessary to accomplish the goal of fundamentally altering the way people view trade. This is a change of at least the same magnitude as that from political isolationism to internationalism and the engineering of that change required a massive program with extremely unpleasant social consequences. The philosophical doubts relate to whether the state in a democratic society has a right to undertake any such program.

Any effort short of such a total effort could backfire. This possibility rests on two findings of research on public opinion and political behavior. The first finding is that support for trade liberalization increases with education and income. The second finding is that political participation increases with education and income. This suggests that the most politically active people tend to support liberal trading relations. While a public relations program could change attitudes, it would also activate people predisposed to oppose liberalization. The total effect is uncertain.[38]

The most promising approach to this problem, which does not fundamentally undermine the democratic relations between the state and society, would involve the creation of institutionalized responses to industrial crisis which do not treat such a crisis as a problem of international trade. Instead of creating a parallel anti-protection mechanism with the political risks mentioned above, the suggestion is to institutionally redefine the politics of industrial crisis to focus more directly on the roots of competitiveness problems: the social relations of production within the economy; and between the state and the

---

37. Finger, "Incorporating the Gains from Trade into Policy," *World Economy* 5:4, p. 376.
38. Bauer, Pool and Dexter, *American Business and Public Policy,* pp. 92-96, make a similar argument.

economy. The fundamental practical problem with this, however, as with the propaganda blitz, is: What political actor is going to implement it?

Appendix: The Data

To directly evaluate the effect of institutions on the structuring of the domestic politics of trade policy and the terms in which the discourse was conducted, we content analysed *New York Times* (NYT) coverage of those politics. Holsti[39] defines content analysis as: "any technique for making inferences by objectively and systematically identifying specified characteristics of messages." In our case, the "messages" were any statements on the condition of the U.S. automobile industry and what (if anything) should be done about it, reported in the *New York Times* between 1 January 1979 and 31 December 1981. The inferential technique used was simple tabulation of statement counts. The results reported in this paper relate primarily to a subset of the larger data base made up of those statements expressing a specific policy position.

Before presenting more detail on the data set itself, an explicit statement of our interpretation of what these data represent may be useful. It is in the nature of the institutions with which we are concerned here that they are readily identifiable, and the first part of the paper presents our analysis of them. What we needed were reliable data on the positions of the major actors in the system. Content analysis provides us with just such information.

The analysis assumes that the statements reported in the NYT represent the positions of the speakers making them, and that the relative frequency of statements within any aggregation period indicates the salience of a position more-or-less accurately. Both of these assumptions require some defense. While the first of these is often a serious problem for content analytic studies, it is not for the research reported here. It is conceivable that some of the actors pursue one set of goals and strategies publicly and another secretly, but there is no evidence in either the academic or journalistic literature that this occurs on a significant scale. This does not mean that actors do not pursue private strategies (ie. "lobbying" in the old fashioned pejorative sense), nor that the public strategies are more or less important than the private

---

39. Holsti, *Content Analysis for the Social Sciences and Humanities* (Reading, Mass.: Addison-Wesley, 1969).

strategies. All it means is that public and private strategies, in the case of the politics of protection, seem to be consistent with one another.

The use of simple counts as a weighting device is considerably more problematic. Bias may enter the data at this point in a number of ways, we will consider two particularly important ones. The first is simple reporting bias. That is, the newspaper may favor some position and systematically over-represent it. This was one reason for selecting the NYT as our source. The NYT is not unbiased (such a thing as lack of bias is unlikely if not impossible), but its bias is clear. As reflected in official editorials in the period covered by the analysis, the NYT has consistently opposed barriers to trade of any kind and generally opposed any form of government intervention in support of the auto industry. Given that one of our central hypotheses is that the system is characterized by a protectionist bias, we would expect the liberal bias of the NYT to impart to the data a bias counter to that of our hypothesis (either by over-reporting statements of anti-protectionist groups or by reporting anti-protectionist arguments more often).

The second source of bias is inherent in using newspapers: no matter what unit of aggregation over time is used, time periods are unlikely to be strictly comparable. This problem is a function of the fact that the "newsworthiness" of any event is relative. Thus, the coverage that a given event receives will depend on what else is happening in the world. We control for this by focussing our analysis on percentages within periods rather than on absolute counts.

Starting with about 1,000 articles on the auto industry relevant to the policy debate, each containing one or more statements by identifiable actors, we isolated the 1,642 statements that constitute our raw data. Each statement was coded for the identity of the speaker in terms of nation (U.S., Japan, and Other), and group affiliation (see Table *4.3).* The content of each statement was coded in terms of statements of policy preference (see Table *4.4),* and arguments supporting some policy preference. The latter category includes statements about the condition of the industry, the cause of that condition, and the effect of a policy-related action. Statements in the latter category need not be made in conjunction with a statement of policy preference, however, the results reported in this paper all relate to the 672 statements of policy preference and the supporting statements contained in them. None of the results reported in the next section are altered by using the full data set and the explicitness of the link between policy position and supporting argument off-sets whatever loss may derive from the reduction in observations.

Chart *4.1* shows how the total of 30 specific categories of supporting statements were reduced to 9 general categories in 3 sets related to the policy process. The first set of statements all relate to the ITC's determination of whether injury exists in the industry and whether or not imports are a substantial cause of injury. A given statement could contain any number (including zero) of statements in this set, each of which were coded as:

1. The condition exists/is getting worse;
2. The condition does not exist/is improving;
3. The condition causes injury, and
4. The condition does not cause injury.

The statements in the second and third sets were coded simply as either accepting (0) or rejecting (1) the argument.

CHAPTER 5
The Origins of The U.S.-Japan Semiconductor Dispute

Jeffrey A. Hart

Introduction

A major trade dispute arose in 1986 between Japan and the United States over the pricing of Japanese semiconductors in U.S. and third-party markets. Japanese firms had begun to challenge the technological superiority of U.S. firms in the late 1970s. By the mid-1980s, they were out-producing and, in a few strategic product areas, out-innovating the U.S. firms. This chapter examines the rise of the Japanese industry and the trade conflicts which ensued.

There was rapid growth in the production of semiconductors in the 1970s and 1980s. This growth was greatly assisted by product innovations in consumer electronics, small computers, and telecommunications equipment and the rapid diffusion of these new information technologies. The lower GNP growth rates of the 1980s, however, together with rising unemployment, focused the attention of policy-makers in the industrialized countries on increased competition in world markets and especially in high technology products and services. Aside from two cyclical downturns in the growth of semiconductor production in 1982 and 1985, the industry continued to grow at a pace approaching twenty percent per year. In certain regions, such as the well known Silicon Valley of Northern California, employment in semiconductor manufacturing was the mainstay of the local economy.

The semiconductor industry was important not just as a source of income and jobs, however, but also as an input to military weaponry. Governments looked upon the industry in the same way they viewed the steel industry in the nineteenth century: as a key to national security. When a trade dispute arose between the leader, the United States, and the challenger, Japan, in the mid-1980s, it was bound to be colored by power political considerations.

Three major factors account for the rapid success of Japanese firms: government assistance in the form of the Very Large Scale Integration (VLSI) Program of 1976–9; the strong demand for advanced components that came from a highly successful consumer electronics industry; and the diversification and vertical integration of Japanese electronics firms, which was in sharp contrast with the mix of integrated and merchant firms in the United States. The VLSI Program was crucial in allowing Japanese firms to move to the technological frontier of semiconductor design and production. Preferential purchasing of Japanese components by the Japanese Ministry of Posts and Telecommunications helped Japanese firms descend their learning curves, just as defense and space program purchases of U.S. components had helped U.S. firms do the same earlier.

U.S. firms were slow to see the extent of the challenge posed by Japanese firms. As in other industries suffering from increased competition, the first response was to compensate for lower wage costs of Japanese manufacturers by using overseas manufacturing facilities. Unlike the Japanese firms, who were early to automate the more labor-intensive parts of the manufacturing process (i.e., bonding, assembly, and packaging), the U.S. firms used their overseas affiliates in Southeast Asia and Latin America to perform these tasks. In addition, Japanese firms were better able to get marginal improvements in their manufacturing of chips by the close cooperation between semiconductor and manufacturing equipment firms made possible by the more integrated nature of Japanese firms and their closer ties to suppliers and contractors. Not only were Japanese firms able to produce existing products less expensively than U.S. firms, but they also began to announce and market new products earlier than their U.S. competitors.

The early 1980s was a period of accelerating rates of investment in semiconductor production. The predictable result was overcapacity. When demand for semiconductors turned down in 1985, there was a very strong temptation for firms that had recently increased capacity, especially the Japanese, to dump their products on overseas markets. Learning curve effects associated with semiconductor production also created incentives to sell below production costs in order to increase volume, thus lowering future production costs. The logic of overcapacity and learning-curve pricing, therefore, led to very low prices and financial losses for almost all the major firms in 1985.

The U.S. merchant firms were hurt the most, and their first recourse was to turn to the U.S. government for shelter. The traditional method for sheltering U.S. firms was to use defense procurement to promote innovation in circuitry by paying premium prices for new devices for military use, allowing a number of firms to finance part of the development costs of these devices. However, by the mid-1980s, the type of circuitry demanded by the military was quite different from that used by civilian technologies. Thus it was not as easy as it had been in the past to subsidize development of new circuits through military procurement. In any case, the problems of the industry were too immediate to be dealt with through procurement policies alone.

Accordingly, the U.S. merchant firms turned to trade policy as a

Table 5.1   World Semiconductor Production, By Region, 1973-86
(in Billions of Dollars)

A. As Estimated by Dataquest

Year	U.S.	Europe	Japan	Rest	World
1973	3.6	1.1	1.3	0.0	6.0
1976	4.5	1.2	1.5	0.2	7.4
1978	5.8	1.7	2.5	0.4	10.4
1981		1.7			14.2
1982		1.9			14.7
1983		2.2			17.5
1985	11.0	2.8	10.7	0.3	24.8
1986	11.7	3.9	15.0	0.5	31.1
1987p	14.2	4.7	16.3	0.7	35.9

B. As Estimated by In-Stat

Year	U.S.	Europe	Japan	Rest	World
1983	7.8	3.0	5.5	2.0	17.3
1984	11.9	4.7	8.1	1.7	26.3
1985	8.1	4.5	7.6	1.2	21.6
1986	8.4	5.8	9.8	1.4	25.4
1987p	8.9	—	—	—	28.9

Note: Dataquest statistics include estimates of captive production of integrated circuits by large computer firms like IBM and In-Stat statistics include only sales of merchant firms.

Sources: a. for Dataquest—Giovanni Dosi, *Technical Change and Industrial Transformation* (New York: St. Martin's Press, 1984), p. 150; *Financial Times* (March 21, 1983), Section IV, p. IV; Special Supplement on Semiconductor Manufacturing and Testing, *Electronic News* (March 9, 1987), p. 5;

  b. for In-Stat—*Electronic News* issues of November 10, 1986 (p. 64), September 30, 1985 (p. 6) and October 1, 1984 (p. 40).

means of obtaining help from the U.S. government. Recent changes in trade law, particularly provisions concerning unfair trade practices and obtaining relief against "injury" caused by rapid increases in imports (whether or not they were the result of unfair trade practices), created a new set of policy instruments for the sheltering of U.S. firms from international competition.[1] U.S. firms were quick to petition the government to use these new instruments in a variety of trade disputes; the semiconductor industry was no exception. The resulting trade dispute, which peaked in 1985-6, is described in detail below, along with some speculations about its overall impact on U.S.-Japanese economic relations.

The Rise of Japan in World Markets

In 1984, world production of semiconductors was estimated to be around $26 billion and of integrated circuits (semiconductor devices which contain entire electronic circuits on a single chip) $9 billion. (See Table 5.1). The share of discrete devices (devices which are *not* integrated circuits) in the overall market for semiconductors has been declining steadily since the invention of integrated circuits in 1971.

Between 1978 and 1983, world production of integrated circuits grew at an annual average rate of 19 percent, despite a recession in 1981-2 (see Table 5.2 below). Between 1983 and 1986, production grew at an annual rate of 26 percent.

The United States accounted for over two thirds of world production of integrated circuits between 1978 and 1985, and more than that in earlier years. According to Dataquest estimates, the United States, Europe and Japan produced more than 93 percent of all semiconductors in the world market for the entire period and more than 98 percent since 1980 (see Table 5.3). Japan has increased its share of world production of integrated circuits from 18 percent in 1978 to 27 percent in 1985. However, if one considers only the open market for semiconductors—sales by merchant producers, excluding consumption of devices by captive producers—the Japanese production share increased from 24 percent in 1978 to 46 percent in 1986. The U.S. share fell from 60 percent in 1978 to 43 percent at the end of 1986.[2]

---

1. For further details, see Stephen Woolcock, Jeffrey Hart and Hans Van der Ven, *Interdependence in the Post-Multilateral Era* (Lanham, MD: University Press of America, 1985), chapter 1.
2. Michael Borrus, *Competing for Control* (Cambridge, MA: Ballinger, 1988).

Table 5.2   *World Production of Integrated Circuits, 1978-83*
              (millions of dollars)

A. As Estimated by Dataquest

Year	USA	Europe	Japan	Rest	Total
1978	4582	453	1195	382	6712
1979	6681	600	1750	675	9706
1980	9055	710	2450	130	12345
1981	8950	790	2590	160	12490
1982	9300	790	3130	160	13380
1983est	10450	855	3910	190	15405

B. As Estimated by In-Stat

Year	USA	Europe	Japan	Rest	Total
1984	6270	2200	4000	430	12900
1984	9800	3400	6500	1200	20900
1984	6610	2990	5650	1250	16500
1986	6850	3050	7300	2100	19300

Note: Dataquest statistics include estimates of captive production of integrated circuits
      by large computer firms like IBM and In-Stat statistics include only sales of
      merchant firms.
Sources: Dataquest—*Trade in High-Technology Products: Industrial Structure and
         Government Policies* (Paris: OECD, 1984), p. 110; In-Stat —Rebecca Day,
         "Worldwide Semiconductor Sales Predicted to Rise Only 15.8% in 1985,"
         Electronic News (September 10, 1984), p. 13; Richard Bambrick, "Semicon
         Sales to Rise Slightly: Buying Patterns in Transition," *Electronic News*
         (January 5, 1987), p. 30.

U.S. production of semiconductors grew at an average annual rate
of over 26 percent between 1955 and 1983 (see Table 5.4 below). The
proportion of semiconductors consumed by military or government
users declined from a high of 50 percent in 1960 to less than 25
percent in 1968; the proportion of integrated circuits consumed by the
military went from 100 percent in 1962 to 9 percent in 1978.[3] The
share of integrated circuits, as opposed to "discrete" devices, in total
semiconductor sales has increased steadily in the U.S. market from 1
percent in 1961 to 70 percent in 1984. The decline in the military
share and the rise of integrated circuits that are too advanced for the

3. Giovanni Dosi, *Technical Change and Industrial Transformation* (New York: St.
Martin's Press, 1984), p. 44.

Table 5.3 *World Production Shares in Integrated Circuits
(in percentages)*

Year	USA	Europe	Japan	Rest
1978	68	7	18	7
1979	69	6	18	7
1980	73	6	20	1
1981	72	6	21	1
1982	70	5	23	2
1983	68	6	25	1
1984	67	5	26	2
1985	67	5	27	1

Note: These production shares are based on Dataquest data. See original data through
1983 in Table 5.2.
Source: *The Semi-Conductor Industry: Trade-Related Issues* (Paris: OECD, 1985), p. 21.

current generation of military hardware makes it more difficult than
it was in the past to use military procurement policies to bolster the
competitiveness of U.S. firms. The leading U.S. firms are no longer
primarily defense contractors.

The increase in the Japanese share of world production is remarkable, but perhaps more important is the domination of markets for
the more advanced integrated circuits and especially the latest generation of random access memories (RAMs). By the end of 1979, the
Japanese firms controlled 43 percent of the U.S. market for 16K
RAM devices.[4] By the end of 1981, they supplied almost 70 percent
of 64K RAM devices in the open part of the US. market.[5] In 1984,
the Japanese firms introduced 256K RAM chips before a number of
major U.S. firms. It was estimated that Japanese firms controlled
over 90 percent of the market in 256K RAMs by 1986.

U.S. firms like Intel, Motorola, Hewlett-Packard and AT&T
(Western Electric) still dominated the market for microprocessors,
however, Japanese firms began to eat into this market as well in the
1984–5 period as they introduced their own "state-of-the-art" microprocessors. NEC and Hitachi were particularly strong in this regard;
NEC displaced Texas Instruments in 1985 as the number one seller of

4. Michael Borrus, James Millstein, and John Zysman, *International Competition in
Advanced Industrial Sectors: Trade and Development in the Semiconductor Industry*
(Washington, DC: Joint Economic Committee of Congress, 1982), p. 106.
5. Gene Bylinsky, "Japan's Ominous Chip Victory," *Fortune*, (December 14, 1981), p.
55.

*Table 5.4 Sales of Semiconductors in the United States, 1961-84
(in millions of dollars)*

Year	ICs	Discrete	Total	Growth
1955			39	
1956			89	128.2
1957			140	57.3
1958			202	44.3
1959			388	92.1
1960			532	37.1
1961	5	533	538	1.1
1962	10	525	535	-.6
1963	20	537	557	4.1
1964	51	617	668	19.9
1965	94	742	836	25.1
1966	173	905	1078	28.9
1967	273	787	1060	-1.7
1968	367	762	1129	6.5
1969	498	858	1356	20.1
1970	524	769	1293	-4.6
1971	534	623	1157	-10.5
1972	718	749	1467	26.8
1973	1421	1335	2756	87.9
1974	1767	1347	3114	13.0
1975	1712	1290	3002	-3.6
1976	2644	1667	4311	43.6
1977	2677	1686	4363	1.2
1978	3538	1973	5511	26.3
1979	4717	2484	7201	30.7
1980	6606	2483	9089	26.2
1981	6976	3333	10309	13.4
1982	7322	3407	10729	4.1
1983	7945	3695	11640	8.5
1984	11275	4725	16000	37.5

Source: *Electronic Market Data Book* (Washington, DC: Electronic Industries Association, 1982), Table 4.2; *Electronic Market Data Book* (Washington, DC: Electronic Industries Association, 1985), p. 134.

semiconductor devices in the world.[6]

Semiconductor Production by Specific Firms

Japanese and U.S. firms dominated the markets for semiconductors and integrated circuits in the early 1980s, as can be seen in Table 5.5 below. Only two European firms ranked among the top ten firms—Philips and Siemens—and those firms did so largely as a result of purchasing U.S. semiconductor firms. While Table 5.5 excludes consideration of captive production of semiconductors, which if included would bring AT&T into the list, nevertheless it gives an indication of shares in the open market for semiconductors and the ranking of firms. It also shows the fall from dominance of Texas Instruments, Motorola and Intel between 1982 and 1986.

Table 5.5 Largest Semiconductor Producers, 1982–6
(Rank Ordered by 1986 Revenues)

Name of Firm	Country	1982	1984	1986
NEC	Japan	1100	2350	2638
Hitachi	Japan	800	2140	2305
IBM (estimate)	USA	—	2000	—
Toshiba	Japan	680	1750	2261
Motorola	USA	1235	1729	2025
Texas Instruments	USA	1422	2390	1820
National Semiconductor	USA	746	1030	1478
Philips/Signetics	Netherleds.	500	1150	1356
Fujitsu	Japan	440	1070	1310
American Micro Devices	USA	358	920	—
Matsushita	Japan	—	920:	1233
Intel	USA	900	1629	991
Siemens	Germany	—	700	—
Gould	USA	318	435	—
Harris Corporation	USA	147	234	—

Note: Data are for fiscal years ending on the column year. There is substantial variation in the fiscal reporting systems used by different firms.
Source: Dataquest estimates.

6. According to Dataquest, NEC sold 1.98 billion dollars worth of semiconductors in 1985 compared with 1.76 billion for Texas Instruments and 1.85 billion for Motorola. See "NEC Tops a List," *New York Times*, (January 16, 1986), p. D4.

The Role of Differences in the Structure of Demand

A fairly large proportion of semiconductor production in the United States is sold on the open market by merchant firms. Generally speaking, a lower proportion of semiconductors is sold on open markets in Japan and Europe because the firms in those two regions tend to be larger and more vertically integrated than the U.S. firms. In addition, the end-use of semiconductors differs considerably among the regions. In the United States, the largest market for semiconductors is the one created by computer manufacturing. In Japan, the largest market for semiconductors, at least until quite recently, was created by consumer electronics. In Europe, consumer electronics and telecommunications equipment are the most important customers of the European semiconductor industry.

The structure of demand for semiconductors was a factor of considerable importance in the initial development of the industry in the three regions. In the early days of the U.S. industry, production was geared to military and space applications. It changed quite drastically when the computer industry displaced government purchasers as the largest source of demand. Computer applications of semiconductors generally required devices that were relatively complex, fast, and ran at cool temperatures. Industrial applications, which figured larger in the early development of the European semiconductor industry, required devices that could handle large amounts of power and that were reliable at high temperatures. Consumer electronics, which were the most important customers for the first Japanese semiconductor producers, generally required devices that used less power than either computer or industrial devices and that had the capacity to handle analog as well as digital signals, i.e., in radios, TVs, and video recorders. As a consequence of the different demand structures, the Europeans did well in power devices, the U.S. did well in developing microprocessors and computer memories with MOS circuitry, and the Japanese did well in CMOS circuits for watches, calculators, and consumer electronics items.[7]

In the mid 1970s, the Japanese perceived that the market was pushing them in the direction of specialization in devices for consumer

---

7. This argument is put forth in Michael Borrus, James E. Millstein, and John Zysman, "Trade and Development in the Semiconductor Industry," in John Zysman and Laura Tyson (eds.), *American Industry in International Competition* (Ithaca, NY: Cornell University Press, 1983); Giovanni Dosi, *Technical Change and Industrial Transformation* (London: Macmillan, 1984); and Franco Malerba, *The Semiconductor Business* (London: Frances Pinter, 1985).

electronics. Worried that production of consumer electronics would shift to the Third World while the U.S. would continue to dominate the world computer industry, a major effort was undertaken by MITI and the Ministry of Posts and Telecommunications (MPT) to promote the development of new devices more suitable for advanced information technology. The VLSI Program of 1976–9 was the result, one of the most successful examples of government promotion of technological development after World War II. A major shift occurred as a consequence of this intervention: the Japanese semiconductor firms were in a much stronger position *vis-a-vis* their U.S. competitors by the late 1970s.

The VLSI project was organized in March 1976 by MITI, MPT and the six major manufactures of semiconductors: Fujitsu, Hitachi, Matsushita, Mitsubishi, NEC, and Toshiba. The project was aimed at developing semiconductor technology for the next generation of computers. There had been an earlier program for large-scale integrated circuits (LSI) which failed because it had not anticipated U.S. innovations in computer technology. Accordingly, the VLSI project focused on manufacturing processes and problems of design of VLSI circuitry.

A substantial portion of the funding was spent for purchasing U.S.-made manufacturing and testing equipment. Much of the initial effort of the project went into "reverse engineering" this equipment so that it could be improved incrementally and produced by Japanese suppliers. Great improvements were made in photolithography and electron-beam technology, both central to the transfer to circuit designs to the surface of silicon wafers. By pooling the research efforts to the major firms and making the results available to all, the VLSI project saved each individual firm millions of dollars in research expenditures.

The savings in research expenditures and the full support and backing of the Japanese government facilitated a major increase in overall investment by Japanese semiconductor firms. Total investment in new plant and equipment rose from 116 million dollars in 1977, to 212 million in 1978, to 420 million in 1979. Japanese producers were further encouraged by a sudden increase in demand for standard memory devices in the United States, an increase which could not be handled by U.S. firms alone, in the beginning of 1978. Japanese integrated circuits jumped from 1 percent of U.S. consumption in 1976 to 8 percent in 1980.[8]

---

8. Michael Borrus, *Competing for Control,* chapter 5.

The Importance of Merchant Firms in the United States

The distinction between integrated and merchant firms in the semiconductor industry is important because of the absence of predominantly merchant firms both in Japan and Europe. All of the major Japanese firms—NEC, Hitachi, Matsushita, Mitsubishi, Sanyo, and Fujitsu—are integrated in the sense that they are major consumers of their own semiconductor production. Sales of semiconductors on the open market account for less than 20 percent of total sales for NEC and less than 10 percent for the rest.[9] NEC, Hitachi, and Fujitsu have become computer manufacturers primarily, while Matsushita, Mitsubishi, and Sanyo remain primarily manufacturers of consumer electronics. All of them are relatively diversified, however, compared to the merchant firms of the United States.

Semiconductors accounted for over 80 percent of total sales for National Semiconductors, AMD and Mostek, and more than 65 percent for Fairchild and Intel. Texas Instruments and Motorola, the two largest firms in the merchant group, were the most diversified in the sense that they both have kept semiconductors in the range of 30-40 percent of total sales. Texas Instruments branched out into consumer products like calculators and personal computers, while Motorola remains a major producer of communications equipment and consumer products. TI tried to break into the mini- and microcomputer markets as well, although the microcomputer effort was a disaster. There is some evidence that other merchant firms, Intel for example, have been trying to integrate downstream into computers, starting with add-on circuit boards for IBM-PCs and PC-clones and with advanced work stations for the computer industry.[10]

Besides downstream vertical integration, most of the important merchant firms began to use foreign subsidiaries, mainly in Southeast Asia, but also in Latin America, to reduce their production costs for portions of the production process. In the late 1970s, Intel, for example, after separating chips on silicon wafers, sent them to its overseas affiliates for insertion in and bonding to the plastic or ceramic packages that protect them from heat and dust. The partially assembled integrated circuits would then be shipped back to the

---

9. Michael Borrus, James E. Millstein, and John Zysman, "Trade and Development in the Semiconductor Industry," in John Zysman and Laura Tyson (eds.), *American Industry in International Competition* (Ithaca, NY: Cornell University Press, 1983), p. 190.

10. See also "Intel may soon compete with its customers," *Business Week*, (March 22, 1982), p. 63.

United States for final assembly and testing. After 1982, most merchant firms added final testing to their overseas activities. The use of overseas affiliates was aimed not just at reducing labor costs but also at making it possible to maintain steady production levels and workforces in the U.S. despite fluctuations in world demand.

This strategy of overseas or "offshore" production was adopted also by AMD, National Semiconductors, Texas Instruments, Motorola and others. The Japanese firms, however, opted for automation of assembly and, for the most part, avoided the use of overseas subsidiaries for manufacturing. It has been argued that this choice of domestic automation over foreign investment was useful to the Japanese firms in the next round of competition, since they were able to gain valuable knowledge about how to improve the overall production process and, in particular, to increase the reliability of their products. But in fact, the U.S. firms that remained competitive with the Japanese were not handicapped by their overseas operations. Their subsequent efforts to automate the production process were carried out within the U.S. initially but with the full intention of applying the new processes overseas as soon as possible.[11]

The Role of IBM and AT&T

The integrated electronics, communications and consumer products firms in the United States have realized that semiconductor production is crucial to their competitiveness. While they continue to purchase a large proportion of their needs from merchant firms, most of them also have developed internal production lines, some of which are also sold on the open market. Hewlett-Packard, a company known for its industrial electronic products as well as for its calculators and small computers, is now one of the leading producers of products and production equipment for very large scale integrated (VLSI) circuits.

Among the mainframe computer manufacturers, IBM stands out as the sole major producer of advanced semiconductor devices. IBM employees claim that IBM semiconductors are second to none in quality and that its production technology is the best, at least in the United States. IBM announced in 1986 that it had begun producing 1 Megabit DRAMs, the first U.S. firm to match the Japanese firms in this area.

---

11. Dieter Ernst, "Automation, Employment and the Third World—The Case of the Electronics Industry," *ISS Working Page No. 29* (The Hague: Institute of Social Studies, November 1985).

But IBM has clearly felt a need to purchase certain devices on the open market, either because they are cheaper or because of fluctuations in internal demand. IBM's purchasing of 64K dynamic RAMs on the open market in 1979 contributed significantly to the rapid growth of Japanese penetration of the world RAM market. U.S. firms were taken by surprise and were not able to increase capacity as rapidly as Japanese firms. Also, because IBM has a philosophy of disarming its critics abroad by appearing to be a "good citizen" in each country in which it operates, the firm frequently purchases components and peripheral devices from national champion firms. Thus, IBM is likely to buy semiconductors from Inmos in Britain, Siemens in Germany, SGS in Italy, Thomson in France, and NEC in Japan. These components are frequently produced under "second source" or licensing arrangement with U.S. firms like Intel. Thus, IBM's desire to avoid political attacks abroad reinforce the tendency of U.S. firms to make second source agreements rather than export their products directly.

Next to IBM, the most important captive producer of semiconductors was AT&T. After the divestment of the regulated regional monopolies in 1984, AT&T began to diversify in the direction of computer and telecommunications equipment manufacturing. Now freed to compete on world markets for computers and telecommunications equipment, AT&T began paying more attention to its semiconductor research and production, which had alway been considered one of the most advanced in the world. It was not without significance that AT&T was one of the first firms in the world along with IBM to announce the successful production of a 1 Megabit DRAM device.[12] For the mid-range future, however, AT&T would probably focus most of its efforts on production and marketing of computers and telecommunications equipment.

The Vulnerability of Merchant Firms to Acquisitions

The U.S. merchant semiconductor firms have become increasingly candidates for acquisition by more diversified or cash-rich firms, and especially European firms. Xerox bought a major stake in Zilog in 1974, Philips purchased Signetics in 1975, Siemens aquired 20 percent of AMD in 1977, Schlumberger bought Fairchild in 1979, IBM

---

12. Michael Schrage, "AT&T Starts Production of Megabit Chip," *Washington Post*, (September 6, 1985), p. 83.

bought a 20 percent stake in Intel in 1982–4, which it sold back to Intel in 1987, and Thomson bought Mostek in 1985 (see Table 5.6). The only two important European merchant firms, Inmos and SGS-ATES, were purchased respectively by Thorn-EMI in 1984 and Thomson in 1987. The attempt by Toshiba to purchase Fairchild semiconductor operations from Schlumberger in 1987 was blocked by the U.S. government. So far this has not reduced the aggressive innovative spirit of the smaller firms, but it may eventually pose such a threat if the trend continues.

Table 5.6 Mergers and Acquisitions in the Semiconductor Industry, 1969–86

Date	Acquirer	Acquired Firm	Price ($mill.)	Equity (%)
1969	Northern Telecom	Monolithic Memories	—	12
1972	Texas Instruments	TI/Sony Japan	—	100
1972	Toyo Electronics	Exar Integrated	—	53
1974	Siemens	Dickson	—	100
1974	Exxon	Zilog	—	80
1975	Philips	Signetics	49	100
1976	Commodore	MOS Technology	1	100
1976	Signal	Semtch	—	23
1977	Commodore	Frontier	—	100
1977	Siemens	Litronix	16	80
1977	Seiko	Micropower Systems	—	100
1977	Siemens	AMD	27	20
1977	Ferranti	Interdesign	4	100
1977	Ferranti	Interdesign	4	100
1977	Thomson-CSF	Sescosem	—	100
1977	Lucas	Siliconix	6	24
1977	Bosch	American Microsystems Inc	14	14
1977	Standard Oil of IN	Analog Devices	—	100
1978	NEC	Electronic Arrays	9	100
1978	Emerson Electric	Western Digital	—	100
1978	Honeywell	Spectronics	3	100
1978	Honeywell	Synertek	2	100
1978	Bourns	Precision Monolithic	—	96
1979	Schlumberger	Fairchild	397	100
1979	Siemens	Databit	25	100
1979	Siemens	Microwave Semiconductor	—	100
1980	VDO Adolf	Solid State Scientific	5	25

Table 5.6 Continued

1980	CIT-Alcatel	Semi Process Inc.	—	25
1980	General Electric	Intersil	11	100
1980	Toshiba	Manuman IC	—	100
1980	Siemens	Threshold Technology	—	100
1980	United Technologies	Mostek	345	93
1981	Gould	American Microsystems Inc	—	100
1981	Olivetti	VLSI Technology	2	8
1981	Olivetti	Linear Technology	2	7
1981	Olivetti	Applied Microcircuit	1	4
1982	United Technologies	Eurosil	—	85
1982	United Technologies	Telefunken Elektronik	35	100
1982	IBM	Intel	250	12
1983	Philips	Vcatec	—	100
1983	Thomson-CSF	Eurotechnique	—	49
1984	Thorn-EMI	Inmos	125	76
1984	IBM	Intel	600	20
1985	Kawasaki Steel	NBK	9	100
1985	AT&T	Synertek	25	100
1985	Thomson-CSF	Mostek	70	100

Sources: Michael Borrus, James Millstein, and John Zysman, *International Competition in Advanced Industrial Sectors: Trade and Development in the Semiconductor Industry* (Washington, DC: Joint Economic Committee of Congress, 1982); Rob van Tulder and Eric van Empel, "European Multinationals in the Semiconductor Industry: Their Position in Microprocessors," unpublished manuscript, University of Amsterdam, Vakgroep vor Internationale Betrekkingen, October 1984; Philippe Delmas, "Le Cow-Boy et le Samourai: Reflexions sur la Competition Nippon-Americaine dans les Hautes Technologies," Ministere des Relations Exterieures, Centre d'Analyse et de Prevision, Paris, January 1984; Business Press.

Trade in Semiconductors: U.S. Deficits, Japanese Surpluses

Analyzing the trade in semiconductors is somewhat complicated by the need to compensate for the fact that many U.S. firms exported semiconductor "parts and accessories" to overseas assembly facilities in Europe, Latin America, and Southeast Asia, and then reimported the assembled devices for sale both in the U.S. and abroad. This practice was encouraged under provisions of the 1974 Trade Act which permitted U.S. firms to import duty free items which had been sent abroad for processing or assembly.

Most U.S. exports of finished integrated circuits went to Britain, France, the Federal Republic of Germany, and Japan. The U.S. had a positive trade balance in semiconductor parts and assembled products of 126 million dollars in 1977. The surplus in semiconductors was around 600 million dollars in 1980. By 1984, it was estimated that semiconductor trade produced a deficit of almost 3 billion dollars.[13]

Japan rapidly went from being a net importer of integrated circuits to a net exporter in 1979 (see Table 5.7 below). Even the United States became a net importer of integrated circuits from Japan, with a deficit in 1984 of $900 million. U.S. firms began to complain loudly about the unfair pricing practices of Japanese firms, as RAM prices dropped faster than anyone had expected. Even though U.S. firms still dominated the markets for certain types of integrated circuits, such as microprocessors and ROMS (read-only memories), the RAM devices were an important source of profits and therefore of research and development funds, especially for the more specialized semiconductor firms. These firms found themselves increasingly squeezed from two directions: loss of market share and inability to put money into developing new types of circuits.

Table 5.7 Japanese Trade in Integrated Circuits, in Billion Yen

Year	Exports	Imports	Balance
1973	2.6	33.2	–40.6
1974	6.7	51.1	–44.4
1975	13.5	40.0	–26.5
1976	22.7	62.7	–40.0
1977	31.6	55.7	–24.1
1978	52.3	61.3	–9.1
1979	108.3	98.5	9.8
1980	183.3	108.9	74.4
1981	199.6	114.3	85.3
1982	285.1	127.4	157.7
1983	418.0	144.0	274.0

Source: 1973–77, Daiwa Securities, as cited in Economic Research Associates, *EEC Protectionism* (Brussels, 1982), p. 222; 1978–83, *Nomura Electronics Handbook 1984* (Tokyo: Nomura Securities Ltd., 1984).

A general downturn in the computer industry in 1984 led to a slashing of semiconductor inventories by 36 percent in 1985. The demand for semiconductors declined sharply and producers responded by cutting

13. Borrus, *et al.*, 1982, p. 49; data from the American Electronics Association as cited in "America's High Tech Crisis," *Business Week*, (March 11, 1985), p. 69.

prices in order to compete for the remaining demand.[14] But besides this general drop in demand corresponding to a cyclical downturn in computers and other types of electronic equipment, a general overcapacity problem had been developing in world markets. One research firm estimated that by late 1985 worldwide demand was approximate 40 percent of production capacity in the semiconductor industry.

Overly ambitious sales projections and government programs designed to aid weaker firms led to an "orgy" of capital spending in the early 1980s. Chipmakers invested 6 billion dollars in plant and equipment in 1984 (remember that total sales during that year were around 21 billion). They invested another 4.5 billion in 1985 despite the turndown in demand.[15] Unless demand recovered in an unprecedentedly spectacular way, there would continue to be a crisis of overcapacity leading to pressures for capacity reduction. The key question politically was where the capacity reductions would occur and who would pay the cost of the reductions.

The Trade Dispute Begins

Important U.S. firms like Intel, Texas Instruments, Motorola, and AMD were losing money and dropping production lines in certain products. Mostek was nearly liquidated before its purchase by Thomson-CSF in 1985. Even the computer and telecommunications equipment manufacturers in the U.S. were beginning to worry. Their interest in being able to buy cheap components had to be weighed against their interest in being assured access to the most advanced devices (particularly worrisome in light of the growing strength of Japanese computer and telecommunications firms).

In 1985, employment at U.S.-based semiconductor companies decreased by 55,000 workers. The industry as a whole suffered a loss of 1 billion dollars.[16] In June 1985, a small firm called Micron Technologies headquartered in Boise, Idaho, filed an anti-dumping suit against Fujitsu, Hitachi, Matsushita, Mitsubishi, NEC, Oki and Toshiba. It asked that countervailing duties of up to 94 percent be

---

14. John Wilson, "The Chips May Not Be Down Much Longer," *Business Week*, (December 16, 1985), p. 26.
15. Bro Uttal, "Who Will Survive the Microchip Shakeout," *Fortune*, (January 5, 1986), p. 82.
16. Intel Corporation, *Annual Shareholders Meeting Report,* (April 16, 1986), Figure 11.

imposed on these firms retroactively for dumping (selling below the cost of production) 64K RAM devices. Although a number of members of the Semiconductor Industry Association (SIA) supported the Micron suit, the SIA as a whole remained neutral.[17]

A few days later, however, the SIA filed a Section 301 complaint against Japan claiming that they had been denied access to the Japanese market, repeating their earlier charges that the Japanese government had targeted the semiconductor industry and that U.S. firms were suffering the consequences. Apparently, the draft version of the Section 301 complaint called for import restrictions against Japan until U.S. firms were granted access to Japanese markets, but IBM and a number of other larger firms opposed this despite the fact that Intel, AMD, Hewlett-Packard and some of the other mechant firms had favored either import restraints or countervailing tariffs, so the final version did not include this demand.[18]

Table 5.8 Unit Prices for 256KD RAMs

Date	Pirce ($)
Jan 84	38.00
Apr 84	21.00
Jul 84	23.50
Oct 84	17.50
Jan 85	14.00
Apr 85	9.75
Jul 85	4.75
Oct 85	2.75
Jan 86	2.10
Apr 86	2.25
Jul 86	2.30
Oct 86	5.00

Source: Dataquest as cited in *Infoworld*, (February 3, 1986), p. 1, and *Infoworld*, (September 22, 1986), p. 1. the October 1986 figure is an estimate.

On September 30, 1985, Intel, AMD and National Semiconductor filed an anti-dumping complaint against eight Japanese firms for dumping EPROMs (eraseable programmable read-only memories).

17. Andrew Pollack, "Japan Seen Target of Chip Plea," *New York Times*, (September 28, 1985), p. 21.
18. Jack Robertson, "SIA Bid to Hit Japan on Trade Disputed," *Electronic News*, (June 24, 1985), p. 1; "The Bloodbath in Chips," *Business Week*, (May 20, 1985), p. 63.

The complainants claimed that the Japanese were selling these devices at 77 to 227 percent below fair value, and that production costs were at least 6 dollars per device while U.S. selling prices were 4-5 dollars.

The International Trade Commission ruled that the U.S. industry had been injured by the trade practices of the Japanese firms in all three cases. The ruling on 64K RAMs was made in August, on EPROMs in November but a ruling on 256K (and above) RAMs was made in January 1986 after an unusual and unprecedented intervention in the process by the President and the Secretary of Commerce. Apparently, the Reagan Administration became convinced of a need to accelerate the process behind the RAM complaint and to change the nature of the complaint somewhat to provide greater bargaining leverage with the Japanese government.

On December 16, 1985, Secretary of Commerce Malcolm Baldridge announced that the Department of Commerce was initiating its own investigation into the possible dumping of 256K RAMs at the request of the President. The Japanese government responded to the changed mood in Washington first by sending MITI officials to meet with industry representatives on January 20, 1986. At this meeting, MITI offered to establish floor prices for devices sold by Japanese firms in the United States. The U.S. firms rejected this offer claiming that it would still allow the Japanese to dump in third country markets and thereby give U.S. equipment firms large incentives to locate their production outside the United States. In addition, they claimed that floor prices would violate antitrust laws. What they wanted, they said, was for Japan to stop dumping on a worldwide basis.[19]

Another Japanese response to the trade dispute was for the firms to raise prices independently.[20] Hitachi also announced a special program to increase imports of electronic components and other items in the United States and to increase contributions to the U.S.-based Hitachi Foundation. But most U.S. observers considered this to be mere window dressing. The appreciation of the yen against the dollar in the first months of 1986 was expected to help somewhat in reducing trade tensions overall, but not much relief could be expected in semiconductors because the underlying source of the dispute was the global overcapacity which resulted from an investment boom in the late 1970s and early 1980s.

19. Jack Robertson, "Japanese Officials Visit IC Cos. on Dumping," *Electronic News*, (January 20, 1986), p. 12.
20. Susan Chira, "Japanese Raising Chip Prices," *New York Times*. (December 4, 1985), p. D1.

On March 14, 1986, Commerce ruled that Japanese firms had indeed dumped 256K RAMs and 1 Megabit RAMs and that the dumping margins for at least two firms, Mitsubishi and NEC, exceeded 100 percent. Commerce had ruled similarly on 64K RAMs in January, so the second ruling was not much of a surprise. Nevertheless, the conversion of the Section 301 complaint into an anti-dumping complaint and the speed with which the two anti-dumping investigations were carried out signalled the intent of the Reagan Administration to make trade in semiconductors a major thrust in its trade diplomacy with Japan.[21]

In late May 1986, the ITC decided to impose countervailing duties on Japanese semiconductor firms, some as high as 35 percent over the current seeling price of certain devices. On May 27, the ITC decided that Micron Technology had suffered economic injury as a result of sales of Japanese 64K dynamic RAMs on the U.S. market because of the severe downward effect of those sales on prices and profits. The six major Japanese producers were named in the ruling.

The U.S.-Japanese Semiconductor Trade Agreement of 1986

In late June 1986, the U.S. Trade Representative and MITI reached a framework agreement on the semiconductor trade issue. The agreement beat the deadline of July 12, after which the USTR would have been forced to impose new penalties and sanctions under the 1974 Trade Act. MITI agreed to adopt measures to raise U.S. firms' share of the Japanese market from 10 to 20 percent in exchange for the dropping of antidumping and Section 301 petitions against Japan. In addition, MITI agreed to help administer a floor-price system based on "fair market value" (FMV). The specifics on the agreement were left to later negotiations. There remained the problem of what to do about the previous antidumping and injury rulings by the ITC and the Department of Commerce.

The U.S. semiconductor industry received the news of this agreement with some skepticism. They were concerned about several issues: 1) the method for establishing fair market value, 2) the treatment of third parties to which semiconductors might be sold at lower than

---

21.Clyde Farnsworth, "U.S. Plans Inquiry on Japanese Chips," *New York Times*, (December 7, 1985), p. 43; Stuart Auerbach, "Tougher U.S. Stance Seen On Chips," *Washington Post*, (December 5, 1985), p. E3; "Cutting Rough with Japan's Chip Makers," *The Economist*, (January 11, 1986), p. 59; Clyde Farnsworth, "New Chip Ruling Goes Against Japan," *New York Times*, (March 14, 1986), p. D2.

FMV, and 3) the includion of other devices besides 64K and 256K RAMs and EPROMs in the agreement. There continued to be conflict between the merchant semiconductor firms and the industrial consumers of semiconductors (mostly computer and electronics firms) about the terms of the agreement. The consumers wanted to maintain their right to purchase devices at low prices and worried that Japanese integrated firms would have an advantage over them if they could not. They were particularly anxious to exclude 1 Megabit DRAMs from the FMV price system.[22]

On July 31, 1986, the U.S. and Japan concluded negotiations for a semiconductor trade agreement. In that agreement, Japan agreed to open its market to further participation by U.S. firms, the FMV price system was to be established and administered by the U.S. Department of Commerce in collaboration with MITI, and the U.S. dropped the antidumping and Section 301 complaints in exchange for guarantees that the Japanese firms would not dump in world markets.

The immediate effect of the agreement was to raise EPROM and DRAM prices dramatically. By late September, 256K DRAM prices had increased from $2.25 to about $5.00 per device.[23] Makers of printed circuit boards for computers and electronic equipment threatened to move their board assembly operations overseas where prices of components could not be so closely monitored. Part of the problem may have been the inaccuracy of the prices established by the Department of Commerce for the FMV system. The American Electronics Association and the Semiconductor Industry Association worked together to provide data to Commerce for the October 15 revisions of the system, so as to bring prices down to more realistic levels. By the end of the first quarter of 1987, 256K DRAMs dropped again to about $4 per unit.

In October 1986, the European Community began to object strenuously to the semiconductor agreement between the U.S. and Japan, claiming that it violated the fair-trade rules of the GATT. On October 8, the European Community requested that the GATT undertake an investigation of the legality of the agreement. The Europeans objected in particular to the provisions for better access to the Japanese market,

---

22. Michael Shrage, "Semiconductor Industry Reacts Warily to Accord with Japan," *Washington Post*, (May 30, 1986), p. F3; Jack Robertson, "Say 6 Mfrs. Seek Inclusion of ASICs in Japan Trade Pact," *Electronic News*, (June 16, 1986), p. 55; Jeff Moad, "Clash of Chip, Systems Vendors Led to Sanctions Compromise," *Datamation*, (June 1, 1987), p. 17.

23. Tom Moran, "Chip Pact Said to Imperil Board Assembly in U.S.," *Infoworld*, (September 22, 1986), p. 1.

suggesting that compliance with the agreement might occur at the expense of European producers.[24]

The Breakdown of the Semiconductor Agreement

Japanese firms began to complain in the fall of 1986 about the relative advantage given to Korean and European firms by the FMV system. They contended that doing the paperwork for administering the FMV system was raising their production costs. They suggested that stabilizing prices would remove incentives to innovate.[25] U.S. firms began to complain about Japanese dumping in third markets and about noncompliance with the FMV system in the U.S. market. In November, the U.S. government warned the Japanese government that dumping in third countries would result in the termination of the July agreement.[26]

In mid March 1987, MITI asked Japanese producers to cut production by 10 percent in an effort to reduce price cutting in third markets. It also tightened up its export licensing system to make it harder to send small batches of semiconductors through third parties. These efforts did not satisfy the SIA or the U.S. government that the Japanese government was serious about living up to the July 1986 agreement. Access to the Japanese market had not improved and third-country dumping continued.[27]

On March 23, 1987, the Senate Finance Committee passed a nonbinding resolution by voice vote calling on the President to retaliate against Japan for failing to live up to the semiconductor trade agreement. On March 27, 1987, President Reagan announced that $300 million in trade sanctions would be imposed on Japanese firms for violating the July 1986 agreement and for restricting access to the Japanese market. The sanctions affected only some Japanese consumer electronic products, power tools, and desktop and laptop personal computers but not semiconductors. U.S. computer and electronics firms wanted to avoid increased input costs for Japanese components and hoped also to avoid direct retaliation against U.S. products in Japan. The SIA agreed to the sanctions to placate the

---

24. "Europeans Protest on Chips," *New York Times*, (October 9, 1986), p. 36.
25. Susan Chira, "Japanese Uneasy on Chip Pact," *New York Times*, (August 2, 1986), p. 17.
26. Clyde Farnsworth, "Japan to Cut U.S. Textile Exports," *New York Times*, (November 15, 1986), p. 17.
27. Jiri Weiss, "Japan Asks Chip Makers for 10% Cut in Production, Tightens Regulations," *Infoworld*, (March 23, 1987), p. 25.

various computer and electronics industry associations, feeling that their message would get across in any case.[28]

The President's move increased the tension in an already strained relationship with Japan. The Japanese government threatened to retaliate if the trade sanctions were actually implemented (the President had given the Japanese government a few weeks to respond).[29] In the end, no agreement could be worked out and sanctions were imposed on April 17.

The Rise of Sematech

One could see the U.S.-Japanese trade dispute of 1985-7 as an initial battle in what might become a much wider trade war. A Defense Science Board Task Force on Semiconductor Dependency was convened in mid February 1986 to assess the "impact on U.S. national security if any leading edge of technologies are no longer in this country." The executive secretary of the Task Force, E.D. (Sonny) Maynard, was also director of the Department of Defense's Very High Speed Integrated Circuits (VHSIC) program. The task force also included representatives from a variety of electronics and defense-oriented firms, a former Undersecretary of Defense, a former Undersecretary of Commerce, and the director of the National Science Foundation.[30] Several of the reports done for the Task Force were so depressing and controversial that they were classified.[31]

The Department of Defense decided, on the basis of these reports, to support a new effort in bolstering U.S. technology called Sematech, short for Semiconductor Manufacturing Technology. Sematech was originally proposed by Charles Sporck, president and CEO of

---

28. Lee Smith, "Let's Not Bash the Japanese," *Fortune*, (April 27, 1987), p. 175; Rachel Parker, "Industry Associations Applaud Sanctions Against Japanese," *Infoworld*, (April 6, 1987), p. 28.

29. Susan Chira, "U.S. Given Warning by Japan," *New York Times*, (April 16, 1987), p. 23.

30. Jack Robertson, "DOD Task Force Eyes Impact of IC Technology Offshore," *Electronic News*, (February 24, 1986), p. 1.

31. One study which remained unclassified was Richard Van Atta, Erland Heginbotham, Forrest Frank, Albert Perrella, and Andrew Hull, *Technical Assessment of U.S. Electronics Dependency* (Alexandria, VA: Institute for Defense Analyses, November 1985).

32. Jeffrey Bairstow, "Can the U.S. Semiconductor Industry be Saved?" *High Techology*, (May 1987), p. 34; David E. Sanger, "Chip Makers in Accord on Plan for Consortium," *New York Times*, (March 5, 1987), p. 29.

National Semiconductor. Sematech would be jointly funded by SIA members and the Department of Defense, and would draw upon the resources of the Semiconductor Research Corporation, an existing research consortium set up by the SIA in North Carolina. The Defense Science Board recommended that the Department of Defense provide $200 million per year over the 1987–92 period, but the actual level of funding for 1988 was to be only $50 million.[32]

In the fall of 1986, Fujitsu announced its intention to acquire 80 percent of the equity of Fairchild Semiconductor, the remaining 20 percent to remain in the hands of Schlumberger. A variety of interests put pressure on the U.S. government to block the sale, on the grounds that it would increase U.S. dependence on Japanese semiconductors.[33] In the end, Fujitsu withdrew its offer. This event was remarkable given the general aversion of the U.S. government to interfere in mergers or acquisitions which do not involve possible violations of antitrust regulations.

Finally, many aspects of the Strategic Defense Initiative (SDI)— and the Strategic Computing program that proceeded it—were clearly aimed at promoting R&D that would have important spinoffs for the semiconductor industry. The European responded to the Japanese VLSI Project, and the American VHSIC program and SDI, with a number of cooperative ventures of their own. Thus, by the mid-1980s, a major information technology subsidy race had begun that raised governmental R&D spending in Japan, the U.S., and Europe. The R&D spending aspect of the race had some of the characteristics of an arms race. It was possible that all the expenditure, which was aimed at achieving competitive *advantages*, might be neutralized by the spending of others.

Summary and Conclusions

The semiconductor industry has been a dynamic industry, both in terms of technological change and in its pattern of economic growth. It has not been immune from the business cycles experienced by other industries, as has been graphically demonstrated by the last two years. It remains, however, one leading contemporary example of the general dynamism of information technology and of the problems created for international economic relations by the sensitivities of

33. Andrew Pollack, "Fujitsu Chip Deal Draws More Flak," *New York Times*, (January 12, 1987), p. 25.

nations to dependence on others for "strategically important" goods. The early lead of the United States in semiconductors provoked responses in Europe and Japan. In Europe, the initial response was to back national champions like Philips, Thomson and Siemens. Now that response is widely perceived to have failed, leading therefore to new efforts at the European level. In Japan, the VLSI Project was the response, and the result was a dramatic improvement in the competitiveness of Japanese firms in international competition.

The organization of production in the United States made it possible for smaller merchant firms to develop alongside larger integrated firms like IBM, AT&T, and Motorola. The conditions that favored the rise and growth of the merchant firms appear to have changed radically. The increasing investment required for the development and production of new devices, the intense competition from integrated electronics firms in Japan and Europe, and the greater ability of large U.S. firms to get access to the capital needed to keep up with that competition seem to have greatly undermined the once nearly unassailable position of the merchant firms.

Their response has been to turn to trade policy remedies to buy time for restructuring. It has also involved an appeal to the Department of Defense for new R&D subsidies in the form of Sematech. This response may not restore the technological edge of U.S. firms nor prevent the trend toward further deterioration of the position of U.S. merchant firms in international markets.

Public policy remains very important in providing sources of assured demand for products, subsidies for R&D and capital investment, and trade policies which insulate the domestic market or provide greater access to foreign markets. But not all countries are equally good at delivering public policies that aid semiconductor producers. The Japanese have been unusually effective compared to both Europe and the United States. Japanese public policy has made it possible for Japanese firms to combine public R&D subsidies with inexpensive capital, and lots of it, to outmanufacture their competitors. The result has been the loss of U.S. technological superiority, global overcapacity, and increasing tension in U.S.-Japanese and Euro-Japanese relations.

CHAPTER 6
The Asian Auto Imbroglio: Patterns of Trade Policy and Business Strategy

James A. Dunn, Jr.

On March 1, 1985 President Reagan, citing the "wisdom of maintaining free and fair trade for the benefit of the world's consumers," announced that the U.S. government would not ask the Japanese Government to extend the "voluntary restraint agreement" (VRA) of 1.85 million units it had imposed on Japanese auto exports to America.[1] The statement was immediately praised as a return to free market principles by the presidents of the Consumers for World Trade and the Imported Automobile Dealers Association. But it was condemned as a "sad day for America" by Lee Iacocca of Chrysler and a return to "one way free trade" by John Danforth, Chairman of the Senate Commerce Committee.[2] Then, on March 5, the Japanese Ministry of International Trade and Industry (MITI) announced that it would continue to restrict Japanese auto exports to America, although at the higher level of 2.3 million units per year.[3] MITI also announced that it would allocate shares of the total quota among the nine different Japanese auto manufacturing firms, and indirectly among the Big Three U.S. auto firms that purchase an increasing share of their subcompact models from Japan.[4] Later, the Japanese government announced that its automakers would continue to be bound by this self-imposed restraint until March 31, 1988.[5]

These developments left commentators somewhat puzzled. Was this a

1. Clyde H. Farnsworth, "U.S. Will Not Ask Japan To Extend Car Export Curbs," *New York Times* (March 2, 1985), 1.
2. *Ibid.*, and "Importers and Association of Car Dealers Hail Move," *New York Times* (March 2, 1985), 33.
3. "Japan seeks to limit auto firms plans for a sharp increase in exports to U.S.," *Wall Street Journal* (March 5, 1985), 3.
4. Susan Chira, "Japan Allotments Set on Cars for U.S.," *New York Times* (April 27, 1985), 31; and "Japanese set car allotments to U.S. Market: domestic makes affiliates get favored treatment," *Wall Street Journal* (April 29, 1985), 3.
5. "Japanese retain auto quota, but critics urge cuts," *Philadelphia Inquirer* (January 28, 1987), E1.

return to free trade? Was it a continuation of protectionism? Did the fact that the VRA could even take place mean that international trade in the auto industry was becoming more protectionist, as many analysts feared? Or did the 1985 removal of the VRA mean a return to more liberal trade after a period of recession-induced protectionism? And what was the significance of the "truly voluntary" export quota that MITI imposed for 1985-88? One thing all observers could agree on, however, was that the V.R.A. demonstrated that political factors will be as important as economic ones in shaping the future of the world auto industry: auto companies will have to adapt their global business decisions to the requirements of politics as well as markets.

This raises a number of important questions about the "rules of the game" for international auto trade and investment, and how changes in government policy effect business strategy and hence the flow of imports, exports and investments. From the Japanese point of view the key concerns are: will the next recession unleash a wave of "Japan-bashing" in the West that will undermine their hard-won market share? What can the Japanese auto companies do that will ease protectionist pressures? Can they alter their corporate investment, manufacturing, and marketing strategies in ways that will reduce the need for more government intervention? From the Western point of view the problem is how to obtain the benefits of Japanese productivity and efficiency while avoiding the problems associated with strict protectionism. Can auto producers with the "Eastasia edge" be accommodated into a globalizing auto industry to the profit of Western auto companies and the benefit of Western consumers without destroying the jobs and industrial capacity associated with a viable domestic auto industry?

This article approaches these questions by focusing on the intersection of politics, diplomacy and commercial strategy. It will examine the way in which trade and investment in the world auto industry have been shaped by the political requirements of governments seeking to promote industrial growth. It will identify political and economic conditions that have permitted international auto trade to flourish, and investigate ways that both governments and companies have responded when the environment for international trade became unfavorable. It will conclude by offering a view of the future evolution of investment and trade in the world auto industry, and especially of how Asian producers are likely to be integrated into a reasonably stable political and economic trade regime in the auto sector.

Regional Dimensions of International Auto Trade and Government Policy

There is little doubt that the rise of Japan (and the promise of the rise of other Asian producers) has brought on major changes in the political economy of the world motor industry. The gravity of the Japanese challenge began to become clear in the mid-1970s. Japan passed Italy to take over 4th place in the ranks of motor vehicle exporting nations in 1968. By 1974 it had leaped past Britain, France, and West Germany to become the world's leading motor vehicle exporter, selling 2,618,100 units abroad. Then between 1974 and 1980, Japan's exports more than doubled again, reaching 5,967,000 assembled motor vehicles.[6] In 1981 Japan passed the U.S. to become the world's largest manufacturer of motor vehicles, and exported more than half of its production abroad. In that same year its auto exports amounted to 41 percent of all motor vehicles traded in the world. Moreover, if trade between the U.S. and Canada, and the exchanges between the Common Market nations of Europe are factored out of the calculation, Japanese exports in 1981 represented 71 percent of total world trade in finished motor vehicles.[7] Such dramatic shifts in the balance of competitive power seemed to some Western observers to portend the destruction of their national auto industries. Small wonder, then, that many voices called for government intervention to save the domestic auto industry.

But it would be a mistake to view Western reactions to the Japanese challenge as heralding the collapse of open trade and the onset of a new and much more protectionist period for auto trade.[8] Instead, the Western efforts to "manage" auto trade are likely to have the paradoxical effect of strengthening and deepening international exchange and cooperation between Western and Asian auto manufacturers. But instead of simply consisting of increasing shipments of assembled vehicles from one region or country to another, much of the future growth of international exchange in the auto sector will take the form of capital investments in local assembly and manufacturing, transfer of technological and management know-how, shipment of original equipment components and replacement parts, complex joint ventures, and cooperative production agreements between Japanese, American, European, and other Asian firms.

---

6. Motor Vehicle Manufacturers Association, *World Motor Vehicle Data,* 1981 edition (Detroit: MVMA, 1982), 10.
7. Calculated from data in ibid. and the European Community Commission, *The European Automobile Industry,* 1981 (Brussels: EC Commission, 1981), annex 8.
8. A scholarly example of this interpretation may be found in: Peter F. Cowhey and Edward Long, "Testing Theories of Regime Change: Hegemonic Decline or Surplus Capacity? *International Organization* 37 (Spring, 1983), 157-183.

Many observers, especially mainstream economists, view the years from the mid-1950s to the late 1970s as a sort of "golden age" of free trade in manufactured products, including motor vehicles. Tariffs were lowered and trade expanded enormously. The conventional assumption is that the former was the cause of the latter. As Susan Strange has demonstrated concerning overall world trade,[9] and as will be shown below in relation to auto trade, it is more plausible to argue that the unprecedented surge in economic growth during those years permitted both increased auto trade and tariff reductions. There was much less genuine global liberalism in auto trade during these years than is commonly supposed. Rather, increased trade in autos was the result of an unprecedented growth in demand for autos in the North Atlantic area coupled with *a carefully managed set of regional political arrangements that permitted trade expansion while at the same time limiting "poaching" in domestic markets by outsiders to tolerable levels.* Indeed, it is far more often the case that advances in fostering international trade occur in an atmosphere in which government trade interventions limit risk and regulate the pace of change than in an environment approximating textbook free trade.

This can be seen by examining the flow of auto trade over the years. Table 6.1 presents data on imports of assembled passenger automobiles according to the region of destination (i.e. the region into which they were imported) and, in some cases, according to their origin (the region from which they were exported).[10] It also shows what proportion of

---

9. Susan Strange, "Protectionism and World Politics," *International Organization* 39 (Spring, 1985), 241.

10. The following should be noted concerning the data in Table 1 on which the interpretations in this section are based. The import figures are drawn from: Automobile International, *World Automotive Market* (New York: Johnson International). This series was chosen for its statistical continuity and because its reporting format of "exports from/ imports to" made it possible to show the regional dimensions of trade. In the years before 1969 the publication was titled: *Global Automotive Market Survey and World Motor Census,* and was published by McGraw Hill, but the same format and categories of reporting were used. There are two drawbacks to the series, however. One is that the Eastern Bloc countries were not covered. The second is that only major exporting countries are covered. The exports of New exporting countries such as Spain, Brazil and Mexico are not given. Thus in the years since the mid-1970s total world trade in autos is increasingly underestimated by this series. By comparing the data in this series with data in the Motor Vehicle Manufacturers Association publication, *World Motor Vehicle Data* it is possible to estimate that the undercounting of free world exports was approximately 800,000 vehicles in 1983. About 560,000 of those exports were cars produced in Spain by U.S. or French multinational auto firms from components manufactured all over Europe and imported into Spain under special arrangements with the Spanish government and then sold in Europe under favorable trade arrangements with the EEC with a view to Spain's immanent accession to Common Market membership. This, then, means that Intra-European trade remains the largest element in world auto trade despite the under-

Table 6.1 Imports of Passenger Cars by Region Selected Years, 1955-1985

Region	Total Imports	% of World Total	Cumula- tive % of World Total
1985			
World	12,208,390	100	100
Intra-European Trade	4,544,340	37.2	37.2
Imports to North America			
from Outside North America	3,237,375	26.5	63.7
U.S.-Canada Trade	1,837,772	15.1	78.8
West. Europe Imports from Japan	1,080,177	8.9	87.7
Rest of World's Imports			
from Japan	920,993	7.5	95.2
Remainder of World's Imports	587,733	4.8	100.0
1970			
World	7,217,415	100	100
Intra-European Trade	3,838,871	53.2	53.2
Imports to North America			
from Outside North America	1,753,739	24.3	77.5
U.S.-Canada Trade	696,702	9.6	87.1
West Europe Imports from Japan	101,136	1.4	88.5
Rest of World's Imports			
from Japan	235,737	3.3	91.8
Remainder of World's Imports	591,230	8.2	100
1960			
World	2,355,730	100	100
Intra-European Trade	956,560	40.6	40.6
Imports to North America			
from Outside North America	608,120	25.8	66.6
U.S.-Canada Trade	27,001	1.1	67.5
West Europe Imports			
from Japan	29	0.0	67.5
Remainder of World Imports	763,930	32.4	100
1955			
World	1,077,004	100	100
Intra-European Trade	480,925	44.6	44.6
Imports to North America			
from Outside North America	92,239	8.6	53.2

Table continues on the following page.

Table 6.1 Continued

U.S.-Canada Trade	27,522	2.5	55.7
West Europe Imports from Japan	6	0.0	55.7
Remainder of World Imports	475,469	44.1	100

Source: Calculated from data in: Automobile International, *World Automotive Market* (New York: Johnson International, various years). See note 10.

total world trade a given region's imports represents. Thus, for example, we can see that in 1985 over 4.5 million passenger cars were exported from European nations, and that this "Intra-European trade" constituted 37.2 percent of total world passenger car trade that year. When the 3.2 million cars imported into North America from other continents are added to the Intra-European percentage of world trade it makes up nearly two-thirds of world auto trade. Adding U.S.-Canada trade and Western European imports from Japan to the cumulative total shows that these flows of imports into and among just two regions make up over 87 percent of the total number of world car imports. Given this dominance of Europe and North America in the flow of world car imports, it is important to examine in more detail the growth of their trade and the political and economic arrangements that enabled them to absorb and adjust to such rapid import growth.

Intra-European Trade

Table 6.1 clearly shows that trade within Western Europe has been the largest single component in international auto trade in the past 25 years. Intra-European trade has consistently accounted for around 40 percent of world imports. In the years around 1970, it made up over 50 percent of total world trade in passenger cars. This certainly represents substantial progress in fostering transnational trade. But the point is that this trade was confined within a single region and was made possible by economic and political preconditions that tend to be found only in specific regional relationships, not on a global or worldwide basis.

What were the conditions that permitted European nations to sustain such long term increases in auto trade? First, growth. From 1950 to 1970 the market for passenger automobiles grew faster in Europe than

counting of exports. Mexican and Brazilian passenger car exports have similar links to government promotion policies and MNCs world marketing strategies, although, as yet, the volume of their exports is much lower than Spain's.

anywhere else in the world.[11] It is always easier to permit imports and lower trade barriers when the size of the total market is expanding. Even firms and nations that were losing market share could experience absolute production increases. For example, Britain's share of total European motor vehicle production fell from 49.6 percent in 1950 to 18.3 percent in 1970, but her absolute level of production increased from 783,672 units to 2,098,498.[12] Second, European nations began the period with similar socio-economic structures and government policies on petrol taxes, ownership fees, highway programs and regulations, etc. This helped auto trade to flourish even before there were major reductions in auto tariffs. In addition, many European nations were involved in or contemplating, major commitments to regional economic integration and political harmonization. The creation and expansion of the EEC, other efforts such as EFTA, and the general intensification of exchanges of all sorts made trade across European frontiers politically less exotic and threatening to established interests than in the past. Thus the opening to imports that was achieved by the European nations was clearly an opening to *European* imports. It is important to point out that during the years from 1950 to 1970 European auto makers did not have any competitors outside their region. The U.S. auto giants had long since accepted the requirement imposed by European governments that they manufacture their products in Europe, and thus did not constitute a threat even in the early postwar years when Detroit could have swamped Europe with vehicles made in the U.S.A. And the Japanese were not ready to begin a major export campaign in Europe until the early 1970s.[13]

Third, the governments of the major continental auto producing nations were prepared to take various official and unofficial steps to protect their auto industries from precipitous decline. These efforts ranged from open subsidies and public ownership of auto firms to off-the-record understandings with foreign auto firms on limiting their

11. Alan Altshuler, et. al., *The Future of the Automobile* (Cambridge, MA: MIT Press, 1984), 22-23, especially Figure 2.3.

12. Motor Vehicle Manufacturers Association, *World Motor Vehicle Data,* 1984/85 edition (Detroit: MVMA, 1985).

13. For a very effective exposition of the point about Europe having no competition in autos and many other industries during these years see Wolfgang Hagar, "Protectionism and Autonomy: How To Preserve Free Trade In Europe," *International Affairs* 58 (Summer, 1982), 413-428; and "Protectionism in the 80's: The Managed Co-Existence of Different Industrial Cultures," in *EEC Protectionism: Present Practice and Future Trends,* vol. 1, by Michael Noelke and Robert Taylor (Brussels: European Research Associates, 1981), 3-34.

sales in times of crisis. Especially after the slowdown in auto growth after the mid-1970s, care was taken to see that rapid competitive changes did not lead to too rapid market share shifts. Even in the 1980s, after 25 years of stepped up trade and lowering of formal trade barriers, each of the big three continental producers still dominated its own domestic auto market. As Table 6.2 shows, in 1983 West German manufacturers still had 72.8 percent of the West German market. French car makers held a 69.0 percent share of their domestic market. Italian auto companies had a 63.5 percent market share in Italy.

Table 6.2 Market Share Held by Domestic and Imported Passenger Cars, Major Producing Nations—1983

Nation	Domestic	All Imports	Japanese Imports
United Kingdom[1]	43.1	56.9	10.7
Italy	63.5	36.5	0.001
France	69.0	31.0	2.7
West Germany	72.8	27.2	10.6
U.S.A.	74.0	26.0	20.9
Japan	98.9	1.1	—

[1] U.K. market shares calculated on basis of total motor vehicles.
Source: Motor Vehicle Manufacturers Association, *World Motor Vehicle Data* (1984/85 edition), passim.

Britain has been unable to develop either a competitive industry or successful trade management techniques and consequently has also been the big loser from liberalized intra-European trade. The U.K. did not enter the EEC until 1973, and all tariffs were not abolished on auto imports from other EEC countries until 1978. In 1972 British firms, including Ford-U.K. and Vauxhaul (GM-U.K.) held a dominant 76.5 percent of their domestic passenger car market. In 1978 their market share had fallen to 56.2 percent by 1983 British-produced cars held only 43.08 percent of the market.[14] Despite massive public subsidies (from

14. Stephen Wilks, *Industrial Policy and the Motor Industry* (Manchester: Manchester University Press, 1984), 70, and Daniel T. Jones, "The European Motor Industry and Government Intervention," Paper prepared for a meeting of the IAP project and EEC Commission (Brussels: 12/13 November, 1979), appendix Table 6, and Motor Vehicle Manufacturers Association, op. cit.

1975 to 1984 2.4 billion pounds of public money was committed to BL alone[15]) and long-term trade protection against Japan (a 1976 "Gentlemen's Agreement" which limits Japanese imports to 10 or 11 percent of the market), the decline of the British motor industry has been termed a "collapse" and "... of a unique order of magnitude." by one of its most careful students.[16] There were, of course, many factors which contributed to the industry's weakness. Everything from poor labor relations to "stop-go" macro-economic policies has been blamed. But the fact is that the proximate agents of the collapse were German, French, and Italian auto firms who invaded the British market with imports and the American owned multinationals who began to manufacture an increasing share of their U.K. sales abroad. The British experience is thus an object lesson on the dangers of mismanaging trade policy in the auto sector. It indicates that liberalizing in the face of stronger foreign competition and a multinationalizing domestic production base can be a recipe for disaster.

Imports to the U.S. from Outside North America: European Imports

The second largest component in world auto trade has been imports into the U.S. market from outside North America. Since the late 1950s these have consistently amounted to about one-quarter of world imports. From the mid-1950s through the early 1970s these imports came overwhelmingly from Europe. During these years North American auto production leveled off as the U.S. market approached "saturation," while European production skyrocketed. By 1970 European production of passenger cars surpassed the American level (11.7 million to 9.4 million).[17] Interestingly, for most of these "golden years" there were very little politicized tensions over auto imports. Why?

In the first place, the U.S. could afford to be broad-minded about imports, since the nation had a substantial surplus in its balance of trade in the automotive sector (measured in dollars spent for all automotive products, not in terms of finished vehicles) until 1968.[18] Second, the heart of the American car market benefited from a substantial degree of "natural protection." With the cheapest gasoline in the world (thanks to the lowest gasoline taxes), the best highway system,

---

15. Wilks, op. cit., 272.
16. *Ibid.,* See also Peter J.S. Dunnett, *The Decline of the British Motor Industry* (London: Croom Helm, 1980).
17. Calculated from data in: Motor Vehicle Manufacturers Association, *World Motor Vehicle Data,* various editions (Detroit: MVMA), passim.
18. U.S. Department of Commerce, *Survey of Current Business* (June, 1970), 48.

and enormous distances to travel, the average American automobile evolved into a machine unlike any in the world. It was larger, heavier, higher-powered, more fuel-hungry than all but a tiny minority of luxury vehicles produced abroad. 85 to 90 percent of American buyers consistently demanded this kind of vehicle and only American manufacturers could supply it. (See Table 6.3) As for the buyers who wanted smaller, more fuel efficient or more off-beat vehicles, U.S. auto companies were tacitly willing to concede this market to imports, since Detroit's own compacts were believed to be unprofitable and more likely to take sales away from their larger models than from imports.

Table 6.3   National Distribution of Passenger Car Production by Engine Displacement—1976

	Percent of cars produced with engine displacement	
Country	below 2,000 cc.	above 2,000 cc.
Japan	93.8	6.2
Italy	99.8	0.2
United Kingdom	65.3[1]	34.7[1]
West Germany	86.1	13.9
France	77.0[2]	23.0[2]
United States	0.1	99.9

[1] Categories for U.K. are "below 1,600 cc." and "above 1,600 cc."
[2] Categories for France are "below 1,500 cc." and "above 1,500 cc."
Source: Thomas G. Whiston, *U.S. Regulatory Control and Innovatory Response in the Automobile Industry—A View from Europe*, (Falmer, England: Science Policy Research Unit, University of Sussex, 1.

A third reason why Detroit was not tempted to resort to political protection against European imports was fear of the predictable reaction of the very European governments whose goodwill was essential to the continued prosperity of U.S.-owned auto companies in Europe. To American auto executives the opportunity for their subsidiaries to compete for a larger share of the rapidly growing European market was much more attractive than the added domestic small car sales they would capture if restrictions were imposed on European imports. This was the way to maximize the companies' market shares and profits *on a worldwide basis,* even if it did entail some loss of sales in the not very profitable small car segment of the American market.

Not until after the effects of the second energy crisis, combined with the advance of federal fuel economy standards mandated by the 1975

Energy Policy and Conservation Act and a dramatic slump in auto demand in 1980 did there seem to be a chance that the small car segment of the U.S. market would be a significant source of future profit for Detroit.[19] By then, however, the Europeans had virtually withdrawn from competition in the imported small car market. Some, like Fiat, simply went home. Volkswagen moved its small car production plant to the U.S. and became a "domestic" car maker. The rest (Volvo, B.M.W., Daimler Benz, etc.) successfully followed a strategy of moving upscale into the luxury car market segments. By 1980, for example, European exports to North America were only 30 percent of Japan's in terms of units, but the dollar value of European auto sales was 54 percent of Japan's.[20]

Japanese Imports

Although Toyota and Nissan had been fringe competitors in the American market since the late 1950s, sales of Japanese cars in the U.S. did not pass the 100,000 mark until 1968. But that year Volkswagen's sales in the U.S. peaked at 580,000 vehicles. The Japanese imports thereafter captured almost all the growth in the low priced import market. Table 6.4 shows the growth of the Japanese market share in the U.S., and how it came to dominate the import market in terms of total units sold.

Indeed, throughout the 1970s, the American market, and to a lesser extent the Western European market, became the engine that drove the growth of the Japanese auto industry. The growth curve of Japan's domestic auto markets had virtually flattened out in the 1970s: 2.9 million new registrations in 1973, 2.85 million in 1980.[21] Exports became the only avenue of expansion, and the only way that smaller Japanese producers such as Honda and Mitsubishi could get around the dominance of Toyota and Nissan in the home market. Thus Japanese companies concentrated an increasing proportion of their rapidly rising export sales in the U.S. market. In 1965, when Japan only exported 100,000 cars, a mere 21.9 percent of her exports went to the U.S. The

---

19. In 1973, for example, only 39.5 percent of the cars sold in the U.S. were classified as small cars. By 1980 64.0 percent were small cars. See Noboru Fujii, "The Road to the U.S.-Japan Auto Crash: Agenda Setting for the Automobile Trade Friction," in Program on U.S.-Japan Relations, *U.S.-Japan Relations: New Attitudes for a New Era* (Cambridge, MA: Harvard University Center for International Affairs), 27.
20. Altshuler, et al., *The Future of the Automobile,* 26.
21. Japan Automobile Manufacturers Association, Inc., *Motor Vehicle Statistics of Japan, 1981* (Tokyo: JAMA, n.d.), 5.

Table 6.4 Growth of Japanese Imports as Proportion of Total Import Share of U.S. Passenger Car Market 1960–1985

Year	All Imports[1]	Japanese Imports[1]	Japanese Imports as % of all Imports
1960	7.5	—	—
1965	6.1	0.2	4.8
1970	15.3	5.3	28.8
1975	18.4	9.4	51.8
1980	28.2	21.3	75.5
1985	27.6	20.1	78.1

[1] Calculated as percent of all passenger cars sold.

Source: Automative News, *Automotive News Market Data Book 1986*, p. 24; and Motor Vehicle Manufacturers Association, *Automobile Fact and Figures 1979*, pp. 20, 85.

American share of Japanese passenger car exports rose to 32.1 percent in 1970 on a total of 725,586 cars exported. In 1975 when Japan sold 1,827,826 cars abroad the American share was 38.9 percent. Five years later in 1980 Americans bought 46.1 percent of a total of 3,947,160 cars exported from Japan.[22] This rise to dominance was financially very profitable, but did entail serious political risks for the Japanese companies. They had become dependent on exports sales. As Dyer and his colleagues remark: "Not only was the rest of the world buying more than 50 percent of the industry's output; even more remarkable, it was buying almost 80 percent of the industry's growth."[23] In America, Japanese cars became virtually synonymous with imports, at least in the lower and moderate price categories. When the great auto sales depression hit in 1980, and it became irresistably popular to blame imports for the industry's desperate plight, the Japanese had no one to share the blame with them.

What was remarkable about the 1981 Voluntary Restraint Agreement was not that it took place, but rather that its terms were so generous. In the midst of the worst recession in postwar history (and the worst year ever for U.S. auto companies), the most that U.S. management, labor, and government could agree on was to ask for a temporary VRA which limited Japanese exports to 1.68 million units for 2 years with an exten-

22. Calculated from data in *ibid.* and *World Motor Vehicle Data,* various years.
23. Davis Dyer, Malcolm S. Salter, and Alan M. Webber, *Changing Alliances: The Harvard Business School Project on the Auto Industry and the American Economy* (Boston, MA: Harvard Business School Press, 1987), 137.

sion for a third year at 1.85 million units. The VRA permitted the Japanese manufacturers to increase their profits by charging premium prices, "loading" every unit they shipped with options to increase the price still further, and moving up-market into the intermediate, sporty, and even the fringes of the luxury segments. For example, in 1984 Mazda's sales volume remained the same as the previous year, but its net income on U.S. operations increased 19 percent.[24] The VRA also actually permitted the Japanese to increase their market share slightly because total car sales in 1981 fell further than anticipated. The VRA proved both so politically valuable in deflecting further pressures for protection and so financially rewarding to Japanese (and American) manufacturers that the Japanese government prudently kept a less stringent version in place even during the recovery years of the mid-1980s when U.S. auto firms rolled up the biggest profits in their history.

U.S.-Canada Trade

Trade between the U.S. and Canada shows a dramatic increase between 1960 and 1970. The conventional view is that this is due to the "free trade" in automobiles that was created between the two nations by the 1965 "Auto Pact." In fact the situation is much more complex than that. The Automotive Products Trade Agreement which was signed between the two governments in January 1965, and the accompanying "letters of undertaking" issues by the Canadian subsidiaries of the Big 3 auto makers to the Canadian government were not textbook free trade at all, but rather a classic example of regional trade management.[25] This is clearly demonstrated by the fact that the U.S. had to seek a waiver from GATT to implement this preferential trade agreement.[26]

In essence the "Pact" constituted by the agreement and the letters provided for a duty free exchange of vehicles and original equipment components among the Big 3 auto firms in America and their Canadian subsidiaries. This would allow them to rationalize production in North

24. *Ibid.,*

25. For details of the Pact and the negotiations see: James F. Keeley, "Cast in Concrete for All Time? The Negotiation of the Auto Pact," *Canadian Journal of Political Science* 16 (June, 1983), 281-298.

26. *An Automotive Strategy for Canada,* Report of the Federal Task Force on the Canadian Motor Vehicle and Automotive Parts Industries to Hon. Edward C. Lumley, M.P., Minister of Industry, Trade and Commerce and Regional Economic Expansion (Ottawa: Minister of Supply and Services, 1983), 18.

America. However, the Canadians insisted that an appropriate share of the production be located in Canada. In the agreement that share was defined as at least equal to 75 percent of the value of a given company's sales in the Canadian market. In practice it has come to mean about 95 percent of the value for automobiles. Thus an auto company, General Motors, for example, must manufacture enough models and parts in Canada for export to the U.S. to offset 95 percent of the value of the models and parts it manufactures in the U.S. for sale in Canada. Canada benefited substantially from the arrangement. Prior to 1965, Canadian value added amounted to 3.7 percent of total North American auto output. By 1970 it had risen to 6.7 percent of North American production, and in 1981 it stood at 6.9 percent. Employment in auto manufacturing in Canada rose from 70,600 in 1964 to a peak of 124,000 in 1979. It fell off sharply due to the recession, but Canada's share of North American auto employment actually reached an all time high of 8.8 percent by 1981.[27]

Thus the Pact was far more than a trade liberalization agreement. It was also, in effect, a sophisticated Canadian local content requirement, with the content balance being measured on total sales rather than on individual vehicles or models. It did indeed remove tariffs and expand trade. It did increase productivity and lead to lower consumer prices, at least in Canada. But it was not laissez-faire trade liberalization. The Canadian government played a very strong role in protecting the Canadian motor manufacturing sector. Unlike the British government, Ottawa was able to manage the trade liberalization process to ensure that Canadians—as workers and members of local auto manufacturing communities—as well as consumers got their fair share of the benefits from trade expansion.

European Imports from Japan

Negligible in 1960, and only 1.4 percent of total world imports in 1970, Western European imports from Japan increased tenfold during the decade of the 1970s. By 1983 they constituted 9.5 percent the world's auto trade. This represents the second large extra-regional trade flow. The fastest growth occurred in European countries without their own national auto industry. Thus Japanese market penetration in the Netherlands rose from 3.2 percent in 1970 to 26.4 percent in 1980. Similarly, by 1980 the Japanese share of the auto market had reached 25.8 percent in Belgium and Luxembourg, over 30 percent in Denmark

---

27. *Ibid.,* 21.

and Ireland, and nearly 50 percent in Greece.[28]

In the major producing countries growth of Japanese imports has been much slower and has been accompanied, in typical European fashion, by a substantial amount of trade management efforts by governments and manufacturers. As Table 2 shows, Italy has virtually excluded Japanese imports. Italy's quota limiting passenger car imports from Japan to no more than 2,200 per year was authorized by the GATT in 1962 as legitimate retaliation for Japan's excluding Italian cars from its market at the time.[29] The French government simply announced in 1977 that it did not wish to see Japanese autos take more than 3 percent of the French market. When Japanese imports threatened to exceed this "official but unwritten" limit, French officials held them on the dock and refused to pass them through customs until Paris was satisfied that the 3 percent cap would be respected.[30] In Great Britain a delegation from the Society of Motor Manufacturers and Traders, accompanied by the Minister of Industry, visited Japan in 1976 and, with the threat of official quotas as an incentive, concluded a "Gentlemen's Agreement" with the Japanese Auto Manufacturers Association to limit Japanese exports to Britain to between 10 and 11 percent of the market.[31] By 1981, in the midst of a serious sales slump, even West Germany, long a bastion of liberalism, was expressing concern about the diversion of Japanese exports from the U.S. to Europe. The Japanese were reported to have unofficially assured the Germans that they would not press their market penetration any further.[32] Even the smaller nations began to ask for restraint. The Belgians, under pressure from Renault which was contemplating new investments in that country, secured a promise from Japan of a 7 percent *reduction* in Japanese imports.[33] The European Community Commission pressed Japan for export restraint and subjected its auto imports to a special statistical surveillance so that trends could be followed from month to month. In February, 1983, the Japanese government publicly renewed a

---

28. European Community Commission, *Commission Activities and EC Rules for the Automobile Industry 1981/1983* (Brussels: EC Commission, 1983), annex.
29. Altshuler, et al., *op. cit.,* 231.
30. *Ibid.,* 232.
31. Dunnett, *op. cit.,* 166-167.
32. The official German position is that there was no "deal" to restrain Japanese exports to Germany. German Economics Ministry and Foreign Ministry Officials told the present author that the Japanese simply gave a "weather forecast" to the German Economics Minister, predicting that Japanese auto exports to Germany would not increase beyond their 1980 levels. In fact the Japanese market share dropped rather sharply below the 10 percent in had reached in 1980.
33. Noelke and Taylor, *EEC Protectionism: Present Practice and Future Trends,* 132.

previously private undertaking that auto exports to the EEC would be "moderate" and extended this moderation to light trucks, motorcycles and forklifts.[34]

These efforts may have slowed down Japan's penetration of the West European market, but they did not halt it. As auto sales recovered in the mid-1980s, the Japanese share of the European market continued to increase. By 1986, Japanese companies sold 11.7 percent of all cars in Western Europe, and some forecasts predicted they would have as much as 18 percent before the end of the decade.[35] This prospect has led the Committee of Common Market Automobile Constructors to ask the European Community Commission to request that the Japanese "voluntarily" hold their market share at about the 10 percent level until the European share of the Japanese domestic car market reaches 5 percent.[36]

Rest of World

As for the rest of the world, neither the levels of economic growth nor the severely protectionist trade policies of most of these nations have as yet permitted them to absorb massive increases of auto imports. Where import growth has occurred, however, the dominance of Japanese exports at the expense of European producers is clear from the data in Table 6.1.

Most industry analysts expect that much of the future growth in demand for autos lies in the newly industrializing countries.[37] But it is precisely these NICs, the Brazils, Koreas and Taiwans of the world, which have learned to use a wide variety of increasingly sophisticated trade management policies in their pursuit of national economic development. Some require major multinational auto firms to locate production plants on their territory as a condition of access to their market, and then add export and sectoral trade balance requirements as further conditions. Brazil's government, for example, has negotiated individual agreements with Ford, Volkswagen, Volvo, GM, Saab-Scania, Alfa Romeo, and Daimler-Benz whereby each company receives permission to import certain components based upon the

---

34. Ibid., 133, and Altshuler, *The Future of the Automobile*, 232.
35. *1987 Ward's Automotive Yearbook* (Detroit, MI Ward's Communications Inc., 67.
36. Ibid.
37. For a summary and comparison of 17 forecasts of annual auto demand in 1990 and/or 2000, as well as for a forecast of the regional distribution of this demand which clearly puts the developing countries in the forefront of auto growth see: Altshuler, et al., *The Future of the Automobile*, 112-117.

company's commitment to export an agreed-on amount and so make a positive contribution to Brazil's payments balance.[38] Korea subsidized and protected auto production by its own national firms until they were competitive on the world scale and ready to export into relatively open Western markets.[39] The governments of third world and newly industrializing countries will bend every effort to become auto producers and auto exporters, thus adding to the potential surplus capacity rather than providing markets for exports of finished vehicles from Western or Japanese producers.

Thus trade expansion in the key regions of Western Europe and North America in the period since 1955 has been sustained by a policy balance which has judiciously mixed liberalizing elements with trade management elements. The mixture has varied between the regions and over time. Europe has lagged behind the U.S. in openness to extra-regional imports. But, compared to other regions, both sides of the North Atlantic have been able to sustain remarkably strong growth in regional and trans-regional imports.

Five Patterns of Government Trade Intervention

Having seen how international trade in autos fall into distinct regional patterns based, to a great extent, on political factors, we can see that there are at least five patterns or postures that governments can adopt when it comes to deciding if, when and how to intervene to alter trade and investment in their auto sector. It is to the international environment formed by the interaction of these patterns of intervention with each other and with market conditions that auto manufacturing firms must adapt in order to do business beyond their borders.

(1) *Promotion of a "National" Industry.* In countries where a combination of population size and per capita income makes domestic auto manufacturing a possibility, some governments will intervene with a broad range of incentives to promote the growth of an auto industry in which "national" capital plays the predominant role. High levels of

38. See: Ronald E. Muller and David H. Moore, *Three Case Studies of Third World Bargaining Power and the Promotion, Control and Structuring of Multinational Corporations: Brazil, Mexico, and Colombia* (New York: United Nations Centre on Transnational Corporations, 1978).

39. Rémy Prud'homme, "Motor Vehicle Use and Production in Less Developed Countries: A Case Study of Korea," Paper prepared for the MIT Future of the Automobile Program International Policy Forum, Stenungsund, Sweden (June, 1983).

auto investment, production and employment are sought while minimizing reliance on "multinational" foreign firms. For policy makers, maximizing total world welfare, producing autos at the lowest possible cost, or adhering strictly to the letter and spirit of free trade are not primary policy objectives, especially where they conflict with the growth of national industry. This pattern not only excludes foreign imports and foreign investments, but, if successful in building a competitive industry, it will place additional pressure on foreign producers as it continues to promote the growth of the industry through exports. Korea is perhaps the foremost example of a practitioner of this pattern at the present. But Japan clearly followed this pattern from its decision to build a domestic auto manufacturing capability in the 1950s until well into the 1970s.

(2) *Access via Investment.* Many governments will require foreign auto firms desiring to make sales and profits in their market to invest in local production facilities, rather than permit them to sell large numbers of imports. European auto producing countries set the precedent for this in the 1920s and 1930s when they obliged American firms to establish new plants (e.g. Ford's integrated facility at Dagenham, England) or to buy local firms (e.g. General Motor's purchase of Adam Open A.G. in Germany). Modern developing nations such as Brazil and Mexico add other conditions such as export requirements as well.

(3) *Fulcrum of Liberalism in Global Auto Trade.* The U.S. adopted this posture during its years of unquestioned supremecy, but even as the competitiveness of its auto manufacturers declines, it remains the nation which sets the pace for open trade and investment. Despite the loss of nearly 30 percent of its market to imports, the United States remains significantly less restrictive of imports than other auto producing countries. The executive branch has taken the lead in lowering tariffs, and has resisted pressures for official non-tariff barriers such as domestic content rules. It has not pressed other nations strongly to abandon such practices as subsidies, protectionist tariffs, investment requirements, restrictions on foreign ownership of auto firms, etc.. Instead, private American auto companies have been expected to take the lead in making the commercial arrangements with foreign interests needed to adapt the investment, production, design and marketing activities of the American auto industry to an international environment in which trade in auto products is largely undirectional. Its market continues to be by far the largest single national market for motor vehicle in the world. No foreign manufac-

turer with even modest share of this lucrative market can afford to lose it because of trade restrictions. Thus, in a crisis, when political pressures for protection seems to be building up too fast, foreign firms and their governments become reluctant partners in working out informal pressure-dampening arrangements such as the VRA with Japan.

(4) *Managed Regional Trade.* Europe is the classic venue of this pattern. Trade between auto-producing countries and European countries with no national auto industry has been encouraged by liberalizing institutions such as the EEC and EFTA. Trade among European auto producing nations themselves has been permitted to rise at more controlled rates with a view to cushioning serious damage to national auto industries (as distinct from individual auto companies). When imports from outside Europe became a major competitive threat, governments of auto producing nations (e.g. France, Italy, Britain) did not hesitate to take effective steps to stem the tide. European Community institutions have not been used to prevent national governments from raising non-tariff barriers, and have even be used to reinforce national concerns about non-European import penetration. The general framework of trade management allows a country such as Germany to strike a very public pose of liberalism, while continuing to benefit from export sales to EC partners that it would probably have lost if the region were totally open to Japanese imports. Despite pledges to create absolutely barrier-free economic exchanges among Common Market nations in the 1990s, it is unlikely that unofficial national restrictions on Japanese imports will be removed simply to honor the principle of free trade. Such controls are likely to be removed only as part of an agreement in which Japan offers a significant *quid pro quo*.

(5) *Superiority and Self-Restraint.* Japan has been under pressure to adopt this posture since the late 1970s. The superiority of the Japanese production system means that the domestic market needs no official protection. Rather, the government intervenes to regulate the exports of its own manufacturers in order to prevent a protectionist backlash among its trading partners. It must be able to ensure compliance with its restraint policies, push its manufacturers toward more production in foreign markets and less reliance on exports of assembled vehicles, be adequately responsive to foreign demands to promote greater imports into the domestic market, and be flexible enough to balance a multitude of divergent economic and political pressures at home and abroad.

Political Dimensions of Firm Strategy

How far are Japanese auto manufacturers willing and able to depart from their export strategy and adopt an investment strategy for penetrating the two largest world auto markets, the U.S. and Europe? To what extent will the major stakeholders in the U.S. and Europe rely on government trade management measures to compel the Japanese to follow the lead of the U.S. companies and compete in foreign markets mainly by local production. Or will they be able to make commercial arrangements that accomplish the same purpose? In a sense, one can say that the prospects for the continuation of a relatively liberal Western trade posture vis-a-vis Japan depends on how fast Japanese manufacturers learn to "play by the rules," that is, to adapt their global investment, production, and marketing strategies to the exigencies of creating enough political allies and economic partners in Western auto industries that effective government protectionist policies can be successfully resisted.

As to the first question, the VRA with the U.S. and the "understandings" with European governments during the great slump of the early 1980s showed that the Japanese government clearly realized that it would have to push its manufacturers to adjust to a more regulated environment in auto trade. Japanese car makers, especially the leading company, Toyota, were initially less than eager to make investments in the U.S. As a former Vice Minister of MITI told a U.S. auto industry conference, "... Japanese car manufacturers... are very concerned about the investment risks. These risks, however, are unavoidable if the transformation into truly international companies is to be effected."[40] Table 6-6 shows that seven Japanese auto manufacturers are now committed to some type of production presence in their biggest overseas market. It is estimated that by the time all of the currently planned U.S. plants are in full operation, Japanese firms will have the capacity to produce over 1.7 million cars per year in America. This will make the Japanese "transplants", taken together, larger than Chrysler.

Initially these Japanese investments will not be fully comparable to U.S. auto investments in Europe. Virtually all of the value added in G.M. and Ford products in Europe is European in origin. For Japanese plants in the U.S. this level of domestic content is beyond reach.

---

40. Shohei Kurihara, "The Japanese Auto Industry: Its Development and Future Problems," in Robert E. Cole, ed., *The American Automobile Industry: Rebirth or Requiem?* (Ann Arbor, MI: Center for Japanese Studies, University of Michigan, 1984), 16.

Japanese plants in America are for stamping and final assembly. Basic manufacturing processes such as casting, forging, machining, and powertrain component production will remain in Japan, as will high-value added operations such as engineering and research and development.[41] For the present, no Japanese manufacturer can attain even 50 percent domestic content. This fits the broad pattern of Japanese overseas investments. As Kiyohiko Fukushima, an economist and manager of the Washington office of the Nomura Research Institute, pointed out, "Recent Japanese investment in the manufacturing industries of advanced countries... is designed to secure markets previously acquired through exports and avoid protectionism."[42] This may not comfort the critics who see Japanese companies as "national multinationals," whose overseas investments stimulate exports and keep high value operations safely in the home islands, rather than "pure multinationals," whose investments flow around the world based on market opportunities.[43]

Nevertheless, this first round of Japanese direct investment in the U.S. auto industry is an important step away from a purely export based strategy. One can expect that when again faced with a major auto recession Japanese auto makers will find themselves obliged to make further highly visible "political" investments in their major overseas markets. Beyond that it is likely that some Japanese firms may decide to make a virtue of necessity and push toward fully "domesticated" production facilities abroad. Smaller firms with less invested in production capacity at home may lead the way toward full multinationalization of their production base. For example, Honda had never been able to challenge Toyota and Nissan's dominant share of the home market. It sought to break out of its impasse by exports. By 1980 it was exporting 77 percent of its passenger car production, compared to Toyota's 50 percent and Nissan's 54 percent. The VRA hit its growth plans especially hard. It was not surprising, therefore, that Honda was the first Japanese auto firm to invest in a U.S. assembly facility. Its plant in Marysville, Ohio has been very successful for the company. Production has increased each year since it opened in 1982. In

41. Robert A. Parkins, "Internationalization of the Japanese Auto Industry: Real Progress or a Snail's Pace?" in Robert E. Cole, ed., *Automobiles and the Future: Compititon, Cooperation, and Change* (Ann Arbor, MI: Center for Japanese Studies, University of Michigan, 1983), 14.
42. Kiyohiko Fukushima, "Japan's Real Trade Policy," *Foreign Policy* (Summer, 1985), 23.
43. Robert B. Reich, *The Next American Frontier* (New York: Penguin Books, 1983), 260.

September, 1987, Honda announced that it would build a second plant in Ohio with capacity of 150,000 vehicles per year. By 1989 Honda will be able to produce a half million vehicles in the U.S., which rivals the 550,000 units it planned to build in Japan that year. In addition, Honda planned to increase the proportion of American-made components in its vehicles to 75 percent by 1991. If these moves are successful, a company spokesman said that Honda might even be in a position *to export some 50,000 cars to Japan each year!*[44] For Honda Motor Co., therefore, the prospect of Japan having to continue some form of voluntary self restraint indefinitely appears less as an obstacle and more as an opportunity.

The answer to the second key question—the mix of official trade actions vs. commercial arrangements adopted in the West to manage auto trade with Japan—is related to the type of investment strategy adopted by the Japanese. Much hinges on whether the Japanese decide to invest independently or to take on a local partner. Independent investment gives the parent corporation greater control over the entire production and marketing process. It enables the Japanese to avoid U.A.W. representation and run their American facility along slightly modified Japanese labor-management lines. But joint ventures with American auto companies have the important political advantages of creating financial incentives for the Americans to oppose stringent types of protectionism, such as domestic content legislation, and of developing a class of internationally hybrid vehicles that cannot be easily labeled either "American" or "Japanese."

Detroit would prefer to see the Japanese come into the U.S. as partners, not independent competitors. In press conferences and congressional testimony U.S. auto executives have often said they wanted to see fewer Japanese imports and more Japanese investments. But it is doubtful if they really want to see substantial, *independent,* Japanese investments in U.S. production facilities. The prospect of a half dozen new multinational competitors with attractive products, state of the art manufacturing facilities located in the south, west, and rural midwest, a hand-picked labor force with little or no U.A.W. representation, run with the much-vaunted Japanese management techniques could hardly thrill them. They would much rather cut their own deals with Japanese manufacturers. As GM's Director of Worldwide Product Planning put

---

44. Philip E. Ross, "Honda Expansion: U.S. Exports a Goal," *New York Times* (September 18, 1987), D1. Data on company export percentages calculated from Japan Automobile Manufacturers Association, Inc., *Motor Vehicle Statistics of Japan 1981* (Tokyo: JAMA, n.d.).

it, "...GM decided that the only sensible interim solution was mutually beneficial business arrangements with the Japanese... These joint ventures will help maintain our distribution organization... and, as in the case with our newest affiliation [Toyota in the Fremont plant] will provide GM with access to new car-building techniques that will accelerate our ability to make small cars competitively."[45] Cooperative ventures not only give U.S. companies a profitable "piece of the action," but enable Detroit to have some degree of control over how far and how fast Japanese manufacturers can push their market penetration. In hard times the companies can count on the U.A.W. to "waive the bloody shirt" of protectionism and thereby strengthen their bargaining position with their Japanese affiliates. In good times the threat of more rapid import penetration strengthens the companies' position in negotiating with their American workers. No wonder then, that GM's John Smith concludes that "Business arrangements ...will prove much more effective in bringing foreign automotive companies to manufacture in the U.S. than any legislation ever could... The forces of the marketplace... have a way of inducing desirable results much faster and more effectively than any law ...because everyone stands to benefit from cooperation..."[46]

Japanese companies are, characteristically, pursuing both the direct investment strategy and the joint venture approach. The pioneer, Honda chose independent investment, as did Nissan. Toyota initially chose a joint venture with GM. Its example was then followed by Toyo Kogyo (Mazda) with Ford and by Mitsubishi with Chrysler. Subsequently Toyota has announced its own independent investment in a plant in Kentucky. Two smaller Japanese producers, Fuji (Subaru) and Isuzu, then followed with a joint Japanese venture to be located in Indiana. (See Table 6.5).

But investments in U.S. plants are not the only kinds of deals the Japanese are making with American auto firms. The U.S. companies are planning to purchase an increasing percentage of their vehicles and components "offshore," i.e. from Japan and other foreign manufacturing locations. U.S. companies have all but conceded the small car market segment to Asian producers. A major share of the components in the GM-Toyota, Mazda-Ford, and Chrysler-Mitsubishi joint ventures will, of course, come from Japan. In addition, both GM and Chrysler have announced plans to step up their imports of assembled

---

45. John F. Smith, Jr., "Prospects and Consequences of American-Japanese Company Cooperation," in Robert E. Cole, ed., *Automobiles and the Future,* 25.
46. Ibid., 27.

Table 6.5 *Japanese Direct Investments in U.S. Auto Assembly Sector*

Company	Plant Site	Start-up Date	Est. Production Capacity
Honda	Marysville, OH	Nov., 1982	350,000
Honda	East Liberty, OH	Fall, 1989	150,000
Nissan	Smyrna, TN	June, 1983 (trucks)	140,000
		Mar., 1985 (cars)	100,000
Toyota-GM	Freemont, CA	June, 1985	250,000
Toyota	Georgetown, KY	Fall, 1988	200,000
Mazda-Ford	Flat Rock, MI	Fall, 1987	240,000
Mitsubishi-Chrysler	Bloomington-Normal, IL	Fall, 1988	180,000
Subaru/Isuzu	Lafayette, IN	Fall, 1989	120,000

Source: John Holusha, "Another Japanese Car Marker," *New York Times* (April 21, 1985), p. E5; Susan Chira, "Chrysler and Mitsubishi to Build Small Cars Jointly in the Midwest," *New York Times* (April 16, 1985), p. 1; "Toyota's Choice," *Time* (December 16, 1985), p. 47, Susan Pastor, "Fuji, Isuzu Plan U.S. Plant," *New York Times* (December 3, 1986), p. D1; Philip E. Ross, "Honda Expansion: U.S. Exports a Goal," *New York Times* (September 18, 1987), p. D1.

vehicles from Japan. GM planned to bring in up to 300,000 sub-compacts from its Japanese affiliates, Suzuki and Isuzu by 1987.[47] Chrysler also planned to increase the number of cars it imports from Mitsubishi by up to 67 percent.[48] Ironically, one of the major factors slowing full realization of these plans was the continuation of the VRA through March, 1988. But the VRA did not restrain U.S. auto markers from bringing in an increasing share of Japanese components to be put in their ostensibly "U.S.-made" vehicles. Chrysler, for example, imported 450,000 V-6 engines from Mitsubishi in 1987.[49]

Japan is not the only Asian source of "offshore" vehicles and components that Detroit will be drawing on. GM plans to import at least 84,000 Pontiac LeMans models built by its Korean partner, Daewoo Motor Company.[50] Ford imported the Festiva, designed with Mazda and built by KIA Industrial Company of Korea for the 1988 model

---

47. John Holusha, "Chevy Turns to the Japanese," *New York Times* (October 6, 1983), D1.

48. Susan Chira, "Chrysler and Mitsubishi to Build Small Cars Jointly in the Midwest," *New York Times* (April 16, 1985), 1.

49. Matt DeLorenzo, "Chrysler to import 150,000 cars," *Automotive News* (October 21, 1985), 4.

50. "G.M., Daewoo To Build Car," *New York Times* (March 7, 1984), D4.

year. Ford-Canada is planning to import the subcompact Mercury Tracer from its affiliate Ford Lio Ho in Taiwan.[51] Some Asian autos will come to the U.S. by more indirect routes. Ford has committed $400 million to build a plant in Mexico to assemble a car for the U.S. market. The car will be designed by Mazda and many of its major components will be manufactured in Japan.[52] Chrysler has arranged to import Dodge Colts from an affiliate of Mitsubishi in Thailand.

Thus both the American and Japanese companies seem to be moving toward commercial strategies and alliances that enhance their ability to manage the political problems associated with Japan's ever-growing presence in the U.S. market. In very bad times, when they feel the need for a more public and political gesture, their clear preference is for the negotiated VRA approach. Local content laws are universal and too rigid. VRA is selective and flexible. It limits the Japanese companies in terms of number of units but not in terms of profits, and allows the American companies to do what they wish in terms of increasing the proportion of imported content in their own fleet.

This will make private capital's role as a buffer and mediator even more important than in the past, as U.S. auto companies attempt to manage and profit from a gradual transition to a more truly internationally-based production and distribution system. Inevitably, the amount of manufacturing value-added in the U.S. auto sector will decline. This will be decried by the U.A.W. and industry critics as "deindustrialization," but accepted as part of the transition to "post-industrialism" by economists and industry executives. As long as the transition remains gradual there should be no need for a more legal-bureaucratic reinforcement of the modest investment presence that Japan has accepted. Japanese manufacturers now realize they must have a more balanced mixture of exports and local production in the U.S. than in the past. They will work hard to preserve a situation where there are no formal investment or local content rules, just a set of mutual expectations of good faith. The aim of this *entente* is definitely not to roll back Japanese imports. On the U.S. side the aim is to make them more politically palatable (to the public) and more profitable (to U.S. auto companies). On the Japanese side the goal is to prepare the way for further market penetration based on a combination of exports of components and assembled vehicles to U.S. auto companies and

---

51. John Holusha, "Auto Industry Adjusting to a Painful New Reality," *New York Times* (December 21, 1986), E4.

52. John Holusha, "Ford Plans New Plant in Mexico: Car Designed in Japan for U.S. Market," *New York Times* (January 10, 1984), D1.

local production under their own names. And it is important to point out that these expectations about investments only apply to Japanese manufacturers. Other foreign producers, whether they be successful and established like Daimler-Benz or newcomers like Hyundai, are free to serve the U.S. market purely by exports.

Bureaucratic trade management measures restricting Japanese imports will not be as controversial in Europe as in the U.S. Policy makers and business leaders in continental nations such as France, Italy and Spain have strong doses of dirigisme in their polical economy traditions that make them less sanguine about the virtues of free trade than Americans. As one European auto executive remarked at an international forum on the auto that this author attended: "Free trade, like alcohol, is stimulating in small doses; in large quantities it is poison." Europeans tend to feel that intra-European competition can be stimulating enough. Since there are more competitors in Europe, the need to challenge the oligopolistic structure of their industry is not as acute as in the U.S. For this same reason both manufacturers and governments tend to be a bit more wary of Japanese investments on European soil, fearing it might simply aggravate the existing over-capacity problem. In addition, Japanese investments in one country do not necessarily provide direct political and economic payoffs to its EC partners. They also threaten to embroil EC members in potentially serious disputes over the European origin of Japanese vehicles produced in one country and exported to another.[53] Nevertheless, direct investments and joint ventures are clearly on the increase in Europe also. Nissan has agreed to joint production agreements with Motor Iberica of Spain and Alfa Romeo of Italy, and Honda is building cars with BL in Britain.[54]

Conclusions

To recapitulate the argument so far: In the 1920s and 1930s governments pursued policies which fall into the first two patterns outlined above, Promotion of a National Industry and Access through Investment, and created a "regime"—a set of national policies and company expectations—for the auto industry that, because of the circumstances of the times, was rather inward-looking. In the more propitious conditions of the 1950s and 1960s there were more no-lose opportuni-

---

53. Altshuler, et al., *The Future of the Automobile,* 231.
54. Dyer, et al., *Changing Alliances,* 141.

ties for expanding trade and the U.S. and most European governments shifted to patterns three and four, Global Liberalism and Managed Regional Trade, respectively. They did so without abandoning the essential stability of maturing national markets as the bedrock for investment and production. In these "golden years" most developing nations adopted variants of the second pattern, Access through Investment, to attract multinational firms to their territory. Japan, however, opted clearly for pattern one, Promoting a "National" industry. The success of Japanese manufacturers resulted in a very competitive industry with an very strong export orientation. When world conditions changed again in the late 1970s and 1980s, there were far fewer no-lose situations in which assembled vehicles could be exchanged across national and regional boundaries. Japan's aggressive pursuit of exports threatened to cause serious problems for the auto sector in some nations. The response was that Japan's major trading partners/competitors adopted a mix of policies, *with respect to Japan,* that were reminiscent of pattern two, Access via Investment. At present we are witnessing the gradual acceptance by Japan of a new pattern based on the idea that her competitive superiority imposes the need for self-restraint. This involves, at a minimum, some adherence to the basic expectation that strong competitors must also manufacture in the major markets where they sell their products.

There are several important differences between the self-restraint expected of Japan today and the kind of self-restraint American auto companies (and the U.S. government) accepted in the days when Detroit was the world's dominant producer. Japan's auto industry today is far more dependent on exports than Detroit ever was. It is being asked to make direct investments in manufacturing in relatively mature U.S. and European markets where it already has won a large segment via exports, and there is some concern that this could lead to the loss of existing jobs in Japan. When U.S. firms had to locate production in foreign markets it was usually with the expectation that those markets would grow very greatly, and that American exports would not be permitted in any case. Even more problematic, perhaps, are the prospects for opening up the Japanese domestic market. Table 6.2 showed how far Japan still has to go in this regard. But it is unlikely that Western nations will completely relax their pressure for Japanese self-restraint until major progress has been made in reducing the disparity in the rate of import penetration among these major auto producing nations, or until their overall trade balance returns to a more favorable equilibrium.

In the past the challenge of trade management efforts for the U.S.

was to accept the fact that its "industry of industries" would not be a major exporter from American soil. Responsibility was "delegated" to auto company executives to negotiate arrangements with foreign governments and manufacturers to assure themselves some share in the profits generated by auto manufacture abroad. In the 1980s the challenge was that U.S. companies were "delegated" to bargain with Japanese and other foreign car makers to serve the U.S. domestic market. The VRA of 1981-85 was not the herald of a breakdown in the auto trade regime, but an important political concommitant of the long term trend toward the increasing integration of the U.S. auto market into an internationalized production system. By signaling the Japanese to begin local production in the U.S., and by giving Detroit the time and money to cut appropriate deals with foreign manufacturers, the VRA actually furthered the process of international integration of the U.S. auto sector. By 1990, the level of international components in the U.S. new vehicle fleet will make recourse to protectionist domestic content legislation virtually impossible to contemplate, so disruptive would it be of American car companies' ability to market products under their own nameplates.[55]

It is evident, then, that recent experience with government trade intervention initiatives in reaction to Japan's export surge does not support the pessimistic prophecies of a collapse of trade into beggar-thy-neighbor protectionism. Rather, it suggests a more positive picture of adaptability on the part of business and government leading to ever more intensive patterns of international exchange and interaction. In the 1950s, except for U.S. foreign investments, the prevalent form of international exchange in the automobile industry was the physical shipment of vehicles to regional neighbors in Europe or to the U.S. Today, this kind of trade has not only grown substantially, but has been overlaid with an increasingly complex and multilateral network of foreign direct investment, joint ventures, foreign sourcing of original equipment components, technology licensing, and other sophisticated forms of exchange. There can be no doubt that the world auto industry is far more integrated and interdependent in the late 1980s than it was in the 1950s—or even the early 1970s. This interdependence flows from

---

55. Estimates of the amount of foreign manufacturing value added which will be found in the American new car fleet in 1990 and beyond vary from the 35 percent range to well above 50 percent. All the estimates agree that the trend is on the increase, and most tend to cluster on the high side. See: John Holusha, "The Disappearing 'U.S. Car,'" *New York Times* (August 10, 1985), 31. See also: John O'Donnell, "Competitive Product Programs and Anticipated Domestic Production and Auto-related Employment for 1990," in Dyer, et al., *Changing Alliances*, Appendix A.

both liberalizing principles (which gave us the Volkswagen, the Toyota, and now the Hyundai), *and* protecting principles (the new Chevy Nova from the Freemont, California joint venture facility, the Nissans from Smyrna, Tennessee, etc.). The world auto industry produced by the regime I have described may not be the maximally efficient use of economic resources portrayed in orthordox economic texts. But it is also far from being the neomercantilism many observers had feared.

CHAPTER 7
Trade Friction and Industrial Adjustment: The Textiles and Apparel in the Pacific Basin

Chung-in Moon

With assistance of *Chul-Ho Chang*

Introduction

The trans-Pacific relationship involving the U.S. and its East Asian allies (i.e., Japan, South Korea, and Taiwan) has traditionally been stable and congenial. America's hegemonic leadership in the region not only created a strong military-strategic alliance system, but also nurtured the economic prosperity of its allies under the shield of a liberal economic order and massive economic aid.

Entering the 1970s, however, this stable patron-client system began to erode. The most visible erosion has stemmed from a quite unanticipated problem: the failure to manage the growing economic interdependence between America and its allies. The liberal economic order under U.S. hegemony provided the conditions for the pursuit of outward-looking economic strategies on the part of its East Asian clients. This outward thrust has gravitated around the American market. Over the last two decades, the economic destinies of these East Asian countries have been critically linked to their ability to sell to the U.S. The U.S. has responded to this flood of exports from the East Asia with protectionist measures, departing from its free market posture. It is this rising tide of protectionism, on one hand, and the failure of the East Asian nations to adjust to it, that has made the trans-Pacific relationship more unstable and discordant.

This study is designed to illuminate the twin problems of trade

We thank Davis Bobrow, Steve Chan, Stephan Haggard, Hyung-kuk Kim and Min-Yong Lee for comments and the University of Kentucky for a Summer Faculty Research Fellowship.

friction and industrial adjustment between the U.S. and its East Asian allies through a case study of the textile and apparel sector.[1] The first section of the paper traces the nature of changing comparative advantage in the textile and apparel sector in the Pacific Basin. The second section examines the pattern of regional textile and apparel trade. The data suggests some surprises. As one would expect, U.S. and Japanese comparative advantage in apparel has declined sharply. But Korea and Taiwan have also experienced an erosion of their competitiveness in this sector. Textiles show an unexpected pattern. Korean and Taiwanese advantage has actually increased while U.S. and Japanese competitiveness has remained surprisingly stable. The third section presents a comparative overview of the adjustment policies of the four countries (U.S., Japan, South Korea and Taiwan) in the wake of changing trade patterns. Again, the findings suggest some surprises. Despite the claims of industrial policy advocates, all of the East Asian countries have experienced difficulties in encouraging exit from the industry. Promotional policies have therefore tended to come to the fore. The final section discusses the political implications of current adjustment efforts.

The Dynamics of Changing Comparative Advantage

As with other late industrializers, the East Asian countries followed a familiar path to industrialization: the textiles and apparel sector was used as a "start-up" industry. In the postwar period, Japan, Korea and Taiwan all singled out textiles and apparel industry as a strategic sector and systematically promoted it. The rationale behind this industrial choice stemmed from characteristics of the industries themselves: low investment requirements, simple processing, a rapid return on investment, labor intensity, and expanding domestic and exports markets.[2]

1. The textiles and apparel industry is chosen because of its salience as a mature industry. The trade friction and industrial adjustment discussed in this sector is likely to present a continuing and general problem, applicable in the future to trade relations with other countries and of other industrial sectors. In addition, trade frictions and industrial adjustment in the Pacific basin have begun to attract scholarly concerns. See Roger Benjamin and Robert Kudrle (eds.), *The Industrial Future of the Pacific Basin* (Boulder, Co.: Westview, 1984) and Lawrence A. Krause and Sueo Sekiguchi (eds), *Economic Interaction in the Pacific Basin* (Washington, DC: Brookings, 1980). John Zysman and Laura Tyson (eds.), *American Industry in International Competition* (Ithaca: Cornell, 1983) also presents several excellent sectoral approaches involving the U.S. and its East Asian trade partners.
2. For a comprehensive overview of the textile industry, see Brian Toyne (ed.), *The*

The U.S. facilitated such a choice by encouraging the growth of the East Asian textile industry through foreign aid programs. These included sales and grants of cotton under PL-480, aid for purchases of machinery through loans from the Export-Import Bank and the Development Loan Fund, procurement of textiles and apparel from East Asian countries by the Commodity Credit Corporation as part of the mutual security program; and technical assistance to the industry under the Mutual Security Act.[3]

The textiles and apparel industry in general is characterized by a distinctive industrial cycle of the embryonic, import substitution, massive exports, maturity and decline stages.[4] In the case of East Asian countries, favorable internal conditions, external incentives, and effective government policies have shortened the phases of this industrial cycle. Beginning in the 1960s, Korea and Taiwan entered the stage of massive exports. Apparel and fabric production have become more sophisticated. The textile complex in these countries continues to diversify its product mixes, and has maintained large net surpluses in the sector.

The textiles and apparel industries in Japan and the U.S. by contrast have reached maturity. Although total output may be increasing through industrial concentration and product and process innovation, the industry as a whole suffers from declining employment (partly because of substitution of capital for labor and partly because of foreign import penetration), an eroding international competitiveness and falling profit margins. This stage does not, however, uniformly apply to both textiles and apparel industries. While Japan and the U.S. have maintained the *status quo* in textiles (especially in man-made fiber), the apparel sector has lost its international competitiveness significantly.

To cope with this eroding comparative advantage, Japanese and American firms began to search for adjustment options. One option was to rely on outward processing, involving offshore production through direct foreign investment and subcontracting. The outward

U.S. Textile Mill Products Industry: Strategies for the 1980s and Beyond (Columbus, SC: Center for Industrial Strategy, Univ. of South Carolina, 1983); F. Clairemonte and J. Cavanagh, *The World in Their Web* (London: Zed Press, 1981); GATT, *Textiles and Clothing in the World Economy* (Geneva: GATT, 1985).
3. Vinod Aggarwal with S. Haggard, "The Politics of Protection in the U.S. Textile and Apparel Industries," in Zysman and Tyson, *American Industry. . .* , p. 261.
4. J. Arpan and B. Toyne, "The Textile Complex and the Textile Mill Products Industry," in Toyne, *The U.S. Textile...*, ch. 2. For a general discussion of the concept, see Kaname Akamatsu, "A Historical Pattern of Economic Growth in Developing Countries," *The Developing Economies*, 1:1 (March-August 1962): 3-25.

Table 7.1 RCA in the Textiles and Apparel Industry (Korea, Japan, Taiwan and the USA)[1]

		1970	1971	1980	1981
Korea	Textiles[2]	1.95	2.30	2.49	2.38
	Apparel[3]	10.67	10.79	4.99	5.14
Japan	Textile	2.46	1.37	0.78	0.78
	Apparel	0.86	0.66	0.11	0.11
Taiwan	Textile	1.67	1.32	1.78	1.82
	Apparel	4.12	3.91	3.62	3.51
USA	Textile	0.56	0.58	0.64	0.67
	Apparel	0.27	0.24	0.10	0.08

1) RCA (Revealed Comparative Advantage) refers to the ratio of a country's share in world exports of an industry group to the country's share in world exports of all non-natural resources based manufactures.
2) SITC 65.
3) SITC 84.
Source: D.S. Cha, S. Y. Park and R. K. Baik, *Hankuk, Il Bon, Taiwan, Singapore ui Suchul Gyungjangryuk Bigyo* (Comparative Analysis of Export Competitiveness of Korea, Japan Taiwan and Singapore) (Seoul: Korea Institute for Economics and Technology, 1984).
United Nations, *Yearbook of International Trade Statistics* (New York: United Nations, 1971, 1972).

processing, which has been relatively concentrated in the East Asian Newly Industrializing Countries (especially Korea, Hong Kong, and Taiwan) and Southeast Asian countries, has expedited the spread of technology, capital, and know-how. Japanese and American corporate strategies to pursue outward processing have enabled Korea and Taiwan to enjoy the global diffusion of textile-related technology, speeding the logic of the product life cycle. Korean and Taiwanese specialization in the labor-intensive textile industry suited their initial factor endowments, but also changed over time in line with the development process.[5]

Shifting comparative advantage can be measured by revealed comparative advantage (RCA).[6] RCA refers to the ratio of a country's share in world exports in a particular sector to its share in world

5. For a succinct discussion of this topic in the East Asian context, see Paul W. Kuznets, "Economic Development, Export Structure, and Shifting Comparative Advantage in the Pacific Region," in Benjamin and Kudrle, *The Industrial Future...*, pp. 35-58.
6. Bela Balassa, "Revealed Comparative Advantage Revisited: An Analysis of the Relative Export Shares of Industrial Countries, 1953-71," *Manchester School* 45:4 (December 1977): 327-344.

exports of all non-natural resource based (NNBR) manufactures. The figure 1.95 in column 1, *Table* 7.1(RCA for Korea's textiles in 1970), for example, means that Korea's share of textile exports is 95 percent more than its share in world exports of all NNRB manufactures. A RCA ratio of 1.00 or more indicates comparative advantage; one of less than 1.00, the absence of comparative advantage.

Table 7.1 presents this RCA ratio in the textiles and apparel sector across the Pacific in four selected years. The comparative advantage of Japan, once a leading powerhorse of textiles and apparel exports, had drastically declined from 2.46 and 0.86 in 1970 to 0.78 and 0.11 in 1981. The rate of decline of the American comparative advantage was less drastic than that of Japan, but nonetheless substantial. As noted before, however, one interesting aspect is that while both countries' advantages in the apparel sector fell beyond any hope of revitalization, their comparative advantage in textiles has steadied, and in the case of the United States, even improved.[7]

Entering the 1970s, on the other hand, Korea and Taiwan began to enjoy a considerable comparative advantage, which is especially impressive in the apparel sector. In 1970, for example, Korea's share in world exports of apparel was almost ten times its share in world exports of all NNRB manufactures. Taiwan's advantage was less than Korea is, but still quite substantial. One interesting phenomenon is that while the comparative advantage of Korea and Taiwan in apparel steadily declined through the 1970s (in the case of Korea almost by half in 1981), advantage in textiles increased slightly. This is largely attributable to the "downstream" linkages of the Korean and Taiwanese textile complexes and to small apparel exporters.[8]

In sum over the last three decades there has been a shift of comparative advantage in favor of Korea and Taiwan and away from Japan and the U.S. This shift was a result of beneficial factor proportions (cheap labor), the spread of technology, and interventionist government policies to promote the textile complex. Market-driven shifts in comparative advantage have also been affected by the industry's politics, however.

---

7. This can be attributed to recent efforts to upgrade productivity through vertical integration, technological innovation, and R&D investment.
8. The backward linking development refers to a pattern of textile industrial development: first came the development of an apparel industry initially designed to replace imports, then the development of a closely linked textile fabric industry, and finally the development of a man-made fiber industry.

Trade Imbalances: Sources of Trans-Pacific Friction

Until the early 1970s, world trade in texiles and apparel was relatively concentrated among the European Community, the U.S. and Japan. In 1970, intra-OECD trade accounted for 33 percent ($74.9 billion) of world total. During the 1970s, however, the pattern of trade concentration began to shift to the East Asian countries. The share of the East Asian Big Three (i.e., Hong Kong, Korea, and Taiwan) of world total exports was 7.6 percent in 1970. But in less than a decade their share almost doubled (13.2 percent), and they began to enjoy an enormous trade surplus in the sector. As of 1981, exports from the Big Three and Japan accounted for more than 20 percent of total world exports.[9]

The destination of exports from East Asian countries was geographically concentrated, however. Their exports were predominantly directed toward the U.S. market. The lion's share of exports went to the American market, with the degree of dependence in apparel especially high. Korea's apparel exports to the U.S. were $138 million in 1970, which represents 39.3 percent of its total apparel exports. Within a decade, the American market share had grown to 49 percent of Korea's apparel exports. Taiwan's dependence on the U.S. has been greater. In 1980, 61.4 percent ($527 million) of its total apparel exports went to the American market, which is a substantial increase from 47.2 percent in 1970. Japan has also been an important export market for Korea and Taiwan. Its share, however, was relatively modest (14.7 percent for Korea and 6.7 percent for Taiwan in 1980). Unlike apparel, the East Asian countries' exports of textiles to the U.S. were less significant. The U.S. market share in their textiles exports has been less than 15 percent throughout the 1970s This can be ascribed largely to U.S. firms' product and process innovation along with their advantage in raw materials, especially in man-made fiber and to protection in the sector.

This export rush by East Asian countries has in turn created excessive import partner concentration in the U.S. As *Table 7.2* illustrates, five East Asian countries (Korea, Japan, Hong Kong, China, and Taiwan) accounted for 63.75 percent of total U.S. textile and apparel imports in 1982. Although their share declined slightly in 1983, they still represented more than half of U.S. imports. In

9. Data in this section are from: Korean Institute for Industrial Economics and Technology, *Segye Sumyu Sanupui Sukub mit Muyukgujo Byunwha* (*Demand and Supply and Changes in the Trade Structure of the World Textile Industry*) (Seoul: KIET, 1984).

apparel, these countries accounted for more than 70 percent in 1982 and 1983. The U.S. is the largest textiles market in the world, absorbing more than 20 percent of total world textile production and is an obvious target for exporters. Moreover, tightened protectionist measures by the European Community, the second largest export market, and other developing countries' growing self-sufficiency in the textiles and apparel sector have limited the scope of geographic diversification, pushing the East Asian countries to concentrate on the U.S..

Table 7.2   The U.S. Textile and Apparel Imports from East Asian Countries[1]

	1982			1983		
	Total Imports	East Asian Five	Ratio (%)	Total Imports	East Asian Five	Ratio (%)
Textile:						
Yarn	495,645.3	114,431.3	23.09	2,035,839.2	223,724.3	10.99
Fabric	1,478,024.8	937,221.9	63.41	1,877,212.8	1,254,235.7	66.81
Miscellenceous	753,865.6	394,783.2	52.37	1,093,400.0	588,874.1	53.86
Apparel	3,383,635.0	2,449,527.1	72.39	3,945,313.1	2,786,525.1	70.63
Total	6,111,170.7	3,895,963.5	63.75	8,951,765.1	4,853,359.2	54.22

Source: U.S. Department of Commerce, *General Imports of Cotton, Wool and Man-made Products* April 1984, pp. 2-001-03.
1) Japan, Korea, Taiwan, Hong Kong, and China.

Import penetration by the East Asian countries has contributed to growing imbalances in the textile and apparel trade of the U.S. As *Table 7.3* demonstrates, the U.S. began to experience trade deficits in textile and apparel as early as 1960. In order to cope with deficits and resulting domestic industrial injuries, the U.S. first resorted to the Short Term Cotton Arrangement (STA), which later evolved into the more restrictive Long Term Arrangement (LTA) and Multifiber Arrangement (MFA) over the 1970s and 1980s. Despite the successive tightening of protectionist measures, culminating in the extension of the MFA III for another four years in 1986, the imbalance in U.S. textile trade has worsened. During the STA ('60-'62) and LTA ('62-'73) period, deficits grew to $1.9 billion and to $3.2 billion respectively. Even under MFA I ('74-'78) and MFA II ('78-'82), trade deficits in the textile and apparel sector were manageable, recording $3.2 billion and $2.2 billion. However, trade deficits grew rapidly during the MFA III ('82-'86), which is regarded as being the most restrictive. In 1982, the first year of the MFA III, the U.S. trade

Table 7.3 U.S. Imports, Exports, and Trade Balances for Textiles and Apparel.

($ millions)

	Textiles			Apparel			Textiles and Apparel		
	Imports	Exports	Trade Balance	Imports	Exports	Trade Balance	Imports	Exports	Trade Balance
1960	na	na	na	na	na	na	886	618	-268
1961	na	na	na	na	na	na	773	578	-195
1962	na	na	na	na	na	na	1,013	580	-433
1963	na	na	na	na	na	na	1,074	583	-491
1964	na	na	na	na	na	na	1,132	681	-451
1965	na	na	na	na	na	na	1,342	640	-702
1966	908	554	-354	608	125	-483	1,516	679	-837
1967	812	531	-281	649	164	-485	1,461	695	-766
1968	963	522	-441	855	176	-679	1,818	698	-1,120
1969	1,019	576	-443	1,106	209	-897	2,125	785	-1,340
1970	1,135	603	-532	1,267	200	-1,067	2,402	803	-1,599
1971	1,392	632	-760	1,521	204	-1,317	2,913	836	-2,077
1972	1,526	779	-747	1,883	240	-1,643	3,409	1,019	-2,390
1973	1,568	1,225	-343	2,168	278	-1,890	3,736	1,503	-2,333
1974	1,621	1,795	+180	2,331	400	-1,931	3,952	2,195	-1,757
1975	1,219	1,625	+406	2,562	403	-2,159	3,781	2,028	-1,753
1976	1,635	1,970	+335	3,634	510	-3,124	5,269	2,480	-2,789
1977	1,772	1,959	+187	4,154	608	-3,546	5,926	2,567	-3,359
1978	2,200	2,225	+25	5,657	677	-4,980	7,857	2,902	-4,955
1979	2,216	3,189	+973	5,876	931	-4,945	-8,092	4,120	-3,972
1980	2,493	3,632	+1,139	6,427	1,202	-45,224	8,920	4,834	-4,086
1981	3,046	3,619	+573	7,537	1,232	-6,305	10,583	4,851	-5,732
1982	2,807	2,784	-23	8,165	953	-7,212	10,972	3,736	-7,236
1983	2,225	2,368	-857	9,583	818	-8,765	12,808	3,186	-9,622
1984	4,874	2,382	-2,492	14,513	807	-13,706	19,387	3,189	-16,198
1985	5,274	2,366	-2,908	16,056	755	-15,301	21,330	3,121	-18,209
1986	6,151	2,570	-3,580	18,554	900	-17,652	24,705	3,469	-21,255

Source: Stanley Nehmer and Mark W. Love, ''Textiles and Apparel: A Negotiated Approach to International Approach,'' in Bruce R. Scott and George C. Lodge (eds.), U.S. Competition in the World Economy (Boston: Harvard Business School Press, 1985), p. 260; American Textile Manufactures Institute, Inc., Textile Hi Lights (Sept. 1987), p. 24.

deficit in the textiles and apparel sector was $7.2 billion, but grew to $21.3 billion in 1986, a three-fold increase in four years. The coincidence of an import surge from East Asian countries and worsening trade deficits began to attract more political attention in Washington in the 1980s than ever before.

Patterns of Industrial Adustment

International trade is seldom dictated by the logic of natural comparative advantage. A few, such as Hong Kong, may adhere to market-conforming behavior with no policy intervention. Most countries, however, pursue a mix of diverse industrial adjustment strategies including positive adjustment (or state-led exit), sectoral promotion and external protection in order to cope with changing internal and external conditions.[10] As we shall discuss below, the textile and apparel industry in the U.S., Japan, South Korea, and Taiwan is no exception. Defying the logic of the dynamics of comparative advantage, each country has protected its own market, while trying to increase its exports through promotional policies. Refusal to adjust postively has in turn deepened trade frictions.

The U.S.: The Mature Industry and Market Protection

The American response to the decline of the textiles and apparel industry has taken the form of "a protectionist snowball growing in size as the years pass."[11] The choice of protection as a policy instrument to cope with domestic surplus capacity, declining productivity, and eroding international competitiveness in the textile complex is attributable to a loosely organized but powerful textile coalition composed of cotton growers, textile manufacturers, and labor.[12]

---

10. On the issue of industrial adjustment, see Chalmers Johnson (ed.), *The Industrial Policy Debate* (San Fransisco: Institute for Contemporary Studies, 1984), Zysman and Tyson, *American Industry. . .*, ch. 1, and Bruce Scott and George C. Lodge (eds.), *U.S. Competitiveness in the World Economy* (Boston: Harvard Business School Press, 1985), pp. 1-70.

11. Aggarwal and Haggard, "The Politics. . .," p. 250.

12. For an extensive discussion of the U.S. protectionism and industrial adjustment, see Vinod Aggarwal, *Liberal Protectionism: the International Politics of Organized Textile Trade* (Berkely, CA: Univ. of California Press, 1985); Aggarwal and Haggard, "The Politics. . .,"; Stanley Nehmer and Mark W. Love, "Textiles and Apparel," in Scott and Lodge, *U.S. Competitiveness...*, pp. 230-262; Brian Toyne, "Strategic Adjustments in Textile Mill Products Industries," in Toyne, *the U.S. Textile. . .*, ch. 8; R. Buford Brandis, *The Making of Textile Trade Policy 1935-1981* (Washington, D.C.: American Textile Manufacturers Institute, 1982).

The first protectionist effort was achieved through the Short Term Cotton Arrangement (STA) of 1961, which cut the total value of cotton-based textiles and apparel imports from $268.7 million in 1960 to $214.5 million in 1961. The STA was soon followed by the Long Term Arrangement (LTA) in 1962 that was renewed until 1974. The LTA limited the growth of cotton textile imports to 5 percent per year for the 19 signatory states. The LTA, nevertheless, did not produce the anticipated outcomes for the industry. It neither helped improve structural adjustment nor enhanced the international competitiveness of the textile industry. U.S. protection slowed the diffusion of labor-displacing innovations and automation, which in turn undermined the modernization of production capacity. Protection expedited geographical, rather than structural, adjustment in terms of industrial relocation to the southern states where labor is cheap. Ironically, the LTA helped East Asian exporters by driving them to product diversification (into man-made fiber and apparel) and to product and process innovation, which in turn allowed them to penetrate in the U.S. market with a more diverse array of products.

As the LTA proved inadequate in the early 1970s, the textile coalition began to seek a more extensive and systematic protectionist mechanism. The result was the creation of the Multi-Fiber Arrangement (MFA) in 1974, which extended protection from cotton to synthetic fibers. The MFA I, which institutionalized bilateral negotiation within the multilateral agreement, allowed major importing countries (mostly the U.S. and the EC) to keep annual increases of exports from developing countries below 6 percent. In 1978, the MFA I was renewed and extended to 1981. The MFA II became more protectionist by including the "reasonable departures" clause, which allowed importing countries to go below the 6 percent export growth rate for sensitive products. The inclusion of the "reasonable departures" clause in turn accommodated American demands to cut back imports from Korea, Taiwan, and Hong Kong by permitting selective treatment of suppliers.

This enforced protectionism via the MFA II notwithstanding, the import surge from East Asia continued and the textiles trade balance worsened (see *Table 7.3*). The U.S. textile lobby, along with that of the EC, regarded MFA II as a failure. As the third round of the MFA began in 1982, the American position became more restrictive. Washington demanded the inclusion of an "anti-surge" provision to prevent a too rapid influx of imports, regardless of quota levels. In addition, it requested the elimination of the "reasonable departures" clause which allowed importers to exercise some discretion in short-term adjustment of quotas. Both demands were incorporated into the MFA III.[13] As shown in *Table 7.2* however, import penetration from East Asian countries increased even *after* the MFA III. 1983 recorded the worst deficit in the U.S. textiles trade balance.

Arguing that the disruptive surge in imports of textiles and textile products during 1981–1984 resulted from the failure of the U.S. to adequately enforce its rights under the MFA, Congressman Jenkins and Senator Thurmond proposed the Textiles and Apparel Trade Enforcement Act in 1985. The bill was ultimately vetoed by President Reagan after a contentious legislative fight. Had it been enacted, however, the Act would have been one of the most protectionist and damaging pieces of trade legislation since the Smoot-Hawley tariff legislation in 1930. Departing from the existing norms, rules, and procedures of the international textiles trading regime (GATT-MFA), the Act would have unilaterally limited the growth of textiles and apparel imports from 12 leading exporters to less than one percent per year. In addition, the Act would have mixed quota restrictions with an import licensing system to control the rate of import penetration. Although the Act was vetoed by President, the U.S. government used it as a bargaining leverage to extend the MFA III for another five years in 1986 and to extract restrictive bilateral agreements with the major East Asian producers.[14]

The protectionist option has neither prevented import penetration nor corrected sectoral trade imbalances. The logical lesson of this failure should have been a policy of internal structural adjustment. The American government has not devised any measures to ease exit, however. No relief or assistance has been extended to the textiles and apparel industry. There were nominal efforts to promote the industry by offering overseas market information and by trying to eliminate foreign trade barriers of American textiles and apparel goods. As with other industries, the United States proved unable to develop sector specific policies.

While the government was relying on protection, there have been corporate-level adjustments. Facing tough competition from abroad, American textiles and apparel firms continued outward processing, offshore production and subcontracting. Internally, they have

---

13. Vinod K. Aggarwal, "The Unraveling of the Multi-Fiber Arrangement, 1981: The Examination of International Regime Change," *International Organization* 37:4 (Autumn 1983), pp. 617-645.
14. Mun Bong Lee, *Seke Sumyu Muyuk Jilsuh Byunhwa wa Urioi Daieung* (*Changes in World Textile Trade Order and Our Countermeasures*) (Seoul: Korea Institute for Industrial Economics and Technology, 1986); The Textiles and Apparel Trade Enforcement Act of 1985 (draft), Feb. 26, 1985.

pursued vertical integration, production innovation, and substitution of capital for labor. These corporate strategies enhanced their competitive position.[15] Nevertheless, their weak coordination with the government in the face of eroding comparative advantage and the search for short-term benefits via market protection reduced the adaptability and competitiveness of the U.S. textile complex. This was only compounded by erratic macroeconomic policy.

Japan: Between Phase-Out and Promotion

As noted before, "center stage" of the trans-Pacific textiles and apparel trade has been the American market. Thus, the pattern of industrial adjustment in East Asian countries has been closely linked to that of U.S. policy. Overall, Japan has pursued both positive adjustment policies of inducing structural changes in the industry and policies of protection and assistance in order to adjust to rising costs, increasing U.S. protection and declining comparative advantage. Unlike the United States, the Japanese government intervened extensively in the industry.[16]

The textile complex was an engine of growth and a primary source of export earnings in the early phase of postwar economic reconstruction. As early as 1946, the Ministry of International Trade and Industry (MITI) established the Three Year Textile Industry Reconstruction Plan in order to promote the industry. The Korean War boosted the Japanese textile industry in the mid-1950s more than the plan had envisioned. Its expansion was soon followed by recession and a sharp downswing, however. Excess capacity and intense competition among small scale enterprises drove the industry to the brink of collapse. The Japanese government intervened by enacting the Temporary Law Governing Textile Industry Equipment in 1956. The law aimed at promoting an orderly growth in exports through the "rationalization" of the industry and a reduction of surplus capacity. To carry out these objectives, the government sought to improve production technology through financial and tax incentives, and to coordi-

---

15. Terence A. Shimp, "Corporate Strategies and Adjustments," in Toyne, *U.S. Textile. . .*, ch. 7.

16. For the Japanese textile industry, see Tsusho Sangyosho (MITI), *Atarashii Jidai no Sen-i Sangyo* (*The Textile Industry in the New Age*) (Tokyo: MITI, 1984); Sueo Sekiguchi and Toshihiro Horiuchi, "Foreign Trade and Industrial Policies: A Review of Japanese Experience," in Benjamin and Kudrle, *The Industrial Future. . .*, pp. 85-87; Brian Ike, "The Japanese Textile Industry: Structural Adjustment and Government Policy," *Asian Survey* 20:5 (May 1980), pp. 532-551; *Sen-i Nenkan: 1984* (Tokyo: Sen-i Shimbunsha, 1985).

nate surplus capacity through tight control of new investment. From 1956 to 1967, when this law was effective, the intention of the Japanese government was to *promote* the industry through effective sector coordination and management.

Toward the end of the 1960s and early 1970s, however, the internal and external market parameters of the Japanese textile complex were changing rapidly. The U.S., then Japan's largest export market, targeted its protectionist policy on Japan through the enforcement of the LTA. In addition, in 1971 the U.S. imposed voluntary export restraints (VER) on Japanese synthetic fibers, which were not covered by the LTA. At the same time, Japan was losing its competitive edge to the East Asian developing countries, which singled out the textile and apparel industry as a leading export sector. Even worse were the effects of Japan's domestic economic conditions. An acute shortage of labor following the industrial restructuring in the 1960s and the appreciation of the Japanese yen further weakened the competitive position of the Japanese textile complex.

The Japanese response to this change was a mixture of domestic protection and promotion. The protection was largely manifested in subsidy and assistance. For example, the Japanese government awarded 48.9 billion yen to textile firms in compensation for injuries incurred from the U.S.-Japan Textile Agreement of 1971, in which Japan accepted U.S. demands for VER. At the same time, the government began to promote the industry selectively by accelerating the disposal of excess machinery, the modernization of production capacity, and the concerntration of production. During the 1967–1974 period, the government provided 329.9 billion yen in low-interest, long-term loans to facilitate its selective promotion of the industry.[17]

These supportive measures could not reverse prevailing market trends, however. The Japanese textile industry continued to lose its competitiveness. Moreover, Japan joined the Generalized System of Preferences in 1971 which allowed a surge of imports from East Asian developing countries, further threatening the Japanese textile complex. At this juncture, the Japanese government began to re-direct its intervention from temporary relief and assistance to structural reorganization of the entire industry. The enactment of the Temporary Law for the Structural Re-organization of the Textile Industry in 1974, which was amended twice (1979, 1984) and extended to June 1986, was a reflection of this government concern. The law identifies three priority policy areas. The first is structural improvement, which

---

17. Ike, "The Japanese Textile. . .," p. 640.

emphasizes the development of new products and technology, promotion of exports through modernization of production capacity, and transition to high value-added products via vertical integration. Second, the law allowed the government to intervene in the industry to eliminate excess capacity and limit new investments in production facilities. Finally, exit and conversion have been strongly encouraged.[18]

These policies have been backed by a variety of tax, finance, R & D, and administrative incentives. Firms are entitled to low-interest (2.6 percent), long-term loans (12 years) for the modernization of production facilities. In the development of new products and technology, firms are entitled to the same interest rate, but for longer term (16 years), plus hefty tax benefits (depreciation, exemption for R & D investments, and local land-tax exemptions). In shutting down existing capacity, reducing the operating ratio, and converting the capacity to other industries, the government provided more comprehensive incentives. They include: financial incentives (low interest, long term loans and guarantee for corporate debts), tax incentives (exemption for new investments, reduction for R&D expenses, corporate tax benefits for converting firms), and employment adjustment fund (for re-training and temporary unemployment relief), management support (marketing information and management guidance for small firms), and national, provincial, and corporate level R&D support.[19]

Results of this policy reorientation remain to be seen. But an interim assessment shows that efforts of both structural reorganization and positive adjustment through exit and conversion into other industrial plants have had limited effects. During the period 1975–1983, a total of 72.1 billion yen was allocated for the concentration of production units and the modernization of production capacity, but that benefitted only 103 projects. On the other hand, phase-out and conversion rates (8.6 percent during 1978–1981) in the textile and apparel sector is higher than that (6.8%) in the manufacturing sector as a whole. But given the urgency attached to the textile program and the high level of government inducements, this differential seems quite low.[20] It was the sagging tempo of adjustment that made it necessary to extend the law to 1989. Meanwhile, the Japanese government has tried to tighten its protectionist position through "orderly import arrangements" and by withdrawing GSP status from selected East Asian countries. Interestingly, Japan has departed from its previous attitude and now

18. MITI, *Atarashii. . .* , pp. 110-142.
19. MITI, *Atarashii. . .* , pp. 162-164.
20. MITI, *Atarashii. . .* , pp. 148-150.

advocates a more restrictive position in recent MFA negotiations. In sum, the Japanese government has been wrestling with the conflicting policy goals of protection, phase-out/conversion, and promotion.

South Korea: From Neglect to Promotion?

The textile and apparel sectors have been one of the dominant manufacturing sectors in Korea. They have consistently been a leading source of export earnings and employment. The expansion of the Korean textile complex was a function largely of shifting comparative advantage, especially cheap labor. The market factor alone does not fully account for this success, however. The Korean government's systematic intervention has also contributed to this success. The consistent theme of government policies has been promotion and assistance, along with protection based on import substituting industralization.[21]

The first textile policy was embodied in the Five Year Emergency Cotton Industry Reconstruction Plan. The Plan was designed to meet surging domestic demand following the Korean War through both supporting domestic cotton manufacturers and rationing American cotton aid. As Korea shifted its pattern of economic development from import substitution to export promotion in 1965, the textiles and apparel industry emerged a strategic export sector. Initially the textile complex benefitted from a wide range of non-discretionary export incentives including preferential finanical and tax treatment. But the textile lobby in Korea argued that given growing domestic and foreign demand and its relative importance in export earnings and employment, the non-discretionary incentive system under the initial export-promotion strategy was not sufficient to guarantee its international competitiveness. Industry demands were met with the enactment of the Temporary Law Governnng Textile Industry Equipment in 1967. The law assisted the industry by improving production capacity through the replacement of old facilities by new ones.

The confluence of changing comparative advantage, the non-discretionary export incentive system, and sector-specific assistance

---

21. For the Korean textile industry, see The Federation of Korean Textile Industries, *Sumyu Sanup Jaidoyakoi Gil: Sumyu Baiksuh* (*Path to the Revitalization of the Textile Industry: A Whie Book*) (Seoul: FKTI, 1985); FKTI, *Migukui Sumyugyuje Siltae wa uri Upgyeui Dangmyun Gwaje* (*The U.S. Textile Import Restriction and the Confronting Tasks of Our Industry*) (Seoul: FKTI, 1984); Yong Mok Sung, *Urinara Sumyu Gongup ui Kukje Gyungjaengryukae gwanhan Nyunku* (*A Study of International Competitiveness of Our Textile Industry*) (Seoul: Sungkyunkwan Univ., unpublished Master Thesis, 1982).

accelerated the expansion of the textile complex. Moreover, the outward processing rush by foreign textile firms and the Korean government's active inducement policy have facilitated the inflow of advanced technology and capital. The creation of free trade zones and the provision of tax holidays, coupled with the government's tight control of labor, attracted foreign investors to the textile sector.

In the mid-1970s, internal and external conditions became less favorable. First, Korea faced increasing protection from major importers. Although Korean textile and apparel exporters effectively coped with this protectionist trend through product diversification and the employment of various countervailing strategies,[22] the government, if not the individual firms, began to foresee the limits on the industry's expansion. Second, the Korean textile complex was gradually losing its comparative advantage to newly emerging Southeast Asian textile exporters. In addition, the textile export boom in the late 1960s and early 1970s led to overinvestment, which in turn created excess capacity and intense competition. Finally, industrial restructuring efforts aimed at creating a heavy industry base that were launched in the mid-1970s caused an acute labor shortage, which, coupled with mounting inflation, drastically increased labor costs. Preoccupation with heavy industry diverted resources from textiles and apparel. These changing conditions led the Korean government to declare the textile complex a sunset industry.

By the end of the 1970s, however, government-sponsored heavy industrialization was revealed to be a costly mistake. The strategic sectors (e.g., heavy machinery, shipbuilding, automobile, petrochemical, and electronics) suffered from capacity underutilization because of overinvestment, duplication, and sagging exports. Ironically, the textile sector filled the gap left by these strategic sectors. Even in the late 1970s, the textile industry still maintained a strong comparative advantage. Korean firms have also managed to adjust surprisingly well to increasing U.S. protectionism. With only a minimum investment (compared to capital intensive heavy industry), the textile lobby argued, the textile complex could generate lucrative export earnings. These factors altered the Korean government's intention to actively phase out the textile sector by reducing support to the industry and allowing competition to operate more freely.

The government shifted from its previous policy position of neglect to one of active promotion. The government's promotional policies were codified through the enactment of the Textile Industry Moderni-

---

22. For these tactics, see David B. Yoffie, *Power and Protectionism* (New York: Columbia Univ. Press, 1983).

zation Act in 1979. The law identifies four operational goals and strategies: 1) the structural reorganization of the textile complex through vertical integration; 2) the differentiation of the industry through product diversification and transition from high volume, low-to-medium priced products to high-value-added and fashion-conscious ones; 3) the modernizatin of production capacity by the inducement of advanced foreign technology, research and development investment, and improvement and replacement of old facilities; and 4) strengthening of production and marketing information.[23]

In order to achieve these goals, the Fund for Textile Industry Modernization was established in 1980. The government was to contribute 60 billion won during the period 1981–1986, half of the total fund. The allocation of the Fund was focused chiefly on improvement and modernization of production facilities and technology development including that of design and fashion. During 1981–1983, 12.7 billion won were allocated for low-interest (8 percent for facilities modernization and 6 percent for technology development), long-term (8 and 5 years respectively) loans. The government also invested an additional 12 billion won in the textile industry (especially small-scale firms) out of the Fund for the Promotion of Small and Medium Scale Enterprises. At the same time, a wide range of tax incentives has been extended: tariff exemption for imported raw materials, tax reduction for investments and depreciation, and special tax provision for small scale textiles and apparel firms.[24]

Korean industrial policy in the textile and apparel sector resembles that of the Japanese in that assistance has been extended in order to revitalize the industry. But they differ in the policy emphasis given to exit and conversion. The Japanese government has attempted to phase out the textile industry and/or convert it to other productive sectors, although with only partial success. Phase-out and conversion options have not been incorporated in the Korean government's textile policy. Several factors explain this. The comparative advantage of the textile complex remains strong, although it is gradually declining. Due to its importance in employment, phase-out or conversion without well-planned adjustment assistance would have more extensive political implications. Given these constraints, the Korean government is rather likely to continue the promotion of the textile complex for

---

23. See Sumyu Sanup Kundaewha Chokjin Bub (The Textile Industry Modernization Act).

24. The Federation of Korean Textile Industries, *Sumyu wa Uiryuae daehan Suchul Jinheung Jungchaek* (*Policies Concerning the Promotion of Textiles and Apparel*) (Seoul: FKTI, 1982), pp. 151-160.

the time being than to positively adjust to mounting protectionism and the increasing challenges from other developing countries.

Taiwan: Revitalizing the Sunset Industry

As in Korea, the textile industry was Taiwan's first major industry and remains one of its leading export items. But Taiwan was ahead of Korea in its development process. The textile industry constituted an integral part of its First Four Year Economic Development Plan in 1953. During the plan period, the textile industry was selected as a strategic import substitution sector. This permitted the intensive promotion of the apparel industry in order to replace imports. As Taiwan made the transition from an inward-looking to an outward-looking economic development strategy in 1961, however, the textile sector became the key export sector. In contrast to Japan and Korea, Taiwan did not implement any sector specific industrial policy. Incentives were largely non-discretionary, including changes in macroeconomic and exchange rate policies. Nonetheless, they were crucial to the industry. The Statute for the Encouragement of Investment (1961) entitled the textile industry to several financial and tax benefits extended to other industries. On the other hand, the enactment and timely amendment of the Statute for Investment by Foreign Nationals (1954) and the Statute for Technical Cooperation (1962) has helped attract foreign capital and technology essential to the expansion and modernization of the industry, including from overseas Chinese. The creation of export processing zones was especially helpful in facilitating the inflow of textile-related foreign capital and technology, mostly Japanese. The expanding export markets and comparative advantage in labor costs made the textile industry a booming sector.[25]

Starting in the late 1960s, however, the Taiwanese textile industry encountered the same constraints that faced the Korean industry: import restrictions, the emergence of ASEAN exporters, and rapid industrial reorganization into high-value-added technology intensive sectors such as electronics and resulting labor shortages that weakened Taiwan's comparative advantage in cheap labor. Despite these market changes, the government did not devise any sector specific industrial policy until the Ten Year Textile Industry Re-vitali-

---

25. A. Arpan, M. Barry, and T. Van Tho, "The Textile Complex in the Pacific Basin," in R. Moxon, T. Roehl, and J. Truitt (eds.) *International Business Strategies in the Asia Pacific Area* (New York: JAI Press, 1984), pp. 101-164; David Ricks, "An Overview of Government Influence," in Brian Toyne and Jeffry S. Arpan (eds.), *The U.S. Textile Mill. . . .*

zation Plan in 1980.

The Ten Year Plan is one of the Taiwanese government's several sector specific industrial policies. The following passage illustrates the background and goals of the Plan:

> In view of domestic economic structural change and the loss of comparative advantage in international markets, the future development of the textile industry shall be oriented toward the integrated development of up-, mid-, and down-stream facilities, renovation of equipment, multi-stage processing. . . , so that productivity may be increased, quality improved, and value-added raised *in order to promote brisk exports.* (emphasis is mine).[26]

Explicit in the preceding passage is that in spite of unfavorable market conditions, Taiwan is not interested in conversion of the industry or its phase-out. On the contrary, the Taiwanese government is attempting to reverse the trend of eroding comparative advantage through promotional policies. The passage also indicates that the textile and apparel industry will remain a key export sector. Taiwan's textile industrial policy has not greatly deviated from that of Korea. It focuses on three elements: 1) structural reorganization through the promotion of vertical integration; 2) differentiation of products into high quality and high-value-added ones through technological innovation and research and development; 3) modernization of product capacity by the improvement and renovation of equipment.

The implementation of the plan has not been smooth. The textile complex in Taiwan is relatively fragmented with the existence of small apparel and textile firms. This fragmentation is held responsible for the weakening the industry's competitiveness. The plan's first target was to consolidate these small firms into larger, vertically integrated groups. Primarily due to resistence by small firms, however, this consolidation process has not been implemented as anticipated. In addition, a package of comprehensive financial, tax, investment, and administrative incentives embodied in the plan has not succeeded in upgrading the quality, developing new products, or improving equipment, as originally planned. These sluggish improvements notwithstanding, it is unlikely that Taiwan will redirect its present textile industrial policy: promotion and assistance.

---

26. Council for Economic Planning and Development, *Four-Year Economic Development Plan for the Republic of China: 1982-1985* (Taipei: Executive Yuan, 1981), p. 62.

Conclusion: Structural Rigidity and Trade Politics

Table 7.4 summarizes industrial adjustment strategies in the U.S., Japan, South Korea, and Taiwan. What is striking is that each of the four countries seeks to protect its domestic market, and at the same time to promote exports. The direction of their promotional policies is strikingly similar: to encourage high-quality, high-value-added, and fashion-conscious products through vertical integration, modernization of production capacity, and technological innovation. All of these countries except Japan are reluctant to phase out the textile industry or convert it to other industrial sectors through conscious policy intervention. Even for Japan, the phase-out and conversion of the industry has only been selectively encouraged, and has met with only partial success. Promotion remains an important element of Japanese policy.

How can a collison among these strategies be avoided? Multilateral coordination through the MFA is an option. The MFA has, however, become increasingly restrictive. The return to GATT principles through the dissolution of the MFA has been a principal developing countries demand to the Uruguay Round of GATT talks, but remains highly unlikely given U.S., European, and increasingly Japanese, interests.[27] U.S. lawmakers from districts with textile and apparel industries introduced new textile legislation before the 100th Congress that would further restrict textile and apparel imports.[28]

The persistence of U.S. protectionism, coupled with failure of East Asian countries to dissolve the MFA and to return to the GATT, forces them to look for alternative options. As David Yoffie points out in this volume, both South Korea and Taiwan have managed to circumvent protectionism by diversifying products, exploiting legal ambiguities, and cheating through transshipment and other measures. Successive rounds of the MFA negotiations have tightened rules and procedures, narrowing the margin of East Asian countries' tactical maneuverability. South Korea and Taiwan have now pursued alternative external adjustment strategies that extend those suggested by Yoffie, including trading partner diversification, off-shore production, and the promotion of trade within developing countries.[29] However,

---

27. See William Cline, *Exports of Manufactures from Developing Countries* (Washington, D.C.: the Brookings Institution, 1984); Martin Wolf, "Managed Trade in Practice: Implications of the Textile Arrangements," in William Cline (ed.), *Trade Policy in the 1980s* (Washington, D.C.: Institute for International Economics, 1983).
28. *Wall Street Journal* February 20, 1987.
29. Lee, *Seke Sumyu. . .* , pp. 170-182.

Table 7.4 Comparative Overview of Industrial Adjustment (Textiles and Apparel Sector)

Type	USA	Japan	Korea	Taiwan
External Adjustment:				
—protection	extensive (MFA IV)	extensive (MFA IV)	extensive	extensive
—outward processing	active/ firm specific	active/ firm specific	no	no
—inducement	no	no	yes, but declining	yes, but declining
Internal Adjustment:				
—phase-out/ conversion	market force/ no policy intervention	extensive state policy (finance/ tax/ retraining/ job compensation)	nominal policy/ market force	nominal/ market force
—relief/ assistance	no sector specific policy	extensive (import relief/ compensation/ supports for re-tooling & reinvestment)	modest (selective bail-out/ re-tooling & re-investment)	modest (selective bail-out/ re-tooling & re-investment)
—promotion	nominal/ low (marketing/ eliminating foreign trade barriers)	sector specific policy/ extensive (high value-added items/ production concentration/ R & D support/ tax & financial incentives)	sector specific/ extensive (high value added items/ R & D and tax incentives/ production concentration)	sector specific/ extensive (high value added items/ R & D incentives/ production concentration/ tax and financial incentives)

growing protectionism by Japan and the EC, the second largest market and continuing shift in comparative advantage to other developing countries weaken the long-term viability of these external options.

Since retaliation by Korea and Taiwan remains unlikely because of the economic and political risks, the dominant strategy is to adjust domestically. Both phase-out and promotion of structural change have their own political costs, however. Internal adjustments raise the political issue of absorbing the economic costs associated with loss of employment. Korea's textile industrial complex, with about 8,000 firms and 750,000 employees, constitutes roughly one fourth of the nation's manufacturing sector. Moreover, the textile labor unions are relatively well organized, and independent labor unions have grown in recent years as a result of the trend to political liberalization. In Taiwan, labor unions owning to the government's tight control are less a problem, but small and medium firms, the core of the Taiwanese textile complex, have been an important source of political support for the government. Political liberalization in both countries thus creates difficult trade-offs for both regimes.

These choices are made more difficult by the growth of U.S. bilateralism described by Haggard and Cheng in this volume. Although the Jenkins bill was vetoed by President Reagan, the bill's selective targeting of East Asian countries had created the popular perception in South Korea and Taiwan that the U.S. is treating them unfairly and that their governments have failed to counter this discrimination.[30] This was particularly true because U.S. protectionism coincided with renewed U.S. pressure for import liberalization by Korea, Taiwan, and Japan, and with the recent decision to withdraw GSP benefits from the East Asian NICs. Student demonstrations in Korea have increasingly linked anti-government slogans with anti-American ones, and textile workers have staged anti-American demonstrations in Seoul. Furthermore, those technocrats who have been negotiating

---

30. As a matter of fact, import surge in the U.S. market over the last several years, the primary concern of the Jenkins bill, came from OECD countries excluding Japan. For example, imports from Canada increased 85 percent between 1983 and 1984 while those from the EC increased 71 percent, or more than double the rate of growth of textiles from all sources. But the bill exempted these countries from the list of import restriction. Even China was dropped from the list in fear of its potential economic retaliation. See the Korean Ministry of Trade and Industry, "The Textile and Apparel Enforcement Act: A Misconceived Policy Prescription" (mimeo for circulation) September 1985; Ablondi and Foster, P.C. (Counsel to the Taiwan Textile Federation), "Opposition of the Taiwan Textile Federation to the Textiles and Apparel Trade Enforcement Act of 1985," (mimeo for circulation), September 20, 1985, pp. 12-13.

with the U.S. have been accused of being pro-American puppets. This anti-American sentiment is less evident in Taiwan largely due to its large trade surplus with the U.S. However, it is undeniable that the traditional Taiwanese-U.S. relationship, which survived past shocks in the wake of diplomatic severance, will also face stress in the future. In both Seoul and Taipei, popular perception of America is shifting from that of a benign hegemonic leader to a hostile spoiler. This changing popular perception, combined with the growth of over anti-American sentiment, not only can undermine Washington's macro-strategic objectives in the region, but also hurt political regimes in both countries, which have pledged their affinity and loyalty to Washington.[31]

The central theme of this paper is that the shifting comparative advantage entails transnational industrial adjustment. Failure to coordinate industrial adjustment leads to the deepening of trade frictions that can also have political costs. The textiles and apparel industry in the Pacific basin illustrates this dilemma *par excellence*. Refusal to adjust harmoniously has led to the perpetuation of a trade friction spreading to other economic sectors. The trans-Pacific alliance ties, which have been traditionally strong and congenial, now suffers from precariousness and uncertainty. To heal this fracture, the countries involved need to come up with a prudent, well planned, and mutually acceptable industrial coordination scheme through either multilateral, regional or bilateral efforts.[32]

---

31. Although it has not been the key determinant of anti-American sentiment, the U.S. textile protectionism served as a catalyst.

32. Equally important is macroeconomic policy coordination. Especially, foreign exchange rate policy is critical in reshaping the nature and direction of textile and apparel trade. Recent sharp appreciation of Korean won and Taiwanese dollar have deprived textiles and apparel exporters in Korea and Taiwan of their compeitive edge significantly.

CHAPTER 8
American-Japanese Cross Investment: A Second Front of Economic Rivalry*

Dennis Encarnation

For much of the 1970s, Japan's trade balance with America was barely sufficient to cover Japan's imports of oil and other raw materials, and its payments for services. But by 1986, that balance reached an unprecedented $53 billion, an increase of over 25 percent from the previous year alone. This infusion of foreign exchange, supplemented by a level of domestic savings that exceeded the capital demanded by local investors, allowed Japan to become a large capital exporter. America, previously a major capital exporter, became a principal recipient.

Japanese investments in America and elsewhere take many forms: bonds, loans, stock portfolios, as well as direct investments. While bonds now account for the largest flow of funds, direct investments attract the greatest attention, since they mix equity ownership with managerial control, and have been growing rapidly.[1] (See Table 1.) Between 1980 and 1986, Japanese direct investments in the United States rose over 33 percent annually, a rate of increase that easily outpaced investments by every other foreign nation.[2] Over those six years, Japan moved ahead of Canada, Germany, and Switzerland to become America's third largest foreign investor, behind Britain and the Netherlands. From the new Japanese factories, assembly plants, and warehouses that have begun to dot the American landscape, Americans purchased an additional $13 billion of "Japanese" goods in 1984. This was the value added by these investments over and above the $57 billion

---

*An earlier version of this paper, subsequently updated and reorganized, appeared in ment by Multinational Firms: A Rivalistic Phenomenon?" *Journal of Post-Keynesian* Press, 1986), pp. 117-149.

---

1. Of Japan's net long-term outflow of $49.6 billion in 1984, for example, $26 billion went into net purchases of foreign, principally government, bonds. Less than $3 billion went into direct investments. *Economist,* September 21, 1985, pp. 79-80.
2. U.S. Department of Commerce, Bureau of Economic Analysis, *Survey of Current Business* 67 (August 1987), p. 90; *Survey of Current Business* 60 (August 1982), p. 36.

of components and finished products imported directly from Japan.[3] Such sales subsequently grew as new projects, typically wholly-owned by the Japanese, proliferated. So, like the original trade competition, this second front of economic rivalry represents an increasingly formidable Japanese challenge to U.S. business—now mounted on America's home ground.

Although less publicized, American business has also been encroaching on the Japanese home market. In 1983, for example, only Canada and Switzerland surpassed Japan as the target for new U.S. direct investment abroad.[4] And by 1984, U.S. multinationals, operating in Japan typically through minority-U.S. or equal-partnership joint ventures, sold $44 billion of locally produced goods. This was the value added by American-affiliated plants over and above the $26 billion of components and other products imported from the United States. Such local sales seem likely to increase with the continued expansion of American investments in Japanese manufacturing, thereby retaining for America its title as Japan's largest foreign investor.

Most cross-investment between Japan and the United States is a recent development, but not entirely. Even before World War II, IBM, General Motors, and other American multinationals operated in Japan, just as Mitsui, Mitsubishi, and other Japanese trading and financial organizations invested directly in the United States. In financial terms, such cross-investment in 1936-37 was roughly equal: $47 million in Japan, $41 million in the United States.[5] After WWII, however, this rough symmetry disappeared, as American investment in Japan quickly spurted, growing nearly 25 percent annually between 1950 and 1970. Despite such growth, Japan still remained a relatively insignificant recipient of American direct investment. Before World War II, it accounted for less than 1 percent of annual American outflows. This figure did not exceed 6 percent for over 20 years after the war.[6] Until the late 1970s, Japan remained an unlikely host for American investors.

The United States was even a less likely host for Japanese investors.

---

3. Kenichi Ohmae, *Beyond National Borders: Reflections on Japan and the World* (Homewood, IL.: Dow Jones-Irwin, 1987), p. 27, Figure 1.
4. *Survey of Current Business* 64 (August 1984), p. 31.
5. Mira Wilkins, "American-Japanese Direct Foreign Investment, 1930-1952," *Business History Review* 56 (Winter 1982), pp. 498-510. U.S. direct investment in Japan probably peaked at $61.4 million in 1930; see Wilkins, Table 2, p. 506. Japanese investment in the U.S. probably peaked at $41.0 million in 1937 following a sharp increase that year; see Wilkins, Table 3, p. 507.
6. For prewar data, see Wilkins, "American-Japanese Direct Foreign Investment," p. 498. For postwar data, see U.S. Department of Commerce, Bureau of Economic Analysis, *Selected Data on U.S. Direct Investment Abroad, 1950-76* (Washington, D.C.: USGPO, 1982), pp. 1, 21.

Much of the postwar period was marked by continued trade deficits and industrial reconstruction, so Japan could ill-afford investments abroad that employed scarce foreign exchange. To the extent that the Japanese invested overseas at all, their attention focused on the developing countries of southeast Asia.[7] In the United States, by contrast, Japanese investments for over twenty years after WWII grew at one-half the rate of U.S. investments in Japan. As late as 1974, the book value of American direct investment in Japan remained ten times larger than the total value of all Japanese investments in the United States. (See Table 8.1)

Subsequently, however, the asymmetry in American-Japanese cross-investment dramatically reversed. By 1981, the stock of all such Japanese investments had surpassed the stock of all U.S. investment in Japan; only five years later, Japanese investment in America stood at nearly twice the value of U.S. investment in Japan. During the 1980s, the United States had replaced Southeast Asia as the principal target for new direct investments,[8] with over two-fifths of all annual investment

Table 8.1 American-Japanese Cross-Investment (FDI) 1974-1986[a]

Year-End	U.S. FDI To Japan	Japanese FDI to U.S.
1974	$ 3.3 billion	$ 0.5 billion
1980	$ 6.2 billion	$ 4.2 billion
1986	$11.3 billion	$23.4 billion

[a] According to the definition employed by the U.S. Commerce Department, direct investment position abroad (FDI) is equal to parent companies' equity (all stock, additional paid-in capital, and parents' shares of undistributed earnings) in, and net outstanding loans (long-term debt, trade credits, and other current liabilities owed to parents) to, foreign affiliates.

Sources: United States Department of Commerce, Bureau of Economic Analysis, *Survey of Current Business,* various issues; *Selected Data on U.S. Direct Investment Abroad, 1950-76,* February 1982, pp. 20-27.

7. Japan, Ministry of Finance, "Japanese Data on Direct Investment Flows," Mimeo, Tokyo, 1984. These data are also reproduced in Japan, Ministry of International Trade and Industry, *White Paper on International Trade, 1985* (Tokyo: MITI, 1985), pp. 88-90. For an excellent review and analysis of these trends, see Hideki Yoshihara, "Multinational Growth of Japanese Manufacturing Enterprises in the Postwar Period," in Akio Okochi and Tadakatsu Inoue, eds., *Overseas Business Activities: Proceedings of the Fuji Conference* (Tokyo: University of Tokyo Press for the Ninth International Conference on Business History, 1984), pp. 95-120.

8. In addition to the sources listed in note 7 above, also see Export-Import Bank of Japan, Research Institute of Overseas Investment, *Exim Review* 5 (March 1985), pp. 37-68.

Table 8.2 Sectoral Distribution of American-Japanese Cross-Investments
(FDI)[a]

a. American FDI in Japan

Sector[b]	Distribution Stocks at Year-End			Compound Annual Growth Rates	
	1974	1980	1986	1974-80	1980-86
Manufacturing	45.8%	47.6%	46.8%	11.8%	10.1%
Trade[c]	8.4	17.9	20.0	25.9	12.6
Finance[d]	1.5	3.0	7.7	24.5	29.5
Petroleum	41.2	25.1	23.1	2.3	8.9
Other	3.1	6.4	2.4	25.8	-6.2
Total Position[a] ($ billion)	$3.3	$6.2	$11.3	13.4%	10.5%

b. Japanese FDI in the U.S.

	1974	1980	1986	1974-80	1980-86
Manufacturing	95.7%	19.8%	12.9%	16.8%	23.8%
Wholesale Trade[e]	-128.1	54.6	55.3	274.3	33.3
Finance[d]	26.2	18.6	13.8	43.4	26.5
Other[e]	-6.2	7.0	18.0	54.9	55.8
Total Position[a] ($ billion)	$0.3	$4.2	$23.4	42.6%	33.1%

[a] Direct investment position abroad, as defined in Table 8-1, note a.

[b] Each foreign affiliate was classified in the major industry group that accounted for the largest percentage of its sales at year end.

[c] Includes both retail and wholesale trade for 1974 and 1980; only wholesale trade for 1986.

[d] Includes banking and insurance.

[e] During 1974, affiliate receivables from their Japanese parents (e.g., credits for exports from American to Japan) exceeded affiliate payables to their Japanese parents. As a result, the net Japanese position in the U.S. wholesale sector was negative that year. Because the column as a whole must net arithmetically to 100%, this distortion also makes the numbers for manufacturing, finance, and other Japanese investments appear to be overstated.

[f] Includes banking.

Sources: United States Department of Commerce, Bureau of Economic Analysis, *Survey of Current Business,* various issues; U.S. Department of Commerce, Bureau of Economic Analysis, *Selected Data on U.S. Direct Investment Abroad, 1950-76,* February 1982, pp. 20-27.

outflows from Japan now destined for America. American-Japanese cross-investment had thus created a new asymmetry—this one dominated by Japan.

Despite such different growth rates for over forty years after WWII, the industrial pattern of Japanese-American cross-investment remained remarkably stable. Indeed, a description of that pattern in 1934 remaines surprisingly accurate the 1970s and far 1980s:

> While American companies in Japan were in the technologically-advanced sectors and prominently in manufacturing, Japanese business in America aimed at aiding Japanese commerce by providing needed financial, insurance, trading, and shipping intermediaries.[9]

Within manufacturing, Americans have favored Japanese investments in machinery, broadly defined, followed by chemicals. The Japanese, by contrast, have long concentrated in the U.S. service sector, where in the 1970s, wholesale trade finally surpassed finance as the largest target of Japanese investment. (See Table 8.2) Those investments continued to grow, so that by 1986, the Japanese presence in U.S. wholesaling alone actually matched the value of all U.S. investments in Japan. Even more importantly, they supported Japan's overall trade strategy.

At one level, such investment may be understood as a natural outgrowth of macroeconomic forces, including the capital surpluses existing at various times in the two countries.[10] But the full story of cross-investment involves a larger tale, that of growing similarities and persistent differences in government policies and industrial structures which have shaped corporate strategies in the two modern economic superpowers.

The American Pattern Explained

The concentration of American multinational investment in the Japanese manufacturing sector imitated the general pattern that U.S. multinationals followed elsewhere in the world.[11] These corporations were among the first to invest in innovative products or manufacturing

---

9. Wilkins, "American-Japanes Direct Foreign Investment," p. 498.
10. For a recent discussion of the macroeconomics of U.S. foreign direct investment, see William H. Branson, "Trends in United States International Trade and Investment Since World War II," in Martin Feldstein, ed., *The American Economy in Transition* (Chicago: University of Chicago Press for the National Bureau of Economic Research, 1980), pp.183-257. For an excellent survey of the macroeconomics of Japanese foreign direct investment, see Makoto Sakurai, "Japanese Direct Foreign Investment: Studies on Its Growth in the 1970s," Yale University, Economic Growth Center, Discussion Paper No. 397, February 1982.

processes which, as they matured, satisfied growing demand in countries less industrialized than the United States. At first, such innovations were diffused through trade. But as trade became threatened, foreign investment accelerated. The principal threat to continued trade were Japanese competitors aided by high transportation costs across the Pacific, cheaper Japanese labor and capital, and the protective policies of the Japanese government (especially the latter). Not until the Japanese liberalized these policies did competitive pressure come to play an equally important role in promoting direct investment. Japanese prefectures then joined the national government in promoting American investment.

Japanese Government Policies

Historically, tariffs and other government restrictions on import competition have proved to be powerful determinants of foreign investments designed to service a local, as opposed to an export, market.[12] Such foreign trade restrictions do not always spur foreign investment, however, since capital controls may limit foreign investment in a newly protected local market. For much of the early history of U.S. investment in Japan, American multinationals had to contend with the opposing forces created by these two sets of government policies. Even after the government liberalized its trade and foreign investment policies, their legacies could still be seen in the pattern of American investment in Japan.

Import restrictions. Well before WWII, U.S. investors employed direct investments to overcome Japanese trade barriers, as the case of American automakers clearly demonstrates.[13] When Ford and General Motors began exporting to Japan both companies seemed positioned to

---

11. For a general overview of this pattern, see Raymond Vernon, *Sovereignty at Bay: The Multinational Spread of U.S. Enterprises* (New York: Basic Books, 1971), especially pp. 60-112.

12. For an early review of, and contribution to this literature, see Grant L. Reuber et al., *Foreign Private Investment in Development* (Oxford: Oxford University Press for the Organization of Economic Cooperation and Development, 1973), esp. pp. 120-32. For a more recent analysis, see Stephen E. Guisinger et al., *Investment Incentives and Performance Requirements: Patterns of International Trade, Production and Investment* (New York: Prager, 1985), esp. pp. 48-54.

13. The discussion of early investments in the Japanese auto industry is based on Wilkins, "American-Japanese Direct Foreign Investment," pp. 499-500; and William C. Duncan, *U.S.-Japan Automobile Diplomacy: A Study in Economic Confrontation* (Cambridge, Mass.: Ballinger, 1973), esp. pp. 55-68; Ira C. Magaziner and Thomas M. Hout, *Japanese Industrial Policy* (Berkeley, Calif.: Institute of International Economics, University of California at Berkeley, 1980), pp. 67-79; Michael A. Cusumano, *The*

satisfy growing Japanese demand by trade alone. In 1930, only 458 automobiles were fully manufactured in Japan, all by Japanese producers. This amounted to 2.5 percent of the Japanese market; GM and Ford supplied the remainder through imports of complete "knocked-down" kits shipped from the United States and assembled in Japan. Next, the Japanese government imposed restrictions on the number of kits that could be imported, and during the 1930s those restrictions tightened. As a result of these import restrictions, Ford and GM could no longer distribute U.S. exports through their Japanese subsidiaries.

Long after the war, U.S. multinationals remained singularly unsuccessful in using their Japanese investments as a beachhead for exports of finished products, sub-assemblies, and manufacturing inputs from the United States. As late as 1985, according to Table 8-3, total exports from the United States to the Japanese affiliates of American companies barely exceeded $3.3 billion (up from $2.5 billion in 1983, and $1.2 billion in 1977), of which 95 percent was shipped by the U.S. parent.[14] Third-country suppliers also played an important role in at least a few industries; again, much of that trade was intra-company, between two subsidiaries of the same U.S. parent.[15] In the semiconductor industry, for example, U.S. subsidiaries in Southeast Asia were important vendors for American-owned manufacturing plants in Japan.[16] Beginning in the 1960s, U.S. semiconductor companies began to move their most labor-intensive assembly operations to Asia; by the 1980s, fully 80 percent of their assemblies would be conducted there or in Latin America. Still, whether from parent or third-country affiliates, imports by U.S. multinationals in Japan remained limited.

Japanese Automobile Industry: Technology and Management at Nissan and Toyota (Cambridge, Mass.: Harvard University Press, 1985), pp. 27-72.

14. For the imports of (nonbank) U.S. affiliates in Japan see, U.S. Department of Commerce, Bureau of Economic Analysis, *U.S. Direct Investment Abroad: Operations of U.S. Parent Companies and Their Affiliates* (Preliminary 1985 Estimates), June 1987, Tables 17 and 18; *U.S. Direct Investment Abroad: Operations of U.S. Parent Companies and Their Affiliates* (Revised 1983 Estimates), June 1985, Tables 17 and 18; *U.S. Direct Investment Abroad, 1977* (Washington, D.C.: GPO, 1981), Table II.I.3, p. 154 and Table II.I.7, p. 156.

15. Third-country suppliers played an important role in food products, pharmaceuticals, paper products, and electronics, including semiconductors; see The American Chamber of Commerce in Japan, *United States Manufacturing Investment in Japan: Executive Summary* (Tokyo: The Chamber, 1980), pp. 17-32.

16. For comparisons of overseas sourcing patterns by U.S. and Japanese semiconductor companies, see the following: Franklin B. Weinstein et al., "Technological Resources," in Daniel I. Okimoto et al., eds., *Competitive Edge: The Semiconductor Industry in the U.S. and Japan* (Stanford, Calif.: Stanford University Press, 1984), pp. 63-5; U S.

Those imports, however, did increase when U.S. parents owned a controlling stake in their Japanese subsidiaries. According to a study commissioned by the American Chamber of Commerce in Japan, in 1978 imports accounted for nearly three-fifths of the material inputs employed in Japan by U.S.-affiliated producers of electrical machinery, electronics (including semiconductors), food products, paper products, and pharmaceuticals.[17] Typically these imports were shipped directly by U.S. parents who owned a majority of the equity in over one-third of their Japanese affiliates. By contrast, in most other Japanese industries, where minority-U.S. or equal partnership joint ventures were more prevalent—transport equipment and parts, for example—the sourcing of material inputs outside of Japan was virtually nonexistent. In these industries, Japanese capital controls restricted U.S. direct investments and, consequently, U.S. exports.

Capital controls. Such restrictions counteract the stimulating effect of trade restrictions on foreign investment. Again, consider U.S. efforts to enter the Japanese automobile industry, beginning before the

Table 8.3 *American Direct Investments and U.S. Trade, 1985*

	($ billion)	
	U.S. Exports to U.S. Affiliates Operating in Japan	U.S. Imports from U.S. Affiliates Operating in Japan
Manufacturing	$1.5	$5.6
Wholesale Trade	$1.7	$0.7
Other	$0.1	$0
Total	$3.3	$6.3

Source: United States Department of Commerce, Bureau of Economic Analysis, *U.S. Direct Investment Abroad: Operations of U.S. Parent Companies and Their Affiliates* (Preliminary 1985 Estimates), June 1987, Tables 17 and 18.

Congress, Joint Economic Committee, *International Competition in Advanced Industrial Sectors: Trade and Development in the Semiconductor Industry* (Washington, D.C.: U.S. Government Printing Office, 1983), p. 50; Robert H. Silin, *The Japanese Semiconductor Industry, 1981-82* (Hong Kong: BA Asia Ltd., 1982), p. 165; Machinery Promotion Association, Economic Research Center, *Survey Report on the Semiconductor Industry in Japan and the United States* (Tokyo: MPAI, 1980), p. 160.

17. American Chamber of Commerce in Japan, *Manufacturing Investment in Japan*, p. 22.

war.[18] Soon after Ford and General Motors first invested in Japan, Japanese buyers of Fords and Chevrolets demanded replacement parts, and many small and medium-sized Japanese companies quickly moved to fill the void with the government's encouragement. Nissan Motors, as one of these, initially supplied parts for Ford. The Japanese suppliers emerged as a visible nucleus of a fledgling automobile industry. Yet their growth remained slow; in 1935, Ford alone sold in Japan two-and-a-half times the number of cars and trucks produced by all Japanese-owned automakers combined. In the following year, however, this pattern changed, when new legislation insisted on greater local production of automobiles. Ford responded by applying for a license to begin manufacturing automobiles in Japan. Subsequent legislation in 1936 restricted such licenses to companies with at least 50 percent Japanese ownership. In fact, only Nissan and Toyota received licenses. Recognizing this, both Ford and GM then sought joint venture partners, initially with little success, since the Japanese could easily copy American products and mimic the technology of mass production. Next, during 1937, the Japanese government blocked remittances by Ford, following a year of good profits. After 1938, the Japanese took a final step by limiting the import permits of Ford and GM, thus curtailing severely the sale of "knocked-down" kits.

During the war these American investments were nationalized, but immediately afterward, Ford and GM expressed interest in re-entering Japan, despite Japanese regulations that blocked foreign investors for the next two decades.[19] First, the Occupation government proscribed American firms from buying "undervalued" Japanese enterprises, and when the Occupation ended, these restrictions were extended and codified in the 1950 Law Concerning Foreign Investment. Reflecting the Japanese government's continuing shortage of foreign exchange as well as deeper fears of losing managerial control over industrial development, the 1950 legislation established a system of prohibitions on all external transactions, including capital transactions. Yen-denominated profits, for example, were usually not convertible into other currencies,

---

18. See note 13 above for citations.
19. For a survey of post-war Japanese policies toward foreign investment, see M.Y. Yoshino, "Japan as Host to the International Corporation," in Charles P. Kindleberger, ed., *The International Corporation* (Cambridge, Mass.: MIT Press, 1970), pp. 345-369; Lawrence B. Krause, "Evolution of Foreign Direct Investment: The United States and Japan," in Jerome B. Cohen, ed., *Pacific Partnership: U.S.-Japan Trade* (Lexington, Mass.: Lexington Books for the Japan Society, 1972, pp. 162-66; Dan F. Henderson, *Foreign Enterprise in Japan: Laws and Policies* (Chapel Hill, N.C.: University of North Carolina Press, 1973), pp. 4-8, 195-290.

and even when IBM and a few other foreign firms were granted convertibility, other restrictions controlled the means and timing of such transactions. Most important, all transactions involving foreign capital outflows and inflows required time-consuming administrative approval.

The 1950 legislation marked a resumption of traditional Japanese policy. Government regulations kept foreign multinationals at bay by insisting on technology licensing decoupled from foreign equity; or, failing to secure the requisite technology, by insisting on joint ventures rather than wholly-owned foreign subsidiaries. These regulations became tighter in 1964, when Japan, as a new member of the OECD, was forced to make the yen convertible. To protect existing foreign reserves in the face of narrow surpluses on the current account, the Ministry of International Trade and Industry (MITI) and the Ministry of Finance (MOF) approved fewer foreign investment applications, and narrowed the acceptable range of technology remittances.

Three years later, in the midst of growing current account surpluses, the Japanese government again reversed policy by decontrolling foreign capital remittances, in the first of what later would be called the "five liberalizations." At this initial hint of liberalization in 1967, Ford and GM returned to Japan (along with Chrysler) to find partners for joint ventures. Their search was complicated, however, by MITI's forced merger of Japan's smaller automakers with either Toyota or Nissan.[20] Luckily for the Americans, Mitsubishi refused to merge, and instead— to the dismay of MITI—sought a partnership with Chrysler. Isuzu and Toyo Kogyo (renamed Mazda) also violated MITI's directives soon thereafter. Isuzu sought out General Motors as a partner, and Mazda pursued Ford. By this process, American automakers indirectly entered the Japanese market. Even more important, smaller Japanese automakers broke the stranglehold exercised by MITI and the industry's two Japanese giants.

American investors in other industries, although faced with constraints on foreign trade and investment comparable to those imposed by the Japanese government on the auto industry, sometimes managed to turn adversity into opportunity. During the late 1960s, this typically was achieved by American companies that were technological leaders, such as Texas Instruments (TI) which entered the Japanese semiconductor market.[21] Like U.S. automakers, TI viewed high Japanese

20. Chalmers Johnson, *MITI and the Japanese Miracle: The Growth of Industrial Policy, 1925-1975* (Stanford: Stanford University Press, 1982), pp. 172-73; Duncan, *Automobile Diplomacy,* pp. 31-52, 83-5.
21. For the history of Texas Instruments and other foreign companies in Japan, see

tariff barriers to trade as an incentive to invest in Japan, because Sony and other Japanese electronics companies, especially those producing calculators, made Japan the second largest and the fastest growing semiconductor market. Moreover, TI's potential Japanese competitors viewed TI's technology as key to their own development of integrated circuits, but those competitors had barely entered semiconductors in 1946 when TI asked for a government license to establish a wholly-owned subsidiary. NEC, for example, was the first Japanese company to enter the semiconductor industry (1963), using less-advanced technology licensed from Fairchild. TI was followed by Hitachi in 1965, and Toshiba and others in 1966. As for TI's foreign, principally American, competitors—tariff barriers would hold them at bay, at least in the short term, while Japan's foreign equity controls would force less technologically advanced competitors to settle for licensing agreements or, at best, joint ventures.

From TI's perspective, the same proprietary technology that had made TI the market leader in the world semiconductor industry would also force MITI to grant it an exception from foreign capital controls. Such dispensation, however, did not come easily. As noted above, the Japanese had just agreed to make the yen convertible in 1964, and they inteded to protect existing foreign reserves through stringent capital controls. Negotiations dragged on for four years, a period designed by MITI to compensate for Japan's late entry into the semiconductor race. By the time MITI and TI reached an agreement in 1968, during the first year of "phased liberalization," Japan's domestic production of semiconductors amounted to less than 10 percent of TI's total world production.[22]

TI won its demand for a wholly-owned subsidiary in Japan, but not without allowing MITI two important concessions.[23] First, TI had to establish a 50-50 joint venture with Sony, one of the major Japanese buyers of semiconductors. Although few knew it at the time, this joint venture was scheduled to last only three years, so by 1971, Sony sold its equity holdings to TI at a prearranged price. TI also acceded to MITI's second demand: that TI make its proprietary technology available to

---

Yasuzo Nakagawa, *The Development of the Japanese Semiconductor Industry* (Tokyo: Diamond, 1982) pp. 154-66; Machinery Promotion Association, *Semiconductor Industry,* p. 115; Silin, *Japanese Semiconductor Industry,* pp. 127-28.

22. Japan Electronics Industry Association, *Thirty-Year History of the Electronics Industry* (Tokyo: JEIA, 1979), pp. 102, 271. Japan's largest producer, then and now, was the country's first entrant, NEC, with 7% of the Japanese market in 1968.

23. For TI's negotiations with MITI, see Nakagawa, *Japanese Semiconductor Industry,* pp. 154-66.

Japanese companies for a 3.5 percent licensing fee—for TI, a unique arrangement and clearly a major concession. By comparison, MITI raised relatively few objections to IBM's plan for building a plant whose production would be completely consumed by other IBM operations in Japan.[24] Aside from TI, however, IBM was the only other foreign semiconductor manufacturer to establish a wholly-owned subsidiary before the near elimination of foreign equity controls in 1973.

In contrast, MITI raised more numerous objections to market entry by foreign-owned semiconductor producers who were not at that time technological leaders, or did not propose to consume most of their own production.[25] Fairchild, for example, the first American producer to license semiconductor technology with a Japanese firm, in 1959, repeatedly failed to secure MITI approval for a wholly-owned transistor and diode assembly plant on the Japanese mainland. As a substitute, Fairchild established a wholly-owned subsidiary in Okinawa, a U.S. territory in 1969, with preferential access to the Japanese market, which it lost when Okinawa reverted to Japanese control in May 1972. At that point, MITI forced Fairchild into a joint venture with TDK, which lasted only until 1977. National Semiconductor, after failing to enter the Japanese mainland, also set up a wholly-owned facility on Okinawa, but unlike Fairchild, National refused to reorganize as a joint venture after 1972. Finally, TI's principal global competitor—number two-ranked Motorola—settled in 1968 for the right to open a Japanese sales office to boost exports and to monitor TI's moves.

The difficulties experienced by American semiconductor and automobile producers were repeated again and again in other industries. By 1974, only one out of every ten foreign manufacturers operating in Japan was fully foreign-owned.[26] Most of the remainder were equal partnership (50/50) joint ventures, each with a single Japanese company. Nowhere in these statistics, of course, are the multitude of prospective investors who stayed away from Japan because of foreign equity controls, or who left Japan after a disappointing experience with a joint venture partner. Indeed, by 1974, Japan was host to only two percent of all new American direct investments entering manufacturing industries overseas. As a result, the total position of U.S. investment was lower in Japan than in Brazil and Mexico, not to mention the six

24. Silin, *Japanese Semiconductor Industry,* pp. 127-28.

25. Silin, *Japanese Semiconductor Industry,* pp. 127-28; Machinery Promotion Association, *Semiconductor Industry,* p. 162.

26. Japan, Ministry of International Trade and Industry, Industrial Location Guidance Division, *Industrial Investment in Japan: 1984* (Tokyo, MITI, 1984), p. 10.

largest EEC countries.[27] Subsequently, however, U.S. investments in Japan grew, stimulated by a further relaxation of Japanese capital controls.

Policy liberalization. In 1973, the government removed a key restriction on foreign entry: the 10 percent ceiling on foreign shareholdings in existing Japanese companies. Still, foreigners had to secure the firm's approval before purchasing larger shares of its equity. The government continued to restrict foreign investment in certain companies and sectors, to regulate the value and timing of selected foreign transactions, and to insist on formal government approval of many transactions. These restrictions were largely eradicated in 1980, when a new era of foreign investment promotion began in Japan. Subsequently, the typical foreign investor simply notified the Bank of Japan about upcoming foreign financial transactions, and then proceeded without further government approval.[28]

In this liberal climate, American investments grew steadily: 11 percent annually between 1974 and 1980, and at a comparable rate over the next six years. (See Table 8.2) While this growth was still significantly slower than the rise of Japanese direct investments in America, nevertheless, annual U.S. outflows to Japan, especially those in manufacturing, began to exceed comparable American direct investments to other countries. At the height of those outflows, during 1983, fully one-fourth of all new U.S. investment overseas went to Japan.[29] Subsequently, however, new American investment in Japan steadily accounted for six to seven percent of all annual outflows—a rate comparable to U.S. direct investments in other industrialized countries. Thus, during the 1980s, Japan had finally become a likely host for American investment abroad.

Despite the influx of new investment following the liberalization of government policies, most U.S. multinationals in Japan still operated minority foreign-owned and equal-partnership joint ventures—not majority U.S.-owned subsidiaries. By 1978, U.S. multinationals able to exploit marketing and technological advantages (in pharmaceuticals, for example) held a majority interest in less than one-half (45%) of their manufacturing subsidiaries in Japan.[30] In a few industries where

---

27. U.S. Department of Commerce, *Selected Data on U.S. Direct Investment Abroad, 1950-76,* pp. 24-5.

28. MITI, *Industrial Investment in Japan,* p. 14.

29. *Survey of Current Business,* 64 (October 1984), p. 38; for subsequent years, see *Survey of Current Business* 67 (August 1987), pp. 63-5.

30. American Chamber of Commerce in Japan, *Manufacturing Investment in Japan,* pp. 17-32 and Exhibit 10.

U.S. firms enjoyed marketing and supplier advantage (food or paper products, for example), they held majority shares in roughly one-third of their manufacturing affiliates. And where such advantages were not generally available, majority-U.S. ownership was even less prevalent, as in transport equipment, where U.S. parents owned majority shares in less than one-tenth of all American affiliates, and in non-electrical machinery, where such majority shares were owned in less than one-quarter. Even by 1981, foreign multinationals owned, on average, a majority of the equity in only about one-third of their Japanese manufacturing affiliates, up from one-sixth in 1974.[31]

To account for the continued high incidence of minority American-owned joint ventures in Japan, even after liberalization, we must consider the timing of these investments. Over one-half of all American stocks owned in 1980 actually flowed into Japan before the 1974 abolition of foreign equity controls. (See Table 8.1) Nevertheless, in the few years following liberalization, American investments doubled, as did the proportion of majority U.S.-owned manufacturing subsidiaries in Japan. Unencumbered by capital controls, American multinationals were now better able to respond to the competition resulting from the characteristics of their separate industries.

Industrial Organization

Among the various competitive forces in an industry that may stimulate foreign investment, the power of buyers figures prominently. Those buyers may reside in the host country or abroad, and may require that their suppliers invest in accordance with the buyers' demands. For example, the price and quality of Japanese-made products have stimulated demand overseas, and have stimulated U.S. investors to consider using Japan as an export platform in their global sourcing network. Moreover, within the Japanese market itself, local buyers may demand that production be customized to meet their peculiar needs, or that products be channeled through decentralized distribution systems. These competitive forces vary widely across industries, and where their importance is pronounced, foreign investment becomes a likely response.

Japanese buyers. Around the world, companies purchasing capital goods often require that machinery and equipment be customized to meet their local needs. Those demands appear to be especially intense in Japan where, according to one long-time observer, "buyers, especially

---

31. ·MITI, *Industrial Investment in Japan,* p. 10.

of capital goods, tend to think local producers are more reliable than importers."[32] Under these conditions, it should come as no surprise that at least one-half of all U.S. investments in Japan have always been concentrated in machinery and equipment, where buyer power constitutes an important determinant of foreign investment.

Pressures for customization were especially pronounced on U.S. manufacturers of the expensive machinery and equipment used to produce and test the semiconductors sold by NEC and other Japanese producers.[33] Of particular concern was the modification of U.S.-designed equipment to the specific requirements of Japanese chipmakers. In response to these pressures, Thermco became the first American semiconductor equipment company to establish a manufacturing plant in Japan in 1968, before the liberalization of foreign equity controls. With an investment in 1984 valued at ¥2 billion and over 50 percent market share in its product lines, TEL-Thermco planned to establish a second manufacturing plant. By then, thirteen American semiconductor equipment companies operated manufacturing facilities in Japan—three times the number of American semiconductor (end-product) vendors operating there. Seven of these equipment manufacturers became operational during 1983-84; they included the wholly-owned subsidiaries of such industry giants as American Machine Tools (¥1.9 billion invested in Japan) and Shipley (¥1.2 billion invested). For all of these manufacturers, having a large and visible local presence demonstrated a market commitment sought by Japanese buyers generally, and especially by purchasers of expensive capital goods such as the equipment used in the manufacture and assembly of semiconductors.

To varying degrees, building a Japanese presence, to quote Peter Drucker, "in service, in market research, in market development and in promotion, mattered in most industries given the behavior of Japanese buyers."[34] Certainly this has been true in the Japanese pharmaceutical industry.[35] The decentralized distribution of drugs through the government-financed health insurance program encouraged foreign companies to establish their own distribution channels and sales forces. Thousands of detail men were required to visit the tens-of-thousands of

---

32. The quotation is from the *Economist,* July 6, 1985, p. 65.
33. "Semiconductor Industry's Global Transition," *Semicon News* 6 (No date), pp. 21-24; Japan Economic Institute, "Semiconductor Equipment: An Example of Growing U.S. Investment in Japan," *JEI Report,* No. 44A, (November 18, 1983), pp. 1-6.
34. The quotation is from the *Wall Street Journal,* July 18, 1985, p. 36.
35. For an overview of government policies and corporate strategies, see Kazuhito Kondo, "The Japanese Pharmaceutical Industry: Its Present and Future," Nomura Research Institute, November 1983, p. 31ff.

doctors and hundreds of hospitals, which together served as the principal purchasers of prescription drugs in the absence of retail pharmacies. The typical foreign pharmaceutical company responded to Japan's national health scheme by licensing their technology to local companies. In fact, between 1978 and 1982, over one-half of all drugs sold in Japan were manufactured using foreign technology.[36] Thus, through technology licensing, multinationals gained initial entry into the Japanese market.

Increasingly, however, direct investments followed technology licensing, especially after the government began in the early 1980s to liberalize its price controls on products new to the Japanese market. By 1983, over 300 foreign pharmaceutical companies had direct investments in Japan, the number having doubled in less than eight years.[37] Most of these multinationals entered into joint ventures, but not all. Beginning in the late-1950s, for example, Pfizer embarked on a long-term plan designed to establish its own independent channels and sales force. Two decades later, others followed Pfizer's lead. By 1983, eight foreign pharmaceutical companies controlled assets in excess of ¥5 billion; for two of these, Merck (U.S.) and Ciba-Geigy (Swiss), assets exceeded ¥20 billion. Merck alone ranked among the three largest pharmaceutical producers in Japan, following its acquisition of Banyu and Torii in the early 1980s. For other companies, however, market entry through acquisition remained uncommon. Indeed, acquisitions have played a miniscule role in the growth of American investment in Japan, as opposed to their greater importance in the comparable growth of U.S. multinationals elsewhere.[38]

---

36. Japanese Federation of Pharmaceutical Manufacturers, Insurance Drug Price Research Council, *Pharmaceutical Industry and Price Standard* (Osaka: JFPM, May 1984), p. 19.

37 Reported in *Pharma Japan,* July 2, 1984, p. 13.

38. According to statistics provided by W.T. Grimm and Company, U.S. entry to the Japanese market through the acquisition of an existing company was virtually non-existent, in marked contrast to U.S. corporate strategy in other markets:

Domicile of Foreign Parent	Number of U.S. Acquisitions	
	1979-83	1984
United Kingdom	169	45
Canada	127	24
West Germany	53	13
France	53	3
Switzerland	25	3
Netherlands	24	3
Brazil	13	3
Mexico	11	0
Japan	8	7

While Merck's acquisitions in Japan remained unique, the rationale for those acquisitions had wide appeal to other American investors.[39] In particular, Banyu's large existing marketing network—including 1600 veteran detail men—made the company an attractive takeover target. Those men had long been familiar with Merck's product line, since Banyu depended heavily on technology licenses with Merck. These already extensive business relations between the two companies helped to accelerate the acquisition, as did dollar-to-yen appreciation during 1983-84, which reduced the dollar cost of investments in Japan. By then, Banyu felt intense pressure from Japanese government regulators who, between 1980 and 1984, had cut by 40 percent the official prices they would pay for drugs with readily-available substitutes. These cost-cutting measures struck Banyu's product line very hard, since the company had failed to invest its large cash balances into ongoing research and development. Merck, by contrast, was poised to introduce into Japan new products directed at Japan's rapidly aging population. This confluence of government policies and demographic factors also fit well into Merck's larger global strategy. Not only would Merck increase its export of products and technology to Japan, but it would also begin marketing in the United States certain antibiotics and dermatological products manufactured by its recently-acquired Japanese subsidiaries.

Japanese sourcing. By using Japan as an export platform for the United States, Merck followed the lead of a few other majority-owned American companies.[40] By 1982, for example, Texas Instruments-Japan exported principally to its parent one-half of the estimated $300 million worth of memory chips it produced. And IBM-Japan, the only IBM subsidiary in the world making XT disk drives, exported 100 percent of its production to its U.S. parent. But these examples still must be regarded as idiosyncratic. During 1985, total exports from U.S. subsidiaries in Japan back to their U.S. parents barely exceeded $6 billion (see Table 8.3 above). The vast majority of American subsidiaries in pharmaceuticals, electronics, and other industries still have sold very little overseas. Most U.S. investors entered Japan to supply the domestic Japanese market, and were encouraged initially to do so by Japanese barriers to trade and the demands of Japanese buyers.

The one notable exception to this general rule can be found in the

---

39. Information in this paragraph was gathered during interviews with industry analysts and company managers held in Tokyo, October 1984.

40. Merck's exports were reported in interviews held in Tokyo in October 1984; TI's exports in John L. Lazlo, *The Japanese Semiconductor Industry: Robust Industry Conditions to Persist Through 1984* (San Francisco: Hambercht and Quist, 1984), p. 17; IBM's exports in *Fortune,* April 8, 1985, p. 40.

automobile industry, where American automakers increasingly integrated Japanese production into their larger global strategies. In fact, by 1983, motor vehicles and parts constituted an estimated three-quarters of total exports by U.S. affiliates in Japan.[41] Subsequently, auto exports by Isuzu and Suzuki to their largest shareholder, General Motors, as well as Mitsubishi sales to Chrysler swelled in 1985 when the Japanese government revised its "voluntary restraints" on car shipments, and allotted much of the increased quota to these captive producers. That relaxation of export restraints contributed to a doubling of total exports of U.S. affiliates in Japan to their parents in America: from $3.6 billion in 1983, to $6.3 billion in 1985.[42]

To insure better coordination of these exports to America with domestic U.S. production, Chrysler increased its minority equity holdings in Mitsubishi, following a strategy charted earlier by GM and Isuzu. By contrast, Ford had no captive imports, but held large, albeit minority equity investments in Mazda and several other Japanese parts suppliers, which have been increasingly integrated into Ford's global operations. Similarly, GM relied on several original equipment manufacturers, who in 1985 established a formal supplier association to discuss issues of mutual concern. Exports from these suppliers were not limited to the United States. According to a 1978 study, more than one-fourth of the total sales of American affiliates operating in the Japanese transport equipment industry were exported to third-country markets, largely in East Asia.[43] Whether for export to the United States or to subsidiaries in third countries, the use of Japan as an export platform for parts, sub-assemblies, and finished products meant expanded investments by American multinationals.

Much of that U.S. investment went into existing automakers and parts suppliers which, like much of Japanese industry, were already concentrated in a 300-mile corridor from Tokyo to Osaka. There, Japanese industry could minimize transportation costs in a country with limited flatlands. And there, Japanese industry could achieve economies of scale and scope conducive to mass production and mass marketing. In short, Japan's strategy of industrial development—constrained by geography—narrowed the scope of industrial invest-

41. For 1983 data, see Dennis J. Encarnation, "Cross-Investment," in Thomas K. McCraw, ed., *America versus Japan* (Boston: Harvard Business School Press, 1986), Table 4-5, p. 141.
42. For 1983 data, see Encarnation, "Cross-Investment," Table 4-5; for 1985 data, see Table 3 above.
43. American Chamber of Commerce in Japan, *Manufacturing Investments in Japan,* pp. 17-32.

ment in rural Japan. To counteract this trend, the Japanese government, joined by rural prefectures, sought to influence the location decisions of U.S. investors who for other reasons—trade restrictions, policy liberalizations, and industrial structures—had already decided to invest in Japan.

The Policies of Japanese Prefectures

Beginning in the late 1960s, the central government tried to lure Japanese and foreign companies out of the congested Tokyo-Osaka corridor and into new, rural, industrial parks. Less expensive real estate was their principal incentive; to this, the central government added infrastructure expenditures, tax breaks, and subsidized loans. Among other objectives, the central government sought to stem the tide of out-migration that had depopulated much of the countryside. Oita prefecture, for example, on the southern-most island of Kyushu, was among the hardest hit. By the early 1970s, its economy was "on the verge of collapse," according to both public and private accounts.[44] Its recovery would serve as a model for the nation.

That recovery would not be based on increased government investment, in marked contrast to the earlier dependence of Oita and other rural prefectures on public subsidies.[45] For much of the post-war period, as urban Japan grew and prospered, jobs in rural economies were generated in part by massive infusions of central government expenditures, notably for public works, funded largely by urban tax revenues. The redistribution of revenues from urban to rural populations had long been a central plank of successive Liberal Democratic governments that depended on continued rural support for their political survival. But the first oil price shock changed all this. A reduction in receipts during the ensuing recession generated a series of budgetary deficits that prompted the central government to curtail spending on public works and other regional development programs. With few other options in the midst of these "fiscal crises," prefectural governments intensified their efforts to attract private investment.

The prefectures first turned to Japanese business, but met with little success. By the recession of 1974, Japanese companies were often more interested in cutting back capacity than in spreading out geographically.

---

44. *Fortune,* May 28, 1984, p. 16; for a survey of the incentives prescribed in "Act on Emergency Measures for Depopulated Areas," see MITI, *Industrial Investment in Japan,* p. 63.
45. The following is consistent with *Fortune,* May 28, 1984, pp. 14-19, and interviews conducted in Japan during June and October 1984.

Even those who thought to invest in rural areas soon discovered that managers and skilled workers refused to move and give up their urban lifestyles. As a result, few Japanese companies moved to the countryside, and the new industrial parks scattered around Japan's 47 prefectures began to look "like industrial wilderness areas."[46] Another option involved the mobilization of local resources, but these were limited in quantity and scope.[47] The final option was to attract foreign direct investment. Here, prefectures became more active.

Their investment incentives, while large in value, were few in number, and their mechanics were well understood.[48] For example, prefectures usually offered newcomers tax holidays and subsidies for such things as recreation centers. They also helped foreigners get loans from local banks and to secure land. Even special loans to foreigners available from the Japan Development Bank, as well as other incentives whose levels were negotiated rather than automatic, were largely based on simple and known criteria. As a result, bidding wars among Japanese prefectures remained almost nonexistent, largely because MITI considered such competition among prefectures to be counterproductive, and negotiated an implicit understanding with prefectural governments to refrain from outbidding each other for foreign investment.[49] To promote accord, MITI exercised further control over the incentive packages assembled by the prefectures. Within each of the regional development authorities, MITI dispatched personnel on two-year assignments ostensibly to aid the investment promotion process. In each of these ways, MITI coordinated regional incentives nationwide.

MITI further limited the competition among Japanese prefectures by controlling their direct "marketing" of investment sites. Much of that marketing, in fact, was carried out by agencies of the central government, the Japan External Trade Organization (JETRO), and the Industrial Guidance Division of MITI. JETRO, for example, provided potential foreign investors with consulting and translation services in its overseas offices; it did not limit itself to promoting Japanese exports. Similarly, MITI established a computerized English-language service to match project proposals with possible investment sites, and to report on available incentives. While MITI and JETRO tried to promote invest-

---

46. *Fortune,* May 28, 1984, p. 16.
47. Of particular note here was the "one-village, one-product" campaign launched by Governor Hiramatsu (Oita Prefecture, Kyushu) and designed to target resources on activities for which a particular community had known expertise.
48. For an excellent survey of these policies see, MITI, *Industrial Investment in Japan,* pp. 62-9, 73-8.
49. The following is based on interviews in Tokyo, June 1984.

ment impartially, central government policies inevitably encouraged some differentiation among rural prefectures. The "Technopolis" program, for example, allowed certain regional governments to target investment in high technology industries while at the same time drawing attention to that region's peculiar advantages.

Some individual prefectures did spotlight the special advantages they could offer foreign investors. For example, Governor Hiramatsu in Oita Prefecture of Kyushu became famous for the services he provided to prospective investors.[50] As a former MITI official, Governor Hiramatsu proved quite adept at moving foreign investment proposals through the bureaucratic labyrinth that still existed during most of the 1970s. Texas Instruments was among the several foreign companies he placed in Oita during the period, helping to transform Kyushu into "Silicon Island." In fact, this new image subsequently bolstered the efforts of other prefectures in Kyushu to attract Fairchild and other American producers of high-technology electronics. In 1982, TI and 14 other majority foreign-owned enterprises established manufacturing plants in Japan. Thirteen of these plants were located in Oita and other economically underdeveloped regions; only two were located in the Tokyo-Osaka corridor.[51]

Certainly, regional incentives contributed in some measure to this geographic dispersal of foreign investment in Japan, but they had far less impact on the initial decision to invest there. Rather, trade policies and prospects, along with the competitive structure of industries proved to be far more important. Yet we should not forget that, in Japan, regional incentives take on added significance because of what they represent symbolically: a fundamental change in the attitudes of government at all levels toward foreign investment.

Here, the career of Governor Hiramatsu of Oita conveniently symbolizes an historic change. As a MITI official in the early 1960s, he was actively involved in the decision to deny Texas Instruments a license to establish a wholly-owned subsidiary in Japan, insisting instead that TI establish an equal-partnership joint venture.[52] Only after Japanese semiconductor suppliers became competitive did Hiramatsu respond to pressure from TI and from Japanese buyers by initialling the deal that satisfied everyone. Then, years later, relations between Governor Hiramutsu and TI became quite different when TI offered employment to residents of Oita prefecture, in exchange for the

---

50. Governor's Hiramatsu's success is detailed in an interview reprinted in *Japan Economic Survey* 8 (October 1984), pp. 6-11.
51. Reported in the *Journal of Japanese Trade and Industry* 1 (1983), p. 11.
52. For an account of this, see *Fortune*, May 28, 1984, pp. 17-8.

governor's promotion of the proposed investment. Hiramatsu's slow evolution from regulator to promoter of foreign investment can be measured in a single life, but the change also represents the shift in Japan's foreign economic policy and its new regional development objectives. At present, with U.S. critics decrying limited foreign access to the Japanese market, active promotion was essential to overcome the lingering spectre of strict regulation, fostered by decades of policy pronouncements by the central government since World War II.

The Japanese Pattern Explained

The same postwar Japanese legislation that tightly regulated foreign investment at home also specified that every Japanese investment located in the United States and elsewhere abroad had to be approved by the Ministry of Finance and the Bank of Japan.[53] Here, Japanese government policies encouraged natural resource projects upstream and trade-related services downstream, to the extent that the encouraged foreign investments at all. Such policies were consistent with Japanese corporate strategies. In the United States, Japanese companies increasingly concentrated their small investments in the wholesale sector, where they exported American raw materials or imported Japanese consumer goods. Thus, the initial objective of Japanese investment in America was to promote international trade, an objective supported by Japanese government policies.[54]

That objective, and the concentration of Japanese investments in U.S. wholesaling, remained largely unaffected by later shifts in Japanese government policy, even after 1969, when a series of liberalizations began, parallel to those affecting U.S. investment in Japan. Even with such policy changes, however, Japanese investment in the United States grew only slowly, at least through 1974, when the Japanese government reimposed restrictions on overseas investment, in response to both the first oil price shock and Japan's first postwar recession. That year, Japanese direct investments in the U.S. wholesale sector were already integral to overall trade strategy. Japanese-affiliated wholesalers exported (principally to Japan) $8.5 billion of

---

53. For further details, see M.Y. Yoshino, *Japan's Multinational Enterprises* (Cambridge, Mass.: Harvard University Press, 1976).

54. For different reasons, Kiyoshi Kojima comes to the same conclusion in his "Direct Foreign Investment Between Advanced Industrial Countries," *Hitotsubashi Journal of Economics* 18 (June 1977), pp. 1-18. For a more recent test of his hyothesis, see Kiyoshi Kojima, "Japanese and American Direct Investment in Asia: A Comparative Analysis, *Hitotsubashi Journal of Economics* 26 (June 1985), pp. 1-35.

American raw materials, and distributed in America $9.3 billion of finished goods imported largely from Japan,[55] thus contributing to the growing trade surplus with the United States.

As U.S. protests over those surpluses grew, Japanese investments in the United States became an integral element in Japan's response. To offset large trade receipts, the Ministry of Finance (MOF) sought to stimulate the outflow of long-term loans and equity investments. Foreign direct investment provided one mechanism for achieving this objective. In the name of "balancing the basic balance," MOF in 1975 removed the last remaining capital controls. MOF quickly changed from being a cautious regulator of Japanese investment overseas to being an enthusiastic promotor, as it initiated, for example, a loan program through the Export-Import Bank of Japan to help Japanese investors going abroad. When internationalization of Japanese companies became a central plank of national foreign economic policy, Japanese direct investment in the United States grew.[56] Again, government policies influenced the pattern of foreign direct investment.

Not until the late 1970s did that investment accelerate, but only after U.S. import restrictions threatened Japanese exports to affiliated wholesalers in America. Those trade restrictions—and later, the appreciation of the yen—emerged as the most powerful determinants of investment strategies upstream from distribution. Into the 1980s, the competitive pressures arising from industrial structures further accelerated the movement of Japanese investors, who then relied on the investment incentives offered by several U.S. states to determine their final location in America. Increasingly, these government policies and industrial structures resembled their counterparts in Japan. But important differences persisted, including the continued importance of foreign trade in determining the pattern of Japanese investment.

Foreign Trade

During the 1980s, Japanese investment and Japanese trade remained intertwined. The U.S. wholesale sector continued to attract Japanese interest. By 1986 it accounted for over one-half of Japanese new investment in America, having grown 33 percent annually since 1980. (See

55. U.S. Department of Commerce, Bureau of Economic Analysis, *Foreign Direct Investment in the United States*, Volume 2, *Report of the Secretary of Commerce, Benchmark Survey, 1974* (Washington, D.C.: GPO, 1976), pp. 57-8, 60.

56. Johnson, *MITI and the Japanese Miracle*, pp. 275-304; *Far Eastern Economic Review*, June 13, 1985, p. 83. According to the *Review*, outstanding loans in this program grew fourfold in seven years, from Y2.4 billion in 1977 to Y258 billion in 1983.

Table 8.2 above.) Just two years earlier, Japan had accounted for nearly 40 percent of the book value of all foreign investments in U.S. wholesale trade, meaning that Japanese companies controlled by 1984 nearly 5 percent of the total stockholders' equity invested in the entire U.S. wholesale sector.[57] As before, these investments remained in the wholesale trade of metals, minerals and farm products. In 1985 they channeled over $17 billion of U.S. commodity exports largely to Japan. (See Table 8.4)

Table 8.4 Japanese Direct Investments and Trade by Japanese Subsidiaries in the United States, 1985

	Position in U.S. ($ billion)[a]	Total Exports from U.S. ($ billion)	Total Imports to U.S. ($ billion)	Imports (% from Japanese Parent)
Wholesale Trade	$11.8	$21.8	$55.1	77.1%
Of which:				
Motor vehicles/parts		c	24.0	72+ [b]
Metals and minerals		9.5	7.6	62.8
Other durable goods		1.1	16.8	98.0
Farm products		7.9	6.0	75+ [b]
Other nondurables		c	0.6	74.7
Manufacturing	$2.7	$1.0	$2.3	77.6%
Other	4.8		$0.2	94.0%
Total	$19.3	$22.8	$57.8	77.2%

[a] Direct investment position abroad, as defined in Table 8-1, note a.

[b] Specific details of intra-company trade suppressed by the U.S. Department of Commerce to avoid disclosing the operations of a few large companies; estimates here are mathematically derived using other data provided.

Source: United States Department of Commerce, Bureau of Economic Analysis, *Survey of Current Business* 66 (October 1986):26. All trade data were supplied separately by the International Investment Division, Bureau of Economic Analysis, U.S. Department of Commerce.

The vast bulk, however, of new Japanese investments went into automobiles and other durable goods—largely to provide American distribution channels for products exported from Japan. (See Table 8.4) By 1985, two-fifths of all imports by Japanese affiliates in the

57. *Survey of Current Business* 64 (October 1984), p. 38.

United States belonged to the wholesale auto sector; and nearly three quarters of these auto imports were shipped by Japanese parents—not trading company intermediaries or third-country affiliates. The same relationship also held for other Japanese exporters, especially of durable goods.[58] As a result, during the decade ending in 1983, exports by all Japanese manufacturers to their U.S. wholesalers grew nearly 16 percent annually.[59] By 1983, such intra-company trade represented over three-fifths of all Japanese exports to the United States during 1983—a proportion that did not change through 1985.[60] Thus, for Japanese exporters, formally affiliated wholesalers downstream in America served as conduits for the export of motor vehicles, sub-assemblies, and spare parts to the United States. But during the 1980s, those exports were being threatened by American trade policies, which forced Japanese investors to move upstream from distribution into U.S. production.

U.S. Import Restrictions

The relationship between trade restrictions and upstream investment showed itself clearly in automobiles. Between 1981 and 1986, "voluntary export restraints" limited Japanese imports to a fixed quota of fully assembled cars and trucks; this did not cover parts or "knocked-down" kits, an obvious omission soon to be exploited.

---

58. By comparison with Europe, intra-company sourcing from Japanese parents has substantially exceeded comparable trade between European multinationals and their American subsidiaries. In 1980, for example, European parents exported $18.8 billion to their U.S. affiliates, compared to $21.9 billion shipped by Japanese parents to their U.S. subsidiaries. Moreover, European intra-company trade was spread across the wholesale (62%) and manufacturing (36%) sectors, while Japanese intra-company trade was concentrated (98%) in the U.S. wholesale sector. These figures reflect the greater concentration (relative to the Japanese) of European investments in U.S. manufacturing, and indicate that the very recent growth of Japanese-owned manufacturing plants in America should be expected to spur export growth in Japanese parts and sub-assemblies, thereby altering the future composition of Japanese exports to America. Between 1974 and 1980, exports by Japanese parents to their U.S. manufacturing subsidiaries grew more than 30 percent annually compared to nearly 20 percent for trade involving the much larger wholesale sector. See, Department of Commerce, *Foreign Direct Investment in the United States, 1980*, Table G-9, p. 149 and Table G-10, p. 150.
59. For 1974 data, see U.S. Department of Commerce, *Benchmark Survey, 1974*, pp. 57-8, 60; for 1983 data, see Encarnation, "Cross-Investment," Table 4-3, p. 126.
60. In 1983 exports (f.o.b.) by Japanese parents to their U.S. subsidiaries in U.S. wholesaling amounted to $26.9 billion [according to Encarnation, "Cross-Investment," Table 4-3], while total Japanese exports (f.o.b.) to the United States equalled $43.3 billion [according to the International Monetary Fund, *Direction of Trade Statistics Yearbook: 1987* (Washington, DC: IMF, 1987, p. 158]. For 1985 data, see Table 8.4 above.

Within that quota, the market shares of Japanese producers remained fixed at the annual average level prevailing during the last three years of free competition. For each of the largest Japanese exporters—Toyota, Honda, and Nissan—this meant a virtual guarantee of a four-to-six-percent share of the U.S. market, to which could be added all sales from American assembly plants. Thus, by directly investing in such a plant, Japanese producers could circumscribe trade barriers on finished autos by exporting parts and knocked-down kits for final assembly in the United States.

Faced with indefinite constraints on exports, Honda and then Nissan quickly established wholly-owned assembly operations in the United States.[61] Next, their largest competitor in Japan, Toyota, came on stream during 1985 in a Japanese-managed joint venture with General Motors. A fourth firm, Mazda, announced that in 1986 it would acquire an existing American facility from its largest shareholder, the Ford Motor Company. Also during 1986, Mitsubishi announced that it would invest directly in the United States, selling most of its production there through its shareholder, Chrysler. Finally, Toyota announced the construction of a wholly-owned assembly plant to be completed during 1988. Collectively, by 1988, these Japanese automakers would be assembling in the U.S. no fewer than one million cars annually, or slightly more than the total production of Chrysler scheduled for that year. Indeed, at that time, Japanese automakers operating in the United States will outnumber American-owned automakers in their home market.

To protect an existing market in the face of restrictive American policies, these Japanese corporations relied on additional foreign investment. Americans called for local content legislation, and the spectre of further trade restrictions in the auto industry grew menacing. So all the proposals from Japanese automakers for new U.S. assembly plants had to include detailed timetables for the phasing-in of local purchases in the United States to replace imports from traditional suppliers in Japan. Honda, for example, was the first Japanese automaker to claim 50 percent American content—parts and labor—in its Ohio cars.[62] So aggressive was Honda's response to U.S. restrictions that, in 1984, it replaced Toyota as the Japanese leader in the U.S. market; and the following year, Honda became America's fourth largest automaker, having exceeded the production of (then-independent)

61. For a recent overview, see *Economist,* December 8, 1984, pp. 75-6. For the Toyota-GM venture, also see *New York Times,* January 30, 1985, p. D1; the special report in *Business Japan,* April 1983, pp. 18-20; and the *Wall Street Journal,* July 9, 1985, p. 2.
62. *Wall Street Journal,* March 29, 1985, pp. 1, 10.

American Motors.[63] Honda's expansion in the United States had to be ambitious largely because that company, among Japanese automakers, stood to lose most from trade restrictions which threatened its largest and fastest growing overseas market, since expansion at home had long been blocked by Toyota and Nissan.

The plight of Japanese automakers signaled exporters in other industries to undertake "preemptive expansion" before the Americans restricted their market. The history of NEC operations in the United States shows both the close linkage among Japanese industries and the effect on foreign investment. As the first semiconductor manufacturer to invest in America, NEC moved rapidly to increase its operations by both acquisition and new construction. At least three factors accounted for the unusual speed of NEC expansion.[64] First was the sheer size of the U.S. market. Even before NEC began to invest substantially in the United States, one-third of its semiconductor sales came from exports, and one-third of those exports were destined for the U.S. market. But, in 1978, increased export sales were threatened by yen-to-dollar appreciation, a second factor that accelerated NEC investment. Such appreciation also reduced the yen cost of that investment, at least temporarily, so NEC opted first for the speedy acquisition of a U.S. company already trading with NEC.

Finally, the spectre of new U.S. trade restrictions (modeled on those in consumer electronics) invited rapid action, beginning in 1975, when American semiconductor manufacturers charged NEC and other Japanese producers with selling capacitors in violation of U.S. antitrust provisions.[65] The following year, these U.S. manufacturers formed the Semiconductor Industry Association (SIA), partly in response to growing Japanese competition. Already, by 1976, Japanese semiconductors accounted for almost 7 percent of total sales in the large U.S. market—roughly comparable to the 12 percent of total sales in the smaller Japanese market supplied by American producers. SIA further anticipated a rapid growth of Japanese imports to fill growing IBM orders for 16K RAMS. American companies could not fill these orders because they had failed to expand capacity after the 1974-75 recession. SIA's fears were justified as the balance of trade in semiconductors shifted to favor the Japanese after 1977, and by 1979, the Japanese had

---

63. *Economist,* December 8, 1984, pp. 75-6; *Fortune,* October 28, 1985, pp. 30-3.

64. Unless otherwise noted, data in this and the next three paragraphs were derived from interviews with industry analysts and company managers in Tokyo, June 1984.

65. *Economist,* December 8, 1984, pp. 75-6. For an overview of the political economy of protectionism in semiconductors, see Daniel I. Okimoto, "Political Context," in Okimoto et al., *Competitive Edge,* esp. pp. 93-4, 100, 105-6, 122-29.

captured 43 percent of the U.S. market for 16K chips.[66] The continued Japanese invasion again prompted the SIA to consider yet another political response.

In such a charged atmosphere, the acquisition of an existing company often seems preferable to building a new plant, in part because it can quickly be accomplished. In 1978, as SIA began to muster political support, NEC acquired its first American company, Electronic Arrays (EA), a financially-troubled firm with which NEC had previously dealt. More generally, however, acquisitions have played a small role in the rapid growth of Japanese direct investment in the United States, in contrast to America's European investors.[67] (By comparison, at home, Japanese acquisitions, while also limited, were nonetheless more common, especially acquisitions of financially distressed suppliers and other affiliated companies.[68]) Instead, new ventures have provided Japanese multinationals with a common form of entry to markets, not only in the United States but elsewhere in the world. Outside the United States, however, joint ventures with local companies became far more prevalent than in America, where the absence of any foreign capital controls allowed Japanese parents to exercise exclusive ownership over their new ventures.[69] NEC demonstrated this broader tendency when it

---

66. The production data in this paragraph are derived from the following sources (in order): Japan, Ministry of International Trade and Industry, *The Semiconductor Industry and Japanese Government Policies* (Tokyo, 1983), p. 3; Joint Economic Committee, *International Competition*, p. 105-06; U.S. Congress, Office of Technology Assessment, *International Competitiveness in Electronics* (Washington, D.C.: U.S. Government Printing Office, 1983), p. 141; Machinery Promotion Association, Economic Research Center, *International Comparison of the Semiconductor Industry in Japan and the U.S.* (Tokyo: MPA, 1981), p. 51.

67. According to statistics provided by W.T. Grimm and Company, the Japanese were among the least likely to gain access to the U.S. market through acquisition:

Domicile of Foreign Parent	Number of Foreign Acquisitions of U.S. Companies	
	1979-83	1984
United Kingdom	288	49
Canada	233	36
West Germany	77	4
France	71	7
Switzerland	43	7
Japan	39	6
Netherlands	37	5

68. James C. Abegglen, *The Strategy of Japanese Business* (Cambridge, Mass.: Ballinger, 1984), pp. 125-39; Sarkis J. Khoury, *Transnational Mergers and Acquisitions in the United States* (Lexington, Mass.: Lexington Books, 1980).

69. Yoshi Tsurumi, *The Japanese Are Coming: A Multinational Interaction of Firms*

constructed a new, wholly-owned American manufacturing plant in 1982. From these new facilities, NEC and other Japanese investors proved better able to respond to the myriad of competitive pressures generated within their industries—unencumbered by foreign capital controls like those imposed on American investors in Japan.

Industrial Organization

While through the mid-1980s U.S. trade restrictions represented the most important determinants of Japanese direct investment downstream from wholesaling, they were not the only predictors. In addition, various competitive forces in an industry accelerated foreign direct investment, among which the enforceable demands of buyers and the gaming strategies of oligopolistic competitors figure prominently.[70] In fact, these two industrial forces can be highly interrelated. A few concentrated buyers may, for example, exert considerable influence over the foreign investment decisions of their multiple suppliers, especially if they demand products that either must be customized to local conditions or can be substituted easily by local suppliers.

In turn, these buyers may actually imitate the behavior of their competitors, especially in moderately concentrated industries with high levels of horizontal and vertical interdependence among oligopolists. From this perspective, companies make foreign investment decisions by following the industry leader and other rivals, or by punishing a rival for an aggressive move made elsewhere. Thus, a risk-averse manager will match a competitor's moves so as to reduce the probability that the

---

and Politics (Cambridge, Mass.: Ballinger, 1976), pp. 71-100, 201-15; Terutomo Ozawa, *Multinationalism, Japanese Style: The Political Economy of Outward Dependency* (Princeton, N.J.: Princeton University Press, 1979), pp. 76-192; Hikoji Katano, *Japanese Enterprises in Asian Countries* (Kobe: Research Institute for Economics and Business Administration, Kobe University, 1981), pp. 33-119 and statistical appendices, pp. 127-223.

70. The gaming strategies of multinationals have been analyzed in at least two ways. First, the so-called follow-the-leader hypothesis is best tested in Frederick T. Knickerbocker, *Oligopolistic Reaction and Multinational Enterprises* (Boston: Graduate School of Business Administration, Harvard University, 1973). Second, the so-called exchange-of-hostage hypothesis is best tested in E.M. Graham, "Transatlantic Investment by Multinational Firms: A Rivalistic Phenomenon?" *Journal of Post-Keynesian Economics* 1 (Fall 1978), pp. 82-99. For an excellent review of these gaming strategies and supporting empirical research, see Richard E. Caves, *Multinational Enterprise and Economic Analysis* (Cambridge: Cambridge University Press, 1982), pp. 97-100, 106-07. Tsurumi, *The Japanese Are Coming,* pp. 88-95, provides empirical support for this logic using data on Japanese foreign investments.

competitor will later jeopardize the future of that manager's own company. According to this reasoning, it is better for both companies to move sequentially, even at the risk of both suffering losses, than for a single company not to move and thereby risk a competitor's net gain.

Oligopolistic competition. Japanese manufacturers have seemed especially prone to such imitative, "follow-the-leader" behavior in their U.S. investments. Consider, again, the semiconductor industry. Since NEC, the Japanese industry leader, initiated the first move in 1978, its Japanese competitors have followed in rapid succession: Hitachi and Fujitsu moved in 1979; Toshiba, in 1980; Hitachi again, in 1981; Mitsubishi, in 1983. Now contrast the timing of Japanese investment with the behavior of U.S. producers in Japan. After IBM's entry in 1971, no other American semiconductor producer established manufacturing operations in Japan until the 1980s, when Motorola and Fairchild reentered Japan. In other words, the liberalization of foreign equity controls by the Japanese government in 1974 had little impact on foreign companies. Nor did the simultaneous liberalization of import controls. And neither did imitative behavior among oligopolistic rivals, so common to Japanese investments in the United States.

To account for the different patterns of American and Japanese investment in semiconductors, existing research suggests that imitative strategies appear most frequently in moderately concentrated industries, not in less concentrated ones, where rivals seem not to recognize interdependence; nor in very concentrated industries, where rivals often collude.[71] Indeed, the semiconductor industry was more highly concentrated in Japan, for example, than in the United States. In 1978 the top four companies—NEC, Hitachi, Fujitsu, and Toshiba—collectively controlled 63 percent of the open market for integrated circuits. NEC alone held 18 percent of the market. The comparable figure in the United States for the top four companies was 49 percent. The U.S. industry was less concentrated in part because of U.S. antitrust policies. IBM and ATT, for example, were both proscribed from entering the open market. Japanese government policies, by contrast, promoted concentration. MITI actually prohibited specialized manufacturers of integrated circuits from entering the industry, thus promoting instead the entry of large, diversified electronics companies that had more in common with each other than did their American rivals.[72]

---

71. Knickerbocker, *Oligopolistic Competition,* Chapter 6. The following estimates of market shares in the Japanese and American semiconductor industries were reported in: Machinery Promotion Association, *Semiconductor Industry,* p. 96; and Semiconductor Industry Association, *International Microelectronic Challenge* (Menlo Park, Calif.: SIA, 1981), p. 35.

72. For the role of Japanese policies in promoting concentration generally, see Richard

Common industry structures have often bred common corporate strategies. For example, the largest Japanese semiconductor manufacturers were themselves horizontally diversified electronics companies of roughly comparable size. The average of semiconductor sales to total sales for the top six Japanese companies was 7 percent in 1979, compared to 70 percent for the top nine U.S. semiconductor companies.[73] Japanese companies, unlike most of their American counterparts, competed in other consumer and industrial product areas as well as in semiconductors. Such comparable diversification heightened pressures for imitative behavior at home and abroad. Sequential foreign investment was one response.

Again, since Japanese production has concentrated in a few diversified conglomerates which consume a small-to-moderate proportion of their own production, cross-purchasing strategies were common among large Japanese companies. In 1979, for example, the top six companies producing over one-half of all semiconductors consumed at least 60 percent of noncaptive domestic production. In the U.S. semiconductor industry, by contrast, a large proportion of noncaptive sales flow from National Semiconductor, Intel, and other full-line semiconductor companies, which themselves consume little of the production of their competitors. With a few captive makers, several full-line companies, and many smaller specialized producers, competitors in the U.S. are far more diverse than those in Japan. Cross-purchasing, like vertical and horizontal integration, contributes to an imitative, follow-the-leader behavior that characterizes Japanese semiconductor producers and other oligopolies in Japan as well.

As in the semiconductor industry, initial investments in U.S. assembly plants by Japanese automakers were made largely in response to trade pressures; but the timing of subsequent moves also at times reflected a pattern of imitative behavior common to oligopolistic rivals.[74]

E. Caves and Masu Uekusa, *Industrial Organization in Japan* (Washington, D.C.: The Brookings Institution, 1976), pp. 141-54; for the semiconductor industry specifically, see Machinery Promotion Association, Economic Research Center (Kikai Shinko-Kyokai Keizai Kendyusho), *Survey Report on the Semiconductor Industry in Japan and the U.S.* (Tokyo: MPA, 1980), p. 122. The only exception to this general statement was Kyodo Electronics Technology Research Center, a joint venture formed by Pioneer, Alps Electric, Tokyo, Japan Chemical Condenser, and Koelen Electric Works.

73. For the internal and external sales of Japanese semiconductor companies reported in this paragraph and the next, see: Joel Stern, "International Structural Differences in Financing," in Semiconductor Industry Association, eds., *An American Response to the Foreign Industrial Challenge in High Technology Industries* (Menlo Park, Calif.: SIA, 1980), pp. 133-34; Office of Technology Assessment, *International Competitiveness in Electronics,* p. 138; Joint Economic Committee, *International Competition,* p. 68.

74. This motive is explored in the *Wall Street Journal,* July 9, 1985, p. 2.

Toyota, for example, accelerated its planned entry into the United States after Nissan had announced that auto production would be added to Nissan's existing truck assembly operations in Tennessee. A Toyota joint venture with GM soon followed; next would come plans for a wholly-owned assembly plant. For the Japanese, follow-the-leader strategies held a strong appeal, one which also influenced their suppliers back home to consider new moves abroad.

Buyer power. Oligopolistic competitors can exert considerable pressure on their existing suppliers to follow their trail overseas, even in the absence of government restrictions on trade. Japanese parts suppliers, for example, have set up American production facilities as a convenience to on-shore buyers engaged in manufacturing.[75] When Honda invested in Ohio, its Japanese-based suppliers of headlights, auto fuel tanks, exhaust pipes, steering wheels, and auto engine parts also established plants within that state. Honda found parts from nearby plants, including American-owned plants, to be cheaper than imports. Having suppliers nearby also facilitated Honda's adoption of the same just-in-the delivery system that it used in Japan. Finally, by utilizing suppliers at hand, whether Japanese- or American-owned, Honda also satisfied its political agenda; such purchases could be classified as "local content" should Congress impose content requirements. For transplanted Japanese suppliers, new sales in America represented one of the few possibilities for growth, given a stagnant home market and U.S. restrictions. Since the proximity of buyers in this industry mattered, foreign investment was a logical step, even for small- to medium-sized companies not otherwise known for their rapid movement overseas.[76]

Even when buyers were not geographically concentrated and markets were more robust, pressures to move became especially intense if suppliers sold only commodity-like products with ready substitutes. For example, vendors of commodity-like semiconductors found that they had to follow existing buyers overseas if they wished to maintain their relationships.[77] So, when Sharp, Matsushita, and other Japanese consumer electronics firms moved to southeast Asia, NEC and other

---

75. *Wall Street Journal,* March 29, 1985, pp. 1, 20; *New York Times,* July 6, 1984, p. D4.

76. According to Kiyoshi Kojima, such small-scale investment was long a hallmark of Japanese overseas operations, in striking contrast to American investments. See his *Direct Foreign Investment: A Japanese Model of Multinational Business Operations* (London: Croon Helm, 1978). With the notable exception of Japanese parts suppliers, his conclusion no longer seems applicable by the 1980s.

77. Kunio Yoshihara, *Japanese Investment in Southeast Asia* (Honolulu: University Press of Hawaii, 1978), pp. 133-78.

Japanese suppliers moved with them. When those same consumer electronics companies moved to the United States as an orderly marketing agreement went into effect in 1977, their Japanese suppliers followed in 1978. The fear of new trade restrictions on commodity-like semiconductors combined with the exercise of buyer power to stimulate foreign investment.

Customization of parts also encouraged suppliers to locate in close geographic proximity to those buyers who required it, a common condition in industrial and communications electronics. Both IBM and ATT, major buyers in the semiconductor segment, demanded customization for the production of computers and telecommunication equipment. The growing need for repetitive and close interaction among engineers from semiconductor suppliers and their buyers—IBM or ATT—added pressure on NEC, as one of those suppliers, to invest in the United States.

New investments in the United States actually began to exceed the value of new investments in Japan, not only for NEC but for Fujitsu and other Japanese electronics companies as well. Until 1984, in fact, electronic producers accounted for the greatest share of Japanese investment in American manufacturing, surpassed that year only by Japanese auto assemblers.[78] Together, in 1986 these machinery and equipment companies accounted for over one-quarter of all Japanese investment in U.S. manufacturing (see Table 8.5). Their investments slowly began to approach U.S. machinery manufacturers in Japan, who also moved abroad in response to the pressures of buyers and the trade restrictions of governments. But unlike these American investors, who remained concentrated in a few industries, Japanese manufacturers in America were more widely diversified and entered industries such as metal fabrication largely unknown to U.S. multinationals operating in Japan.

Not only were these Japanese investments in U.S. manufacturing more widely diversified, they also rapidly approached the size of American investments in the Japanese manufacturing sector. For example, the ratio of American to Japanese cross-investment in manufacturing was 5 to 1 in 1974, just under 4 to 1 in 1980, and less than 2 to 1 in 1985. (see Table 8.5) This growth of Japanese manufacturing in the United States was part of a much larger process. By 1981, foreign direct investment contributed 5 percent to America's gross fixed capital formation, the highest level in modern U.S. history.[79] The

---

78. Economist Intelligence Unit, *Japanese Overseas Investment: The New Challenge,* Special Report No. 142 (London: Economist Intelligence Unit, 1983), pp. 109-14, 123-28.
79. U.S. Department of Commerce, International Trade Administration, *International*

comparable figure in 1971 was one percent. With that investment came employment, especially as manufacturing and assembly operations became increasingly more attractive to Japanese investors—to the delight of many American states and localities keen to replace American investors who earlier had moved abroad.

Table 8.5 *American-Japanese Cross-Investments in Manufacturing*[a]

Industry[b]	1974	1980	1986
Machinery[c]			
In Japan[d]	59.4%	54.0%	52.8%
In USA[e]	f	21.0	28.9
Chemicals			
In Japan[d]	21.5	23.6	30.8
In USA[e]	12.7	27.1	9.6
Metals			
In Japan[d]	1.2	2.8	2.2
In USA[e]	f	28.7	18.3
Other			
In Japan[d]	17.9	19.6	14.2
In USA[e]	34.5	23.2	43.2
Total Cross-Investment in Manufacturing ($ billion)			
In Japan[d]	$1.5	$3.0	$5.3
In USA[e]	$0.3	$0.8	$3.0

[a] Distribution of direct investment position at year-end in each country's manufacturing sector.

[b] Listed in the order of importance to U.S. investors.

[c] Includes electrical and nonelectrical machinery, and transport equipment.

[d] American direct investments in Japan, as defined in Table 1, note a.

[e] Japanese direct investments in America, as defined in Table 1, note a.

[f] Data supressed by the U.S. Commerce Department to avoid the disclosure of individual company operations.

Sources: United States Department of Commerce, Bureau of Economic Analysis, *Survey of Current Business*, various issues; U.S. issues); *Selected Data on U.S. Direct Investment Abroad, 1950-76,* February 1982, pp. 20-27.

Direct Investment: Global Trends and the U.S. Role (Washington, D.C.: GPO, 1984), Table 7, p. 48.

The Policies of U.S. States and Localities

For Japanese investors who had already decided to invest somewhere in the United States, state and local governments tried to influence their final choice of a plant location through combinations of fiscal incentives, infrastructure expenditures, and marketing practices.[80] Competition among American states and localities often proved intense, especially since Ohio and other industrialized states of the midwest and northeast looked to foreign investment as one way to overcome economic stagnation. Japanese prefectures were also looking to foreign investors to provide more jobs, but bidding wars there remained almost nonexistent. Simple and predictable packages of incentives, limited marketing activities, and the active involvement of the national government to keep a lid on competitive bidding—all so common to investment promotion by Japanese prefectures—seldom prevailed in America.

To the contrary, in the United States, competition among the several states often became quite heated—and costly.[81] Ohio, for example, beat out several contenders by offering Honda an incentive package which included: a $2.5 million grant to develop the site; a substantial reduction in annual property taxes; designation of the site as a foreign trade subzone with reduced duties; a guarantee that the federal government would make railroad improvements; and free English tutoring for Japanese expatriates and their children at a nearby state university. At roughly the same time, Ohio's neighbor and competitor, Tennessee, proposed a comparable incentive package to Honda's rival, Nissan.[82] Among many inducements, Tennessee spent $12 million for new roads to the Nissan plant, and $7 million to help train plant employees, while the county government reduced property taxes by another $1 million over the first 10 years of the project.

In all, Ohio and Tennessee, like other governments, tailored a package of incentives to the needs of each investor.[83] Usually, among

80. For a recent review of these policies and their effects on investment see Guisinger, *Investment Incentives,* pp. 48-54.

81. For a survey of the strategies pursued by governments competing for investment, see Dennis J. Encarnation and Louis T. Wells, "Sovereignty En Garde: Negotiating with Foreign Investors," *International Organization* 39 (Winter 1985), pp. 47-48; Encarnation and Wells, "Competitive Strategies in Global Industries: A View from Host Countries," in Michael E. Porter, ed., *Competitive Strategies in Global Industries,* forthcoming.

82. The competitive bids of Ohio and Tennessee are reviewed in Economist Intelligence Unit, *Japanese Overseas Investment, pp. 109-14.*

83. See John Rees, "Government Policy and Industrial Location in the United States," in U.S. Congress, Joint Economic Committee, *Special Study on Economic Change,* Vol.

comparably-sized firms, the first Japanese entrant in each industry received the most lucrative package. And across industries, those that offered the greatest employment potential typically received the greatest assistance. Aggressive bidding strategies were adopted by most states, but California, by far the favorite site for Japanese investors, remained a notable exception. Indeed, that state's unitary tax, a subject of vitriolic debate, was often cited as a disincentive to Japanese investment there.

Bidding with incentives has not been the only competitive strategy used by state and local governments. They also seek investors by creating images of attractive business climates, a strategy that parallels efforts used by private firms to differentiate their products from those of competitors. For example, governments in the United States have introduced industrial parks and foreign trade zones, which appeal to only a narrow segment of Japanese investors. Governments also "market" possible location sites through media advertisements, promotional and "sales" offices, and overseas missions sent in search of new investors. These efforts at differentiation complement aggressive bidding and occasionally even substitute for that bidding. State governments, in particular, have placed great emphasis on differentiating their jurisdictions. Again, Ohio stands out. The Ohio Development Board represents the marketing arm of the state government, which not only grants incentives but also conducts complete marketing operations, with offices in Japan and Europe as well as at home.[84] By the mid-1980s, their job became much easier, as the spectre of increased U.S. trade restrictions loomed ever-larger, and as the relative value of the U.S. dollar declined significantly.

Foreign Exchange Appreciation

Beginning in March 1985, the value of the dollar began to fall precipitously: From a high of ¥263, it plummeted over the next 20 months (into November 1987) to a post-war low of ¥134. That 50 percent appreciation of the yen greatly accelerated Japanese direct investments in the United States. Indeed, at the end of 1984, in the midst of an upward revaluation of the dollar, those investments stood as $7.9 billion. Two years later, in the throes of the dollar's devaluation, the

---

7, *State and Local Finance: Adjustments in a Changing Economy* (Washington, D.C.: G.P.O., 1980), pp. 128-79.
84. This paragraph summarizes interviews with government officials, industry analysts, and company managers held in Tokyo, June 1984.

book value (stock) of all Japanese direct investments in U.S. factories, assembly plants, and warehouses had grown over 40 percent, to reach $11.3 billion. Such growth, at an annual rate of 20 percent between 1984 and 1986, was unprecedented in the history of Japanese-American cross-investment.

Looking ahead, Japanese direct investments will undoubtedly grow, and their sectoral distribution will probably change. Not only have the yen-prices of dollar-denominated assets in the United States fallen since March 1985, but so have the profits associated with Japanese exports. Both factors stimulated investments, but not instantaneously. Rather than let the dollar-price of their exports rise apace with the yen, most Japanese exporters initially tried to maintain their dollar prices by cutting profit margins. This strategy, however, proved to be unsustainable as the dollar continued to plummet. Rather than continue exporting from Japan, these exporters have feverishly sought to produce offshore. In search of lower wages, Japanese exporters have moved rapidly to the countries of east and southeast Asia, and to Mexico—following a pattern reminiscent of U.S. multinationals seeking to avoid the ravages of a strong dollar during the late 1970s and into the 1980s. But unlike their American counterparts at an earlier time, Japanese exporters today have also moved in far greater numbers to invest in their principal overseas markets—most notably, in the United States—where they hope to add more local assembly and manufacturing operations to feed their existing wholesale network. This heightened form of economic rivalry, now mounted on America's home ground, represents an increasingly formidable challenge to U.S. public policies and corporate strategies.

Conclusions

Government policymakers and business managers have long recognized that foreign direct investment is a formidable weapon in commercial competition. Indeed, early fears of American industrial prowess prompted Japan's erection of foreign trade barriers to encourage fledgling competitors. But by themselves these barriers were insufficient to reduce American competition, since tariffs and quaotas in several industries served as strong inducements for American companies to invest in Japan. Recognizing this, the Japanese government instituted capital controls to limit foreign investors in the domestic market. Not only was the volume of American investment diminished by these regulations, but these restrictions also minimized American managerial control over the occasional joint ventures permitted by the Japanese government. In such a hostile environment, few American multinationals invested in Japan, fewer than in any other major overseas market. Those that did invest were able to exploit in a newly-industrializing Japan the same marketing and technological innovations that earlier had given them formidable competitive advantages in U.S. manufacturing, especially in machinery and in chemicals.

Early Japanese investors in America enjoyed few of the same advantages. Rather than diversify horizontally, as the Americans in Japan were doing with their assembly and manufacturing plants, the Japanese integrated vertically, unencumbered by any American version of foreign equity controls. Early investments upstream in the wholesale trade and finance of food products, metals, and minerals satisfied Japan's growing demand for America's natural resources. Later investments downstream in the wholesale trade of motor vehicles and other durables complemented Japan's export strategy. With few marketing and technological advantages to exploit early on, with few impediments to continued trade with America, and finally, with Japanese capital controls limiting large outflows of still-scarce foreign exchange, Japan's direct investments in the United States remained miniscule for nearly 30 years after World War II. In 1974, those investments were one-tenth the size of comparatively small American investments in Japan.

But during the late 1970s, this changed as Japan's current account surpluses grew. By 1984, the book value of Japanese investment in the U.S. wholesale sector alone, mainly in selling autos and other durables, exceeded the value of all American investments in Japan. Japanese-affiliated wholesalers distributed at least 70 percent of all Japanese exports to America, and gave Japanese manufacturers an unusual degree of control over the marketing of their exports to America. From the perspective of Japanese foreign economic policy, these direct investments in America also provided a mechanism for recycling ever-growing export receipts, in an effort to quiet U.S. critics of Japanese trade policies. Somewhat paradoxically, new Japanese investments, most of which entered the U.S. wholesale sector, stimulated even more Japanese exports.

The composition of those Japanese exports gradually changed, however, as sub-assemblies and "knocked-down" kits began to be shipped to new Japanese assembly and manufacturing plants in America. While the book value of these new plants paled in contrast to the total value of Japanese investments in U.S. wholesaling, Japanese investments in

U.S. manufacturing nevertheless experienced rapid growth following America's erection of new barriers to Japanese exports. Japanese manufacturers, like their earlier American counterparts, sidestepped these barriers by investing in the protected market. But unlike the Americans, several Japanese investors could exploit their earlier beachhead in U.S. wholesaling by integrating backward into assembly and manufacture. Japanese manufacturers often moved sequentially to America, following their industry leaders; and when Japanese manufacturers moved, they tended to bring their suppliers along. Once both buyer and supplier had decided to invest in the United States, their subsequent plant-location decisions were often shaped by the proposals of American state and local governments, competing with one another for employment-generating investment. Given this constellation of forces between 1974 and 1986, Japanese investments in U.S. manufacturing grew rapidly; in fact, twice as fast as U.S. manufacturing investments in Japan.

Still, American direct investments in Japan, most of which remained in manufacturing, did nearly quadruple in the twelve years following Japan's 1974 liberalization of capital controls. Similarly, the proportion of wholly-owned foreign subsidiaries in Japan doubled after this liberalization; and in high-technology industries, these proportions were even greater. Japanese buyers of American goods, and especially capital goods, often demanded an increased American presence. Like many Japanese investors in America, some U.S. manufacturers in pharmaceuticals and elsewhere discovered that increased sales were possible only when they controlled their own distribution channels downstream. But unlike the Japanese, few U.S. multinationals used their Japanese investments to expand their exports from America. To the extent that they traded at all, these U.S. subsidiaries, and especially those in the automobile sector, supplied their American parents and third-country affiliates with parts and equipment manufactured in Japan. Whether they were built for export or not, the location of U.S. manufacturing plants inside Japan was increasingly influenced by the pleas of Japanese prefectures in those economically backward regions that actively sought employment-generating investment.

The strategies of local governments competing for foreign investment in both Japan and the United States illustrate a larger theme that emerges from this chapter: the growing similarity in the patterns of American-Japanese cross-investments. These patterns are reflected in the corporate strategies of Japanese and American investors; they can be explained in part by declining variation, not only in government policies, but also in product markets and in the competitive structures of industries. Even the widely touted reluctance of Japanese enterprise to invest in manufacturing and assembly plants overseas—a reluctance typically explained in terms of cultural and organizational factors— began to disappear when Japanese exporters came face to face with many of the same conditions that once sent American multinationals to Japan: ever-larger capital surpluses at home, government-induced barriers to trade overseas, the appreciation of their currency relative to their competitors, and, to a lesser extent, the demands of local buyers and the competitive pressures of concentrated industries.

In explaining the growing "multinationalization" of Japanese enterprise (to use a phrase popular in Japan today) one other consideration should also be added the increasing comparability of Japanese and American policies toward investment outflows and inflows. Policy now seeks to promote, rather than regulate, direct foreign investment. In the coming years, these and other similarities of corporate .strategy and government policy should become even more prominent, as American-Japanese cross-investments continue to grow.

Because of its newness, this investment weapon will undoubtedly be exploited in many different ways by business and government in both countries, as one nation maneuvers to counter the other. The persistence of important differences in the government policies and corporate strategies adopted in America and Japan provides a second theme of this chapter. Some of these differences are legacies of the past. For example, to account for the higher incidence of minority American-owned joint ventures, compared to the preponderance of wholly-owned Japanese subsidiaries in America, one must take into consideration the timing of these investments. Over one-quarter of all American-owned stocks flowed into Japan before the 1974 abolition of foreign equity controls. Those controls had no counterparts in America. Moreover, American regulators did not inhibit foreign companies from entering the U.S. wholesale sector, in contrast to the actions of Japanese policy-makers prior to liberalization.

After liberalization, of course, American investors could have expanded rapidly in manufacturing and wholesale trade if they had actively acquired existing Japanese companies, an investment strategy common to the rapid expansion of American multinationals in other foreign markets. But standard business practices in Japan, and Japanese government policies as well, simply denied foreign investors any opportunity to enter Japan through mergers and acquisitions. These business practices have also shaped Japanese investments in America. The dearth of American acquisitions in Japan has been matched by the relative infrequence of Japanese acquisitions in the

United States. This divergence in the functioning of Japanese and American capital markets illustrates the continued variation of corporate strategies in the two countries even after government-induced barriers to change have dissipated.

The persistence of important differences in corporate strategies and government policies also shows itself in recent developments. Over time, pressures for change in the two countries have pushed them in opposite directions, thus increasing their differences. Japanese manufacturers, for example, have become even more concerned about controlling the marketing of their autos and other durable exports to America. Correspondingly, they have increased the proportion of their total investments going into U.S. wholesaling, from one-half of all Japanese investments in 1979 to two-thirds in 1984. With fewer manufacturing exports to Japan to worry about, American investors, by contrast, either have viewed Japan as an export platform (in auto parts, for example) or, more generally, have sought to establish a visible presence in industries (electrical machinery and customized semiconductors for example) where close interaction with a limited number of buyers has been important. Pressures for changing government policies in the two countries have also pushed in opposite directions. In trade policy the threat of greater U.S. protectionism increased just as Japan's long-existing import restrictions began to wane. And again in foreign investment policy, the slow decline in Japanese resistance to foreign investors was paralleled by a growing concern in the United States that Japan might someday control America's most productive assets.

At this juncture in the history of American-Japanese cross-investments, the greatest threat to bilateral relations is that the growing asymmetry evident in 1986—with Japanese investments in the United States now over two times larger than American investments in Japan—would become another source of tension. That new "investment gap," like the already present "trade gap," adds further credence to the claim that Japan has moved much too slowly in opening its markets to Americans. Fearing this criticism, Japanese attempts to promote the inflow of American investments have been carefully orchestrated by the national government, in contrast to the benign neglect still practiced by U.S. federal agencies. While Japanese promotional efforts have so far met with limited success in attracting American investors, they have stymied for the moment those U.S. critics who earlier pointed to Japanese constraints on market entry through foreign direct investment.

Still other American critics, pointing to the flood of Japanese investment into U.S. industry, claim Japanese companies are financing their expansion at America's expense. Using profits from exports to the United States, the Japanese can build new state-of-the-art plants, which will generate new profits to be wholly repatriated to Japan. Others complain that their new Japanese rivals are misrepresenting themselves by putting the "Made in America" label on products merely assembled here, after most of the work was done in Japan with lower labor costs and with the benefit (up until 1985) of favorable exchange rates. Subsequent yen-to-dollar appreciation has not totally stymied this criticism, even though it has forced many Japanese assemblers and manufacturers to shift purchases to low-cost local suppliers. Among these local suppliers we now find newly-arrived Japanese parts manufacturers whose long-standing relations with Japanese buyers intensify local competition with American suppliers. In defense of these Japanese investments, several American states have argued that Japanese companies (both buyers and suppliers) contribute jobs, tax revenues, and even dividends to their American stockholders—all benefits to the society from which they are reaping profits. Perhaps the most unlikely defenders of Japanese investments in America can be found in the U.S. auto industry, where both labor and management have long wanted to bring the Japanese to our shores. In a 1980 speech, Chrysler Corporation Chairman Lee A. Iacocca succinctly explained: "That way, we draw from the same pool, pay the same taxes, meet the same regulations and contribute to the health of the economy we all compete in."[85] His may be the best argument yet for continuing unimpeded cross-investments between America and Japan.

---

85. *Fortune*, October 28, 1985, p. 30.

CHAPTER 9
Trade Conflicts and the Common Defense: The United States and Japan*

Stephen D. Krasner

For the United States, providing for the common defense has always had a specific and distinctly American meaning. With few exceptions American policy makers have had a Manichean view of the world, a perspective that reflects the profound and pervasive influence of Lockean liberalism on foreign policy.[1] Other countries have been classified as either good or evil, as friends or enemies. The United States has found it difficult to identify common interests with enemies or irreconcilable conflicts with friends. In dealing with enemies American policy makers have been very alert to any alteration in the relative distribution of power and unconcerned with absolute gains. In contrast, in dealing with friends American policy makers have been almost exclusively concerned with absolute gains and have been insensitive to relative shifts in power capabilities.

The precept that a country has permanent interests but not permanent friends is profoundly antithetical to the American perspective. U.S. policy makers have paid little attention to the possibility that a loss of power vis-à-vis friends could present serious and unforeseen difficulties, either because friends can become enemies or because managing the international system may be more difficult in a world in which power is more evenly distributed.

Nowhere has this obliviousness to the potential importance of relative gains among friends been more apparent than in American foreign economic policy in the postwar period. In the years'

---

*This article is reprinted with permission from *Political Science Quarterly* Volume 101, Number 5, (1986): 787-806.
1. The classic statement of this argument is Louis Hartz, *The Liberal Tradition in America* (New York: Harcourt Brace, 1955), esp. chap. XI.

immediately after World War II, the basic American concern was that economic breakdown would lead to political collapse and the "loss" of Europe and Japan. Economic prosperity was thought to be a prophylactic against communism. American policy makers promoted European economic unification even though this was, in the long term, bound to place American products at some comparative disadvantage; and they tolerated explicit Japanese discrimination against American exports and direct foreign investment. Their basic objective was to increase the level of absolute well-being in the western alliance. They were unconcerned with relative power except with regard to the Soviet Union.

In pursuit of their goal of promoting absolute well-being, they fostered a set of norms for the international economic system in which reciprocity, the balancing of concessions, played a central role. But they understood reciprocity in a specific way. American leaders endorsed diffuse rather than specific reciprocity. Diffuse reciprocity implies that negotiations and agreements should be multilateral. Rules are designed to regulate procedures rather than actual outcomes. Specific balancing is eschewed in favor of promoting long-term reductions in trade barriers which, it is inferred, will benefit all countries. Some free riding is tolerated because the costs of coercion and policing outweigh the benefits provided by greater openness.

In contrast, specific reciprocity requires careful equilibration of benefits. Negotiations and agreements are limited to specific countries. Rules are designed to achieve particular behavioral outcomes. Free riding is not tolerated; tit for tat is a legitimate strategy. Rules can vary across countries and issue areas. Different problems are treated in different ways.[2]

In many ways, diffuse reciprocity was a stunning success for the United States. Formal trade barriers have fallen across the industrialized world. Trade has grown dramatically since the conclusion of World War II. Economic growth rates have been unprecedented. Communist movements, whether legitimate or illegitimate, have been frustrated. Diffuse reciprocity reinforced American conceptions of the common defense, which denied profound disagreements among allies.

---

2. Robert Keohane, "Reciprocity in International Relations," *International Organization* 40 (Winter 1986), develops the distinction between diffuse and specific reciprocity and analyzes in general terms the costs and benefits of these two approaches.

The Problem

In recent years, however, the economic consequences for the United States of a global regime based upon multilateral accords and diffuse reciprocity have become more problematic. Nowhere is this more apparent than in American relations with East Asia, especially Japan. Japan has challenged the United States even in the highest of high-technology goods. Moreover, it has been extremely difficult for American producers to secure access to the Japanese market despite the removal of virtually all non-agricultural tariff and nontariff barriers in Japan. The difficulty American producers have had in penetrating the Japanese market can no longer be attributed to violations of the General Agreement on Tariffs and Trade (GATT) rules. The rules are being followed, but the expected consequences have not materialized.

One explanation for this disjunction between policy change and behavioral continuity is that the structure of Japan's domestic political economy defies external penetration. Japanese institutions, both public and private, are linked in a dense network of reciprocal obligations. Relations are, in Oliver Williamson's term, hierarchies rather than markets.[3] It is extremely difficult for new actors to pierce this network, whether they be Japanese or foreign; given the past system of formal closure, almost all foreign actors are bound to be new.

If domestic political-economic structures vary, then similar universal rules, such as those codified in the GATT, can have very different behavioral outcomes. Tariff reductions in a market-oriented system like the United States will offer more opportunities to foreign producers than similar reductions in Japan, because buyers are more likely to consider only the costs and benefits of a specific transaction rather than to also incorporate assessments about past and future relationships with prospective suppliers. Diffuse reciprocity will not work even in the absence of conscious efforts at exploitation by the Japanese. The differences between the domestic structures of these two states guaranteed that a universal open system base upon diffuse reciprocity will leave the United States with the "sucker's" payoff.[4]

---

3. See Oliver Williamson, *Markets and Hierarchies* (New York: Free Press, 1975).

4. For an elaboration of payoff matrices in prisoner's dilemma, see Robert Axelrod, *The Evolution of Cooperation* (New York: Basic Books, 1984). Analyses that emphasize direct state intervention in the control of the Japanese economy would lead to an even more pessimistic conclusion about the prospects for an effective system of diffuse reciprocity. If there is direct state intervention in Japan, it is now at such a

Given this situation there is a strong incentive for the United States to adopt a policy of specific reciprocity. Such a policy would be economically beneficial for the United States. It would be more sustainable in the long run than the current policy based on diffuse reciprocity. And it could decrease the economic tensions generated by misperceptions in both countries arising from differing understandings of the implications of the GATT regime.

The adoption of such a policy will, however, be difficult for the United States, because its conception of the common defense is so strongly oriented toward enemies rather than friends. The notions that Japan could pose a threat to the United States simply by virtue of its growing power capabilities and that American policies could be contributing to that growth are difficult to assimilate into a Manichean view of the world. Even if Japan is not perceived as a potential military enemy, the creation of an environment in which an effective Japanese-American condominium could be established, or even a transition to Japanese leadership effected, has not been seriously addressed.

In some areas Japan is gaining rapidly on the United States. If present trends continue, Japan's manufacturing output would surpass America's in about twenty years. Technological progress is making the absence of raw materials less important. While Japan's relatively small population, at least in comparison with the United States, makes Japanese hegemony unlikely even in the long term, the need to share power is already apparent in some issue areas such as international finance. Past experience as well as collective choice analysis should not make us sanguine about the ease with which a Japanese-American condominium could be established.[5]

---

subtle level that it cannot adequately be dealt with through the GATT framework, or even through some of its ancillary arrangements such as the Government Procurement Code.

5. Bruce Russett has correctly pointed out that despite a decline since World War Ii, the power position of the United States remains extraordinary. See his "The Mysterious Case of Vanishing Hegemony; Or, Is Mark Twain Really Dead," *International Organization* 39 (Spring 1985); for a discussion of the importance of having one leading power, see Charles P. Kindleberger, *The World In Depression, 1929–1939* (Berkeley: University of California Press, 1973).

Japan's Political Economy

Japan cannot fail to arrest the attention of policy makers and analysts. For the last three decades it has been the world's most dynamic economy. Growth has been coupled with political stability. Japan has become an exemplar, a model for other states to follow. But what remains at issue is what exactly it exemplifies.

Two conflicting interpretations of Japan's economic success have been presented in the United States. The first, which can be termed the market-oriented model, argues that Japanese attainment can be explained in conventional neoclassical economic terms. The most important institution is the private corporation. Japanese firms have vigorously pursued market opportunities. The hand of the Japanese state has been light, particularly at the sectoral and specific company levels. State intervention is only apparent at the macroeconomic level, where government policy has promoted a high rate of savings and maintained a stable overall environment for private business.

The second line of argument, which can be labelled the developmental state model, places great emphasis on the network of relationships between public and private actors and among private firms. Indeed, it questions the appropriateness of even labelling Japanese firms as "private," as if they were equivalent to corporations in the United States or Europe, because their ties with other actors in the polity and their definition of their own objectives are so heavily influenced by nonmarket considerations. Public institutions are seen as playing an important role in Japanese economic success through their intervention at the sector and firm, as well as the macroeconomic, levels.[6]

---

6. The term developmental state is taken from Chalmers Johnson, *MITI and the Japanese Miracle: The Growth of Industrial Policy, 1925–1975* (Stanford, Calif.: Stanford University Press, 1982). In this presentation, however, I have placed somewhat more emphasis on the network of relations between the public and the private sector rather than the directive role of central state institutions such as MITI (Ministry of International Trade and Industry) and the Ministry of Finance. Daniel Okimoto has called this conceptualization the network state. See Daniel I, Okimoto, *Between MITI and the Market* (Stanford, Calif.: Stanford University Press, 1982), esp. chap. 3. I have not tried to resolve the difference between Johnson and Okimoto in this article since my objective is to provide an alternative to the market-oriented model. Both Johnson's more statist perspective and Okimoto's more corporatist one stand in clear contrast with the market-oriented approach, which sees government activity restricted to macroeconomic measures and market structure dominated by strongly competitive firms.

The Market Model

Many scholars, and certainly almost all economists, have interpreted Japan's economic success from a conventional neoclassical theoretical perspective. Japan's performance is not seen as the result of state intervention, or a peculiar configuration of state-private interaction, or unique cultural values, but rather as the product of factor endowments, including managerial skills, residing in the private sector. In what is still one of the most influential interpretations of contemporary Japan, Hugh Patrick and Henry Rosovsky state that "We gently suggest that Japanese growth was not miraculous: it can be reasonably well understood and explained by ordinary economic causes."[7] In particular, they point to Japan's highly skilled labor force, the existence of widespread dualism in the labor market after World War II that provided a pool of workers, and "substantial managerial, organizational, scientific, and engineering skills capable of rapidly absorbing and adapting the best foreign technology."[8] Coupling these labor endowments, which reflected in part the level of industrial accomplishment that Japan had reached before the war, with capital and technology produced rapid economic growth.

The state does not disappear entirely from the market-oriented model, but its role is distinctly secondary. It acts as a facilitator for the private sector. Adherents of the market model generally accept the claim that the government gave private business the first claim on investible resources. This was particularly important after the war, when capital and foreign exchange were both in short supply. The state adopted policies that encouraged high levels of savings. Expenditures on social welfare and tax rates were both kept low in comparison with other advanced industrialized states. (In fact, Japan could hardly be classified as an advanced country in the early 1950s when its per capita gross national product (GNP) was lower than that of Brazil and Chile.) The following quote, again from Patrick and Rosovsky's 1976 study, captures the essential thrust of the market-oriented model.

> Our view is that, while the government has certainly provided a favorable environment, the main impetus to growth has been private—business investment demand, private saving, and industrious and skilled labor

7. Hugh Patrick and Henry Rosovsky, "Japan's Economic Performance: An Overview" in Hugh Patrick and Henry Rosovsky, eds., *Asia's New Giant: How the Japanese Economy Works* (Washington, D.C.: Brookings Institution, 1976), p. 6.
8. *Ibid.*, 12.

operating in a market-oriented environment of relative prices. Government intervention generally has tended (and intended) to accelerate trends already put in motion by private market forces—the development of infant industries, the structural adjustment of declining industries, and the like.[9]

Analysts utilizing a market-oriented perspective see Japan's political system as pluralistic. Power is divided among a number of different groups. There is intense competition within the private sector. Although the Liberal Democratic Party (LDP) has won every election since 1950, it is internally split into factions and is better seen as a loose coalition of quasi-independent political baronies rather than a coherent and unified political party. Government agencies are at times forced to capitulate to particular private pressures. While bureaucracies do often develop close relationships with firms in specific sectors of the economy, this does not lead to coherent and centralized direction, because different bureaus have different goals and objectives. Philip Tresize concludes his assessment of Japan's political system with the observation that

> the tidier the model of Japan's political economy, the less it conforms to reality. . . . The bureaucracy has indeed has an extensive and active part in economic life, but it has been constantly subject to conflicting pressures, ranging from narrow loyalty to individual ministries to the demands of competing special interests groups. And politics in Japan has hardly been more orderly a process of bargain and compromise than in any other democracy.[10]

For adherents of the market model, economic conflict between the United States and Japan reflects either specific interests group pressures or the failure to adequately coordinate macroeconomic policy. Even with equilibrium exchange rates, Japan would have a large bilateral surplus with the United States: it would run a large deficit with primary commodity producing areas (especially oil exporting states) which it would offset by running surpluses with other areas of the world.[11] The widely-noted lack of manufactured imports into Japan can be explained from the market-oriented perspective by factor endowments. Given an absence of raw materials, high rates of saving, and a skilled labor force, Japan would

---

9. *Ibid.*, 47.
10. Philip H. Tresize, "Politics, Government, and Economic Growth in Japan's in Patrick and Rosovsky, eds., *Asia's New Giant*, 810.
11. C. Fred Bergsten and William R. Cline, *The United States-Japan Economic Problem* (Washington, D.C.: Institute for International Economics, October 1985), 40.

not be expected to import many manufactures.[12] Periods of economic tension between the United States and Japan have coincided with high American bilateral trade deficits that are related to the dollar/yen exchange rate, which is in turn heavily influenced by macroeconomic policy choices. Greater coordination could lessen or even eliminate conflict.[13]

The Developmental State Model

The developmental state model stands in contrast with the market-oriented model. Adherents of this approach reject the notion that the basic impetus, the root cause of Japan's economic performance has been the private market. They argue that market transactions in Japan are conditioned by non-economic factors. The market itself cannot be understood without reference to the larger social context within which it is embedded. Direct action by the state is seen as fundamental for understanding Japan's development. The strongest advocates of the developmental state model contend that the state is the primary cause of Japanese growth. Alternative versions of this argument focus on the involvement of public and private organizations in a dense policy network that blurs the boundary between the state and civil society.[14]

At the core of the developmental state system are a group of interlocked private, political, and public institutions that share core values emphasizing the central importance of aggregate economic performance—large industrial and financial enterprises, the Liberal Democratic party, and the Ministry of Finance and Ministry of International Trade and Industry (MITI). The strength of these various actors has changed over time. The power of the major ministries—Finance, International Trade and Industry—has slipped in relation to the LDP and the major corporations, because party members have

12. Gary Saxonhouse has developed this argument at some length. See, for instance, his "Japan's Intractable Trade Surpluses in a New Era," Seminar Discussion Paper No. 178, Research Seminar in International Economics, University of Michigan, n.d.
13. One formulation is Ronald McKinnon's call for the establishment of stable exchange rates between the United States, Europe, and Japan. See, for instance, his *An International Standard for Monetary Stability* (Washington, D.C.: Institute for International Economics, 1984).
14. The two versions of this argument correspond with statist approaches that see public institutions as separate and distinct from civil society and with corporatist approaches that place more emphasis on the network of relationships between the public and private sphere. Two major texts that reflect this distinction are Chalmers Johnson's *MITI* and Daniel Okimoto's *Between MITI and the Market*.

become more knowledgeable about specialized bureaucratic affairs and firms have more direct control over financial and other resources. More actors, with conflicting interests, have become involved in policy making. Nevertheless, in comparison with the United States and most other industrialized countries, the most powerful institutions in Japan are very closely bonded through shared values and personal networks.

The ability to arrive at a shared consensus about basic goals has been a hallmark of Japan's political economy. The survival of the community is the fundamental objective, rather than the interests of particular groups or individuals. At least through the early 1970s Japanese policy makers agreed that perpetuation of the community required rapid economic growth. The national interest was interpreted as recovery from the devastation left by World War II. This goal took precedence over other possible objectives, such as equity, and was shared by the Liberal Democratic Party, the bureaucracy, large corporations, and most of the electorate.[15]

In the private sector, the significance of community survival is reflected in the choice of market share and long-term growth as the most important corporate goals. Firms are linked by joint stock ownership, which not only reinforces mutuality of interests but also acts as a barrier to outside or hostile takeovers. Seventy percent of the stock of Japan's major companies is owned by other companies and is never traded on the open market.[16] Companies are joined through *keiretsu* (corporate group) ties, among which the provision of finance from the group's bank may be particularly important. Labor unions have been excluded from the policy network in Japan, reducing concern with short-term questions of equity and distribution.[17] Coordination within firms is accomplished through the sharing of information among incumbent employees who, at least in large firms, enjoy permanent tenure.[18]

---

15. See, for instance, Kozo Yamamura, "Caveat Emptor: The Industrial Policy of Japan" in Paul R. Krugman, ed., *Strategic Trade Policy and the New Internatonal Economics* (Cambridge, Mass.: MIT Press, 1986), 169-170; also Johnson, *MITI*.

16. Chalmers Johnson "How to Think About Economic Competition Coming From Japan," paper presented at the West Coast Forum on the Japanese Political Economy, University of Washington, n.d. (c. 1986).

17. See T.J. Pempel's discussion of the exclusion of labor in his "Japanese Foreign Economic Policy: The Domestic Bases for International Behavior" in Peter J. Katzenstein, ed., *Between Power and Plenty* (Madison: University of Wisconsin Press, 1978).

18. Masahiko Aoki, "The Japanese Bureaucracy in Economic Administration: A Rational Regulator or Pluralist Agent?" CEPR Publication No. 68, Center for Economic Policy Research, Stanford University (March 1986), 34-35.

Decision making requires consensus. No major conclusion is arrived at without extensive discussion among interested parties. Leaders are obligated to adopt an inclusionary style. Any bargaining situation is encumbered with long-term calculations of mutual obligations that have been accumulated in the past and are assumed to extend into the future.[19]

It is difficult to accommodate new actors within established networks, even if they are Japanese. Foreigners are even more problematic, since there is no reason to suppose that they will share national objectives and their sense of commitment even to local objectives (such as those in a specific industrial sector) are likely to be suspect if only because they have competing interests outside of Japan. In a system based on diffuse reciprocity, such as Japan's policy networks, the admittance of new actors would open the danger of exploitation even with the best intentioned participants.

For the developmental state model, public organizations— especially the central economic ministries, the Ministry of Finance, and MITI— are critically important. Daniel Okimoto argues that in Japan "the presumption is that the state is there to do whatever is appropriate and necessary to promote industrial growth and prosperity." Interesting contrasts between American and Japanese attitudes are sometimes provided by small incidents. In March 1986, American and Japanese semiconductor manufacturers met in Los Angeles to discuss their disagreements related to international trade. The Japanese delegation included representatives from MITI. American government officials did not participate. S. Bruce Smart, the Under Secretary of Commerce for international trade, stated that "we don't think we should put ourselves in the middle of private transactions, and secondly we were concerned that our presence there somehow would be interpreted as some form of repeal or absolution from antitrust laws."[20]

While the leverage of the major policy-making ministries has eroded in recent years, their power is still formidable. Even without formal authority in an area, they can exercise influence over firms and sectors through the use of administrative guidance. Ministries do have sanctions that can be used to punish those actors in the private

19. I.M. Destler, *Managing an Alliance: The Politics of U.S.-Japanese Relations* (Washington, D.C.: Brookings Institution, 1976), 108-111. See also Stephen D. Cohen, *Uneasy Partnership: Competition and Conflict in U.S.-Japan Trade Relations* (Cambridge, Mass.: Ballinger, 1984), 7.

20. Daniel I. Okimoto, "Domestic Political Configurations and Trade," mimeo, Department of Political Science, Stanford University, April 1983, 25; *New York Times*, 14 March 1986.

sector that ignore administrative guidance. For instance, a ministry could reject a firm's application for approval of a specific kind of transaction. The ministry might refuse to cooperate with a recalcitrant firm in other issue areas. The ministries retain control over issuing ordinances, licensing, and registration. Even without sanctions, their considerable prestige makes administrative guidance a potent instrument.[21]

The state not only provides a general economic environment that is conducive to economic growth by, for instance, sustaining stable macroeconomic conditions, it also may intervene at the level of the sector or the firm. Japanese policy makers have rejected the dictates of static comparative advantage, which in the 1950s suggested that Japan ought to concentrate on labor-intensive products. Instead they have attempted to identify industrial sectors that are characterized by increasing returns to scale, high income elasticities of demand, attractive exports markets, and extensive technological linkages.[22]

The developmental state model does not reject the importance of the market. Indeed, Japan's focus on export-oriented, as opposed to import-substituting, growth compelled Japanese firms to operate in environments that were completely outside the control of their own government. But the state actively encouraged the redeployment of resources to make Japanese firms more competitive. Japanese economic development took precedence over all other objectives.

Chalmers Johnson has argued that the very conception of the market in Japan is different from that which exists in other polities, especially the United States. For the Japanese the basic function of the market is to promote economic growth. The state acts to create conditions that will elicit the maximal entrepreneurial effort from the private sector. In contrast, Americans see the market as a form of social organization that maximizes efficiency. The only fully legitimate

---

21. See the discussion in Chalmers Johnson, "MITI, MPT, and the Telecom Wars: How Japan Makes Policy for High Technology," Department of Political Science, University of California, n.d. (c. 1986); Allan D. Smith, "The Japanese Foreign Exchange and Foreign Trade Control Law and Administrative Guidance: the Labyrinth and the Castle," *Law and Policy in International Business* 16 (1984): 426.

22. Stephen D. Cohen *Uneasy Partnership*, 64; and Bruce R. Scott, "National Strategies" in Bruce R. Scott and George C. Lodge, eds., *U.S. Competitiveness in the World Economy* (Boston: Harvard Business School Press, 1985), 96-100. The most thorough exposition of Japan's industrial policy can be found in Chalmers Johnson, *MITI*, esp. chap. 3. For an anlysis of the concept of created comparative advantage see the work of the Berkeley Roundtable on the International Economy (BRIE); for example, John Zysman and Laura Tyson, eds., *American Industry in International Competition: Government Policies and Corporate Strategies* (Ithaca, N.Y.: Cornell University Press, 1983).

role for the state is regulatory; public action should be limited to promoting conditions that will approximate as closely as possible those of the perfectly competitive market. Other forms of state intervention are counterproductive, even illegitimate, because they distort prices which offer the best guide for allocating resources.[23]

Japan has employed several strategies. Through the rapid growth period that ended in the early 1970s, Japan followed a pattern that Kozo Yamamura has described as "protection and nurture." Foreign products were kept out of the domestic market. Oligopoly pricing was tolerated within Japan. The Ministries of Finance and International Trade and Industry encouraged investment in sectors where declining costs were possible because of the application of foreign technology. Products were sold in foreign markets, especially the United States, at prices far below those in Japan. In more recent years MITI has emphasized the importance of high-technology industries and has organized a number of joint research projects. The role of state bureaucracies in steering the economy or directly promoting the activities of major firms has not been questioned.[24]

Some analysts working with the developmental state model have also been particularly attentive to path-dependent patterns of industrial development, where choices made at a particular moment can have a dramatic effect on the future trajectory of a particular sector or firm. State intervention at a specific juncture can have a decisive impact over the long term. Even if basic economic variables such as interest rates, labor costs, and exchange rates return to some earlier set of values, the effect on production and consumption will not be the same because of earlier choices. With declining cost curves, for instance, there are long-term advantages to seizing larger market shares even if this means reduced short-term profits. By moving quickly down the learning curve, a firm may be able to eliminate its competition for successive generations of products. If economic processes are characterized by path-dependent development and irreversibilities, then future competitive advantages will outweigh the opportunity costs of profits foregone in the present.[25]

One example of the consequences of irreversibilities is offered by the case of computer chips known as 256 k DRAMS. In March 1986

23. Chalmers Johnson, "How to Think About the Economic Competition Coming from Japan," mimeo, 12-15.

24. Yamamura, "Caveat Emptor," 174-179 and 192-200.

25. For a discussion of path dependent development, although not one related to long-term profitability or the nature of Japan's political economy, see Paul David, "Clio and the Economics of QWERTY," *American Economic Reivew* 75 (May 1985); for an

the Commerce Department found that the Japanese had been dumping this product in the American market and imposed substantial antidumping duties. By the time of the decision, however, most American manufactures had withdrawn in the face of intense Japanese competition. A spokeswoman for Intel stated that "Our decision to exit the DRAM market is irreversible.[26]

If the developmental state model is an accurate depiction of Japan's political economy, then the appropriateness for Japan's trading partners of a policy based on diffuse reciprocity and generalized rules is questionable. With different economic structures, different policies will not necessarily have the same effect. The removal of tariff barriers would increase imports in a market-oriented system, assuming foreign products were competitive. But it would not necessarily have much impact in a developmental state, either because nonstate actors resisted penetration or because the policy network composed of private and public institutions frustrated new entrants, or because state actors themselves imposed formal or informal barriers.

Evidence

The debate about the nature of Japan's political economy will not be easily resolved. Adherents of the market-oriented and developmental state models do not always look at the same evidence or even understand the nature of actors in the same way. For instance, the market-oriented model assumes that corporations in Japan, like those in the United States, are private organizations with autonomously generated preferences; while the developmental state model views these same entities as so entwined with the state and so imbued with general societal objectives that they cannot simply be understood as part of an autonomous private sphere of activity.[27] The empirical evidence does not weigh all in one direction, hardly surprising given the fact that well-informed analysts have arrived at very different conclusions about the nature of Japan's political economy. Adherents

anlysis of Japan's strategy based on declining costs in the electronics industry in particular, see Michael Borrus, et al., *U.S. Japanese Competition in the Semiconductor Industry: A Study in International Trade and Technological Development* (Berkeley, Calif.: Institute of International Studies, Policy Papers in International Affairs, No. 17).

26. *San Francisco Chronicle,* 14 March 1986.

27. See, for instance, Chalmers Johnson, "The Institutional Foundations of Japanese Industrial Policy," *California Management Review* 27 (Summer 1985).

of the developmental state and market-oriented models have frequently interpreted the same empirical data in different ways. This problem arises in its most acute form when trying to assess the significance of changes in Japanese policy. Adherents of the market-oriented model see increasing liberalization as evidence of the basic commitment of Japanese policy makers to an open market-oriented system. Adherents of the developmental state model see Japanese leaders implementing more liberal policies only to assuage pressure from foreign trading partners; should pressure cease, concessions would be eliminated. From one reading liberalization supports the market-oriented approach, from another the developmental state.

Finally, the nature of Japan's political economy appears to be a moving target. Some forms of behavior have changed, and changed rapidly. Many have argued that the 1970s marked a watershed because Japan caught up with the West. The power of the Ministry of Finance and MITI has been eroded by the ability of Japanese firms to tap international capital markets, and the consensus on growth as the overriding objective has been weakened by prosperity.[28]

It is, however, important to recognize that the interpretation of change depends critically on how Japan's initial situation is described. If the developmental state model is taken to mean that the key to understanding Japan's political economy is the directive role of the central bureaucracies, then most commentators would agree that change has taken place. The Ministry of Finance and MITI can do less now than in the past, although their power is still formidable in comparison with bureaucracies in the United States. But if a corporatist interpretation of the developmental state model is employed, in which the defining characteristic is not the directive role of state bureaucracies but rather a dense network of ongoing relationships involving both public and private actors, then it is more difficult to conclude that the basic nature of the system has changed. The specific actors involved in policy networks and their relative standings may have altered, but the formulation of policy within a closed circle places a high emphasis on consensus decision making, the maintenance of long term relationships, and shared goals.

There is substantial evidence for the developmental state model. Japan has a very sectorally skewed pattern of trade. It has relatively little intrasectoral trade and imports relatively few manufactured goods in comparison with other major states. In 1980, manufactures accounted for 2.9 percent of Japan's GNP in contrast with 5.7 for the

---

28. See, for instance, Yamamura, "Caveat Emptor," 185-192.

United States, 12.7 for Italy, and 15.0 for Germany. During the 1970s Japan's manufacturing imports increased less, as a percent of gross domestic product (GDP), then those of any other major industrialized country. Through 1984 there was no change in this pattern.[29]

Adherents of the market-oriented model explain Japan's unique pattern of trade in terms of factor endowments. Since Japan is resource poor, it is compelled to import raw materials and export manufactures. This same line of reasoning, however, would suggest that other resource-poor areas, such as western Europe, would have the same pattern of trade. But this is not the case.

If factor endowments alone do not explain Japan's pattern of sectoral imports, then perhaps a combination of geographical position and factor endowments might salvage the market-oriented model. Japan's relatively small proportion of manufactured imports could be seen as a function of barriers imposed by distance (both geographical and cultural) not just factor endowments. A comparison of Japanese trade patterns with those of South Korea sheds some light on this claim. South Korea is also poorly endowed with resources and in the same geographic and cultural area as Japan. It is the largest of the newly industrializing countries (NICs) with a population of about 40 million in 1984, compared with 120 million for Japan. While it is smaller than Japan it cannot be simply viewed as an entrepôt like Hong Kong or Singapore. In 1983, however, 47.3 percent of South Korea's imports were accounted for by manufactures compared with 21.1 percent for Japan. Similar disparities have typified trade for these two countries at least since 1955.[30]

One common assertion, consistent with a market-oriented approach, is that Japan's low level of manufactured imports can be attributed to the lack of competitiveness of products from other countries. This proposition is questionable in light of the ability of non-Japanese producers to compete effectively against Japanese products in the rest of the world. American engineering products have sold well in the rest of the world except Japan, even where there is Japanese com-

---

29. Derived from figures in World Bank, *World Tables*, 3rd ed., Comparative Economic Data, Table 6 and country pages, Economic Data, Sheet 1, using current prices to determine imports as a percent of GDP. For recent trade figures see GATT, *International Trade 1984/85* appendix tables for Japan, the United States, and the European Community.
30. See figures in United Nations, *Yearbook of International Trade Statistics 1957, 1962, 1963, and 1966*; and United Nations Conference on Trade and Development, *Handbook of International Trade and Development Statistics 1985 Supplement*, Table 4.2.

petition. But the same American products have not done as well in Japan as comparable Japanese products have done in the United States.[31]

Direct foreign investment is another activity that can shed some light on the relative merits of the market-oriented and developmental state interpretations of Japan's political economy. Throughout the postwar period, the Japanese government has actively promoted the transfer of technology to Japan through licensing rather than direct foreign investment. Public agencies effectively intervened between foreign and domestic firms. State agencies prevented competitive bidding among Japanese companies. Chalmers Johnson offers the following observation:

> Before the capital liberalization of the late 1960s and 1970s, no technology entered the country without MITI's approval. . . . no patent rights were ever brought without MITI's pressuring the seller to lower the royalties or to make other changes advantageous to Japanese industry as a whole; and no program for the importation of foreign technology was ever approved until MITI and its various advisory committees had agreed that the time was right and that the industry involved was scheduled for 'nurturing' (*ikusei*).[32]

Japan's policy of promoting the transfer of technology through licenses, patents, and other agreements between Japanese and American firms, rather than direct foreign investment, was effective. After World War II, Europe and Japan were technologically backward relative to the United States. They both had skilled labor and other factors that facilitated the absorption of foreign technology. The form of this diffusion, however, differed substantially. Diffusion may occur as a result of information contained in technical literature, the reverse engineering of products, direct foreign investment, or the purchase of patents or other rights by national firms. Japan secured relatively more technology through licensing than did Europe. There has been much less direct foreign investment in Japan than in Europe.[33]

There is, then, a plausible body of evidence that supports the

31. For a detailed statistical presentation see Stephen D. Krasner and Daniel I. Okimoto, "Japan's Trade Posture: From Myopic Self-Interest to Liberal Accommodation?" mimeo, Department of Political Science, Stanfor University, February 1985.

32. Johnson, *MITI*, 17; see also Michael Borrus, et al., *U.S.-Japanese Competition in the Semiconductor Industry*, 68; and Raymond Vernon, *Two Hungry Giants* (Cambridge, Mass.: Harvard University Press, 1983), 88.

33. See Stephen D. Krasner, "Technology Transfer and Institutional Structures in Japan and the United States," mimeo, Department of Political Science, Stanford University, May 1985, Tables 6 and 7.

developmental state argument. Whether as a result of formal mechanisms in the 1950s and 1960s or less formal, and perhaps even private, mechanisms in more recent years, it has been difficult for foreign manufactures to penetrate the Japanese market, especially in those areas that have been targeted by the central economic ministries. *Keiretsu* and other long-term ties among horizontally linked firms, visions of the industrial future that are shared by both public and private officials, and close working relationships between the state and private sector create a network of relationships that cannot be penetrated by new actors, especially those from outside Japan.

It is, however, necessary to recognize that the available evidence does not weigh in one direction. The level of direct foreign investment in Japan has increased substantially since 1980. Many Amrican firms have operated successfull in Japan. Since the elimination of capital controls there has been little disparity betwen Euro-yen interest rates and rates in short-term domestic markets. Such disparties were significant before the mid 1980s; this change does suggest that in the area of finance, changes in policy have changed behavior. Japanese capital markets are no longer insulated from world markets.[34]

This evidence cannot be entirely dismissed by adherents of the developmental state model. The most obvious counter, however, is that direct foreign investment and the elimination of capital controls do not have a negative impact on the fundamental objective of promoting Japanese industrial competitiveness. Such policies may even have positive consequences. By securing access to global capital markets, liberalization may increase the resources available to Japanese firms. Direct foreign investment may make it easier to obtain some types of foreign technology. The dangers of foreign domination are much less than they would have been in the 1960s and even the 1970s because of the growing strength and competitiveness of the Japanese economy and its major firms. Japanese officials may feel compelled to make some accommodations to foreign pressures lest trading partners close their home markets.

I do not want to claim that these arguments for the developmental state model are decisive. Such finality is unlikely. Relations between the United States and Japan are in flux. Japanese policy is changing. Greater involvements in international financial markets may loosen

---

34. For a discussion of the impact of the removal of capital controls see Jeffrey A. Frankel, *The Yen/Dollar Agreement: Liberalizing Japanese Capital Markets* (Washington, D.C.: Institute for International Economics, 1984), 22.

domestic ties. Stories can be offered by proponents of many persuasions. But there is enough evidence, both institutional and behavioral, to support the plausibility of the developmental state model.

Policy Implications

If the developmental state model is an accurate description of Japan's political economy, then an international economic strategy based on diffuse reciprocity, universal procedural rules, and the assumption that costs and benefits will balance out in the long term will not be optimal for the United States. Similar policies will not have similar behavioral outcomes in a market-dominated as opposed to developmental state political economy. Decreasing formal trade barriers may have more of an impact on imports in the United States than Japan. American technology may be more available, because a larger proportion of innovations are made by smaller companies that need foreign capital or foreign sales. Market access through direct foreign investment, especially through the takeover of established firms, is more easily accomplished in the United States than Japan. Marketing networks in the United States are more accessible because they are not encumbered by traditional ties. American officials have rarely been sensitive to the benefits that might accrue by intervening on behalf of industries with declining costs; for such an enterprise a small initial advantage can guarantee market dominance as the industry matures. Existing international rules are simply not drawn to cope with most of these situations; and even if attempts were made, it is difficult to believe that general stipulations could have similar behavioral effects in different countries. Given substantial asymmetries in national institutions, specific reciprocity, which is based upon securing roughly equivalent advantage for a delimited set of transactions, is a more attractive option.

Specific reciprocity does not, however, sit comfortably with the basic beliefs of most American policy makers. The United States has been and continues to be the strongest supporter of diffuse reciprocity especially with regard to international trade. This commitment to diffuse reciprocity has a number of sources, one of the most important being the lessons that were drawn from the 1930s.[35] To quote from

---

35. Judith Goldstein has developed this argument. See, for instance, "The Political Economy of Trade: Institutions of Protection," *American Political Science Reivew* 80 (March 1986).

one recent book, the keynote volume for a new set of studies being conducted by the Council on Foreign Relations: "The experience of the 1930s shows that bilateral agreements cannot provide a stable, consistent and expanding trading system."[36] In a highly uncertain world, vivid historical exemplars have a force that surpasses specific argumentation.[37]

Specific reciprocity is also difficult to incorporate within a view of the common defense, which ignores power shifts among friends. If friends cannot have fundamental disagreements, then the power relations among them will in the last analysis be irrelevant. If everyone is becoming wealthier, as diffuse reciprocity suggests they should, then issues of relative gain are not critical.

However, while the general principles and commitments of American policy makers have not changed, both external and internal pressures have led to the adoption of policies that are based more on specific than diffuse reciprocity. Nowhere is this more apparent than in American economic relations with Japan. There is a long history of sector specific arrangements involving the United States and Japan. In 1955 the United States successfully secured Japanese acquiescence to an accord that limited the sale of Japanese cotton textiles in the United States. This settlement ultimately evolved into the MultiFibre Agreement. Other settlements involving major products such as steel and automobiles were negotiated in subsequent years. In the early 1980s the United States and Japan began a series of negotiations known as MOSS (Market Opening Sector Specific). These discussions dealt with several specific commodities including telecommunications equipment, electronic devices, wood and paper products, and medical equipment and pharmaceuticals. The United States has focused its attention on procedural issues such as technical standards, certification requirements, access to MITI advisory committees, and the processing of patent applications. Such initiatives flow naturally from policies based on the norm of diffuse reciprocity, which has informed most American trade policy. But in some cases the United States has also press for explicit market shares, an objective much more consistent with a policy of specific reciprocity.[38] The MOSS negotiations

---

36. C. Michael Aho and Jonathan David Aronson, *Trade Talks: America Better Listen!* (New York: Council on Foreign Relations, 1985), 129.
37. In the terminology of recent work in cognitive psychology such exemples may be particularly "available" to decision makers. See Daniel Kahneman, Paul Slovic, and Amos Tversky *Judgment Under Uncertainty: Heuristics and Biases* (New York: Cambridge University Press, 1982), part IV.
38. A listing of the issues dealt with in the MOSS negotiations can be found in Japan

implicitly recognize that given the variations in institutional arrangements between the United States and Japan, general rules do not provide either mutually satisfactory outcomes or adequate policy guidance. Different sectors in Japan may require different initiatives to secure access for American producers.

There has been a growing disparity between American principles and American policy. The principles espoused by American top governmental decision makers still reflect diffuse reciprocity. They are based on general rules that can be applied to all countries. They assume that similar policy measures will more or less have similar effects. At the same time the force of circumstances—sometimes domestic protectionist pressures, sometimes glaring difficulties at the international level—have led American policy makers toward specific reciprocity.

Should principle follow practice (in which case American policy makers would begin to espouse specific reciprocity as a first best option) or should some current pratices be regarded as unfortunate departures from the ideal of free trade liberalism? This is a central question for American foreign economic policy. The 1930s seemed to show that specific reciprocity was unwise and that diffuse reciprocity was the only avenue to global economic stability. But the developments of the last two decades, especially in American-Japanese relations, suggest that diffuse reciprocity is not without its difficulties. Persistent high visibility and politically charged trade conflicts often involving heads of state have generated bitterness and resentment. Central decision makers in both countries believe that their counterparts are not playing by the rules.

The sense of injustice experienced on both sides of the Pacific is an inevitable consequence of the tensions arising from the universal principles advocated by the United States, differences between domestic political-economic institutions in the United States and Japan, and

---

Economic Institute, *JEI Report* No. 2B, 17 January 1986, 3-6. American negotiators did suggest a target level of purchases in the MOSS telecommunications negotiations. See *JEI Report*, 21B, 7 June 1985, 8. Some specific import levels were also informally mentioned in a settlement worked out between Japan and the United States to resolve several anti-dumping suits that had been brought by American firms against Japanese exporters of micro-chips. Both American and Japanese negotiators said that they had discussed a target of 20 percent of the Japanese market for American firms by 1991, but no formal commitments were made. In the mid-1980s American producers accounted for only 8 percent of the Japanese market. See *New York Times*, 1 August 1986. A Commerce Department official who played a major role in negotiations with Japan expressed great skepticism about the impact of the 1986 agreement. See the statement of Clyde Prestowitz in *Business Week*, 16 August 1986, 63.

Japan's dynamic growth. If Japan had not experienced such dramatic economic achievement, American decision makers would not have been particularly concerned about Japan's domestic practices. If Japan's domestic political economy was like that of the United States, the procedural changes associated with diffuse reciprocity, which have been at the core of American initiatives, would have had more impact. But the ineffectiveness of policies based on diffuse reciprocity and continued Japanese success led American policy makers to adopt specific reciprocity in an ad hoc fashion. And this in turn led Japanese leaders to conclude that their American counterparts were not playing by the rules that had been endorsed by the United States.

If diffuse reciprocity is problematic, can specific reciprocity work? Is it possible to create mutually acceptable rules that vary across countries and issue areas? This is not an easy task. First, it is hard to establish exact equivalents. Second, there is an incentive to increase barriers for bargaining purposes. Third, sequential interdependent decision making can invalidate earlier agreements; for instance, the benefits of a particular tariff concession may be compromised if reductions are later extended to other countries. Fourth, specific reciprocity complicates administrative enforcement and monitoring because different issues are treated in different ways.[39]

These are serious problems, but for the United States and Japan they may be less severe than would be the case for some other bilateral relationships. The problem of equivalencies is mitigated by the large amount of trade between the two countries. This increases the possibilities for trade-offs among a limited number of specific issue areas. The problem of strategic behavior cannot be eliminated, but it also arises in many situations other than bilateral interactions among major economic powers. Even in a multilateral setting governed by diffuse reciprocity there is an incentive for major actors to behave strategically, since their conduct will be noticed and they may be able to influence the choices made by other states. The problems posed by sequential decision making are lessened again by the sheer size of trade between the two major Pacific economies. In areas where the United States and Japan are each other's dominant supplier, subsequent agreements with third parties may not have much impact on prior concessions. Finally, the problem of monitoring is also lessened by size. Given the large volume of transactions between the United States and Japan, the use of different rules and procedures in different issue areas need not create insurmountable administrative

39. Keohane, "Reciprocity," 17-19.

inefficiencies or pose overwhelming monitoring problems. If officials had to deal with many small and variegated transactions, each covered by different rules, such problems would be much more substantial.

In a recent eassy on Japan's role in the world Hugh Patrick and Henry Rosovsky write: "The central issue is whether the systemic differences between Japan and other modern industrial nations are so deep, profound, intransigent, and unchanging (on both sides), creating an incompatibility so great, that either a new international economic system will have to be created or special rules will have to be devised for dealing with Japan.[40] The analysis in this article suggests that the latter option is the most viable. Specific reciprocity is no panacea. But it does offer the possibility of establishing a more stable relationship between the United States and Japan.

Adopting specific reciprocity as a basic norm would eliminate the growing disparity between principle and policy in the United States by ending even rhetorical adherence to universal procedural rules. This could lead to a more coherent American policy by providing general guidelines that could steer American decisions rather than basing them in an ad hoc fashion on immediate pressures. It would make American policy more effective by making it easier for central decision makers to anticipate needed changes rather than simply reacting to pressures from civil society. In a rapidly changing technological environmental such anticipation may be critical, because delay can severely damage American industries. Specific reciprocity as a basic norm would lessen the fragmentation in American policy making that has frequently pitted the executive against the legislature. Greater coherence could make American policy more efficacious by facilitating the use of decisive administrative interventions rather than lengthy judicial proceedings, which must usually be initiated by the private sector and which may provide financial remedies only long after domestic productive capacity has been mortally weakened.

And, although it may appear to be paradoxical, adopting specific reciprocity as the norm guiding American foreign economic policy could reduce the hostility that has been generated by conscious or unconscious American hypocrisy. Japan has conformed to the rules of the GATT system, yet it continues to be criticized by the United States. These criticisms can only be interpreted for what they are—attacks on the basic nature of Japan's political economy, on the way

---

40. Hugh Patrick and Henry Rosovsky, "The End of An Era?" mimeo, paper prepared for the Japanese Political Economy Research Committee (JPERC), 27.

in which the Japanese have interacted among themselves and with the outside world. It may be far easier for public and private decision makers in both countries to agree on desirable outcomes such as specific market shares than it is for them to agree on procedural norms.[41]

Conclusion

The international economic policies adopted by the United States at the conclusion of the World War II were guided more by concerns for the common defense, or more accurately a vision of how international political and economic life should be ordered, than by narrow considerations of American economic self-interest or relative power capabilities. American leaders wanted to create a liberal international system that mirrored political and economic arrangements in the United States. They embraced this vision, because they believed that its realization would create a stable and peaceful international environment.

The war and its aftermath had ruptured older patterns of international economic relations. European colonial empires and the preferential trade arrangements that they entailed disappeared. Many of Japan's ties with the mainland of Asia, especially China, were broken. The creation of the Communist bloc severed long-standing trading relationships in the center of Europe.

American leaders wanted to establish a new order. They supported the creation of international organizations—the International Monetary Fund, the World Bank, and GATT. They facilitated recovery in Europe with the Marshall Plan. They opened American markets to foreign products but accepted some protectionism in other countries. They tolerated subsidy and dumping practices that were suspect under American law. Some of the policies did have short-term economic benefits for the United States, but many did not. They were understood as part of a long-term effort to create a liberal international economy within which democracy would flourish and individuals would prosper, and outside of which the Soviet bloc might wither.

American leaders were not much concerned with relative, as opposed

---

41. For a similar conclusion see Gilbert R. Winham and Ikuo Kabashima, "The Politics of U.S.-Japanese Auto Trade" in I.M. Destler and Hideo Sato, eds., *Coping with U.S.-Japanese Economic Conflicts* (Lexington, Mass.: Lexington Books, 1982), 17.

to absolute, gains within the western alliance. The exceptional postwar power capabilities of the United States elevated them above such conventional considerations, except with regard to the Soviet Union. Yet, for a hegemonic leader there is likely to be a trade-off between maximizing absolute gains by providing a disproportionate share of the collective goods needed to maintain international economic regimes and maximizing relative power capabilities by increasing national economic returns.[42] Relative capabilities might even be enhanced by adopting policies that lower absolute economic benefits, so long as the dominant power experiences relatively less decline.

The western world has prospered under these American policies, and none has prospered more mightily than Japan. It is too early to say that Japan or East Asia more generally will emerge as the dynamic center of international economic and political life in the twenty-first century. But in the ebb and flow of world history that center has not remained in the same place. Political and economic leadership moved from northwestern Europe to North America at the beginning of the twentieth century. It is likely to move again in the future. The power of the United States has already begun to wane. The international economic policies that were predicated on that power will change. Managing the transition to a new center of world power, or to an American-Japanese condominium, or even an American-Japanese-European triumvirate will be a major challenge for the United States—one that cannot be adequately met by pursuing international economic policies based on diffuse reciprocity and global (at least for the non-Communist part of the globe) liberalism.

---

42. Arthur A. Stein, "The Hegemon's Dilemmas; Great Britain, the United States, and International Economic Order," *International Organization* 38 (Spring 1984).

CHAPTER 10
Trade Adjustment and Foreign Direct Investment: Japan in the United States

Harry I. Chernotsky

Foreign direct investment (FDI) in the United States has grown substantially over the past decade. This investment is dispersed throughout the country and is targeted toward virtually all key sectors of the economy. Still, the bulk of FDI continues to originate from those few countries which have traditionally been most active in this regard. By 1985, firms from a mere seven countries—Canada, France, Germany, Japan, Netherlands, Switzerland and the United Kingdom—accounted for approximately 80 percent of the total.[1]

One of the more notable features of recent FDI is the expanding activity of the Japanese. Between 1977 and 1985, the average annual rate of Japanese FDI growth was a stunning 37.3 percent, far beyond that of any other country. Although still responsible for only 10.4 percent of total FDI, Japan now ranks third among all U.S. investors.[2]

This paper seeks to uncover the factors underlying this recent surge of Japanese FDI in the United States. As such, it will address the forces contributing to the increasing allure or "pull" of the United States, as well as those which have served to "push" Japanese investors in this direction. This investment will also be analyzed in the context of Japan's overall overseas business strategy, with particular reference to its relationship to shifting trade conditions.

---

1. U.S. Department of Commerce, *Survey of Current Business* (August 1986), p. 85.
2. *Ibid.,* p. 79.

Table 10.1 FDI in the United States

	($ Mill.) 1977	% of Total	($ Mill.) 1985	% of Total	Average Annual Growth 1977-85
JAPAN	1,755	5.1	19,116	10.4	37.3
Canada	5,650	16.3	16,678	9.1	14.2
France	1,800	5.2	6,295	3.4	18.7
Germany	2,529	7.3	14,417	7.9	24.8
Netherlands	7,830	22.6	36,124	19.7	22.5
Switzerland	2,651	7.7	11,040	6.0	19.7
United Kingdom	6,397	18.5	43,766	23.9	25.9
		(82.7)		(80.4)	
U.S. (All Countries)	34,595		182,951		22.5

Source: U.S. Dept. of Commerce, Survey of Current Business (August 1979, August 1985, August 1986).

FDI Motives

Although long recognized as an attractive and potentially profitable market, the United States did not acquire significant amounts of FDI until the late 1960s. In 1959, total FDI was a mere $6.6 billion; by 1974, this figure had reached a respectable $26.5 billion. This rapid expansion was attributable to a number of developments, perhaps the most significant of which was the progressive increase in the size and competitiveness of corporations from other industrial countries. More firms were capable of operating effectively in the U.S. market, whose attraction was enhanced by the disarray of the international monetary system and subsequent devaluation of the dollar. Additional incentives were provided by the increasingly unfavorable politico-economic circumstances elsewhere, as well as the need to secure direct access to energy and other critical raw materials.[3]

The pace of FDI continued to escalate, reaching $182.9 billion by 1985. Particularly impressive were the approximately 25 percent annual increases during the 1978–81 period. This stemmed from the steady depreciation of the dollar which served to lessen the costs of establishing or expanding U.S. operations, the relatively slower rate of growth in U.S. production costs, sharp price increases in fuels and other resources and continuing political instability in many other prospective host countries.[4]

More recently, however, the rate of FDI has fluctuated. Increasing by only 14.7 percent in 1982 and 9.9 percent in 1983, FDI was affected adversely by the persistence of high interest rates and the greater expense involved in utilizing foreign currencies whose values declined sharply relative to the dollar since 1981. The surge in the U.S. stock market, moreover, served to raise the cost of acquisitions while stimulating further the diversion of capital inflows from FDI to portfolio investment in U.S. stocks and bonds. This brief hiatus notwithstanding, FDI in 1984 picked up considerably (20.1 percent), owing largely to the strength of the U.S. economic recovery and the low rate of inflation.[5]

The rise of FDI in the United States might be best understood in the context of broader theories designed to explain the existence of

---

3. U.S. Department of Commerce, *Foreign Direct Investment in the United States: Report to the Congress*, Vol. 1, 5 (1976), pp. 97-111.
4. U.S. Department of Commerce, *Survey of Current Business* (August 1982), p. 32 (October 1984), p. 26 (August 1986), p. 74.
5. U.S. Department of Commerce, *Survey of Current Business* (August 1983), p. 33 (October 1984), p. 27 (August 1986), p. 74.

FDI in the international economy. While developed largely in response to the earlier advances of American-based firms abroad, a number of these theoretical approaches also appear quite applicable to the more recent "reverse investment" phenomenon. When taken as a whole, they provide the perspective necessary for understanding both the internal and external conditions which have stimulated this flow of foreign capital.

A scanning of the literature suggests that no single theory can account fully for FDI. Considerable insight is provided, however, through those works focusing on the operation of the market and its imperfections. In this context, FDI is seen as a transitory phenomenon arising only when there is a departure from perfect competition. Among the more salient disequilibria affording opportunities for greater returns in a foreign setting are those arising in foreign exchange, capital, labor and/or technology markets. Thus, incentives to invest abroad may be greatest under conditions where firms possess distinct advantages over local competitors and the market for the sale of that advantage is imperfect.[6]

The potential role of government policies in the creation of market distortions cannot be underestimated. Any number of government-imposed regulations affecting business conditions in one country relative to another (e.g., exchange controls, anti-trust legislation, tax provisions, wage or price restraints) may provide important FDI inducements. Especially in those areas offering sizeable export opportunities, any pressure to restrict trade through erection of tariff and/or non-tariff barriers could result in expanded FDI.[7] As will be suggested, trade-based concerns appear particularly instrumental in facilitating the flow of Japanese FDI into the United States.

Departures from purely market-determined prices, brought about by the existence of monopolistic or oligopolistic conditions, suggest additional insights into the role of the market as a determinant of FDI. Viewed as an outgrowth of industrial organization, FDI may be understood as a strategy undertaken by advantaged firms to preserve their competitive edge. The restricted, yet fierce competition often prevalent in oligopolistic settings might offer further stimuli. This is detected in the product life cycle thesis tracing FDI to threats of losing markets as a product matures, the so-called "follow the leader" view of FDI as a reaction to expansionary moves by another

---

6. A.L. Calvet, "A Synthesis of Foreign Direct Investment Theories and Theories of the Multinational Firm", *Journal of International Business Studies* (Spring/Summer 1981), p. 44.

7. *Ibid.*, p. 46.

firm within a given industry and the "exchange of threats" hypothesis accounting for the activities of oligopolists in one another's markets.[8] Together, these approaches focusing on industrial organization call attention to some of the more significant benefits deriving from FDI by firms seeking to exploit existing and future advantages.

Some more recent efforts to project the direction and composition of FDI flows have centered directly on the investing agent—the multinational corporation. The propensity to invest abroad may be attributed to the ownership-specific endowments implied in the aforementioned industrial organization perspective. In addition, however, attention is directed toward a host of location-specific attributes to account for the particular value of a given investment site. Decisions to invest, according to this view, may be predicated largely on desires to capitalize on opportunities presented by prevailing market, capital, raw material, technology or labor conditions. This allure might be accentuated in certain instances, moreover, through the offering of extraordinary financial inducements.[9] It is interesting to note that any number of these factors have been cited among the primary ingredients contributing to the FDI appeal of the United States.[10]

The various approaches sketched above illustrate the multiplicity of considerations that might account for the increasing activity of foreign investors in the United States. In the remaining sections of this paper, an effort will be made to identify the most salient organizational, push and pull factors underlying FDI in the Japanese case, with particular emphasis on the progressive shift in Japan's overall overseas business strategy. This shift, evidenced most directly in the vast expansion of American-based activity, is best seen in the context of Japan's broader FDI objectives.

---

8. *Ibid.*, pp. 46-47.
9. *Ibid.*, pp. 49-51.
10. Riad A. Ajami, and David A. Ricks, "Motives of Non-American Firms Investing in the United States", *Journal of International Business Studies* (Winter 1981), p. 31; Harry I., Chernotsky, "Selecting U.S. Sites: A Case Study of German and Japanese Firms", *M,anagement International Review*, Vol. 23, No. 2 (1983), p. 53; Jane Sneddon Little, Locational Decisions of Foreign Direct Investors in the United States, *New England Economic Review* (July/August 1978), p. 58; Hsin-Min Tong and C. K. Walter, "An Empirical Study of Plant Location Decisions of Foreign Manufacturing Investors in the United States", *Columbia Journal of World Business* (Spring 1980), p. 71.

Japanese Overseas Investment: Development and Evolution

Appreciable FDI by Japanese firms is a relatively recent phenomenon. To be sure, it was not until the 1960s, when Japan's post-war recovery was sufficiently in place, that full-scale international involvement was feasible. Up to that point, FDI was rather sporadic and limited to natural resources and select manufacturing sectors. This relative absence of activity was also attributable somewhat to a basic Japanese distrust of long-term commitments involving foreigners, as reflected in the substantial government restrictions on both inward and outward investment.[11]

Perhaps most noteworthy in this regard was Japan's Foreign Exchange and Foreign Trade Control Law (FEFTCL, 12/1/49) designed to monitor and regulate all foreign business dealings. Providing for extensive government oversight in the review of pending international transactions, this law was intended to provide for the control of foreign exchange, foreign trade and other foreign transactions, necessary for the proper development of foreign trade and for the safeguarding of the balance of international payments and the stability of the currency, as well as the most economic and beneficial use of foreign currency funds, for the sake of the rehabilitation and the expansion of the national economy.[12]

Subsequent revisions of the FEFTCL and other relevant legislation, mandated by Japan's global economic interests, were to pave the way for greater FDI by Japanese firms. Continuing efforts to curtail foreign presence in the Japanese economy is evident, however, by the absence of equally extentive liberalization of inward investment policy. Most government initiatives in this area have been accompanied by elaborate loopholes to protect domestic companies and to limit foreign involvement in critical economic sectors. The 1973 law permitting up to 100 percent FDI in both new and existing enterprises, for example, did not extend this right to agriculture, forestry, fishing, mining, petroleum and leather industries or to cases where foreign takeovers were against the wishes of affected local firms.[13] Thus, Japan remains unique among the more active investing countries in that it does not provide for commensurate levels of FDI within its own borders. Although by far the most dominant, American business

---

11. H. Robert Heller and Emily E. Heller, *Japanese Investment in the United States* (New York: Praeger Publishers, 1974), p. 35.
12. H. Robert Heller and Emily E. Heller, *Ibid.,* p. 125.
13. U.S. Department of Commerce, *Overseas Business Reports* 80-21 (June 1980), p. 21.

still accounted for a mere $8.4 billion in Japan by 1984.[14]

This reluctance to encourage inward investment notwithstanding, Japanese investment abroad has proceeded in accordance with underlying national and corporate objectives. The critical role of the Japanese government is suggested by the congruence of FDI activity and the country's overall international payments position. Japanese FDI did not truly accelerate until 1968, when ample exchange reserves and a substantial payments surplus were evident. The 32 percent average annual growth of FDI between 1968 and 1971 was considerably beyond that of the previous period marked by periodic shifts in Japan's official settlements balance. Interestingly enough, the greatest surge in FDI to that point occurred in 1972 and 1973 (50 percent average annual growth), owing largely to the government's need to reduce burgeoning exchange reserves.[15]

Although turning somewhat sluggish in the mid 1970s due to the adverse effects of the oil crisis, FDI rebounded strongly toward the latter part of the decade—reaching $31.8 billion in 1979 and expanding to $61.2 billion by 1983. Particularly impressive have been the $8 billion annual additions since 1981. Japan's 7 percent share of the world's overseas investments ranks far above its 2.5 percent share in 1970. With few exceptions, recent FDI in lesser developed regions has remained fairly constant. This is in sharp contrast to the rapid growth (averaging approximately 20 percent annually in the 1980s) in advanced industrial countries.[16]

Japanese FDI is perhaps best understood as a defensive measure to relieve pressures on the domestic economy and to assist in the realization of key economy policy goals. It is also an integral part of Japan's enduring "trade or die" philosophy. As such, it seems predicated largely on the need to acquire direct access to raw material supplies to support industrial activity, to secure established and additional markets for the sale of Japanese products and to enhance further the country's worldwide competitiveness.[17] These elements are reflected in the composition and direction of FDI activity.

---

14. U.S. Department of Commerce, *Survey of Current Buiness* (November 1984), p. 24.
15. H. Robert Heller and Emily E. Heller, *Japanese Investment in the United States* (New York: Praeger Publishers, 1974), p. 32.
16. Japan Economic Journal, *Industrial Review of Japan* (1981), p. 38; "Japan's Growing Stake in America", *Fortune*, Vol. 110 (August 20, 1984), p. 51; Focus Japan, *Japan Scence: Fiscal 1985 JETRO White Paper (Investment)*, Vol. 12, No. 4 (April 1985), p. 1.
17. U.S. Department of Commerce, *Foreign Direct Investment in the United States: Report to the Congress*, Vols. 1, 5 (1976), p. G-276.

Until the mid 1970s, the vast majority of Japanese FDI was targeted toward Asia and other lesser developed regions. By 1973, for example, approximately 64 percent of FDI occurred in these areas while approximately 63 percent of all Japanese overseas investments was directed toward mining and manufacturing.[18] Both encouraged and supported by the Japanese government, resource oriented FDI was designed to preserve the security of critical raw material supplies. It was also intended to protect pre-existing export markets in countries that had traditionally fulfilled the bulk of Japanese needs in this regard.

Japan's manufacturing FDI in these areas was also of a defensive character, in that it was often undertaken to avert potential losses of export outlets. This was particularly the case with that FDI launched in response to expanding import substitution drives or threats of competition from U.S. firms. Many of these investments were rather small and were designed initially to perform limited production functions. They were often in the form of joint ventures and involved extensive participation by trading companies.[19] .

In other instances, Japanese manufacturing FDI appeared to stem from a variety of labor considerations. Japan's mounting labor shortages and escalating wage rates stood in stark contrast to prevailing conditions in a number of neighboring Asian countries. Not only did Japanese companies seek to capitalize on these favorable labor circumstances to limit production costs. They were also forced to respond to the increasing competitive threats posed by these countries themselves, as well as by American businesses exploiting these opportunities through the establishing of offshore facilities. FDI was somewhat constrained initially by the Japanese government's policy of discouraging large firms from producing for the home market through offshore activities. It was to become an even more attractive option with the revaluations of the yen and other emerging threats to Japanese exports the difficulties of expanding operations in a progressively overcrowded Japan marked by the existence of costly environmental controls and the offer of often extraordinary incentives by governments seeking to acquire new industries.[20]

---

18. H. Robert Heller, and Emily E. Heller, *Japanese Investment in the United States* (New York: Praeger Publishers, 1974), pp. 43-48.
19. M. Y. Yoshino, "Japanese Foreign Direct Investment", Isaiah Frank (ed.) *The Japanese Economy in International Perspective* (Baltimore, Maryland: The Johns Hopkins University Press, 1975), pp. 261-262.
20. M. Y. Yoshino, *Ibid.,* pp. 263-266; Yoshi Tsurumi, The Multinational Spread of Japanese Firms and Asian Neighbors' Reactions", David E. Apter and Louis Wolf

Thus, in many respects, Japanese FDI during this period can be viewed as an effort to exploit the comparative advantage of recipient countries and to assist the labor and resource-scarce Japanese economy. It was also a means to insure access to established and prospective markets otherwise served through exports. While in many cases an alternative to exporting, some FDI proved to generate exports of component parts necessary to sustain foreign operations. Nearly half of the total investment was directed toward just three industries—textiles, timber/pulp and steel. It is also interesting to note that the bulk of FDI was from small to medium size Japanese parent companies capitalized at less than $3.4 million.[21]

The success of Japanese investors in these areas rested largely on their ability to transfer production techniques without significant prior local experience, as well as their pre-empting of indigenous competition.[22] Also quite helpful was the Japanese willingness to accept minority ownership status. By 1972, for example, only 41 percent of manufacturing subsidiaries in Asia involved majority parent ownership.[23] This was in response to the somewhat greater leverage of host country governments, given the local market orientation of the bulk of Japanese enterprises. Yet, it also served the interests of many Japanese parents intent on spreading the risk for their activities.

By the mid 1970s however, majority ownership had become increasingly important to Japanese overseas operations. This coincided with an emphasis on integrating subsidiaries producing related products within a given country and the concurrent need for greater production control to insure uniform sales and export strategies. Unfortunately, this shift was to pose difficulties for a number of Japanese concerns—particularly in Asia, where memories of pre-war expansionism remained intact. This situation was complicated further by the high visibility of many Japanese firms involved in the production of simple industrial and consumer goods directed mainly to the domestic markets of these host countries.[24]

Goodman (eds.), *The Multinational Corporation and Social Change* (New York: Praeger Publishers, 1976); pp. 128-129.

21. Masaaki Kotabe, "Changing Roles of the Sogo Shoshas, the Manufacturing Firms, and the MITI in the Context of the Japanese Trade or Die Mentality", *Columbia Journal of World Business* (Fall 1984), pp. 35-36.

22. Yoshi Tsurumi, "The Strategic Framework for Japanese Investments in the United States", *Columbia Journal of World Business*, Vol. 13, No. 4 (Winter 1973), pp. 19-25.

23. Yoshi Tsurumi, "The Multinational Spread of Japanese Firms and Asian Neighbors' Reactions", David E. Apter and Louis Wolf Goodman (eds.) *The Multinational Corporation and Social Change* (New York: Praeger Publishers, 1976), p. 41.

24. Yoshi Tsurumi, *Ibid.*, pp. 138-142.

It was also during this period that resistance to Japan's aggressive exporting practices intensified in advanced industrial countries. Thus, many Japanese firms were forced to assume a more comprehensive posture toward FDI in the context of their overall business strategies. Although still somewhat defensive in nature, FDI came to be viewed more as an opportunity to improve production and marketing efficiency. Perhaps most importantly, it was seen as a means for expanding markets while serving to offset import barriers and manage exchange rate uncertainty.[25]

As can be seen in Table 10.2, lesser developed regions continue to provide important outlets for Japanese FDI. In certain instances, this activity has been prompted by the need to sustain access to markets in which commitments have been made or to discourage competitors from entering these markets.[26] Recent FDI in these areas has been directed mainly toward a handful of more secure Asian NIC's (Newly Industrializing Countries), owing largely to their growth potential and efforts to attract foreign capital for 'high tech' and export-oriented industries.[27] The mounting risk and vulnerability accompanying FDI in most other regions, however, have spawned additional efforts to diversify the base of Japan's overseas operations. This has encouraged further the progressive targeting of industrial country markets, the most significant of which has been the United States. Attention now turns to identifying those factors most responsible for this expanding American penetration.

Sources of Japanese FDI in the United States

Japanese FDI in industrial countries remains somewhat distinct from that in developing regions, in that it is highly concentrated in the service sector. This is certainly evident in the United States, where considerable sums have been placed in commercial and financial outlets to support the sale and distribution of Japanese products. Since 1983, for example, approximately two-thirds of all Japanese capital inflows have been in wholesale trade; two-thirds of the overall

---

25. Mary Saso and Stuart Kirby, *Japanese Industrial Competition to 1990* (Cambridge, Mass.: Abt Books, EIU Series 1, 1982), p. 123.
26. M. Y. Yoshino, "Japanese Foreign Direct Investment", Isaiah Frnak (ed.), *The Japanese Economy in International Perspective* (Baltimore, Maryland: The Johns Hopkins University Press, 1975), p. 266.
27. Japan External Trade Organization, *Overseas Direct Investment Helps Industrial Cooperation: White Paper of JETRO,* New Release (March 18, 1985), p. 7.

Table 10.2 Cumulative Japanese Overseas Investments

	1951–79		1982	
	Value ($ Bill.)	Percent	Value ($ Bill.)	Percent
North America	8.2	25.8	15.2	28.7
(U.S.)	(7.4)	(23.3)	(14.0)	(26.3)
Europe	3.9	12.3	6.2	11.6
Asia/Oceania	10.7	33.6	17.9	33.7
Latin America	5.6	17.6	8.8	16.6
Middle East/Africa	3.4	10.7	5.0	9.4
TOTAL	31.8	100.0	53.1	100.0

Source: Japan Economic Journal, *Industrial Review of Japan* (1981), p. 38; *Fortune,* (*August 20, 1984*), p. 51.

increase of FDI in this sector, moreover, have been the result of Japanese activity.[28]

This is not to underestimate the growing importance of manufacturing in the American case. By 1983, 334 Japanese manufacturing firms were operating in the United States, 309 of which involved controlling Japanese interests. These 309 companies were operating 479 plants and employing approximately 73,000 workers. Among the product lines with the largest number of manufacturing facilities were color televisions, microwave ovens, integrated circuits, communications equipment, bearings and machine tools. Thus, even in this instance, the need to preserve markets for Japanese goods in the face of rising trade protectionist pressures appears a primary motivation.[29] The U.S. market for Japan's heavy industries remains quite limited, thereby resulting in the targeting of the Middle East and other developing areas for the bulk of Japan's FDI in this sector.[30]

Japan's FDI strategy in the United States is also quite flexible. There still appears to be a marked preference for majority ownership. Yet, a growing number of Japanese companies have entered the U.S. market through joint ventureships with American firms. The purchase of a 50% share of National Steel Corporation by Nippon Kokan K.K. provide National new capital with which to revitalize its plants while offering NKK the opportunity to gain a strong foothold in the

28. U.S. Department of Commerce, *Survey of Current Business* (October 1984), pp. 29-30; (August 1986), p. 80.
29. Susan MacKnight, "Japan's Expanding U.S. Manufacturing Presence: 1983 Update", *JEI Report,* Japan Economic Institute, No. 15A (April 1984), pp. 3-4.
30. "Japan's Growing Stake in America", *Fortune,* Vol. 110 (August 20, 1984), p. 48.

American market. Other Japanese firms have been content to confine their activity to production agreements with U.S. businesses. This is exemplified by the accord between National Semiconductor Corporation and Oki Electric Industry Company, under which Oki supplies production know-how in return for completed computer chips which the Japanese company sells in the United States.[31]

Another notable feature of Japanese FDI is its disbursement throughout the United States. A number of states have actively recruited

Table 10.3 Japanese FDI Distribution 1979 (Percent)

	Resource	Manufacturing	Service
Total (Average)	25.5	35.9	38.5
North America	9.7	24.5	65.8
Europe	3.4	18.1	78.6
Oceania	40.8	31.1	28.0
Asia	39.3	44.3	16.4
Latin America	18.9	52.0	29.0
Middle East	45.6	48.2	6.3
Africa	44.5	7.4	48.2

Source: Japan Economic Journal, *Industrial Review of Japan* (1980), p. 38.

Table 10.4 Distribution of FDI in U.S. (%)

	All Countries		Japan	
	1977	1985	1977	1985
Petroleum	19.0	15.4	2.7	0.2
Manufacturing	40.6	33.2	18.7	13.7
Trade	20.9	18.7	45.7	61.9
Finance	6.4	8.9	29.0	15.1
Insurance	6.7	6.1	2.2	0.6
Real Estate	2.5	10.1	1.7	5.5
Other	3.9	7.6	—	3.0

Source: U.S. Dept. of Commerce, *Survey of Current Business* (August 1979, August 1985, August 1986).

---

31. "Japan's Growing Stake in America", *Fortune*, Vol. 110 (August 20, 1984), pp. 43-44.

Japanese businesses. Yet, FDI is by no means restricted to any particular region of the country. It is worth nothing, however, that approximately 70 percent of manufacturing investment is located in a mere eight states (California, Georgia, Illinois, Washington, New Jersey, New York, Tennessee and Texas). Despite its unitary tax provision, California continues to host the greatest number of Japanese facilities—concentrated largely in the industrial and consumer electronics field.[32]

Organizational Capacity

Although not a sufficient explanation for the recent surge of Japanese FDI in the United States, the dramatic rise in the organizational strength of Japanese firms may be seen as a pre-requisite for full-scale entry into the American market. The internal capabilities of Japanese companies are a product of that country's unique industrial and managerial structure marked by extensive government regulation and intervention.[33] Their ability to compete internationally, as evidenced by resounding export successes, is similarly rooted in Japanese government policy initiatives to enhance economic growth and development.

As was noted earlier, Japan's FDI strategy has evolved largely in response to circumstances affecting exports of her products and access to necessary resources. Thus, it is understandable that most government efforts to facilitate FDI should occur in the context of possible disruptions in securing these objectives. This was particularly evident in the series of measures commencing in the early 1970s, when Japanese exports to the United States were jeopardized by global monetary disarray and the revaluation of the yen. The reaction to this threat was to provide additional financing for overseas ventures and to ease licensing and foreign exchange restrictions.[34] Such steps were to prove especially useful in subsequent years, as mounting U.S./Japanese trade imbalances were to raise the spectre of American trade protectionism. FDI as a means to guarantee market position has

32. Susan MacKnight, "Japan's Expanding U.S. Manufacturing Presence: 1983 Update", *JEI Report*, Japan Economic Institute, No. 15A (April 1984), p. 4; Japan External Trade Organization, *Overseas Direct Investment Helps Industrial Cooperation: White Paper of JETRO*, News Release (March 18, 1985), p. 10.
33. Jane Sneddon Little, "Locational Decisions of Foreign Direct Investors in the United States," *New England Economic Review* (July/August 1978), pp. 23-28.
34. H. Robert Heller and Emily E. Heller, *Japanese Investment in the United States* (New York: Praeger Publishers, 1974), pp. 72—78.

hinged largely on this government-supported growth of financing and international experience.[35]

Yet, any discussion of the organizational advantages accruing to Japanese investors must take special note of the role of trading companies in enhancing overall competitiveness. These trading companies have been most instrumental in furthering Japanese trade and investment worldwide; by 1975, for example, they occupied the top five positions among leading Japanese investing companies.[36] Owing largely to their prior export-related activities and familiarity with the American market, trading companies have also been heavily involved in Japan's FDI in the United States, particularly in the primary sector and certain manufacturing areas. This has derived from their superior capabilities in procuring resources, financing, distribution and information processing.[37]

It should be noted, however, that the role of trading companies has declined somewhat with the shift of investments toward technology-intensive and high value added product areas. With this thrust has come the need for greater attention toward such matters as customer service and marketing research which are not especially well handled by trading companies. Given their growing international experience, increasing reliance on wholly owned subsidiaries, financial strength and marketing skills, many firms have begun operating more independently of the trading companies in pursuing their American objectives.[38] Nevertheless, trading companies remain quite active in many large-scale projects involving basic materials firms (e.g., petrochemical and aluminum refining) and have served businesses in other areas through their performance of vital residual tasks. In addition to absorbing some of the risk, they often prove instrumental in assembling project consortia and acting as an intermediary in arranging for necessary credit.[39]

---

35. U.S. Department of Commerce, *Foreign Direct Investment in the United States: Report to the Congress*, Vols. 1, 5 (1976), p. 273.

36. Sueo Sekiguchi, *Japanese Direct Foreign Investment* (Montclair, New Jersey: Allenheld Osmun Co., 1979), p. 140.

37. U.S. Department of Commerce, *Foreign Direct Investment in the United States: Report to the Congress*, Vols. 1, 5 (1976), p. G-285.

38. Masaaki Kotabe, "Changing Roles of the Sogo Shoshas, the Manufacturig Firms, and the MITI in the Context of the Japanese Trade or Die Mentality", *Columbia Journal of World Business* (Fall 1984), pp. 36-38.

39. Mary Saso, and Stuart Kirby, *Japanese Industrial Competition to 1990* (Cambridge, Mass.: Abt Books, EIU Special Series 1, 1982), p. 3; U.S. Department of Commerce, *Foreign Direct Investment in the United States: Report to the Congress*, Vols. 1, 5 (1976), p. G-285.

This brief review serves to underscore the importance of effective industrial organization as a precondition for viable overseas activities. In many respects, the Japanese system-designed to support such international involvement through an intricate network of incentives and supports—can be viewed in the context of classic oligopolistic behavior. The advantages provided Japanese firms, while intended mainly to promote their export drives, have left them well positioned to engage in FDI to defend against possible encroachments from competing companies or from alterations in host market conditions. This latter element has proven especially significant in the American case, where FDI has become an increasingly essential device to sustain market access in the face of growing bilateral trade friction.

Foreign Market Push

These organizational features notwithstanding, the selection of the United States by Japanese investors cannot be explained without more direct reference to circumstances in Japan and other prospective investment locations which have pushed them in this direction. A number of constraints undermining corporate expansion within Japan itself have been identified in the previous discussion of Japan's evolving worldwide investment strategy in the 1970s. These included the limited availability of sites for additions to existing facilities or construction of new plants, growing labor shortages accompanied by considerable escalation of domestic wage rates and the additional production costs mandated by the need to adhere to strict environmental standards.[40] Such bottlenecks appear especially pronounced when considered in light of the relatively favorable U.S. cost-related environment during the period of most sustained Japanese FDI growth.

Still, the direct push to the United States can be seen more fully in contrast to the diminishing allure of other potential investment outlets —particularly those in lesser developed regions. Fewer opportunities to redirect investments to more stable climates may be available to primary and processing industries, whose decisions are often dictated by the existence of desired resources. Yet, Japanese investors, like all others, have been forced to become increasingly sensitive to country risk considerations in devising overseas strategies.

---

40. U.S. Department of Commerce, *Foreign Direct Investment in the United States: Report to the Congress*, Vol. 1, 5 (1976), p. G-282.

As was suggested earlier, Japanese FDI in Asia remains quite strong. Dynamic growth, coupled with the frequent offering of financial inducements, has resulted in a flurry of Japanese activity throughout the area. This interest has even extended to China, where Japanese firms had previously encountered serious difficulties due to political uncertainties and frequent government policy shifts. By 1984, seventeen joint Chinese/Japanese ventures had been undertaken.[41]

Nevertheless, the recent investment climate in most other developing areas has been quite bleak. Most Japanese investors have shied away from involvement in the Middle East, owing largely to the decline in oil prices and the continuing political instability which threatens the safety of any such endeavors. Nor has there been much attention directed toward Africa, where rising debts and slumping raw material prices have limited the availability of foreign exchange. Adverse economic conditions have also curtailed FDI in Latin America, even in Brazil and Mexico where Japanese firms have had an ongoing stake and where recent activities have been confined mainly to merely supporting existing enterprises.[42]

It should be noted that recent FDI in Europe has been rather extensive—especially in manufacturing, finance and insurance. By 1984, 170 Japanese manufacturing plants and some 1100 non-manufacturing firms were operating in the region. Yet, given its relative geographic proximity, critical market and more extensive economic and financial ties, the United States remains a more attractive site.[43]

This is not to imply the elimination of all risk when investing in the United States. Particularly vexing to the Japanese have been the increasing efforts of American labor unions to organize within their plants and the difficulties arising in certain cases when seeking to adapt standard management practices.[44] Yet, such threats have proven far less intimidating when compared to those likely to be encountered in inherently unstable political and economic locales. Neither have they had much impact in discouraging investments by Japanese concerns faced with losses of American export outlets.

41. Japan Economic Journal, *Industrial Review of Japan* (1981), p. 55; Japan External Trade Organization, *Overseas Direct investment Helps Industrial Cooperation: White Paper of JETRO,* New Release (March 18, 1985), pp. 2-7.

42. Japan External Trade Organization, *Overseas Direct Investment Helps Industrial Cooperation: White Paper of JETRO* , News Release (March 18, 1985), pp. 2-7.

43. Focus Japan, *Japan Scene: Fiscal 1985 JETRO White Paper (Investment)*, Vol. 12, No. 4 (April 1985), p. 2.

44. Yoshi, Tsurumi. The Best of Times and the Worst of Times: Japanese Management in America'', *Columbia Journal of World Business*, Vol. 13, No. 1 (Spring 1978), pp. 56-61.

U.S. Market Pull

As has been suggested throughout, this need to preserve access to the lucrative American market can account for much of this recent "discovery" of the United States as a host for Japanese FDI. While influenced by organizational and foreign market push considerations, Japanese FDI in the United States has been prompted largely by shifts in prevailing market conditions. These alterations have accentuated the overall FDI appeal of the United States but have been especially alluring to Japanese firms.

Inasmuch as Japanese investors have been pulled toward the United States to insure continuing sales opportunities, attention must focus on those forces threatening to undermine this vital need. The potential ramifications of any sizeable loss of U.S. export markets are suggested in Table 10.5. Japan has come to rely upon the United States for more than one-fourth of its total exports, far greater than that of virtually all other major investing countries. Of course, the Japanese still remain considerably less dependent than the Canadians, whose inordinate trade concentration has been dictated largely by geographic proximity. Yet, not only is the U.S. market critical in sustaining the volume of Japanese exports. It is also one in which Japan has realized substantial financial gain.

The disproportionate benefits accruing to the Japanese in their trade with the United States are reflected most graphically in the extraordinary surpluses that have been generated. These Japanese surpluses have risen steadily and reached an unprecedented $21.6 billion in 1983, $36.8 billion in 1984 and $49.7 billion in 1985.[45] In many respects, Japanese exporters have been the victims of previous marketing and production successes which permitted them to offer competitively priced, high quality products to American consumers. Yet, charges of unfair practices, coupled with the reluctance of Japan to ease its own import barriers, have led to a number of steps to limit Japanese export accessibility. This exceedingly hostile environment has left many Japanese companies with little choice— either risk the loss of the American market or invest directly to insure sales of existing and new product lines.

The linkage between Japan's trade and FDI strategy in the United States is evident in the heavy targeting of those sectors most directly subject to protectionist measures. Particularly noteworthy is the automobile industry where Japanese firms have been forced to respond to

---

45. International Monetary Fund, *Direction of Trade Statistics* (1986), p. 401.

Table 10.5 Trade with the United States 1977-85

	US Exports % Total	Avg. Annual Trade Balance ($ Mill.)
JAPAN	29.5	21,252
Canada	69.0	11,226
France	6.4	196
Germany	7.9	4,447
Netherlands	4.3	-4,623
Switzerland	9.6	-232
United Kingdom	12.4	885

Source: IMF, *Direction of Trade Statistics* (1983, pp. 3, 397; 1985, pp. 3, 399; 1986, pp. 3, 399; 1986, pp. 3, 401).

"voluntary" restraints (until very recently) and the prospects of future restrictions if imports exceed acceptable levels. The shape of this response is exemplified by decisions of major Japanese auto manufacturers to expand their American-based operations. Honda has installed car production facilities in Ohio adjacent to its existing motorcycle plant, Nissan is now assembling trucks and cars in Tennessee, Mazda is preparing its plant in Michigan while Toyota and Mitsubishi proceed with their joint automobile ventures with General Motors and Chrysler respectively.[46]

Direct investment has become quite pronounced in other trade-sensitive areas. Anti-dumping investigations in the late 1970s played a major role in the decisions of Matsushita, Sanyo, Sharp and Toshiba to produce microwave ovens in this country. Similarly, mounting trade tensions in such industries as semi-conductors and bearings have also been accompanied by a wave of FDI by a number of Japanese companies involved in these product areas.[47]

The experiences of the color television industry are especially instructive. By the mid-1970s, Japanese producers had gained a considerable share of the market, thereby prompting American firms to seek appropriate government redress. The subsequent 1977 orderly marketing agreement (OMA) resulted in a wave of Japanese FDI. Vulnerability to possible restraints had already forced some Japanese firms to establish U.S. bases. Although initiated primarily to merely

46. Japan Economic Journal, *Industrial Review of Japan* (1981), p. 55.
47. Susan MacKnight, *Japan's Expanding Manufacturing Presence in the United States: A Profile* (Washington, D.C.: Japan Economic Institute 1981), p. 2.

relieve U.S. demand pressure, the American operations of Sony were expanded considerably in response to these increasing threats. Similarly, Matsushita moved to acquire Motorla's color television division while retaining the Quasar brand name and Sanyo gained control over Warwick through the purchase of Whirlpool's interests. Following the imposition of the OMA, Mitsubishi, Toshiba, Sharp and Hitachi set up their own U.S. production facilities.[48]

These activities highlight the importance of the trade/FDI link, inasmuch a they evolved in response to escalating tensions threatening to deny Japanese companies access to the crtical, lucrative U.S. market. The outgrowth of sustained and intense relief efforts by U.S. manufacturers, the OMA mandated a 40 percent reduction in Japanese imports; similar agreements with Taiwan and Korea precluded the option of exporting to the United States from these bases.[49] The strategy is quite explicit—to maintain and expand market share through volume production and a broadening of product line. Production facilities have been supplemented by field service and sales networks, thereby enhancing market response capabilities. By 1979, only 9.2 percent of the U.S. market was accounted for by imports. However, more than 40 percent of that market served by thirteen companies had been captured by those nine which were foreign owned. With few exceptions (most notably, Dutch-owned Magnavox), these operations could be traced to Japan.[50]

This case also serves to illustrate the conditions under which FDI becomes a desirable alternative for surmounting anticipated and/or actual trade disruptions. To be sure, FDI remains directly linked to the perceived importance of a given market. Many color television manufacturers, for example, expanded their U.S. operations despite severe price and product competition which resulted in little or no return during the initial phases of activity.[51] Still, the feasibility of FDI appears greatest when there are additional incentives operating to enhance the allure or "pull" of a prospective location. Such inducements emanating directly from alterations in prevailing market conditions certainly seem to have added to the viability of Japan's FDI strategy. Thus, in an important sense, decisions to move into the United States may be seen as expected responses to opportunities presented by critical market disequilibria.[52]

---

48. Jack Baranson, *The Japanese Challenge to U.S. Industry* (Lexington, Mass.: Lexington Books, D.C. Heath and Company, 1981), pp. 78-80.

49. Ibid., p. 82.

50. Ibid., pp. 75-78.

51. Ibid., p. 91.

52. A. L. Calvet, "A Synthesis of Foreign Direct Investment Theories and Theories of

Perhaps most significant in this regard were the disequilibria evident in foreign exchange markets. As can be seen in Table 10.6, yen/dollar exchange rates shifted dramatically during this period of mounting Japanese FDI. It is somewhat difficult to measure precisely the direct impact of exchange rate alterations, given the lag time that may be involved before such changes are reflected in FDI flows. Some evidence to this effect is suggested by the steady growth of Japanese FDI despite the depreciation of the yen from its peak in 1978. Nevertheless, the value of the yen throughout this period remained considerably above its previous level.

The relative appreciation of the yen provided major cost-related advantages, as Japanese firms had the equivalent of a subsidy for their investments in the United States. The establishment of facilities with overvalued yen became less expensive, while a larger quantity of inputs could be purchased with the same amount of money.[53] As indicated in Table 10.7, moreover, the yen fared far better than the currencies of other chief investing countries during this period of rather extreme exchange rate realignments. Particularly noteworthy was the rebounding of the yen in 1983 when the dollar demonstrated considerable strength relative to all other leading currencies. Thus, both in absolute and relative terms, Japanese investors operated within a favorable exchange rate environment as they moved toward expanding and solidifying their presence in the United States. The even more dramatic recent realignments, which have pushed the yen to around the 150 level, should provide additional inducements in upcoming years.

Additional evidence of the allure of the U.S. market is apparent in Table 10.8, which reflects the transformation of the United States into a country of lower relative production costs. Once again, the incentives for Japanese investors appear somewhat greater than those for most other nationalities. This is due largely to the previously noted labor conditions in Japan which, in addition to alterations in exchange rates, have affected this relative competitive status. From a purely cost-related perspective, then, FDI in the United States seemed to be an especially effective means to insure continuing market accessibility.

Finally, the pervasive pull of the United States is detected in the opportunities afforded Japanese firms to realize acceptable returns on

---

the Multinational Firm'', *Journal of International Business Studies* (Spring/Summer 1981), pp. 44-45.

53. H. Robert Heller and Emily E. Heller, *Japanese Investment in the United States* (New York: Praeger Publisher, 1974), p. 8.

Table 10.6 *Yen/Dollar Exchange Rate*

Year	Yen/Dollar
1965–70 (Avg.)	360
1971–75 (Avg.)	303
1976–85 (Avg.)	240
1976	297
1977	268
1978	210
1979	219
1980	227
1981	221
1982	249
1983	237
1984	238
1985	239

Source: OECD, *Economic Outlook* (Dec. 1983, p. 165; Dec. 1984, p. 179; Dec. 1985, p. 185; Dec. 1986, p. 169).

Table 10.7 *Exchange Rates/Dollar*

	Net % Change: 1977–85
JAPAN	17.5
Canada	–33.4
France	–71.1
Germany	–20.8
Netherlands	–28.2
Switzerland	– 4.8
United Kingdom	–37.6

Source: OECD, *Economic Outlook* (Dec. 1985, p. 185; Dec. 1986, p. 169).

their investments. The overall investment climate of any country will be measured against the likelihood of deriving appreciable profits. For the Japanese, the United States has proven quite lucrative. As suggested in Table 10.9, Japanese concerns have fared well-ranking second only to the Dutch, whose American holdings are attributable largely to massive investments by relatively few firms.[54] Japanese affi-

54. Jeffrey S. Arpan and David A. Ricks, "Foreign Direct Investments in the United States," *Journal of International Business Studies* (Winter 1979), pp. 85-88.

liates have rebounded strongly, following disappointing performances in 1982 owing to the prolonged impact of the global recession and the increased costs of borrowing funds at inflated interest rates.[55] Since 1983, the rates of return have exceeded those of all others.

Table 10.8 Competitive Positions
       Relative Unit Labor Costs in Manufacturing (1970 = 100)

	Average 1977–84
JAPAN	125.0
Canada	100.3
France	101.1
Germany	108.0
Netherlands	101.3
Switzerland	152.1
United Kingdom	118.5
United States	73.2

Source: OECD, *Economic Outlook* (Dec. 1985, p. 168).

Table 10.9 Rate of Return

	Average 1976–84	1984	1985
JAPAN	12.5	13.8	9.7
Canada	5.0	2.3	4.3
France	2.8	−2.9	−2.3
Germany	2.9	6.9	3.8
Netherlands	12.9	9.9	7.7
Switzerland	6.8	7.2	6.5
United Kingdom	9.0	6.5	5.9
All Countries	8.4	6.1	4.6

Source: U.S. Dept. of Commerce, *Survey of Current Business* (August 1979–83, 1985, 1986).

55. U.S. Department of Commerce, *Survey of Current Business* (August 1986), p. 77.

The earnings generated by existing operations will affect decisions to sustain activities through reinvestment, particularly when circumstances mitigate against massive inflows of new capital. Once again, the Japanese demonstrated considerable inclination to pursue this strategy as a means for expanding their U.S. involvement (Table 10.10). Although less severe than for other key investing countries, the substantial decline in reinvestment earnings in 1982 was quite pronounced in light of extraordinarily high previous levels. This may be traced largely, however, to the adverse business conditions responsible for the diminished returns during the year. Despite this temporary lull, Japan's ongoing commitment is reflected in the more sizeable reinvestment accompanying the respectable returns since 1983.

Table 10.10 Reinvested Earnings Ratio

	U.S. Total	Japan	Europe	Canada
1976	.69	.67	.70	.69
1977	.65	.84	.65	.78
1978	.75	.85	.74	.78
1979	.78	.92	.75	.86
1980	.80	.91	.71	.96
1981	.65	.90	.60	*
1982	*	.36	.12	*
1983	.03	.59	.33	*
1984	.45	.72	.61	*
1985	.21	.65	.44	*

*Negative Reinv. Earnings.·
Source: U.S. Dept. of Commerce, *Survey of Current Business* (August 1979-83, 1985, 1986).

Conclusion

Japanese FDI in the United States is perhaps best understood as both a means to capitalize on existing opportunities and a hedge against emerging threats to underlying national and corporate interests. As such, the impetus to secure an expanded base for operations within the country is not all that difficult to determine. Nor need it be viewed in complete isolation. In large measure, recent Japanese investment behavior appears as an expected response to those organizational and market imperatives commonly assumed to foster flows of capital across the world economy. At the same time, however, it is the product of certain features unique to the bilateral American/ Japanese relationship—most importantly, intensified efforts to deny Japanese exporters access to their most critical market.

In certain respects, Japan's FDI in the United States might be viewed as an extension of the pattern evident during the earlier expansion in lesser developed countries—that is, to avert market losses stemming from import substitution drives and to capitalize on cheaper labor costs. This latter element was critical during the late 1970s when the depreciated dollar lessened relative production costs, while the former remains quite prominent in light of continued threats to limit Japanese exports. This FDI also serves to enhance Japanese competitiveness in the United States by permitting more efficient servicing and greater market response capabilities.

Japan's global business success has elicited expected responses from the United States and other major partners faced with expanding trade deficits and progressive losses of market shares. In those instances where FDI is not deemed justified, tensions have often been resolved through technology exports to affected U.S. companies. In other cases, Original Equipment Manufacturers Contracts (OEM's) enabling Japanese firms to sell through a U.S. partner under the latter's trademark and through its distribution channel have proven useful. This is well illustrated by Kyosera's exports of portable computers to Tandy Corporation and sold under the Radio Shack label. Yet, more and more, the route has been FDI. Thus, somewhat contrary to popular belief, Japanese firms are far more than narrow minded exporters in their quest to preserve and expand U.S. market opportunities.[56]

Despite considerable successes within the United States, Japanese investors are not without their difficulties. As reflected in the slowing of the rate of FDI growth in 1982 and 1983, they have not remained immune from the vestiges of global recession or the pressures of exchange rate alterations. Additional impediments have surfaced within Japan itself, as concern has grown over the need to insure appropriate levels of domestic investment to protect Japanese jobs. Serious questions have also been raised as to the so-called boomerang effect of overseas investment—that is, its likelihood of displacing future Japanese exports.[57]

A number of Japanese firms have encountered problems, as well,

---

56. Kiyohiko Fukushima, "Japan's Real Trade Policy", *Foreign Policy*, No. 59 (Summer 1985), pp. 22-26.

57. Japan Economic Journal, *Industrial Review of Japan* (1981), p. 95.

in adapting management practices to the American business environment. In certain instances, efforts to infuse Japanese organizational patterns and decision making styles have been met with resistance by American managerial staffs. Among the more salient tensions are those pertaining to the defining of formal lines of authority, the integration of Japanese and American executives within the unit, the communicating of tasks and responsibilities and the methods of resolving conflicts over the setting of goals and priorities. To minimize these obstacles, many Japanese companies are now moving more slowly in implanting management systems unfamiliar to American personnel.[58] Interestingly enough, relations between Japanese executives and their rank and file American employees tend to be more consensual. Yet, unionization drives by organized labor—perceived as noncooperative and little concerned with the situations of individual companies—have also proven disruptive.[59]

Mention must also be made of the hindrances resulting from the imposition of unitary tax systems in a number of American states. Under this method, tax liabilities are based on the overall activity of the parent and its world-wide subsidiaries, rather than merely on the profits made in the state. This has not prevented Japanese firms from locating in those states where unitary taxation is in force. Yet, it has proven quite disconcerting to many of the "high tech" industries naturally drawn to California and has been the focus of considerable Japanese lobbying for repeal. In addition to impacting severely on affected Japanese businesses, the unitary tax may prove increasingly debilitating to the future recruitment efforts of these states.[60]

These obstacles and threats notwithstanding, it seems reasonable to expect Japanese FDI in the United States to continue at relatively healthy levels. Even in 1982, the total number of Japanese transactions climbed to 149 (from 129 in 1981). Since 1983, moreover, the rate of growth of Japanese FDI has exceeded that of all other countries (17.1 percent in 1983, 41.5 percent in 1984, 19.1 percent in 1985).[61] Continuing trade tensions, coupled with the depreciation of

---

58. "Japan's Growing Stake in America", *Fortune,* Vol. 110 (August 20, 1984), p. 56.
59. Richard T. Johnson, "Success and Failure of Japanese Subsidiaries in America", *Columbia Journal of World Business*, Vol. 12, No. 1 (Spring 1977), pp. 30-37; Yoshi Tsurumi, "The Best of Times and the Worst of Times: Japanese Management in America," *Columbia Journal of World Business*, VOL. 13, No. 1 (Spring 1978), pp. 56-61.
60. "Japan's Growing Stake in America", *Fortune*, Vol. 110 (August 20, 1984), pp. 57-58.
61. "Foreign Investment in USA Fell in 1982", *USA Today* (February 24, 1984), p. 1-

the dollar and the incentives offered by many states and localities, should augur well for future activity.[62]

More recent Japanese investments also reflect a willingness to tap any new opportunities that might be presented. This is particularly evident in the U.S. real estate market, where Japan has aggressively pursued the purchase of office buildings (including such landmarks as ARCO Plaza in Los Angeles and Manhattan's EXXON Building). Rising prices in Japan stand in contrast to the relatively lower U.S. purchase prices stemming from the appreciation of the yen. Prospects for stabilized exchange rates and the differences in tax laws and land availability in the two countries have also contributed to the allure of such acquisitions. Unlike the case in Japan, most of the purchase price of U.S. real estate is represented by the building. Inasmuch as land is not depreciable for tax purposes in either country and U.S. depreciation schedules remain relatively more favorable, FDI offers a significant cost advantage.[63]

FDI has become an important tool in the pursuit of Japan's global economic objectives. As such, it is likely to proceed at a respectable pace to secure necessary supplies of raw materials and natural resources and to retain access to critical markets. The United States has come to occupy a key role in the advancing of this strategy. The ever increasing efforts to limit Japan's imports into the country leave Japanese policy makers with little other choice.

B; "Japanese Invest More in the USA", *USA Today* (September 13, 1985), p. 1-B; Susan MacKnight, "Japan's Expanding U.S. Manufacturing Presence: 1983 Update", *JEI Report*, Japan Economic Institute, No. 15A (April 1984), p. 1; U.S. Department of Commerce, Survey of Current Business (August 1986), p. 85.

62. Cedric L. Suzman, *The Costs and Benefits of Foreign Investment from a State Perspective*, The Southern Center for International Studies (Atlanta, Georgia: August 1982), pp. 7-29.

63. U.S. Department of Commerce, Survey of Current Business (May 1987), p. 31.

Reference

Ajami, Riad A. and David A. Ricks, "Motives of Non-American Firms Investing in the United States," *Journal of International Business Studies*, Winter 1981, pp. 25-34.

Arpan, Jeffrey S. and David A. Ricks, "Foreign Direct Investments in the U.S.," *Journal of International Business Studies,* Winter 1979, pp. 85-88.

Arpan, Jeffrey S., et al., "Foreign Direct Investment in the United States: The State of Knowledge in Research," *Journal of International Business Studies*, Spring/Summer 1981, pp. 137-154.

Baranson, Jack. *The Japanese Challenge to U.S. Industry,* Lexington, Mass., Lexington Books, D.C. Heath and Company, 1981.

Calvet, A.L., "A Synthesis of Foreign Direct Investment Theories and Theories of the Multinational Firm," *Journal of International Business Studies*, Spring/Summer 1981, pp. 43-59.

Chernotsky, Harry I., "Selecting U.S. Sites: A Case Study of German and Japanese Firms," *Management International Review*, Vol. 23, No. 2,1 983, pp. 45-55.

Conference Board, "Foreign Manufacturing Investment in the U.S.: A Temporary Lull," *World Business Perspectives*, No. 71, February 1983, pp. 1-4.

Focus Japan, *Japan Scene: Fiscal 1985 JETRO White Paper (Investment)*, Vol. 12, No. 4, April 1985.

"Foreign Investment in USA Feel in 1982," *USA Today,* February 24, 1984, p. 1-B.

Fukushima, Kiyohiko, "Japan's Real Trade Policy," *Foreign Policy*, No. 59, Summer 1985, pp. 22-39.

Haendel, Dan. *Foreign Investments and the Management of Political Risk*, Boulder, Colorado, Westview Press, 1979.

International Monetary Fund. *Direction of Trade Statistics*, 1983, pp. 3, 397; 1984, pp. 3, 395; 1985, pp. 3, 399; 1986, pp. 3, 401.

Japan Economic Journal. *Industrial Review of Japan*, 1981.

Japan External Trade Organization. *Overseas Direct Investment Helps Industrial Cooperation: White Paper of JETRO,* News Release, March 18, 1985.

"Japan's Growing Stake in America," *Fortune*, Vol. 110, August 20, 1984, pp. 39-63.

"Japanese Invest More in the USA," *USA Today*, September 13, 1985, p. 1-B.

Johnson, Richard T., "Success and Failure of Japanese Subsidiaries in America," *Columbia Journal of World Business*, Vol. 12, No. 1, Spring 1977, pp. 30-37.

Kojima, Kiyoshi. *Japan and a New World Economic Order*, Boulder, Colorado, Westview Press, 1977, pp. 75-119.

Kotabe, Masaaki, "Changing Role of the Sogo Shoshas, the Manufacturing Firms, and the MITI in the Context of the Japanese Trade or Die Mentality," *Columbia Journal of World Business*, Fall 1984, pp. 33-42.

Little, Jane Sneddon, "Locational Decisions of Foreign Direct Investors in the United States," *New England Economic Review*, July/ August 1978, pp. 43-63.

MacKnight, Susan. *Japan's Expanding Manufacturing Presence in the United States: A Profile,* Washington, D.C., Japan Economic Institute, 1981, pp. 1-25.

MacKnight, Susan, "Japan's Expanding U.S. Manufacturing Presence: 1983 Update," *JEI Report*, Japan Economic Institute, No. 15A, April 1984, pp. 1-5.

Organization for Economic Cooperation and Development. *Economic Outlook*, December 1981-86.

Ozawa, Terutomo, "Japan's New Industrial Offensive in the United States," *MSU Business Topics*, Vol. 21, No. 4, Autumn 1973, pp. 23-28.

Saso, Mary and Stuart Kirby, *Japanese Industrial Competition to 1990*, Cambridge, Mass., Abt Books, EIU Special Series 1, 1982.

Sekiguchi, Sueo. *Japanese Direct Foreign Investment*, Montclair, New Jersey, Allanheld, Osmun & Co., 1979.

Suzman, Cedric L. *The Costs and Benefits of Foreign Investment from a State Perspective*, Atlanta, Georgia, The Southern Center for International Studies, August 1982.

Takitani, Kenji, "A Prototype for Japanese Investment in the United States," *Columbia Journal of World Business,* Vol. 8, No. 4, Winter 1973, pp. 31-33.

Tong, Hsin-Min and C.K. Walter, "An Empirical Study of Plant Location Decisions of Foreign Manufacturing Investors in the United States," *Columbia Journal of World Business,* Spring 1980, pp. 66-73.

Tsurumi, Yoshi. *Multinational Management: Business Strategy and Government Policy*, Cambridge, MA, Ballinger Publishing Co., 1977, pp. 1-150.

Tsurumi, Yoshi, "The Best of Times and the Worst of Times: Japanese Management in America," *Columbia Journal of World Business*, Vol. 13, No. 1, Spring 1978, pp. 56-61.

Tsurumi, Yoshi, "The Multinational Spread of Japanese Firms and Asian Neighbors' Reactions," *The Multinational Corporation and Social Change*, David E. Apter and Louis Wolf Goodman (eds), New York, Praeger Publishers, 1976, pp. 118-147.

Tsurumi, Yoshi, "The Strategic Framework for Japanese Investments in the United States," *Columbia Journal of World Business*, Vol. 13, No. 4, Winter 1973, pp. 19-25.

U.S. Department of Commerce, *Foreign Direct Investment in the United States: Report to the Congress*, Vols. 1, 5, 1976.

U.S. Department of Commerce, *Foreign Economic Trends*, 82-031 (May 1982), 83-093 (October 1983).

U.S. Department of Commerce, *Overseas Business Reports*, 80-21 (June 1980).

U.S. Department of Commerce, *Survey of Current Business*, August 1979, pp. 38-51; August 1980, pp. 38-51; August 1981, pp. 40-51; August 1982, pp. 11-41; August 1983, pp. 31-41; October 1984, pp. 26-48; November 1984, p. 24; May 1985, pp. 18-23; August 1983, pp. 47-66; August 1986, pp. 74-88; May 1987, pp. 36-51.

Yoshino M.Y., "Japanese Foreign Direct Investment," *The Japanese Economy In International Perspective*, Isaiah Frank (ed.), Baltimore, Maryland, The Johns Hopkins University Press, 1975, pp. 248-272.

CHAPTER 11
The New Bilateralism:
The East Asian NICs in American
Foreign Economic Policy*

Stephan Haggard and Tun-jen Cheng

The export-oriented industrialization of East Asia has produced a profound shift in the world's economic geography. In 1964, Japan and the East Asian newly industrializing countries (NICs)—Korea, Taiwan, Singapore and Hong Kong—accounted for 23 percent of the United States' imports. By 1985, this had risen to almost 44 percent, surpassing Europe's share (Table 11.1).

The macroeconomic cycles of the 1980s exacerbated these secular trends. The United States recovered more quickly and vigorously from the recession of the early 1980s than the rest of the world and until 1985 the dollar was unusually strong. As a result, the United States absorbed a large share of the developing world's manufacturing exports, over half from East Asia. In 1985, the United States took 68 percent of all manufactured exports from the Third World, up from 52 percent in 1981. Europe's share dropped to 24 percent in 1985 compared with 40 percent in 1981, while Japan's stayed constant at 7 percent.[1]

These changes in the pattern of world trade naturally had political consequences. One component of the U.S. response to the growing presence of the NICs in world trade was multilateral.

*This paper is a substantially revised version of Ch. 7 of Tun-jen Cheng and Stephan Haggard, *Newly Industrializing Asia in Transition: Policy Reform and American Response* (Berkeley: The Institute of International Studies, 1987). We would like to thank Don Babai, Hee-jung Shin, Peter Cowhey, Wayne Edisis, Tamara Fish, Chung-in Moon, Raymond Vernon, Louis T. Welles, Jr. and Peter Wollitzer for their comments and assistance on this paper. Funding for this research was provided by the Institute for the Study of World Politics, Harvard University, and the Graduate School of International Relations and Pacific Studies, University of California, San Diego.

1. Henry Nau, "Bargaining and the New Round: The NICs and the United States," paper prepared for the Quadrangular Forum, July 1986, p. 10.

Table 11.1 U.S. Imports of Manufactures from Japan and the East Asian NICs (as a share of total manufactured imports, various years)

	1964	1973	1980	1985
Japan	19.5	21.4	25.0	28.0
Korea	.2	2.2	3.4	4.1
Taiwan	.5	3.9	5.7	6.8
Singapore	.1	.9	1.3	1.5
Hong Kong	2.8	3.1	3.8	3.5
Total, East Asian NICs	3.6	10.2	14.3	15.9
Total, East Asia	23.1	31.7	39.8	43.9

(Totals may not add due to rounding. Source: Organization for Economic Cooperation and Development, Trade Series C.)

The U.S. actively encouraged NIC participation in the Uruguay Round of GATT negotiations, the first round of international trade talks in which the developing countries "graduated" to a central role. Korea, Hong Kong and Singapore were among the group of small trading nations with a strong interest in seeing the trade talks progress. The distinguishing feature of American policy toward the East Asian NICs during the second Reagan administration, however, was its bilateralism. The origins of the new policy were, in the first instance, political, driven by efforts to forestall protectionist legislation emanating from Capital Hill. The new bilateralism also reflected a longer term trend in American foreign economic policy, however: a drift away from multilateralism and rule-conforming behavior toward a more aggressive use of market power to achieve economic goals.

This new bilateralism was visible in three areas.[2] First, the United States played the role of *demandeur* in widening the trade policy agenda to include new issues such as services, the protection of intellectual property and trade-distorting investment measures. Though these issues constituted the heart of the new round of trade

2. Useful summaries of the issues before the new round are contained in C. Michael Aho and Jonathan David Aronson, *Trade Talks: America Better Listen!* (New York: Council on Foreign Relations, 1985); Robert E. Baldwin and J. David Richardson, *Current U.S. Trade Policy: Analysis, Agenda and Administration* (Cambridge MA: National Bureau of Economic Research, 1986); "U.S. Objectives in the New Round of Multilateral Trade Negotiations," summary of testimony by Clayton K. Yeutter, U.S. Trade Representative, before the Senate Subcommittee on International Trade, May 14, 1986 (USTR mimeo); and Congressional Budget Office, *The GATT Neogitations and U.S. Trade Policy* (Washington, D.C.: U.S. Government Printing Office, 1987).

talks, they were also the subject of bilateral negotiations with Korea, Taiwan and, to a lesser extent, Singapore. Second, the United States became more aggressive in retaliating against perceived violations of existing international commerical law. The reasons, again, were political. The banner of "reciprocity" and "fair trade" provided the rationale for increased pressure on the NICs to open their markets, and even to balance their trade with the U.S. on a bilateral basis. To fend off legislated measures, the Reagan administration launched a comprehensive attack on unfair trade practices in 1985. The NICs were major targets of this plan. Very little thought was given to the concessions that might be offered in return, beyond a promise to keep the American market open. Even this promise could not be made explicit, since protectionist pressures were strongest precisely in sectors in which the East Asian NICs have been most competitive, such as footwear, textiles and apparel, and steel. This posed a profound political dilemma for the NICs. While under extreme pressure to make concessions, there was very little likelihood of gaining any new concessions from the United States in return.

The third set of issues centered on alternative means for managing trade imbalances. Two received attention in the early and mid-1980s. The first was to address trade problems through multilateral action on exchange rates and macroeconomic policy. This strategy was pursued with some results among the U.S., Japan, Germany, France and Britain, the so-called Group of Five. The multilateralism of the Plaza and Louvre accords on exchange rates and macroeconomic policy was altogether absent in U.S. dealings with the East Asian NICs, however. As with trade issues, the U.S. resorted to bilateral pressure.

An alternative approach advanced during the Reagan years was to work toward a regional approach to trade and economic conflicts. Proposals for Pacific Basin cooperation had a long pedigree, but political cleavages within the region blocked the formation of an organization that could effectively address economic conflicts.

The East Asian NICs thus provide an interesting case study of the new reciprocity. Some of the reforms sought by the U.S. already had supporters in the NICs themselves. In these cases, the balance of internal and external pressures in generating reform is difficult to gauge. The accumulation of surpluses by Taiwan, for example, gave that country an economic as well as political motivation for import liberalization. Nonetheless, it is plausible in such cases of partly convergent preferences that external pressure strengthened the domestic political hand of reformers. Where disagreements remained over the legitimacy of certain policy stances but the issue was not broadly politicized, such as the degree to which intellectual property should be protected, the NICs sought to limit unwanted concessions by agreeing in principle, but delaying in implementation. In some cases this probably constituted a self-conscious strategy; in others, it may simply have reflected the inability to surmount bureaucratic opposition to policy change or difficulties in monitoring compliance domestically. This class of cases produces particular problems of compliance that the proponents of the new reciprocity have not fully resolved. The transparency of domestic regulation is much less than for actions taken at the border, and the stringency or laxness of domestic enforcement is difficult to monitor from outside.

Finally, in several cases, particularly import liberalization in Korea, American pressure became the subject of open political controversy. The transformation of the political systems of the East Asian NICs in a more plural, if not fully democratic, direction makes national leaderships more sensitive to the costs of adjustment. This category of cases poses an additional set of problems to the new reciprocity. Naturally, opponents of reform may seek to use the existence of domestic political resistance to strengthen their bargaining position with the U.S. On the other hand, a strategy of "getting tough" runs the risk of politically isolating potential allies and even spilling over into broader political relations. In such cases, compensation and concession are critical, but it is precisely in outlining such concessions that the new bilateralism has been weakest.

New Issues on the Agenda

Services

Trade in services was a top U.S. priority for the Uruguay Round, though through 1988, there was little sense of how the complex menu of service-related issues could be negotiated multilaterally.[3] Figure 11.1 suggests the scope of the service sector, and also some of the problems in negotiating the liberalization of service "trade." Many services are not really "traded" at all; the real issue is the right of establishment and national treatment. Negotiations over services necessarily concern national regulatory policies governing the operations and entry of foreign firms and banks.

---

3. Jonathan David Aronson and Peter Cowhey, *Trade in Services: A Case for Open Markets* (Washington: American Enterprise Institute, 1984), p. 3.

Figure 11.1 Types of Services

---

1. Travel, Transportation, Tourism and Leisure Services
   Lodging, recreational and cultural services
   Shipping, rail and air transport
   Tourism and passenger transport

2. Return on Capital
   Direct and other investment income
   Licensing, including royalties, license fees, copyrights
   Rental

3. Support Services
   Accounting
   Advertising
   Education
   Personal consumer services
   Professional services, including consulting, legal, economic, and
       technical services
   Maintenance

4. Construction and Engineering Services

5. Telecommunications, Information and Data Processing
   Data processing and computer treatment of information bases
   Motion pictures, printing and artwork
   Telephone, telegraph, television and tele-data transmission

6. Banking, Brokerage, Financial and Insurance Services

7. Management Services
   Employment
   Franchising and chartering
   Health services and hospital management
   Leasing
   Wholesaling and retailing

(Adopted from Jonathan David Aronson and Peter F. Cowhey, *Trade in Services: A Case for Open Markets* (Washington D.C.: American Enterprise Institute, 1984), pp. 4-5.)

Even prior to the launching of the new GATT round, the United States was involved in disputes with Korea and Taiwan over services. Under Section 301 of the 1974 Trade Act, which deals with unfair foreign practices, the USTR was allowed to initiate its own trade actions. The 1984 Trade Act incorporated several "reciprocity" provisions, including the grant of authority to the President to negotiate the reduction of services barriers, and an extension of Section 301 to allow retaliation against barriers to trade in services. In his "trade policy action plan" announced in September 1985, President Reagan promised to use the new Section 301 authority "vigorously" to open particular markets for U.S. firms.

Taiwan has generally been more forthcoming than Korea in these disputes. Though the services market has been relatively closed in Taiwan, the overall posture toward foreign investment has been more liberal, in part because the country's political isolation argued for the maintenance of liberal trade and investment relations. The most important factor in forcing liberalization, however, has been Taiwan's curious balance of payments situation. While Korea generally ran payments deficits prior to 1986, by 1987, Taiwan had accumulated reserves second in size only to those of Japan. Large surpluses made it extremely difficult for Taiwan to resist pressure for liberalization, and provided a political opening for liberal technocrats within the bureaucracy. In August 1985, in line with the new Reagan initiative, the American Institute in Taiwan (AIT), the unofficial United States embassy, submitted proposals to liberalize barriers to banking, insurance, shipping, and motion-picture distribution and leasing.[4] The AIT banking proposal requested permission for foreign banks to establish more than one branch office in Taiwan, to conduct foreign exchange business with companies in the Export Processing Zones and to gain broader access to local currency funding. Some of these steps were implemented in 1986, but reforms that would give both local and foreign banks greater freedom to set rates and would have permitted banks to accept deposits and engage in trust business have been held captive to the broader issue of banking and financial market reform. Strongly conservative forces in the Ministry of Finance and Central Bank have slowed progress in this area. The most dramatic financial market reform to date has been the result of accumulating surpluses rather than American pressure. In July 1987, the government eased long-standing foreign exchange controls, creating greater opportunities for outward investment. Through 1987, however, only five local banks and investment companies were authorized to invest in foreign securities, and then only on a trust fund basis.

In August 1986, agreement was reached on a wide-ranging insurance pact that would grant free entry and full national treatment to American companies meeting certain criteria. Initially, only two firms were allowed

---

4. This and other details in this section on pending bilateral negotiations can be found in *Annual Report of the President of the United States on the Trade Agreements Program 1984-85* (Washington: USGPO, 1984), p. 14. (Hereafter cited as *USTR Annual Report,* and in the useful Annual Report on National Trade Estimates 1985 (Washington: USTR, n.d.).

to operate in Taiwan, and only in a restricted range of business. The agreement allowed four foreign insurance firms (two for property insurance, two for life insurance) to enter the Taiwan market every year, but excluded European firms. Leasing has also been liberalized to allow foreign companies to own 80 percent of a leasing firm, but cargo transport was only partly opened to foreign participation. American firms could manage air cargo warehouses, but not inland container shipment.

Korea's stance toward liberalization has historically been cautious. The economy is less open than Taiwan's and fear of Japanese domination of the economy remains a recurrent political theme. In addition, political liberalization has resulted in open opposition to reform initiatives and added to the ruling party's caution. As a result, bilateral negotiations have been more contentious and politicized. In the fall of 1985, the USTR initiated a series of 301 investigations concerning insurance, film distribution and intellectual property rights. Korea was also threatened with a suspension of privileges under the Generalized System of Preferences, of which it has been a prime beneficiary. A general agreement was reached in July 1986 that eased entry for insurance firms and lifted some restrictions on film distribution. The actions were not considered satisfactory by the American business community, however, which pointed out that Korea consistently dragged its feet in implementing negotiated agreements and used other instruments to restrict entry.[5] Among the restrictions limiting foreign participation in the Korean services market were highly discretionary foreign exchange controls. Of fifty current invisible transactions listed in the *OECD Code of Liberalization and Current Invisible Operations*, the Ministry of Finance restricted foreign payments for twenty-nine. Plans had been developed to liberalize virtually all of the remaining items of the OECD list by 1989, but a variety of other restrictions still remained. Licensing and registration requirements limited foreign participation in construction and engineering. Access to the advertising market was completely closed.

Banking was a major point of contention. In May 1984, the Ministry of Finance announced a two-year liberalization of banking that would give foreign branches more equal treatment with Korean banks. Between 1983 and 1986, however, the foreign bank share of total won deposits was virtually unchanged despite the reforms.[6] Foreign banks

5. American Chamber of Commerce in Korea, *United States-Korean Trade Issues* (Seoul, June 1987). On the Korean insurance agreement see White House Press Release and Fact Sheet, July 21, 1986. On Korean banking reforms, see *National Trade Estimates 1985*.

6. American Chamber of Commerce, *United States-Korean Trade Issues,* p. 29.

complained about rules limiting each bank to two branches, restrictions of funding sources and investments, inability to exercise mortgage rights of foreclosure, and restrictions on the introduction of new products and services.

Korea contended that the service sector was hamstrung by past government intervention; to internationalize services prior to domestic de-regulation and liberalization would result in unfair advantages to foreign competitors. The government's stance toward liberalization is stated clearly in a study of Korea's bargaining position in the GATT and suggests a strategy of selective compromises and gradualism:

> Industrial engineering, banking, insurance and advertising all appear to be Korea's infant service sectors. This means two things. It would be undesirable to admit foreign operations in infant sectors on an MFN basis and extend full national treatment to foreign firms. On the other hand, it would be desirable to have some foreign presence to stimulate innovation and learning. Thus, foreign operations may have to be admitted to a limited degree and / or there may have to be qualified national treatment to admitted ones.[7]

Protecting Intellectual Property

The U.S. also succeeded in placing the protection of intellectual property on the GATT agenda, though as with services, Section 301 was employed to force the issue bilaterally. Counterfeiting and trademark abuse was one visible target of U.S. concern. The U.S. International Trade Commission estimated in 1986 that as much as $6 billion in trade entering the United States each year was counterfeit. Taiwan and Korea were considered among the worst violators, but Singapore was accused of audio and videotape piracy as well.[8] The city-states' free port status made them regional centers for the distribution of counterfeit goods, including those made in Taiwan and Korea.

The intellectual property issue extended beyond counterfeiting to questions of the investment climate, particularly the degree of protection afforded patents, copyrights, and trade secrets. The U.S. argued that without such protection, firms would be dissuaded from licensing technology, or even copyrighting and patenting innovations at all. The U.S. also claimed that the lack of protection extended to intellectual

---

7. Soogil Young, "Trade Policy Problems of the Republic of Korea and the Country's Objectives at the Uruguay Round," paper prepared for the Rockefeller Trade Policy Project, Korean Development Institute, July 1987, p. 48.

8. Aho, *Trade Talks*, pp. 51-2; *National Trade Estimates 1985*, pp. 183-4.

property affected firms' willingness to invest. There was little empirical evidence to back these claims. The likelihood of technology being copied is not increased substantially by foreign investment, nor are patents alone likely to halt the diffusion of technology. Nonetheless, the United States pressed the issue vigorously. In late 1987, for example, the U.S. opened bilateral neogtiations with Taiwan on copyrights—the first in a series of bilateral negotiations with the East Asian NICs on the issue—by demanding thirty-year retroactive compensation!

The NICs generally justified their restrictive policies on developmental grounds, arguing that they accelerated rather than retarded technological diffusion. Until recent changes, for example, Taiwan limited patent protection for chemicals and pharmaceuticals to process patents and Korea excluded foodstuffs, chemicals, pharmaceuticals and methods of using chemicals from being patented altogether. Trademark licenses in Korea were granted only in the context of a technical assistance agreement or joint venture. Taiwan lacked clearly defined codes that permitted successful prosecution of trademark and patent infringement. Under the July 1986 agreement with the U.S., Korea consented to join the Universal Copyright Convention, the Geneva Phonograms Convention and even the Budapest Treaty on microbe patents! New legislation was introduced to reform copyright, patent and trademark laws. In the past, however, enforcement of those laws that did exist in the NICs was extremely lax. In 1987, a visitor to Taipei could still buy pirated hardcover versions of the entire *New York Times* bestseller list for about a quarter of their cost in the U.S., and pirated records and tapes were easily available in Hong Kong. To enforce compliance of such laws demanded complex monitoring procedures not only of traded goods, but within the domestic markets of the targeted countries. Such monitoring itself had significant costs.

Trade-distorting Investment Laws

The third new issue the U.S. advanced was trade-distorting investment laws, specifically export requirements and domestic content legislation.[9] As with services and intellectual property, the 1984 Trade

---

9. For a review of these issues, see Harvey E. Bale, Jr., "Trade Policy Aspects of International Direct Investment Policies," in Robert E. Baldwin ed., *Recent Issues and Initiatives in U.S. Trade Policy* (Cambridge MA: National Bureau of Economic Research, 1984); USTR Press Release, July 24, 1986, "Statement on Trade-Related Investment Requirements"; and Dennis J. Encarnation and Louis T. Wells, Jr.,

Act authorized the USTR to engage in consultations to remove foreign export requirements. The East Asian NICs are not as important as sites for foreign investors as Brazil and Mexico, nor are they on the whole as restrictive as other LDCs. Nonetheless, both countries have regulated the trade behavior of multinationals. On a case-by-case basis, Taiwan has tied foreign investment approvals to minimum export requirement ranging from 5 percent to 50 percent of output. Generally, products made locally faced export requirements while those without local production would not, but the rationale for such measures was more complex than protection for domestic firms. Export requirements were intended in part to offset biases that existed toward local production and to counter the efforts by multinationals to impose territorial limits on the sales of their subsidiaries. Under pressure from the United States, Taiwan reduced export requirements for firms in the export processing zones and dropped export requirements imposed on a Toyota branch plant that was planning to export to the United States. USTR claimed these actions as precedents, and Taiwan promised to forego such requirements in the automobile sector, but whether Taiwan's acquiesence in this particular case represents a new policy line remains to be seen.

Taiwan has imposed local content requirements on the production of color televisions, video recorders, heavy sedans and motorcycles. These restrictions were designed to foster linked parts and component industries. Korea developed more wide-ranging local content requirements over the 1970s in line with its heavy and chemical industrialization plan. Korea announced a series of measures liberalizing the rules governing foreign investment in the early 1980s, but investors complained that the old practice of exercising broad discretion over the terms of entry remained in place. As with other "new issues," monitoring compliance and achieving transparency posed serious challenges.

Industrial Targeting

A final issue of relevance to the East Asian NICs has been U.S. concern over foreign industrial targeting. In 1983 the International Trade Commission conducted a study of foreign targeting practices that included analyses of industrial policies in Mexico, Brazil, Korea

---

"Evaluating Foreign Investment," in Theodore Moran, ed., *Investing in Development: New Roles for Private Capital?* (New Brunswick: Transaction Books for the Overseas Development Council, 1986).

and Taiwan.[10] The 1984 Trade Act also called on the USTR and other agencies to conduct a study of the effects of foreign industrial targeting on U.S. industry and whether existing law, particularly Section 301, was adequate to deal with them.[11] Trade legislation also called on the President to retaliate against foreign targeting practices.[12]

The findings of the various studies proved ambiguous, in part because established economic theory remained undecided on the efficacy of targeting practices. On the one hand, the USTR report found that targeting only worked against the backdrop of a "natural" comparative advantage in the product, that financial and tax subsidies were less significant than commonly thought, that the U.S. engaged in its own industry-specific programs and that targeting introduced macroeconomic distortions. As a result, "because targeting is so often unsuccessful, the appropriate policy response to it is to take offsetting action only if it has burdened, or is likely to burden, U.S. commerce."[13] At the same time, however, the USTR report found that targeting could work in industries characterized by a short product-life cycle, steep learning curves and high R&D expenditures. Home market protection and R&D programs may be "potent" targeting practices that have "long-lived effects on the competitiveness of the recipient industry." Most troubling for trade policy was the fact that targeting practices had delayed effects against which it is difficult to retaliate:

> Once firms are able to reduce variable costs by increasing their volume, they can become quite competitive in world markets and the protection can be eliminated. Hence, the effects of targeting policies may not be felt in international markets until long after the policies have been eliminated.[14]

---

10. The ITC studies are *Foreign Industrial Targeting and Its Effects on U.S. Industry Phase I: Japan* (Washington, D.C.: USITC, 1983); *Foreign Industrial Targeting and Its Effects on U.S. Industry Phase II: The European Community and Member States* (Washington, D.C.: USITC, 1984); *Foreign Industrial Targeting and Its Effects on U.S. Industry Phase III: Brazil, Canada, The Republic of Korea, Mexico and Taiwan* (Washington, D.C.: USITC, 1985).

11. The final USTR report is contained as Appendix O to *USTR Annual Report 1984–85*. The sector studies were: *The Effects of Foreign Targeting on the U.S. Automobile Industry*, done by Booz-Allen for the USTR and submitted in 1985, which contained material on Japan, Brazil, Korea, France and West Germany; *The Impact of Foreign Industrial Practices on the U.S. Computer Industry*, done by the Futures Group and submitted in 1985, which contained material on Japan, France, Brazil and Singapore; and *An Analysis of the Effects of Targeting on the Competitiveness of the U.S. Semiconductor Industry*, done by Quick Finan, and containing material on Japan, France, Great Britain, West Germany, Korea and Taiwan.

12. *Wall Street Journal*, May 23, 1986.

13. Appendix O, *USTR Annual Report 1984–85*, p. 159.

14. Appendix O, *USTR Annual Report 1984–85*, p. 163.

The USTR concluded that trade law should be amended to authorize the U.S. to offer concessions in return for the elimination of practices which, while fair under the GATT, are likely to enhance the long-term competitiveness of the industry in question. "Market-opening, sector-specific" (MOSS) negotiations with Japan have been conducted in a number of "hi-tech" sectors, including semiconductors; it is probable that such negotiations will open with the NICs at some point in the future. It is unlikely, however, that the targeting issue can be addressed effectively in a multilateral setting. At the GATT Ministerial Meeting in 1982, the U.S. asked for work on high-technology industries, but did not succeed in convincing the other parties that these sectors could be treated differently from others. According to Aho and Aronson, "the initiative was so poorly defined that LDC representatives asked how high-technology discussions could be related to transfer of technology."[15] The whole issue was eventually dropped.

Two conclusions emerge from this review of new issues. First, despite U.S. interest in placing these new issues on the GATT agenda, they are already being managed through bilateral negotiations. Given their trade-dependence, Korea and Taiwan are particularly vulnerable to these bilateral pressures, though Korea has been more reluctant to comply. Even were agreement to be reached on a broader set of principles governing these issues, and even if the framework for such an agreement was posing substantial difficulties to negotiators through the end of 1987, bilateral negotiations would continue.

A second conclusion concerns the difficulty of negotiating service issues. Trade policies have been the subject of bilateral discussions with the NICs throughout the post-war period; the U.S. wrangled with Korea and Taiwan over them in the 1950s. Unlike trade policies, however, discussions of service liberalization touch more closely on policies usually considered purely "domestic." These discussions are difficult since there are underlying disagreements on the legitimacy of certain interventions. Negotiating *commitments* is therefore easier than guaranteeing *compliance*. As a result, bilateral negotiations are likely to be followed by a growing web of bilateral consultative and surveillance mechanisms designed to insure implementation.

---

15. Aho and Aronson, *Trade Talks*, p. 44.

The East Asian NICs and the Traditional GATT Agenda

Until the 1970s, the core of the multilateral trade talks was tariff liberalization. The Tokyo Round negotiations added the issue of non-tariff barriers. Both areas have been of obvious importance to the trade-oriented East Asian NICs. Their political position is particularly weak, however. First, it is unlikely that the new GATT round will reverse the trend toward bilateral protectionist arrangements, such as selective safeguards and voluntary export restraints (VERs). Second, the new round will be the occasion for increased pressure on the NICs to make concessions of their own and to abandon those preferences they now have, an expectation summarized in the concept of "graduation."

The External Environment: Continuing Restrictions

The NICs face a variety of restraints on their exports, including discriminatory quantitative restrictions and voluntary export restraints, nondiscriminatory relief measures and unfair trade practice remedies. The last two categories of restraints are not, strictly speaking, protectionist. Nonetheless, the vigorous prosecution of illegal trade practices became a central pillar of U.S. trade policy under Reagan.

Given their growing share of LDC exports, it is not surprising that the East Asian NICs have figured in a number of anti-dumping, subsidy/countervailing duty, and escape clause actions over recent years.[16] In 1984, for example, anti-dumping duties were imposed on twenty-two products. Seven of these cases involved Korea and Taiwan. Through August 31, 1985, eighty new anti-dumping cases had been filed. Twenty-eight of these involved the newly industrializing countries; ten of them were against the East Asian NICs. In 1984, countervailing duties were imposed against fourteen products. While none of them went against the East Asian LDCs, eleven of the fourteen were imposed against the large Latin American NICs, Argentina, Brazil and Mexico. There have been fewer escape clause actions, though the East Asian NICs figured in four of six investigations completed between January 1984 and August 1985. In two important cases, however, certain steel products and nonrubber footwear, President Reagan refused to grant relief.

The newly industrializing countries tend to view these arrangements as disguised forms of protection. A study on Korea's goals in the

---

16. The following is extracted from the *USTR Annual Report 1984–85*.

Uruguay Round notes, for example, that while unfair trade practices are formally consistent with the GATT:

> They are founded on concepts such as dumping, subsidies, material injury, industry, etc., which defy clean-cut definitions, let alone objective measurement. Accordingly, there is a great deal of room for arbitrary implementation and even the abuse of these concepts. . . . Furthermore, many national provisions have been known to have features that allow employment of petition to be used as an instrument for harassing foreign competitors.[17]

Historically, the United States has been reluctant, or politically unable, to bring its trade policy into line with international norms. Recent trade legislation has "strengthened" both the escape clause and unfair trade practice laws in ways that make it easier for petitioners to file, while reducing the discretion of the President in denying relief.

The core of the new protectionism, however, remains the network of sectoral arrangements based on quantitative restrictions. These include the negotiation of voluntary export restraints (VERs) among industry groups designed to forestall protectionist action. In the first half of 1987, for example, Korea negotiated nine new VERs covering a range of products from TV sets and videocasette recorders to stuffed toys and pianos.

By far the most important restraint agreement for the NICs, however, is the Multi-Fiber Arrangement.[18] Though the United States stated its willingness to consider textiles in the upcoming round, through the end of 1987, a textile negotiator had not been appointed. Election year politics and the strength of the textile lobby made it highly unlikely that any progress would be made in reintegrating textile trade into the GATT framework. Nor was it entirely clear that the "Big Three" textile exporters—Korea, Taiwan and Hong Kong—had an interest in scrapping the MFA altogether, since it guaranteed market share against the looming threat of massive Chinese entry into world textile and apparel trade. Nonetheless, progress on textiles in the new round could be an important political test for the LDCs of the developed countries' sincerity.

A brief overview of the MFA suggests that each renegotiation has involved new restrictions and an increase in the discretion of the importers. Following the economic disruption associated with the oil crisis, the 1977 renegotiation contained a safety clause that permitted

---

17. Soogil Young, "Trade Policy Problems..." p. 21.
18. See Vinod Aggarwal, *Liberal Protectionism: The International Politics of Organized Textile Trade* (Berkeley: University of California Press, 1985).

importers to severely limit, and even reduce, annual export growth, the so-called "reasonable departure" clause. In 1977, the United States used this clause to eliminate export growth for the "Big Three." Under pressure from the textile-apparel coalition, and seeking to push the legislation ratifying the Tokyo Round through Congress, the Carter Administration developed a textile industry plan in 1979 that included new restrictions on the major East Asian exporters. The 1981 renegotiation of the MFA marked an even sharper turn in a protectionist direction by giving the developed countries further leeway in their negotiation of bilaterals with exporters. Primarily due to the united front of the exporting countries during the negotiation, the "reasonable departure" clause was dropped from MFA III. However, the U.S., Canada and the European Community reduced quota base levels for major suppliers so as to make room for new producers, a tactic interpreted as a political effort to split the exporters. For the first time, the United States instituted a "call system" that allowed it to *unilaterally* suspend the import of a product category from a particular country.

American rules of origin were also tightened in 1984, aimed in large part at Hong Kong's apparel industry. Hong Kong businessmen had moved knitting operations to China's Special Economic Zones, re-importing partly finished goods to be finished and exported under Hong Kong's larger quota. Rules of origin dictated that the finishing work done in Hong Kong did not constitute a "substantial trans-formation." These restrictions were estimated to affect as much as $300,000,000 in trade; 16 percent of Hong Kong's exports to the U.S., 6 percent of total textile exports and 2 percent of total exports.[19]

By 1985, these various restrictions were deemed inadequate by the American industry. The initial version of the highly popular Jenkins Bill would have reduced textile imports from by as much as 40 percent, in complete contravention of the already-restrictive MFA III. The final textile quota bill, passed by the House in December 1985 on a vote of 255-161, would have rolled back current import levels an average of 30 percent.[20] The bill was vetoed by President Reagan, but there was a genuine threat that the veto would be overturned. As part of its Con-gressional strategy, the USTR negotiated extremely restrictive textile bilaterals with the Big Three prior to the final vote on the veto override. The announcement of the restrictive Korean bilateral came two days before the veto override vote, on which the administration's vigorous

---

19. Data supplied by U.S. Consulate, Hong Kong.
20. *Wall Street Journal*, August 4, 1986.

lobbying proved effective. While all of the bilaterals allow for annual increases in import growth rates, they will dramatically curtail the Big Three's textile exports to the U.S. Taiwan's allows for an average of 0.5 percent growth over three years; Korea's 0.8 percent over four years, Hong Kong's about 1 percent annually through 1991.[21]

The domestic political climate also affected the third renegotiation of the Multi-Fiber Arrangement, MFA IV, which was concluded July 31, 1986. As with previous renegotiations, this one proved more restrictive than its predecessor, extending the coverage to additional fibers, granting more discretion to importers in handling import surges and demanding exporters' cooperation in prosecuting cases of fraud and circumvention.[22] While MFA IV prohibits any cutback of quota, it permits an importing country to unilaterally renew and freeze a quota for one year.

Steel provides a second example of a restrictive agreement taking slightly different form. In September 1984, President Reagan announced a steel program with a somewhat confusing rationale. On the one hand, it was aimed at the growth of *unfairly* traded steel products into the U.S. market through the vigorous pursuit of countervailing and antidumping duties. At the same time, countries were "invited" to negotiate steel agreements with the United States limiting the exporter's market share to a fixed percent regardless of their trade practices. This allowed the administration to announce an overall ceiling on imports of 18.5 percent. In return for the negotiation of these bilaterals, the industry agreed not to file new unfair trade actions during the life of the agreement. Korea was an important target, negotiating an agreement limiting its market share to 1.9 percent until 1989. As a result, the value of Korea's steel exports to the United States, which grew 7 percent in 1984, shrank by 9 percent in 1985.[23]

It is difficult to predict how negotiated protection will evolve in the future. Vinod Aggarwal, Robert Keohane and David Yoffie have developed a model that distinguishes types of negotiated protection on the basis of certain industry characteristics, including barriers to entry and exit and the size of the domestic industry.[24] They argue that protectionist accords are likely to become institutionalized, as they have in textiles, only in instances where barriers to entry are low, the

21. See *Wall Street Journal*, July 1 and August 5, 1986.
22. Initial details on MFA IV are contained in White House Press Release, August 1, 1986, "Press Briefing by U.S. Trade Representative Clayton Yeutter."
23. For a summary of the Reagan steel plan, see *USTR Annual Report 1984–85*, p. 70.
24. Vinod K. Aggarwal, Robert Keohane and David Yoffie, "The Dynamics of Negotiated Protectionism," *American Political Science Review*, 81, 2 (June 1987).

industry is large, and exit is difficult. Where the domestic industry is small and exit easy, protection is likely to be temporary and not particulary restrictive. Restrictions on footwear and televisions reached with the NICs in the late 1970s are examples of agreements that were allowed to lapse. New efforts to secure protection in the 1980s were not successful. Finally, where barriers to entry are high, as in steel and autos, protectionism is likely to be sporadic and periodically re-negotiated. David Yoffie has also noted that quantitative restrictions are relatively easy to circumvent, and while they have slowed trade from a theoretical free trade trajectory, they have not necessarily been as restrictive as is commonly thought, and may even provide incentives for the NICs to upgrade.[25]

These findings should not necessarily be taken as a source of optimism for the NICs, however. First, the NICs have industrial capabilities across all three types of industries noted by Aggarwal, Keohane and Yoffie, and thus will continue to face negotiated protection regardless of the outcome of the multilateral trade talks. In addition, if industries are not securing the levels of protection required to solve their adjustment difficulties—and such adjustment may simply be impossible given the inherent cost advantages of the NICs—then they are likely to continue to press for bilateral, product-specific restrictions.[26]

Pressures to Liberalize

While confronting an inauspicious external trade environment, Korea and Taiwan have come under strong pressure to liberalize their own domestic markets. Despite the common image of the East Asian NICs as liberal, market-oriented economies, Korea and Taiwan have had bifurcated trade regimes; low barriers or duty free entry for inputs to export industries, coupled with relatively high levels of protection on a range of final products.[27] Because of mounting surpluses and wide bilateral imbalances in trade with the United States, Taiwan has been more vulnerable to U.S. pressure. Since 1986, Taiwan has accelerated its trade liberalization program, and has even gone so far as to propose the negotiation of a free trade agreement that would parallel those negotiated with Israel in 1984 and Canada in 1987.[28] A

---

25. See David Yoffie, *Power and Protectionism* (New York: Columbia University Press, 1983).
26. I am indebted to David Yoffie for this point.
27. This is a theme of Tun-jen Cheng and Stephan Haggard, *Newly Industrializing Asia in Transition: Policy Reform and American Response* (Berkeley: The Institute of International Studies, 1987).
28. For the Taiwan government's view of its trade policy efforts, see "How the

fairly dramatic lowering of tariffs has taken place since 1980, even though some discretionary control over imports is allowed for industrial policy purposes and to limit import dependence on Japan. In 1981, Taiwan's average effective tariff rate was 9.1 percent. In 1985, it was 7.6 percent, and the goal was to reduce it to 5 percent by 1990. In 1987, however, two new packages of concessions were offered that would lower average tariff rates to 3.6 percent, roughly equal to American tariff levels. The new concessions on manufacturers were likely to favor Japan more than the United States, however, and in late 1987, the U.S. renewed pressure for the liberalization of agricultural imports. Since 1978, Taiwan has also sought to reduce bilateral imbalances and protectionist sentiment by sending "Buy America" missions to the United States. These have been responsible for over $10 billion in purchases.

The Korean system has been substantially more protectionist than the Taiwanese one, reflecting a more interventionist style of economic management.[29] Liberalization has also been the subject of substantial political controversy. Import licensing is administered by the Ministry of Commerce and Industry, which has used trade restrictions for industrial policy purposes. Under a provision for the diversification of import sources, imports coming largely from Japan have also been restricted. A variety of other controls have also operated, though some are now slated for elimination, such as a "surveillance" system that limited import surge, and variety of special laws that gave various ministries and even industry associations approval over imports. Liberalization of restrictive licenses, i.e., removing commodities from the list of products requiring prior import approval, has been the key ingredient in the import liberalization program launched in 1978. In 1977, approximately half of all commodities demanded import licenses. Following the completion of a five-year schedule launched in 1984, 96 percent of imports will be free of these restrictions. This so-called "import liberalization ratio" is often used by Korean authorities to point to the dramatic liberalization that has occurred, though in fact, this ratio does not reflect quantitative restrictions under special, sector-specific laws or the import-source diversification scheme.

Nor does it take account of tariff rates, which remain high by developed, it not developing, country standards. A five-year program

Republic of China is Working to Promote Better Trade Relations with the U.S.A.," Taipei, Government Information Office, May 1986; and *Taiwan Industrial Panorama,* Taipei, Industrial Development and Investment Center, various issues.

29. For contending views of Korea's liberalization efforts, see Soogil Young, "Trade Policy Problems..." and American Chamber of Commerce, *United States-Korean Trade Issues.*

of tariff cuts was initiated in 1984. Under this program, the average general tariff rate was brought down from 23.7 percent in 1983 to 18.1 percent in January 1988. The recent import liberalization measures have been limited largely to manufactured products. The agricultural sector in Korea remains heavily protected; on a weighted average, it is more protected than the Japanese market. As in Japan, liberalizing agricultural trade involves a complex of policy issues that include promoting structural change, generating off-farm income, land-use policy, productivity improvements and considerations of equity.

American pressure has probably been more important in Korea's liberalization efforts than in Taiwan, where the pressure of accumulated surpluses was undoubtedly a central factor. As in Taiwan, liberal technocrats have seen import liberalization as one aspect of rationalizing the system of economic management, but they have been strongly opposed by a loose coalition of farmers, opposition leaders and politicians, students and potentially affected industries. The industrial policy segments of the bureaucracy itself have also argued for gradualism in liberalizing. Since Reagan's visit to Seoul in November 1983, the United States has been aggressive in demanding specific market opening measures, making six such requests through 1987. By the middle of 1987, the number of commodities for which licensing liberalization was requested totalled 4.8 percent of all commodities. Korea complied with 82 percent of these requests. On the whole, U.S. pressure has been strongest in agriculture. About 40 percent of the requests for both licensing and tariff liberalization have involved agricultural products, the sector in which liberalization will prove the most difficult politically.

The review of the core GATT agenda—trade in manufactures—suggests that the U.S. has become more aggressive in prosecuting unfair trade practices and in pressing liberalization, even while continuing to engage in bilateral restraint agreements. The extent to which these bilateral actions will slow market penetration or reduce the trade deficit is highly exaggerated, however. There are a shrinking number of identifiable unfair trade practices for the U.S. to pursue. Trade deficits are a function of many factors, including macroeconomic policy, and will not be affected substantially by sector-level trade policies nor by success in getting trade partners to liberalize. Hong Kong and Singapore are already free ports, and even if the United States is wholly successful in securing liberalization in Korea and Taiwan, it would not forestall protectionist pressures.

On the other hand, the picture for the NICs is not altogether bleak.

As other studies in this volume show, close attention must be given to the market consequences of restrictionist actions. Quotas provide perverse incentives to exporters to upgrade the quality of restricted items and to move into new products altogether.[30] Protectionism has been a major force driving the NICs' efforts to geographically diversify trade and to move into new sectors.

Second, the making of trade policy in the U.S. is characterized by cross-cutting interests. Though a number of sectors in the United States face import competition, not all are likely to respond with calls for protection, particularly if they have extensive multinational ties or are themselves extensively engaged in trade. As I.M. Destler and John Odell have shown, there has been a significant rise in anti-protectionist lobbying by American exporters, leading retail chains, users of industrial imports and lobbyists representing foreign countries, including Taiwan, and sometimes embarrassingly, Korea.[31] Certain sectors have been altogether opposed to protection. Until recently, the farm belt voiced reservations, fearing retaliation in their export markets. Bankers worry about how trade restrictions will affect the capacity of heavily indebted LDCs to service their financial obligations. A number of retail chains, such as K-Mart and Zayre's, have extensive buying operations in the East Asian NICs and face difficulties in substituting American-made products for imports. Mobilizing these advocates of free trade may not permit the "rollback" of protection that the LDCs seek, but it will act as a brake.

Protection provides incentives for exporters to invest directly as well, both as a means of evading quotas and as a political strategy to ease trade tensions. In a number of important sectors, U.S. firms have welcomed foreign investment and marketing links; witness American and Canadian auto dealers' enthusiastic reception of the Hyundai Excel. As Dunn and Hart show, a variety of consortia-like agreements have emerged among firms in the automobile and semiconductor sectors for the purpose of offshore sourcing and cross-licensing of technology.[32] Foreign investment and the formation of inter-firm pacts create new trade ties and thus provide an additional political counterweight to protectionist forces.

---

30. See David Yoffie, *Power and Protectionism*.

31. See I.M. Destler and John Odell, *Anti-Protection: Changing Forces in United States Trade Politics* (Washington, D.C.: Institute for International Economics, 1987). See also Helen Milner *Resisting Protectionism: Global Industries and the Politics of International Trade* (Princeton: Princeton University Press, 1988).

32. Stephan Haggard, "The International Politics of East Asian Industrialization," *Pacific Focus* 1:1 (Spring 1986), pp. 97-124.

Finally, U.S. pressure on Korea and Taiwan may have a salutary effect on their economic policies over the long run. In Korea, protection has been but one component of a dirigiste industrial policy that, while successful overall, has resulted in a number of distortions, including high prices to consumers. External pressure provides an opportunity for liberalizers to force adjustments that many see as being in the country's long-term interests.

Macroeconomic Policy and Exchange Rate Coordination

Given the weakness of trade policy instruments in solving the fundamental imbalances in the international economy, policy attention has gradually shifted to macroeconomic and/or exchange rate coordination. At the height of Congressional concern over trade in September 1985, Treasury Secretary Baker reversed the U.S. policy of benign neglect toward the dollar, and called a meeting at the Plaza Hotel in New York with his counterparts from Japan, Germany, Britain and France. The aim of the meeting was to steer the dollar lower, particularly against the yen. The agreement suffered from an important limitation. While the currencies of Japan and the European countries rose against the dollar, the currencies of a number of other trading partners, including the East Asian NICs, did not; this was one reason why, despite the apparently large fall in the dollar, the U.S. trade deficit failed to correct itself. (Table 11.2) The Federal Reserve Board's Trade-Weighted U.S. Dollar Index fell more than 30 percent between February 1985 and September 1986, but Mexico and the East Asian NICs were not included in the index.[33]

Subsequently, the U.S. sought to address these continuing misalignments with the NICs on a bilateral basis. Initially, upward movements in the currencies of Korea and Taiwan in the 15-20 percent range were sought.[34] Of the four NICs, Taiwan permitted its currency to appreciate the most, though this was probably a result of accumulated surpluses as much as U.S. pressure.[35] The New Taiwan dollar appreciated 36 percent between the Plaza agreement and December 1987, though this was only half the rise of the Japanese yen and German mark over the same period. In 1987, the appreciation ac-

33. On background on the exchange rate issue, see Jeffrey Frankel, *The Yen/Dollar Agreement: Liberalizing Japanese Capital Markets* (Washington, D.C.: Institute for International Economics, 1984); *New York Times,* September 19, 1984.
34. *Wall Street Journal,* August 4, 1986.
35. *New York Times,* November 27, 1987.

celerated, however, with the New Taiwan dollar rising by 19 percent, compared to 15 percent for the mark and 18 percent for the yen. Korea has pursued perhaps the most mercantilist exchange rate policy. Between the Plaza accord and December 1987, the Korean won rose only 12 percent despite very large trade surpluses in 1986 and 1987. The Singapore dollar rose 7.5 percent over the same period and the Hong Kong dollar remained pegged to the U.S. dollar, but the trade surpluses the two city-states ran with the United States were almost fully offset by deficits with other countries.

Table 11.2 Origins of U.S. Imports and Bilateral Currency Realignments

Country	U.S. Import Share 1985	Change in Dollar (2/28/85–7/26/87)
Canada	19.3%	0
Europe	20.2	
U.K.	4.3	−27.4%
Germany	5.8	−37.2
Italy	2.9	−29.3
France	2.8	−32.8
Japan	20.0	−40.4
Asian NICs	13.1	
Korea	3.0	5.1
Taiwan	5.0	−2.8
Singapore	1.2	−2.3
Hong Kong	2.5	0.1

Source: *Wall Street Journal*, August 18, 1986, citing Wharton Econometric Forecasting Associates, Inc.

A Regionalist Alternative?

A different option for managing trade conflict was regionalism.[36] The Reagan administration claimed to give new attention to the Pacific Rim as part of its overall foreign policy. In 1984, Secretary of State Schultz appointed a special ambassador with exclusive responsibility for Pacific regional cooperation. The administration also endorsed the formation of a U.S. Committee for Pacific Cooperation to be composed of members of Congress, academics, business leaders and government officials acting in a non-official capacity. A number

---

36. On the idea of a Pacific Basin Community, see Robert L. Downen and Bruce J. Dickson, *The Emerging Pacific Community: A Regional Perspective* (Boulder: Westview Press, 1984); and the special issue of *Asian Survey* 23:12 (December 1983).

of political figures, including most prominently Gary Hart, sought to transform the Pacific Rim concept into an electoral issue. The number of seminars and conferences devoted to Pacific Basin co-operation increased exponentially, though without a corresponding sharpening of focus. Expectations for formal regional cooperation on economic issues far outran the underlying political realities. Too many axes of potential conflict existed within the region—North-North; North-South; South-South; and increasingly, East-West—for them to be easily encapsulated in a regional organization.

The idea of a regional cooperative mechanism is not new. Inspired by the EEC, Prof. Kiyoshi Kojima proposed a free trade zone for the Pacific in 1965. In 1980, an agreement between Japanese Prime Minister Masayoshi Ohira and Australian Prime Minister Malcolm Fraser led to the first Pacific Economic Cooperation Conference (PECC), a broad regional forum including business, academic and non-official government participation. The PECC was a trans-societal platform rather than a formal inter-governmental organ. The barriers to the emergence of the latter were multiple. First, and ironically, security and economic conditions were less conducive to close inter-governmental cooperation than they had been in Europe in the immediate postwar period, or even in Southeast Asia following the fall of Saigon in 1975. The EEC and the OECD were formed against the backdrop of the Cold War and the dire economic circumstances that lent them urgency. ASEAN received a political boost from the Communist victory in Vietnam. Regional schemes for the Pacific, by contrast, were proposed during a period of relative security, the absence of an external threat and economic prosperity. Second, there were problems of leadership. Both Japan and the U.S. seemed unwilling to take the lead in organizing a regional organization, probably because of the fears such a scheme might arouse among smaller countries of unwanted hegemony. ASEAN (the Association of Southeast Asian nations: the Philippines, Malaysia, Thailand, Singapore, Indonesia and more recently, Brunei) in particular feared that the creation of any regional organization would swamp its efforts, would be dominated by Japan and the U.S., and would inevitably draw ASEAN into the triangular geopolitics among the United States, China and the Soviet Union. ASEAN did establish a joint conference between the ministers of the ASEAN six and five advanced industrial states: Australia, Canada, Japan, New Zealand and the United States. The first so-called "6 + 5" meeting took place in July 1984. But the underlying question of membership plagues the "6 + 5" meetings, as it is likely to plague any official regionalist efforts. ASEAN is wary of including the more economically advanced NICs, including Korea. Hong Kong's status precludes its effective participation, and Taiwan's inclusion would probably block the participation of China, and vice versa. China has opened itself to external economic forces and actors to an unprecedented degree and has a strong interest in a smoothly functioning regional economy. It is unlikely to enter any organization or forum which might suggest more formal alignment with the West, however. Because of both its strategic and economic importance, the PRC will be able to fare quite well pursuing a bilateralist economic diplomacy.

There is also an inherent tension between closer regional cooperation in the Pacific and the world trading system, as the political problems surrounding the formation and growth of the EEC should suggest. Pacific cooperation led by the U.S. and Japan would have a profound systemic impact, freeing trade within the region at the expense of outsiders. Such a development would strain U.S.-European relations, particularly given that many Europeans already interpret U.S. Pacific policy as a way of pressuring Atlantic allies to be more forthcoming in multilateral discussions.

Conclusion: A Multi-tiered Foreign Economic Policy

The limitations on both multilateralism and regionalism suggest a future that looks somewhat like the present. GATT negotiations and consultations through the emerging network of regional organizations will continue, but the United States will pursue outstanding issues with the East Asian NICs on a sectoral and bilateral basis. A sectoral and bilateral approach to the management of economic issues does not necessarily mean an abandonment of multilateralism. New momentum on some issues will only be achieved through a more disaggregated approach. Nor should the political drawbacks of a principled multilateralism be overlooked. Without the gains negotiable at the sectoral and bilateral level, domestic pressure for "drastic action" is likely to mount.

There is a limit on what bilateral and sectoral efforts aimed at liberalization can achieve, however. Singapore and Hong Kong are already virtually free ports. With the exception of some relatively minor investment and intellectual property issues with Singapore, there is little to negotiate unless the decision is taken to follow the example set with Israel and move toward the negotiation of a Free Trade Zone. In such a negotiation with Singapore, most of the concessions would have to come from the United States. In Korea and Taiwan the range of government interventions and domestic political changes make bilateral and sectoral negotiations more complex. But even were Korea and Taiwan to fully liberalize their markets and adjust their exchange rates—to remake themselves in America's desired image—the adjustment problems created by their underlying competitiveness would continue. If there is truth to neo-classical arguments, they would become even more competitive than they now are! In addition, the bilateral approach, particularly in the heavy-handed way it has been pursued in Korea, threatens to create a political backlash.

Perhaps the greatest danger of the new bilateralism is in fostering the illusion that the problems facing American industry can be solved by forcing others to adjust. Without supporting measures that confront underlying fiscal imbalances, stabilize exchange rates, stimulate commercial R&D, ease adjustment costs for workers and address problems of retraining, American industry will continue to face competitive problems. But these are questions that go far beyond U.S. relations with the East Asian NICs.

CHAPTER 12
The Newly Industrializing Countries and the Political Economy of Protectionism*

David B. Yoffie

Political scientists are frequently caught by surprise when "weak" actors succeed in influencing "stronger" actors. Particularly in international relations, there is a tendency to proclaim a paradox every time a country with few resources turns its adversity into advantage. From the time of Thucydides (1951: 331), scholars have predicted that "the strong do what they can and the weak suffer what they must." Nonetheless, examples abound of weak states performing feats that are apparently beyond their means. During World War II, several small European nations managed to avoid the total devastation that engulfed their continent (Fox, 1958); tiny Malta has bargained successfully with the more powerful Britain (Wriggins, 1976); countries in Africa have discovered ways to extract bargaining concessions from the European Economic Community (EEC; Zartman, 1971); and "small allies" have had a "big influence" on American policies (Keohane, 1971). Regardless of the issue, traditional stereotypes about the weak and the strong rarely seem to hold.

This article aims to explore similar asymmetrical relationships in the context of the world trading system. It is widely accepted in the vast literature on North-South relations and dependency that developing nations are at a disadvantage in confrontations with the more developed countries, especially in the realm of international trade (Strange, 1979; Knorr, 1975; Caporaso, 1978). On one hand, most newly industrializing countries depend on their exports of manufactured goods for their economic growth and well-being. On the other hand, the United States and Europe control the access to a large percentage of the world's markets. Since importers in these circumstances usually have the greater bargaining capabilities, we are told to expect

---

*This article is reprinted with permission from *International Studies Quarterly,* 25:4 (December 1981): 569-599.

the "weaker" developing nations to be more or less at the mercy of the "stronger" industrial ones.[1]

If trade operated on the principles of Ricardian liberalism, politics and asymmetrical power relationships might be unimportant. Weak and powerful countries could simply exchange goods. But unfortunately, protectionism has always been a problem in the modern international economy, a problem that has had foreboding implications for the developing countries in recent years. Although tariffs in 1980 are at their lowest point in more than three decades, a more disturbing and potentially more dangerous form of trade barriers has emerged. No longer are tariffs and global quotas the primary tools for limiting imports. Instead, the industrial nations have discovered a "new protectionism" that discriminates against the developing states. The concept of most favored nation, which was the cornerstone of the "old protectionism" and assured the equal treatment of all nations, is not upheld in this new system. Rather, the United States and Europe avoid restricting each other while they simultaneously single out the most successful developing countries for protection. The result is that trade has become an urgent problem of politics and power, pitting the weak directly against the strong. According to the United Nations Conference on Trade and Development (UNCTAD, 1978: iii), the proliferation of these "restrictionist regimes" is a "grave threat to the trade and development of developing countries and would undermine their reliance on international trade as an engine for growth."

The most important policy tools of the new protectionism are voluntary export restraints (VERs) and orderly marketing agreements (OMAs). Both have become so widespread that they now rank equally with tariffs and quotas as the most widely used foreign commercial policies (Bergsten, 1975: 239). The Americans alone have used these arrangements in textiles and apparel since 1957, in automobiles since 1981, in color TVs and footwear from the late 1970s to the early 1980s, and in steel from the late 1960s to the early 1970s. OMAs and VERs are agreements between importing and exporting countries that limit quantities of exports at their source. Either through a formal or a tacit arrangement, the exporting nation or its industry "agrees" to restrict the shipments of a particular good to a level lower than it might otherwise expect in a competitive market.[2]

---

1. Only when there is a scarcity or restricted supply of a needed good will exporters command oligopolistic powers and bargain on a relatively equal level with importers (Strange, 1979).

2. VERs are virtually identical to OMAs except in their degree of formality. OMAs are usually explicit, legal contracts between importing and exporting governments. VERs,

Analysts view VERs and OMAs as especially worrisome to developing countries because they have unique discriminatory qualities. First, they are usually directed at labor-intensive, low-priced goods, products such as textiles and footwear, where a developing country has the greatest comparative advantage. Second, they put the burden of implementation on the exporting country. Third, they are negotiated accords. Unlike tariffs and quotas, which are implemented unilaterally, VERs and OMAs may lead to political tensions as a result of bargaining. Fourth, they involve bilateral negotiations that isolate the weak, developing country in a one-on-one confrontation against a powerful, developed state or the EC.

From the perspective of organizations such as UNCTAD, this final feature is the most disconcerting. Since the industrializing nations are in a "substantially disadvantageous bargaining position," "unable to retaliate," and "fearful of more severe restrictions," they are supposedly doomed to suffer in their foreign trade (UNCTAD, 1978: 10). UNCTAD fears that the consequences of this new protectionism may be "irreparable damage to the world economy and in particular to the developing countries" (UNCTAD, 1978: 19).

Despite the dire predictions, protectionism has not always been translated into disaster. An increasing number of rapidly developing nations, the so-called newly industrializing countries (NICs), have found ways to avoid the damaging effects of import barriers. Some nations have, unfortunately, suffered. Countries ranging from India to Colombia have had difficulties responding to trade restrictions. At the same time, however, others, such as Korea and Hong Kong, have rebounded resourcefully and minimized the damage of protectionism. As Table 12.1 illustrates, some countries have increased their quotas far beyond most expectations.

The reason for this apparent anomaly is that there are political and economic weaknesses in the structure of modern protectionism that a handful of NICs have learned to exploit. Ad hoc and short-run American policies have produced a structurally deficient form of protection; also, a limited number of exporters have discovered a formula that dissipates the effect of these U.S. trade barriers. The politics of protectionism have provided an opportunity for the "weak" to beat the "strong" at their own game.

---

however, range from industry-to-industry agreements to agreements between an exporting industry and an importing government to tacit arrangements between two governments. For a review of these various types of VERs and OMAs, see Smith (1973) and Yoffie (1981b).

Table 12.1. *U.S. Textile Quotas for Selected Exporters, 1956–1976*

(in millions of square-yard equivalents)

	First U.S. VER[a]	First Multifiber VER[b]	Quantitative Increase	% Increase
Japan	245	1737	1492	609%
Taiwan	56	562	506	903%
Korea	26	403	377	1450%
Mexico	75	278	203	270%
Philippines	45	189	144	320%
Pakistan	55	181	126	229%
India	79	152	73	92%
Colombia	24	91	67	279%

[a] Cotton textiles only, negotiated in 1956 for Japan, and between 1963 and 1965 for the other nations.
[b] Cotton, synthetic, and wool textiles, negotiated in 1971 for Japan, Korea, and Taiwan, and in 1975 or 1976 for the other nations.

Source: USITC, 1978.

In the first part of this article, I shall examine some of the strategies that NICs can employ to overcome modern trade barriers imposed by industrial nations and particularly by the United States. Taking examples from the most persistent objects of protectionism (textiles, apparel, and footwear), I will illustrate how Japan in the 1950s and 1960s, and Hong Kong, Korea, and Taiwan through 1980, have generally increased their export earnings and maintained a stable bilateral trading relationship with the United States, despite VERs, OMAs, and elaborate market-sharing arrangements. The second section will then focus on whether these strategies, or variants of them, are still viable for the NICs of the 1980s. To a large extent, exporters' success has been predicated on the accommodating nature of American protectionist policies. Hegemonic objectives and bureaucratic politics have constrained U.S. bargaining and implementation during most of the last twenty-five years. Yet as interest groups and decision makers in the United States learn from their experiences in individual sectors, an accommodating America can no longer be totally assured.

Exporting Strategies and the Weakness in Modern Protectionism

NICs have extended their export earnings in the presence of VERs and OMAs by recognizing the political basis of trade, and by following two core policies and several supplemental tactics. The essence of a successful exporting strategy is that a NIC give priority to long-term rather than immediate political and economic interests, and that it use the bargaining opportunity created by the new protectionism to negotiate for short-run needs. Within this general policy framework, an exporting country can also try to supplement its profits by linking issues, cheating on restrictive arrangements, and mobilizing transnational and transgovernmental allies on its behalf. When assembled as a coherent unit, these policies will exploit the political and economic weaknesses in modern protectionism and spell victory for the exporting country.

Prerequisite: Understanding the Political Basis of Trade

An exporting country must understand the political basis of trade. Protectionism, and particularly American protectionism, is the product of complex political choices. When the United States employs VERs and OMAs in a sector, it is usually an attempt to reconcile hegemonic interests in free trade with particularistic demands for import barriers. Therefore, the decision for a bilateral arrangement generally reflects pressures by domestic groups and competing government departments rather than a concerted effort to combat "unfair competition," "market disruptions," or a real threat to a domestic industry. As Bauer et al. (1963: 1) have noted, "in foreign trade issues, foreign policy becomes domestic politics."

Trade is inherently political, and there is little that a NIC can do to alter this situation. But grasping the nature of the game is an important step toward manipulating the rules to one's own advantage. When the sole purpose of import restrictions is political appeasement, the exporter may be able to accommodate that purpose without reducing its rate of export growth. The importance of political symbols for American and other policy makers may allow the NIC to exchange deference on politically salient questions for tangible economic benefits in less sensitive areas. The political nature of trade constitutes the first loophole in modern protectionism.

Core Policy I: Pursuing Long-Run Gains[3]

Politically motivated protectionism is essentially a refusal to adjust to change (Blackhurst et al., 1977). An industrialized country imposes trade restrictions when a loss in comparative advantage creates difficulties for domestic industries and labor. Protectionism of this sort, however, is filled with contradictions. It may provide immediate benefits to domestic groups, yet it is a "short-run and ultimately self-defeating alternative to the needed adjustment" (Blackhurst et al., 1978: 1). Protectionism alone does not foster increased productivity or movement of domestic firms into more dynamic sectors: its objective is to give domestic producers a breathing spell, a chance to operate under conditions of restricted competition. Furthermore, protectionist policies often come very late, with import barriers being implemented only after a market has become stagnant or has gone into decline. As a consequence, protectionism is fraught with dangers for an industrialized nation such as the United States. Instead of facilitating adjustments, protectionism may actually slow the process by "insulating the industry and [labor] from the very factors which encourage it" (Morton and Tullock, 1977: 182).

The very short-run nature of the industrialized state's policy has its benefits for the developing country. Protectionism is paradoxical: an immediate loss of trading gains for a developing nation can produce even larger long-term benefits if adjustments are made.[4] Firms in the newly industrializing countries are frequently short-run profit maximizers that fail to take into account changing costs and long-run market trends. Moreover, many foreign exporters operate under conditions of imperfect information. This helps to explain why NICs do not always produce an optimal product mix, and why they often saturate relatively slow growth markets in the industralized nations. Protectionism by the importing nations in these market segments, however,

---

3. "Long run" and "short run" are obviously relative terms. For the purposes of this analysis, they will be defined as they are in the business and economic literature: The short run will refer to a period of time in which certain types of inputs cannot be varied (a period of usually less than one year); the long run will refer to five or more years in the future, when all factors of production can be varied.

4. In his classic work, *National Power and the Structure of Foreign Trade,* Albert Hirschman (1945: 20) demonstrated this relationship between adjustment and trading gains. Hirschman showed that powerful states could limit the growth of developing countries if they could prevent adjustments and provide incentives for the exporting nations to focus on "urgent demand" and short-term gains from trade. The logic of Hirschman's argument is identical to the argument being made here; that is, the country that makes adjustments to the market is ultimately the greatest gainer.

can prod NIC governments to institute policies that correct for these externalities. Jolted by politically motivated, myopic protectionism, an exporting government and its producers may therefore be able to enhance profits if they focus on *long-run gains from trade*—that is, if they pursue a course of economic and political adjustment in order to facilitate a more efficient allocation of resources and political stability. If the exporting country can be flexible and make adjustments to slowly growing markets while the industrialized nation becomes entrenched in the status quo, the power of the industrialized country to force others to change can ultimately be turned back on itself.

The Political Component. Politically, the best strategy is one of short-term compromise and a willingness to make concessions. During the past twenty years, rapidly developing countries have successfully minimized large-scale trading problems with the United States when they made specific sectoral concessions; they have exacerbated political tensions almost every time they refused to compromise. Maintaining broad market access is critical for NICs if they wish to continue their reliance on export trade. The executive branch of the U.S. government has normally hesitated to implement strong protectionist measures, but it has required cooperation from the exporting country to avoid a choice between severe political dissatisfaction at home and overt protectionism. Since the latter would be inconsistent with international obligations or at least with a general policy of trade liberalism, NICs potentially have some leverage.

In 1956, for instance, American textile and apparel manufacturers almost sabotaged Japan's burgeoning trade with the United States. Complaining about unfair competition, the textile industry made a forceful entrance into American politics in the early 1950s. It organized boycotts of Japanese goods, filed petitions with the U.S. Tariff Commission for import barriers, and flooded Congress with a plethora of requests for restrictive quotas (Bauer et al., 1963). If the Japanese had not made a short-run sacrifice to appease the U.S. textile lobby, Congress may have taken more severe action, something inconsistent with Eisenhower's free trade orientation.[5] Once the Japanese signed a VER in 1957, however, economic peace was restored. The boycotts were dropped, no new petitions were filed, and the President was able to veto the restrictive tariffs recommended by the Tariff Commission without worrying about a congressional override.

---

5. On June 28, 1956, for example, a Senate motion to impose rigid export quotas on textiles failed by only two votes (Lynch, 1967). A full account of U.S. trade conflict with the Far East in textiles and footwear can be found in Yoffie (1981a).

While this policy is not meant to achieve desirable new objectives, the alternative—to reject protectionism forthrightly—can have much worse results. This strategy promotes confrontation, which is likely to stimulate action by domestic protectionist groups. Japanese resistance to American demands for textile export restraints in the late 1960s provided a good example of the costs of such confrontation. For almost three years, the Nixon administration insisted that Japan (as well as Hong Kong, Korea, and Taiwan) limit exports of synthetic and wool textiles. At times, the issue even dominated the agenda of U.S.-Japanese relations. It took a close defeat of a massive protectionist program (the 1970 Trade Act), a Nixon threat to invoke the "Trading with the Enemy Act," and implicit linkages to America's reversion of Okinawa to Japan before the Japanese finally came to terms (Destler et al., 1980).

The Economic Component. If confrontation were necessary to preserve essential economic interests, it might be worthwhile for the exporting country. But this is generally not the case. Indeed, accepting certain forms of protectionism may in the long run be *beneficial* for the economic development of a NIC. Myopic protectionism, oriented toward short-term political problems, by the importing state can have a twofold impact on bilateral trading patterns: First, it can impede adjustment in the importing country by encouraging production in slow growth markets; second, it can serve as a signal to the exporter to allocate resources more efficiently, diversify markets, and upgrade product lines. In the absence of trade barriers, a profitable NIC manufacturer will normally continue production in established product lines until it has fully exploited the available productivity gains. There is little incentive to shift production in the short run.

The calculus of profitability, however, changes with protection. If quantitative restrictions are imposed, exporters will need to sell fewer goods for more money, and will therefore need to improve the quality of their exports. "Trading up" is standard wisdom in international trading circles, particularly for those exporters that produce low-quality, labor-intensive goods. This effect will be reinforced if controls are imposed on slowly growing market segments, but not on more dynamic sectors. In this case, NIC producers will have even stronger incentives to upgrade their exports rapidly, which will, in the long run, benefit their growth. NIC governments, meanwhile, will be able to promote this upgrading by pointing to the protectionist measures as making these steps necessary in the short run as well as desirable in the long run. Although exporting governments may have planned to upgrade before protectionism, the external stimulus can help NIC leaders overcome any domestic resistance to adjustments as well as accelerate the diversification process.

Cases abound in which producers have adjusted successfully to restrictions by following such a strategy. The best illustration is the pattern of exports that followed the implementation of the Long-Term Arrangement Regarding International Trade in Cotton Textiles (LTA). The common assumption about the LTA is that it was a very restrictive regime that severely damaged developing countries' economies. Upon closer scrutiny, however, a more complex picture emerges.

The world cotton textile market was growing relatively slowly as early as the mid-1950s. Between 1955 and 1963, consumption of natural fibers increased only by 1.8% a year, while the rate of growth for synthetic fibers was 22.5% (Lynch, 1967: 178). Even though the LTA restrained (perhaps unfairly) exports in cotton products (the slow growth market), it created an incentive for exporting countries to shift to synthetic fibers (the dynamic market segment). As a result, the imports of such fibers into the United States increased more than tenfold between 1962 and 1970, with Japan, Taiwan, Korea, and Hong Kong accounting for almost 90% of the non-European total (USITC, 1978: vi). It is interesting to note that Hong Kong was among the slower of these four to enter the synthetic market. This can largely be attributed to Hong Kong's substantial cotton quota and a relative lack of incentive to diversify. Thus, despite the supposed restrictiveness of the LTA, overall textile exports from these four countries increased

Table 12.2 U.S. Imports of Japanese, Taiwanese, Korean, and Hong Kong Textiles and Apparel Under the LTA

	Japan		Hong Kong		Taiwan		Korea	
	Cotton	Synthetic	Cotton	Synthetic	Cotton	Synthetic	Cotton	Synthetic
1962	351	110	269	n.a.	85	n.a.	11	n.a.
1963	304	126	257	n.a.	35	n.a.	35	n.a.
1964	324	164	264	11	47	14	34	3
1965	404	310	293	19	52	25	26	18
1966	412	445	353	39	61	33	24	28
1967	376	352	354	74	69	60	30	64
1968	391	435	401	99	71	123	37	137
1969	395	585	413	145	61	238	36	212
1970	330	775	377	188	66	350	39	254

(in millions of square-yard equivalents)

Source: USITC, 1978.

dramatically in response to the restrictions (see Table 12.2).

The pattern of upgrading product lines and increasing gains in response to export restraints has also been exhibited in the shoe industry. In June 1977, the United States negotiated OMAs with Taiwan and Korea in "nonrubber" footwear. For the year immediately preceding the agreements, Taiwan exported $395.2 million worth of shoes to the United States, while Korea shipped exports worth $321.1 million. In the year ending December 1978, Taiwanese footwear exports rose to $589.9 million, and Korean shoes were valued at $437.1 million (Brimmer & Co., 1979). In other words, within eighteen months Taiwan had increased the overall value of its exports by 49% and Korea by 36%, despite restrictions. Part of the increase can be attributed to inflation and rising wage costs, but as Table 12.3 illustrates, a great deal of upgrading has also occurred. The Korean government, in particular, took an active role in pushing up the value of its exports. When the restrictions were about to be implemented, they set up an "incentive system" to upgrade their existing product lines. Simultaneously, they established an "export check price formula" to endure minimum prices on all shoes exported to the United States, and they have increased these floor prices almost every three months (*Korea Herald,* January 11, 1977).

It is important to mention that this strategy of focusing on long-term gains from trade is not without risks. First, agreeing to accept protectionism should come only when restrictions are inevitable. An importing country may go through the motions of seeking protection to appease domestic interest groups without ever intending to pursue it. Therefore, concessions by the developing nation under these circumstances would be premature. Second, accepting protectionism in one sector can establish a precedent for restricting other sectors in the future. To increase overall gains, developing states want to avoid a spillover into new sectors. Moreover, if restrictions become more comprehensive, the economic benefits diminish over time: there will no longer be the prospect of diversifying. This has happened in textiles, where restraints have gradually been extended to all fibers and most product types.

These potential dangers, however, do not outweigh the benefits of this approach. It is critical that NICs maintain good relations with major importers. The costs to the exporting nation of the collapse of trade regimes are extremely high. Furthermore, over the long run NICs must be prepared to give up their market shares in labor-intensive protected sectors. If their economic growth is to continue, they cannot indefinitely rely on textiles, footwear, and similar products in their foreign trade.

Table 12.3 Percentage of Total U.S. Nonrubber Footwear Imports

Price Brackets	Taiwan July 1977	Taiwan Dec. 1978	Korea July 1977	Korea Dec. 1978	World July 1977	World Dec. 1978
Up to $1.25	39.9%	0.8%	11.5%	0.05%	24.1%	9.6%
$1.26-$2.50	30.9%	23.2%	13.1%	1.06%	19.2%	15.7%
$2.51-$5.00	22.5%	66.6%	62.3%	14.9%	29.6%	33.5%
$5.01-$8.00	6.2%	8.3%	9.8%	59.6%	13.1%	21.4%
Over $8.00	0.02%	1.1%	3.2%	24.4%	13.7%	19.8%

Source: Volume Footwear Retailers of America Publication, Bulletin #7, February 23, 1979.

Core Policy II: Bargaining

The essence of a long-term strategy is to avoid overspecialization, encourage export diversification, and simultaneously minimize political tensions. This is consistent with Schumpeter's conception of industrial growth as involving "creative destruction." Economies must continually adjust, supplanting the old with the new. Yet long-run plans must be consistent with short-term needs. Economies need time to adjust to these changes. The more severe are the short-run dislocations, the harder is the long-run adjustment. The dilemma for the NICs is how to meet short-run requirements without jeopardizing long-run objectives.

This dilemma can be resolved for a NIC if an importing state is willing to bargain. Ironically, this is another "advantage" of the new protectionism. In the past, tariff and quota barriers to trade were implemented unilaterally, leaving an exporter with little choice. Today, even when a NIC decides that it has to accept a VER or an OMA, it can try to obtain its short-run objectives through the bargaining process. Bargaining allows an exporter to influence the nature of an agreement and therefore reduce the severity of its short-run impact. Moreover, protectionism can be more restrictive in form than in substance, and effective negotiating can help to assure this result.

The problem, of course, is that, as weak states, NICs have relatively little leverage. They will often have the least control over the most basic issues in contention, such as the aggregate levels of imports or the scope of protection. Exporters may occasionally be able to stall the negotiations to provide time to raise their export levels and/or stockpile future supplies. But in most instances aggregate numbers are highly visible and are the most obvious target of attack for dissatisfied domestic interests.

This does not mean, however, that industrializing countries are helpless. Within these constraints, two bargaining tactics can be employed, separately or in combination, to reduce the restrictiveness of protection and meet an exporter's short-term requirements. The most important of these negotiating policies is trying to foster ambiguity in the agreement as well as bargain for flexibility. To supplement this approach, a NIC may be able to link unrelated issues to the negotiations tiations at hand to compensate for possible losses.[6]

Figure 12.1   Welfare Effects of Protection

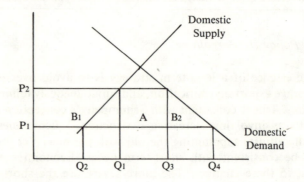

Q₁Q₃ = imports after quotas
Q₂Q₄ = imports before quotas
P₁   = price of imports before quotas
P₂   = price of imports after quotas
A   = scarcity rent as a result of quotas (or tariff of P₂P₁)
B₁,B₂ = deadweight losses

Source: C. Fred Bergsten, Towards a New World Trade Policy, 1975.

---

6. There is a third element of the bargaining strategy that is technical and relates specifically to OMAs and VERs. Through the bargaining process, an exporting country can try to assure that it captures any scarcity gains that might be generated by protection. Under an export restraint arrangement and other quantitative restrictions, prices usually increase because of forced or anticipated reductions in supply. Unlike the situation under a tariff, windfall profits are not automatically absorbed by the importing government. Who captures the scarcity gains will partly depend on the relative concentration of market power between exporters and importers, partly on the market-sharing arrangements the exporting nation can bargain for, and partly on any abnormal shifts in supply and demand that may occur.

Figure 1 depicts the situation in formal terms in partial equilibrium. A tariff of (P₂–P₁) would produce a net transfer equal to area A to the importing government. A quota of Q₁Q₃ will, given the domestic supply and demand curves, have the same effect on the price of the goods, but the scarcity rent, A, will not accrue to the government. If the

Flexibility and Ambiguity. The key to a weak state's bargaining success is to negotiate for loopholes. Appearances can be deceiving in politics. A restrictive arrangement on paper may be an unworkable arrangement in practice. Therefore, the optimal agreement for an exporting state to seek is one that is impermeable and restrictive in form (thus satisfying potential political problems) but porous and unrestrictive in substance——that is, an agreement so filled with loopholes that it is impossible to implement.

In protectionist accords, ambiguity and flexibility are the most important loopholes desired by an exporting country. Ambiguity is the ally of a restricted party because an ambiguous agreement, like ambiguous legislation, is difficult to implement. And unlike negotiating over numbers, bargaining for ambiguity is much easier for an exporting country, particularly in agreements with the United States. Sometimes ambiguity will be mutually agreed upon, and at other times it will be an unintended consequence of the bargaining. But if NICs make ambiguity a priority, many issues, particularly technical and complex ones, can be easily clouded in the negotiation process. Furthermore, definitions of products can provide sizable loopholes for exporters. U.S. customs definitions are often so antiquated and/or complicated that they verge on being useless. Thus, classifying a shoe as rubber or nonrubber, a fabric as cotton or synthetic, or a television set as a subassembly or less than subassembly can be as important to the eventual exports figures as base levels, growth rates, and numerous other clauses that a NIC cannot control.

In both textiles and footwear, product definitions have played a critical role in determining exporters' success. Under the LTA, for example, textiles had to be at least 50% cotton by weight to be restricted. As mentioned earlier, this allowed for substantial gains, far above what could have been realized with higher cotton quotas. Similar loopholes were also apparent in the recent footwear OMAs. Taiwan and Korea were careful to limit the restrictions to nonrubber shoes. By adding rubber to footwear soles or making other minor alterations, both countries found that exported shoes could be classified as rubber and thus

foreign producers do not fear a loss of market shares from other suppliers, they can be expected to raise prices to absorb the rent for themselves. If exporters fear such a loss from other foreign suppliers, and the U.S. importers are relatively concentrated, the importers will absorb the rent. Last, if both foreign producers and importers are worried about a loss of market shares from the possible entry of domestic firms, prices would remain low. If a government such as Taiwan or Korea takes an active role in fostering exporters' concentration or bargaining with importers, this will tend to transfer more of the windfall gains to the exporters, even in a market characterized by a high degree of importers' concentration. See William Cline (1979).

legally outside of the quota's jurisdiction.

Flexibility provisions can have advantages similar to ambiguity. Trade agreements usually include some limitations on switching between categories, borrowing against future quotas, and carrying over unused quotas to future years. When these flexibility clauses are too restrictive, an exporting country may be unable to meet aggregate quotas. If market conditions change quickly, rigid restraints will inhibit effective responses. But generous flexibility not only will overcome these problems; it also may allow an exporter to compensate for low aggregate limits. By borrowing against future years and exploiting various other technicalities, an exporting country can sometimes overship allotments legally and find a great deal of freedom to maneuver. Moreover, like ambiguity, bargaining for flexibility is feasible for a weaker country. In many cases, flexibility provisions are politically less salient than issues such as overall quotas.

In the footwear OMAs, for example, both Korea and Taiwan were able to negotiate for significant flexibility in their accords. American compromises included a pipeline clause (which allowed each country an additional 27% of the first year's restraint to enter the United States without being considered in violation of the quota), shorter duration for the treaties, and late starting dates for enacting the agreements, above and beyond borrowing and carryover clauses. The pipeline and carryover provisions, plus an ambiguity concerning the definition of the exporting dates, severely lessened the quotas' short-run impact.

A few statistics will illustrate the point. In 1976, the Koreans exported 44 million pairs of nonrubber shoes to the United States, and they stated in January 1977 that their total export capacity for that year was 60 million pairs. The United States sought to roll back Korea's 1976 total as far as possible and, in fact, reduced Korea's quota level to 33 million pairs for the year ending June 1978. However, by using their various flexibility clauses, the Koreans were able to export 58 million pairs of shoes during 1977 (U.S. Department of Commerce, 1978: 26-27).

The point of this discussion is that a variety of opportunities for an industrializing country are built into the bargaining process. The greater the success of the NIC in blurring the clarity and consistency of a protectionist accord, the less likely is the country to suffer short-term losses. And bargaining, even very tenacious bargaining, does not jeopardize long-run goals. It is part of the rules of the game.

The biggest danger of this strategy is that the importing state will later close loopholes, leaving an exporting nation with low quotas and no other compensation. Earlier concessions on overall limits in exchange for flexibility and ambiguity could therefore have a devastating impact. Yet, fortunately for the NICs, the technical complexity of most manufacturing trade is too great to plug all the loopholes. Even if agreements become complex and comprehensive in labor-intensive sectors, more loopholes are likely to open up. It is rare, indeed, to find a truly restrictive agreement, completely devoid of some exploitable ambiguity.

Supplemental Policy I: Issue Linkage

Pursuing long-run objectives and bargaining for ambiguity and flexibility are the core of a weak state's strategy, because in combination they should always lead to an increase in gains from trade, preservation of the bilateral trade relationship, and a minimally restrictive protectionist accord. Yet an exporting country need not stop here. Under certain conditions, a NIC may be able to link issues, cheat, and mobilize transnational and transgovernmental allies to make additional profits. Although these strategies entail higher risks than the core policies and they cannot be used at all times, they can still be useful in supplementing a long-run-oriented, loophole-seeking approach.

Issue linkage is a profitable tactic when bargaining for loopholes does not suffice to yield an acceptable accord. If the NIC can broaden the arena of negotiation to include other issues, it may be able to use leverage from other areas to counterbalance weakness in trade. Yet linkage strategies are risky for a NIC. Once an exporter broadens the bargaining agenda, a Pandora's Box is potentially opened: both sides have the opportunity to introduce new issues into the negotiations. Thus, linkages can only be profitably pursued under certain conditions. All other things being equal, the more broadly the importing country defines its interests, outside of narrow protectionist ones, the greater is the developing state's potential leverage. When the importing country is a hegemonic state, for example, it usually identifies its interests with the interests of the trading system (Krasner, 1976; Kindleberger, 1973). Unlike an importing country that is worried exclusively about its domestic industry or nationalistic goals, a hegemonic state is more likely to give side payments to cushion the impact of a particularistic policy.

The best linkage opportunities exist when the importing state has strong bilateral ties to the exporting country. The United States has military bases, intelligence operations, and other important stakes all over the world. If a NIC has a relatively narrow range of interests, and controls one of these vital assets, its bargaining power is enhanced. The more "powerful" country is more constrained in using its overall capabilities for narrow gains, while the exporting country has additional bargaining chips for advancing its own interests.

Several types of linkages, many of them implicit, have been employed historically by NICs confronted with protectionist demands. During the 1960s, for example, one of the reasons that the United States was unwilling to push Hong Kong harder than it did on textile limits was the American need for the intelligence post used to monitor the People's Republic of China. The most successful known case of linkage was devised by the Koreans in 1971. As part of the ROK's agreement to limit multifiber textiles, the United States promised Korea over $700 million in P.L. 480 agricultural aid. This was an explicit tradeoff to offset Korea's projected costs. According to the principal negotiator, David Kennedy, some P.L. 480 aid was planned before the Korean linkage, but the amount was nonetheless increased. When there was a chance a few years later that all of the aid would not be delivered, the Koreans reversed the linkage back to textiles. They insisted that their quotas be readjusted upward if the promise could not be fulfilled.[7]

Although linkages can have highly lucrative payoffs, it must be remembered that linkages are hazardous. In most instances, they require a developing state to operate on the edge of brinkmanship. Moreover, the conditions for linkages do not remain constant over time. If the global and bilateral commitments of a hegemonic state shrink, there is a greater likelihood that the most powerful state will use its own linkage strategies if the NIC helps to create the opportunity. Hence, linkages are best used selectively and only as a last resort.

Supplemental Policy II: Cheating

Once the bargaining has been completed (and assuming a VER or OMA is signed), importers and exporters move into the phase of implementation. At this point, exporting countries can pursue an option rarely mentioned in the political science or economic literature: they can cheat! Cheating in this context refers to the evasion of regulations: the achievement of some outcome through a process not explicitly permitted by the rules of the game. Operationally, cheating extends along a spectrum from legal evasion, or circumvention, to

---

7. This information was received through the Freedom of Information Act in documents released by the Department of Agriculture. Former Secretary of the Treasury and Ambassador-at-Large, David Kennedy, confirmed most of the details in an interview on November 11, 1979.

illegal evasion, actions clearly specified as improper by an agreement or statute.[8] Although cheating is an integral part of an exporter's strategy, and despite Machiavelli's advice many years ago that "men from obscure conditions... have succeeded by fraud alone," cheating has remained one of the most neglected arts in the study of politics (Machiavelli, n.d.: 319).

Circumventing restrictions is an ancient art in international trading circles. Exporters and importers have found numerous ways to sidestep bilateral quota restrictions dating at least back to the interwar period (Heuser, 1939). One of the easiest forms of evasion is transshipment (exporting to an uncontrolled third party en route to the final destination). Another common method is to substitute a freely exportable or half-finished good for the restricted product. When a cotton product is protected, a producer simply adds a little wool and it is no longer "cotton." Dutch, Swiss, and Belgian exporters were well known for such tactics in the 1930s.

In textiles, apparel, and footwear, transshipment has become the classic quota dodge. Under global quantitative restrictions or tariffs, transshipments would be difficult or unimportant. With a string of bilateral arrangements, however, they are virtually impossible to prevent. For obvious reasons, hard data on the extent of transshipment are unavailable. Yet, in a series of articles recently published by the U.S. textile and apparel industry newspaper, it was suggested that hundreds of thousands of dozens of various categories of apparel, worth millions of dollars, are shipped every year from Asia in direct violation of U.S. quotas (*Daily News Record*, October 29, 1979). The cost of buying false papers in Hong Kong, for example, is about one-third the cost of buying legitimate quotas. Goods can then be shipped in massive quantities to Indonesia or Sri Lanka, which are quotafree ports, and relabeled before they are sent off to the United States. Transshipment was also immediately apparent after the footwear OMAs. Taiwanese companies shipped footwear parts to Hong Kong for assembly, which resulted in Hong Kong's shoe exports jumping 225% in 1978.

Under some conditions, exporters may have little incentive to circumvent quotas. Prices generally increase under a quota, and the net results may benefit exporters, particularly if these firms have other markets available or can switch into more sophisticated products not covered by import restrictions. Incentives to cheat arise when losses

8. Even illegal and legal circumvention are not always easy to distinguish: what is legal or illegal is often a matter of perception. The legality of a particular action in politics is easy to identify only at the extremes.

from quotas exceed price benefits. Low quotas may lead to underutilized capacity in the short run, inducing firms to cheat temporarily while restructuring production. If there is concern about elasticity of supply, exporters may not receive high enough prices as a result of quotas; in this case, both importers and exporters will have a strong incentive to cheat. Finally, when restrictions become comprehensive, cheating may be the only way to avoid surplus capacity. As one observer commented, it is "economic forces, and not an aversion to the principles of international law," that are at the root of cheating (Fisher, 1971: 433).

Cheating can also be encouraged or tacitly approved by the restricting party. The importing country may realize that strict enforcement is oppressive and unfair. Cheating can thereby satisfy legitimate grievances of the restricted nation and simultaneously keep the protectionist system largely intact. During the Vietnam War, for example, the United States purposely allowed textile and apparel overshipments to fill surplus demand (*Daily News Record,* March 14, 1966).

While some cheating is necessary and beneficial to both the restricted and the restricting nations, cheating in large doses can be dysfunctional. If a NIC attempts to circumvent quotas too often or too much, a breakdown in the system is likely to occur. Moderate circumvention can keep a protectionist arrangement within politically acceptable bounds. But excessive cheating can lead to such rapid losses that the United States has little choice but to crack down. Instead of reinforcing short-term profits, too much circumvention can result in a more restrictive system. In 1972, Hong Kong threatened to sabotage its recently negotiated multifiber agreement when it overshipped synthetic-fiber knit textiles.[9] This led to a mini-crisis in U.S.-Hong Kong trade relations, which was not settled for several months. Treaties and laws no longer function as a guide to behavior when circumvention goes beyond tolerable levels.

The greatest practical problem with circumventing quotas is that the efficacy of the strategy declines over time. The textile lobbies in the United States have put increasing pressure on the American administration to prevent transshipment and to tighten loopholes. The Special Trade Representative's Office responded to this pressure in March 1979 with a statement now known as the Textile White Paper. Under the heading of "Law Enforcement," the government

---

9. This information comes from the *Daily News Record,* May 5, 1972; interviews with Ron Levin and Don Foote, Department of Commerce, Textile Division, October 12, 1979; and an interview with Stanley Nehmer, former Deputy Assistant Secretary of Commerce for Resources, October 11, 1979.

stated its intention to "dramatically improve the administration's enforcement of all our textile agreements" (STR Press Release 302, 1979: 3-4). This included improved monitoring of transshipments, stronger regulations on certificates of origin, and fewer exceptions to allowing overshipments. The ability of the United States to achieve these goals is questionable; but its renewed efforts in this realm indicate some of the dangers associated with continued cheating.

Supplemental Policy III: Transnational and
Transgovernmental Ties

In its struggle to shape the form of protectionism, a NIC is not limited to formal diplomatic channels. This is particularly true in the United States, since the American government is rarely united on protectionist issues, and the U.S. bureaucracy is relatively open to transnational and transgovernmental contacts. The State Department and Treasury will usually argue for free trade; Labor and Commerce normally insist on important restraints; and when the Special Trade Representative's office is involved, it often takes a middle ground. Strong lobbying with sympathetic agencies can reinforce bureaucratic splits. Such lobbying can be particularly effective if the NIC forms coalitions with organized domestic interest groups such as trading associations, or with multinational corporations (Keohane and Nye, 1974).

Transnational and transgovernmental allies can provide numerous services for a weak state. First, transnational alliances can constrain American protectionist policies by exacerbating Congressional-executive ties. Coalitions with free traders in Congress can affect legislation or limit an administration's freedom of action. Second, transgovernmental ties can exploit differences within the American bureaucracy. If bureaucrats disagree with their government's negotiating position, they can help the exporter by leaking bargaining instructions, arguing for bargaining concessions, and the like.

In 1977, for instance, the chief negotiator for the footwear agreements, Stephen Lande of the Special Trade Representative's Office, allegedly leaked the U.S. negotiating instructions to the American Importers Association, which passed them on to Korea and Taiwan. Lande had hoped that knowledge of the American position by the NICs would make the negotiations easier. In fact, the evidence seems to suggest that this allowed both countries to push the U.S. negotiators to the limit of their instructions, and even beyond on certain points.[10]

10. This information about Stephen Lande was conveyed to the author by Thomas

Finally, transnational and governmental connections can ease some of the burdens of implementation. Friends in "high places" can warn of an impending crackdown, thus warding off costly embargoes. The American Textile Manufacturers Institute, the most prominent textile lobby in Washington, has accused the State Department of notifying Asian governments that their quotas are about to reach their limits. The State Department then supposedly makes an arrangement for these nations to submit requests for quota adjustments that are to be reviewed "favorably."[11]

Transnational and transgovernmental lobbying efforts, however, cannot be pushed too far. Dramatic efforts, such as those pursued by Japan, Hong Kong, and others late in the 1971 textile negotiations, are likely to politicize an issue, strengthening the resolve of higher-level officials. If the importing state's decision makers perceive the foreign government directly "meddling" in its internal affairs, it can unite an otherwise divided government. A transnational and transgovernmental approach is best viewed as a supplementary means to other tactics for subtly influencing America's domestic policy process.

The NICs, Protectionism, and the 1980s

The ideal typical strategy for increasing NIC's trade earnings under VERs and OMAs is briefly summarized in Table 12.4. East Asian exporters did not always follow this long-term-oriented, loophole-seeking approach; but when they did, they usually exploited the political and economic weaknesses in textile, apparel, and footwear protectionism. The question remains, however, of whether these strategies can continue to work for the "gang of four" and other rapidly developing countries confronted by protectionism in the decade ahead. In the past, the United States followed a less than maximizing set of protectionist policies, which was a critical element in NIC success. If American interests become less free-trade-oriented, and interest groups within individual sectors learn from their previous mistakes, future profits for the NICs may not look at bright.

Graham, former Deputy General Counsel, STR, and negotiator of the OMAs with Taiwan and Korea, September 4, 1979. Mr. Graham has subsequently differed with the interpretation offered here, but he did not deny the essential facts. Mr. Graham felt that the United States could have extracted additional concessions from Taiwan and Korea, and that Lande was a first-rate negotiator.

11. Interview with officials in the American Textile Manufacturers Institute, June 22, 1979.

Table 12.4 Summary of Exporting Strategy

Context	Strategies	Policies	Benefits	Risks
Prerequisite	Political sensitivity to the rules of the game	Understanding the political basis of trade	Foundations for manipulating the game in one's own favor	Setting precedents
Core Policies	Pursuing long-run gains	Making short-run sacrifices and adapting	Minimizes political tensions, avoids over-specialization, encourages export diversification	
	Bargaining	Negotiating for ambiguity and flexibility	Minimizes hort-run dislocations, provides loopholes and protectionist "image"	Low quotas, no compensation if loopholes closed
Supplemental Policies	Cross-issue linking	Manipulating inter-dependence, security and patronage	Provides added long-term and/or short-term compensation	Reversed linkage
	Cheating	Evading regulations	Minimizes short-run dislocations, provides added profits	Sanctions
	Creating transnational ties	Reinforcing bureaucratic splits	Fosters importing state's shortsightedness	Fosters importing state coherence

The Calculus of American Protectionism

OMAs and VERs have been adopted by the U.S. government as short-run policies designed to satisfy domestic political pressures. In essence, these bilateral arrangements began as little more than ad hoc compromises. Free traders, on one hand, wanted VERs to preserve a semblance of open borders; protectionists, on the other hand, were interested in using these agreements to defend domestic industries. Since these competing groups have rarely been able to agree on restrictive trade barriers, the U.S. negotiating position has invariably had some degree of flexibility. Countries that then pursued the strategies outlined in the first part of this article could exploit the weaknesses in the American approach.

As long as the United States wants to play a leading role in the international trading system, *and* as long as domestic lobbies are willing to accept OMAs and VERs, exporters may be able to beat the United States at the protectionist game. But this calculus of protectionism is not a constant. American decision makers and interest groups learn from the failures of their protectionist policies. This does not lead them toward freer trade, but toward more effective restrictions.

Within an individual sector, firms do not remain quiescent when import penetration continues to rise. .The greater is the success of exporting countries in overcoming restrictions, the greater is the effort by lobbyists to have the loopholes filled. The evolution of the international regime in textiles is a case in point: As international arrangements and bilateral accords have been renegotiated over the past twenty-five years, they have become more restrictive and sophisticated. In 1957, textile restrictions were negotiated with only one country (Japan) in thirty-two cotton-product categories (Hunsberger, 1964). In 1980, the U.S. had agreements with more than twenty countries, sometimes covering all fibers with several hundred product distinctions.[12] Today, textiles and apparel manufacturers want to go one step further. They have been calling for the dismantling of the bilateral agreement system (known as the Multifiber Agreement), to be replaced by more comprehensive quotas (*Daily New Record*, March 3, 1980).

Furthermore, the United States in the 1980s may be less willing to absorb the costs of maintaining an open trading environment (Krasner, 1979). Principles such as nondiscrimination and reciprocity may be less important to a United States that seeks to promote *particularistic* rather than global interests. More and more, industries are at a com-

---

12. For a review of the present textile quota system, see USITC (1978).

petitive disadvantage with imports in the late 1970s and early 1980s as opposed to any other period in the postwar era. Textiles and steel were the only major sectors to receive protection before 1970. In the past few years, however, footwear, color TVs, and now automobiles have been added to the list (Yoffie, 1981c). To date, each industry has been protected in isolation from the others: There has been no "system of protectionism." But as similar problems begin to recur across industries, the likelihood seems to grow that the United States will pursue more restrictive trading strategies.[13]

What predictions can one make from this evolution in United States policy? Are the NICs doomed to suffer in their foreign trade as UNCTAD thought they would in the 1970s? While it will certainly be more difficult for exporting nations to repeat their remarkable successes of the past, one need not despair. Despite a lesser commitment to free trade and narrower global interests, the United States remains constrained from adopting truly ironclad restrictions. Its interdependence with the rest of the world, commitments to allies, a vast network of transgovernmental and transnational ties, and the persistance of bureaucratic politics still make it difficult for any administration to close all the loopholes that the NICs can exploit.[14]

The changing political environment, however, will require the industrializing countries to pay even more attention to certain parts of the "success formula." NICs must continue to recognize the political basis of trade and to emphasize long-term gains. They must also bargain for as much ambiguity as possible. Their diplomats, even more than in the past, will need to lobby effectively with various branches of the U.S. government while maintaining a low public profile to avoid politicized confrontations. The "search for loopholes" must go on.

At the same time, the magnitude of NIC exports, especially to the U.S., ensures that they will increasingly be targets of political attack in the importing countries. Between 1963 and 1976, the share of world exports in manufactured goods held by Korea, Taiwan, and

---

13. In 1979 and 1980, discussions were taking place within the government about the effects of OMAs and VERs across industries. There does not appear to have been a definitive conclusion, but the discussions themselves represent a partial awakening of the U.S. bureaucracy to the problems of the new protectionism.

14. At the outset of the Reagan administration, these bureaucratic differences continue to plague policy discussions. For example, different agencies took different positions on the question of VERs in the automobile industry (*Wall Street Journal,* March 5, 1981). Furthermore, despite pressure from the footwear industry, Congress, and some bureaucratic agencies, Reagan has allowed the OMAs on shoes to lapse.

Hong Kong tripled from 1.35% to 4.1%, and in selected sectors such as footwear, their market shares are much greater (OECD, 1979: 19). In 1977, Taiwan and Korea alone accounted for 61.2% of all shoe imports into the United States, close to 30% of the entire American footwear supply (U.S. Department of Commerce, 1978). This size and concentration of NIC exports will make it increasingly difficult for countries such as Korea and Taiwan to cheat on agreements without facing retaliation, or to attempt linkage strategies without having the linkages reversed.

Conclusion

For most of the last 25 years, the exporting nations of East Asia have managed to increase their gains from trade and maintain a relatively stable bilateral trade relationship with the United States. Despite VERs, OMAs, and elaborate market-sharing arrangements such as the Long-Term Arrangement Regarding International Trade in Cotton Textiles, these relatively weak, rapidly industrializing countries have generally prospered in textile, apparel, and footwear trade. This success was accomplished by capitalizing on the weaknesses in modern protectionism and taking advantage of the United States' accommodating approach to trade barriers. Exporters did not always find the right formula, and the political obstacles to success have recently become more severe. Nonetheless, as long as the United States continues to be constrained from pursuing a rigid protectionist path, NICs can beat the protectionist game. Countries that recognize the political basis of American policy and combine a long-run orientation with short-run bargaining tactics, occasional cheating, linkages, and mobilizing transnational and governmental allies will realize the greatest gains.

On a more theoretical level, the success of the newly industrializing countries in trade raises some serious questions about the simple use of the concepts of "weak" and "strong" in international relations.[15] Less powerful nations are generally assumed to be at a disadvantage in world politics because they do not possess many tangible resources. This is why organizations such as UNCTAD believe that developing nations have little choice when confronted with protectionist demands from industrial countries. Power, according to this view, is like a vector diagram in classical mechanics. Outcomes are predicted by weighing the potential resources of the actors and finding a resultant (March, 1966).

Unfortunately, what this notion of "weak" and "strong" ignores is *politics*.[15] Weak countries do not need to overpower their opponents to achieve their objectives. Who gets what in international relations is a function of bargaining, implementation, and the use of symbols, as well as coercion. Therefore, in trade as well as in other issue areas, nations with few perceivable assets can overcome their "objective" weaknesses if they are skillful in negotiation, can manipulate political symbols, and are cunning enough to circumvent restrictions. What the NICs have been able to show us is that states that emphasize bargaining rather than coercion, adjustment rather than intransigence, and the substance of international agreements rather than their forms, can capitalize on their strengths and turn adversity into advantage.

References

Bauer, R., I. de Sola Pool, and L. A. Dexter (1963) American Business and Public Policy: The Politics of Foreign Trade. New York: Atherton.

Bergsten, C.F. (1975) "On the non-equivalence of import quotas and 'voluntary' export restraints," in C. F. Bergsten (ed.) Towards a New World Trade Policy: The Maidenhead Papers. Lexington, MA: D.C. Heath.

Blackhurst, R., N. Marian, and J. Tumlir (1978) "Adjustment, trade, and growth in developed and developing countries." GATT Studies in International Trade 6.

_____ (1977) "Trade liberalization, protectionism and interdependence." GATT Studies in International Trade 5.

Brimmer & Co. (1979) Trends in the Demand for the Supply of Non-Rubber Footwear. New York: Author.

Caporaso, J. (1978) "Dependence and Dependency in the Global System." International Organization 32 (Winter), special issue.

Cline, W. (1979) Imports and Consumer Prices: A Survey Analysis. American Retail Federation.

Destler, I. M., H. Fukui, and H. Sato (1979) The Textile Wrangle: Conflict in Japanese-American Relations, 1969–71. Ithaca, NY: Cornell Univ. Press.

Fisher, B. (1971) "Enforcing export quota commodity agreements:

---

15. The concepts of "weak" and "strong" should not be confused with Peter Katzenstein's (1976) definition, which refers to the structure of a country's policy network. "Weak" refers to relatively less powerful nations with few resources; "strong" refers to countries that possess many tangible capabilities.

the case of coffee.'' Harvard International Law J. 12 (Summer): 401-435.

Fox. A. B. (1958) The Power of Small States. Chicago: Univ. of Chicago Press.

Heuser, H. (1939) Control of International Trade. Philadelphia: P. Blakiston's Sons.

Hirschman, A. (1945) National Power and the Structure of Foreign Trde. Berkeley: Univ. of California Press.

Hunsberger, W. (1964) Japan and the United States in World Trade. New York: Harper & Row.

Katzenstein, P. (1976) ''International relations and domestic structures: foreign economic policies of advanced industrial states.'' International Organization 30 (Fall): 1-47.

Keohane, R. (1971) ''The big influence of small allies.'' Foreign Policy 2 (Spring): 161-182.

_____ and J. Nye (1974) ''Transgovernmental relations and international organizations.'' World Politics 27 (October): 39-62.

Kindleberger, C. (1973) The World in Depression 1929–1939. Berkeley: Univ. of California Press.

Knorr, K. (1975) The Power of Nations: The Political Economy of International Relations. New York: Basic Books.

Krasner, S. (1979) ''The Tokyo round: particularistic interests and prospects for stability in the global trading system.'' International Studies Quarterly 1. 23 (December): 491-530.

_____ (1976) ''State power and the structure of international trade.'' World Politics 28 (April): 317-347.

Lynch, J. (1967) Towards an Orderly Market: An Intensive Study of Japan's Voluntary Quota in Cotton Textile Exports. Tokyo: Sophia Univ. Press.

Machiavelli, N. (1950) The Prince and the Discourses. NY: Modern Library.

March, J. (1966) ''The power of power,'' pp. 39-70 in D. Easton (ed.) Varieties of Political Analysis. Englewood Cliffs, NJ: Prentice-Hall.

Morton, K. and P. Tullock (1977) Trade and Developing Countries. New York: John Wiley.

Organisation for Economic Cooperation and Development [OECD] (1979) The Impact of the Newly Industrializing Countries on Production and Trade in Manufactures. Paris: OECD.

Smith, M. (1973) ''Voluntary export quotas and U.S. trade policy—a new nontariff barrier.'' Law and Policy in International Business 5, 10: 10-54.

Special Trade Representative's Office [STR] (1979) press release

302: 3-4.

Strange, S. (1979) "The management of surplus capacity: how does theory stand up to protectionism, seventies style?" International Organization 33 (Summer): 303-334.

Thucydides (1951) The Peloponnesian War. New York: Modern Library.

United Nations Conference on Trade and Development [UNCTAD] (1978) "Growing protectionism and the standstill on trade barriers against imports from developing countries." Geneva, TD.B.D.2/194 (May).

U.S. Department of Commerce (1978) Footwear Industry Revitalization Program: First Annual Progress Report.

U.S. International Trade Commission [USITC] (1978) The History and Current Status of the Multifiber Arrangement. USITC Publication 850 (January).

Wriggins, W. H. (1976) "Up for auction: Malta bargains with Great Britain," in I. W. Zartman (ed.) The 50% Solution. New York: Anchor.

Yoffie, D. B. (1981a) "The advantages of adversity: weak states and the political economy of trade." Ph.D. dissertation, Stanford University.

———— (1981b) "Orderly marketing agreements as an industrial policy: the case of the footwear industry." Public Policy 29 (Winter): 93-119.

———— (1981c) "Reagan's mythical auto restraint agreement." Wall Street Journal (May 18).

Zartman, I. W. (1971) The Politics of Trade Negotiations Between Africa and the EEC: The Weak Confront the Strong. Princeton, NJ: Princeton Univ. Press.

CHAPTER 13
Conclusion: A Dissenting View on the Pacific Future

Chung-in Moon

Economic dynamism, strategic interactions, and cultural vitality have contributed to the growing importance of the Pacific Basin. But the emergence of the Pacific as a new center of economic activity portends conflict as well as cooperation. Increasing intra-regional economic interdependence, shaped and nurtured by U.S. hegemony, has brought countries in the region closer, and produced new forms of consultation. On the other hand, as several authors in this volume have demonstrated, rapidly expanding economic transactions have bred trade frictions and adjustment difficulties. Discord over economic issues threatens to spill over into security relations which have otherwise been strong and congenial. The region is torn between harmony and discord and between universalistic and particularistic interests. The tension raises fundamental questions regarding the future of the Pacific economy.

In the introductory chapter, Stephan Haggard projects a cautious, but optimistic outlook for the Pacific relations. Haggard concurs with the other contributors that the protectionist mood is increasing in the U.S. and that trade frictions have been on the rise. He argues, however, that these adjustment difficulties are not insurmountable and that economic interdependence will only deepen. This optimistic projection stems from three factors. First, high politics favors continued close cooperation among nations in the region. The strategic ties that bind the U.S. and its East Asian allies are still alive and well, providing the basic political structure for continuing economic cooperation. Second, protectionist coalitions are not the dominant social force in shaping the nature and direction of trade policy. Liberal forces are still well entrenched in the U.S., and are growing in Japan, Korea and Taiwan. The internationalization of business has contributed to this trend on both sides of the Pacific. Finally, American policy initiatives are

I would like to thank Stephan Haggard for his comments.

creating a new momentum for the institutionalization of a more liberal economic regime in the region. U.S. bilateral trade pressure, based on the principle of specific reciprocity, has facilitated liberalization of trade in goods and services as well as investment in its East Asian trading partners. Bilateralism runs certain risks, as Haggard and Cheng (chapter 11) argue, but it need not mean a decline into retaliatory trade wars. The rhetoric of trade politics must be separated from the reality of continued and deepening trade, investment and financial ties.[1]

I agree with the main thrust of his projection. Certainly the Pacific economy has become internationalized more than ever before and economic liberalization may well be irreversible. I argue, however, that growing internationalization and liberalization do not necessarily bring about cooperation, and that the Pacific economy faces formidable political challenges. First and foremost, the newly emerging pattern of sectoral competition, coupled with segmented corporate strategies and industrial policies, diminishes the room for cooperation among nations in the region and draws the state more closely into industrial competition. Domestic structural rigidities, resulting from both institutional inertia and the dynamics of national politics, will pose stumbling blocks to positive adjustment, and place countries in the region on a collision course. Moreover the increasing political weight of distributional coalitions in the wake of recent democratization in the Asia-Pacific countries will reinforce economic nationalism, and undermine the commitment to a liberal economic order. Chances for collective management and coordination of economic differences on the regional level are also limited. The Asia-Pacific region is too fragmented and heterogenous to reach a regionalist solution. Finally, a new regional politics, characterized by waning U.S. hegemony, weakening intra-regional security ties, and the assertive entry of new powers will preclude the continuation of past patterns of cooperation between the U.S. and its Pacific partners.

---

1. See Mann-Kyu Kim, (ed.), *The Pacific Century: Trade, Development and Leadership* (Inchon: Center for International Studies, Inha Univ., 1988); Robert Downen and Bruce J. Dickson (eds.), *The Emerging Pacific Community* (Boulder: Westview, 1984); Staffan B. Linder, *The Pacific Century* (Stanford: Stanford Univ. Press, 1986); P.N. Nemetz, (ed.), *The Pacific Rim: Investment, Development and Trade* (Vancouver: University of British Columbia Press, 1983); Roger Benjamin and Robert T. Kurdle, (eds.), *The Industrial Future of the Pacific Basin* (Boulder: Westview Press, 1984); Lawrance Krause and Sueo Sekiguchi (eds.), *Economic Interaction in the Pacific Basin* (Washington, D.C.: Brookings, 1980); James W. Morley, (ed.), *The Pacific Basin:New Challenges for the U.S.* (New York: Academy of Political Science, 1986).

Sectoral Conversion, Industrial Disorder and New Competition

In the traditional model of an international capitalist division of labor, economic transactions take place at the level of the firm. International, regional, and national political factors can be viewed as ecological variables influencing market interactions. Thus, the core of the Pacific adjustment problems lie at the sectoral and microeconomic level and involve corporate strategies and industrial policies.

Sectoral adjustment difficulties in the Pacific region are a relatively new phenomenon. The past pattern of regional economic development was rather smooth and complementary. While the U.S. was the innovator and the pace setter, Japan followed the American lead, exploiting lower labor costs and emulating its technological and process innovations. As Gereffi and Wyman (chapter 2) illustrate, the East Asian NICs followed in Japan's footsteps. The region as a whole developed through cascading "follow the leader" policies. This "flying geese" pattern of regional development not only fostered intraregional interdependence, but also prevented sectoral frictions. Shifting comparative advantage operated relatively smoothly to allocate investment activities.[2]

The traditional division of labor does not appear to work as smoothly any longer, however. Two factors account for this change: the increased speed of the product cycle as a result of technological change; and the increasing involvement of the state in industrial restructuring.[3] The "flying geese" model implies that regional industrial hierarchy was shaped largely by technology. The regional division of labor based on the differential levels of technological development has increasingly become blurred, however. Easy access to advanced technology and the integration of international financial markets have shortened the duration of the product life cycle. Furthermore, the relatively free flow of knowledge, coupled with a rapid improvement of research and development infrastructure in the Asian-Pacific countries, has helped them graduate rather quickly from the stage of imitation and

2. The "flying geese" model was coined by a Japanese economist, Kaname Akamatsu who differentiated it from a vertical division of labor in which an unequal exchange takes place between raw material and manufactured goods exporting countries. See Saburo Okita, "Pacific Development and its Implication for the World Economy," in James Morley (ed.), *The Pacific Basin: New Challenges for the U.S.,* pp. 26-27; Kaname Akamatzu, "A Historical Pattern of Economic Growth in Developing Countries," in *The Developing Economies* 1:1 (March/August 1962):3-25.
3. For an excellent discussion of sectoral competition involving the U.S., Japan, and the East Asian NICs, see John Zysman and Laura Tyson (eds.), *American Industry in International Competition* (Ithaca: Cornell Univ. Press, 1984).

assimilation toward innovation of technological frontiers. This increasing technological parity has resulted in conversion of interests in the development of certain cutting-edge, high-tech industries and thus deepened competition across the Pacific.[4]

Technology and related market forces alone have not been the only factor affecting the existing division of labor, however. The creation of "arbitrary comparative advantage" through a concerted orchestration of industrial, trade, and science and technology policies in the East Asian countries has also facilitated this process.[5] Ultimately, it is the pursuit of comparative advantage by states through aggressive trade and industrial policies that threatens the regional industrial order and fosters a new type of mercantile competition.

Technological change and state intervention have resulted in a new form of competition among firms and nations. All the countries in the region, regardless of level of development, have attempted to move into more value-added, capital- and technology-intensive industries. As the chapters on individual industries show, each country has promoted new industries such as semiconductors and computers (Hart, chapter 5), while protecting sunset industries such as steel and textiles (Moon with Chang, chapter 7). Industries in transition such as automoblile (Dunn, chapter 6) and consumer electronics have also received government supports. To paraphrase Chairman Mao, "walking on two legs," i.e., maintaining both sunrise and sunset industries, has become a new industrial motto in the Asia-Pacific countries. As a result, the logic of conventional comparative advantage has weakened, and surplus capacity has become a new reality of the Pacific economy. In contrast to the vertical, "flying geese" pattern, we now have a horizontal "swarming sparrow" pattern of development, which makes trans-Pacific economic adjustment much more complicated.

As Encarnation and Chernotsky (chapters 8 and 9) show, trade frictions, protectionism, and exchange rate instability have encouraged adjustment through cross-investment. In the past, the flow of capital in the region was from developed countries, the U.S. and Japan, to developing countries. Now, capital flows in a variety of directions. Japan and the East Asian NICs have made new investments in the U.S. and the NICs are investing in Southeast Asian countries.[6]

---

4. Peter Drucker describes this sectoral conversion as an "outflanking" syndrome, but argues that Japan and the East Asian NICs are still in the stage of imitation. See his "Japan's Choices," *Foreign Affairs* 65:5 (Summer 1987), p. 928.

5. On the concept of "arbitarary comparative advatange," see Zysman and Tyson, *American Industry*... ch. 1, and William Cline (ed.), *Trade Policy in the 1980s* (Washington, D.C.: Institute of International Economics, 1983), pp. 155-158.

6. As Chernotsky (chapter 10) notes, this new investment is a result of multiple factors:

The new investment is not without problems. Host countries are increasingly reluctant to absorb foreign capital because of fears of economic dependence. The American public has become increasingly hostile to foreign direct investment in the U.S.,[7] and the East Asian countries have proven even more resistant to the intrusion of foreign capital. As Encarnation shows, an intricate web of cozy corporate ties, coupled with public policy discrimination, limits the access of foreign investors in Japan.[8] South Korea has liberalized foreign investment to a considerable degree, but institutional constraints such as policy disincentives and anti-foreign sentiment among labor continue to limit foreign investment, which is less than in either Singapore or Hong Kong.

Public opinion in investing countries has also been divided about new investment. A new argument has surfaced in Japan, South Korea, and Taiwan that parallels the U.S. concern about "runaway shops."[9] The outflow of capital and technology through direct investment is accused of creating "donut effects" and a "hollowing out," in which the core of national economic vitality is shipped out to foreign competitors.[10] Reflecting this sentiment, the Japanese Ministry of International Trade and Industry has reinforced safeguard measures to minimize the domestic repercussions of the "boomerang effects" associated with technology transfer to competitors.[11] Jeffrey Hart (chapter 5) shows that the U.S. is also taking a more cautious and conservative posture regarding both the transfer of advanced technologies to Japan and the East Asian NICs and their direct investment in U.S. high-tech firms. Both commercial and national security interests

exchange rates, labor costs, and protectionism. The U.S. removal of the Generalized System of Preference from the East Asian NICs has driven them to invest in ASEAN.

7. This anti-foreign sentiment can be attributed in part to foreign investment in real estates in major cities of the U.S.

8. A recent episode of T. Boone Pickens, an American investor, illustrates this point well. He failed to control the board of Koito Manufacturing, Inc., a Japanese automotive issue.

9. See Robert Gilpin, *U.S. Power and the Multinational Corporations: The Political Economy of Foreign Direct Investment* (New York: Basic Books, 1975).

10. For recent debates on Japanese investment in the U.S., see Kozo Yamamura (ed.), *Japanese Investment in the U.S.: Should We Be Concerned* (Seattle: Society for Japanese Studies, 1989). "Hollowing out" effects were first raised by Yukuo Ajima, a labor union leader in Japan. See his contribution to *Nihon Keizai Shinbum* September 15, 1986. Taiwanese and South Korean mass media have also recently drawn public attention to this issue.

11. The term "boomerang effect" was coined by Miyohei Shinohara. See his *Industrial Growth. Trade and Dynamic Patterns in the Japanese Economy* (Tokyo: Univ. of Tokyo Press, 1982), pp. 71-86.

dictate this posture. Ths U.S. government has shown political sensitivity to foreign, especially Japanese, acquisitions in sectors related to defense, and have expressed a growing concren over boomerang effects that might occur as a result of competitors' commercial application of American military technology.[12]

Frictions are not limited to trade and investment. Reflecting the phenomenal growth in trade in manufacturing and a new role of foreign investment in the Asia-Pacific region, the service sector has become a new focal point of economic disputes across the Pacific (see Encarnation, chapter 8; Haggard and Cheng, chapter 11). The U.S. has pushed for the liberalization of services by placing the issue on the agenda of the Uruguay Round negotiations, and by exercising bilateral pressures on East Asian trading partners. The U.S. has an obvious reason to push for such liberalization because of its comapartive advantage. Nevertheless, the liberalization of services is not likely to be smooth. Lack of consensus on the definition of the service sector portends protracted diplomatic negotiations. At the same time, political and institutional obstacles to liberalization are considerable in East Asia. Compared with the manufacturing sector, the East Asian service sector is still in its infancy, which has been used as a justification for heavy state intervention. It is not easy to dismantle this intervention, especially in the areas of financial services and insurance which have been closely linked to macroeconomic, industrial and financial policies. Finally, state protection of the service sector has created a strong political constituency. Despite an official commitment, the Japanese government was unable to allow the participation of American firms in the Osaka Airport construction project because of intense political opposition by domestic construction firms. In a similar vein, American demand for the liberalization of insurance, financial services, and tourism markets in Korea faced immense political opposition, causing significant delays.[13]

---

12. See Takehiko Yamamoto, "Technology and National Security: The Case of Japan," *Pacific Focus* 3:2 (Fall 1988); Takashi Inoguchi, "Trade, Technology and Security: Implications for East Asia and the West," in Robert O'Neill (ed.), *The West and International Security* (London: Macmillan, 1987), pp. 197-213.

13. For an overall discussion of the service sector, see Jonathan Aronson and Peter F. Cowhey, *Trade in Services: A Case for Open Market* (Washington, D.C.: American Enterprise Institute, 1984). In the Asia-Pacific context, see Chung H. Lee and Seiji Naya (eds.), *Trade and Investment in Services in the Asia-Pacific Region* (Inchon and Boulder: Center for International Studies, Inha University, and Westview Press, 1988).

National Politics and Structural Rigidity

The economic changes outlined above—rapid technological change, cross-investment, the rise of the service sector—have created a new generation of policy problems. Yet it is national politics and institutional inertia that make these changes salient; the growing gap between the speed of market changes and the ability of affected groups to accept adjustment.

This is most clear in the U.S. In the post-war period, the U.S. broke away from protectionism and isolation, and established a liberal economic regime in the GATT and Breton Woods system. However, we are witnessing the revival of the ghosts of "Smoot-Hawley" in the 1980s. The executive branch has lost some of its control over foreign economic policy, and as Nelson argues, both legislative and administrative protectionism have become key features of the American trade policy system. Despite the lobbying efforts by free trade groups, protectionist coalitions continue to exercise enormous power and influence.[14]

The executive has responded to this pressure by taking a more aggressive trade policy stance. In 1983, the Reagan administration announced a new trade policy based on the principle of specific reciprocity and began to enforce a selective, sectoral protectionism.[15] Its rationale is to ensure "fairness" in trade by taming freeriders and spoilers through bilateral leverage, and eventually to induce greater economic liberalization. As Krasner (chapter 9) and Haggard and Cheng (chapter 11) argue, there are limits to this strategic protectionism. It may force "cooperation" by the East Asian trading partners and bring about short-term improvements in the trade balance, but these effects are exaggerated and the long-term consequences are likely to be harmful. Apart from the danger of bilateral retaliation, the application of specific reciprocity implants uncertainty and emotional animosity in the process of adjustment, and diminishes the chances of cooperation in the long term. In addition, as Yoffie (chapter 12) notes, such bilateral pressures make the East Asian countries more competitive and induce them to move away from the U.S. through trading partner diversification. Specific reciprocity is a politically inexpensive solution

---

14. See I.M. Destler, *Trade Politics* (Washington, D.C.: Institute of International Economics, 1986); I.E. Destler and John Odell, *Anti-Protection: Changing Forces in U.S. Trade Politics* (Washington, D.C.: IIE, 1987).

15. On the concept of specific reciprocity, see Robert Keohane, "Reciprocity in International Relations," *International Organization* 40:1 (Winter 1986), pp. 1-27 and Krasner's chapter in this volume.

on the part of the strong, but one that could undermine the multilateral foundation of the international trading system.

The is not to deny that the U.S. bilateral initiative has produced positive outcomes. As a result of the U.S. offensive, East Asian countries have opened their markets considerably. Such opening appears to be based on a fragile political and institutional foundation, however. Recent trade liberalization in the East Asian NICs was undertaken at bureaucratic initiative, possible largely because of the strength of the East Asian states vis-à-vis their societies. The state in the East Asian NICs has traditionally penetrated and steered the economy, and insulated economic decision-making machinery from contending social forces. Consequently, liberalization could be formulated and implemented in a rather speedy and flexible fashion.[16] But recent domestic political changes cloud the future of this liberalization. Democratization and the proliferation of distributional coalitions impose fundamental societal constraints on state action. Democratization in South Korea, for example, has produced a new balance of power between state and society. The popular sector comprising labor, farmers, radical students, and intellectuals, has become increasingly politicized, and has actively opposed import liberalization of farm products and the opening of the service sector to the U.S. Moreover, the protectionist tone of the National Assembly, now dominated by opposition political parties, complicates the situation further.[17] In Taiwan, increasing labor activism in the wake of democratization has weakened the government position toward economic liberalization.

In Japan, as Van Wolfren argues in an influential article in *Foreign Affairs*, liberalization might be more fictional than real.[18] Japan is structurally protectionist. Both Nakasone and Takeshita argued for a more open national economy, but their efforts to accommodate "outside pressure" (*gaiatsu*) were limited by reactions from a ruling coali-

---

16. On the topic of "strong state" in the East Asian development, see Chalmers Johnson, *MITI and the Japanese Miracle* (Stanford: Stanford Univ. Press, 1981); Gordon White and Robert Wade (eds.), *Developmental States in East Asia* (Brighton: Institute of Development Studies, 1985); Frederic Deyo (ed.), *The Political Economy of the New Asian Industrialism* (Ithaca: Cornell Univ. Press, 1987); Stephan Haggard and Chung-in Moon, "The Korean State in the International Economy: Liberal, Dependent or Mercantile?" in John Ruggie (ed.), *Antinomies of Interdependence* (New York: Columbia Univ. Press, 1983).

17. For economic consequences of distributional coalition in South Kore, see Stephan Haggard and Chung-in Moon, "Institutions, Policy and Economic Growth: Theory and a Korean Case," *World Politics* (forthcoming).

18. Karel G. van Wolfren, "The Japan Problem," *Foreign Affairs* 65:2 (Winter 1986/87), p. 292.

tion composed of powerful bureaucrats, political cliques, and clusters of industrialists. Contrary to conventional wisdom, the Japanese political leadership is relatively weak and ineffective in implementing economic changes facing strong opposition; Noble (chapter 3) shares this nicely.[19] The domestic political backlash produced by liberalization efforts is likely to make the Japanese policy machinery more fractious, decentralized, and directionless. Import liberalization of farm products has cost the LDP's rural support, and the imposition of a sales tax as part of a fiscal reform has weakened urban middle class support. Coupled with the Recruit and sex scandals, these attempted reforms are weakening the political power base of the ruling Liberal Democratic Party, further limiting the maneuverability of the Japanese political leadership. In short, rigidities stemming from the rise of political pluralism or divisive domestic political arrangements hinder the state's capability to formulate and implement liberalization policy in East Asia.

Equally troublesome is the disjuncture between policy and actual outcomes. In recent years, the East Asian countries announced a series of policy reforms for liberalization as a result of increasing foreign pressures. As a growing number of foreign firms complain, however, they are not being effectively translated into market outcomes. Resistance to foreign penetration is still high, and invisible barriers are extensive. This disjuncture is by and large an outcome of mercantile mindsets and extramarket institutional arrangements.[20] The protection of enfant industry has been the backbone of the East Asian "catchup" strategy, and has created a collective mercantile mindset. While the consumption of domestic products was equated with virtue and partiotism, those who prefer foreign products were branded as traitors. This group psychology remains strong and alive. Japan imports California rice, but its importers are scorned by the mass media and public. Korea opened its tobacco market, but those who smoke American

---

19. Refuting Chalmers Johnson's interpretation of Japan as a "strong, developmentalist state," several scholars began to advance a revisionist view of state and society in Japan. See Noble's excellent survey in this volume; Karel Van Wolfren, "The Japan Problem"; T.J. Pempel, "The Unbundling of 'Japan, Inc.'": The Changing Dynamics of Japanese Policy Formation," in Kenneth B. Pyle (ed.), *The Trade Crisis: How Will Japan Respond* (Seattle: Society for Japanese Study, 1987), pp. 117-152.
20. For a theoretical and succinct discussion of extramarket institutional barriers, see Stephan Krasner and Daniel Okimoto, "Japan's Evolving Trade Posture," in Akira Iriye and Warren I. Cohen (eds.), *The U.S. and Japan in the Postwar World* (Lexington: Univ. of Kentucky Press, 1989), pp. 142-143. Shumpei Kumon's "Dilemma of a New Phase: Can Japan Meet the Challenge?" in Pyle, *The Trade Crisis...*, pp. 229-240 neatly elucidates the nature of Japanese mercantile mindset.

cigarettes are considered unpatriotic.

Apart from these psychological legacies of enfant industry protection, extramarket institutions pose serious barriers to liberalization. In the East Asian countries, distribution networks are so highly concentrated that foreign firms cannot find local distributors easily. Stiff product standards delay production processes. Equally troublesome is bureaucratic redtape from local level to central government. Complex procedures, myriad of legal barriers, and often adversarial attitude of bureaucrats hinder smooth business operation. Moreover, as Encarnation shows, discrimination in terms of public policies undermines the competitiveness of foreign firms. Foreign firms cannot find their own niche in this business climate.[21] Deprogramming mercantile mindset and dismantling parochial institutional template is not an easy task.

In view of these barriers, the full institutionalization of a liberal economic order across the Pacific remains elusive. The aggressive corporate and national strategies that heighten trans-Pacific friction are backed by a mercantile "ethos" deeply rooted in the fabric of East Asia societies.

Regional Fragmentation and Limits of Collective Management

Such national differences might be resolved or at least amerliorated, if collective coordination could be institutionalized on the regional level. Inspired by the formation and continuing success of the European Economic Community, nations in the region have periodically attempted to form an Asia-Pacific economic bloc. Professor Kojima proposed the Pacific Free Trade Area (PAFTA) in 1966, and Professors Peter Drysdale and Hugh Patrick further developed the idea into a more expanded regional system, the Organization for Pacific Trade and Development (OPTAD) in the late 1970s. This forum for scholarly discussion developed a theoretical rationale for regional cooperation.[22] Japan has been most enthusiastic about the idea of regional economic integration. Since Prime Minister Ohira proposed Pacific Basin Cooperation (PBC) in 1979, Japan has played a leading role in this regard.

---

21. This is what Drucker terms "adversarial trade" as opposed to fair, competitive trade. Druker, "Japan's Choices," p. 928.
22. See Kiyoshi Kojima, *Japan and a Pacific Free Trade Area* (Berkeley: University of California Press, 1971); Peter A. Drysdale and Hugh Patrick, "Evaluation of a Proposed Asian-Pacific Regional Economic Organization," in Congressional Research Service, *An Asian-Pacific Regional Economic Organization: An Exploratory Concept Paper* (Washington, D.C.: USGOP, 1979).

South Korea has been an active advocate too. In 1982, president Chun Doo Hwan proposed a Pacific Summit Conference.

Japanese and Korean initiatives have met with lukewarm responses from other nations in the region, however. The U.S. showed an interest in the idea of Pacific integration in the late 1970s, and the U.S. Congress held extensive hearings on the matter. But the Reagan administration made it clear that the private sector, not the government, should lead any regional economic schemes. Very recently, Senators Alan Cranston and Richard Lugar introduced a bill on the establishment of a Pacific Basin Trade Forum, which would open the way for American legislators to begin talks with leaders in the Pacific Rim countries and to initiate annual summit meetings in order to establish a permanent secretariat. In a similar vein, Secretary of State James Baker also called for the establishment of a "Pan-Pacific Alliance" involving the U.S., Japan, and the East Asian NICs.

Despite such legislative and executive initiatives, the overall domestic and international climate is not conducive. ASEAN has shown little interest in the scheme and is generally wary of regional efforts. Perhaps the most active debates on the formation of the Pacific economic bloc have taken place on the private level. The Pacific Basin Economic Council (PBEC), a non-governmental organization established in 1967, has been most active. In 1982, the PBEC initiated a proposal on the Pacific Economic Community under the principles of cooperation, communication, and consultation but such efforts are unlikely to have major influence.[23]

Despite the proliferation of ideas and the feast of political rhetoric, the formation of a regional economic organization remains remote. There are several barriers impeding the advent of a new regionalism. First and foremost is regional heterogeneity. The Asia-Pacific region is composed of extremely diverse political, ideological, cultural, and historical entities. Different levels of economic development may lead to economic complementarities, but actually weaken the possibility of formal economic cooperation; ASEAN's reluctance demonstrates this This heterogeneity has blocked consensus on the fundamental issue: the membership and scope of cooperation. The bilateral initiatives of the U.S.—free trade zones with Canada and Israel, but bilateral pressure on East Asian trading partners—have deepened divisions across the Pacific. Though far from complete, ASEAN has formed its own economic bloc. The South Pacific economies have also created a more

---

23. For an excellent survey of divergent views of the Pacific community, see Japan Center for International Exchange, *The Pacific Community Concept: Views from Eight Nations* (Tokyo: JCIE, 1980).

explicit form of economic subbloc through a free trade arrangement. Japan has long cultivated its own zone of influence in Southeast Asia through bilateral cooperation accords; South Korea and Taiwan have followed suit.[24] The rise of economic subblocs and bilateralism diminishes the necessity to establish a regional economic integrative scheme.

Apart form heterogeneity, bitter historical memories and fear of dominance/dependence relations are yet to be solved. Who will assume the costs of leadership in a new integrative system? Who are the beneficiaries and losers?; in the words of Lynn Mytelka, the "salience of gain" matters.[25] Japan and the East Asian NICs have everything to gain, but others, especially ASEAN, do not see any immediate gains from regional arrangements. The U.S. also sees no immediate and pressing need for such an arrangement.[26] As long as this perception of unequal return of benefits prevails, the regionalist alternative remains a dim possibility.

The Regional Consequences of Declining Hegemony

The Asia-Pacific region confirms the realist assertion that economics cannot be separated from politics. As Haggard argues in the introduction, the rise and decline of hegemonic power helps account for the changing pattern of economic interactions in the region, ranging from the Japanese imperial design of a Greater Asian Co-prosperity Sphere to the liberal economic order imposed by the U.S. American hegemonic commitment to the region has buttressed economic cooperation in the post-war period through economic aid, policy intervention for export promotion, the provision of export markets, and tying East Asian countries to liberal international economic institutions.

Since the 1970s, however, the strategtic outlook has evolved in a quite different direction. The growing disjunction between military and economic power has created new possibilities for political conflict. Ongoing debates on defense burden sharing and the transfer of military technology are the clearest evidence that the mismatch of military and

---

24. For the survey of existing economic institutions in the region, see Congressional Research Service, *An Asian-Pacific...,* Appendix 4.

25. Lynn K. Myltelka, "The Salience of Gain in Third-World Integrative Systems," *World Politics*, 25:2 (January 1973), pp. 236-46.

26. For an analysis of pay-off structure of regional integration by nations and sectors, see Mann-Kyu Kim, Chung-in Moon and Kwang-il Baek, *A Study of Pacific Cooperation in the 21st Century* (Inchon: Center for International Studies, 1988), pp. 52-74 in Korean.

economic power is breeding new tensions in traditional security alliances. Calls for burden-sharing by alliance leaders inevitably lead to demands for power sharing by alliance partners. No matter how disproportionate, once power is shared, leading powers cannot retain their position of dominance. Recent American efforts to realign burden sharing with East Asian clients in the name of "strategic reciprocity" weaken the level of support and loyalty from them.[27]

Eroding security relations are not limited to burdensharing. Recent frictions over military technology have added new bilateral tensions. As noted before, U.S.-Japan relations have become strained as a result of American reluctance to transfer advanced military technology to Japan. Moreover, recent political debates on the FX program, although it was approved, have left deep scars on the bilateral relations. South Korea has also echoed a strong disenchantment with the U.S. in this area. The U.S. has been reluctant to allow South Korean participation in the SDI despite the latter's strong desire, and has strengthened its supplier control to prevent South Korea's third country sales of arms which are produced under U.S. technical assistance. Frictions emanating from military technology are not serious enough to dissolve security relations, but contribute to undermining American credibility in East Asia.[28]

This changing political context naturally raises the question of Japan's new international role. As Paul H. Kreisberg writes, "many

27. For a most recent overview of U.S. security relations with the Asia-Pacific nations, see Ronald McLaurin and Chung-in Moon, *The U.S. and the Defense of the Pacific* (Boulder: Westview Press, 1989). On U.S.-Japan relations, see Edward Olsen, *U.S.— Japan Strategic Reciprocity* (Stanford: Hoover Press, 1985).

28. See Yamamoto, "Technology..." for the U.S. regulation of transfer of advanced defense technology to Japan. For U.S.-Korean bilateral friction over military technology, see Mann Woo Lee, Ron McLaurin and Chung-in Moon, *Alliance under Tension: The Evolution of U.S.-South Korean Relations* (Boulder: Westview Press, 1988), chapter 4.

29. Paul H. Kreisberg, "The Decline of the West in the East," *Washington Post*, January 1, 1989, p. B2.

30. There is no clear consensus on the Japanese military role. Mike M. Mochizuki describes four basic schools: military realism, Japanese Gaulism, political realism, and unarmed neutralism. Of these, only Japanese Gaulism insists on a more independent military role. All other approaches, except unarmed neutralism, emphasize the importance of security ties with the U.S. "Japan's Search for Strategy," *International Security* 8:3 (Winter 1983-4), pp. 152-179. Nevertheless, some American scholars fear the rise of Japanese military power. See John W. Dower, "Japan's New Military Edge," *The Nation*, July 3, 1989, p. 1.

Asians are looking for some new ark which will keep them afloat, and most look toward Japan.''[29] Can Japan then become a new ark? Given existing institutional constraints, the revival of Japan as a military superpower appears remote.[30] Over the last decade, however, Japan has emerged as an economic and financial superpower. Japan is the leading creditor nation in the capitalist system, and the number one aid donor in the world. Japan is also a great source of technology. Along with this changed international economic position, Japan has gradually broken its traditional silence, and projected a leadership role in the international setting. The issues at stake are not simply Japan's economic penetration of the U.S., but the prospect that Japan might form an economic or trade group in which it would be the dominant economic power. Its intention is not yet clear, but Japan is *capable* of charting a new regional economic order with or without the U.S.[31]

Declining U.S. hegemony has coincided with a new security environment in East Asia. Domestic reforms have resulted in new regional strategies in both the Soviet Union and China. The spirit of "perestroika" and "glasnost" has been strongly echoed in the Pacific theater. Gorvachev has declared the Soviet Union a Pacific power, and while such a claim is somewhat premature, the Soviet Union is making enormous effort to develop a Pacific connection, primarily for economic reasons.

China has also drastically changed its posture toward Pacific economic interactions. In searching for a new niche in the capitalist division of labor, China has aggressively realigned its relations with the East Asian countries. Shanghai was designated as the platform of China's long reach into the Pacific economy, and economic cooperation with the Pacific region is being actively sought. Sino-Japanese economic relations are now more than a decade old. Beijing is planning to exchange trade representatives with South Korea despite political barriers. More surprising is the mainland's attitude toward Taiwan. Over the past two years, economic relations between the two Chinas have flourished. In the name of detente, countries in the region have shown a favorable response to these intiatives by the two socialist giants. While Japan is more cautious to this change, South Korea and Taiwan are rushing to exploit the new opportunity.

To the extent that there is an economic opening to the socialist

31. See Takashi Inoguchi and Daniel Okimoto (eds.), *The Political Economy of Japan: Vol. 2- The Chaning International Context;* Kenneth B. Pyle, ''In Pursuit of a Grand Design: Nakasone Betwixt the Past and the Future,'' in Pyle, *The Trade Crisis*, pp. 5-32; ''Japan's New Internationalism: Struggling with the Burden of History,'' in Yamamura, *Japanese Investment...*, pp. 265-289.

powers in the region, it could lead to a reduction of key tensions, including the Taiwan-China question and the problems on the Korean peninsula. This will result in somewhat diversified trade relations. It will probably not fundamentally reorient trade patterns, however, since the Soviet market is limited, and China remains economically cautious. The events in China in 1989 will make business even more cautious, at least in the short-run. In the longer-term, however, the conversion of economic interests among Japan, the East Asian NICs, China and the Soviet Union could produce a different kind of inter-nationalization in the region in which the role of the U.S. will be substantially diluted, changing the traditional pattern of trans-Pacific cooperation and consulation.[32]

Conclusion

The challenges facing the Pacific economy are deep and far-reading. National and corporate interests diverge at a number of points. Het-erogenous political structures and institutional arrangements weaken the foundation of a common ideology. Liberal forces are on the rise, but mercantilist forces are also strong. The effects of American pressure remain unclear. The strategic ties that have strengthened trans-Pacific security alliances are gradually eroding, opening new space for political and economic maneuver by countries in the region.

Current Pacific dynamics appear to confirm the realist assertion that despite interdependence, international politics are still dictated by the anarchic structure of the international system. Multilateral regimes have become too weak to enforce the "rules of the game" in the period of rapid industrial change. Intraregional fragmentation and unequal gains to trade and investment diminish the chances for regi-onalist alternatives. Tough bilateral bargaining and management are likely to proliferate in the Pacific basin.

Bilateralism may provide a positive alternative, however. It does not necessarily signal the erosion of a liberal, multilateral order. First, bilateral pressures have, ironically, increased demands for multilateral and regional solutions. Second, bilateral initiatives by the U.S. have provided a new external impetus for the domestic transformation of the East Asian countries, which are otherwise likely to maintain mer-centile inertia.

---

32. Immanuel Wallerstein makes an interesting speculation on this changing realign-ment. See his *The Politics of the World Economy* (London: Cambridge Univ. Press, 1984), pp. 58-68.

The new bilateralism is not only inevitable; it could also be an effective way of navigating the Pacific's future direction. But bilateralism can be only a short-term, strategic remedy. If "beggar-thy-neighbor" policies are to avoided, and common prosperity of the region is to be secured in the long run, *domestic* changes, both institutional and social, must take place first.

About the Editors and Contributors

Editors

Stephan Haggard is Associtate Professor of Government and an Associate of the Center for International Affairs, Harvard University. He received his B.A. and Ph.D. from the University of California. He has written on industrialization in the East Asian and Latin American newly industrializing countries, American trade policy and on the politics of stabilization and structural adjustment in developing countries. His articles have appeared in *World Politics, International Organization, Pacific Focus, The Journal of Concerned Asian Scholars, Latin American Research Review* and a number of edited collections. He is the author with Tun-jen Cheng of *Newly Industrializing Asia in Transition: Policy Reform and American Response* (Berkeley: Institute for International Studies, 1987) and the forthcoming *Pathways from the Periphery: The Politics of Growth in the Newly Industrializing Countries.*

Chung-in Moon is Associate Professor of Political Science, University of Kentucky. He received his Ph.D. from University of Maryland. His current research interests comprise comparative industrial policy, democratic changes and coalition politics, and state and society in third world national security. Over two dozens of his articles have appeared in edited volumes and in such scholarly journals as *International Interactions, International Studies Quarterly, Journal of Developing Societies, Journal of Northeast Asian Studies, Middle East Journal, Millennium,* and *World Politics.* His books include *The Third World National Security, Alliance under Tension, The United States and the Defense of the Pacific,* and *Supplier Control and Recipient Autonomy; The Dilemma of Defense Industrialization in the Third World.*

Contributors ————————————————————————————

Chul-Ho Chang is a candidate for JD and MBA at Duke University. He received his BA from Williams College. His research interests include international corporate and investment law and international management.

Harry I. Chernotsky is Associate Professor of Political Science at the University of North Carolina at Charlotte. He received his Ph.D. from Rutgers University. He has written a number of articles on foreign direct investment in the United States. He has also written on international studies education, and is currently serving as Chair of the International Education Section of the International Studies Association.

Tun-jen Cheng is Associate Professor in the Graduate School of International Relations and Pacific Studies, University of California, San Diego. He received his Ph.D. from the University of California, Berkeley. His research interests include comparative industrial policy, agrarian politics and the connection between economic development and political change. His articles have appeared in a number of edited collections, including Fred Deyo, *The Political Economy of the New Asian Industrialism* (Cornell University Press, 1987) and Gary Gereffi and Don Wyman, *Development Strateqies in East Asia and Latin America* (Princeton University Press, forthcoming). He is the author with Stephen Haggard of *Newly Industrializing Asia in Transition; Policy Reform and American Response* (Berkeley: Institute for International Studies, 1987).

James A. Dunn, Jr., is Associate Professor of Political Science and Public Policy at Rutgers University, Camden. He received his Ph.D. from the University of Pennsylvania. He was a Research Associate with the MIT Future of the Automobile Program, and has written numerous articles on public policy and transportation. He is the author of *Miles to Go: European and American Transportation Policies* (MIT Press, 1981).

Dennis Encarnation is Assistant Professor at the Harvard Business School where he specializes in th management of international business. He received his Ph.D. from Duke University. He has written a number of articles on bargaining relations among multinationals, national governments, and local enterprises in developing countries, and is the author of a forthcoming book on multinationals in India. His current project examines the motivations and consequences of cross-investment between American and Japanese firms.

Gary Gereffi is Associate Professor of Sociology at Duke University. His current research forcuses on development issues in Latin America and East Asia. He is the author of *The Pharmaceutical Industry in the Third World* (Princetion University Press, 1983) and editor with Donald Wyman of *Development Strategies in East Asia and Latin America* (Princeton University Press, forthcoming). His articles have appeared in *International Organization, Latin American Research Review* and the *International Journal of Comparative Sociology*.

Jeffrey A. Hart is Professor of Political Science at Indiana University in Bloomington. He received his Ph.D. from the University of California, Berkeley. From 1973 through 1979, he taught at Princeton University. In 1980, he was a professional staff member of the President's Commission for A National Agenda for the Eightees. He joined the faculty of Indiana University in 1981. In 1982-3, he was the Paul Henri Spaak Fellow for U.S.-European Affairs at the Center for International Affairs, Harvard University. He was a fellow at the Office of Technology Assessment of the U.S. Congress in 1985-86, and co authored the report on *International Competition in Services* (1987). His other major publications include *The New International Economic Order* (St. Martins, 1983); *Interdependence in the Post-Multilateral Era* (University Press of America, 1985) and the forthcoming *Atlantic Riptides.*

Stephen Krasner is Professor of Political Science and chairman of the Department of Political Science at Stanford University. He has also taught at Harvard University and the University of California, Los Angeles. He is the author of *Defending the National Interest* (Princeton University, 1978); *Structural Conflict* (University of California Press, 1985); and *Asymmetries in Japanese-American Trade* (Institute for International Studies, Berkeley, 1987) and the editor of *International Regimes* (Cornell University Press, 1985). His articles have appeared in a number of journals, including *World Politics, International Organization,* and *Comparative Politics.* He is currently the editor of *International Organization.*

Doug Nelson is currently visiting Assistant Professor of Political Economy at the School of Business at Washington University, St. Louis. He has previously worked in the Office of Trade Research of the Treasury Department and for various departments of the World Bank, including the *World Development Report* research staff, the International Economic Research Division and the Development Strategy Division. He has also taught in the Department of Political Science at Rutgers University and the International Management Studies program of the University of Texas, Dallas. His articles have appeared in *The American Economic Review, The American Journal of Political Science* and in the World Bank's *Staff Paper* series. He is currently finishing a monograph entitled *Social Choice in the Tropics: Microanalytic approaches to the Political Economy of Development Policy.* His research interests include the political economy of development policy.

Until his untimely death in 1987, **Donald Wyman** was Associate Dean of the Graduate School of International Relations and Pacific Studies at the University of California, San Diego. He received his Ph.D. in history from Harvard. His research interests included development in the Pacific Basin Countries and the debt crises and Taiwan, including an article in *Asian Survey,* and is currently writing a book comparing the formulation and implementation of industrial policy in Taiwan and Japan.

Greg Noble is Assistant Professor of Political Science, University of California, San Diego. He received his Ph.D. from Harvard University in 1988. He has written on industrial policy in financial institutions in Mexico. He edited *Mexico's Economic Crisis: Challenges and Opportunities* (San Diego: Center for U.S. Mexican Studies, 1985) and is co-editor with Gary Gereffl of *Development Strategies in Latin America and East Asia* (Princeton University Press, forthcoming).

David B. Yoffie is Associate Professor at the Harvard Business School in the Business, Government and Competition area. He received his B.A. summa cum laude from Brandeis University and his Ph.D. from Stanford University. He is the author of *Power and Protectionism; Strategies of the Newly Industrializing Countries* (Columbia University Press, 1983). His articles have appeared in *The American Political Science Review, International Studies Quarterly, Public Policy, Journal of Forecasting, California Management Review, International Management and the Harvard Business Review,* and he has authored over thirty case studies and teaching notes on issues in international strategy and management. His present research focuses on theories of industrial organization and their application to the political economy of international trade.

Index